ARE

2014

Structural Systems

Study Guide

Executive Director of Architecture Education: Brian S. Reitzel, PE

STRUCTURAL SYSTEMS ARE STUDY GUIDE 2014

Published by Kaplan Architecture Education

1-877-884-0828

www.kaplanarchitecture.com

Printed in the United States of America.

14 15 16 10 9 8 7 6 5 4 3 2 1

ISBN: 978-1-4277-4535-4 / 1-4277-4535-8
PPN: 3200-3346

CONTENTS

PART III: ARE STRATEGIC PLANNING

LESSON EIGHTEEN

WELCOME

Thank you for choosing Kaplan AE Education for your ARE study needs. We offer updates annually to keep abreast of code and exam changes and to address any errors discovered since the previous update was published. We wish you the best of luck in your pursuit of licensure.

ARE OVERVIEW

Since the State of Illinois first pioneered the practice of licensing architects in 1897, architectural licensing has been increasingly adopted as a means to protect the public health, safety, and welfare. Today, the United States and Canadian provinces require licensing for individuals practicing architecture. Licensing requirements vary by jurisdiction; however, the minimum requirements are uniform and in all cases include passing the Architect Registration Examination (ARE). This makes the ARE a required rite of passage for all those entering the profession, and you should be congratulated on undertaking this challenging endeavor.

Developed by the National Council of Architectural Registration Boards (NCARB), the ARE is the only exam by which architecture candidates can become registered in the United States or Canada. The ARE assesses candidates' knowledge, skills, and abilities in seven different areas of professional practice, including a candidate's competency in decision making and knowledge of various areas of the profession. The exam also tests competence in fulfilling an architect's responsibilities and in coordinating the activities of others while working with a team of design and construction specialists. In all jurisdictions, candidates must pass the seven divisions of the exam to become registered.

The ARE is designed and prepared by architects, making it a practice-based exam. It is generally not a test of academic knowledge, but rather a means to test decision-making ability as it relates to the responsibilities of the architectural profession. For example, the exam does not expect candidates to memorize specific details of the building code, but requires them to understand a model code's general requirements, scope, and purpose, and to know the architect's responsibilities related to that code. As such, there is no substitute for a well-rounded internship to help prepare for the ARE.

4.0 Exam Format

The seven ARE 4.0 divisions are outlined in the table below.

DIVISION	QUESTIONS	VIGNETTES
Building Design & Construction Systems	85	Accessibility/ Ramp Roof Plan Stair Design
Building Systems	95	Mechanical & Electrical Plan
Construction Documents & Services	100	Building Section
Programming, Planning & Practice	85	Site Zoning
Schematic Design	-	Building Layout Interior Layout
Site Planning & Design	65	Site Design Site Grading
Structural Systems	125	Structural Layout

ARCHITECTURAL HISTORY

Questions pertaining to the history of architecture appear in all of the multiple-choice divisions. The prominence of historical questions will vary not only by division but also within different versions of the exam for each division. In general, however, history tends to be lightly tested, with approximately three to seven history questions per division, depending upon the total number of questions within the division. One aspect common to all the divisions is that whatever history questions are presented will be related to that division's subject matter. For example, a question regarding Chicago's John Hancock Center and the purpose of its unique exterior cross bracing may appear on the Structural Systems exam.

Though it is difficult to predict how essential your knowledge of architectural history will be to passing any of the multiple-choice divisions, it is recommended that you refer to a primer in this field—such as Kaplan's *Architectural History*—before taking each exam, and that you keep an eye out for topics relevant to the division for which you are studying. It is always better to be overprepared than taken by surprise at the testing center.

The exam presents multiple-choice questions individually. Candidates may answer questions, skip questions, or mark questions for further review. Candidates may also move backward or forward within the exam using simple on-screen icons. The vignettes require candidates to create a graphic solution according to program and code requirements.

Actual appointment times for taking the exam are slightly longer than the actual exam time, allowing candidates to check in and out of the testing center. All ARE candidates are encouraged to review NCARB's *ARE Guidelines* for further detail about the exam format. These guidelines are available via free download at NCARB's Web site (*www.ncarb.org*).

Exam Format

It is important for exam candidates to familiarize themselves not only with exam content, but also with question format. Familiarity with the basic question types found in the ARE will reduce confusion, save time, and help you pass the exam. The ARE contains three basic question types.

The first and most common type is a straightforward multiple-choice question followed by four choices (A, B, C, and D). Candidates are expected to select the correct answer. This type of question is shown in the following example.

Which of the following cities is the capital of the United States?

A. New York

B. Washington, D.C.

C. Chicago

D. Los Angeles

The second type of question is a negatively worded question. In questions such as this, the negative wording is usually highlighted using all caps, as shown below.

Which of the following cities is NOT located on the west coast of the United States?

A. Los Angeles

B. San Diego

C. San Francisco

D. New York

The third type of question is a combination question. In a combination question, more than one choice may be correct; candidates must select from combinations of potentially correct choices. An example of a combination question is shown below.

Which of the following cities are located within the United States?

I. New York

II. Toronto

III. Montreal

IV. Los Angeles

A. I only

B. I and II

C. II and III

D. I and IV

The single most important thing candidates can do to prepare themselves for the vignettes is to learn to proficiently navigate NCARB's graphic software. Practice software can be downloaded free of charge from their Web site. Candidates should download it and become thoroughly familiar with its use.

Recommendations on Exam Division Order

NCARB allows candidates to choose the order in which they take the exams, and the choice is an important one. While only you know what works best for you, the following are some general considerations that many have found to be beneficial:

1. The Building Design & Construction Systems and Programming, Planning & Practice divisions are perhaps the broadest of all the divisions. Although this can make them among the most intimidating, taking these divisions early in the process will give a candidate a broad base of knowledge and may prove helpful in preparing for subsequent divisions. An alternative to this approach is to take these two divisions last, since you will already be familiar with much of their content. This latter approach likely is most beneficial when you take the exam divisions in fairly rapid succession so that details learned while studying for earlier divisions will still be fresh in your mind.
2. The Construction Documents & Services exam covers a broad range of subjects, dealing primarily with the architect's role and responsibilities within the building design and construction team. Because these subjects serve as one of the core foundations of the ARE, it may be advisable to take this division early in the process, as knowledge gained preparing for this exam can help in subsequent divisions.
3. Take exams that particularly concern you early in the process. NCARB rules prohibit retaking an exam for six months. Therefore, failing an exam early in the process will allow the candidate to use the waiting period to prepare for and take other exams.

EXAM PREPARATION

Overview

There is little argument that preparation is key to passing the ARE. With this in mind, Kaplan has developed a complete learning system for each exam division, including study guides, question-and-answer handbooks, mock exams, and flash cards. The study guides offer a condensed course of study and will best prepare you for the exam when utilized along with the other tools in the learning system. The system is designed to provide you with the general background necessary to pass the exam and to provide an indication of specific content areas that demand additional attention.

In addition to the Kaplan learning system, materials from industry-standard documents may prove useful for the various divisions. Several of these sources are noted in the "Supplementary Study Materials" section on the following page.

Understanding the Field

The subject of structures may fall under the direct responsibility of the architect or under the responsibility of a structural consultant, depending on the scale and complexity of the project. In either case, however, properly designed building structures are critical in protecting the safety of building occupants. This significant role means that the subject of structural systems must be thoroughly understood in order for architects to properly integrate structural elements into their designs and permit constructive interaction with other members of the building design team.

Understanding the Exam

The Structural Systems exam is among the most daunting of the ARE exams due to the technical nature of the subject and the wide array of problems that may be presented. Candidates will find questions covering statics, wood construction, concrete, steel, foundations, long span structural systems, and connections frequently on the exam. Many candidates, however, allow the breadth of the exam to intimidate unnecessarily.

Candidates must be familiar with the *Manual of Steel Construction* published by the American Institute of Steel Construction, but it is not necessary to memorize any part of this volume. Rather, candidates should be aware of how to utilize the charts and tables found in the book. Tables and charts needed to solve problems within the ARE will be provided during the exam. Additionally, candidates should be aware that a series of equations, tables, and formulas are available within the ARE software by clicking on the References button. Therefore, candidates should not waste time memorizing formulas, but rather should focus their studies on practicing their use.

Many have also found that the question-and-answer books and mock exams published by Kaplan are especially useful for this exam as they provide additional opportunities to practice working through structural problems.

Preparation Basics

The first step in preparation should be a review of the exam specifications and reference materials published by NCARB. These statements are available for each of the seven ARE divisions to serve as a guide for preparing for the exam. Download these statements and familiarize yourself with their content. This will help you focus your attention on the subjects on which the exam focuses.

Prior CAD knowledge is not necessary to successfully complete vignettes. In fact, it's important for candidates familiar with CAD to realize they will experience significant differences between CAD and the drawing tools used on the exam.

Though no two people will have exactly the same ARE experience, the following are recommended best practices to adopt in your studies and should serve as a guide.

Set aside scheduled study time.

Establish a routine and adopt study strategies that reflect your strengths and mirror your approach in other successful academic pursuits. Most importantly, set aside a definite amount of study time each week, just as if you were taking a lecture course, and carefully read all of the material.

Take—and retake—quizzes.
After studying each lesson in the study guide, take the quiz found at its conclusion. The quiz questions are intended to be straightforward and objective. Answers and explanations can be found at the back of the book. If you answer a question incorrectly, see if you can determine why the correct answer is correct before reading the explanation. Retake the quiz until you answer every question correctly and understand why the correct answers are correct.

Identify areas for improvement.
The quizzes allow you the opportunity to pinpoint areas where you need improvement. Reread and take note of the sections that cover these areas and seek additional information from other sources. Use the question-and-answer handbook and online test bank as a final tune-up for the exam.

Take the final exam.
A final exam designed to simulate the ARE follows the last lesson of each study guide. Answers and explanations can be found on the pages following the exam. As with the lesson quizzes, retake the final exam until you answer every question correctly and understand why the correct answers are correct.

Use the flash cards.
If you've purchased the flash cards, go through them once and set aside any terms you know at first glance. Take the rest to work, reviewing them on the train, over lunch, or before bed. Remove cards as you become familiar with their terms until you know all the terms. Review all the cards a final time before taking the exam.

Practice using the NCARB software.
Work through the practice vignettes contained within the NCARB software. You should work through each vignette repeatedly until you can solve it easily. As your skills develop, track how long it takes to work through a solution for each vignette.

Supplementary Study Materials

In addition to the Kaplan learning system, materials from industry-standard sources may prove useful in your studies. Candidates should consult the list of exam references in the NCARB guidelines for the council's recommendations and pay particular attention to the following publications, which are essential to successfully completing this exam:

- International Code Council (ICC) *International Building Code*
- American Institute of Steel Construction *Manual of Steel Construction: Allowable Stress Design,* Ninth Edition

Test-Taking Advice

Preparation for the exam should include a review of successful test-taking procedures—especially for those who have been out of the classroom for some time. Following is advice to aid in your success.

Pace yourself.
Each division allows candidates at least one minute per question. You should be able to comfortably read and reread each question and fully understand what is being asked before answering. Each vignette allows candidates ample time to complete a solution within the time allotted.

Read carefully.
Begin each question by reading it carefully and fully reviewing the choices, eliminating those that are obviously incorrect. Interpret language literally, and keep an eye out for negatively worded questions. With vignettes, carefully review instructions and requirements. Quickly make a list of program and code requirements to check your work against as you proceed through the vignette.

Guess.
All unanswered questions are considered incorrect, so answer every question. If you are unsure of the correct answer, select your best guess and/or mark the question for later review. If you continue to be unsure of the answer after returning to the question a second time, it is usually best to stick with your first guess.

Review difficult questions.
The exam allows candidates to review and change answers within the time limit. Utilize this feature to mark troubling questions for review upon completing the rest of the exam.

Reference material.
Some divisions include reference materials accessible through an on-screen icon. These materials include formulas and other reference content that may prove helpful when answering questions in these divisions. Note that candidates may *not* bring reference material with them to the testing center.

Best answer questions.
Many candidates fall victim to questions seeking the "best" answer. In these cases, it may appear at first glance as though several choices are correct. Remember the importance of reviewing the question carefully and interpreting the language literally. Consider the following example.

> Which of these cities is located on the east coast of the United States?
>
> **A.** Boston
>
> **B.** Philadelphia
>
> **C.** Washington, D.C.
>
> **D.** Atlanta

At first glance, it may appear that all of the cities could be correct answers. However, if you interpret the question literally, you'll identify the critical phrase as "on the east coast." Although each of the cities listed is arguably an "eastern" city, only Boston sits on the Atlantic coast. All the other choices are located in the eastern part of the country, but are not coastal cities.

Style doesn't count.
Vignettes are graded on their conformance with program requirements and instructions. Don't waste time creating aesthetically pleasing solutions and adding unnecessary design elements.

ACKNOWLEDGMENTS

Many thanks to David M. Berg, PE, and Robert Marks, PE, authors of the first editions of *General Structures* and *Lateral Forces.* While titles relying so heavily on building codes necessarily undergo significant change over the years, it says much about the quality of the original authors' endeavors that the organizational structure of these books endures, and that passages addressing the basic nature of structures, earthquakes, and wind forces remain largely untouched.

Contributing editor David A. Fanella, PhD, SE, PE, FASCE. A licensed structural and professional engineer, David is an Associate Principal with Klein and Hoffman in Chicago, and earned a PhD in structural engineering from the University of Illinois at Chicago. He has designed buildings for more than 20 years, and has written multiple books on earthquake and wind loads. David also served as primary author of two reports analyzing the collapse of the Twin Towers for the National Institute of Standards and Technology. He is an active member of numerous national technical committees, including the American Society of Civil Engineers' Minimum Design Loads for Buildings and Other Structures.

Special thanks to Suzanne Galletti, BSE, MArch, architect, for reviewing this update. Her contributions have helped create a more reader-friendly, accurate, and current title.

Thanks also to David Eckmann and Stephanie Hautzinger for their contributions to previous updates.

This introduction was written by John F. Hardt, AIA, MArch. John is a senior project architect and senior associate with Karlsberger, an architecture, planning, and design firm based in Columbus, Ohio. He is a graduate of Ohio State University (MArch).

Tables and charts from the International Building Code are reproduced courtesy of the International Code Council (ICC). For information about ICC codes and related services, contact ICC customer service at 1-800-786-4452.

ABOUT KAPLAN

Thank you for choosing Kaplan AE Education as your source for ARE preparation materials. Whether helping future professors prepare for the GRE or providing tomorrow's doctors the tools they need to pass the MCAT, Kaplan possesses more than 50 years of experience as a global leader in exam prep and educational publishing. It is that experience and history that Kaplan brings to the world of architectural education, pairing unparalleled resources with acknowledged experts in ARE content areas to bring you the very best in licensure study materials.

Only Kaplan AE offers a complete catalog of individual products and integrated learning systems to help you pass all seven divisions of the ARE. Kaplan's ARE materials include study guides, mock exams, question-and-answer handbooks, video workshops, and flash cards. Products may be purchased individually or in division-specific learning systems to suit your needs. These systems are designed to help you better focus on essential information for each division, provide flexibility in how you study, and save you money.

To order, please visit *www.kaplanarchitecture.com* or call 800-420-1429.

ACKNOWLEDGMENTS

Kaplan AE is grateful for the cooperation of many experts, professionals, organizations, and associations in the creation of this book. The following images are reproduced with the permission of the copyright holders.

Title	*Page*
Simple Beam—Uniformly Distributed Load	32
Simple Beam—Concentrated Load at Center	32
Simple Beam—Two Equal Concentrated Loads Symmetrically Placed	50
W Shapes—Selection by Plastic Section Modulus Table	65–67
Maximum Total Uniform Load—W Shapes	68
W Shapes—Available Moment vs. Unbraced Length	72
Available Critical Stress for Compression Members	75
Available Strength in Axial Compression—W Shapes	76
Available Strength in Axial Compression—Pipe	77
Available Strength in Axial Compression—Square HSS	78
Six Idealized Conditions	81
Simple Beam—Load Increasing Uniformly to One End	115
Available Shear Strength of Bolts	131
Slip-Critical Connections—Available Shear Strength	132
Available Bearing Strength at Bolt Holes Based on Bolt Spacing	134
Available Bearing Strength at Bolt Holes Based on Edge Distance	135
All-Bolted Double-Angle Connections	142
Workable Gages in Angle Legs	228
Available Strength in Axial Compression—Double Angles—LLBB	231

Title	Page
Table 1607.1—Minimum Uniformly Distributed Live Loads and Minimum Concentrated Live Loads	161
Table 1604.3—Deflection Limits	381
Table 1806.2—Presumptive Load-Bearing Values	386
Table 601—Fire-Resistance Rating Requirements for Building Elements	386

Title	Page
Section Properties of Standard Dressed (S4S) Sawn Lumber	47
Section Properties of Western Species Structural Glued Laminated Lumber	57
Table 11A—Bolts: Reference Lateral Design Values (Z) for Single Shear (two-member) Connections	123
Table 11F—Bolts: Reference Lateral Design Values (Z) for Double Shear (three-member) Connections	124
Table 4A—Reference Design Values for Visually Graded Dimension Lumber	382
Table 4D—Reference Design Values for Visually Graded Timbers (5" × 5" and larger)	383
Table 5A—Expanded-Reference Design Values for Structural Glued Laminated Softwood Timber Combinations	384

Reprinted courtesy of American Forest & Paper Association, Washington, D.C.

ASTM Standard Reinforcing Bars 90

Reproduced with permission. Copyright ASTM International.

Table 5.4-1—Number of Rows of Top Chord Bridging (K-Series) 189

Table 104.5-1—Max. Spacing of Lines of Top Chord Bridging (LH and DLH Series Joists) 189

ASD Standard Load Table for Longspan Steel Joists, LH-Series 191, 192

ASD Joist Girder Weight 194

Reprinted from *Standard Specifications for Steel Joists*, copyright the Steel Joist Institute. All rights reserved.

Table 9.5(a)—Minimum Thickness of Nonprestressed Beams or One-Way Slabs Unless Deflections Are Calculated 385

Reprinted from Building Code Requirements for Structural Concrete and Commentary, copyright the American Concrete Institute.

SYMBOLS & ABBREVIATIONS

The following abbreviations and symbols are used in this course and are generally understood in structural design practice.

Abbreviation/Symbol	**Reads as**
ft. or '	foot
ft.^{2} or sq.ft.	square foot
ft.^{3} or cu.ft.	cubic foot
ft.-kip or 'k	foot-kip
ft.-lb. or '#	foot-pound
in. or "	inch
in.^{2} or sq.in.	square inch
in.^{3} or cu.in.	cubic inch
in.-kip or "k	inch-kip
in.-lb. or "#	inch-pound
kip or k	kip (1 kip = 1 kilo pound or 1,000 pounds)
ksi	kips per square inch
lb. or #	pound
lb./cu.ft. or #/ft.^{3} or pcf	pounds per cubic foot
plf or #/' or #/ft.	pounds per lineal foot
psf or #/ft.^{2}	pounds per square foot
psi or #/in.^{2}	pounds per square inch
yd^3 or cu.yd.	cubic yard
Σ	summation of
$>$	is greater than
$<$	is less than

Part I

The Multiple-Choice Exam

LESSON ONE

1

STATICS

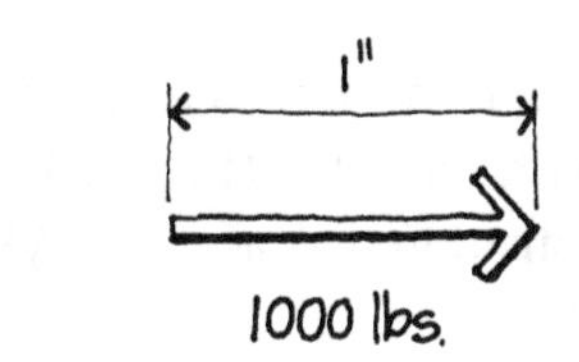

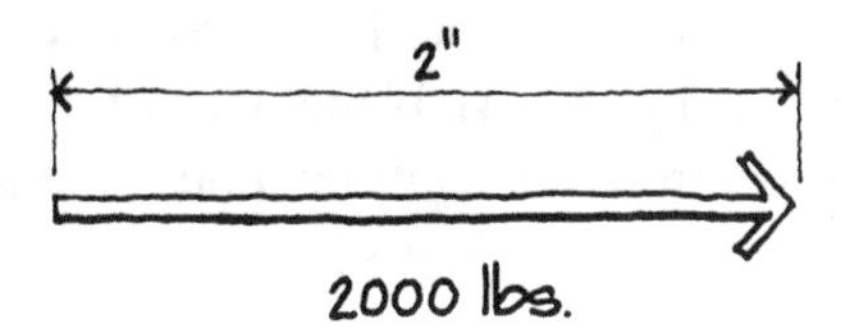

Scale : 1" = 1000#

Figure 1.1

FORCES

The essential function of a building's structure is to resist forces. It is therefore logical to begin our study of structures by reviewing the nature of forces.

What is a force? Simply stated, it is a *push* or *pull exerted on an object.* The description of a given force includes its magnitude, direction, and point of application. Forces are drawn as arrows showing direction, with the length indicating magnitude. For example, if one inch represents 1,000 pounds of force, then two inches would represent a force of 2,000 pounds (Figure 1.1).

Forces are measured in units of weight, such as pounds or kips (1 kip = 1,000 pounds). In metric (SI) units, forces are measured in newtons (N), where 1 lbf (pound force) = 4.44822 N.

A force applied to a body is called an external force or a *load.* The resistance of the body to the load is called an internal force or a *stress* (Figure 1.2).

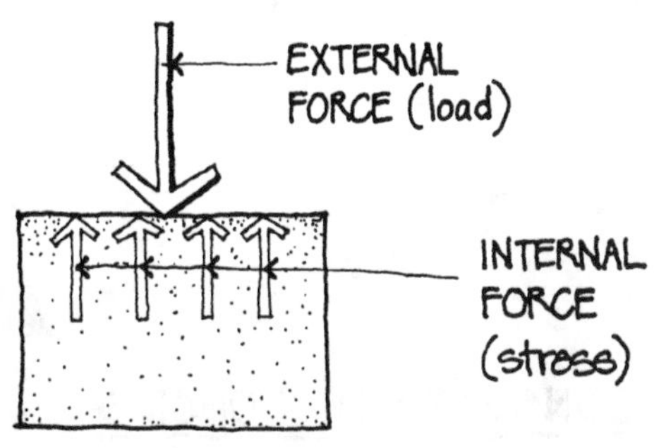

Figure 1.2

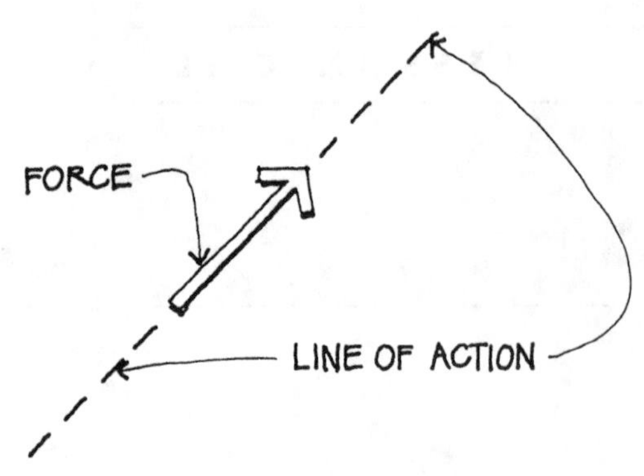

Figure 1.3

The *line of action* of a force is a line parallel to and in line with the force (Figure 1.3). The force may be considered as acting anywhere along its line of action.

If the lines of action of several forces pass through a common point, the forces are called *concurrent* (Figure 1.4). If the lines of action do not pass through a common point, the forces are *non-concurrent* (Figure 1.5).

Adding Forces

The resultant is one force that will produce the same effect on a body as two or more other forces. If the forces have the same line of action, they may be added directly to produce the resultant (Figure 1.6).

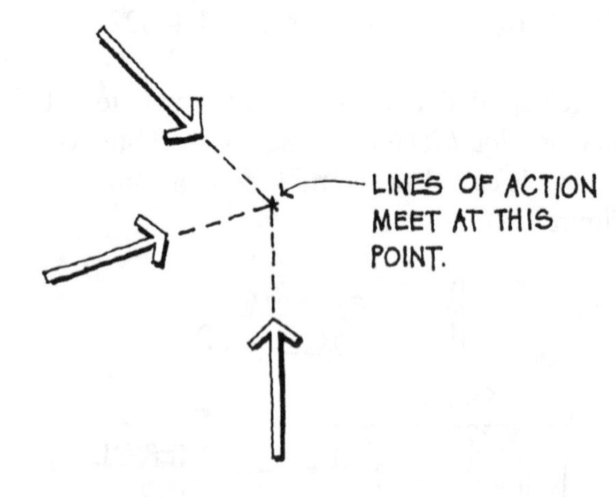

Figure 1.4

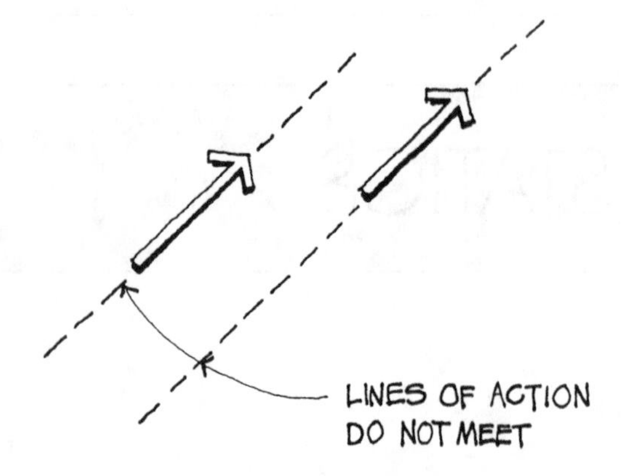

Figure 1.5

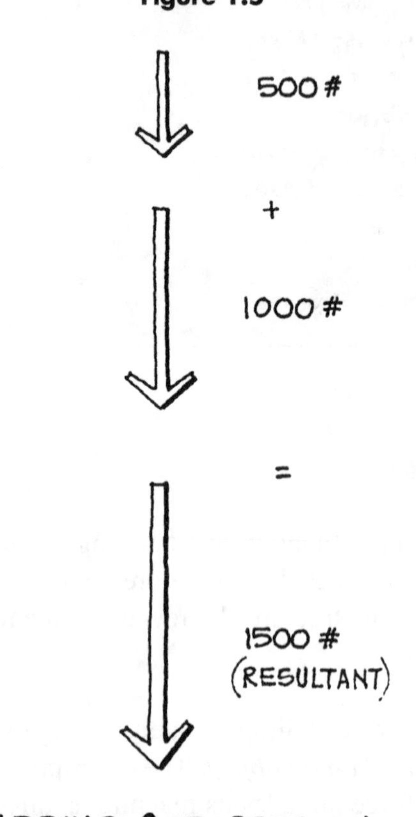

Figure 1.6

Concurrent forces cannot be added directly, but must be added vectorially, a procedure that is best illustrated by the following examples.

Example #1

Find the resultant of the two forces shown in Figure 1.7.

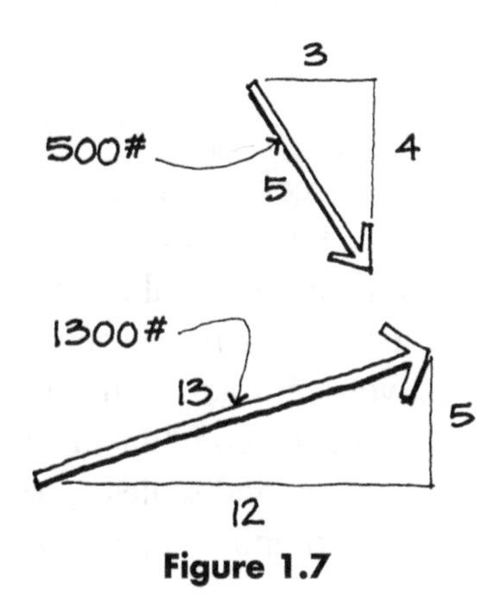

Figure 1.7

Solution:

Extend the lines of action of the two forces as shown in Figure 1.8. From their intersection, lay off the two forces to scale, in this case 1/8" = 100 pounds. Add parallel lines to form a parallelogram. The resultant, in both magnitude and direction, is the diagonal of the parallelogram, from the point of intersection of the lines of action to the diagonally opposite corner. In this example, we scale the value of the resultant as *1,503 pounds.*

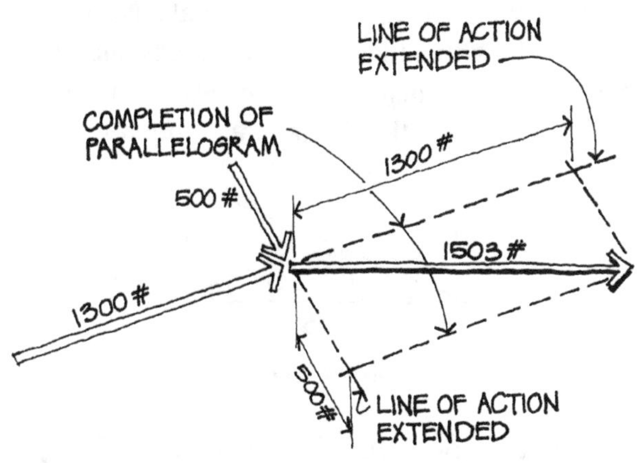

Figure 1.8

The following example illustrates the procedure for finding the resultant of more than two concurrent forces.

Example #2

Find the resultant of the three forces shown in Figure 1.9.

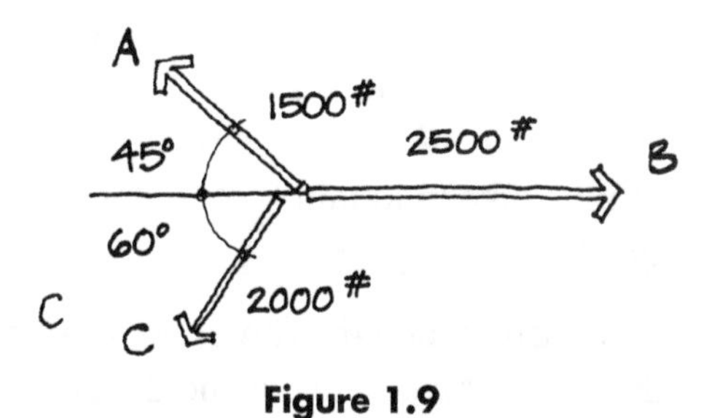

Figure 1.9

Solution:

From any point (in this case, point O), lay off one of the forces to scale and in the correct direction (Figure 1.10). From its arrowhead end, lay off another of the forces in the same manner, and so on until all of the forces have been included. In this case, we followed the order A-B-C in laying off the forces. Finally, draw a line from the starting point (O) to the arrowhead end of the final force (C). This line is the resultant, in both magnitude and direction. In this example, the resultant scales *802#.* This diagram is called a *force polygon.*

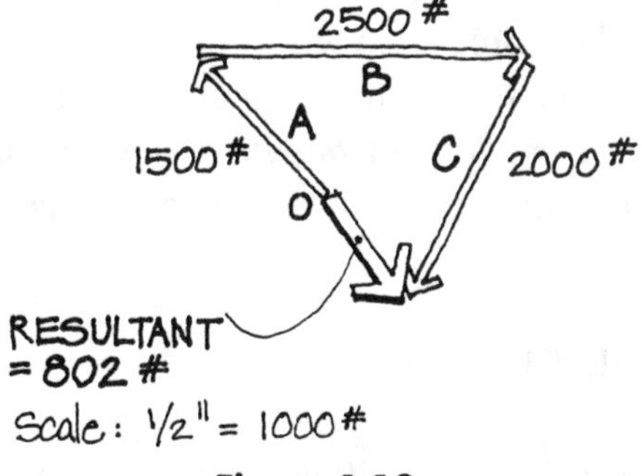

Figure 1.10

Let us look again at the original three forces along with the resultant that we have just determined.

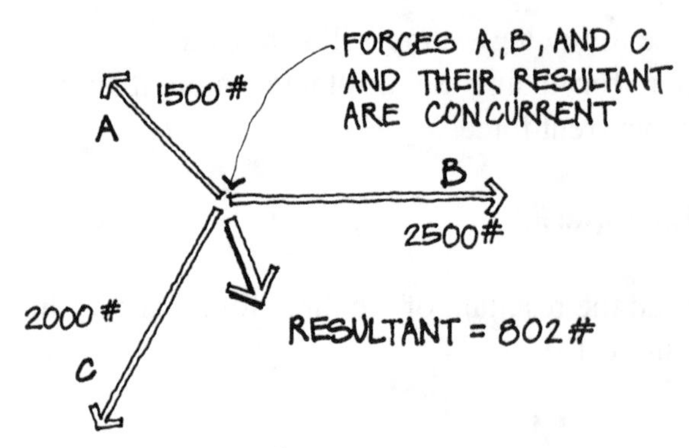

Figure 1.11

Three things to keep in mind when determining the resultant are:

1. The order in which the forces are drawn in the force polygon makes no difference; the resultant will be the same in both magnitude and direction regardless of the order followed.
2. The resultant is directed *away* from the starting point.
3. The resultant is *concurrent* with the original forces (their lines of action pass through a common point); see Figure 1.11.

A force equal in magnitude to the resultant, but opposite in direction and on the same line of action as the resultant is called the *equilibrant*.

Resolving Forces

It is sometimes convenient in the analysis of structures to replace one force with two or more other forces that will produce the same effect on a body as the original force. These forces are called *components* of the original force, and the procedure is called *resolving forces*.

Example #3

Resolve a force of 1,300# acting at 23° to the horizontal into two components, one horizontal and the other vertical (See Figure 1.12).

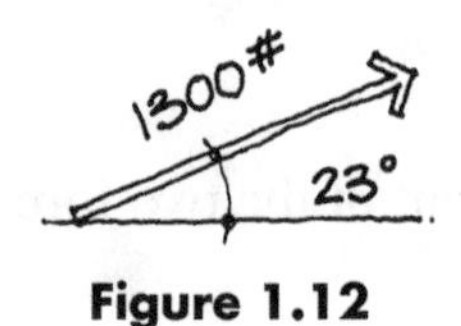

Figure 1.12

Solution:

Draw the 1,300# force to scale at 23° to the horizontal (Figure 1.13). From beginning point O draw horizontal and vertical lines. From the arrowhead end of the force (point C) draw horizontal and vertical lines. Points of intersection A and B define the components: OB is the horizontal component and scales *1,200#*, while OA is the vertical component and scales *500#.*

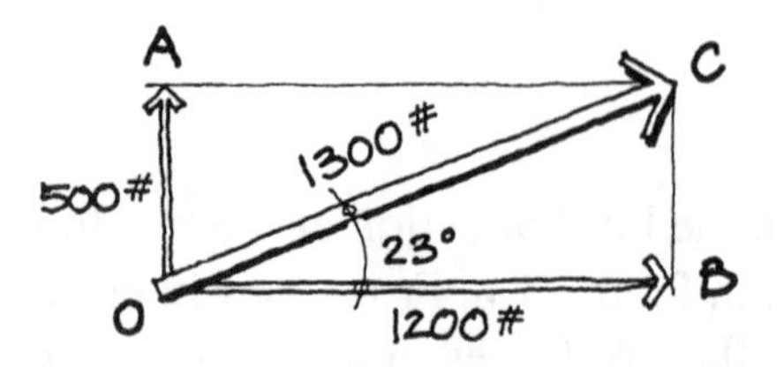

Scale: 1/8" = 100#

Figure 1.13

Analytical Method

Thus far we have used graphical methods for adding forces to obtain the resultant, as well as for resolving forces into components. Now we will see how to add and resolve forces analytically. First we must review some basic trigonometry, such as the Pythagorean theorem.

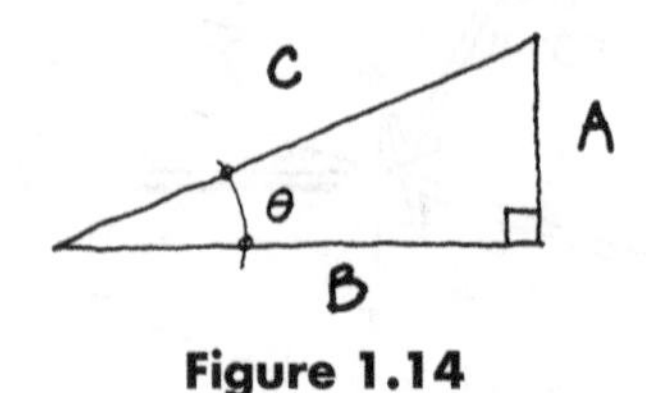

Figure 1.14

In right triangle ABC, $\sin \Theta = A/C$, $\cos \Theta = B/C$, and $\tan \Theta = A/B$. Thus, $A = C \sin \Theta$ and $B = C \cos \Theta$ (Figure 1.14). Also $C^2 = A^2 + B^2$, from which C = $C = \sqrt{A^2 + B^2}$.

These basic rules may easily be used to determine the vertical and horizontal components of a force. For example, the horizontal component of the 1,300# force of Example #3 is OB, which is equal to 1,300 cos 23° = 1,300 × .920 = 1,196#. The vertical component is OA, which is equal to CB, which is 1,300 sin 23° = 1,300 × .390 = 507#. These values are in close agreement with those determined graphically in Example #3.

Values of sines and cosines of angles may be found in many standard reference books, or may easily be obtained by using an electronic calculator which has trigonometric functions.

To obtain the resultant of two or more concurrent forces analytically, each force must first be resolved into its horizontal and vertical components, as described above. You must be consistent with the signs of the forces: the horizontal components are positive (+) when they act towards the right and negative (–) when they act towards the left. Vertical components are positive (+) when they act upward and negative (–) when they act downward.

Total the horizontal components, being careful to add (+) values and subtract (–) values. The total of the horizontal components is called ΣF_x.

Add the vertical components in the same manner; the total of the vertical components is called ΣF_y.

The resultant $R = \sqrt{(\Sigma F_x)^2 + (\Sigma F_y)^2}$ and the angle Θ that the resultant makes with the horizontal is such that $\tan \Theta = \dfrac{\Sigma F_y}{\Sigma F_x}$.

Example #4

Find the magnitude and direction of the resultant of the four forces shown in Figure 1.15.

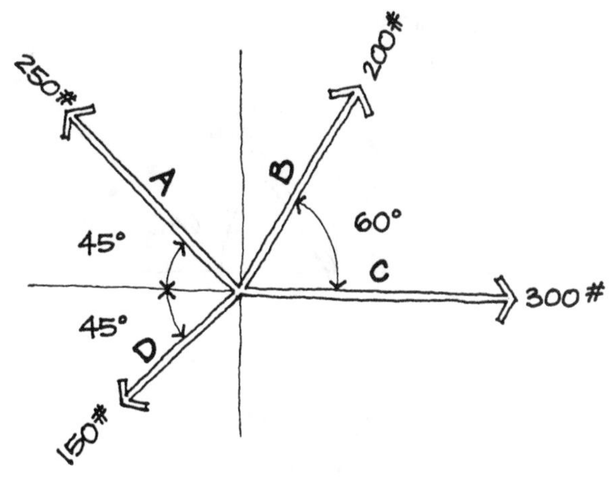

Figure 1.15

Solution:

Set up a table with each force and its horizontal and vertical components.

Force	Value	Horiz. Comp.	Vert. Comp.
A	250	–250 cos 45 = –176	250 sin 45 = +176
B	200	200 cos 60 = +100	200 sin 60 = +173
C	300	300 cos 0 = +300	300 sin 0 = 0
D	150	–150 cos 45 = –106	–150 sin 45 = –106
Total		$\Sigma F_x = +118$	$\Sigma F_y = +243$

Table 1.1

The resultant $R = \sqrt{(\Sigma F_x)^2 + (\Sigma F_y)^2} = \sqrt{118^2 + 243^2} = 270\#$ (see Figure 1.16)

$$\tan \Theta = \frac{\Sigma F_y}{\Sigma F_x} = \frac{243}{118} = 2.06$$

Using a table of trig functions or a calculator, we determine that $\Theta = 64.1°$ *with the horizontal.*

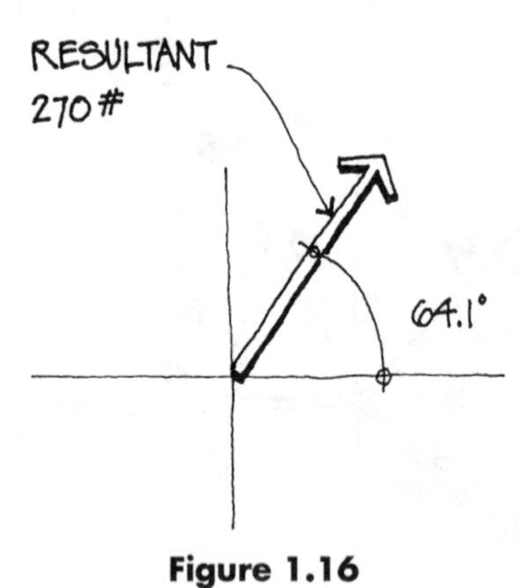

Figure 1.16

Note that the horizontal components of forces A and D are negative, since they act towards the left; the horizontal component of force B is positive, since it acts to the right; and the horizontal component of force C is also positive, since it acts to the right, and is equal to force C, since the force is horizontal. Similarly, the vertical components of forces A and B are positive, since they act upward; the vertical component of force D is negative, since it acts downward; and force C has no vertical component.

Since both ΣF_x and ΣF_y are positive, the resultant acts *upward and to the right.*

MOMENTS

A *moment* (M) is defined as the tendency of a force to cause rotation about a given point or axis.

The point is called the *center of moments* or *axis of rotation* and the distance, called the *moment arm* or *lever arm*, is measured in a direction perpendicular to the line of action of the force.

The magnitude of the moment is the magnitude of the force (in pounds or kips) multiplied by its distance from the given point (in feet or inches). If the line of force intersects the given point, the distance equals zero, so then M = 0.

The units of moment are therefore foot-pounds, inch-pounds, foot-kips, or inch-kips. In metric (SI) units, moments are measured in newton meters (N·m), where 1 ft·lb (foot pound) = 1.35582 N·m.

The value of the moment shown in Figure 1.17, about point A, is F × e.

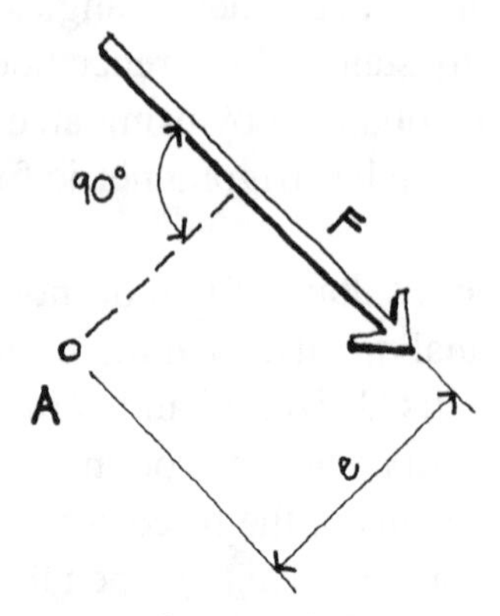

Figure 1.17

It is sometimes convenient to compute a moment by resolving the force into its horizontal and vertical components (Figure 1.18). The moment of the force is then equal to the moment of its horizontal component plus the moment of its vertical component.

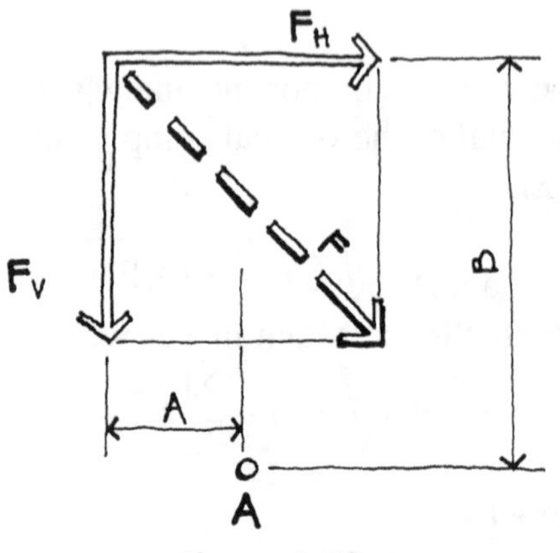

Figure 1.18

If we resolve force F into its horizontal and vertical components F_H and F_V, the moment about point A caused by F_H is clockwise, while that caused by F_V is counterclockwise. If we call clockwise moments positive and counterclockwise moments negative, the moment is $F_H \times B - F_V \times A$. This moment has the same magnitude as the moment F × e previously indicated.

Moment is not rotation itself, but the tendency to rotate. Consider Figure 1.19.

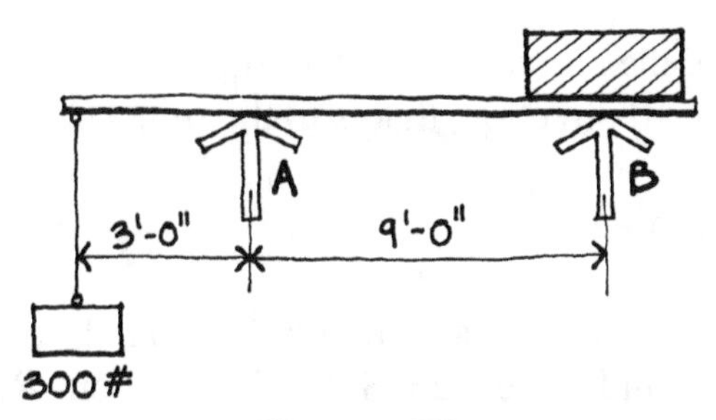

Figure 1.19

The load at the end of the lever causes a moment about point A of 300 × 3 = 900 foot-pounds (counterclockwise). The lever does not rotate because the load at point B prevents that end from moving up. The load required at B to prevent movement is 300 × 3/9 = 100 pounds.

Two forces equal in magnitude, but opposite in direction, and acting at some distance from each other, form a *couple.*

The moment produced by a couple is equal to the value of one force multiplied by the distance between the two forces (Figure 1.20). This moment is the same regardless of the axis of rotation.

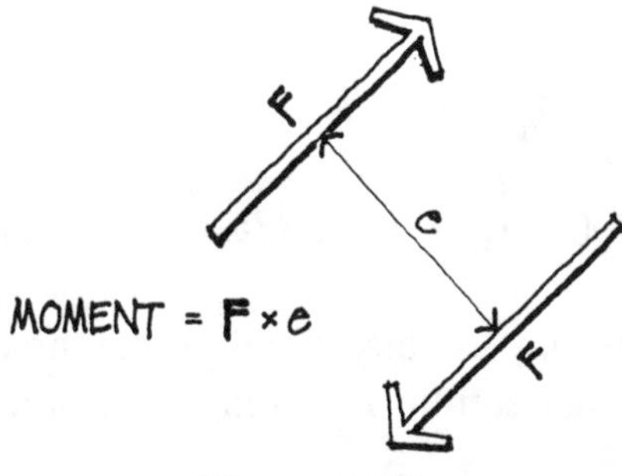

Figure 1.20

For example, when a lug wrench is used to remove a nut, one hand pushing up and the other pushing down with a force of 40 pounds produces a moment of 40 × 14 = 560 inch-pounds (Figure 1.21). The two forces form a couple.

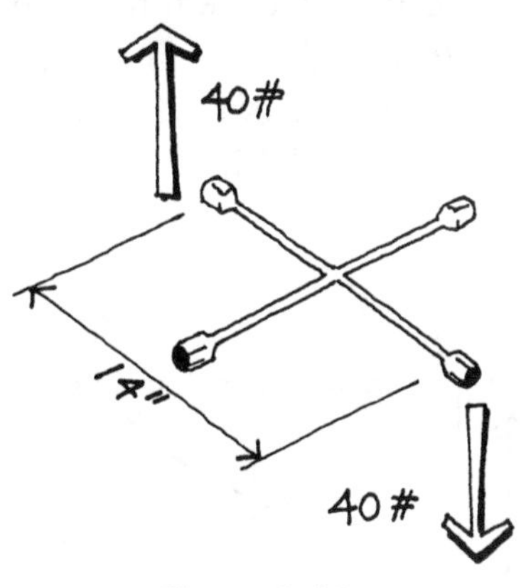

Figure 1.21

EQUILIBRIUM

Architects deal with buildings, which are objects at rest (in equilibrium). For an object to be in equilibrium, it must have no unbalanced force acting on it, or it would move. It must also have no unbalanced moment acting on it, or it would rotate. The three conditions of equilibrium may be stated as follows:

1. The summation of all the horizontal forces acting on the body must equal zero ($\Sigma F_x = 0$).
2. The summation of all the vertical forces acting on the body must equal zero ($\Sigma F_y = 0$).
3. The summation of all the moments acting on the body must equal zero ($\Sigma M = 0$).

Another way of expressing the conditions of equilibrium is that the sum of all the forces acting to the right must equal the sum of all the forces acting to the left, the sum of all the downward forces must equal the sum of all the upward forces, and the sum of all the clockwise moments must equal the sum of all the counterclockwise moments.

Example #5

A horizontal force of 200# to the left and a vertical force of 400# downward act at the end of the structure as shown in Figure 1.22. The forces acting at supports A and B are called *reactions*. Reactions are the forces acting at the supports that hold the beam in equilibrium. What is the magnitude of the horizontal reaction at A (A_H) and at B (B_H)?

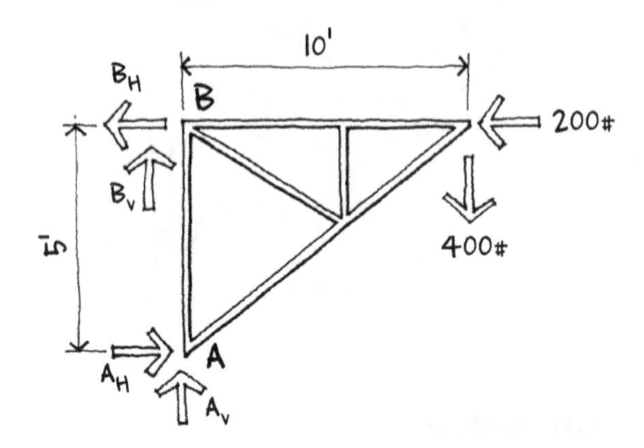

Figure 1.22

Solution:

Take moments about A and apply the equilibrium equation $\Sigma M_A = 0$.

The lines of action of A_V, B_V, and A_H all pass through point A, and their moments about point A, therefore, are zero. The only reaction having a moment about A is B_H.

$\Sigma M_A = 0$
$(400\# \times 10\text{ft.}) - (200\# \times 5\text{ft.}) - (B_H \times 5\text{ft.}) = 0$
$4{,}000 - 1{,}000 - B_H \times 5 = 0$
$3{,}000 = 5B_H$
$B_H = 600\#$

Note that the 400 pound force produces a clockwise moment about point A, which we call positive, and the 200 pound force produces a counterclockwise moment, which we call negative. Reaction B_H, which acts to the left, causes a counterclockwise moment about A, which is negative. If we had assumed B_H acting to the right, our answer for B_H would have been –600#, which would simply mean that our assumed direction was wrong and that B_H actually acts to the left.

Now we apply the equation $\Sigma H = 0$.

$A_H - B_H - 200\# = 0$
$A_H - 600\# - 200\# = 0$
$A_H = 800\#$

Since the answer is positive, A_H acts in the assumed direction, that is to the right.

Example #6

A beam supports a load of 5,000 pounds as shown in Figure 1.23. Solve for reactions R_1 and R_2, neglecting the weight of the beam.

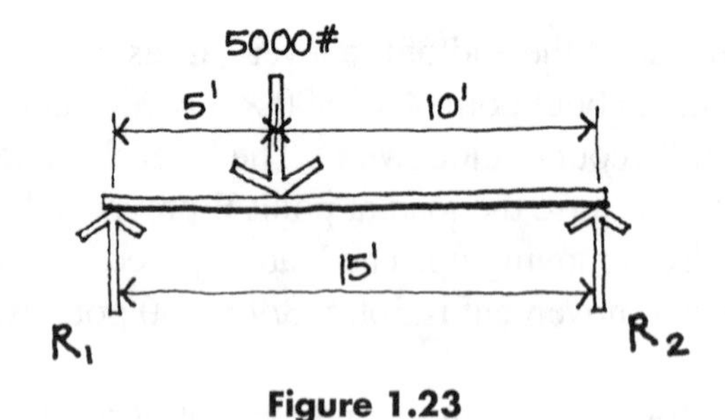

Figure 1.23

Solution:

Take moments about R_1.

$\Sigma M_{R_1} = 0$
$(5{,}000\# \times 5 \text{ ft.}) - (R_2 \times 15 \text{ ft.}) = 0$

$$R_2 = \frac{5{,}000 \times 5}{15} = 1{,}667\#$$

$\Sigma F_y = 0$
$-5{,}000\# + R_1 + R_2 = 0$
$R_1 + 1{,}667 = 5{,}000$
$R_1 = 5{,}000 - 1{,}667 = 3{,}333\#$

Since we may use any, or all, of the three equations of equilibrium and in any order, let's solve the problem differently, by taking moments about R_2.

$\Sigma M = 0$
$-(5{,}000\# \times 10 \text{ ft.}) + (R_1 \times 15 \text{ ft.}) = 0$

$$R_1 = \frac{5{,}000 \times 10}{15} = \mathit{3{,}333\#}$$

$\Sigma V = 0$
$-5{,}000\# + R_1 + R_2 = 0$
$3{,}333 + R_2 = 5{,}000$
$R_2 = 5{,}000 - 3{,}333 = \mathit{1{,}667\#}$

Therefore the results are the same, whichever approach we use.

Example #7

A bar of negligible weight is 20 feet long. It supports a load of 250 lbs. at one end and a load of 50 lbs. at the other end (see Figure 1.24). The bar is to be supported at one point only (the *fulcrum*). Where should the fulcrum be located so the bar is balanced?

Solution:

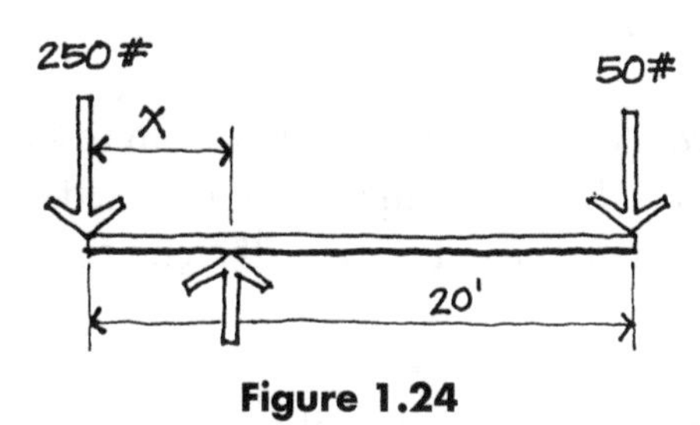

Figure 1.24

If we call the distance from the fulcrum to the 250 lb. load x, then the distance from the fulcrum to the 50 lb. load will be 20 – x.

Then, taking moments about the fulcrum, and applying $\Sigma M = 0$

$-250x + 50(20 - x) = 0$
$-250x + 1{,}000 - 50x = 0$
$300x = 1{,}000$
$x = \mathit{3.33\ ft.}$

FREE BODY DIAGRAMS

To determine the internal forces in a structure, we make an imaginary cut through the structure, and apply the equations of equilibrium to the remaining portion, called the *free body diagram.*

Example #8

What is the internal force and moment for the beam of Example #6, at a point 8 feet from the left support (see Figure 1.25)?

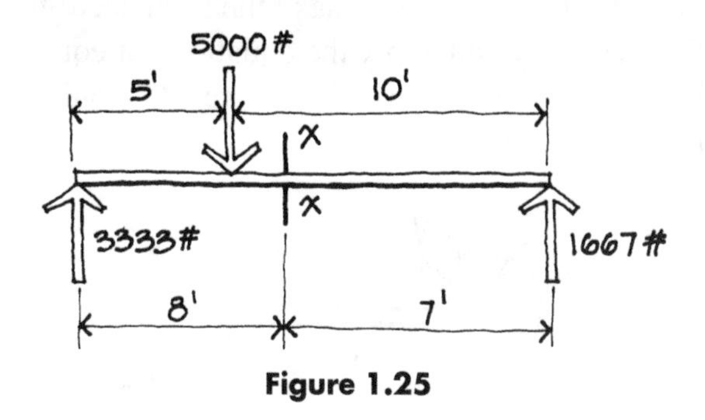

Figure 1.25

Solution:

Imagine cutting the beam at section x-x, eight feet from the left support as shown in Figure 1.26. Make the left portion the *free body diagram* and apply the equations of equilibrium.

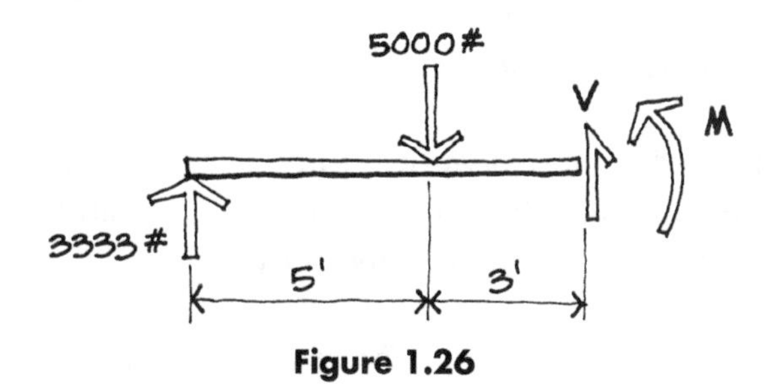

Figure 1.26

We call the internal vertical force in the beam V, and the internal moment M.

$\Sigma V = 0$
$-5{,}000\# + 3{,}333\# + V = 0$

V = *1,667#*. Since the answer comes out positive, our assumption for the direction of V is correct: the force acts *up* on the cut section. Now take moments about section x-x.

$\Sigma M = 0$
$(3{,}333\# \times 8 \text{ ft.}) - (5{,}000\# \times 3 \text{ ft.}) - M = 0$
$26{,}664 - 15{,}000 - M = 0$
$M = \mathit{11{,}664\ ft. - \#}$

Again, our assumption for the direction of moment M is correct.

For additional practice, make the right section the free body and apply the equations of equilibrium to it (Figure 1.27). Your result should be:

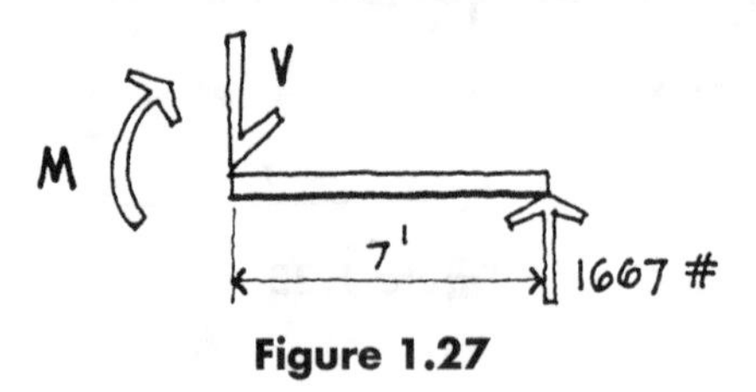

Figure 1.27

$\Sigma V = 0$
$1{,}667\# - V = 0$
$V = \mathit{1{,}667\#}$

$\Sigma M = 0$
$-(1{,}667\# \times 7 \text{ ft.}) + M = 0$
$M = \mathit{11{,}669\ ft.–\#}$

Thus, the internal vertical force and moment are the same whether we make the left section or the right section the free body.

Also note that internal forces are sometimes shown with half an arrowhead, as in Figure 1.27.

PROPERTIES OF AREAS

Up to now, this lesson has dealt with forces and the moments caused by forces. Now we depart from these topics to review the properties of areas. Later we will see how the properties of areas are used in structural analysis and design.

Two-dimensional shapes, such as rectangles, circles, and squares, possess geometric properties, such as area, centroid, and moment of inertia.

The area of a shape is a familiar concept and requires no discussion here.

The *centroid* of an area is equivalent to the center of gravity of the area: all of the area may be considered concentrated at the centroid without affecting the moment of the area about any axis. For symmetrical shapes such as rectangles, the centroid is the geometric center of the shape: a point midway between the upper and lower boundaries of the area and between the left and right boundaries of the area (see Figure 1.28).

The *statical moment* of an area with respect to an axis is defined as the area multiplied by the perpendicular distance from the centroid of the area to the axis.

To compute the location of the centroid of an unsymmetrical shape, divide the shape into two or more simple shapes, compute the statical moments of the simple shapes about any axis, add them to obtain the total, then divide by the total area.

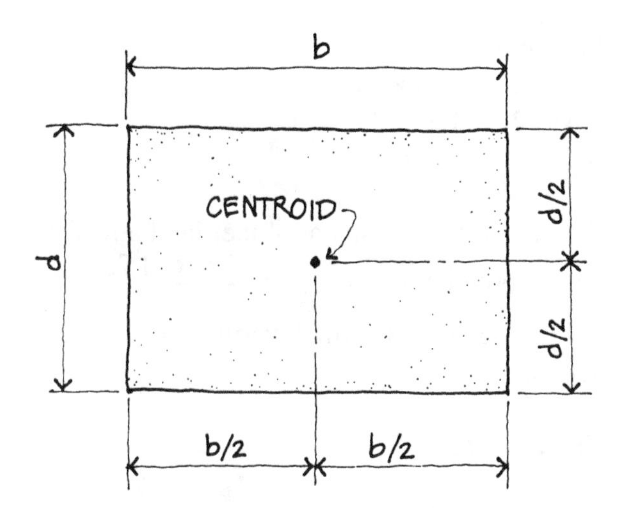

Figure 1.28

Example #9

Find the centroid of the area shown in Figure 1.29.

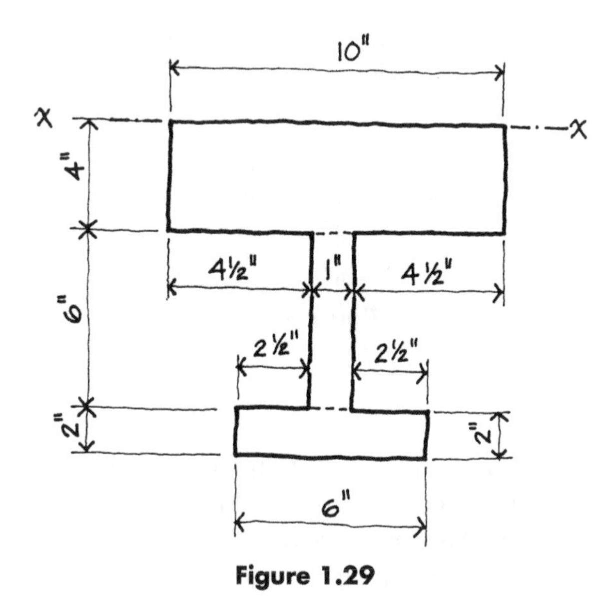

Figure 1.29

Solution:

Divide the area into three rectangles: A, B, and C, as shown in Figure 1.30. Calculate the area of each rectangle, and the statical moment of each rectangle about axis x-x, the top of the area. Calculations are shown in Table 1.2.

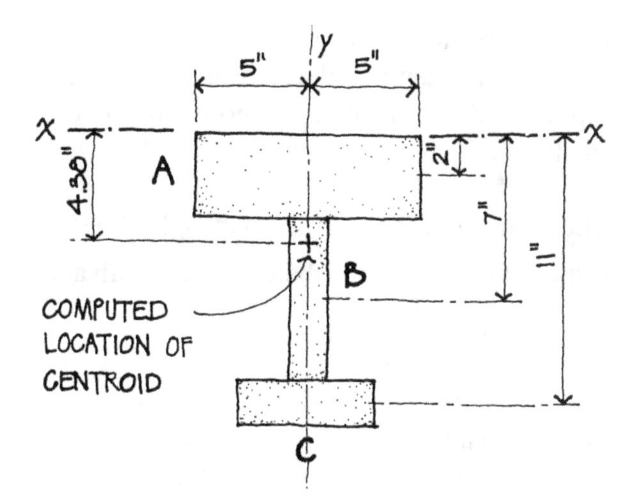

Figure 1.30

	Area	y	Area × y
A	4 × 10 = 40	2	80
B	6 × 1 = 6	7	42
C	6 × 2 = 12	11	132
Total	58		254

Table 1.2

$\bar{y}$ = the distance from axis x-x to the centroid of the area = $\frac{254}{58}$ = 4.38".

Because of symmetry, the centroid is 5" from the left or right edge of rectangle A. If the area were unsymmetrical about the vertical axis, the horizontal distance from its centroid to a given axis could be computed in the same way that we computed $\bar{y}$.

For additional practice, try calculating the distance from the *bottom* of the area to the centroid. Your result should be the same as above: 12.0 – 4.38 = 7.62" from the bottom.

The *moment of inertia* of an area about an axis is defined as the sum of the products obtained by multiplying all of the infinitesimal elements of area by the square of their distances from the axis. Unless otherwise specified, when we refer to the moment of inertia of an area, it is with respect to the centroidal axis of the area, and is represented by I. Its units are normally inches[4] (inches to the fourth power).

The moment of inertia I of a rectangle about its centroidal axis is $\frac{bd^3}{12}$ where b = the width and d = the depth of the rectangle.

The moment of inertia of an area about any axis is equal to the moment of inertia about its centroidal axis plus the area multiplied by the square of the distance between the axes.

Example #10

What is the moment of inertia of the area shown in Figure 1.31 about its horizontal centroidal axis? About its base?

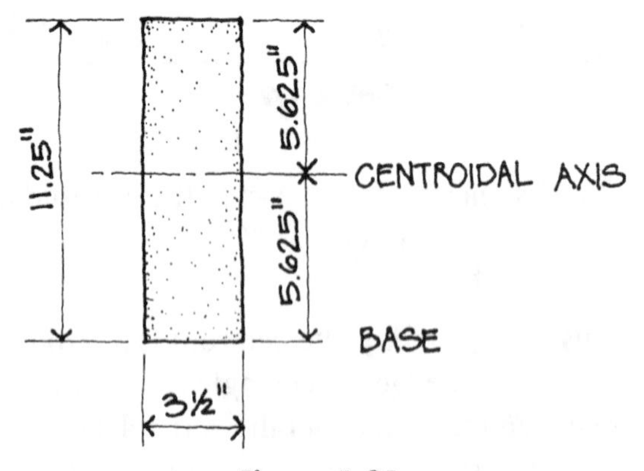

Figure 1.31

Solution:

The moment of inertia of a rectangle about its centroidal axis = $\frac{bd^3}{12} = \frac{3.5 \times 11.25^3}{12} =$ *415.28 inches⁴*.

The moment of inertia about the base = moment of inertia about centroidal axis + Ay^2, where A = area and y = the distance from the centroidal axis to the base.

I about base
$= 415.28 + (3.5 \times 11.25) \times 5.625^2$
$= 415.28 + 1{,}245.85$
$=$ *1,661.13 inches⁴*

Incidentally, the moment of inertia of a rectangle about its base = $\frac{bd^3}{3} = \frac{3.5 \times 11.25^3}{3}$ = 1,661.13 inches⁴, which verifies our answer above.

Example #11

Find the location of the centroidal axis and the moment of inertia of the girder shown in Figure 1.32 about its centroidal axis.

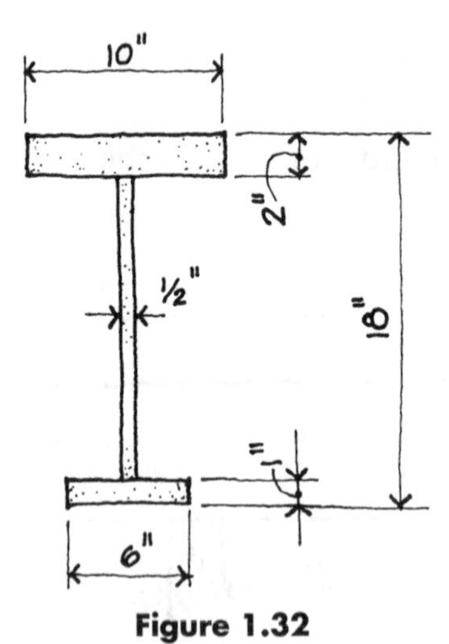

Figure 1.32

Solution:

Divide the cross-section into three rectangular areas: 1, 2, and 3, as shown in Figure 1.33. Calculate the area of each rectangle, and the statical moment of each rectangle about axis x-x, the top line (see Table 1.3).

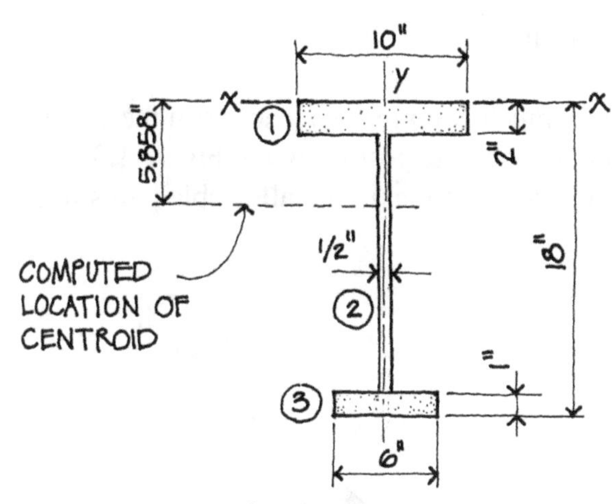

Figure 1.33

	Area	y	Area × y
1	10 × 2 = 20	1	20
2	1/2 × 15 = 7.5	9.5	71.25
3	6 × 1 = 6	17.5	105
Total	33.5		196.25

Table 1.3

$\bar{y}$ = the distance from axis x-x to the centroid = $\frac{196.25}{33.5} = 5.858"$.

Now set up another table, Table 1.4, as follows:

	Area	I_o	y'	$(y')^2$	Ay'^2	$I_o + Ay'^2$
1	20	6.66	4.858	23.6	472	478.6
2	7.5	140.63	3.642	13.26	99.5	240.1
3	6	0.5	11.642	135.5	813	813.5
Total	33.5					1532.2

Table 1.4

I_o = moment of inertia of each area about its own centroidal axis = $bd^3/12$

y' = distance from centroid of each area to centroidal axis of girder

The moment of inertia I of the girder about its centroidal axis = *1,532.2 in*^{4}.

The properties of common geometric shapes may be found in many standard reference books, including the *AISC Manual of Steel Construction.*

STRESSES

A *stress* in a body is an internal resistance to an external force.

Total stress is the total internal force on a section and is measured in pounds or kips (newtons in metric units).

Unit stress is the stress per unit of area of the section and is measured in pounds per square inch (psi) or kips per square inch (ksi) (kilopascals (kPa) in metric units).

The three kinds of stress that most concern us in the design of buildings are *tension, compression*, and *shear* (see Figures 1.34 and 1.35).

In every case, the unit stress (f) is equal to the load (P) divided by the cross-sectional area (A).

Example #12

The cable shown in Figure 1.36 has a diameter of 3/4 inch and supports a load of 10,000 pounds. What is the unit tensile stress in the cable?

Solution:

The unit stress (f) is equal to the load (P) divided by the cross-sectional area (A).

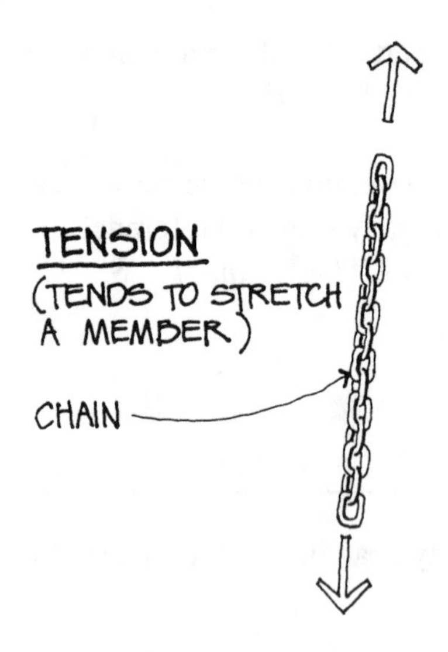

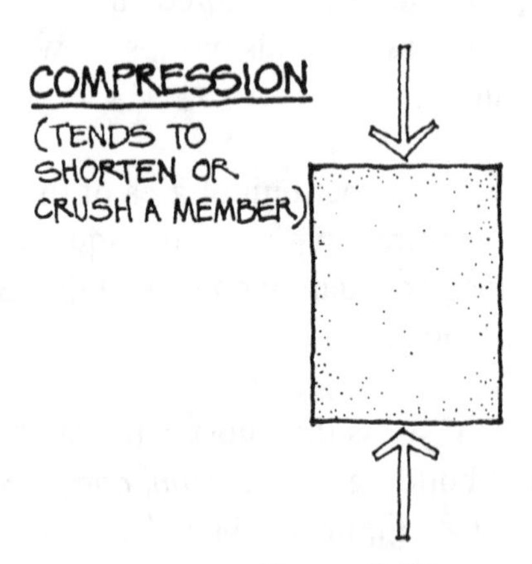

Figure 1.34

A = Area of bar = $\pi/4 \times (\text{diameter})^2$ = $0.7854(0.75)^2 = 0.4417$ square inches

P = 10,000#

$$f = \frac{P}{A} = \frac{10{,}000\#}{0.4417 \text{ in}^2} = \textit{22,640 psi}$$

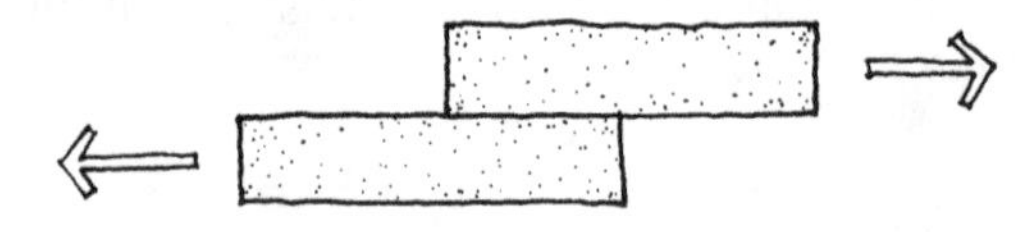

SHEAR
(2 MEMBERS TEND TO SLIDE PAST EACH OTHER.)

Figure 1.35

Example #13

How much compressive load P can the 10-inch-square concrete post shown in Figure 1.37 support, if the maximum allowable unit stress is 1750 psi?

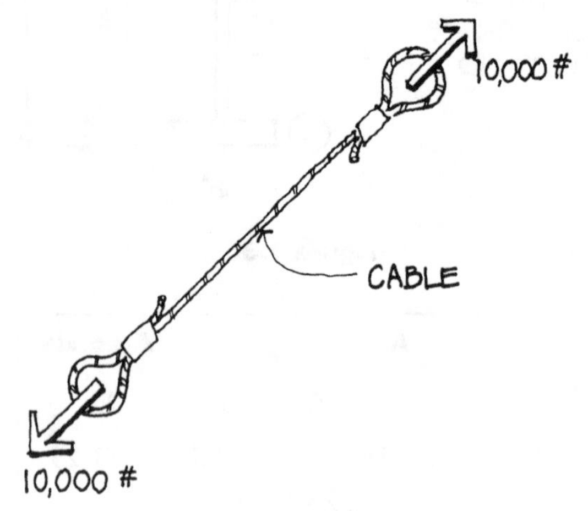

Figure 1.36

Solution:

Since f = P/A, P = fA
P = 1,750 psi × 10" × 10" = *175,000#*

Example #14

Two angles are connected by two 5/8-inch-diameter bolts (see Figure 1.38). The load is 6,000 pounds. What is the unit shear stress in the bolts?

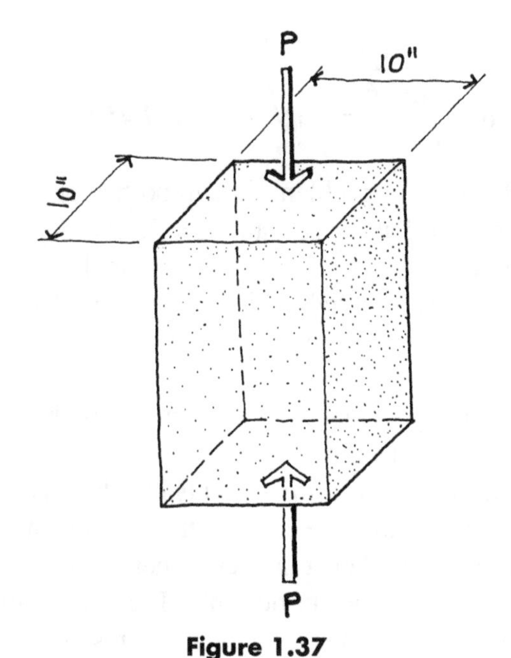

Figure 1.37

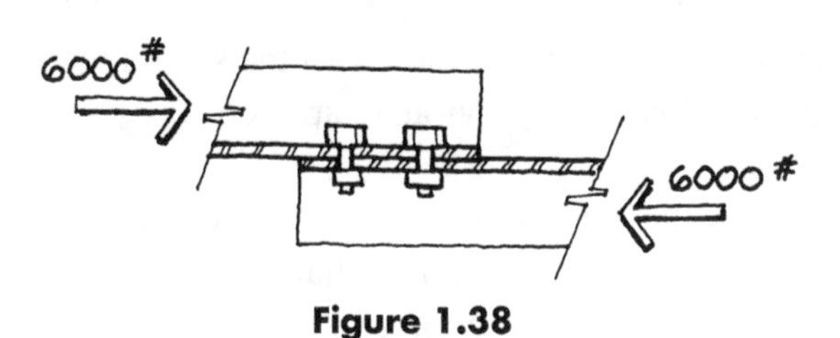

Figure 1.38

Solution:

The area subject to shear stress is the cross-sectional area of the bolts.

$$\text{A per bolt} = \frac{\pi d^2}{4} = 0.7854(0.625)^2 = 0.306 \text{ sq.in.}$$

$$\text{Total A} = 2(0.306) = 0.612 \text{ sq.in.}$$

$$f = \frac{P}{A} = \frac{6{,}000\#}{0.612 \text{ in}^2} = \mathit{9{,}804\ psi}$$

STRAIN

Strain is the deformation, or change in size, of a body caused by external loads. Tensile loads stretch a body, while compressive loads shorten it.

Total strain is the total elongation or shortening of a body, and is represented by Δ.

Unit strain ($\in$) is the total strain (Δ) divided by the original length (L). Thus $\in = \Delta/L$.

Hooke's Law states that up to a certain unit stress, called the *elastic limit*, unit stress (P/A) is directly proportional to unit strain (Δ/L). In other words, for a body of constant cross-section, if a load of 10,000 pounds causes it to stretch x inches, then a load of 20,000 pounds will stretch it 2x inches.

Since unit stress is directly proportional to unit strain, within the elastic limit, the unit stress divided by the unit strain has a constant value. This constant value, for a given material, is called the *modulus of elasticity* or *Young's modulus* and is represented by capital letter E.

$$E = \frac{\text{unit stress}}{\text{unit strain}} = \frac{P/A}{\Delta/L} = \frac{PL}{A\Delta} \text{ or } \Delta = \frac{PL}{AE}$$

Although E is expressed in pounds per square inch (psi), it is not a stress. Rather, it is a measure of the *stiffness* of a material—that is, its resistance to deformation. The greater the E value of a material, the more stress it takes to deform it a given amount.

The construction material with the greatest value of E is steel. Although the other properties of steel may vary, its E value is constant at about 29,000,000 psi. The E value of wood varies, depending on species and grade. For example, Douglas fir, which is very widely used for framing lumber, has an E value varying from 1,300,000 to 1,900,000 psi. The E value of normal weight concrete is approximately 57,000 $\sqrt{f'_c}$ where f'_c is its ultimate strength at 28 days. Depending on strength, this comes out to about 3,000,000 psi to 5,000,000 psi.

The E value for steel is very accurate, while those for wood and concrete are only approximate, since the stress-strain relationships of these materials do not precisely obey Hooke's Law.

Example #15

A 30-inch-long steel bar 1-1/2 inches square is subject to a tensile force of 50,000 pounds. How much does it stretch? Use E = 29,000,000 psi.

Solution:

$$\Delta = PL/AE = \frac{50{,}000 \times 30''}{(1.5\text{ in.})^2 \times 29{,}000{,}000\text{ psi}}$$

$= 0.0230''$

The unit stress in the bar is $P/A = \frac{50{,}000\#}{(1.5\text{ in.})^2} =$ 22,222 psi

Example #16

A 24-inch-long round steel bar resists a tensile force of 10,000 pounds.

(a) What is the required diameter if its unit stress may not exceed 24,000 psi?
(b) What is the required diameter if its elongation is not to exceed 0.005"?

Solution:

(a) $f = 24{,}000$ psi

$P = 10{,}000\#$

$A = \pi d^2/4$

$f = P/A$

$$24{,}000\text{ psi} = \frac{10{,}000\#}{\pi d^2/4}$$

$$d^2 = \frac{10{,}000}{\pi/4(24{,}000)} = 0.5305$$

$$d = \sqrt{0.5305} = 0.72836''$$

(b) $$A = PL/\Delta E = \frac{10{,}000\# \times 24''}{0.005'' \times 29{,}000{,}000\text{ psi}}$$

$= 1.66$ sq.in. $= \pi d^2/4$

$$d = \sqrt{\frac{4(1.66)}{\pi}} = \sqrt{2.114} = 1.454''$$

In this type of problem, it is important to remember that the formula $\Delta = PL/AE$ applies only when the unit stress is less than the elastic limit of the material, which is about 50,000 psi for the most common type of structural steel.

Let us imagine that we are testing a steel bar in a testing machine. Up to the elastic limit, an increase in unit stress results in a directly proportional increase in unit strain as shown in Figure 1.39. What happens when the unit stress exceeds the elastic limit? The unit strain then increases faster than the unit stress—in other words, for each increment of load, the bar stretches more than it did when the same increment of load was previously applied.

As we continue to test the bar, we reach a point where the bar yields, that is, it continues to stretch with no increase in load. The unit stress at which this stretch occurs is called the *yield point,* and its value is slightly greater than the elastic limit. Later we will discuss the significance of the yield point in structural design. Ductile materials, such as steel, have well-defined elastic limits and yield points. On the other hand, non-ductile materials, such as concrete, have no yield point and elastic limits which are not well-defined (see Figure 1.39).

If we continue to increase the load on our steel bar in the testing machine, its unit stress and unit strain increase until it reaches the maximum unit stress which can be developed before it fractures. This stress is called the *ultimate strength.* The ultimate strength of most steel used in buildings is 65,000 to 80,000 psi. Concrete in compression has an ultimate strength of about 3,000 to 6,000 psi, although higher strengths are obtainable. Wood has an ultimate strength of about 2,000 to 8,000 psi.

Because there are uncertainties with respect to the properties of structural materials, the loads actually to be imposed on the structure, and the methods used to calculate stresses, it is necessary that the maximum unit stress permissible in a structural member be designed to be considerably less than the ultimate strength. This maximum permissible unit stress is called the *allowable stress.*

The term *factor of safety* is often used to denote the ratio of the ultimate strength of a material to its working stress. This is a misleading term, since failure of a structural member may be considered to start when it is stressed to the yield point, which is considerably less than the ultimate strength.

Since materials like wood and concrete are less uniform and predictable than steel, their working stresses must be considerably lower than their ultimate strengths, i.e., they have high factors of safety. Ductile materials, such as steel, have the ability to resist unexpected overloads or high localized stresses without failing; while brittle materials, such as masonry or concrete, are more likely to rupture suddenly when overloaded.

THERMAL STRESSES

Materials expand when heated and contract when cooled. The ratio of unit strain to temperature change is called the *coefficient of thermal expansion* and is constant for a given material. For steel, its value is 0.0000065.

Example #17

A steel bar 2.5 square inches in cross-section and 20 feet long has its temperature decreased 50°F.

(a) How much does it contract?

(b) If it were built into unyielding supports at each end, what would be the resulting unit stress in the bar?

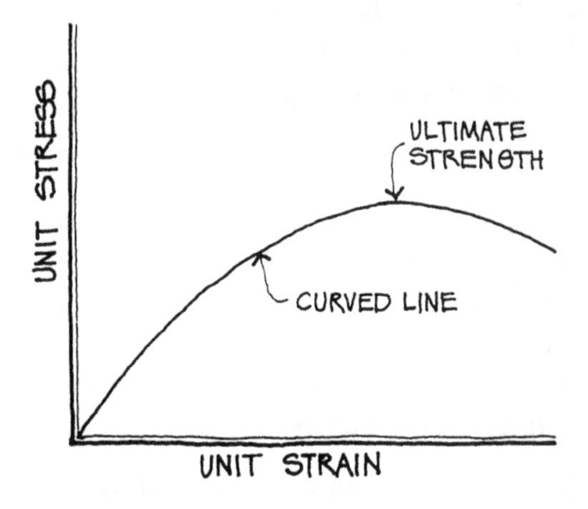

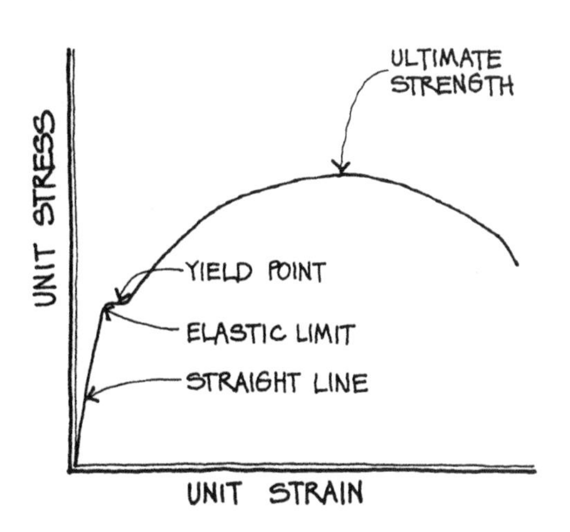

Figure 1.39

Solution:

(a) n = coefficient of thermal expansion = 0.0000065

Δ = change in length

L = original length

t = temperature change (°F)

$$n = \frac{\Delta/L}{t}$$

$$\Delta = nLt$$

$$= (0.0000065)(20 \text{ ft.}) \times (12 \text{ in./ft.})(50°)$$

$$= 0.078''$$

Note that we used the factor 12 to convert the length from feet to inches.

(b) If the bar were built into unyielding supports, it could not contract. You can think of it in this way: the bar contracts 0.078" and then the support applies a force to it to stretch it 0.078", back to its original position. The problem then becomes: what is the unit stress in the bar when it is stretched 0.078".

Since $\Delta = PL/AE$, $P/A = \Delta E/L$

$$= \frac{(0.078'')(29{,}000{,}000 \text{ psi})}{(20 \text{ ft.} \times 12 \text{ in./ft.})} = 9{,}425 \text{ psi}$$

Since the support effectively stretches the bar, the stress is *tension*. Note that the unit stress is independent of the cross-sectional area of the bar.

LESSON 1 QUIZ

1. Forces whose lines of action pass through a common point are called

   A. concurrent forces.

   B. nonconcurrent forces.

   C. components.

   D. vectors.

2. External loads acting on a body cause it to elongate or shorten. This change in size is known as ____________________.

3. A material's resistance to deformation is determined by its

   A. factor of safety.

   B. moment of inertia.

   C. modulus of elasticity.

   D. yield point.

4. 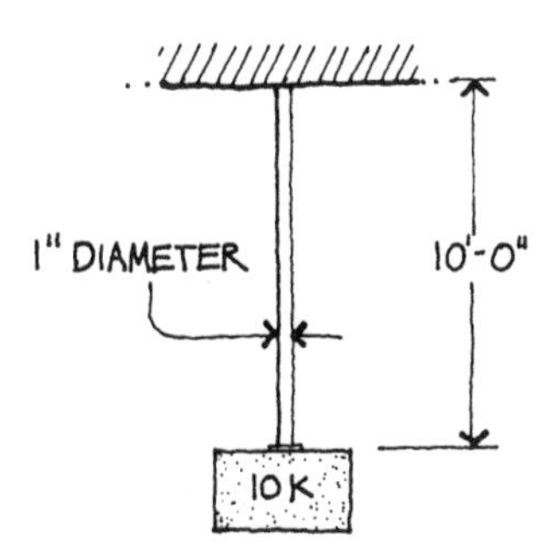

   A load of 10 kips is supported by a one-inch-diameter rod as shown above. If the modulus of elasticity of steel is taken as 29,000,000 psi, how much does the rod stretch?

   A. 0.013 inch   C. 0.053 inch

   B. 0.004 inch   D. 0.044 inch

5. After cutting a free body diagram of a structure, if the calculated internal forces are negative, then the

   A. structure is not in equilibrium.

   B. internal forces act in the opposite direction from that assumed.

   C. material has reached its yield point.

   D. loads should act in the opposite direction.

6. If a load acts through a body's center of gravity, then the body

   I. has no tendency to rotate.

   II. tends to translate in the direction of the applied force.

   III. develops no stress.

   IV. tends to spin about its axis of rotation.

   A. I, II, III, and IV   C. II and III

   B. IV only   D. I and II

7. A 10-inch-square concrete post supports a compressive load of 20,000 pounds. What is the unit stress in the post?

   A. 20,000 psi

   B. 2,000 psi

   C. 200 psi

   D. Depends on the strength of the concrete.

8. The yield point is the

A. maximum unit stress that can be developed in the material.

B. maximum permissible unit stress used to design a member.

C. unit stress beyond which unit strain increases faster than unit stress.

D. unit stress at which the material continues to deform with no increase in load.

9. Hooke's Law states that

A. the ratio of unit stress to unit strain is a measure of the stiffness of the material.

B. up to the elastic limit, unit stress is in direct proportion to unit strain.

C. the total elongation or shortening of a body is equal to PL/AE.

D. the strain, or change in size of a body, is caused by external forces applied to that body.

10. The moment of inertia of the section shown below about its centroidal axis (x-x) is

A. 512 inches^4.

B. 13,824 inches^4.

C. 192 inches^4.

D. 1,152 inches^4.

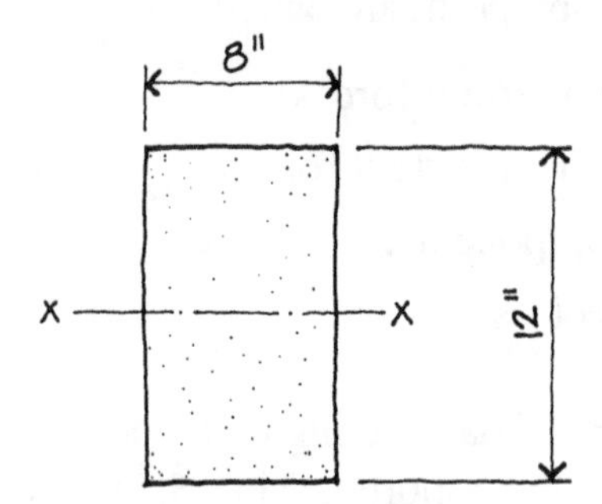

LESSON TWO

BEAMS AND COLUMNS

TYPES OF BEAMS AND LOADS

Beams comprise one of the most basic structural components used in buildings. It is therefore vital that candidates preparing for the Architect Registration Examination have a clear understanding of how beams work.

A beam is a member that supports loads perpendicular to its longitudinal axis. There are several types of beams, as described here and on the following pages.

1. A *simple beam,* which is depicted in Figure 2.1, is one that rests on a support at each end, and whose ends are free to rotate.

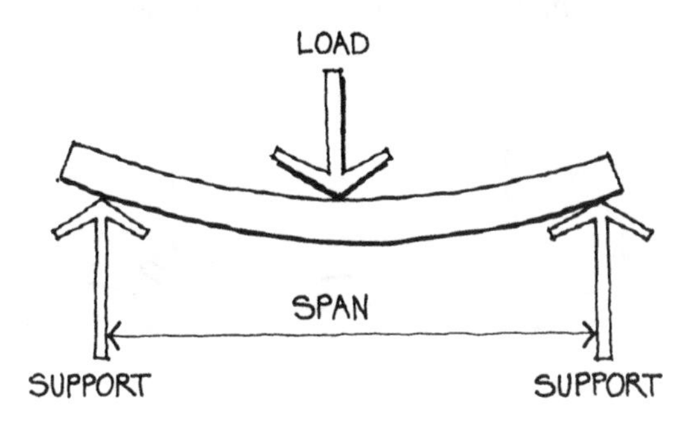

Figure 2.1

2. A *cantilever beam* is one that is supported at one end only, and which is restrained against rotation at that end (see Figure 2.2).

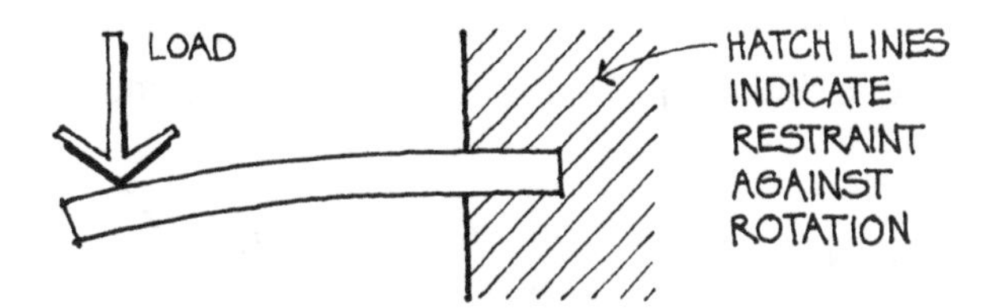

Figure 2.2

3. An *overhanging beam* is one that rests on two or more supports and has one or both ends projecting beyond the support (see Figure 2.3).

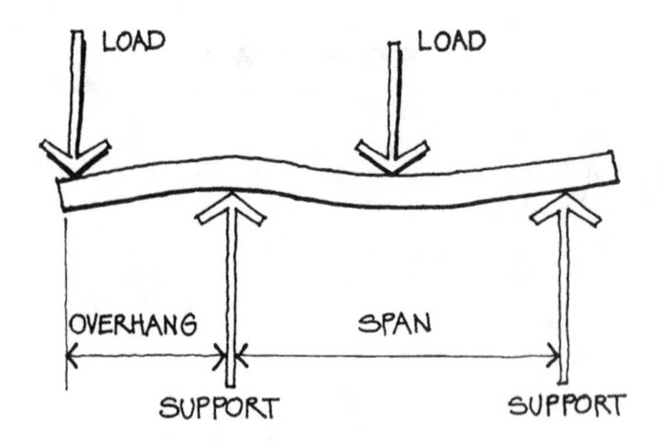

Figure 2.3

4. A *continuous beam* is one that rests on more than two supports (Figure 2.4).

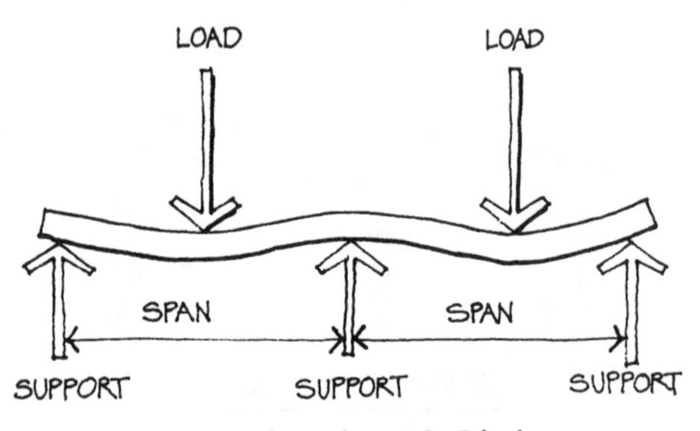

Figure 2.4

5. A *fixed end beam* is one that is restrained (fixed) against rotation at its ends as shown in Figure 2.5. Some beams are fixed at one end and simply supported at the other end.

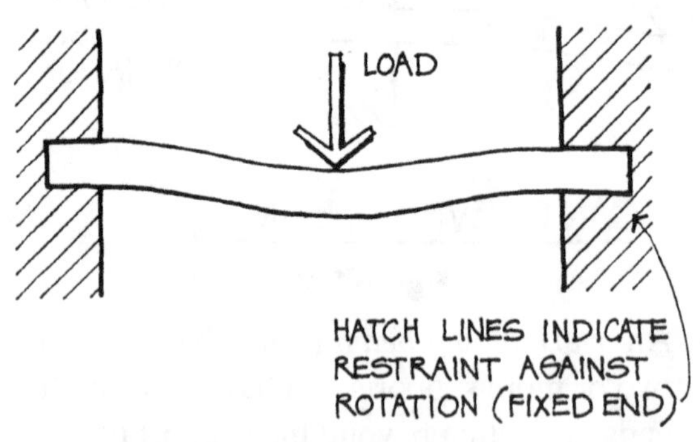

Figure 2.5

The two kinds of loads that commonly act on beams are *concentrated* and *distributed*. A concentrated load acts at one point on the beam, while a distributed load acts over a length of the beam. If the load per unit of length of the beam is constant, it is called a *uniformly distributed load*, or simply a *uniform load*.

A concentrated load is usually indicated as an arrow and called P (see Figure 2.6).

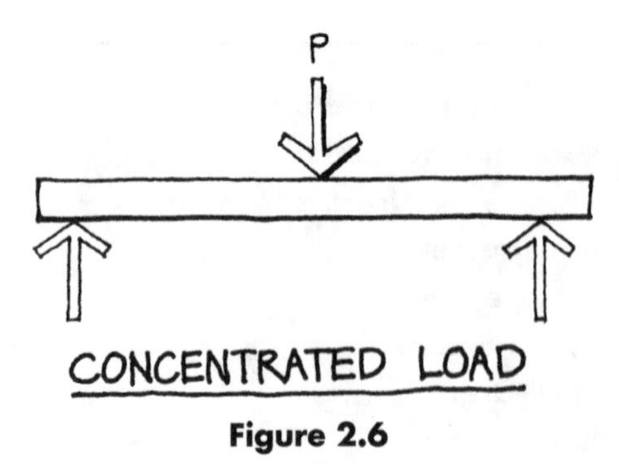

Figure 2.6

A uniformly distributed load is usually indicated as shown in Figure 2.7 and called W (total load) or w (load per foot).

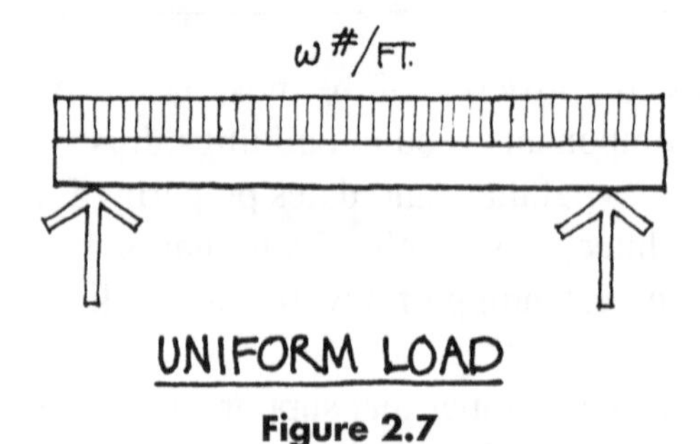

Figure 2.7

Beams for which the reactions can be found from the equations of equilibrium ($\Sigma H = 0$, $\Sigma V = 0$, $\Sigma M = 0$) are called *statically determinate beams*. Simple and cantilever beams and overhanging beams that rest on two supports are statically determinate.

Beams whose reactions cannot be found from the equations of equilibrium only, but require additional equations, are called *statically indeterminate beams.* Continuous and fixed end beams are statically indeterminate.

This lesson deals mainly with determining the reactions and stresses in statically determinate beams.

Example #1

Determine the reactions R_1 and R_2 for the beam shown in Figure 2.8, neglecting the weight of the beam.

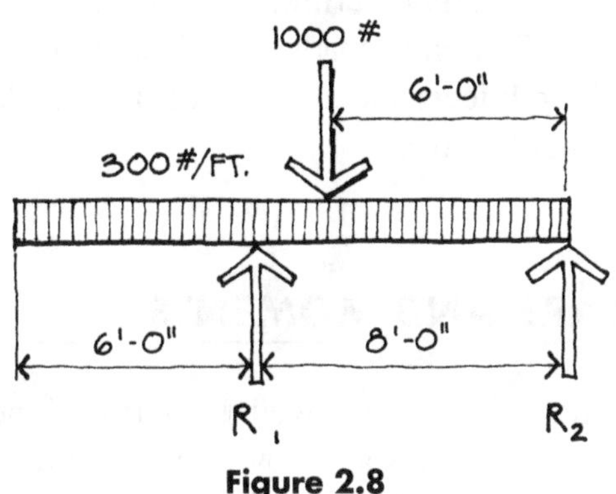

Figure 2.8

Solution:

This is an overhanging beam with both concentrated and uniformly distributed loads. The reactions are determined from the equations of equilibrium.

Take moments about R_2.

$\Sigma M = 0$

$$(-300\#/\text{ft.}) \times (8 + 6)\text{ft.} \times (8 + 6)\text{ft.}/2$$
$$-(1{,}000\# \times 6\text{ft.}) + (R_1 \times 8\text{ft.}) = 0$$
$$-29{,}400 - 6{,}000 + 8R_1 = 0$$
$$8R_1 = 35{,}400 \text{ ft.}$$

$R_1 = \mathit{4{,}425\#}$

$\Sigma V = 0$

$$R_1 + R_2 - 1{,}000\# - (300 \times 14)\# = 0$$
$$4{,}425\# + R_2 - 5{,}200\# = 0$$

$R_2 = \mathit{775\#}$

Keep the following in mind when solving for beam reactions:

1. Be consistent with the signs of moments and forces. In the previous example, we called clockwise moments positive and counterclockwise moments negative, upward forces positive and downward forces negative.
2. The calculated reactions must be the same regardless of which point is used as the center of moments. In this example, we took moments about R_2. As an exercise, take moments about R_1 and see if your calculated reactions come out the same.
3. In solving for reactions, consider that a uniformly distributed load is equivalent to a concentrated load of the same magnitude acting at the midpoint of the distributed load. In the example above, the uniformly distributed load (300#/ft. × 14ft.) acts at its midpoint (14ft. ÷ 2 = 7 feet left of R_2).

Example #2

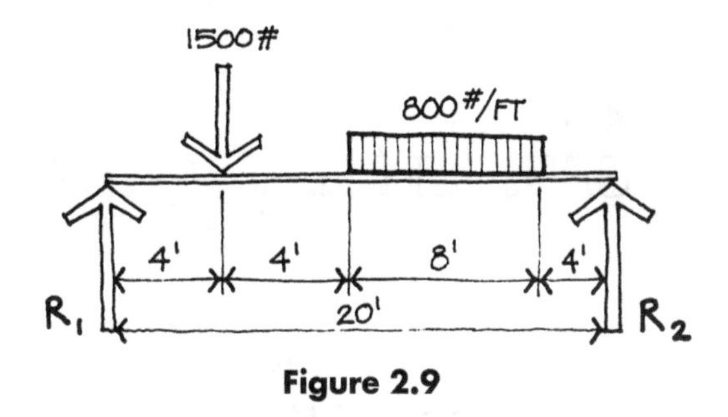

Figure 2.9

A beam of negligible weight supports the loads shown in Figure 2.9. Solve for reactions R_1 and R_2.

Solution:

This simple beam supports a concentrated load and a uniformly distributed load acting over a portion of the beam. Consider the uniformly distributed load acting at its midpoint and take moments about R_2.

$\Sigma M = 0$
$-1{,}500\#(4 + 8 + 4)\text{ft.} -800\#/\text{ft.}(8\text{ft.})$
$(8\text{ft.}) + R_1(20\text{ft.}) = 0$
$20R_1 = 24{,}000 + 51{,}200 = 75{,}200$

$$R_1 = \frac{75{,}200}{20} = \mathit{3{,}760\#}$$

$\Sigma V = 0$
$-1{,}500\# - 800\#/\text{ft.}(8\text{ft.}) + R_1 + R_2 = 0$
$R_1 + R_2 = 1{,}500 + 6{,}400 = 7{,}900$
$R_2 = 7{,}900 - R_1 = 7{,}900 - 3{,}760 = \mathit{4{,}140\#}$

As a check, we can take moments about R_1

$\Sigma M = 0$
$1{,}500(4) + 800(8)\ (4 + 4 + 8/2) - R_2(20) = 0$
$20R_2 = 6{,}000 + 76{,}800 = 82{,}800$

$$R_2 = \frac{82{,}800}{20} = \mathit{4{,}140\#},$$

which verifies our previous result.

Example #3

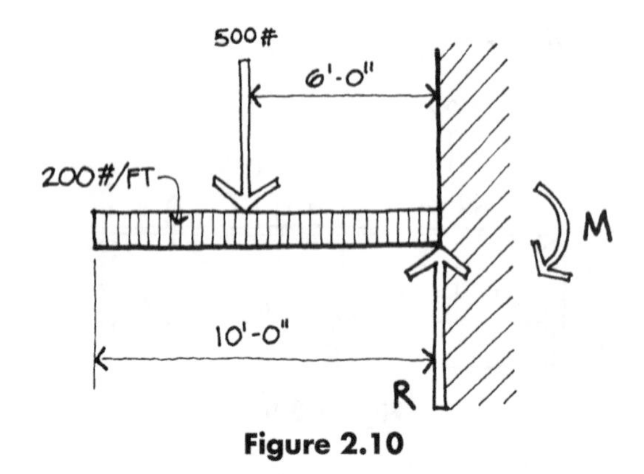

Figure 2.10

Determine the reaction R and the moment M at the support of the cantilever beam shown in Figure 2.10, neglecting the weight of the beam.

Solution:

$\Sigma V = 0$
$-200\#/\text{ft.}(10\text{ft.}) - 500\# + R = 0$
$R = 2{,}000 + 500 = \mathit{2{,}500\#}$

Take moments about the right end of the beam.

$\Sigma M = 0$
$-500\#(6\text{ft.}) - 200\#/\text{ft.}$
$(10\text{ft.})(5\text{ft.}) + M = 0$
$M = 3{,}000 + 10{,}000 = \mathit{13{,}000'\#}$

Note that the reaction at the support is equal to the sum of all the loads on the beam. This is always true of cantilever beams.

You should be able to determine the reactions for any statically determinate beam resisting any combination of loads. This is a basic skill that all architects and all architectural candidates should have.

SHEARS AND MOMENTS

Now that we have reviewed the various kinds of beams and loads and how beam reactions are computed, we will investigate the nature of the internal forces (stresses) in the beam. We will begin by defining a few terms.

Vertical shear (V) at any section of a beam is the algebraic sum of the forces that are on one side of the section. We may use the forces on either side of the section, but for convenience we usually use the forces to the *left* of the section. Remember that the forces include the applied loads and the reactions, and that upward forces are considered positive and downward forces negative. Vertical shear is expressed in the same units as the forces, usually pounds or kips.

The *shear diagram* is a graphic representation of the value of the vertical shear at any point along the beam.

Bending moment (M) at any section of a beam is the algebraic sum of the moments about the section of the forces on one side of the section. For convenience, we usually use the moments of the forces to the *left* of the section. Remember to include the moments of the applied loads and the reactions, and that clockwise moments are considered positive and counterclockwise moments negative. Since bending moment is obtained by multiplying forces by distances, its units are foot-pounds or foot-kips.

Both vertical shear and bending moment are calculated by using free body diagrams, as described in Lesson One. The *moment diagram* is a graphic representation of the value of the bending moment at any point along the beam.

Example #4

Compute the reactions and draw the shear diagram for the beam shown in Figure 2.11, neglecting the weight of the beam.

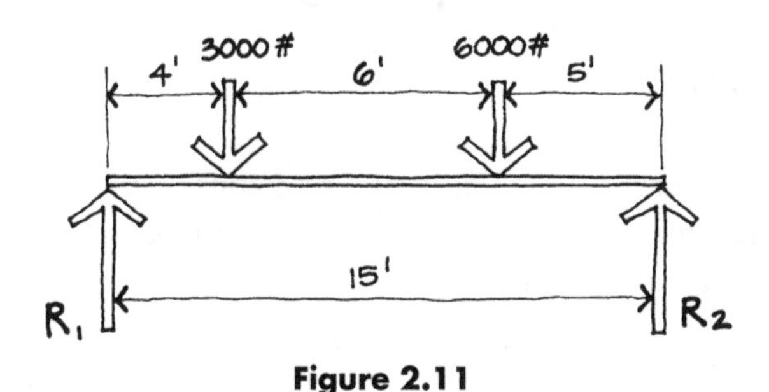

Figure 2.11

Solution:

First we compute the reactions as previously described. Take moments about R_2.

$\Sigma M = 0$
$-6{,}000\#(5\text{ft.}) - 3{,}000\#(6 + 5)\text{ft.} +$
$R_1(15\text{ft.}) = 0$
$15R_1 = 30{,}000 + 33{,}000 = 63{,}000$

$$R_1 = \frac{63{,}000}{15} = 4{,}200\#$$

$\Sigma V = 0$
$R_1 + R_2 - 3{,}000 - 6{,}000 = 0$
$R_2 = 9{,}000 - R_1 = 9{,}000 - 4{,}200 = 4{,}800\#$

Now we will calculate the value of V at various points along the beam. What is the value of V just to the right of the left support (see Figure 2.12)?

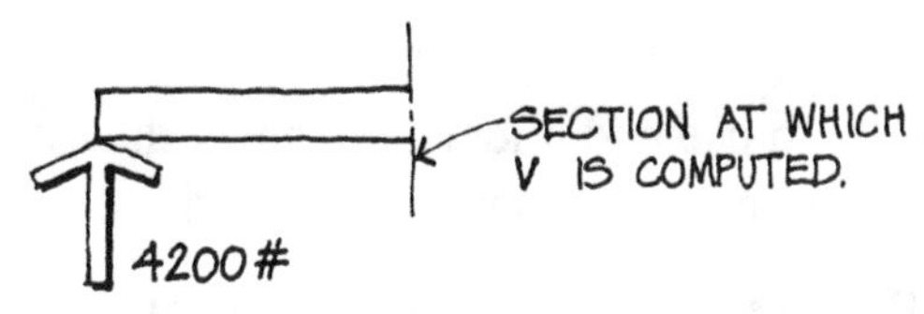

Figure 2.12

V is the algebraic sum of the forces left of the section. Since the only force left of the section is the reaction of 4,200# acting up, V is equal to *+4,200#*.

The value of V is +4,200# up to the 3,000# concentrated load, since there are no intervening loads.

What is the value of V just to the right of the 3,000# load (see Figure 2.13)?

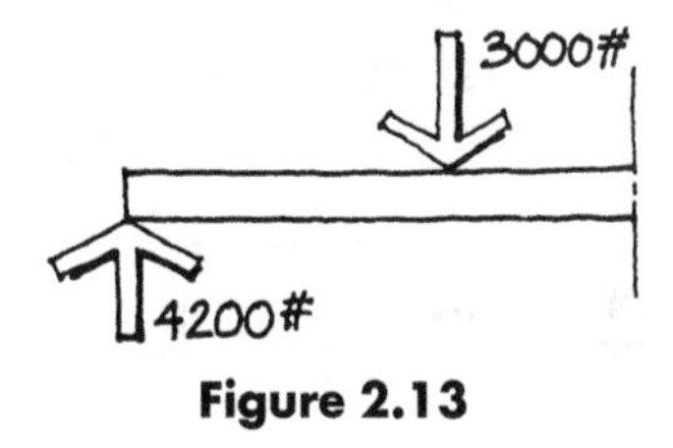

Figure 2.13

V is the algebraic sum of the forces left of the section. V = 4,200 – 3,000 = *+1,200#*

The value of V remains +1,200# up to the 6,000# concentrated load, since there are no intervening loads.

What is the value of V just to the right of the 6,000# load (see Figure 2.14)?

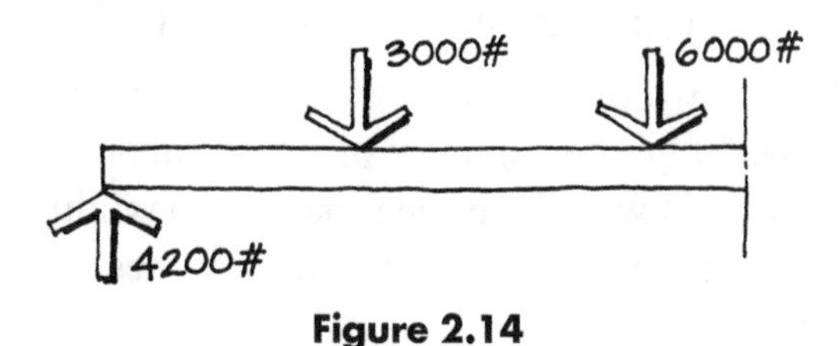

Figure 2.14

$V = 4{,}200 - 3{,}000 - 6{,}000 = -4{,}800\#$ and remains at this value up to the right reaction.

We can now draw the shear diagram (see Figure 2.15).

In order to use and understand shear diagrams, the following points should be noted:

1. Positive values of shear are shown above the base line and negative values below the base line.

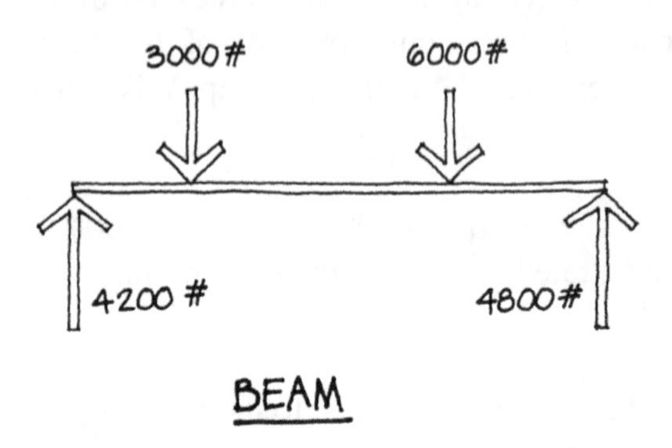

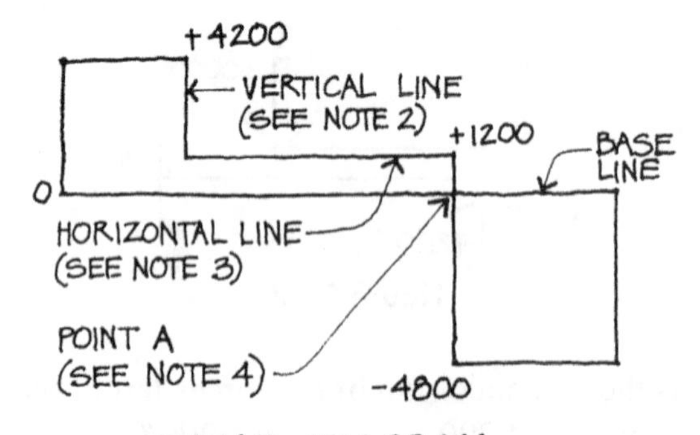

Figure 2.15

2. The value of shear changes abruptly at concentrated loads and reactions and is indicated by a vertical line.
3. Where the value of shear is constant, the shear diagram is horizontal. In Example #4, this occurs between concentrated loads.
4. The two points on the shear diagram that are of greatest interest in structural design are: (1) where the shear has the maximum value, and (2) where the shear passes through zero; that is, where it changes from positive to negative or vice versa. In Example #4, the maximum value of shear is 4,800#, and the shear passes through zero at point A.

Example #5

A beam of negligible weight supports a 3,000 lb. concentrated load and a uniform load of 400 lbs. per foot as shown in Figure 2.16.

Draw the shear diagram.

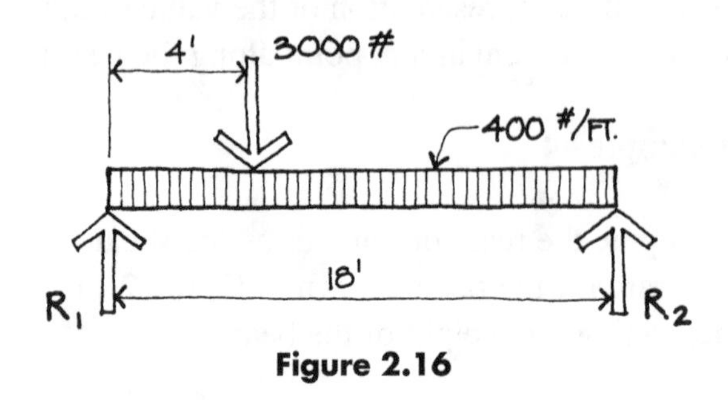

Figure 2.16

Solution:

We start by calculating the reactions. Call the left reaction R_1 and the right reaction R_2 and take moments about R_2.

$\Sigma M = 0$
$-3{,}000\#(14\text{ft.}) - 400\#/\text{ft.}(18\text{ft.})$
$(9\text{ft.}) + R_1\,(18\text{ft.}) = 0$
$18R_1 = 42{,}000 + 64{,}800 = 106{,}800$

$$R_1 = \frac{106{,}800}{18} = 5{,}933.3\#$$

$\Sigma V = 0$
$R_1 + R_2 - 3{,}000 - 400(18) = 0$
$R_1 + R_2 = 3{,}000 + 7{,}200 = 10{,}200$
$R_2 = 10{,}200 - R_1 = 10{,}200 - 5{,}933.3 =$
$4{,}266.7\#$

Now we will calculate the value of shear (V) at various points along the beam.

V just to the right of the left reaction is *+5,933.3#*

V just left of the 3,000# concentrated load = 5,933.3 – 4(400) = *4,333.3#*

V just right of the 3,000# concentrated load = 5,933.3 – 4(400) – 3,000 = *+1,333.3#*

V just to the left of the right reaction = *–4,266.7#*

In addition to the previous comments, note that the value of shear changes at a uniform rate at uniformly distributed loads, and therefore the corresponding shear diagram is a line of constant slope (see Figure 2.17).

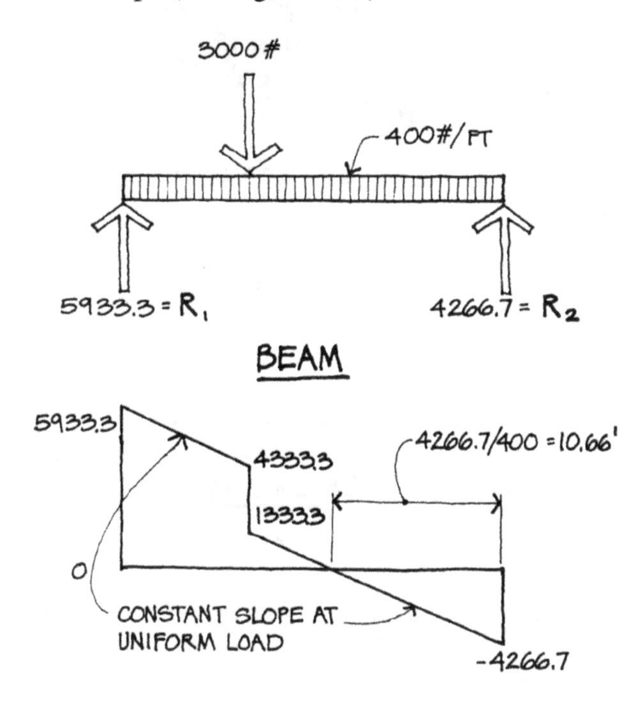

Figure 2.17

Example #6

Draw the moment diagram for the beam of Example #4.

Solution:

The value of the moment (M) at the supports of a simple beam is zero.

Next we calculate the value of M just to the right of the 3,000# load (see Figure 2.18).

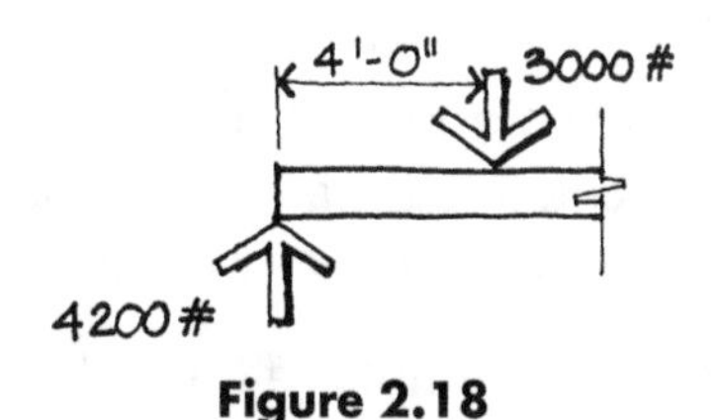

Figure 2.18

M is the algebraic sum of the moments left of the section.

M = 4,200#(4ft.) – 3,000#(0)
= 4,200(4) = *16,800'#*

What is the value of M just to the right of the 6,000# load (see Figure 2.19)?

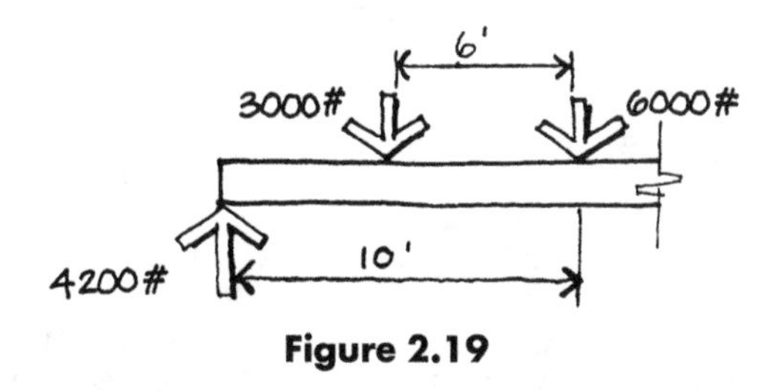

Figure 2.19

M = 4,200#(10ft.) – 3,000#(6ft.) – 6,000#(0)
= 42,000 – 18,000 – 0 = *24,000'#*

We can now draw the moment diagram, which is depicted in Figure 2.20.

Please note the following:

1. Positive values of moment are shown above the base line and negative values, if any, below the base line.
2. The slope of the moment diagram is equal to the value of the shear diagram at the same point on the beam. Thus, where the

shear is constant, the shear diagram is horizontal and the moment diagram is a line of constant slope. At concentrated loads, the shear changes abruptly, and therefore the slope of the moment diagram changes abruptly.

3. *The bending moment is maximum where the shear passes through zero.*
4. The change of moment between any two points is equal to the area of the shear diagram between those two points.

In the beam of Examples #4 and #6, the change of moment from the left end of the beam to the 3,000# concentrated load is equal to the area of the shear diagram between the left end and the 3,000# concentrated load, which is 4,200 × 4 = 16,800'#, as previously determined.

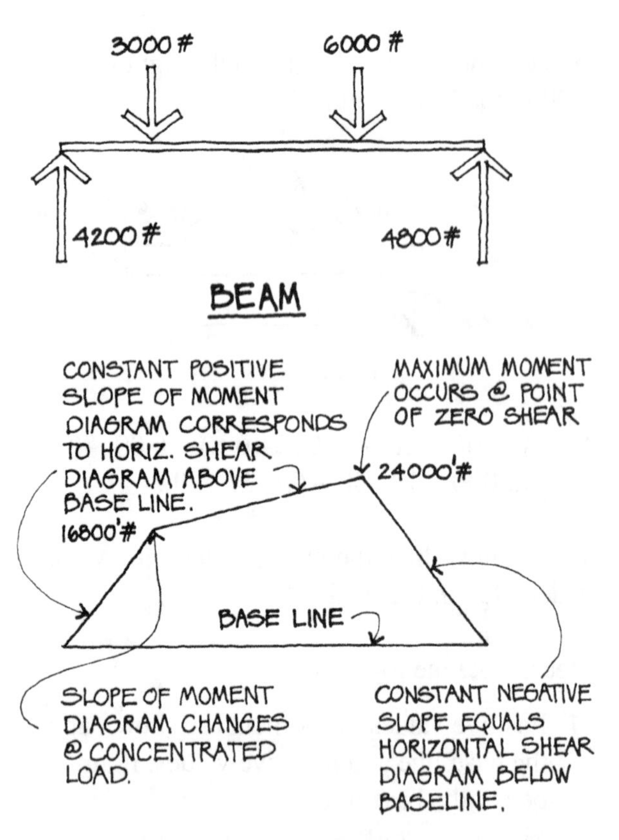

Figure 2.20

Example #7

Draw the moment diagram for the beam of Example #5.

Solution:

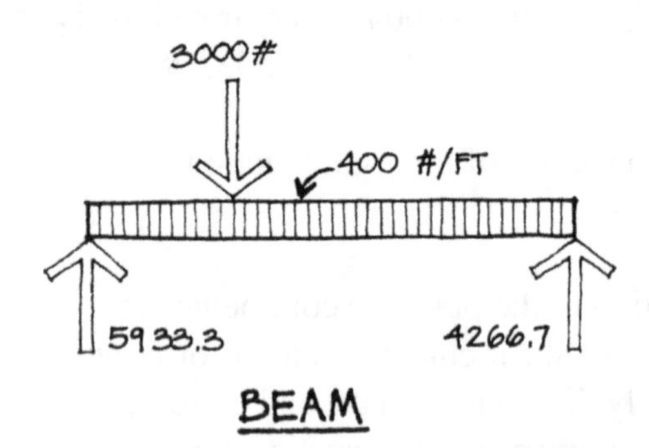

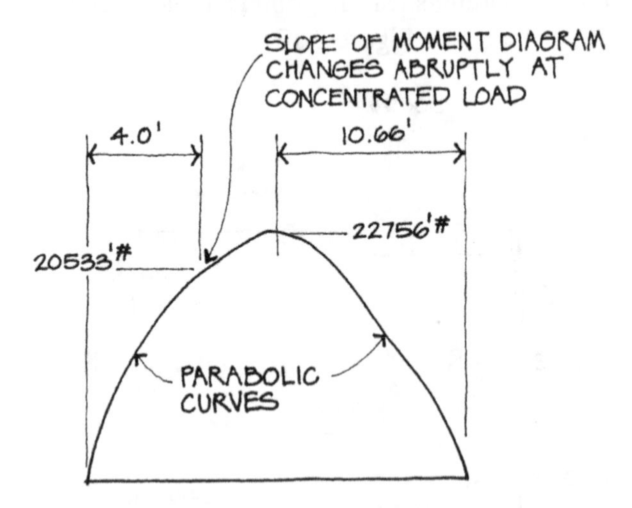

Figure 2.21

The value of M is zero at the supports, as shown in Figure 2.21.

The value of M at the 3,000# concentrated load is 5,933.3#(4ft.) – 400#/ft.(4ft.)(2ft.) = 23,733 – 3,200 = *20,533'#*

The maximum moment occurs at the point of zero shear. To locate this point, we divide R_2 by the value of the uniform load = 4,266.7/400 = 10.66 feet from the right support.

The maximum moment M
= 4,266.7#(10.66ft.) – 400#/ft.(10.66ft.)(5.33ft.)
= *22,756'#*

In addition to the previous comments, note that the shape of the moment diagram at a uniformly distributed load is a parabolic curve.

In a similar manner, we can draw the shear and moment diagrams for the beam of Example #1 on page 25. These diagrams appear in Figure 2.22.

In this case, the shear passes through zero at two points: at the left support and at a point 2.58 feet left of the right support. The moment at each point of zero shear must be computed in order to determine which has the greater value.

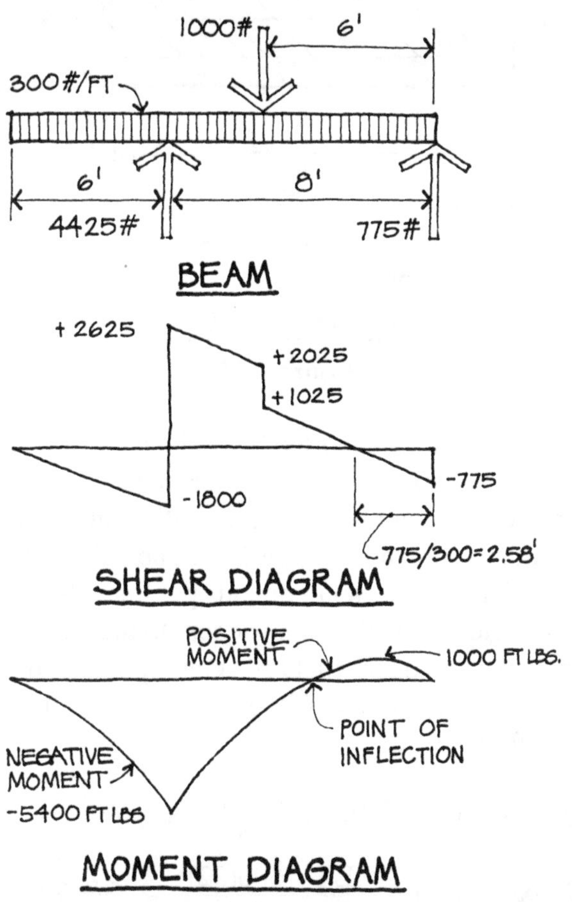

Figure 2.22

This beam has both negative and positive moments (terms that are explained on page 34). The point where the moment changes from positive to negative is called the *point of inflection.*

And similarly, we can draw the shear and moment diagrams for the beam of Example #3 on page 26. These diagrams appear in Figure 2.23.

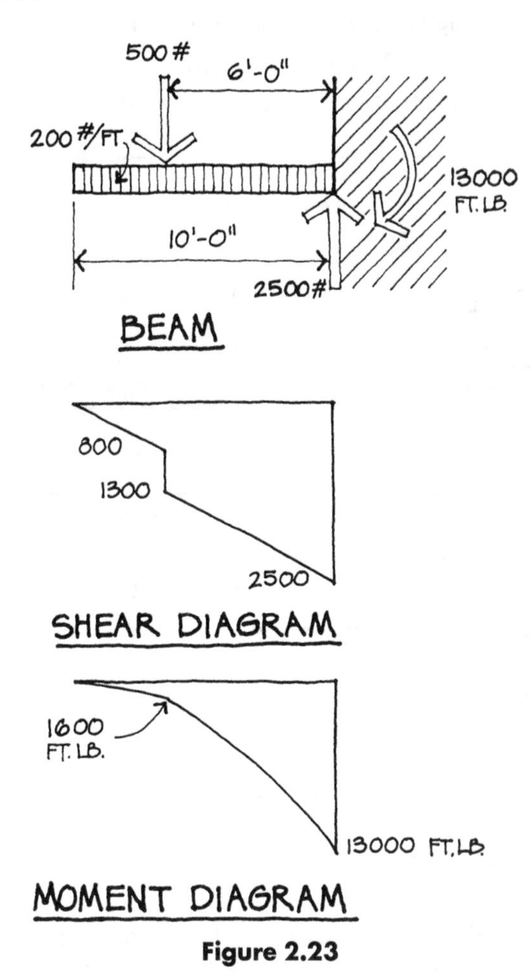

Figure 2.23

The methods of computing shears and moments and drawing shear and moment diagrams described in this lesson may be used for all types of beams and all loading conditions.

A number of beam and load types and their corresponding shear and moment diagrams are shown in the *AISC Manual of Steel Construction.* We suggest that you become familiar with this material in case you need to refer to it.

Two conditions occur so frequently in building structures that their formulas and shear and moment diagrams are worth remembering. These conditions are: (1) a simple beam with a uniformly distributed load over the entire span, and (2) a simple beam with a concentrated load at midspan.

The beam diagrams and formulas for these two conditions are shown in Figure 2.24, reprinted from the AISC Manual by permission of the American Institute of Steel Construction.

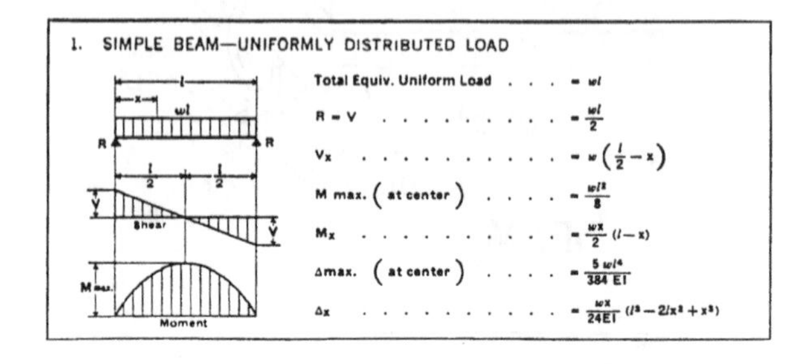

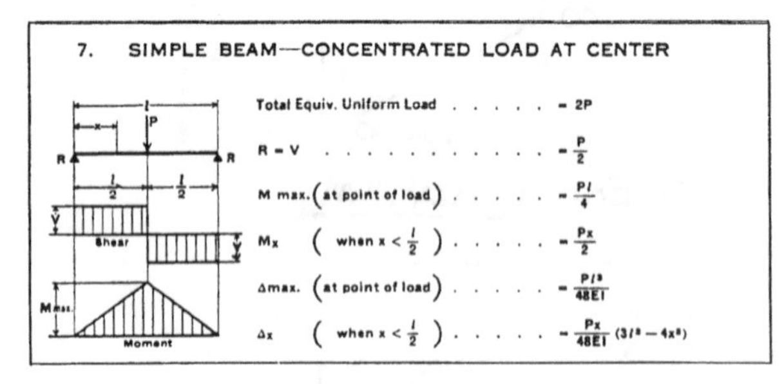

Figure 2.24

Example #8

A simple beam 20 feet long has a concentrated load of 5,000 lbs. at midspan. What is the maximum shear (V) and the maximum bending moment (M)?

Solution:

From the second diagram in Figure 2.24,

$$\text{Maximum V} = \frac{P}{2} = \frac{5{,}000}{2} = \mathit{2{,}500\#}$$

$$\text{Maximum M} = \frac{PL}{4} = \frac{5{,}000(20)}{4} = \mathit{25{,}000'\#}$$

Example #9

The framing system for a floor consists of beams spaced at six feet on center which span 32 feet between girders (see Figure 2.25). The beams support a uniform load of 100 pounds per square foot. What is the maximum shear (V) and the maximum bending moment (M) for each floor beam?

Solution:

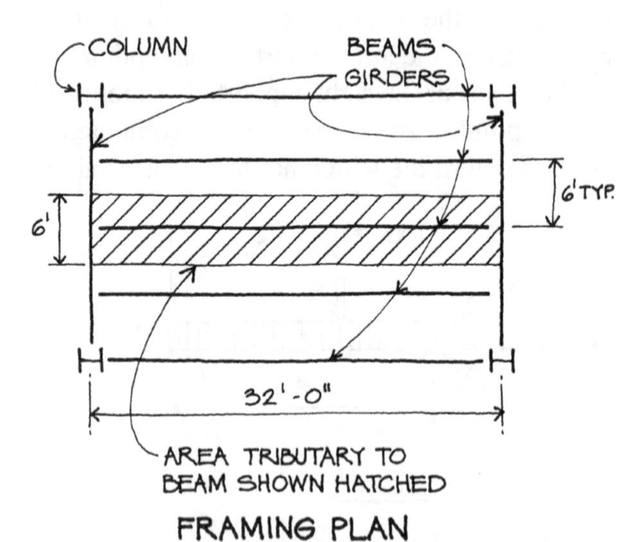

Figure 2.25

The floor area supported by each beam, known as the beam's *tributary area*, extends halfway to each adjacent beam and the full length of the beam, as shown in Figure 2.25. The uniformly distributed load (w) in pounds per lineal foot supported by each beam is equal to the load in pounds per square foot (100) multiplied by the beam spacing in feet (6), or 100 × 6 = 600 lbs. per foot.

$$\text{Maximum V} = \frac{wL}{2} = \frac{600(32)}{2} = \mathit{9{,}600\#}$$

$$\text{Maximum M} = \frac{wL^2}{8} = \frac{600(32)^2}{8} = \mathit{76{,}800'\#}$$

The uniformly distributed load is sometimes expressed as the total load (W), instead of the load per foot (w). Since W = wL, the maximum bending moment can be written WL/8 instead of $wL^2/8$. The two formulas are equivalent.

In Example #9, if the uniformly distributed load had been given as 19,200# total (100#/sq.ft. × 6 ft. o.c. × 32 ft.) instead of 100# per sq. ft. or 600# per ft., the maximum moment would be calculated as:

$$\frac{WL}{8} = \frac{19{,}200(32)}{8} = 76{,}800'\#,$$

the same result as before.

One way that the examiners can test your familiarity with shear and moment diagrams is to see if you can determine the shape of the moment diagram from a given shear diagram, or vice versa.

Example #10

The shear diagram for a beam is shown in Figure 2.26. Draw the moment diagram and calculate the change in moment between points 1 and 2.

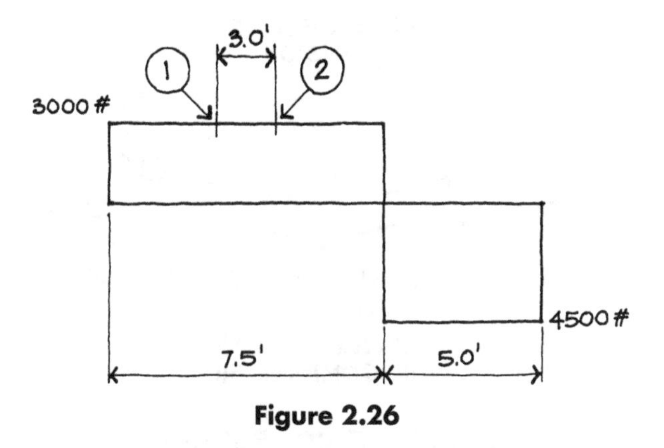

Figure 2.26

Solution:

From the left support to 7.5 feet past the left support, the shear is constant and positive and therefore the slope of the moment diagram is constant and positive.

At 7.5 feet past the left support, the value of the shear changes abruptly and passes through zero. Therefore, at this point the moment is maximum, and the slope of the moment diagram changes abruptly. From 7.5 feet past the left support to the right support, the shear diagram is constant and negative, and therefore the slope of the moment diagram is constant and negative.

The maximum moment is equal to the area of the shear diagram between the left support and 7.5 feet to the right of the left support.

M = 3,000# × 7.5 ft. = *22,500'#*

We can now draw the moment diagram, which is shown in Figure 2.27.

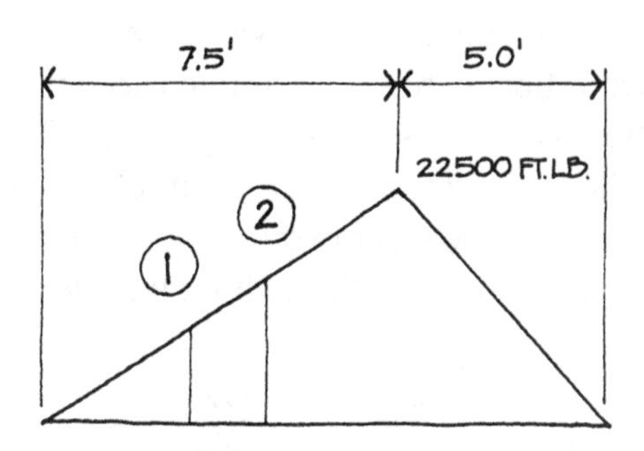

MOMENT DIAGRAM

Figure 2.27

The change in moment between points 1 and 2 is equal to the area of the shear diagram between the two points = 3,000# × 3 ft. = *9,000'#.*

Positive and Negative Moments

In designing beams for buildings, as well as for the exam, it is helpful to visualize the shape of the beam under load. See pages 23 and 24 for descriptions of the different types of beams.

1. Simple beams are bent concave upward, as shown in Figure 2.28; the upper beam fibers are stressed in compression, while the lower fibers are in tension. This is called *positive moment*. Simple beams supporting downward vertical loads have positive moment across the entire span.

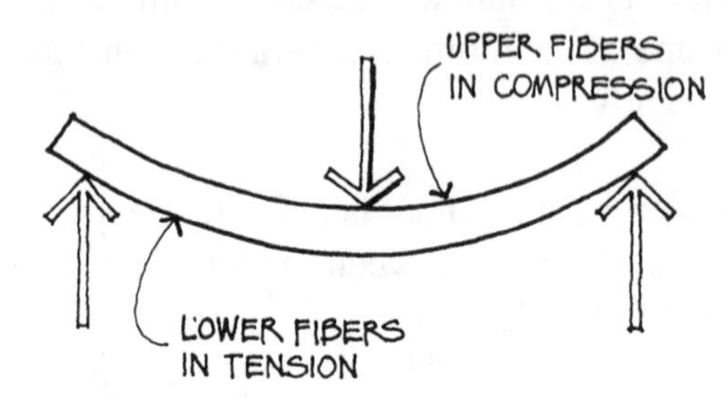

Figure 2.28

2. Cantilever beams, as shown in Figure 2.29, are bent concave downward; the upper beam fibers are stressed in tension, while the lower fibers are in compression. This is called *negative moment*. Cantilever beams resisting downward vertical loads have negative moment throughout their length.

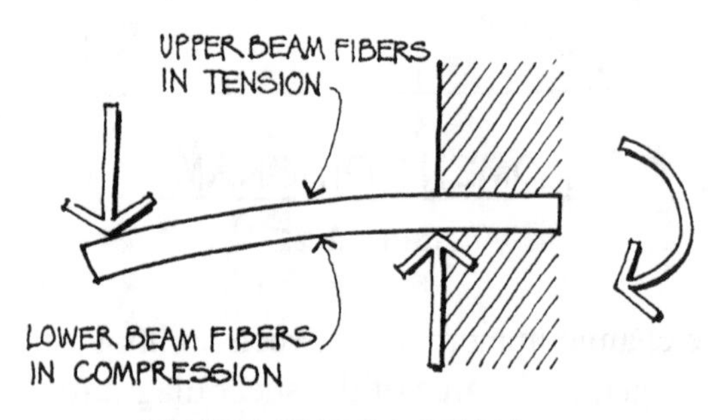

Figure 2.29

3. Overhanging beams may have both positive and negative moment, such as the beam of Example #1 on page 25.
4. Continuous beams supporting downward vertical loads have negative moment over the interior supports, positive moment between supports, and no moment at the exterior supports (see Figure 2.30).

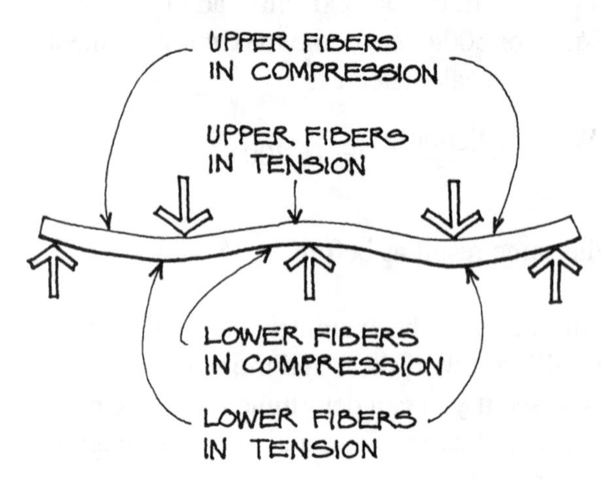

Figure 2.30

5. Fixed end beams supporting downward vertical loads have negative moment at their ends and positive moment between the ends (see Figure 2.31).

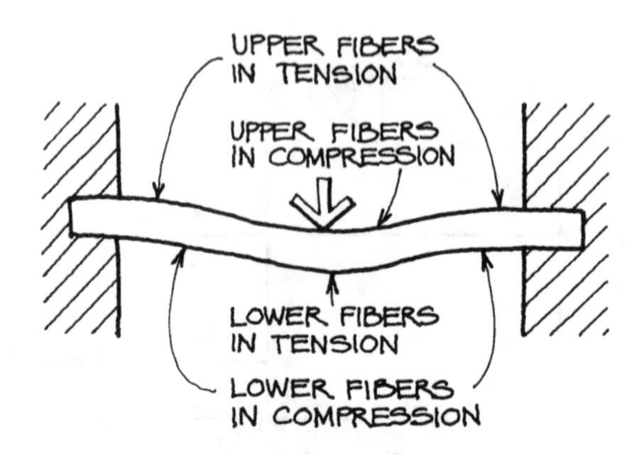

Figure 2.31

In general, a simple beam will have greater maximum moment and more deflection than either a continuous beam or a fixed end beam supporting the same load on the same span.

BEAM STRESSES

We have discussed shears and moments in beams, and we will now consider the internal forces (stresses) that result from those shears and moments.

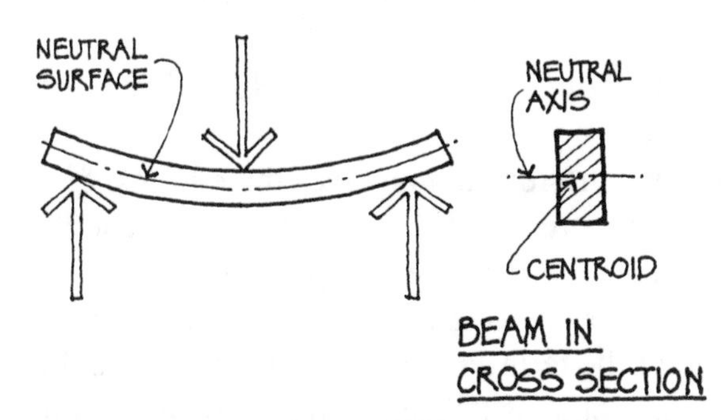

Figure 2.32

Let's say that a simple beam supports a concentrated load, as shown in Figure 2.32. The beam is bent concave upward, which we call positive moment. The upper beam fibers shorten and are thus stressed in compression, while the lower beam fibers elongate and are stressed in tension. The plane that has neither compressive nor tensile stress and whose length remains the same is called the neutral surface. The intersection of the neutral surface with the beam cross-section is a line called the *neutral axis* (N.A.). For a homogeneous beam (one made of only one material), the neutral axis passes through the centroid of the beam cross-section.

Flexure Formula

The flexure formula is used in one form or another in the design of all homogeneous beams.

The formula is $f = \frac{M\bar{y}}{I}$

where

- f = the flexural (bending) unit stress in tension or compression, in psi
- M = bending moment, in inch-lbs.
- $\bar{y}$ = the distance from the neutral axis to the fiber under consideration, in inches
- I = the moment of inertia of the beam cross-section about the neutral axis, in inches^4

The further the beam fiber is from the neutral axis, the higher the unit stress. The maximum unit stress therefore occurs at the outermost fiber, whose distance from the neutral axis is called c.

Therefore, the maximum unit bending stress formual is $f = \frac{Mc}{I}$. This is the usual way of expressing the flexure formula.

The term I/c is called the *section modulus* (S). Thus, another way of writing the flexure formula is f = M/S.

Example #11

A homogeneous rectangular beam 4" wide and 12" deep is subject to a positive bending moment of 80,000 inch-pounds. What is the maximum flexural unit stress in tension? What is the maximum flexural unit stress in compression? What is the unit stress 3" below the top of the cross-section? What is the value of the section modulus of the cross-section?

Solution:

We apply the basic flexure formula: $f = \frac{Mc}{I}$

M = 80,000"#

$c = \frac{d}{2} = \frac{12}{2} = 6"$

I for a rectangular cross-section = $\dfrac{bd^3}{12}$
(see page 14)

$= \dfrac{4(12)^3}{12} = 576 \text{ in.}^4$

$f = \dfrac{80{,}000" \times 6"}{576 \text{ in.}^4} = 833.3 \text{ psi}$

Since the section is symmetrical, the c distance is 6" for both the upper part of the beam (compression) and the lower part of the beam (tension), and the maximum flexural unit stresses in tension and compression are equal.

The maximum flexural unit stress in tension = *833.3 psi*

The maximum flexural unit stress in compression = *833.3 psi*

Unit stress 3" below top of cross-section

$= \dfrac{M\bar{y}}{I} = \dfrac{80{,}000" \times (6" - 3")}{576 \text{ in.}^4}$

= *416.7 psi in compression*

The section modulus $S = \dfrac{I}{c} = \dfrac{576 \text{ in}^4}{6 \text{ in.}} = 96 \text{ in.}^3$

The variation of flexural stress is shown in Figure 2.33.

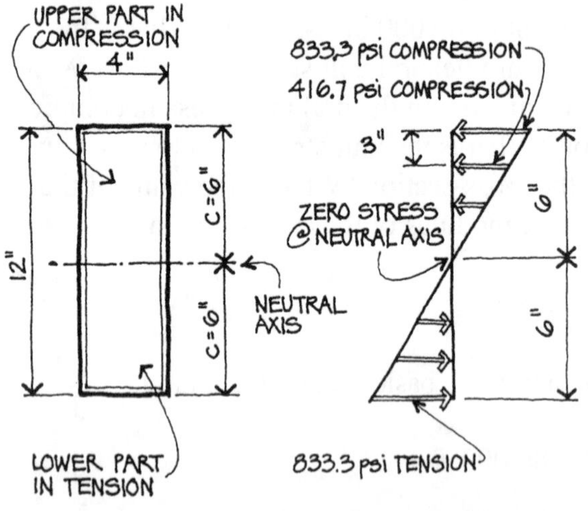

Figure 2.33

Example #12

What is the moment capacity of a wood beam whose section modulus is 102.41 in.³ if the allowable stress is 1,350 psi?

Solution:

We apply the flexure formula:

$f = \dfrac{Mc}{I}$ from which $M = \dfrac{fI}{c} = fS$

$= 1{,}350\#/\text{in}^2 \times 102.41 \text{ in}^3 = 138{,}253.5"\#$

The flexure formula involves three terms: f (flexural stress), M (bending moment), and I/c or S (section modulus). In problems involving the use of the formula, you are generally given two of the terms and asked to solve for the third term. Problems thus come down to one of three types:

1. Knowing the moment and the section modulus, what is the flexural stress? Use the formula f = Mc/I or f = M/S, as in Example #11.
2. Knowing the allowable stress and the section modulus, what is the allowable moment? Use the formula M = fI/c, or M = fS, as in Example #12.
3. Knowing the allowable stress and the moment, what is the required section modulus? Use the formula I/c = M/f, or S = M/f.

Thus, the flexure formula may be expressed in several different ways, which are all equivalent:

f = Mc/I	f = M/S
M = fI/c	M = fS
I/c = M/f	S = M/f

In using the flexure formula, you should be aware of the following:

1. The units must be consistent. Generally this means that moment is in inch-pounds, section modulus is in inches^3, and unit stress is in psi.
2. Unless stated otherwise, the moment of inertia of a beam cross-section is with respect to its neutral axis. For a rectangular section $I = \frac{bd^3}{12}$, and $S = I/c = \frac{bd^3/12}{d/2} = \frac{bd^2}{6}$
3. For a symmetrical cross-section, the values of c for both tension and compression are equal, as in Example #11. For an unsymmetrical cross-section, the values of c for tension and compression are not equal, and the maximum unit stresses in tension and compression are therefore different.

Shear Stresses

Wherever vertical shear exists in a beam, there is also horizontal shear. Looking at a very small area of the beam, the vertical shear stresses along each edge form a couple, which is held in equilibrium by another couple composed of the horizontal shear stresses at each edge (see Figure 2.34). The vertical and horizontal shear unit stresses at any point are of equal magnitude and are perpendicular to each other.

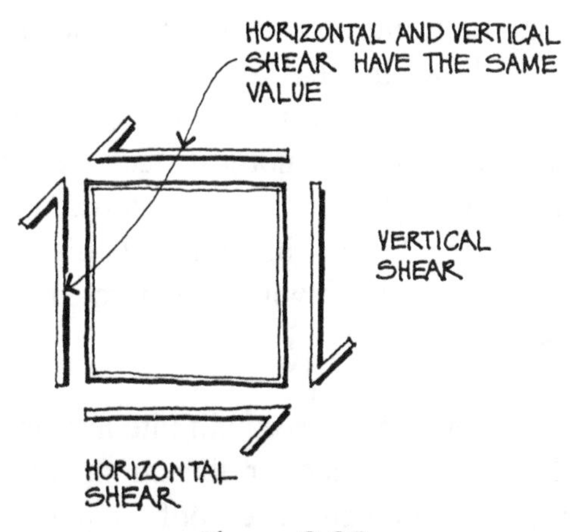

Figure 2.34

The horizontal shear stress $v = \frac{VQ}{Ib}$

where

v = horizontal shear stress, in psi
V = vertical shear at the section under consideration, in pounds
Q = statical moment about the neutral axis of the area above the plane under consideration, in inches^3
I = moment of inertia of the beam cross-section about the neutral axis, in inches^4
b = thickness of the beam at the plane under consideration, in inches

Shear stress in a beam varies parabolically, from zero at the outermost fibers to a maximum value at the neutral axis (see Figure 2.35).

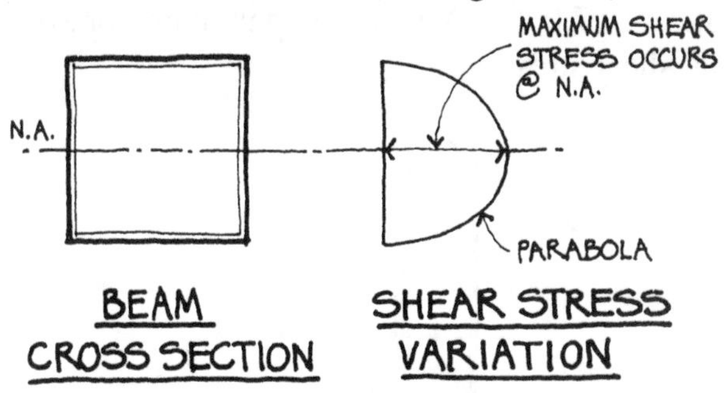

Figure 2.35

One way to visualize horizontal shear is to consider two planks supporting a load and not connected to each other as depicted in Figure 2.36.

The planks slide past each other, and the total bending resistance is the sum of the bending resistances of the two planks acting independently.

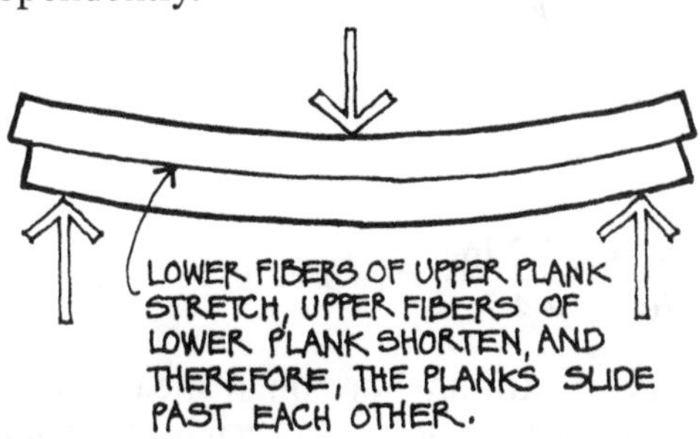

Figure 2.36

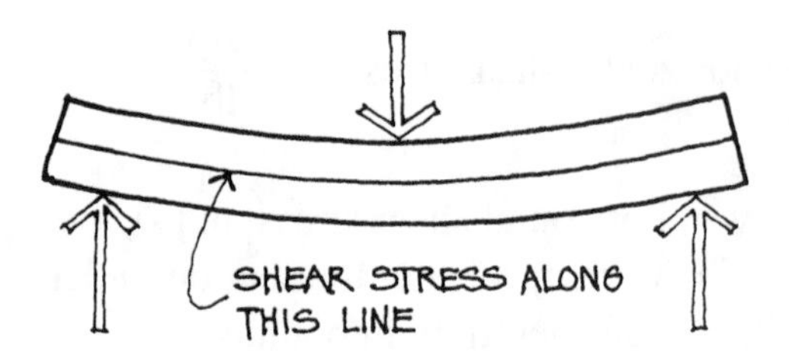

Figure 2.37

If the planks are connected along their line of contact so as to develop adequate shear strength, the two planks can be made to act as one beam.

Example #13

The beam shown in Figure 2.38 has a vertical shear of 7,000#. What is the maximum horizontal shear stress?

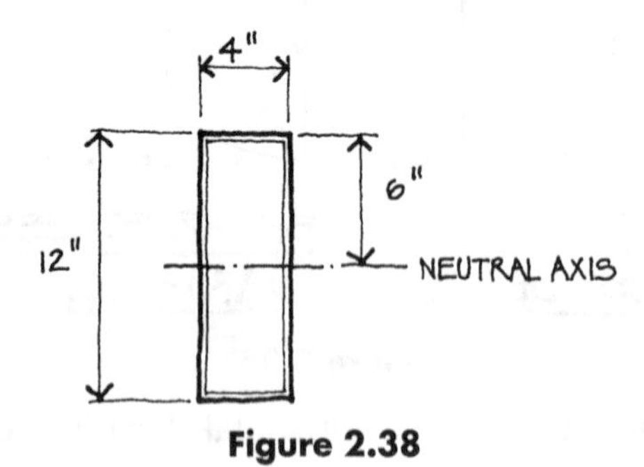

Figure 2.38

Solution:

Maximum horizontal shear stress $v = \dfrac{VQ}{Ib}$

$V = 7{,}000\#$

$Q = 4 \times 6 \times 6/2 = 72 \text{ in.}^3$

$I = \dfrac{bd^3}{12} = \dfrac{4(12)^3}{12} = 576 \text{ in.}^4$

$b = 4 \text{ in.}$

$v = \dfrac{7{,}000\# \times 72 \text{ in.}^3}{576 \text{ in.} \times 4 \text{ in.}} = \mathit{218.75\ psi}$

This is the maximum horizontal shear stress and occurs at the neutral axis. There is also a vertical shear stress perpendicular to the horizontal shear stress, whose magnitude at the neutral axis is also 218.75 psi.

For beams of rectangular cross-section, the expression for the maximum horizontal shear stress (v) can be simplified from $\dfrac{VQ}{Ib}$ to $\dfrac{3V}{2bd}$ where b = beam width and d = beam depth.

This formula is frequently used to compute the horizontal shear stress in rectangular wood beams.

Example #14

Calculate the maximum horizontal shear stress in the beam of Example #13 using the simplified formula for rectangular cross-sections.

Solution:

$v = \dfrac{3}{2}\dfrac{V}{bd} = \dfrac{3}{2} \times \dfrac{7{,}000}{4 \times 12} = \mathit{218.75\ psi}$, the same as our previous result.

Deflection

Deflection is the movement of a beam from its original location when load is applied to it. A beam should have sufficient stiffness so that its deflection does not exceed the values in IBC Table 1604.3.

Deflection formulas for various beam loads and types are shown in the *AISC Manual of Steel Construction*. These formulas may be used for beams of any material, not just steel. Deflection formulas for the two most common beam conditions (simple beam with uniformly distributed load and simple beam with concentrated load at midspan) are shown in Figure 2.24.

Camber, which is a curve built into a structural member to compensate for deflection, is often provided for wood and steel roof members.

COLUMNS

A column is a member primarily subject to an axial compressive load (Figure 2.39). Sometimes a column also resists bending moment; for example, if loads are applied off center (eccentrically), or perpendicular to its length, such as wind loads, as shown in Figure 2.40.

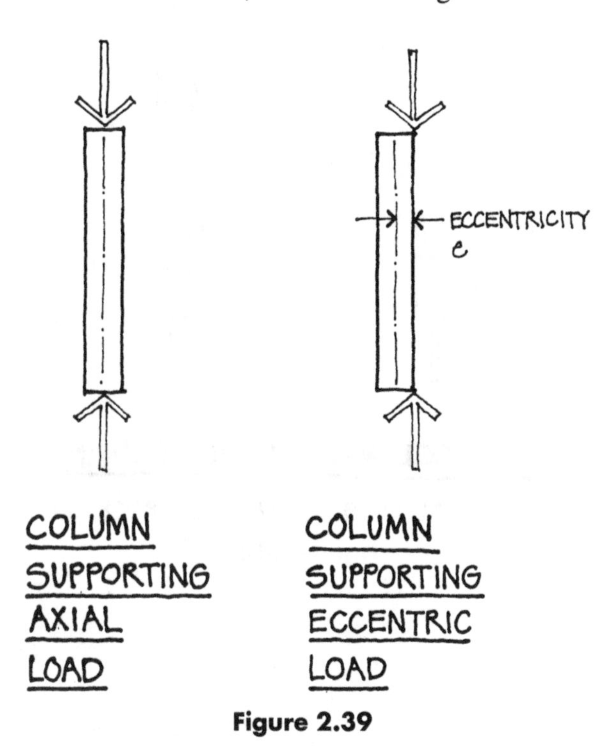

Figure 2.39

When a column is subject to combined bending and axial compression, the compressive stress is P/A, and the flexural stress at the extreme fiber is Mc/I (tensile on one face and compressive on the other). On the compressive face, the flexural stress Mc/I *adds* to the compressive stress P/A to obtain the maximum combined stress of P/A + Mc/I. On the tensile face, the flexural stress Mc/I *subtracts* from the compressive stress P/A to obtain the minimum combined stress of P/A – Mc/I. These stresses are depicted in Figure 2.41.

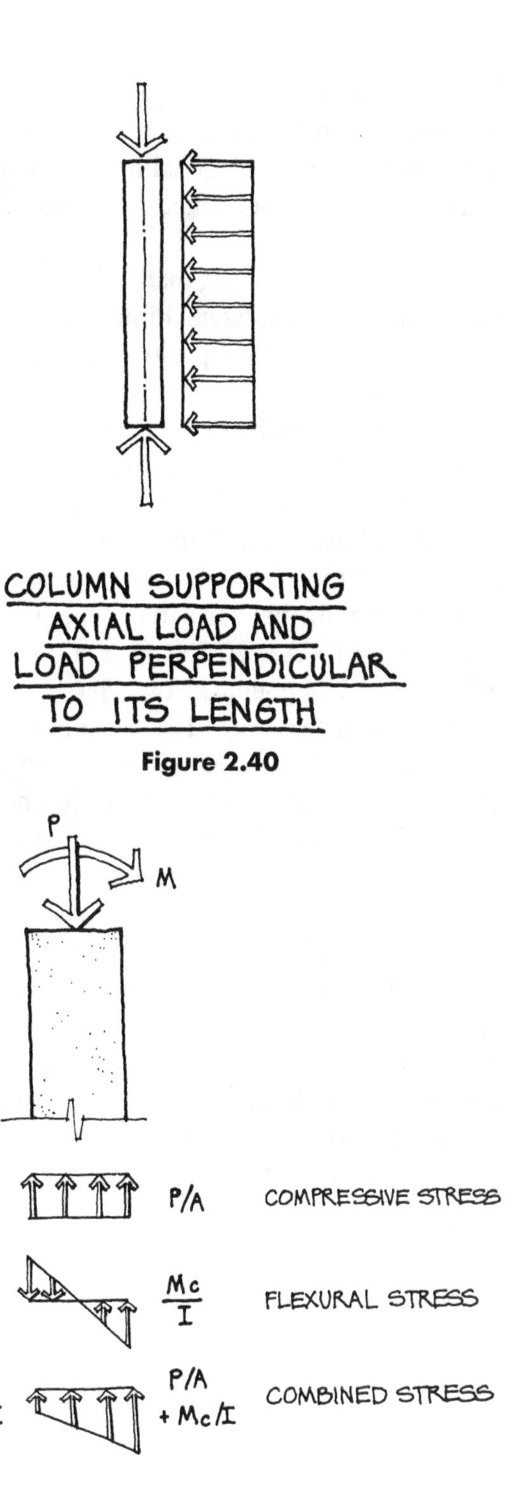

Figure 2.41

Building columns are usually vertical, but some members that resist axial compressive load are not vertical, such as diagonal bracing members and the compression members of trusses.

If a column is very short, that is, if the ratio of its length to its least lateral dimension is small, then its load-carrying capacity is limited by the crushing strength of its material (see Figure 2.42). On the other hand, long compressive members tend to fail by buckling; they will buckle before they crush. For example, consider a steel bar one inch in diameter and one inch high. If the allowable compressive stress is 20,000 psi, the bar can safely support an axial load of $f \times A = 20{,}000 \times \pi d^2/4 = 20{,}000 \times \pi(1)^2/4 = 15{,}708$ pounds. Now what happens if the bar is four feet long? The bar can no longer safely support 15,708 pounds because it is so slender compared to its length that it tends to buckle. Therefore, slenderness must be considered in the design of columns.

The maximum stress that a slender column can resist without failing by sudden buckling is determined from *Euler's equation*, and depends on the column's *slenderness ratio* l/r, where l is the length of the column and r is the radius of gyration of the column cross-section, which is $\sqrt{I/A}$.

We will discuss the design of columns of various materials in later lessons.

In this lesson, we have discussed the basic theory of beams and columns, and especially statically determinate beams. We cannot stress too highly the importance of this subject!

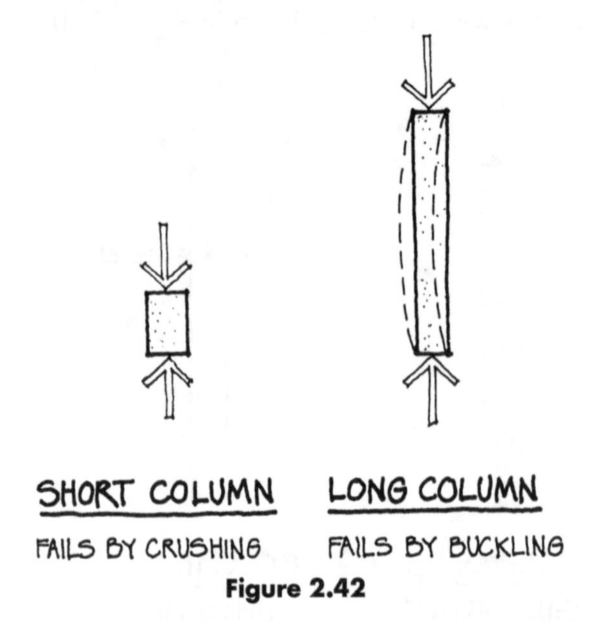

Figure 2.42

LESSON 2 QUIZ

1. The maximum stress that a column can resist without failing by buckling depends on the column's

   I. unbraced length.

   II. section modulus.

   III. radius of gyration.

   A. I and II
   B. II and III
   C. I and III
   D. I, II, and III

2. Where is the bending moment in a simply supported beam maximum?

   A. At midspan
   B. At the point of zero shear
   C. At the point of inflection
   D. Where the shear is maximum

3. The flexural stress in a beam is a function of which of the listed items? Check all that apply.

   A. Bending moment
   B. Reactions
   C. Section modulus of the beam
   D. Span

4. The shear stress in a beam is

   A. maximum at the neutral axis.
   B. uniform over the beam's cross-section.
   C. maximum where the moment is zero.
   D. maximum at midspan.

5. The point where the bending moment in a beam changes sign is called the

   A. neutral surface.
   B. end support.
   C. centroid.
   D. point of inflection.

6.

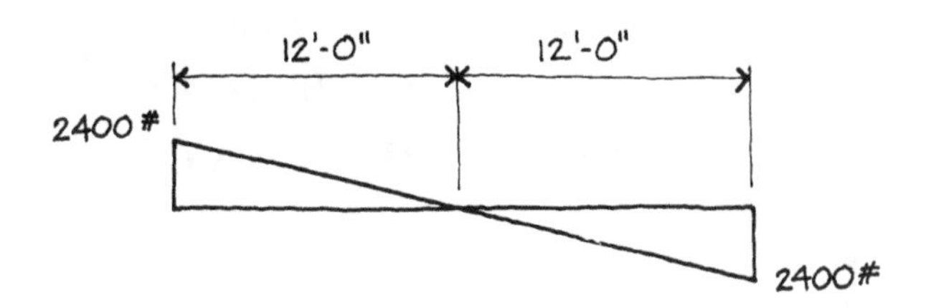

   A simply supported uniformly loaded beam has a shear diagram as shown above. What is the uniform load on the beam?

   A. 1.2k/ft.
   B. 0.2k/ft.
   C. 2.4k/ft.
   D. 0.4k/ft.

7. A simple beam supporting a uniformly distributed load over a 20-foot span deflects 1/2 inch. If a beam with the same cross-section were to span 25 feet supporting the same uniform load per foot, how much would it deflect?

   A. 0.7813 inches
   B. 0.2048 inches
   C. 0.9211 inches
   D. 1.2207 inches

8. Select the CORRECT statements about a continuous beam having two equal spans supporting a uniformly distributed load over both spans.

I. The moment over the middle support is negative.

II. The point of inflection occurs over the middle support.

III. The beam is statically indeterminate.

IV. The moment at the end supports is zero.

V. The moment diagram for the two spans is comprised of parabolic curves.

A. I, II, and IV
B. II, III, and V
C. I, III, IV, and V
D. I, II, and V

9.

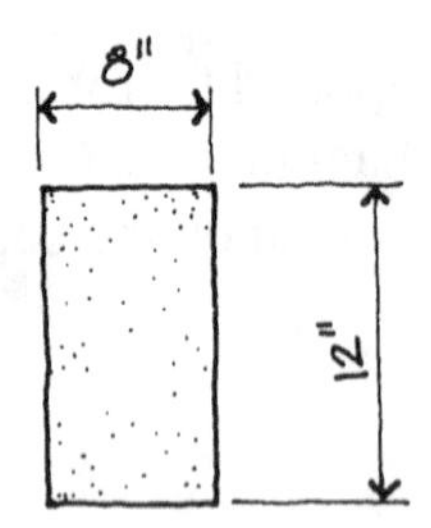

The beam shown above has a maximum vertical shear of 6,000 pounds. What is the maximum horizontal shear stress? The maximum vertical shear stress?

A. Horizontal shear = 62.5 psi, vertical shear = 62.5 psi.

B. Horizontal shear = 93.8 psi, vertical shear = 62.5 psi.

C. Horizontal shear = 93.8 psi, vertical shear = 93.8 psi.

D. Horizontal shear = 62.5 psi, vertical shear = 93.8 psi.

10. The flexural stress in a beam

I. varies parabolically over the beam's cross-section.

II. is uniform over the beam's cross-section.

III. is maximum at the neutral axis.

IV. is maximum at the extreme fibers.

V. is a function of the beam's section modulus.

A. I, III, and V
B. IV and V
C. I and IV
D. II and V

LESSON THREE

WOOD CONSTRUCTION

INTRODUCTION

The structure of a building consists essentially of three major elements: horizontal framing (joists, beams, and girders), vertical members (columns and walls), and the foundation. A bracing system to resist horizontal wind or earthquake forces must also be provided.

Only four basic materials are commonly used in structural systems: wood, structural steel, reinforced concrete, and masonry. In this lesson we will cover wood framing, while subsequent lessons will deal with systems utilizing other structural materials.

PROPERTIES OF WOOD

Species of trees are divided into two classes: *hardwoods*, which have broad leaves, and *softwoods*, which have scalelike or needlelike leaves. Hardwoods are generally deciduous, shedding their leaves annually. Most softwoods, also called *conifers*, are evergreen. The terms hardwood and softwood are misleading, since they do not refer directly to the hardness or softness of the wood; in fact, some hardwoods are softer than certain softwoods. Most of the lumber used for structural purposes is softwood.

The moisture content of timber varies with the temperature and humidity of the surrounding atmosphere. Dry wood in a humid space will absorb moisture; conversely, wet wood loses moisture in a dry environment. Between zero moisture content and the *fiber saturation point* (about 30 percent), wood swells as it absorbs moisture and shrinks as it loses moisture. The amount of the shrinkage or swelling is directly proportional to the amount of moisture lost or gained below the fiber saturation point. Above the fiber saturation point, no further swelling takes place.

Wood in the living tree contains a large amount of water. Ideally, wood members should be seasoned (dried) before installation until they reach the moisture content they will experience in use. In practice, this does not always occur. Consequently, construction details should allow for some shrinkage of wood members. The greatest shrinkage occurs perpendicular to the grain, while the longitudinal shrinkage is usually negligible. In structural lumber, one should anticipate a shrinkage of about 1/64" per inch of width when the wood dries from *green* (having the moisture content of the living tree) to the average moisture content in use. As the wood dries, it not only shrinks, but it also becomes stronger and stiffer (its modulus of elasticity increases).

Like other solids, wood expands on heating and contracts on cooling. However, its coefficient of thermal expansion is very low compared to that of other common building materials and therefore can be neglected in most structural designs.

Wood is one of the most widely used construction materials in the United States, largely because of its ready availability and ease of fabrication and installation. Some of its other qualities include high strength-to-weight ratio, an attractive appearance, and low heat conductivity. For many structural purposes, wood is the most economical material that can be used.

However, wood also has several distinct disadvantages: it shrinks and swells when exposed to moisture changes, it is combustible, it is subject to attack by fungi and certain insects, and it can easily be damaged because of its softness.

Some of these disadvantages may be partially overcome. For example, proper seasoning and good construction detailing should minimize problems caused by shrinkage and swelling. Reasonable care in shipping and handling of wood will usually prevent its being damaged. Even the combustibility of wood, one of its most serious deficiencies, can be offset to some extent by the use of fire-retardant chemical treatment.

The other major destroyers of wood are decay and insects, principally termites. Decay is caused by fungi, which require food (wood cellulose), moisture, air, and favorable temperatures to develop. Therefore, decay may be prevented by installing wood members in weather-protected locations and isolated from the ground or other moisture sources. Wood continually submerged in water will not decay either, since fungi require air. Intermittent wetting, however, provides a very favorable environment for decay. Where there is the possibility of decay, wood should be a decay-resistant species, such as redwood, or pressure treated with a preservative.

Since termites require an environment similar to that needed by fungi, they require similar preventive measures. Buildings should have proper drainage, good ventilation, adequate crawl space, and impervious concrete foundations. In addition, all wood should be kept well separated from the ground. As with decay, the most positive protection is pressure treatment of the wood with a preservative.

REFERENCE DESIGN VALUES

Most of the structural lumber used in the United States is manufactured from softwoods, principally Douglas fir and Southern pine. Obviously, the local availability and cost largely determine the species and grades used for any specific building project.

There are many factors that determine the reference design values (previously referred to as allowable stresses) for lumber. These include the natural properties of the species, the size and location of knots and other defects, slope of grain, degree of density, and the condition of seasoning.

Most structural lumber is stress graded—that is, graded for its ability to resist stresses, and appropriately marked. Each grade is given a designation (such as No. 1, select structural, etc.) to which a schedule of design values are assigned. These values include extreme fiber stress in bending (F_b), tension parallel to grain (F_t), compression parallel to grain (F_c), compression perpendicular to grain ($F_{c\perp}$) and horizontal shear (F_v).

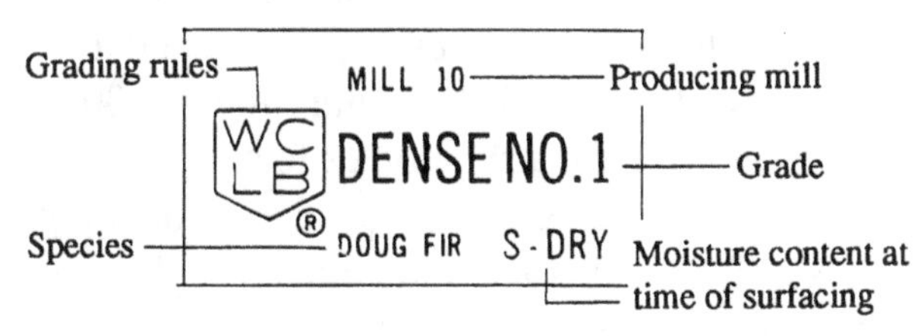

Figure 3.1

Unlike steel and concrete, the strength of wood is not the same in every direction; for example, it is much stronger in compression parallel to the grain (F_c) than perpendicular to the grain ($F_{c\perp}$).

Stress graded lumber is divided into categories by size, such as *2" to 4" thick 2" and wider, beams and stringers,* and *posts and timbers.* Beams and stringers are usually 5" and thicker with a width more than 2" greater than thickness. Posts and timbers are generally 5" × 5" and larger with a width not more than 2" greater than thickness. By these criteria, a 6 × 6 and 6 × 8 are *posts and timbers*, while a 6 × 10 is a *beam and stringer.*

Grading is done either visually by experienced graders, or by machine (machine stress-rating), under rules established by approved lumber grading agencies. Machine rating will probably become more widespread in the future. A typical grade stamp is depicted in Figure 3.1.

The 2009 IBC references the 2005 edition of the National Design Specification for Wood Construction for the design of wood members. Tables 4A and 4D in the appendix are reproduced in part from the Supplement to the National Design Specification. Table 4A contains design values for visually graded dimension lumber; Table 4D provides the same for timbers.

The National Design Specification permits wood members to be designed either by Allowable Stress Design (ASD) or Load and Resistance Factor Design (LRFD). This lesson will focus on the ASD method.

SIZES OF LUMBER

A piece of structural lumber is designated by its nominal cross-sectional dimensions (breadth and depth). For example, a 4 × 10 (four by ten) is a piece of lumber having a nominal breadth of four inches and depth of ten inches. The actual dimensions of this timber after being surfaced on four sides (S4S) are 3-1/2" × 9-1/4". Since most structural lumber is surfaced, it is necessary to use the actual sizes in structural design rather than the nominal sizes. The dressed (surfaced) sizes in Figure 3.2, from the National Design Specification Supplement, conform to the American Softwood Lumber Standard PS 20-99. The tabulated areas (A), moments of inertia (I), and section moduli (S) are all based on the actual sizes. The weight of structural lumber averages about 35 pounds per cubic foot, and the weights of standard sizes of lumber are available in tables.

DESIGN OF WOOD BEAMS

Flexural members are referred to by a bewildering and not altogether precise assortment of names: joists (closely spaced repetitive members supporting floor, roof, or ceiling loads), rafters (closely spaced repetitive members, generally sloping, supporting roof loads), purlins (regularly spaced roof beams that span between trusses or girders), planking (flat members spanning between beams), beams (main members), stringers (beams supporting stairways), lintels (members over openings in walls), headers (members supporting the ends of joists or rafters at an opening), and girders (main members that support the ends of joists or beams and span between columns or bearing walls). Whatever their name, their structural function is identical: they are all beams.

Almost all wood beams are simple beams and are designed by the methods described in Lesson Two. The design values for wood members may be increased for short-time loading as follows:

- 15% for two months' duration, as for snow
- 25% for seven days' duration, as for construction loads or roof live loads
- 60% for wind or earthquake
- 100% for impact

The increases are not cumulative. For combined duration of loadings, the structural members may not be smaller than required for the longer duration of loading.

Design for Bending Stress

Use the flexure formula $F_b = \frac{Mc}{I} = \frac{M}{S}$

First calculate the maximum bending moment (M). Then determine the design value for bending (F_b) from a table, such as the one in the appendix, based on the species and grade of lumber to be used. Calculate S, the required section modulus. Select a standard rectangular beam from a table, such as the one in Figure 3.2.

Variations on this procedure include (a) for a given moment and member size, solve for the maximum flexural stress, and (b) given the member size and the design flexural stress, determine the moment capacity of the member.

Example #1

A wood roof beam has a uniformly distributed dead load of 100 pounds per foot and a live load of less than 7 days' duration of 100 pounds per foot on a span of 22 feet. What size beam should be used, considering bending stress only, using No. 1 and better Douglas fir-larch and assuming that the weight of the beam is included in the dead load?

Solution:

The moment from dead load = $M_{dl} = \frac{wL^2}{8} = \frac{100 \times 22^2}{8} = 6{,}050$ ft.-lbs.

The moment from total load = $M_{tl} = \frac{wL^2}{8} = \frac{200 \times 22^2}{8} = = 12{,}100$ ft.-lbs.

S_{dl} required $= \frac{M}{F_b} = \frac{6{,}050 \times 12}{1{,}200} = 60.5$ in.^{3}

S_{tl} required $= \frac{M}{F_b} = \frac{12{,}100 \times 12}{1{,}200 \times 1.25} = 96.8$ in.^{3}

Use 4 × *14.* Actual S, is 102.4 in.^{3}.

TABLE 1B SECTION PROPERTIES OF STANDARD DRESSED (S4S) SAWN LUMBER

Nominal Size b×d	Standard Dressed Size (S4S) b×d inches×inches	Area of Section A in.²	X-X Axis: Section Modulus S_{xx} in.³	X-X Axis: Moment of Inertia I_{xx} in.⁴	Y-Y Axis: Section Modulus S_{yy} in.³	Y-Y Axis: Moment of Inertia I_{yy} in.⁴	Approximate weight in pounds per linear foot (lb./ft.) of piece when density of wood equals: 25 lb./ft.³	30 lb./ft.³	35 lb./ft.³	40 lb./ft.³	45 lb./ft.³	50 lb./ft.³
1 × 3	3/4 × 2-1/2	1.875	0.781	0.977	0.234	0.088	0.326	0.391	0.456	0.521	0.586	0.651
1 × 4	3/4 × 3-1/2	2.625	1.531	2.680	0.328	0.123	0.456	0.547	0.638	0.729	0.820	0.911
1 × 6	3/4 × 5-1/2	4.125	3.781	10.40	0.516	0.193	0.716	0.859	1.003	1.146	1.289	1.432
1 × 8	3/4 × 7-1/4	5.438	6.570	23.82	0.680	0.255	0.944	1.133	1.322	1.510	1.699	1.888
1 × 10	3/4 × 9-1/4	6.938	10.70	49.47	0.867	0.325	1.204	1.445	1.686	1.927	2.168	2.409
1 × 12	3/4 × 11-1/4	8.438	15.82	88.99	1.055	0.396	1.465	1.758	2.051	2.344	2.637	2.930
2 × 3	1-1/2 × 2-1/2	3.750	1.563	1.953	0.938	0.703	0.651	0.781	0.911	1.042	1.172	1.302
2 × 4	1-1/2 × 3-1/2	5.250	3.063	5.359	1.313	0.984	0.911	1.094	1.276	1.458	1.641	1.823
2 × 5	1-1/2 × 4-1/2	6.750	5.063	11.39	1.688	1.266	1.172	1.406	1.641	1.875	2.109	2.344
2 × 6	1-1/2 × 5-1/2	8.250	7.563	20.80	2.063	1.547	1.432	1.719	2.005	2.292	2.578	2.865
2 × 8	1-1/2 × 7-1/4	10.88	13.14	47.63	2.719	2.039	1.888	2.266	2.643	3.021	3.398	3.776
2 × 10	1-1/2 × 9-1/4	13.88	21.39	98.93	3.469	2.602	2.409	2.891	3.372	3.854	4.336	4.818
2 × 12	1-1/2 × 11-1/4	16.88	31.64	178.0	4.219	3.164	2.930	3.516	4.102	4.688	5.273	5.859
2 × 14	1-1/2 × 13-1/4	19.88	43.89	290.8	4.969	3.727	3.451	4.141	4.831	5.521	6.211	6.901
3 × 4	2-1/2 × 3-1/2	8.750	5.104	8.932	3.646	4.557	1.519	1.823	2.127	2.431	2.734	3.038
3 × 5	2-1/2 × 4-1/2	11.25	8.438	18.98	4.688	5.859	1.953	2.344	2.734	3.125	3.516	3.906
3 × 6	2-1/2 × 5-1/2	13.75	12.60	34.66	5.729	7.161	2.387	2.865	3.342	3.819	4.297	4.774
3 × 8	2-1/2 × 7-1/4	18.13	21.90	79.39	7.552	9.440	3.147	3.776	4.405	5.035	5.664	6.293
3 × 10	2-1/2 × 9-1/4	23.13	35.65	164.9	9.635	12.04	4.015	4.818	5.621	6.424	7.227	8.030
3 × 12	2-1/2 × 11-1/4	28.13	52.73	296.6	11.72	14.65	4.883	5.859	6.836	7.813	8.789	9.766
3 × 14	2-1/2 × 13-1/4	33.13	73.15	484.6	13.80	17.25	5.751	6.901	8.051	9.201	10.35	11.50
3 × 16	2-1/2 × 15-1/4	38.13	96.90	738.9	15.89	19.86	6.619	7.943	9.266	10.59	11.91	13.24
4 × 4	3-1/2 × 3-1/2	12.25	7.146	12.51	7.146	12.51	2.127	2.552	2.977	3.403	3.828	4.253
4 × 5	3-1/2 × 4-1/2	15.75	11.81	26.58	9.188	16.08	2.734	3.281	3.828	4.375	4.922	5.469
4 × 6	3-1/2 × 5-1/2	19.25	17.65	48.53	11.23	19.65	3.342	4.010	4.679	5.347	6.016	6.684
4 × 8	3-1/2 × 7-1/4	25.38	30.66	111.1	14.80	25.90	4.405	5.286	6.168	7.049	7.930	8.811
4 × 10	3-1/2 × 9-1/4	32.38	49.91	230.8	18.89	33.05	5.621	6.745	7.869	8.993	10.12	11.24
4 × 12	3-1/2 × 11-1/4	39.38	73.83	415.3	22.97	40.20	6.836	8.203	9.570	10.94	12.30	13.67
4 × 14	3-1/2 × 13-1/4	46.38	102.4	678.5	27.05	47.34	8.051	9.661	11.27	12.88	14.49	16.10
4 × 16	3-1/2 × 15-1/4	53.38	135.7	1034	31.14	54.49	9.266	11.12	12.97	14.83	16.68	18.53
5 × 5	4-1/2 × 4-1/2	20.25	15.19	34.17	15.19	34.17	3.516	4.219	4.922	5.625	6.328	7.031
6 × 6	5-1/2 × 5-1/2	30.25	27.73	76.26	27.73	76.26	5.252	6.302	7.352	8.403	9.453	10.50
6 × 8	5-1/2 × 7-1/2	41.25	51.56	193.4	37.81	104.0	7.161	8.594	10.03	11.46	12.89	14.32
6 × 10	5-1/2 × 9-1/2	52.25	82.73	393.0	47.90	131.7	9.071	10.89	12.70	14.51	16.33	18.14
6 × 12	5-1/2 × 11-1/2	63.25	121.2	697.1	57.98	159.4	10.98	13.18	15.37	17.57	19.77	21.96
6 × 14	5-1/2 × 13-1/2	74.25	167.1	1128	68.06	187.2	12.89	15.47	18.05	20.63	23.20	25.78
6 × 16	5-1/2 × 15-1/2	85.25	220.2	1707	78.15	214.9	14.80	17.76	20.72	23.68	26.64	29.60
6 × 18	5-1/2 × 17-1/2	96.25	280.7	2456	88.23	242.6	16.71	20.05	23.39	26.74	30.08	33.42
6 × 20	5-1/2 × 19-1/2	107.3	348.6	3398	98.31	270.4	18.62	22.34	26.07	29.79	33.52	37.24
6 × 22	5-1/2 × 21-1/2	118.3	423.7	4555	108.4	298.1	20.53	24.64	28.74	32.85	36.95	41.06
6 × 24	5-1/2 × 23-1/2	129.3	506.2	5948	118.5	325.8	22.44	26.93	31.41	35.90	40.39	44.88
8 × 8	7-1/2 × 7-1/2	56.25	70.31	263.7	70.31	263.7	9.766	11.72	13.67	15.63	17.58	19.53
8 × 10	7-1/2 × 9-1/2	71.25	112.8	535.9	89.06	334.0	12.37	14.84	17.32	19.79	22.27	24.74
8 × 12	7-1/2 × 11-1/2	86.25	165.3	950.5	107.8	404.3	14.97	17.97	20.96	23.96	26.95	29.95
8 × 14	7-1/2 × 13-1/2	101.3	227.8	1538	126.6	474.6	17.58	21.09	24.61	28.13	31.64	35.16
8 × 16	7-1/2 × 15-1/2	116.3	300.3	2327	145.3	544.9	20.18	24.22	28.26	32.29	36.33	40.36
8 × 18	7-1/2 × 17-1/2	131.3	382.8	3350	164.1	615.2	22.79	27.34	31.90	36.46	41.02	45.57
8 × 20	7-1/2 × 19-1/2	146.3	475.3	4634	182.8	685.5	25.39	30.47	35.55	40.63	45.70	50.78
8 × 22	7-1/2 × 21-1/2	161.3	577.8	6211	201.6	755.9	27.99	33.59	39.19	44.79	50.39	55.99
8 × 24	7-1/2 × 23-1/2	176.3	690.3	8111	220.3	826.2	30.60	36.72	42.84	48.96	55.08	61.20
10 × 10	9-1/2 × 9-1/2	90.25	142.9	678.8	142.9	678.8	15.67	18.80	21.94	25.07	28.20	31.34
10 × 12	9-1/2 × 11-1/2	109.3	209.4	1204	173.0	821.7	18.97	22.76	26.55	30.35	34.14	37.93
10 × 14	9-1/2 × 13-1/2	128.3	288.6	1948	203.1	964.5	22.27	26.72	31.17	35.63	40.08	44.53
10 × 16	9-1/2 × 15-1/2	147.3	380.4	2948	233.1	1107	25.56	30.68	35.79	40.90	46.02	51.13
10 × 18	9-1/2 × 17-1/2	166.3	484.9	4243	263.2	1250	28.86	34.64	40.41	46.18	51.95	57.73
10 × 20	9-1/2 × 19-1/2	185.3	602.1	5870	293.3	1393	32.16	38.59	45.03	51.46	57.89	64.32
10 × 22	9-1/2 × 21-1/2	204.3	731.9	7868	323.4	1536	35.46	42.55	49.64	56.74	63.83	70.92
10 × 24	9-1/2 × 23-1/2	223.3	874.4	10270	353.5	1679	38.76	46.51	54.26	62.01	69.77	77.52

Figure 3.2

Please note the following:

1. The factor 12 in the numerator is used to convert ft.-lbs. to inch-lbs.
2. The design value for bending (F_b) of 1,200 psi is determined from Table 4A in the appendix, size classification *2" to 4" thick 2" and wider*, for *No.1 and better Douglas fir-larch*, as shown in Figure 3.3.

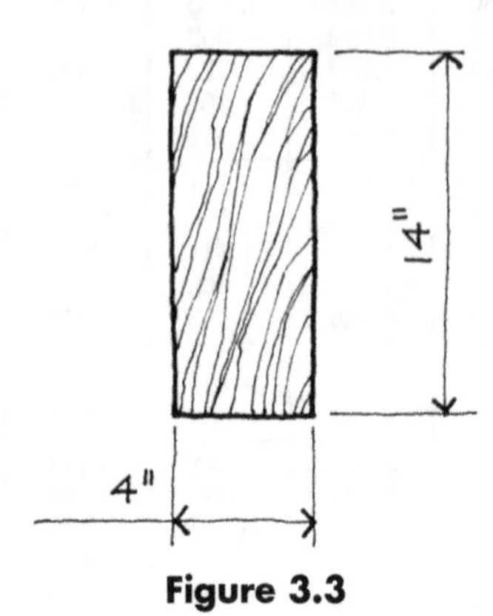

Figure 3.3

Since the member selected is a 4 × 14, the size classification we used is correct.

3. The factor 1.25 in the denominator of the second calculation for S provides for a 25 percent increase in the design stress, which is permitted for a wood member supporting roof loads. Note that where members are subject to loads of different durations, each load combination should be checked to determine the required section modulus.
4. This example illustrates design for bending strength only. In actual practice, horizontal shear and deflection would also be considered, as explained on the following pages.

Example #2

A 2 × 12 wood joist is loaded as shown in Figure 3.4. What is the maximum moment and the maximum fiber stress?

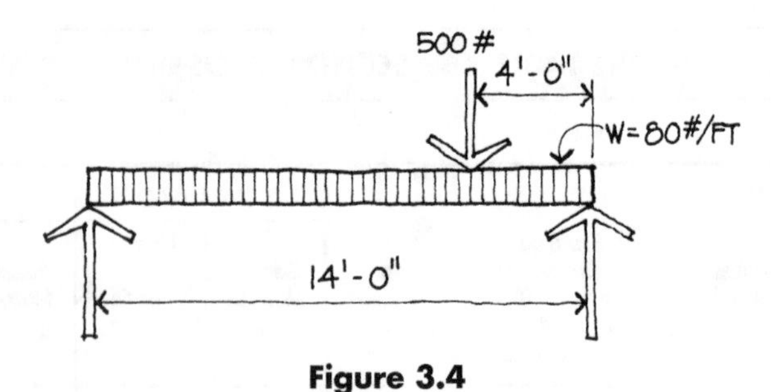

Figure 3.4

Solution:

As explained in Lesson Two, calculate the end reactions. As an aid in calculating the moment, draw the shear diagram to determine the point of zero shear, where the moment is maximum (see Figure 3.5).

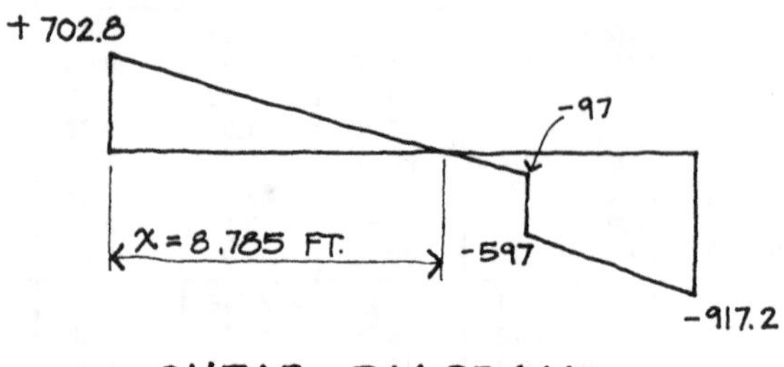

Figure 3.5

To determine the point of zero shear, divide the reaction at the left by the uniform load.

$$x = \frac{702.8\#}{80\ \#/\text{ft.}} = 8.785\ \text{ft., then}$$

$$M = 702.8\# \times 8.785\ \text{ft.} - 80\ \#/\text{ft.} \times 8.785\ \text{ft.} \times \frac{8.785\ \text{ft.}}{2}$$

$= \mathit{3{,}087\ ft.\text{-}lbs.}$

From the table in Figure 3.2, S for a 2 × 12 is 31.64 in^3.

The maximum fiber stress f = M/S

$$= \frac{3{,}087\ \text{ft.-lbs.} \times 12\ \text{in./ft.}}{31.64\ \text{in.}^3} = \mathit{1{,}170\ psi}$$

Again the factor 12 is used to change ft.-lbs. to in.-lbs., in order to keep the units consistent.

Design for Horizontal Shear Stress

From Lesson Two, we have the formula $f_v = \frac{3}{2}\frac{V}{bd}$ for rectangular sections.

Example #3

Calculate the maximum shear stress in the joist of Example #2.

Solution:

The maximum shear is the right reaction of 917.2 lbs.

$$f_v = \frac{3}{2}\frac{V}{bd} = \frac{3}{2} \times \frac{917.2\#}{16.88 \text{ in.}^2} = \mathit{81.5\ psi}$$

Note that we used the area (bd) of a 2 × 12 (16.88 in.^{2}), which we determined from the table in Figure 3.2.

Based on the calculated bending stress of 1,170 psi and shear stress of 81.5 psi, and referring to Table 4A in the appendix, an acceptable species and grade would be select structural Douglas fir-larch, which has design values F_b of 1,500 psi and F_v of 180 psi.

Horizontal shear stress should be checked for all wood flexural members and is most critical for short span beams with large loads.

In wood construction it is permissible to neglect loads within a distance from either support equal to the depth of the beam.

Design for Deflection

It is desirable to limit the deflection of a wood beam for a number of reasons: to minimize the possibility of plaster cracking, to limit the vibration or springiness of a floor, and to prevent unsightly sag.

The maximum deflections permitted by the IBC for a roof member supporting plaster or a floor member are shown in Table 1604.3 in the appendix. For longer spans, these deflection criteria will often determine the required size of the member.

Example #4

What is the total deflection of the 4 × 14 beam in Example #1? Assuming that the beam supports plaster, does it meet the IBC criteria?

Solution:

The value of E, determined from Table 4A in the appendix, is 1,800,000 psi.

The value of I, determined from Figure 3.2, is 678.5 in.^{4}.

The formula for the deflection Δ of a simply supported uniformly loaded simple beam is

$$\Delta = \frac{5}{384}\frac{wL^4}{EI} \quad \text{(see Figure 2.24)}$$

In order to keep the units consistent and have Δ expressed in inches, we can rewrite the formula as

$$\Delta = \frac{5}{384}\frac{wL \times L^3}{EI}$$

$$= \frac{5}{384} \times \frac{(200\#/\text{ft.} \times 22 \text{ ft.}) \times (22 \text{ ft.} \times 12\text{in./ft.})^3}{(1.8 \times 10^6\#/\text{in.}^2) \times (678.5 \text{ in.}^4)}$$

$$= 0.863 \text{ in.}$$

Note that we convert the span (22 ft.) to inches in the numerator by multiplying by 12, then cube the entire term $(22 \times 12)^3$.

Since the live and dead loads are each equal to half of the total load, the live load deflection is equal to

$$\frac{0.863}{2} = \mathit{0.4315''}.$$

The IBC limits are: For live load only, L/360 = 22 × 12/360 = 0.733". For live load plus dead load, L/240 = 22 × 12/240 = 1.10". *The 4 × 14 beam satisfies both criteria.* Its live load deflection of 0.4315" is less than the allowable of 0.733", and its live plus dead load deflection of 0.863" is below the allowable of 1.10".

Unless we are given specific information to the contrary, *sawn wood members are generally assumed to be unseasoned* for the purpose of determining the maximum allowable deflection.

Beam Design Procedure

We have isolated the three criteria that determine the size of a wood beam: bending stress, horizontal shear stress, and deflection. In an actual problem, all three criteria should be satisfied, by using the following procedure:

1. Calculate the required section modulus (S) using the flexure formula $S = \frac{M}{F_b}$.

   Select a beam whose value of S is at least equal to that required.
2. Check the selected size for horizontal shear using the formula $f_v = \frac{3}{2}\frac{V}{bd}$.

   If necessary, use a larger beam.
3. Calculate the deflection of the selected beam, if required by the governing code or the problem statement, and verify that it is less than the allowable. If the deflection exceeds the allowable, increase the size of the beam. For this purpose, it is more efficient to increase the beam's depth (d) than its breadth (b). Deflection formulas for common loading conditions may be found in the AISC Manual. Deflection formulas for the two most common beam conditions are shown in Figure 2.24.

Example #5

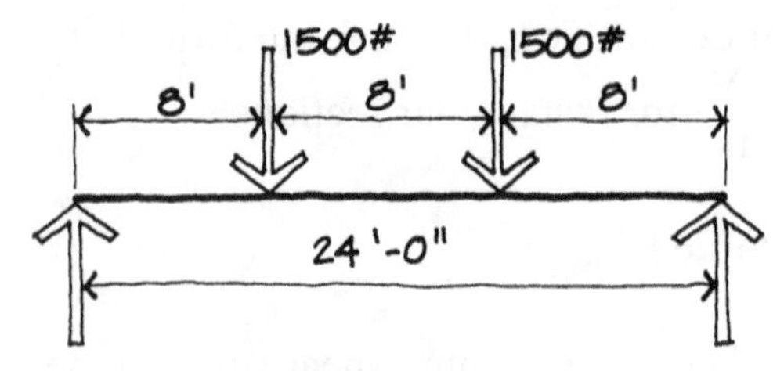

Figure 3.6

What size wood beam should be used for the span and load shown in Figure 3.6, neglecting the weight of the beam? Use select structural Douglas fir-larch and limit the deflection to 0.80". The diagrams and formulas for this loading condition are reproduced in Figure 3.7 from the AISC Manual by permission of the American Institute of Steel Construction.

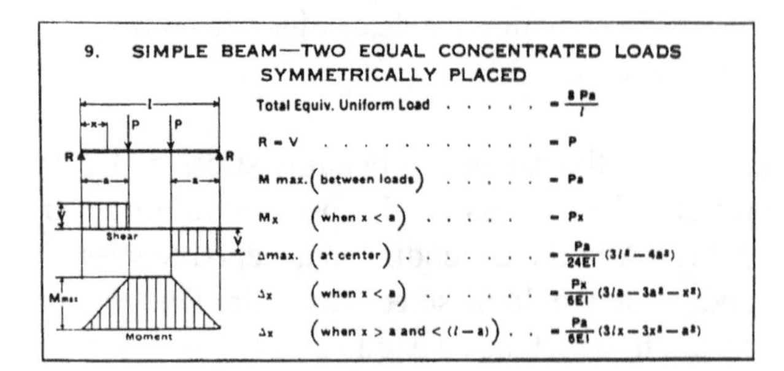

Figure 3.7

Solution:

1. From the formulas in Figure 3.7, the maximum bending moment (M) = Pa = 1,500# × 8 ft. = 12,000 ft.-lbs. Next determine the design value for bending (F_b). From Table 4D in the appendix, for the categories *Select Structural* and *Beams and Stringers*, we read F_b = 1,600 psi, F_v = 170 psi, and E = 1,600,000 psi.

   We compute the required section modulus (S) = M/F_b = 12,000 ft.-lbs. × 12 in./ft. ÷ 1,600 psi = 90 in.^{3}. We do not apply any increase to the design value, since the problem statement does not give any indication of the nature of the load.

Using Figure 3.2, we tentatively select a 6 × 12. Its tabulated properties are:

$A = 63.25$ sq. in., $S = 121.2$ in.^{3},

$I = 697.1$ in.^{4}.

2. We check the maximum horizontal shear stress $f_v = \frac{3}{2}\frac{V}{bd}$. The maximum vertical shear is equal to either reaction = 1,500#. The area (bd) of a 6 × 12 is 63.25 sq. in.

$$f_v = \frac{3 \times 1{,}500\#}{2 \times 63.25 \text{ in.}^2} = 35.5 \text{ psi}$$

This is well under the allowable F_v of 170 psi as determined from Table 4D in the appendix.

3. We calculate the deflection from the formula

$$\Delta = \frac{Pa(3L^2 - 4a^2)}{24\ EI}$$

$$= \frac{1{,}500 \times 8 \times (3 \times 24^2 - 4 \times 8^2) \times 1{,}728}{24(1.6 \times 10^6)(697.1)}$$

$= 1.14$"

Note that we used the factor 1,728 in the numerator, which is 12^3, to convert feet^3 to in.^{3}.

Since the problem statement limits the deflection to 0.8", we must use a beam with a greater moment of inertia (I).

By ratio, the required $I = \frac{1.14}{0.80} \times 697.1$ in.^{4} $= 993$ in.^{4}

We choose a 6 × 14 from Figure 3.2 ($I = 1{,}128$ in.^{4}).

Although this problem is longer than any of the problems you are likely to encounter on the exam, it serves to illustrate the three beam design criteria: bending, shear, and deflection.

Size Factor

For wood beams deeper than 12 inches, the design value for bending F_b is multiplied by the size factor C_F, which has the effect of slightly decreasing the value of F_b. The notes preceding Table 4D in the National Design Specification supplement gives information on how to determine C_F.

Notching of Beams

Notching of wood beams should be avoided if possible. If unavoidable, a beam notched on its lower side at the end should be checked for horizontal shear by the formula:

$$V_r' = \left[\frac{2}{3}F_V bd_n\right]\left[\frac{d_n}{d}\right]^2$$

where

V_r' = adjusted design shear force in pounds
F_v = design value for shear in pounds per square inch
b = breadth of beam in inches
d_n = depth of member remaining at a notch in inches
d = total depth of beam in inches

The maximum calculated shear force V from the analysis must be less than or equal to V_r'.

Example #6

An 8 × 12 beam is notched 2-1/4" on its lower side at its end support (see Figure 3.8). If the shear force is 3,000#, check the shear at the notch, given $F_v = 175$ psi.

Solution:

A nominal 8 × 12 has actual dimensions of 7-1/2" by 11-1/2" (see the table in Figure 3.2). The actual depth at the notch is 11-1/2 minus 2-1/4 = 9-1/4".

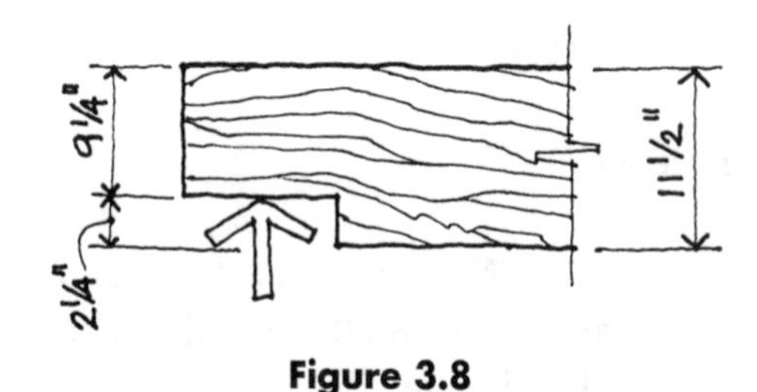

Figure 3.8

Use the formula

$$V_r' = \left[\frac{2}{3}F_V b d_n\right]\left[\frac{d_n}{d}\right]^2$$

$$= \left[\frac{2}{3}(175)(7.5)(9.25)\right]\left[\frac{9.25}{11.5}\right]^2$$

$$= 5{,}237\#$$

Since 5,237# is greater than 3,000#, the beam is adequate for shear at the notch.

Some additional comments about the notching of wood beams:

1. A gradual change in cross-section compared with a square notch increases the shearing strength nearly to that computed for the actual depth above the notch.
2. A beam notched or beveled on its upper side at the ends causes a less severe condition from the standpoint of stress concentration.
3. When notches are near the middle of a beam span, the net depth is used to determine bending strength.
4. The deflection of a beam is practically unaffected by small, widely spaced notches.

Bearing

The load on a wood beam compresses the wood fibers where the beam rests on its supports. The area of the beam bearing on the supports must therefore be large enough to transfer the load without overstressing the wood fibers. The required bearing area is determined by dividing the beam reaction V by the design values in compression perpendicular to grain $F_{c\perp}$, for the species and grade of lumber used.

Example #7

An 8 × 10 Dense No. 1 Douglas fir-larch beam has an end reaction of 8,000 pounds. What is the required bearing area?

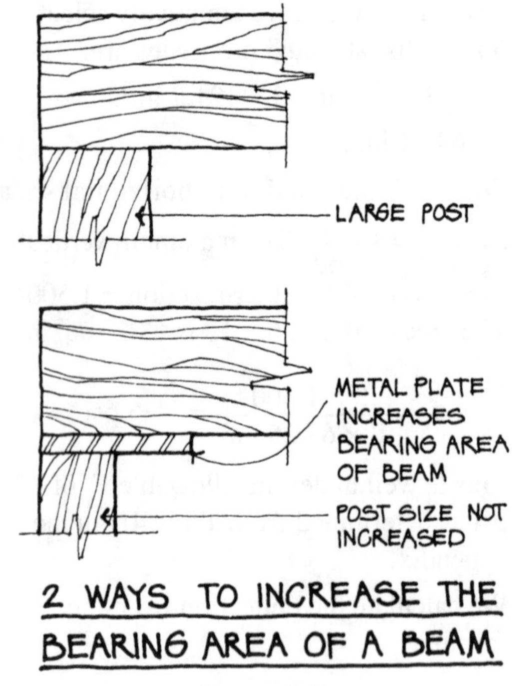

Figure 3.9

Solution:

The required bearing area $= \frac{V}{F_{c\perp}} = \frac{8{,}000\#}{730 \text{ psi}}$

$= 10.96$ *sq. in.*

Note that the value of $F_{C\perp}$ (730 psi) is determined from Table 4D in the appendix.

If the required bearing area exceeds the actual area of the beam bearing on its supporting post, two ways to increase the bearing area are (1) using a larger post, or (2) using a metal plate between the beam and the post (see Figure 3.9).

Joists

Joists are small, closely spaced beams used to support floor, ceiling, or roof loads. They are usually 2 × 8, 2 × 10, or 2 × 12, and spaced 12, 16, or 24 inches apart.

Since they are actually beams, joists may be designed by the methods described in this lesson. However, to simplify the selection of joists supporting uniformly distributed loads, a number of tables have been devised for various spans, joist sizes and spacings, and loads.

These tables take into account the factors previously discussed in this lesson: bending stress, horizontal shear stress, and deflection.

Because joists are very slender, it is usually necessary to laterally stabilize them by using solid blocking between the joists at the supports, and cross bridging or solid blocking between the joists at intervals of eight feet or less, as shown in Figure 3.10.

Joists are generally supported by bearing directly on the supporting beam or wall, or by being hung from the supporting wood or steel beam with metal hangers (see Figure 3.10).

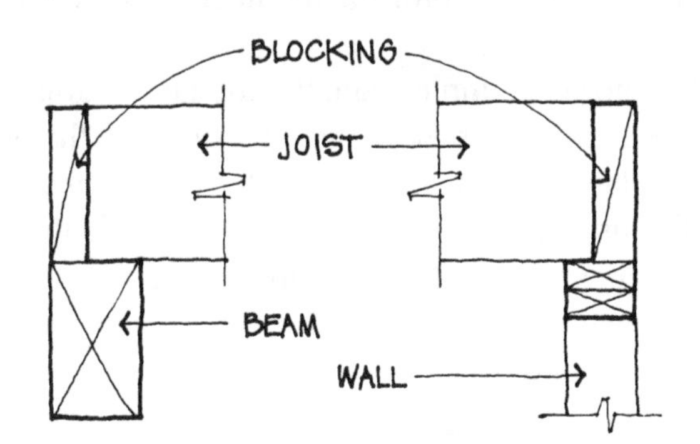

Figure 3.10

Planking

Wood planking laid flat and spanning between supporting beams frequently provides an economical roof or floor framing system. The planks are generally tongue-and-groove, either sawn or glued-laminated, and two, three, or four inches thick in nominal dimension.

Planks are usually designed as simple beams supporting a uniformly distributed load; as with joists, tables are available which facilitate their design.

The advantages of planking include simplicity (one material functions as finish floor, ceiling, and insulation), good fire resistance, attractive appearance, and easy installation.

However, planking systems frequently are uneconomical in the use of material, they provide no space in which to conceal mechanical and electrical ducts, and they may deflect excessively on longer spans.

GLUED LAMINATED BEAMS

Glued laminated beams comprise an assembly of wood laminations in which the grain of all laminations is approximately parallel longitudinally, with the laminations bonded with adhesives. Almost without exception, *glulam* members are manufactured at a plant, under exacting quality control.

The individual laminations are 1" or 2" in nominal thickness, seasoned and properly joined, and shop glued together. Using proven adhesives, the joints are durable and as strong as the wood itself. The species generally used is either Douglas fir or Southern pine, although the species and stress grades of the laminations that comprise a beam may vary.

For a number of reasons, glued laminated members have design values that are greater than those for sawn timbers: the individual laminations are thin and easily seasoned before fabrication, the laminations may be selected free of defects normally found in larger timbers, and any defects in the beam will be small and dispersed. In addition, high grade laminations may be placed in the outer portions of a beam, where their high strength may be effectively used, and lower grade laminations in the inner portion, where their low strength will not greatly affect the overall strength of the member.

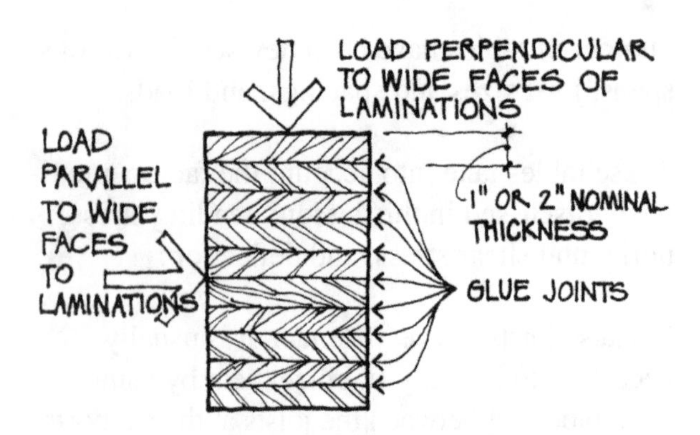

Figure 3.11

Glued laminated beams are usually loaded perpendicular to the wide faces of the laminations although occasionally, the load is applied parallel to the wide faces (see Figure 3.11). In that case, the design values are usually reduced. This can be seen in Table 5A of the National Design Specification Supplement, referenced in the 2009 IBC (see appendix).

The design values given in the table are listed not by species but by combination symbol. A description of the columns in the table follows:

Column 1 The combination symbol consists of two digits, the letter F, a dash, the letter V, and a number. The first two digits are the design tensile bending stress in hundreds of psi, the letter F designates stress rating, and the letter V followed by a number indicates visual grading. If the letter were E instead of V, it would designate machine stress rating.

Column 2 Species, etc., indicates the lumber used for outer and core laminations. DF stands for Douglas fir-larch, HF for hem-fir, AC for Alaska cedar, SP for Southern pine, and SW for softwood species.

Columns 3 & 4 Extreme fiber stress in bending F_{bx}. The design values given are for tension and compression for the member in bending.

In a normal simple span, the top laminations of a beam are stressed in compression and the bottom laminations in tension. Under certain conditions it is possible that the upper laminations, which are normally in compression, might be in tension. The design values in column 4 apply to this condition.

Columns 5 & 6 Compression perpendicular to grain

Columns 7–14 Self-explanatory

Columns 15–17 Design values and modulus of elasticity for laminated wood members used primarily for axial load, such as columns, posts, and truss members

Columns 18 & 19 Specific gravity of the timber for use in fastener design

The table also includes suggested uses for various combinations, as well as a number of footnotes.

In general, dry-use adhesive (casein) should be used for interior locations, and wet-use (waterproof) adhesive for exterior locations.

The design values for glued laminated members are modified by a number of factors, including the following:

1. Curvature Factor

For the curved portion of members, the design value in bending is modified by multiplication by the curvature factor

$$C_c = 1 - 2{,}000\left(\frac{t}{R}\right)^2$$

where
t = thickness of lamination in inches
R = radius of curvature of the inside face of lamination in inches

2. Volume Factor

When the size of a glued laminated beam is other than 5-1/8" wide or 12" deep, or if its span is other than 21 feet, F_b is multiplied by the volume factor C_v, which has the effect of slightly decreasing the value of F_b (see National Design Specification Section 5.3.6 for more information).

3. Beam Stability Factor

When the depth of a beam exceeds its breadth, the design values in bending may require reduction, unless the beam is adequately supported laterally. Refer to the National Design Specification Section 5.3.5 for more detailed information.

4. Short-Time Loading

The design values for glued laminated members may be increased for short-time loading the same as for sawn timbers, as described in National Design Specification Section 2.3.2.

Standard Sizes of Glued Laminated Members

The standard finished widths of glued laminated members are as follows:

Nominal Width	Net Finished Width
4"	3-1/8"
6"	5-1/8"
8"	6-3/4"
10"	8-3/4"
12"	10-3/4"
14"	12-1/4"

Net Depth in Inches

The standard depths of glued laminated members are based on an exact number of laminations; the net depth of a 2" nominal lamination is 1-1/2", and the net depth of a 1" nominal lamination is 3/4".

Number of Laminations	1" Nominal Laminations	2" Nominal Laminations
4	3"	6"
5	3-3/4"	7-1/2"
6	4-1/2"	9"
etc.	etc.	etc.

While it is not difficult to compute the section properties of glued laminated beams, it is a time-consuming exercise. Tables of section properties, which are available from various sources, greatly facilitate the selection of glulam members. The table in Figure 3.12 is from the National Design Specification Supplement. Glulam members may be fabricated in virtually any size or shape, and theoretically to any length. Camber (curvature opposite to the anticipated deflection) can easily be built in, to eliminate ponding of water and avoid the appearance of sag. Compared to sawn timbers, glued laminated members are more dimensionally stable, less subject to shrinkage and warping, and have greater strength and generally a more attractive appearance. An additional advantage to glued laminated members is that the architect has the opportunity to select an *appearance grade* consistent with the use of the structure—*industrial* where appearance is not of primary concern, *architectural* where appearance is an important requirement, and *premium* where the finest appearance is required. These grades are for appearance only and are not related to strength.

Example #8

A glued laminated beam 5-1/8" wide made from 1-1/2" net laminations spans 30 feet and supports the following loads:

Dead load including its own weight: 300 plf
Roof live load: 600 plf

Use combination 24F-V1, which has design values F_b of 2,400 psi and F_v of 300 psi.

What size beam should be used? Do not consider deflection or the volume or beam stability factor.

Solution:

Case 1 Maximum dead load moment =

$$\frac{wL^2}{8} = \frac{300 \times 30^2}{8} = 33{,}750'\#$$

Case 2 Maximum dead + live load moment =

$$\frac{(300 + 600)30^2}{8} = 101{,}250'\#$$

Case 1 Section modulus required (S) =

$$\frac{M}{F_b} = \frac{33{,}750 \times 12}{2{,}400} = 168.75 \text{ in.}^3$$

Case 2 $S = \dfrac{101{,}250 \times 12}{2{,}400 \times 1.25} = 405.0 \text{ in.}^3$

Note that in both cases, the factor 12 is used to change foot-lbs to inch-lbs. to maintain consistent units. In case 2, the factor 1.25 in the denominator provides for a 25 percent increase in the design value for bending, which is permitted for a wood member supporting roof live load.

Comparing the two values of S required indicates that case 2 governs. Using Figure 3.12, we now tentatively select a 5-1/8" × 22-1/2" deep beam, which has a section modulus S of 432.4 in.^{3}.

We now check the horizontal shear stress

$$f_v = \frac{3}{2}\frac{V}{bd} = \frac{3}{2}\frac{(300 + 600) \times 15}{115.3} = 175.6 \text{ psi}$$

Note that 115.3 is the area (bd) of a 5-1/8" × 22-1/2" beam.

The design value for shear F_v is 300 psi. We can increase this by 25 percent, since the member supports roof live load.

$$F_v = 300 \times 1.25 = 375 \text{ psi}$$

Since the calculated shear stress f_v of 175.6 psi is less than the F_v of 375 psi, *the 5-1/8" × 22-1/2" beam is satisfactory.*

WOOD COLUMNS

Axial Compression

Solid wood columns are generally square or rectangular and can be classified as short, intermediate, or long, each of which has a different mode of failure at ultimate load. For short columns, which have a small *slenderness ratio* l/d (length to least lateral dimension),

Depth	Area	X-X Axis			Y-Y Axis	
d (in.)	A (in.^{2})	I_x (in.^{4})	S_x (in.^{3})	r_x (in.)	I_y (in.^{4})	S_y (in.^{3})
		5-1/8 in. Width			**(r_y = 1.479 in.)**	
6	30.75	92.25	30.75	1.732	67.31	26.27
7-1/2	38.44	180.2	48.05	2.165	84.13	32.83
9	46.13	311.3	69.19	2.598	101.0	39.40
10-1/2	53.81	494.4	94.17	3.031	117.8	45.96
12	61.50	738.0	123.0	3.464	134.6	52.53
13-1/2	69.19	1051	155.7	3.897	151.4	59.10
15	76.88	1441	192.2	4.330	168.3	65.66
16-1/2	84.56	1919	232.5	4.763	185.1	72.23
18	92.25	2491	276.8	5.196	201.9	78.80
19-1/2	99.94	3167	324.8	5.629	218.7	85.36
21	107.6	3955	376.7	6.062	235.6	91.93
22-1/2	115.3	4865	432.4	6.495	252.4	98.50
24	123.0	5904	492.0	6.928	269.2	105.1
25-1/2	130.7	7082	555.4	7.361	286.0	111.6
27	138.4	8406	622.7	7.794	302.9	118.2
28-1/2	146.1	9887	693.8	8.227	319.7	124.8
30	153.8	11530	768.8	8.660	336.5	131.3
31-1/2	161.4	13350	847.5	9.093	353.4	137.9
33	169.1	15350	930.2	9.526	370.2	144.5
34-1/2	176.8	17540	1017	9.959	387.0	151.0
36	184.5	19930	1107	10.39	403.8	157.6
		5-1/2 in. Width			**(r_y = 1.588 in.)**	
6	33.00	99.00	33.00	1.732	83.19	30.25
7-1/2	41.25	193.4	51.56	2.165	104.0	37.81
9	49.50	334.1	74.25	2.598	124.8	45.38
10-1/2	57.75	530.6	101.1	3.031	145.6	52.94
12	66.00	792.0	132.0	3.464	166.4	60.50
13-1/2	74.25	1128	167.1	3.897	187.2	68.06
15	82.50	1547	206.3	4.330	208.0	75.63
16-1/2	90.75	2059	249.6	4.763	228.8	83.19
18	99.00	2673	297.0	5.196	249.6	90.75
19-1/2	107.3	3398	348.6	5.629	270.4	98.31
21	115.5	4245	404.3	6.062	291.2	105.9
22-1/2	123.8	5221	464.1	6.495	312.0	113.4
24	132.0	6336	528.0	6.928	332.8	121.0
25-1/2	140.3	7600	596.1	7.361	353.5	128.6
27	148.5	9021	668.3	7.794	374.3	136.1
28-1/2	156.8	10610	744.6	8.227	395.1	143.7
30	165.0	12380	825.0	8.660	415.9	151.3
31-1/2	173.3	14330	909.6	9.093	436.7	158.8
33	181.5	16470	998.3	9.526	457.5	166.4
34-1/2	189.8	18820	1091	9.959	478.3	173.9
36	198.0	21380	1188	10.39	499.1	181.5

Figure 3.12

failure is by crushing. At intermediate values of l/d, failure is by a combination of crushing and buckling. At large l/d values, long wood columns fail by lateral deflection or buckling. The maximum allowable l/d value is 50.

Sometimes a column is braced in one direction and not the other. In that case, there are two different values of unbraced length 1. The ratio l/d must be evaluated in both directions, and the larger value is used in the design of the column.

The adjusted compressive stress in a wood column F_c' may not exceed F_c, the design value for compression parallel to the grain (see Table 4A or 4D in the appendix), which may be adjusted for short-time loading as noted on page 46.

Figure 3.13 graphically shows the relationship between F_c' and l/d for short, intermediate, and long wood columns.

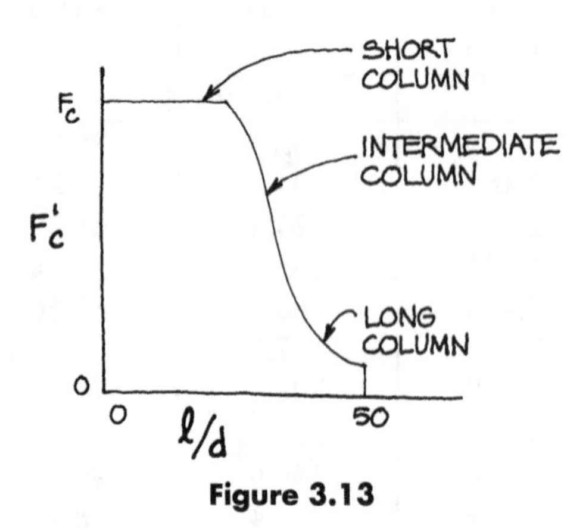

Figure 3.13

Combined Bending and Axial Compression

Wood columns are sometimes required to resist bending moment as well as axial compression. The bending moment may be caused by lateral load, such as wind, or by eccentric load. For such cases, an interaction formula is used, which is also subject to the duration of load adjustments shown on page 46.

Since the axial stress is P/A and the bending stress is Mc/I, the stress under combined bending and axial compression is equal to P/A + Mc/I (see page 39).

Column Base Details

There are a number of methods commonly used to anchor the base of a wood column. One typical detail, shown in Figure 3.14, utilizes a steel U-strap cast into the concrete and bolted to the wood column. A steel plate or other moisture barrier should be placed between the column and the concrete. This detail can resist both horizontal forces and uplift, while downward forces are resisted by direct bearing.

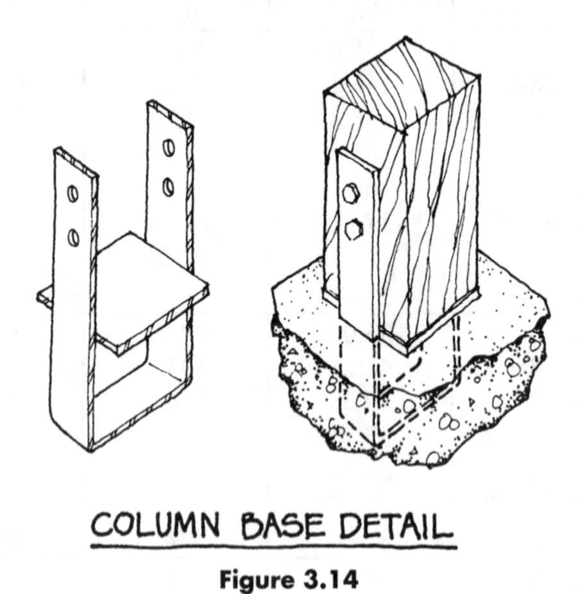

Figure 3.14

LESSON 3 QUIZ

1. What is especially critical for wood beams with short spans and large loads?

   A. Bending moment **C.** Flexural stress
   B. Deflection **D.** Horizontal shear

2. Which of the following are generally ignored in the design and detailing of structural wood members? Check all that apply.

   A. Cross-grain shrinkage
   B. Longitudinal shrinkage
   C. Thermal expansion
   D. Thermal contraction

3. Most lumber used structurally is

   I. softwood.
   II. hardwood.
   III. from coniferous trees.
   IV. from deciduous trees.

   A. I and III **C.** I and IV
   B. II and IV **D.** II and III

4. The value of F_b for a glued laminated member is 2,200 psi. If the beam has a 6-3/4" × 24" rectangular cross-section, what is the maximum moment it can safely develop? Assume the volume factor is 1.

   A. 1,188.0 in.-kips **C.** 1,069.2 in.-kips
   B. 1,425.6 in.-kips **D.** 1,555.2 in.-kips

5. Why is camber often built into a glued laminated beam?

   I. To reduce flexural stress
   II. To avoid the appearance of sag
   III. To allow the use of lumber with a lower stress rating
   IV. To eliminate the ponding of water
   V. To compensate for deflection
   VI. To minimize deflection

   A. II, IV, and V **C.** I, III, and VI
   B. II, V, and VI **D.** II, IV, and VI

6. A roof beam has a maximum moment of 30,000 foot-pounds. What is the required section modulus for a No. 1 Douglas fir-larch member to resist this moment, if the tabulated F_b value is 1,000 psi?

   A. 177.8 in.^{3} **C.** 222.2 in.^{3}
   B. 213.3 in.^{3} **D.** 288.0 in.^{3}

7. Select the CORRECT statements concerning the notching of wood beams.

I. The stress concentration is the same whether the beam is notched on its upper or lower side.

II. A notch near the middle of a beam's span has practically no effect on its deflection.

III. The loss of shear strength caused by a gradual change in a beam's cross-section is not as great as that caused by a square notch.

IV. The maximum depth of notch permitted by the IBC is 20 percent of the beam depth.

V. When notches are small and widely spaced, they do not affect flexural strength.

A. I, II, and IV
C. III only
B. II and III
D. II, III, and V

8. Select the CORRECT statement about glued laminated beams.

A. They are normally loaded parallel to the wide faces of the individual laminations.

B. The individual laminations are thin and therefore difficult to season before fabrication.

C. Since the glue is the weakest part of the beam, stresses at the glue line are critical.

D. The individual laminations are not required to be of the same species and stress grade.

9. For a given load and span condition, a 4 × 10 beam has a flexural stress of 1,250 psi. What is the flexural stress for a 4 × 12 beam with the same span and loading?

A. 1,041.7 psi
C. 694.8 psi
B. 845.0 psi
D. 1,500 psi

10. A sawn timber floor beam spans 22 feet. What is the maximum deflection permitted by the IBC for live load only, and for total load?

A. 1.10 inches for live load, 1.10 inches for total load

B. 0.55 inch for live load, 0.73 inch for total load

C. 0.73 inch for live load, 1.10 inches for total load

D. Deflection is not limited by the IBC

STEEL CONSTRUCTION

PROPERTIES OF STEEL

Structural steel is one of the most important structural materials used in buildings. Its widespread use can be attributed to its desirable properties, its availability, and its generally competitive price.

Some of its desirable qualities include great strength and stiffness, durability, workability, and reliability. Steel is subject to closer quality control during its manufacture and fabrication than any other structural material in general use. In addition, steel is incombustible, unaffected by fungi or insects, and dimensionally stable.

However, there are some disadvantages associated with steel, principally its susceptibility to corrosion and its lack of fire resistance.

Corrosion requires the presence of both moisture and oxygen. There are a number of protective coatings that are used to resist corrosion, including paint, zinc (galvanizing), and concrete. Another method widely used to improve the corrosion resistance of steel is alloying, which is the addition of other elements, such as chromium, copper, and nickel.

Although steel is incombustible, it lacks fire resistance. When steel is exposed to a severe fire, it does not burn, but it loses so much strength that it may fail or deform excessively. Therefore, to be fire resistant, steel generally needs a protective coating of concrete, plaster, gypsum wallboard, or an approved sprayed-on material. Fire ratings of from one to four hours may be attained in this manner.

All steel, regardless of its strength or other properties, has the same value of modulus of elasticity (E)—about 29,000,000 psi. This is higher than any other structural material in general use, and indicates that steel is very stiff, that is, very resistant to deformation.

Steel has another property that makes it particularly advantageous in earthquake-resistant construction: ductility. In this context, ductility means the ability to absorb energy in the inelastic range, when the material is stressed above its yield point, without failure.

The chemical composition of structural steel is: iron (over 98 percent), carbon (about one quarter of 1 percent), and small quantities of other elements. The most important factor affecting the properties of steel is usually its carbon content; a small amount of carbon generally results in an increase in strength and a decrease in ductility. Most steels have a carbon content between 0.12 and 1.8 percent.

Most of the structural steel shapes used in this country are formed by rolling. Rolled shapes are designated by a standard nomenclature consisting of a group symbol, the nominal depth in inches, and the weight in pounds per lineal foot, in that sequence. The most widely used groups are designated W (wide flange), M (miscellaneous), S (American standard beams), C (American standard channels), and L (angles). Inch and pound symbols are not used, nor is reference made to any specific manufacturer. Table 4.1 illustrates some of the standard designations.

Group	Example	Profile
Wide Flange Shapes (W)	W 24 x 76	I
Miscellaneous Shapes (M)	M 8 x 6.5	I
American Standard Beams (S)	S 24 x 100	I
American Standard Channels (C)	C 12 x 20.7	[
Angles (L)	L 6 x 4 x 3/8	L

Table 4.1

Thus, W24 × 76 indicates a 24-inch-deep wide flange section which weighs 76 pounds per foot.

There are a number of other standard shapes, including pipe columns, tubing, plates, and tees, which are used for specific applications in buildings. However, the W (or wide flange) is the most common shape used for beams, girders, and columns. It is symmetrical about both axes, and is a very efficient bending member because most of its material is in the flanges, at the greatest possible distance from the neutral axis (see Figure 4.1).

Structural steel shapes are available in the United States with different yield strength, F_y, and ultimate strength, F_u, as established by the American Society for Testing and Materials (ASTM). W shapes are ASTM A992, with $F_y = 50$ ksi and $F_u = 65$ ksi. ASTM A36 steel, widely used for many years, is currently being replaced with steel having higher yield points. M, S, C, L shapes and bars and plates are still selectively available in A36 steel, $F_y = 36$ ksi, and $F_u = 58$ ksi, although many of these shapes will be A572 Grade 50, $F_y = 50$ ksi and $F_u = 65$ ksi, or A992.

The properties and dimensions of all standard shapes are tabulated in the AISC Manual. Candidates are urged to obtain a copy of this mandatory reference of steel construction and become familiar with its scope and format.

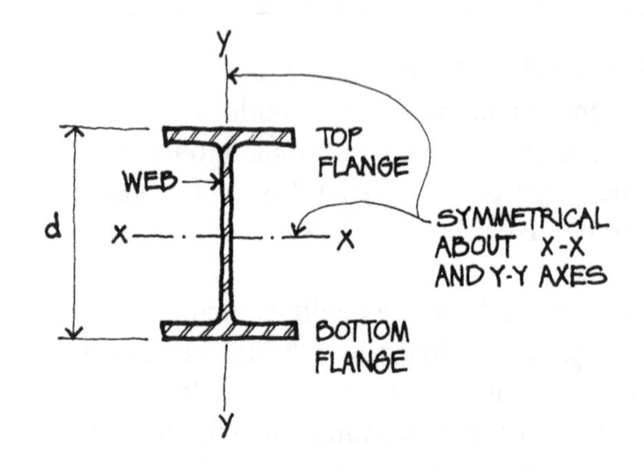

Figure 4.1

DESIGN REQUIREMENTS

The design requirements for structural steel are established by the American Institute of Steel Construction (AISC) Specification, which is part of the AISC Manual. This specification is adopted by reference in the IBC.

The AISC Specifications provide design requirements for both Load and Resistance Factor Design (LRFD) and Allowable Stress Design (ASD). In the LRFD method, sometimes called *limit states design*, the various loads (dead, live, etc.) are multiplied by their respective *load factors*; these factored loads are used to determine the required strength R_u that the member must be able to resist. The nominal strength R_n, which is specified in Chapters B through K of the Specifications, is multiplied by a resistance factor φ, which is less than 1. The design strength of the member φR_n must be greater than or equal to the required strength R_u.

In the ASD method, the required strength R_a is determined from the actual service loads that act on a member. The nominal strength R_n is divided by the safety factor Ω, which is also specified in Chapters B through K. The allowable strength R_n/Ω must be greater than or equal to the required strength R_a. In this lesson, we will focus on designing steel members using the ASD method.

The nominal strength R_n is generally a function of F_y, the specified minimum yield point of the grade of steel used, in kips per square inch (ksi). For ASTM A36 steel, $F_y = 36$ ksi, and for ASTM A992 steel, $F_y = 50$ ksi.

The following is a summary of nominal flexural strength M_n for ASTM A992 steel. All values are for tension and compression caused by bending, unless noted otherwise.

For compact, adequately braced, symmetrical members, $M_n = 50Z_x$, where Z_x is the plastic section modulus of the section about the x-x axis. According to the User Note in Section F2 of the Specification, most wide flange and channel sections fall in this category.

In the case of I-shaped and channels bent about their weak axis, $M_n = 50Z_y$, but less than or equal to $80S_y$, where Z_y and S_y are the plastic and elastic section moduli of the section, respectively, about the y-y axis.

Equations of M_n are also given in the Specifications for many other steel shapes, including tubes, tees, double angles, and single angles, to name a few.

For the vast majority of currently available W sections, the nominal shear strength V_n of unstiffened or stiffened webs is $30A_w$, where A_w is the web area, which is equal to the overall depth d times the web thickness t_w.

To summarize, then, the nominal flexural strength M_n for most symmetrical, laterally braced members is $50Z_x$. For W sections that are laterally unbraced along their compression flanges, the charts on pages 3-96 through 3-131 of the AISC Manual are used. Later in this lesson we will illustrate the use of these charts and discuss lateral support of beams.

STEEL BEAM DESIGN FOR FLEXURE

The procedure used to design a steel beam for flexure (bending) is as follows.

1. Determine the maximum bending moment (M).
2. Compute the required plastic section modulus (Z) by using the flexure formula $Z = M\Omega/F_y$, where for flexure, the safety factor $\Omega = 1.67$.

3. Refer to the tables of properties of steel sections, pages 1-10 through 1-110 of the AISC Manual, or the plastic section modulus table on pages 3-11 through 3-19 of the Manual, and select a beam with a plastic section modulus Z greater than or equal to that required. In general, the lightest section is the most economical. For laterally supported beams supporting uniformly distributed loads, an alternate procedure is to use the tables on pages 3-33 through 3-95 of the AISC Manual.

Example #1

Select the most economical structural steel beam to support the load shown in Figure 4.2. Assume full lateral support and use ASTM A992 steel.

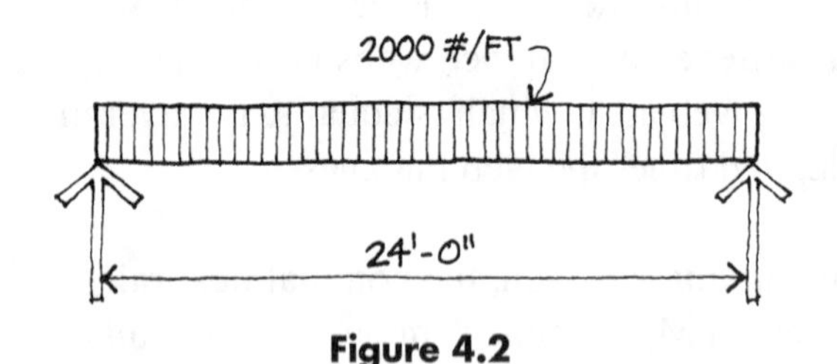

Figure 4.2

Solution:

There are three alternative procedures that may be used to select the most economical steel beam, and we will illustrate the use of all three methods.

Compute the maximum bending moment $M = \frac{wL^2}{8} = 2{,}000(24)^2/8 = 144{,}000'\#$

Compute $Z = M\Omega/F_y = 144{,}000 \times 12 \times 1.67/50{,}000 = 57.7 \text{ in.}^3$

Note that the factor 12 in the numerator is used to convert ft.-lbs. to inch-lbs.

Method 1: Enter the table on page 3-18 of the AISC Manual, a portion of which is reproduced in Figure 4.3, and select *W18 × 35*, which has a plastic section modulus of 66.5 in.^{3}. In this table, the lightest shape in each group is shown in bold-face type. The resisting moment may also be read from this table; its value of 166 ft.-kips is greater than the actual moment of 144 ft.-kips.

Method 2: The W18 × 35 beam may also be selected from page 1-19 of the AISC Manual. Portions of pages 1-18 and 1-19 are reproduced in Figure 4.4.

Method 3: Use the table on page 3-59 of the AISC Manual, part of which is reproduced in Figure 4.5. Where the line representing the 24-foot span intersects the column headed W18 × 35, we read the value of 55.3 kips, which is the *total* allowable uniform load. The allowable uniform load per foot = 55.3 kips/24 feet = 2.3 kips per foot. Since this is greater than the actual load of 2.0 kips per ft., the W18 × 35 is adequate.

LATERAL SUPPORT

In the preceding example, we used the term *lateral support*, and now we will explain that concept more fully. When a beam resists bending moment, one flange is in compression while the other flange is in tension. For simple beams supporting downward vertical loads, the top flange is in compression and the bottom flange is in tension, as shown in Figure 4.6.

For cantilever and overhanging beams supporting downward vertical loads, the top flange is in tension and the bottom flange is in compression (see Figure 4.7).

Z_X

Table 3-2 (continued)
W Shapes
Selection by Z_x

F_y = 50 ksi

Shape	Z_x	M_{px}/Ω_b	$\phi_b M_{px}$	M_{rx}/Ω_b	$\phi_b M_{rx}$	BF		L_p	L_r	I_x	V_{nx}/Ω_v	$\phi_v V_{nx}$
		kip-ft	kip-ft	kip-ft	kip-ft	kips	kips				kips	kips
	in.^{3}	ASD	LRFD	ASD	LRFD	ASD	LRFD	ft	ft	in.^{4}	ASD	LRFD
W18×35	**66.5**	**166**	**249**	**101**	**151**	**8.07**	**12.1**	**4.31**	**12.4**	**510**	**106**	**159**
W12×45	64.2	160	241	101	151	3.83	5.75	6.89	22.4	348	80.8	121
W16×36	64.0	160	240	98.7	148	6.19	9.31	5.37	15.2	448	93.6	140
W14×38	61.5	153	231	95.4	143	5.39	8.10	5.47	16.2	385	87.4	131
W10×49	60.4	151	227	95.4	143	2.44	3.67	8.97	31.6	272	68.0	102
W8×58	59.8	149	224	90.8	137	1.70	2.56	7.42	41.7	228	89.3	134
W12×40	57.0	142	214	89.9	135	3.66	5.50	6.85	21.1	307	70.4	106
W10×45	54.9	137	206	85.8	129	2.59	3.89	7.10	26.9	248	70.7	106
W14×34	**54.6**	**136**	**205**	**84.9**	**128**	**5.05**	**7.59**	**5.40**	**15.6**	**340**	**79.7**	**120**
W16×31	**54.0**	**135**	**203**	**82.4**	**124**	**6.76**	**10.2**	**4.13**	**11.9**	**375**	**87.3**	**131**
W12×35	51.2	128	192	79.6	120	4.28	6.43	5.44	16.7	285	75.0	113
W8×48	49.0	122	184	75.4	113	1.68	2.53	7.35	35.2	184	68.0	102
W14×30	**47.3**	**118**	**177**	**73.4**	**110**	**4.65**	**6.99**	**5.26**	**14.9**	**291**	**74.7**	**112**
W10×39	46.8	117	176	73.5	111	2.51	3.77	6.99	24.2	209	62.5	93.7
W16×26^v	**44.2**	**110**	**166**	**67.1**	**101**	**5.96**	**8.96**	**3.96**	**11.2**	**301**	**70.5**	**106**
W12×30	43.1	108	162	67.4	101	3.92	5.89	5.37	15.6	238	64.2	96.3
W14×26	**40.2**	**100**	**151**	**61.7**	**92.7**	**5.32**	**7.99**	**3.81**	**11.1**	**245**	**70.9**	**106**
W8×40	39.8	99.3	149	62.0	93.2	1.64	2.47	7.21	29.9	146	59.4	89.1
W10×33	38.8	96.8	146	61.1	91.9	2.39	3.59	6.85	21.8	171	56.4	84.7
W12×26	**37.2**	**92.8**	**140**	**58.3**	**87.7**	**3.61**	**5.42**	**5.33**	**14.9**	**204**	**56.2**	**84.3**
W10×30	36.6	91.3	137	56.6	85.0	3.08	4.62	4.84	16.1	170	62.8	94.2
W8×35	34.7	86.6	130	54.5	81.9	1.62	2.43	7.17	27.0	127	50.3	75.5
W14×22	**33.2**	**82.8**	**125**	**50.6**	**76.1**	**4.75**	**7.14**	**3.67**	**10.4**	**199**	**63.2**	**94.8**
W10×26	31.3	78.1	117	48.7	73.2	2.90	4.36	4.80	14.9	144	53.7	80.6
W8×31^f	30.4	75.8	114	48.0	72.2	1.58	2.37	7.18	24.8	110	45.6	68.4
W12×22	**29.3**	**73.1**	**110**	**44.4**	**66.7**	**4.65**	**6.99**	**3.00**	**9.17**	**156**	**64.0**	**96.0**
W8×28	27.2	67.9	102	42.4	63.8	1.66	2.50	5.72	21.0	98.0	45.9	68.9
W10×22	**26.0**	**64.9**	**97.5**	**40.5**	**60.9**	**2.68**	**4.02**	**4.70**	**13.8**	**118**	**48.8**	**73.2**
W12×19	**24.7**	**61.6**	**92.6**	**37.2**	**55.9**	**4.27**	**6.43**	**2.90**	**8.62**	**130**	**57.2**	**85.7**
W8×24	23.1	57.6	86.6	36.5	54.9	1.59	2.39	5.69	19.0	82.7	38.9	58.3
W10×19	**21.6**	**53.9**	**81.0**	**32.8**	**49.3**	**3.17**	**4.77**	**3.09**	**9.72**	**96.3**	**51.2**	**76.8**
W8×21	20.4	50.9	76.5	31.8	47.8	1.86	2.79	4.45	14.8	75.3	41.4	62.1

ASD	LRFD
$\Omega_b = 1.67$	$\phi_b = 0.90$
$\Omega_v = 1.50$	$\phi_v = 1.00$

^{f} Shape exceeds compact limit for flexure with F_y = 50 ksi.
^{v} Shape does not meet the h/t_w limit for shear in Specification Section G2.1a with F_y = 50 ksi, $\Omega_v = 1.67$, $\phi_v = 0.90$.

Figure 4.3

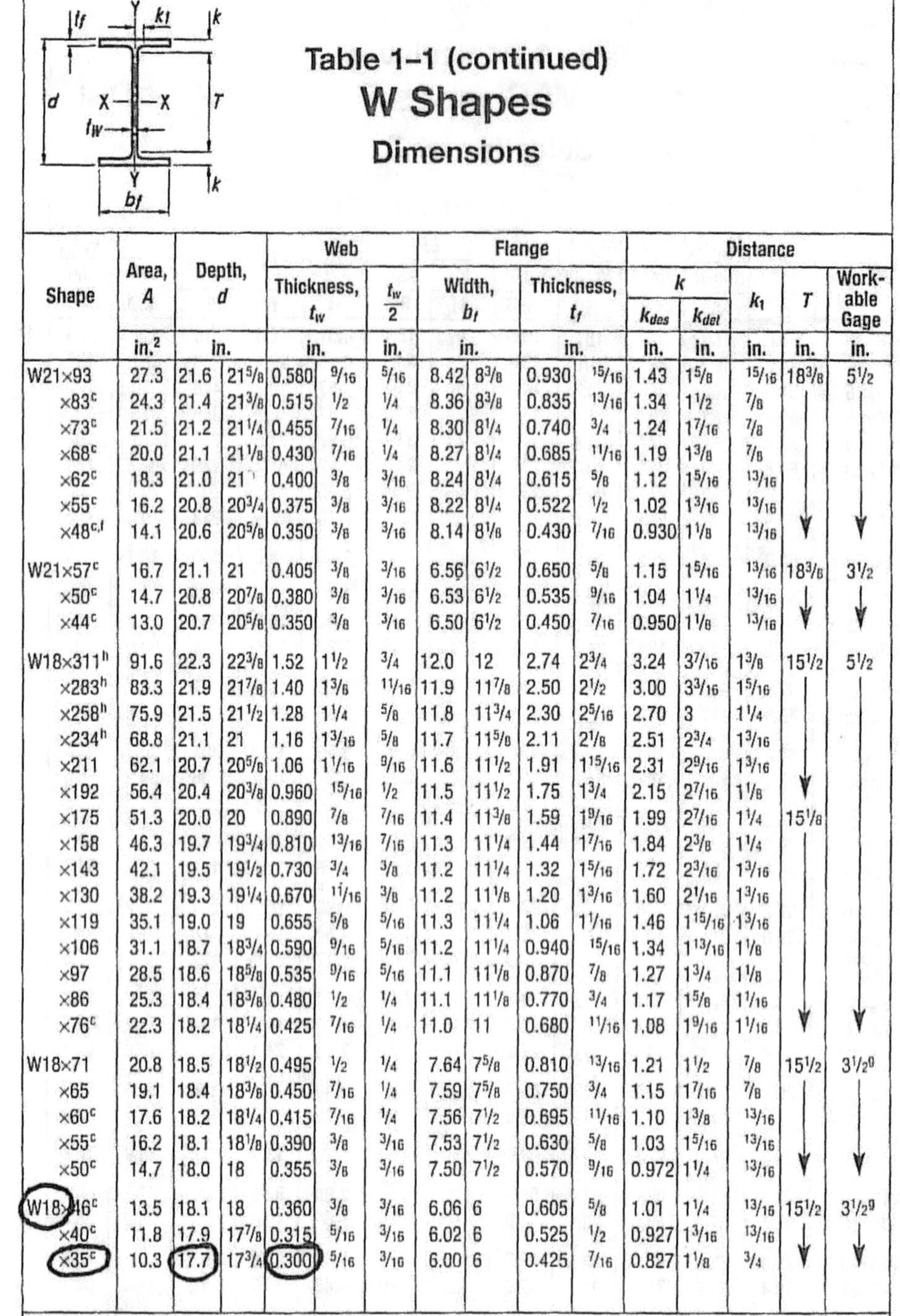

Table 1-1 (continued)
W Shapes
Dimensions

Shape	Area, A	Depth, d		Web Thickness, t_w		Web $\frac{t_w}{2}$	Flange Width, b_f		Flange Thickness, t_f		Distance k k_{des}	Distance k k_{det}	k_1	T	Workable Gage
	in.²	in.		in.		in.	in.		in.		in.	in.	in.	in.	in.
W21×93	27.3	21.6	21 5/8	0.580	9/16	5/16	8.42	8 3/8	0.930	15/16	1.43	1 5/8	15/16	18 3/8	5 1/2
×83[c]	24.3	21.4	21 3/8	0.515	1/2	1/4	8.36	8 3/8	0.835	13/16	1.34	1 1/2	7/8		
×73[c]	21.5	21.2	21 1/4	0.455	7/16	1/4	8.30	8 1/4	0.740	3/4	1.24	1 7/16	7/8		
×68[c]	20.0	21.1	21 1/8	0.430	7/16	1/4	8.27	8 1/4	0.685	11/16	1.19	1 3/8	7/8		
×62[c]	18.3	21.0	21	0.400	3/8	3/16	8.24	8 1/4	0.615	5/8	1.12	1 5/16	13/16		
×55[c]	16.2	20.8	20 3/4	0.375	3/8	3/16	8.22	8 1/4	0.522	1/2	1.02	1 3/16	13/16		
×48[c,f]	14.1	20.6	20 5/8	0.350	3/8	3/16	8.14	8 1/8	0.430	7/16	0.930	1 1/8	13/16		
W21×57[c]	16.7	21.1	21	0.405	3/8	3/16	6.56	6 1/2	0.650	5/8	1.15	1 5/16	13/16	18 3/8	3 1/2
×50[c]	14.7	20.8	20 7/8	0.380	3/8	3/16	6.53	6 1/2	0.535	9/16	1.04	1 1/4	13/16		
×44[c]	13.0	20.7	20 5/8	0.350	3/8	3/16	6.50	6 1/2	0.450	7/16	0.950	1 1/8	13/16		
W18×311[h]	91.6	22.3	22 3/8	1.52	1 1/2	3/4	12.0	12	2.74	2 3/4	3.24	3 7/16	1 3/8	15 1/2	5 1/2
×283[h]	83.3	21.9	21 7/8	1.40	1 3/8	11/16	11.9	11 7/8	2.50	2 1/2	3.00	3 3/16	1 5/16		
×258[h]	75.9	21.5	21 1/2	1.28	1 1/4	5/8	11.8	11 3/4	2.30	2 5/16	2.70	3	1 1/4		
×234[h]	68.8	21.1	21	1.16	1 3/16	5/8	11.7	11 5/8	2.11	2 1/8	2.51	2 3/4	1 3/16		
×211	62.1	20.7	20 5/8	1.06	1 1/16	9/16	11.6	11 1/2	1.91	1 15/16	2.31	2 9/16	1 3/16		
×192	56.4	20.4	20 3/8	0.960	15/16	1/2	11.5	11 1/2	1.75	1 3/4	2.15	2 7/16	1 1/8		
×175	51.3	20.0	20	0.890	7/8	7/16	11.4	11 3/8	1.59	1 9/16	1.99	2 7/16	1 1/4	15 1/8	
×158	46.3	19.7	19 3/4	0.810	13/16	7/16	11.3	11 1/4	1.44	1 7/16	1.84	2 3/8	1 1/4		
×143	42.1	19.5	19 1/2	0.730	3/4	3/8	11.2	11 1/4	1.32	1 5/16	1.72	2 3/16	1 3/16		
×130	38.2	19.3	19 1/4	0.670	11/16	3/8	11.2	11 1/8	1.20	1 3/16	1.60	2 1/16	1 3/16		
×119	35.1	19.0	19	0.655	5/8	5/16	11.3	11 1/4	1.06	1 1/16	1.46	1 15/16	1 3/16		
×106	31.1	18.7	18 3/4	0.590	9/16	5/16	11.2	11 1/4	0.940	15/16	1.34	1 13/16	1 1/8		
×97	28.5	18.6	18 5/8	0.535	9/16	5/16	11.1	11 1/8	0.870	7/8	1.27	1 3/4	1 1/8		
×86	25.3	18.4	18 3/8	0.480	1/2	1/4	11.1	11 1/8	0.770	3/4	1.17	1 5/8	1 1/16		
×76[c]	22.3	18.2	18 1/4	0.425	7/16	1/4	11.0	11	0.680	11/16	1.08	1 9/16	1 1/16		
W18×71	20.8	18.5	18 1/2	0.495	1/2	1/4	7.64	7 5/8	0.810	13/16	1.21	1 1/2	7/8	15 1/2	3 1/2[g]
×65	19.1	18.4	18 3/8	0.450	7/16	1/4	7.59	7 5/8	0.750	3/4	1.15	1 7/16	7/8		
×60[c]	17.6	18.2	18 1/4	0.415	7/16	1/4	7.56	7 1/2	0.695	11/16	1.10	1 3/8	13/16		
×55[c]	16.2	18.1	18 1/8	0.390	3/8	3/16	7.53	7 1/2	0.630	5/8	1.03	1 5/16	13/16		
×50[c]	14.7	18.0	18	0.355	3/8	3/16	7.50	7 1/2	0.570	9/16	0.972	1 1/4	13/16		
W18×46[c]	13.5	18.1	18	0.360	3/8	3/16	6.06	6	0.605	5/8	1.01	1 1/4	13/16	15 1/2	3 1/2[g]
×40[c]	11.8	17.9	17 7/8	0.315	5/16	3/16	6.02	6	0.525	1/2	0.927	1 3/16	13/16		
×35[c]	10.3	17.7	17 3/4	0.300	5/16	3/16	6.00	6	0.425	7/16	0.827	1 1/8	3/4		

[c] Shape is slender for compression with F_y = 50 ksi.
[f] Shape exceeds compact limit for flexure with F_y = 50 ksi.
[g] The actual size, combination, and orientation of fastener components should be compared with the geometry of the cross-section to ensure compatibility.
[h] Flange thickness greater than 2 in. Special requirements may apply per AISC Specification Section A3.1c.

Figure 4.4

Table 1–1 (continued)
W Shapes
Properties

W21 – W18

Nom-inal Wt.	Compact Section Criteria		Axis X-X				Axis Y-Y				r_{ts}	h_o	Torsional Properties	
	$\frac{b_f}{2t_f}$	$\frac{h}{t_w}$	I	S	r	Z	I	S	r	Z			J	C_w
lb/ft			in.^{4}	in.^{3}	in.	in.^{3}	in.^{4}	in.^{3}	in.	in.^{3}	in.	in.	in.^{4}	in.^{6}
93	4.53	32.3	2070	192	8.70	221	92.9	22.1	1.84	34.7	2.24	20.7	6.03	9940
83	5.00	36.4	1830	171	8.67	196	81.4	19.5	1.83	30.5	2.21	20.6	4.34	8630
73	5.60	41.2	1600	151	8.64	172	70.6	17.0	1.81	26.6	2.19	20.5	3.02	7410
68	6.04	43.6	1480	140	8.60	160	64.7	15.7	1.80	24.4	2.17	20.4	2.45	6760
62	6.70	46.9	1330	127	8.54	144	57.5	14.0	1.77	21.7	2.15	20.4	1.83	5960
55	7.87	50.0	1140	110	8.40	126	48.4	11.8	1.73	18.4	2.11	20.3	1.24	4980
48	9.47	53.6	959	93.0	8.24	107	38.7	9.52	1.66	14.9	2.05	20.2	0.803	3950
57	5.04	46.3	1170	111	8.36	129	30.6	9.35	1.35	14.8	1.68	20.4	1.77	3190
50	6.10	49.4	984	94.5	8.18	110	24.9	7.64	1.30	12.2	1.64	20.3	1.14	2570
44	7.22	53.6	843	81.6	8.06	95.4	20.7	6.37	1.26	10.2	1.60	20.2	0.770	2110
311	2.19	10.4	6970	624	8.72	754	795	132	2.95	207	3.53	19.6	176	76200
283	2.38	11.3	6170	565	8.61	676	704	118	2.91	185	3.47	19.4	134	65900
258	2.56	12.5	5510	514	8.53	611	628	107	2.88	166	3.42	19.2	103	57600
234	2.76	13.8	4900	466	8.44	549	558	95.8	2.85	149	3.37	19.0	78.7	50100
211	3.02	15.1	4330	419	8.35	490	493	85.3	2.82	132	3.32	18.8	58.6	43400
192	3.27	16.7	3870	380	8.28	442	440	76.8	2.79	119	3.28	18.6	44.7	38000
175	3.58	18.0	3450	344	8.20	398	391	68.8	2.76	106	3.24	18.5	33.8	33300
158	3.92	19.8	3060	310	8.12	356	347	61.4	2.74	94.8	3.20	18.3	25.2	29000
143	4.25	22.0	2750	282	8.09	322	311	55.5	2.72	85.4	3.17	18.2	19.2	25700
130	4.65	23.9	2460	256	8.03	290	278	49.9	2.70	76.7	3.13	18.1	14.5	22700
119	5.31	24.5	2190	231	7.90	262	253	44.9	2.69	69.1	3.13	17.9	10.6	20300
106	5.96	27.2	1910	204	7.84	230	220	39.4	2.66	60.5	3.10	17.8	7.48	17400
97	6.41	30.0	1750	188	7.82	211	201	36.1	2.65	55.3	3.08	17.7	5.86	15800
86	7.20	33.4	1530	166	7.77	186	175	31.6	2.63	48.4	3.05	17.6	4.10	13600
76	8.11	37.8	1330	146	7.73	163	152	27.6	2.61	42.2	3.02	17.5	2.83	11700
71	4.71	32.4	1170	127	7.50	146	60.3	15.8	1.70	24.7	2.05	17.7	3.49	4700
65	5.06	35.7	1070	117	7.49	133	54.8	14.4	1.69	22.5	2.03	17.6	2.73	4240
60	5.44	38.7	984	108	7.47	123	50.1	13.3	1.68	20.6	2.02	17.5	2.17	3850
55	5.98	41.1	890	98.3	7.41	112	44.9	11.9	1.67	18.5	2.00	17.5	1.66	3430
50	6.57	45.2	800	88.9	7.38	101	40.1	10.7	1.65	16.6	1.98	17.4	1.24	3040
46	5.01	44.6	712	78.8	7.25	90.7	22.5	7.43	1.29	11.7	1.58	17.5	1.22	1720
40	5.73	50.9	612	68.4	7.21	78.4	19.1	6.35	1.27	10.0	1.56	17.4	0.810	1440
35	7.06	53.6	510	57.6	7.04	66.5	15.3	5.12	1.22	8.06	1.52	17.3	0.506	1140

Figure 4.4 (continued)

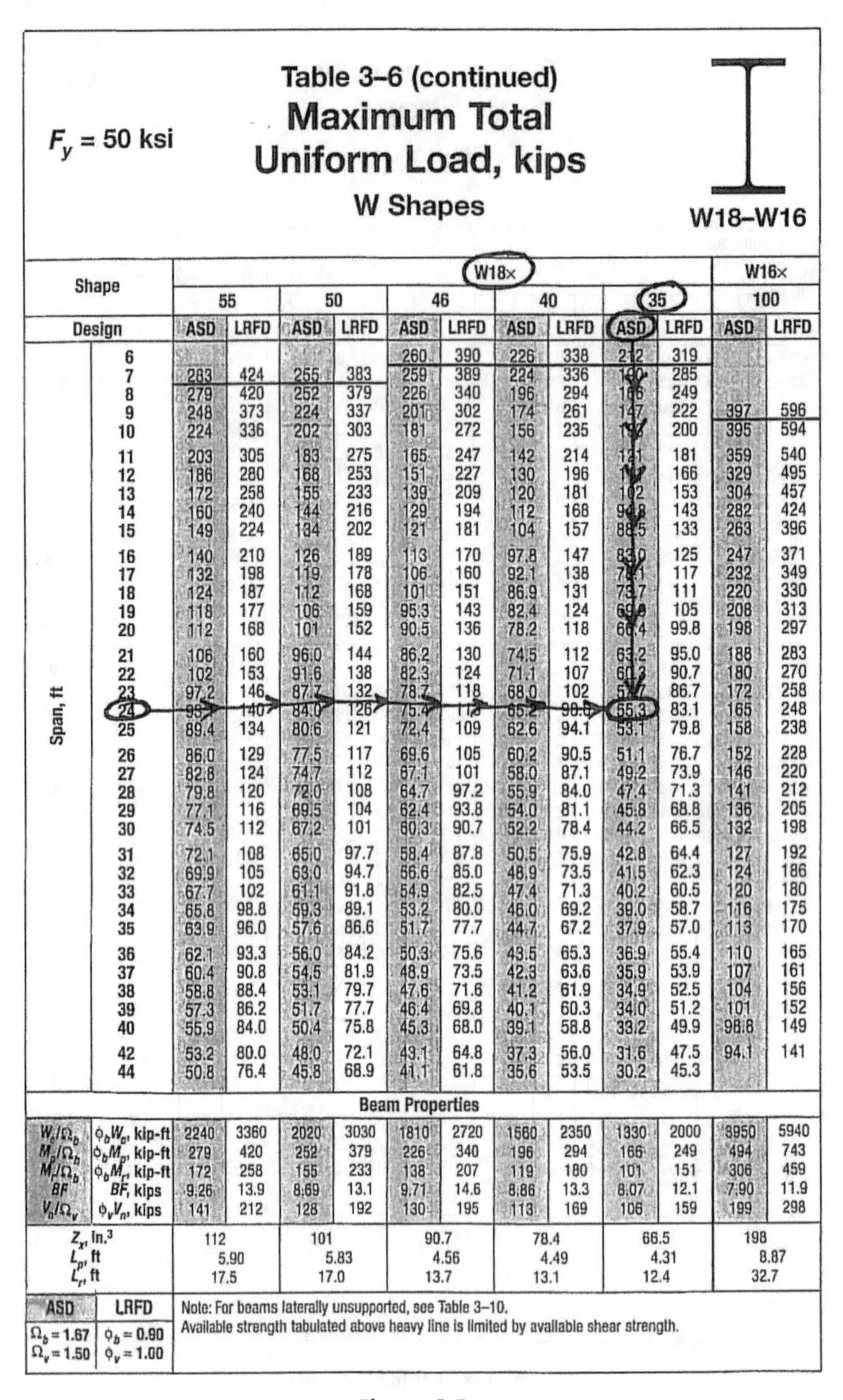

Table 3-6 (continued)
Maximum Total Uniform Load, kips
W Shapes

F_y = 50 ksi

W18–W16

Shape	W18×										W16×	
	55		50		46		40		35		100	
Design	ASD	LRFD	ASD	LRFD	ASD	LRFD	ASD	LRFD	ASD	LRFD	ASD	LRFD
Span, ft 6					260	390	226	338	212	319		
7	283	424	255	383	259	389	224	336	190	285		
8	279	420	252	379	226	340	196	294	166	249		
9	248	373	224	337	201	302	174	261	147	222	397	596
10	224	336	202	303	181	272	156	235	[illegible]	200	395	594
11	203	305	183	275	165	247	142	214	121	181	359	540
12	186	280	168	253	151	227	130	196	[illegible]	166	329	495
13	172	258	155	233	139	209	120	181	102	153	304	457
14	160	240	144	216	129	194	112	168	94.8	143	282	424
15	149	224	134	202	121	181	104	157	88.5	133	263	396
16	140	210	126	189	113	170	97.8	147	83.0	125	247	371
17	132	198	119	178	106	160	92.1	138	78.1	117	232	349
18	124	187	112	168	101	151	86.9	131	73.7	111	220	330
19	118	177	106	159	95.3	143	82.4	124	[illegible]	105	208	313
20	112	168	101	152	90.5	136	78.2	118	66.4	99.8	198	297
21	106	160	96.0	144	86.2	130	74.5	112	63.2	95.0	188	283
22	102	153	91.6	138	82.3	124	71.1	107	60.3	90.7	180	270
23	97.2	146	87.7	132	78.7	118	68.0	102	57.7	86.7	172	258
24	[illegible]	140	84.0	126	75.4	[illegible]	65.2	[illegible]	55.3	83.1	165	248
25	89.4	134	80.6	121	72.4	109	62.6	94.1	53.1	79.8	158	238
26	86.0	129	77.5	117	69.6	105	60.2	90.5	51.1	76.7	152	228
27	82.8	124	74.7	112	67.1	101	58.0	87.1	49.2	73.9	146	220
28	79.8	120	72.0	108	64.7	97.2	55.9	84.0	47.4	71.3	141	212
29	77.1	116	69.5	104	62.4	93.8	54.0	81.1	45.8	68.8	136	205
30	74.5	112	67.2	101	60.3	90.7	52.2	78.4	44.2	66.5	132	198
31	72.1	108	65.0	97.7	58.4	87.8	50.5	75.9	42.8	64.4	127	192
32	69.9	105	63.0	94.7	56.6	85.0	48.9	73.5	41.5	62.3	124	186
33	67.7	102	61.1	91.8	54.9	82.5	47.4	71.3	40.2	60.5	120	180
34	65.8	98.8	59.3	89.1	53.2	80.0	46.0	69.2	39.0	58.7	116	175
35	63.9	96.0	57.6	86.6	51.7	77.7	44.7	67.2	37.9	57.0	113	170
36	62.1	93.3	56.0	84.2	50.3	75.6	43.5	65.3	36.9	55.4	110	165
37	60.4	90.8	54.5	81.9	48.9	73.5	42.3	63.6	35.9	53.9	107	161
38	58.8	88.4	53.1	79.7	47.6	71.6	41.2	61.9	34.9	52.5	104	156
39	57.3	86.2	51.7	77.7	46.4	69.8	40.1	60.3	34.0	51.2	101	152
40	55.9	84.0	50.4	75.8	45.3	68.0	39.1	58.8	33.2	49.9	98.8	149
42	53.2	80.0	48.0	72.1	43.1	64.8	37.3	56.0	31.6	47.5	94.1	141
44	50.8	76.4	45.8	68.9	41.1	61.8	35.6	53.5	30.2	45.3		
Beam Properties												
W_c/Ω_b / $\phi_b W_c$, kip-ft	2240	3360	2020	3030	1810	2720	1560	2350	1330	2000	3950	5940
M_p/Ω_b / $\phi_b M_p$, kip-ft	279	420	252	379	226	340	196	294	166	249	494	743
M_r/Ω_b / $\phi_b M_r$, kip-ft	172	258	155	233	138	207	119	180	101	151	306	459
BF / BF, kips	9.26	13.9	8.69	13.1	9.71	14.6	8.86	13.3	8.07	12.1	7.90	11.9
V_n/Ω_v / $\phi_v V_n$, kips	141	212	128	192	130	195	113	169	106	159	199	298
Z_x, in.^{3}	112		101		90.7		78.4		66.5		198	
L_p, ft	5.90		5.83		4.56		4.49		4.31		8.87	
L_r, ft	17.5		17.0		13.7		13.1		12.4		32.7	

ASD	LRFD
$\Omega_b = 1.67$	$\phi_b = 0.90$
$\Omega_v = 1.50$	$\phi_v = 1.00$

Note: For beams laterally unsupported, see Table 3–10.
Available strength tabulated above heavy line is limited by available shear strength.

Figure 4.5

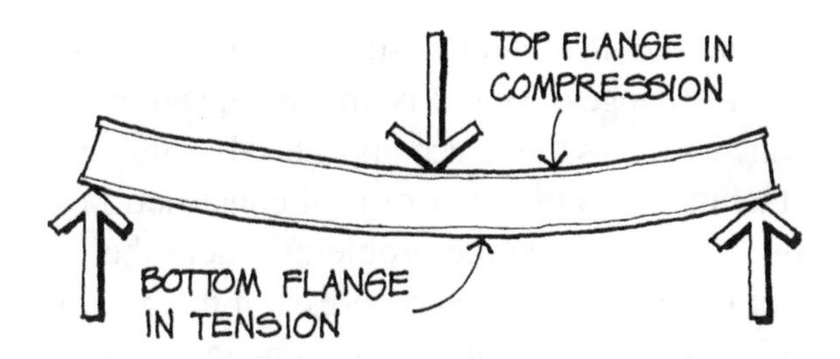

SIMPLE BEAM

- BENT CONCAVE UPWARD
- POSITIVE MOMENT

Figure 4.6

The compression flange of a steel beam is similar to a column. Its compressive stress tends to make it buckle, just as a column tends to buckle under compressive load, as shown in Figure 4.8.

If lateral support is provided, the beam cannot buckle laterally, and the full allowable bending stress may be used. However, if the beam is laterally unsupported, its tendency to buckle reduces its ability to resist moment.

Most simple beams used in building framing have continuous lateral support along the top (compression) flange provided by the floor or roof construction. Some examples are shown in Figures 4.9 and 4.10.

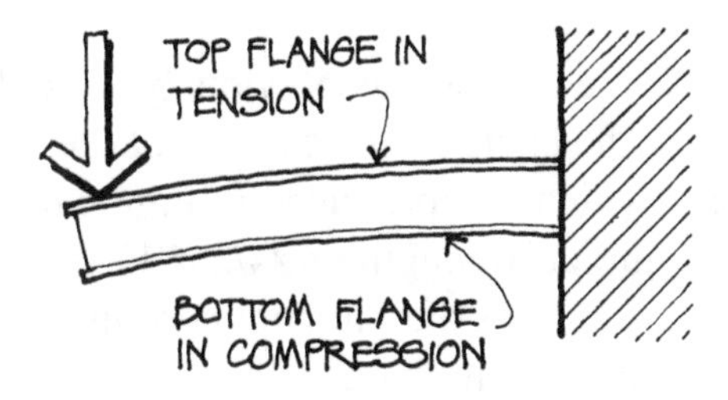

CANTILEVER BEAM

- BENT CONCAVE DOWNWARD
- NEGATIVE MOMENT

Figure 4.7

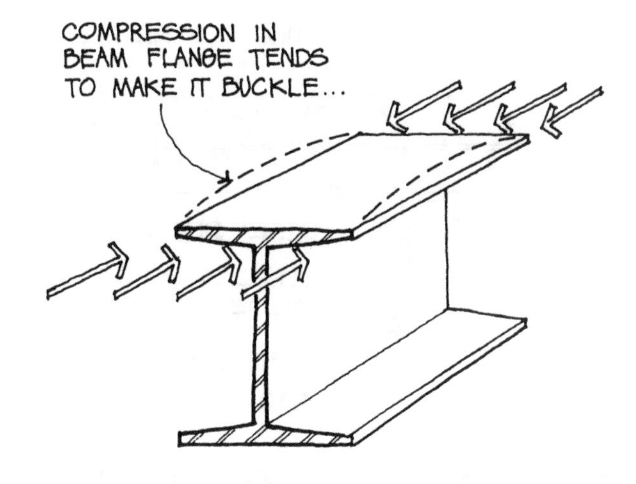

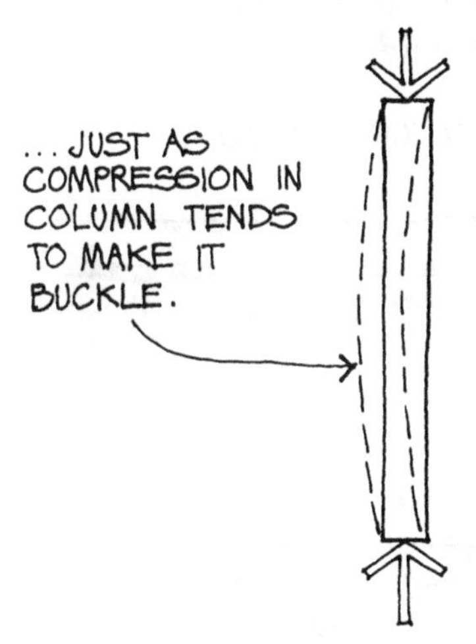

Figure 4.8

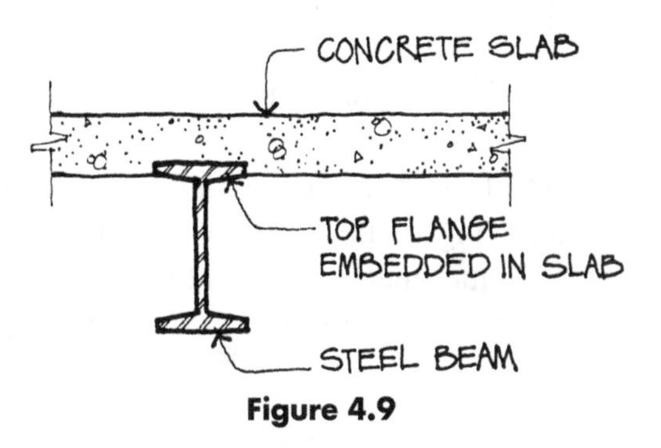

Figure 4.9

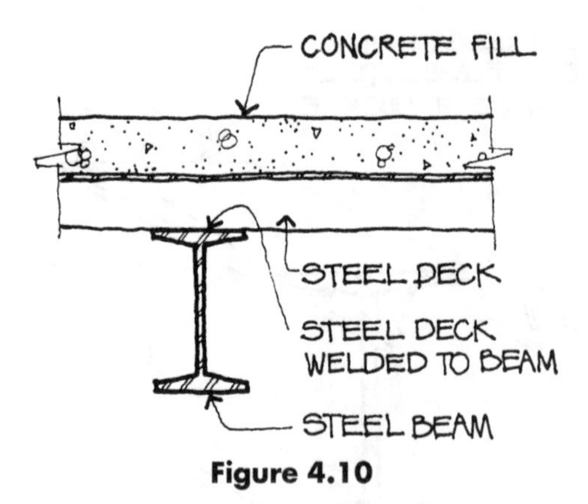

Figure 4.10

Some beams or girders are laterally supported at concentrated loads only, such as the floor girder shown in Figure 4.11.

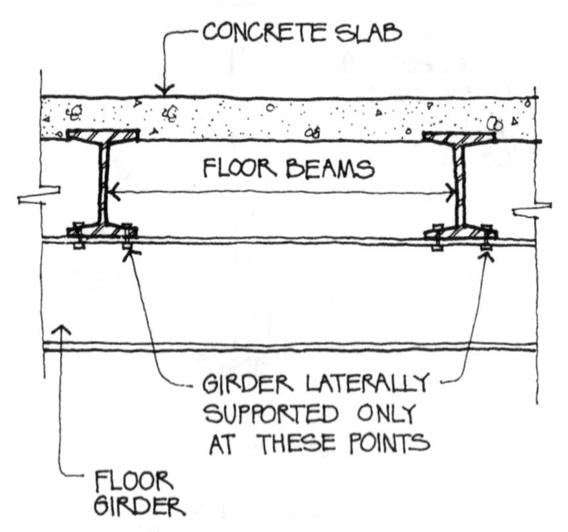

Figure 4.11

In the case of overhanging or cantilever beams, the bottom flange is in compression. Knee braces are sometimes connected to the bottom flange to laterally support it, as shown in Figure 4.12.

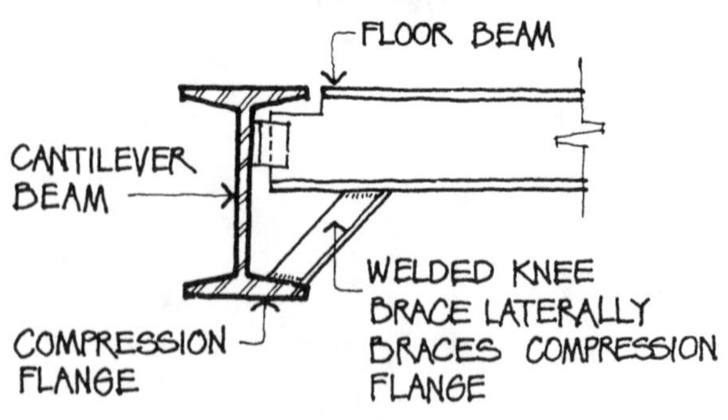

Figure 4.12

We can see that laterally supporting the compression flange of beams is important, and that in many instances this is accomplished by simple attachments to the floor or roof construction. But how do we handle problems where the lateral supports of the compression flange are widely spaced? In the tables on pages 3-33 through 3-95 of the AISC Manual, such as that in Figure 4.5, L_p is the maximum unbraced length of the compression flange for the limit state of yielding—that is, where the nominal flexural strength $M_n = F_y Z_x$. Also, L_r is the unbraced length for the limit state of inelastic lateral-torsional buckling. The equations for the nominal flexural strength M_n are very complicated once the unbraced length of the compression flange is greater than L_p.

The most practical approach to the design of such beams is the use of the charts on pages 3-96 through 3-139 of the AISC Manual, as illustrated by the following example.

Example #2

For the beam of Example #1, select the most economical steel beam if its compression (top) flange is laterally supported only at its ends and at midspan.

Solution:

Compute the moment as previously = 144'k, and enter the chart on page 3-125 of the AISC Manual, which is reproduced in Figure 4.13, with an unbraced length of 24/2 = 12 ft. Locate the intersection of the moment coordinate (144'k) and the length coordinate (12 ft.). Any beam whose curve lies above and to the right of this intersection satisfies the allowable bending stress requirement, and the nearest such curve that is a solid line represents the lightest weight section that is satisfactory.

In this example, the first solid line curve above and to the right of the intersection is that of *W14 × 43*; this is therefore the lightest section satisfying the problem criteria. Note that this beam is eight pounds per foot heavier than the W18 × 35 of Example #1, where full lateral support was provided.

SHEAR

Most steel beams designed for flexure are adequate to resist shear, and consequently are not checked for shear. Two exceptions are (1) short, heavily-loaded beams and (2) beams with large loads near the support. For these, shear may be critical and should therefore be checked.

For the vast majority of rolled I-shaped members, the nominal shear strength V_n is obtained from the formula $V_n = 0.6F_yA_w$, where A_w is the overall depth of the section d times the web thickness t_w. The safety factor for shear Ω is equal to 1.5, so the maximum permitted shear force on an I-shaped member is $V_n/\Omega = 0.6F_yA_w/1.5 = 0.4F_yA_w$.

Example #3

Check the W18 × 35 beam of Example #1 for shear.

Solution:

$V = 2000(24/2) = 24{,}000\#$

$d = 17.7''$, $t_w = 0.3''$ from AISC Manual page 1-18, reproduced in Figure 4.4

$V_n/\Omega = 0.4F_yA_w = 0.4(50{,}000)(17.7)(0.3) = 106{,}200\# > 24{,}000\#$

Note that V_n/Ω can also be obtained from AISC Manual page 3-18, reproduced in Figure 4.3.

As expected, the *W18 × 35 is not critical for shear.*

DEFLECTION

As with wood beams, steel beams must not only be able to safely resist bending and shear stresses, but they must also not deflect excessively. Excessive deflection may cause plaster cracking, unsightly sag, bouncy floors, ponding of rain water on roofs, and even damage to the structure.

The maximum deflections permitted by the code for a roof member supporting plaster or a floor member are shown in Table 1604.3 of the IBC, reproduced in the appendix. *The deflection limit for steel members is L/360 for live load only.* For longer spans, this deflection limit will often govern the selection of the member. Generally, floor beams having a depth at least 1/22 of the span length will not deflect excessively. This is an approximation useful for preliminary design only and should not be used in lieu of accurate calculations.

Example #4

What is the total deflection of the W18 × 35 beam of Example #1? If half of the load is live load, does the beam meet the deflection limits of the IBC? I for the beam is 510 in.^{4} (see Figure 4.4) and E = 29,000,000 psi.

Solution:

$$\text{Deflection} = \frac{5}{384}\frac{wL^4}{EI} = \frac{5}{384}\frac{wL \times L^3}{EI}$$

(see Figure 2.24)

$$= \frac{5}{384} \times \frac{2{,}000 \times 24 \times (24 \times 12)^3}{29{,}000{,}000 \times 510}$$

$$= 1.0''$$

Live load deflection = 1/2 × 1.0
= 0.5"

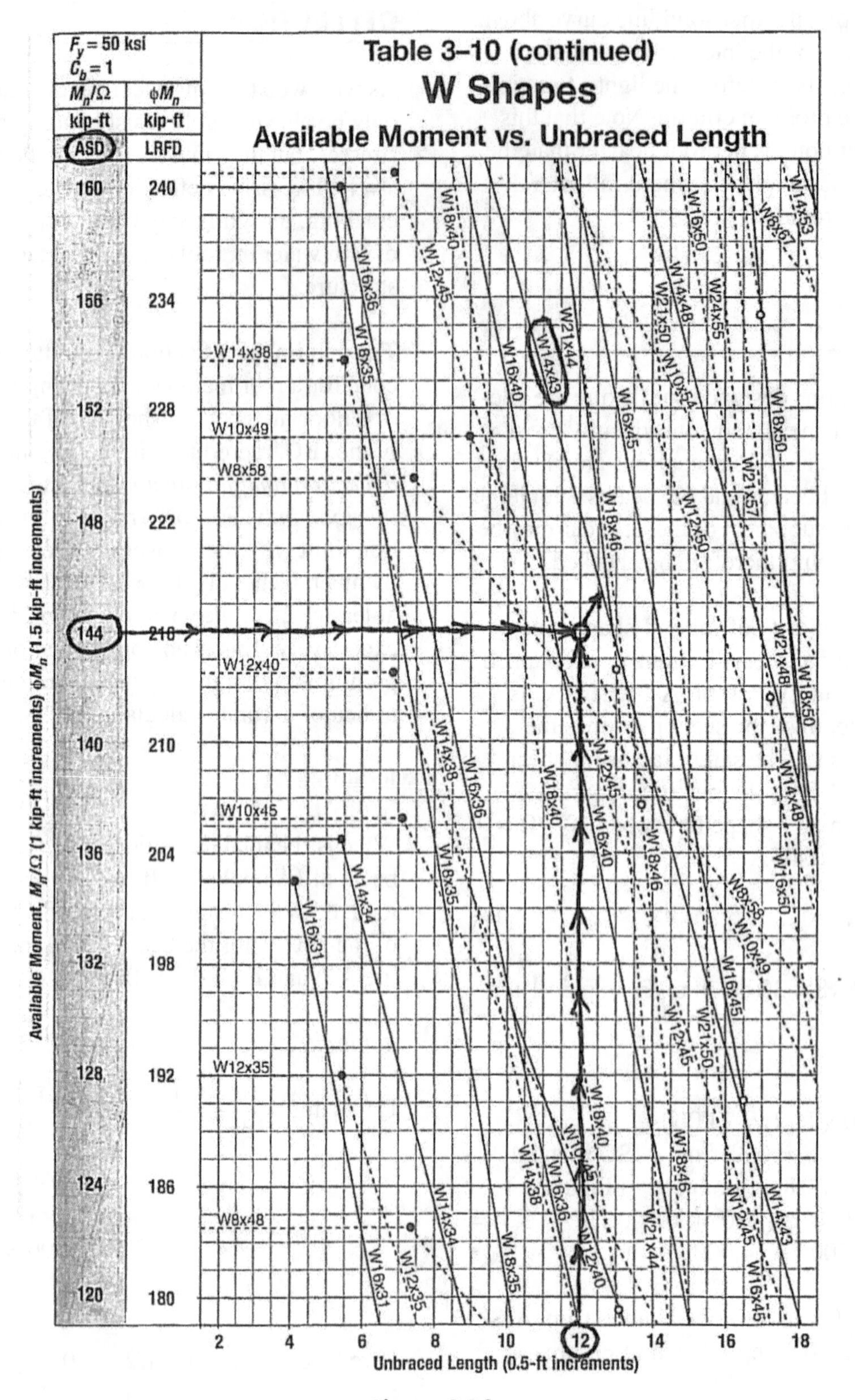
F_y = 50 ksi
C_b = 1
M_n/Ω
ϕM_n
kip-ft
kip-ft
ASD
LRFD
Table 3–10 (continued)
W Shapes
Available Moment vs. Unbraced Length
Available Moment, M_n/Ω (1 kip-ft increments) ϕM_n (1.5 kip-ft increments)
160 240
156 234
152 228
148 222
144 216
140 210
136 204
132 198
128 192
124 186
120 180
2 4 6 8 10 12 14 16 18
Unbraced Length (0.5-ft increments)
W14x38
W10x49
W8x58
W12x40
W10x45
W12x35
W8x48
W16x36
W18x35
W18x40
W12x45
W16x40
W21x44
W14x43
W16x45
W18x46
W16x50
W14x48
W21x50
W24x55
W10x54
W8x67
W14x53
W18x50
W21x57
W12x50
W21x48
W14x34
W16x31
W12x35
W14x38
W16x36
W12x40
W10x45
W21x44
W12x45
W21x50
W16x45
W8x58
W10x49
W16x50
W14x43

Figure 4.13

Deflection limit = L/360 for live load only
= 24 × 12/360 = 0.8"

Since 0.5 in. < 0.8 in., *the W18 × 35 beam meets the IBC deflection limits.*

In the expression for deflection (Δ), note that we convert the length L (24 feet) from feet to inches by multiplying by 12, then cube the entire term $(24 \times 12)^3$.

Since the moment of inertia (I) is in the denominator of all deflection formulas, the deflection varies inversely with I: the greater the value of I, the lower the deflection. Therefore, if the deflection of a steel beam is excessive, we must select a different beam section having a greater moment of inertia (I). For this purpose, we use the tables of properties of steel sections on pages 1-10 through 1-110 of the AISC Manual. An example is the table reproduced in Figure 4.4.

As an alternative means of selection, the moments of inertia are listed on pages 3-20 through 3-21 of the AISC Manual. In this table, as in the plastic section modulus table in Figure 4.3, the beam in bold type is the lightest shape in its group.

COMPOSITE DESIGN

The term *composite beam* refers to a steel beam and a concrete slab that are connected so that they act together as a single structural unit to resist bending stresses (Figure 4.14). In composite construction, the concrete slab becomes part of the top flange and resists compressive bending stresses. Composite construction is most efficient with heavy loading, relatively long spans, widely-spaced beams, and slab thicknesses of 4 to 5-1/2 inches. A composite beam is much stiffer than a non-composite beam having the same depth, load, and span: the composite beam will deflect about 1/3 to 1/2 less than the non-composite beam. However, since composite design usually allows us to use shallower beams, deflections should be checked.

The shear stresses between the steel beam and the concrete slab are resisted by shear connectors, such as studs, welded to the top flange of the beam and embedded in the concrete.

Light gage steel decking covered with a concrete slab is often used for floors and roofs in steel frame buildings (Figure 4.15). The decking serves as a working platform as well as a permanent form for the slab. In such cases, composite design may be used in either or both of the following ways:

1. The concrete slab and the steel beam may act together as a *composite beam* if they are adequately connected the same as if the concrete slab were formed.

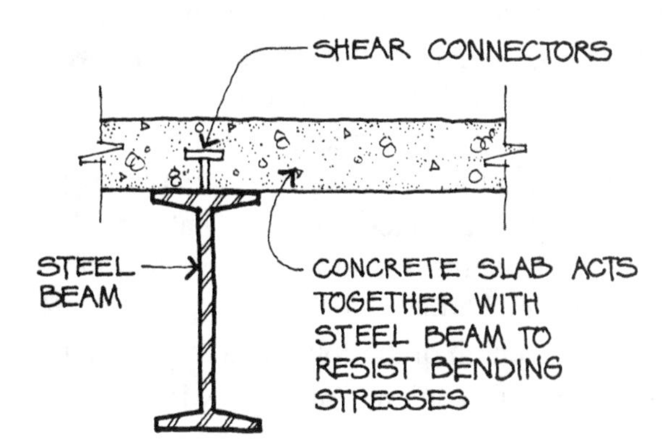

Figure 4.14

2. The steel decking and the concrete slab may act together as a *composite deck* to span between beams. Decking used for this purpose must have sufficient indentations and protrusions to bond to the concrete slab and resist the shear stresses between the slab and the steel deck.

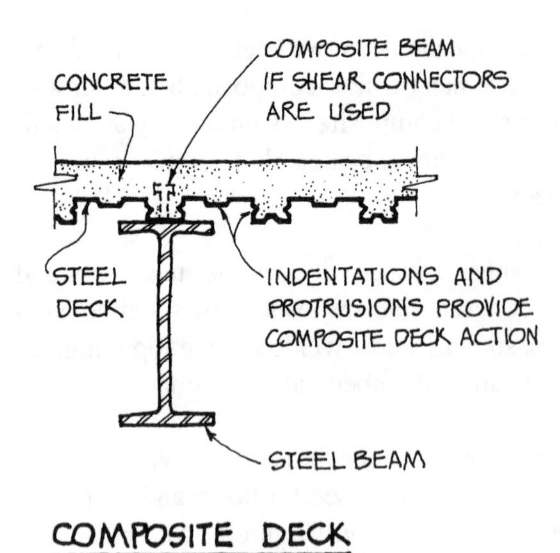

Figure 4.15

STRUCTURAL STEEL COLUMNS

Axial Compression

The AISC Specification includes design requirements for members subject to compression (columns). In general, the nominal compressive strength P_n divided by the safety factor $\Omega = 1.67$ must be greater than or equal to the total applied compressive force on the column. Formulas for P_n are expressed in terms of a critical stress F_{cr} and the slenderness ratio Kl/r, where K is the effective length factor (see Figure 4.25), l is the unbraced length of the column in inches, and r is the radius of gyration in inches $= \sqrt{I/A}$.

The formulas to determine F_{cr} are cumbersome to use. Fortunately, the AISC Manual tabulates F_{cr} for values of Kl/r up to 200 (the preferable maximum allowed) and for various values of F_y. These tables are given on pages 4-318 through 4-322 of the AISC Manual. The table on AISC page 4-320 is reproduced in Figure 4.17.

Most steel columns have two different values of r, the larger value with respect to the x-x axis and the smaller value with respect to the y-y axis. These values are tabulated in the AISC Manual. For a steel column whose unbraced length is the *same* with respect to its x-x and y-y axes, we use the *smaller r value* in order to obtain the *larger Kl/r value* to use in designing the column.

If a steel column's unbraced length with respect to the x-x and y-y axes are *different*, as shown in Figure 4.16, the slenderness ratio Kl/r is calculated for each direction, and the *larger value of Kl/r* is used to design the column.

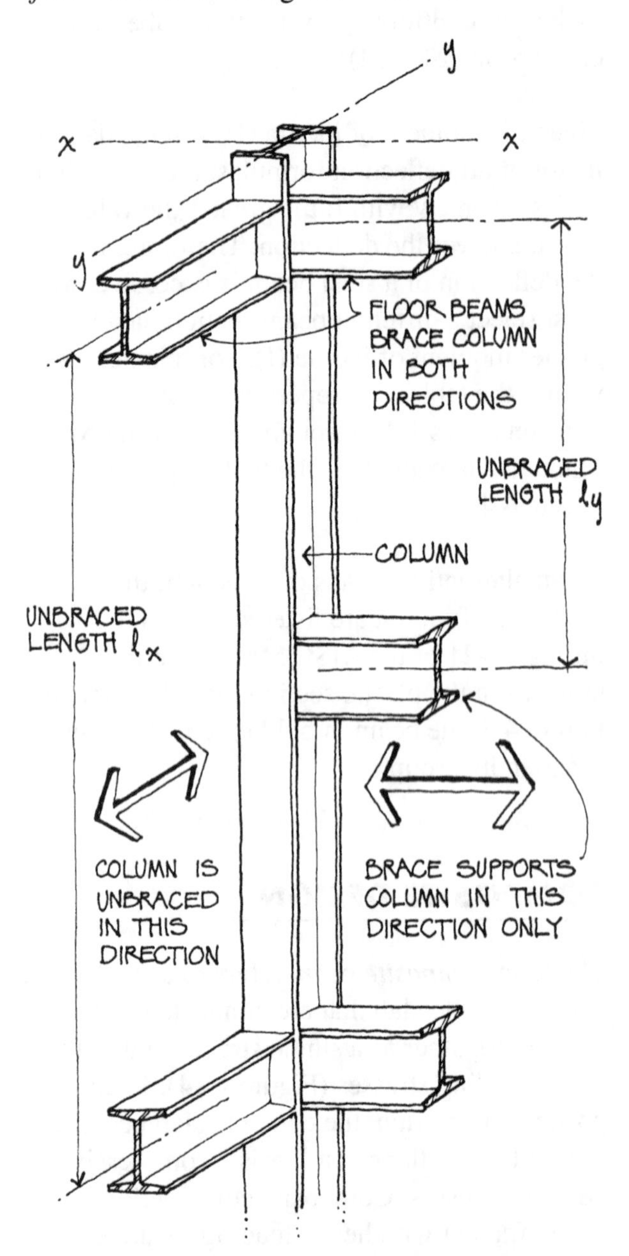

Figure 4.16

Table 4–22 (continued)
Available Critical Stress for Compression Members

$F_y = 35$ksi			$F_y = 36$ksi			$F_y = 42$ksi			$F_y = 46$ksi			$F_y = 50$ksi		
$\frac{Kl}{r}$	F_{cr}/Ω_c ksi ASD	$\phi_c F_{cr}$ ksi LRFD	$\frac{Kl}{r}$	F_{cr}/Ω_c ksi ASD	$\phi_c F_{cr}$ ksi LRFD	$\frac{Kl}{r}$	F_{cr}/Ω_c ksi ASD	$\phi_c F_{cr}$ ksi LRFD	$\frac{Kl}{r}$	F_{cr}/Ω_c ksi ASD	$\phi_c F_{cr}$ ksi LRFD	$\frac{Kl}{r}$	F_{cr}/Ω_c ksi ASD	$\phi_c F_{cr}$ ksi LRFD
81	15.0	22.5	81	15.3	22.9	81	16.8	25.3	81	17.7	26.6	81	18.5	27.9
82	14.9	22.3	82	15.1	22.7	82	16.6	25.0	82	17.5	26.3	82	18.3	27.5
83	14.7	22.1	83	15.0	22.5	83	16.5	24.8	83	17.3	26.0	83	18.1	27.2
84	14.6	22.0	84	14.9	22.3	84	16.3	24.5	84	17.1	25.8	84	17.9	26.9
85	14.5	21.8	85	14.7	22.1	85	16.1	24.3	85	16.9	25.5	85	17.7	26.5
86	14.4	21.6	86	14.6	22.0	86	16.0	24.0	86	16.7	25.2	86	17.4	26.2
87	14.2	21.4	87	14.5	21.8	87	15.8	23.7	87	16.6	24.9	87	17.2	25.9
88	14.1	21.2	88	14.3	21.6	88	15.6	23.5	88	16.4	24.6	88	17.0	25.5
89	14.0	21.0	89	14.2	21.4	89	15.5	23.2	89	16.2	24.3	89	16.8	25.2
90	13.8	20.8	90	14.1	21.2	90	15.3	23.0	90	16.0	24.0	90	16.6	24.9
91	13.7	20.6	91	13.9	21.0	91	15.1	22.7	91	15.8	23.7	91	16.3	24.6
92	13.6	20.4	92	13.8	20.8	92	15.0	22.5	92	15.6	23.4	92	16.1	24.2
93	13.5	20.2	93	13.7	20.5	93	14.8	22.2	93	15.4	23.1	93	15.9	23.9
94	13.3	20.0	94	13.5	20.3	94	14.6	22.0	94	15.2	22.8	94	15.7	23.6
95	13.2	19.9	95	13.4	20.1	95	14.4	21.7	95	15.0	22.6	95	15.5	23.3
96	13.1	19.7	96	13.3	19.9	96	14.3	21.5	96	14.8	22.3	96	15.3	22.9
97	13.0	19.5	97	13.1	19.7	97	14.1	21.2	97	14.6	22.0	97	15.0	22.6
98	12.8	19.3	98	13.0	19.5	98	13.9	21.0	98	14.4	21.7	98	14.8	22.3
99	12.7	19.1	99	12.9	19.3	99	13.8	20.7	99	14.2	21.4	99	14.6	22.0
100	12.6	18.9	100	12.7	19.1	100	13.6	20.5	100	14.1	21.1	100	14.4	21.7
101	12.4	18.7	101	12.6	18.9	101	13.4	20.2	101	13.9	20.8	101	14.2	21.3
102	12.3	18.5	102	12.5	18.7	102	13.3	20.0	102	13.7	20.6	102	14.0	21.0
103	12.2	18.3	103	12.3	18.5	103	13.1	19.7	103	13.5	20.3	103	13.8	20.7
104	12.1	18.1	104	12.2	18.3	104	12.9	19.5	104	13.3	20.0	104	13.6	20.4
105	11.9	17.9	105	12.1	18.1	105	12.8	19.2	105	13.1	19.7	105	13.4	20.1
106	11.8	17.7	106	11.9	17.9	106	12.6	19.0	106	12.9	19.4	106	13.2	19.8
107	11.7	17.5	107	11.8	17.7	107	12.4	18.7	107	12.8	19.2	107	13.0	19.5
108	11.5	17.3	108	11.7	17.5	108	12.3	18.5	108	12.6	18.9	108	12.8	19.2
109	11.4	17.2	109	11.5	17.3	109	12.1	18.2	109	12.4	18.6	109	12.6	18.9
110	11.3	17.0	110	11.4	17.1	110	12.0	18.0	110	12.2	18.3	110	12.4	18.6
111	11.2	16.8	111	11.3	16.9	111	11.8	17.7	111	12.0	18.1	111	12.2	18.3
112	11.0	16.6	112	11.1	16.7	112	11.6	17.5	112	11.8	17.8	112	12.0	18.0
113	10.9	16.4	113	11.0	16.5	113	11.5	17.3	113	11.7	17.5	113	11.8	17.7
114	10.8	16.2	114	10.9	16.3	114	11.3	17.0	114	11.5	17.3	114	11.6	17.4
115	10.7	16.0	115	10.7	16.2	115	11.2	16.8	115	11.3	17.0	115	11.4	17.1
116	10.5	15.8	116	10.6	16.0	116	11.0	16.5	116	11.1	16.7	116	11.2	16.8
117	10.4	15.6	117	10.5	15.8	117	10.8	16.3	117	11.0	16.5	117	11.0	16.5
118	10.3	15.5	118	10.4	15.6	118	10.7	16.1	118	10.8	16.2	118	10.8	16.2
119	10.2	15.3	119	10.2	15.4	119	10.5	15.8	119	10.6	16.0	119	10.6	16.0
120	10.0	15.1	120	10.1	15.2	120	10.4	15.6	120	10.4	15.7	120	10.4	15.7

ASD	LRFD
$\Omega_c = 1.67$	$\phi_c = 0.90$

Figure 4.17

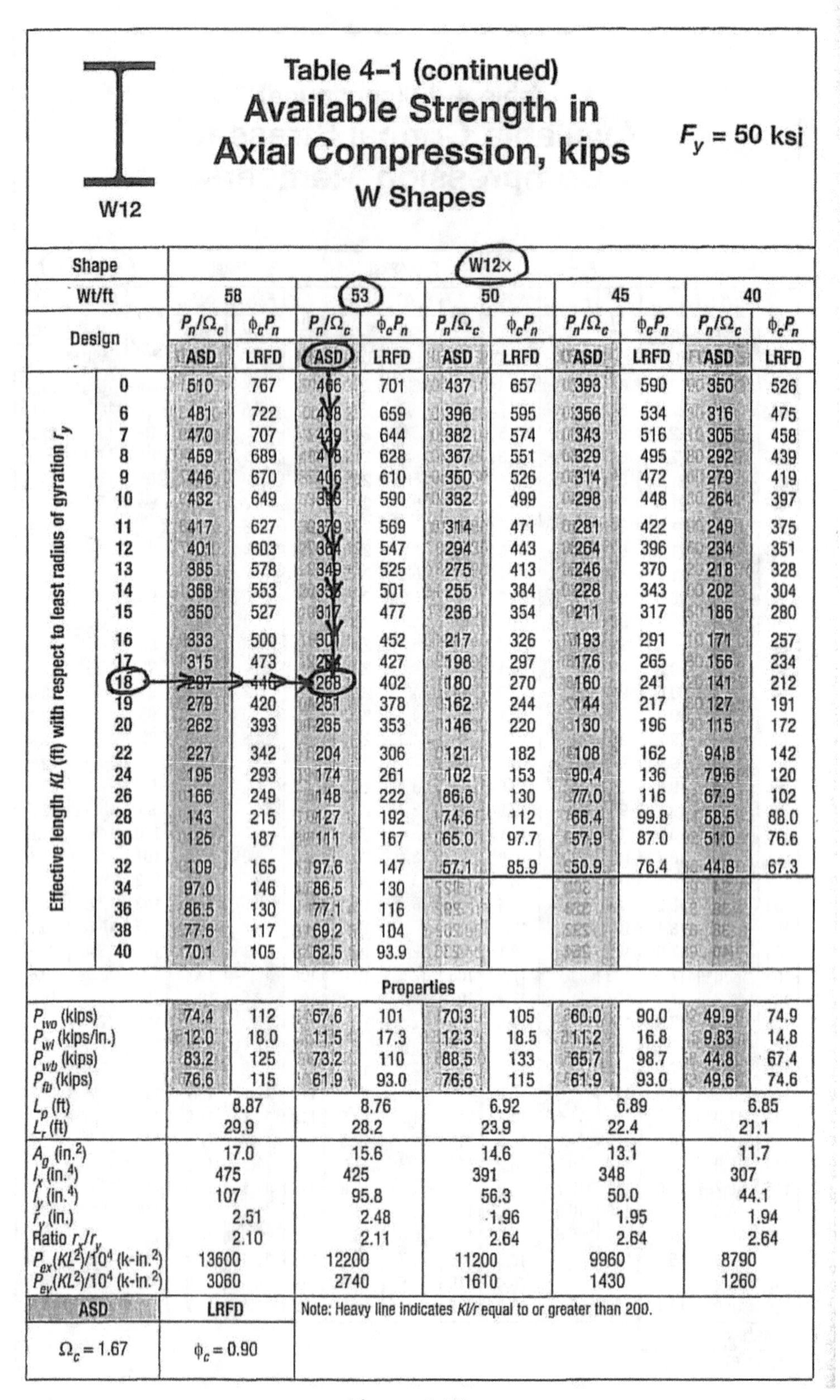

Table 4–1 (continued)
Available Strength in Axial Compression, kips
W Shapes

W12 — F_y = 50 ksi

Shape	W12×									
Wt/ft	58		53		50		45		40	
Design	P_n/Ω_c	$\phi_c P_n$	P_n/Ω_c	$\phi_c P_n$	P_n/Ω_c	$\phi_c P_n$	P_n/Ω_c	$\phi_c P_n$	P_n/Ω_c	$\phi_c P_n$
	ASD	LRFD	ASD	LRFD	ASD	LRFD	ASD	LRFD	ASD	LRFD
Effective length *KL* (ft) with respect to least radius of gyration r_y										
0	510	767	466	701	437	657	393	590	350	526
6	481	722	438	659	396	595	356	534	316	475
7	470	707	429	644	382	574	343	516	305	458
8	459	689	418	628	367	551	329	495	292	439
9	446	670	406	610	350	526	314	472	279	419
10	432	649	393	590	332	499	298	448	264	397
11	417	627	379	569	314	471	281	422	249	375
12	401	603	364	547	294	443	264	396	234	351
13	385	578	349	525	275	413	246	370	218	328
14	368	553	333	501	255	384	228	343	202	304
15	350	527	317	477	236	354	211	317	186	280
16	333	500	301	452	217	326	193	291	171	257
17	315	473	284	427	198	297	176	265	156	234
18	297	446	268	402	180	270	160	241	141	212
19	279	420	251	378	162	244	144	217	127	191
20	262	393	235	353	146	220	130	196	115	172
22	227	342	204	306	121	182	108	162	94.8	142
24	195	293	174	261	102	153	90.4	136	79.6	120
26	166	249	148	222	86.6	130	77.0	116	67.9	102
28	143	215	127	192	74.6	112	66.4	99.8	58.5	88.0
30	125	187	111	167	65.0	97.7	57.9	87.0	51.0	76.6
32	109	165	97.6	147	57.1	85.9	50.9	76.4	44.8	67.3
34	97.0	146	86.5	130						
36	86.5	130	77.1	116						
38	77.6	117	69.2	104						
40	70.1	105	62.5	93.9						
Properties										
P_{wo} (kips)	74.4	112	67.6	101	70.3	105	60.0	90.0	49.9	74.9
P_{wi} (kips/in.)	12.0	18.0	11.5	17.3	12.3	18.5	11.2	16.8	9.83	14.8
P_{wb} (kips)	83.2	125	73.2	110	88.5	133	65.7	98.7	44.8	67.4
P_{fb} (kips)	76.6	115	61.9	93.0	76.6	115	61.9	93.0	49.6	74.6
L_p (ft)	8.87		8.76		6.92		6.89		6.85	
L_r (ft)	29.9		28.2		23.9		22.4		21.1	
A_g (in.^{2})	17.0		15.6		14.6		13.1		11.7	
I_x (in.^{4})	475		425		391		348		307	
I_y (in.^{4})	107		95.8		56.3		50.0		44.1	
r_y (in.)	2.51		2.48		1.96		1.95		1.94	
Ratio r_x/r_y	2.10		2.11		2.64		2.64		2.64	
$P_{ex}(KL^2)/10^4$ (k-in.^{2})	13600		12200		11200		9960		8790	
$P_{ey}(KL^2)/10^4$ (k-in.^{2})	3060		2740		1610		1430		1260	

ASD	LRFD	Note: Heavy line indicates *Kl/r* equal to or greater than 200.
$\Omega_c = 1.67$	$\phi_c = 0.90$	

Figure 4.18

Table 4-6
Available Strength in Axial Compression, kips
Pipe

F_y = 35 ksi

PIPE 12–PIPE 8

Shape	Pipe 12				Pipe 10				Pipe 8			
	XS		Std		XS		Std		XXS		XS	
t_{design}, in.	0.465		0.349		0.465		0.340		0.816		0.465	
Wt/ft	65.5		49.6		54.8		40.5		72.5		43.4	
Design	P_n/Ω_c	$\phi_c P_n$	P_n/Ω_c	$\phi_c P_n$	P_n/Ω_c	$\phi_c P_n$	P_n/Ω_c	$\phi_c P_n$	P_n/Ω_c	$\phi_c P_n$	P_n/Ω_c	$\phi_c P_n$
	ASD	LRFD	ASD	LRFD	ASD	LRFD	ASD	LRFD	ASD	LRFD	ASD	LRFD
Effective length KL (ft) with respect to least radius of gyration r_y												
0	376	565	285	428	315	473	233	350	419	630	250	375
6	371	557	281	422	309	464	229	343	405	609	242	364
7	369	555	280	420	306	461	227	341	400	601	239	360
8	367	551	278	418	304	457	225	338	394	593	236	355
9	364	548	276	415	301	452	223	335	388	583	233	350
10	362	544	274	412	298	448	221	332	381	573	229	344
11	359	539	272	409	294	442	218	328	373	561	225	337
12	356	534	270	405	291	437	215	324	365	549	220	331
13	352	529	267	401	287	431	213	320	357	536	215	323
14	348	524	264	397	282	424	209	315	348	522	210	316
15	345	518	261	393	278	418	206	310	338	508	205	308
16	340	512	258	388	273	410	203	305	328	493	199	300
17	336	505	255	383	268	403	199	299	318	478	194	291
18	331	498	252	378	263	395	196	294	307	462	188	282
19	327	491	248	373	257	387	192	288	297	446	182	273
20	322	484	244	367	252	379	187	282	286	430	176	264
21	317	476	241	362	246	370	183	276	275	413	169	254
22	311	468	237	356	241	362	179	269	264	397	163	245
23	306	460	233	350	235	353	175	263	253	380	157	235
24	300	452	229	343	229	344	170	256	242	363	150	226
25	295	443	224	337	222	334	166	249	231	347	144	216
26	289	434	220	331	216	325	161	243	220	330	138	207
27	283	425	216	324	210	316	157	236	209	314	131	197
28	277	416	211	317	204	306	152	229	198	298	125	188
29	271	407	206	310	197	296	148	222	188	282	119	179
30	265	398	202	303	191	287	143	215	177	266	113	170
32	252	379	192	289	178	268	134	201	157	236	101	152
34	240	360	183	275	166	249	124	187	139	209	89.9	135
36	227	341	173	261	153	230	115	173	124	187	80.2	120
38	214	322	164	246	141	212	106	160	111	167	71.9	108
40	201	303	154	232	129	194	97.7	147	101	151	64.9	97.6
Properties												
A_g (in.^{2})	17.9		13.6		15.0		11.1		20.0		11.9	
I (in.^{4})	339		262		199		151		154		100	
r (in.)	4.35		4.39		3.64		3.68		2.78		2.89	

ASD	LRFD
Ω_c = 1.67	ϕ_c = 0.90

Figure 4.19

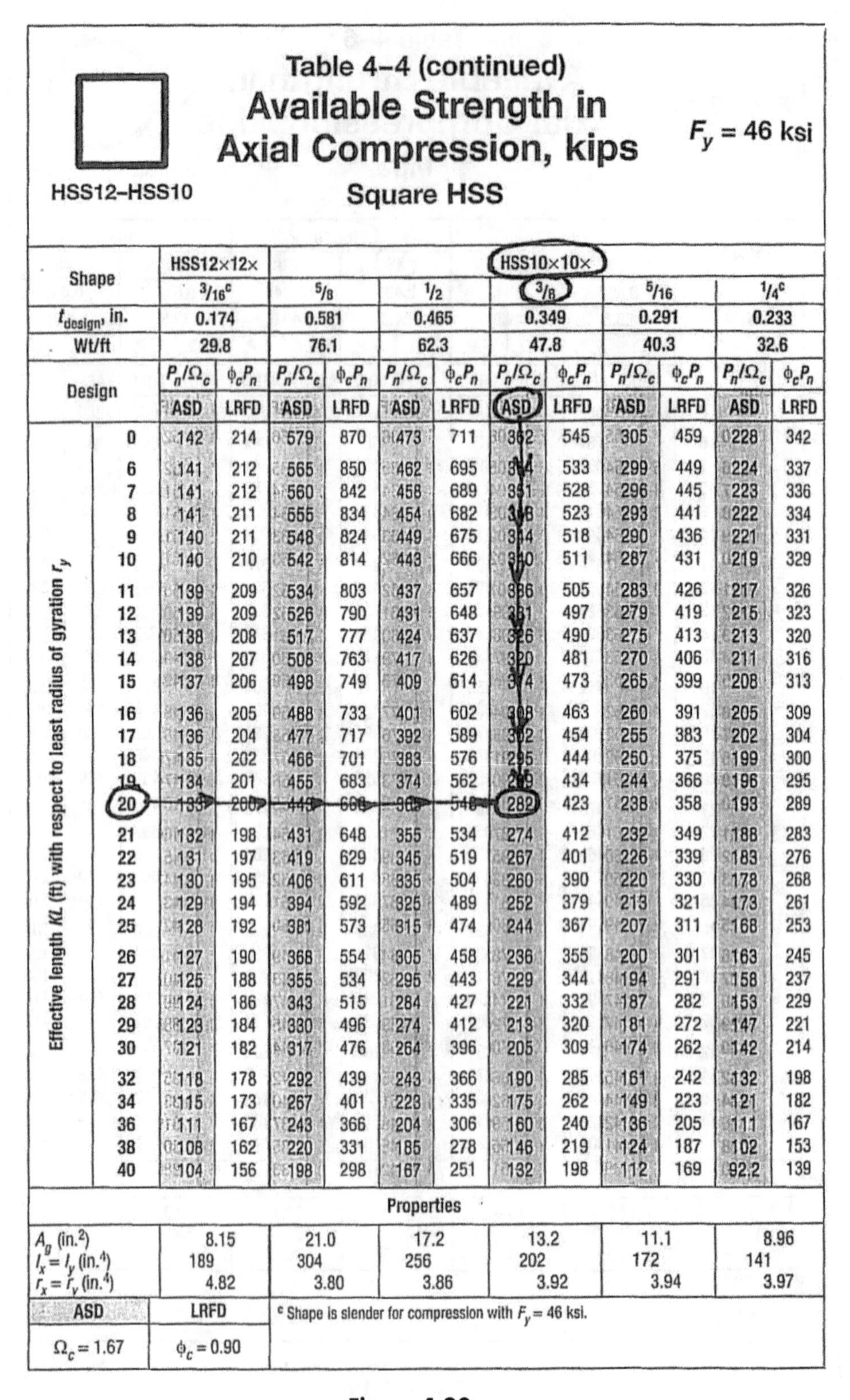

Table 4-4 (continued)

Available Strength in Axial Compression, kips

HSS12–HSS10 — Square HSS — F_y = 46 ksi

Shape	HSS12×12×		HSS10×10×									
	3/16^c		5/8		1/2		3/8		5/16		1/4^c	
t_{design}, in.	0.174		0.581		0.465		0.349		0.291		0.233	
Wt/ft	29.8		76.1		62.3		47.8		40.3		32.6	
Design	P_n/Ω_c	$\phi_c P_n$	P_n/Ω_c	$\phi_c P_n$	P_n/Ω_c	$\phi_c P_n$	P_n/Ω_c	$\phi_c P_n$	P_n/Ω_c	$\phi_c P_n$	P_n/Ω_c	$\phi_c P_n$
	ASD	LRFD	ASD	LRFD	ASD	LRFD	ASD	LRFD	ASD	LRFD	ASD	LRFD
0	142	214	579	870	473	711	362	545	305	459	228	342
6	141	212	565	850	462	695	354	533	299	449	224	337
7	141	212	560	842	458	689	351	528	296	445	223	336
8	141	211	555	834	454	682	348	523	293	441	222	334
9	140	211	548	824	449	675	344	518	290	436	221	331
10	140	210	542	814	443	666	340	511	287	431	219	329
11	139	209	534	803	437	657	336	505	283	426	217	326
12	139	209	526	790	431	648	331	497	279	419	215	323
13	138	208	517	777	424	637	326	490	275	413	213	320
14	138	207	508	763	417	626	320	481	270	406	211	316
15	137	206	498	749	409	614	314	473	265	399	208	313
16	136	205	488	733	401	602	308	463	260	391	205	309
17	136	204	477	717	392	589	302	454	255	383	202	304
18	135	202	466	701	383	576	295	444	250	375	199	300
19	134	201	455	683	374	562	289	434	244	366	196	295
20	133	200	443	666	365	548	282	423	238	358	193	289
21	132	198	431	648	355	534	274	412	232	349	188	283
22	131	197	419	629	345	519	267	401	226	339	183	276
23	130	195	406	611	335	504	260	390	220	330	178	268
24	129	194	394	592	325	489	252	379	213	321	173	261
25	128	192	381	573	315	474	244	367	207	311	168	253
26	127	190	368	554	305	458	236	355	200	301	163	245
27	125	188	355	534	295	443	229	344	194	291	158	237
28	124	186	343	515	284	427	221	332	187	282	153	229
29	123	184	330	496	274	412	213	320	181	272	147	221
30	121	182	317	476	264	396	205	309	174	262	142	214
32	118	178	292	439	243	366	190	285	161	242	132	198
34	115	173	267	401	223	335	175	262	149	223	121	182
36	111	167	243	366	204	306	160	240	136	205	111	167
38	108	162	220	331	185	278	146	219	124	187	102	153
40	104	156	198	298	167	251	132	198	112	169	92.2	139

Effective length KL (ft) with respect to least radius of gyration r_y

Properties

	HSS12×12×3/16	HSS10×10×5/8	HSS10×10×1/2	HSS10×10×3/8	HSS10×10×5/16	HSS10×10×1/4
A_g (in.^{2})	8.15	21.0	17.2	13.2	11.1	8.96
$I_x = I_y$ (in.^{4})	189	304	256	202	172	141
$r_x = r_y$ (in.^{4})	4.82	3.80	3.86	3.92	3.94	3.97

ASD	LRFD
$\Omega_c = 1.67$	$\phi_c = 0.90$

^c Shape is slender for compression with F_y = 46 ksi.

Figure 4.20

Example #5

A W12 × 53 column of A992 steel has an unbraced length of 18 feet. The column has pinned ends (K = 1.0, see Figure 4.25). What is the maximum allowable compressive load? The properties of the column are as follows: $r_{x\text{-}x}$ = 5.23 inches, $r_{y\text{-}y}$ = 2.48 inches, A = 15.6 square inches.

Solution:

l = 18 × 12 = 216 inches
Kl/r = 1.0 × 216/2.48 = 87.1

Notice that we use the *smaller* value of r in computing the slenderness ratio Kl/r. Enter the table on page 4-320 of the AISC Manual, which is reproduced in Figure 4.17, with Kl/r = 87 (the whole number closest to the calculated value of 87.1), and read the F_{cr}/Ω value of 17.2 ksi.

$P_n/\Omega = F_{cr}/\Omega \times A = 17.2 \times 15.6 =$ *268.3 kips*

As you can see, this is a relatively simple procedure. However, an even simpler and more direct way to determine the allowable axial load on any column section of any unbraced length is to use the tables on pages 4-10 through 4-316 of the AISC Manual. In this case, we enter the table on page 4-18 of the AISC Manual, a portion of which is reproduced in Figure 4.18, with the member (W12 × 53) and the effective length Kl (18 feet), and read *268 kips*, compared with *268.3 kips* computed above.

These tables may also be used to *select* a column section, if the axial load and column length are given.

Example #6

Select the lightest pipe column and the lightest square tubular column that can safely support an axial load of 250 kips. The unsupported length is 20 feet and K = 1.0.

Solution:

To solve this problem, we scan the tables in the AISC Manual and select a *10" extra strong (XS) pipe column* and a *10 × 10 × 3/8 tubular (square HSS) column*, since these are the lightest sections that will satisfy the requirements of the problem. A portion of pages 4-81 and 4-50 are reproduced in Figures 4.19 and 4.20, by permission of the American Institute of Steel Construction.

For columns supporting light to moderate loads, pipe columns and square tubular columns are often lighter than wide flange (W) columns. The reason is that W columns have a greater tendency to buckle in the weak direction ($r_{y\text{-}y}$ is low and therefore Kl/r is high; see Figure 4.21). Consequently, W columns have a lower available strength and require more area and hence more weight than pipe columns or square tubes.

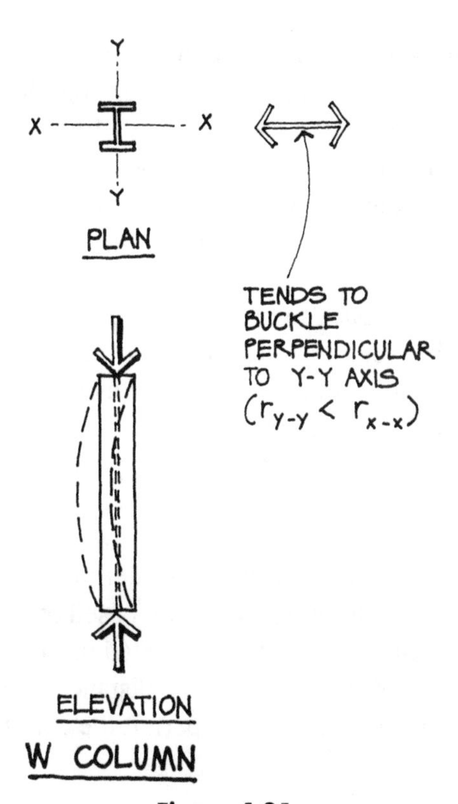

Figure 4.21

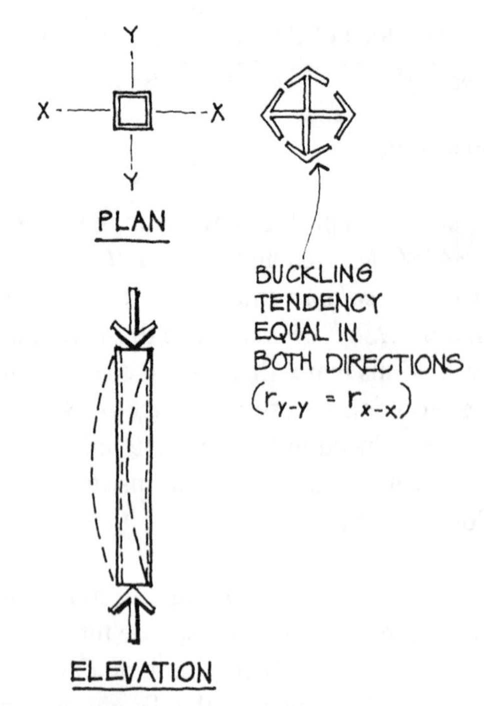

Figure 4.22

In theory, the ideal column is one for which the buckling tendency is equal in both directions; i.e., $r_{x\text{-}x} = r_{y\text{-}y}$ and $I_{x\text{-}x} = I_{y\text{-}y}$. These conditions are met by pipe columns and square steel tubes (Figure 4.22). However, their use in actual practice is limited because of the difficulty of making good beam connections. Most steel columns are therefore W sections; where the loads are unusually great, plates are sometimes welded to the flanges for additional strength.

Two angles back to back with about a 3/8" space between are often used for the compression members of trusses, as well as for bracing members (see Figure 4.23). Their design is expedited by using the tables on pages 4-118 through 4-156 of the AISC Manual.

What is the factor K in the slenderness ratio Kl/r? The column formulas are based on a pin-ended column, which is illustrated in Figure 4.24, where the length l is the distance between the pinned ends. The pins permit the column to rotate, but not to translate (move laterally). Actual building columns, however, cannot be categorized so neatly; their ends may permit rotation (pin end), or may permit no rotation (fixed end), or may be somewhere in between. Likewise, they may be fixed against translation, or free to translate, or somewhere between.

To make the column formulas usable for all conditions, we use the factor K. This factor, which may be greater or less than 1.0, is multiplied by the actual unbraced length l to arrive at the effective column length. Six idealized conditions are shown in Figure 4.25, reproduced from the AISC Manual by permission of the American Institute of Steel Construction.

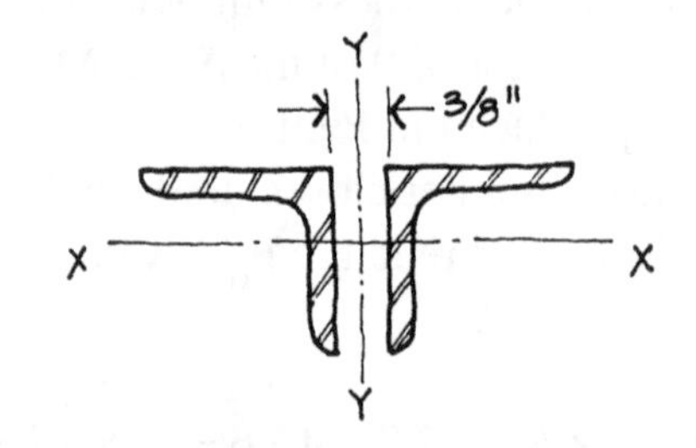

Figure 4.23

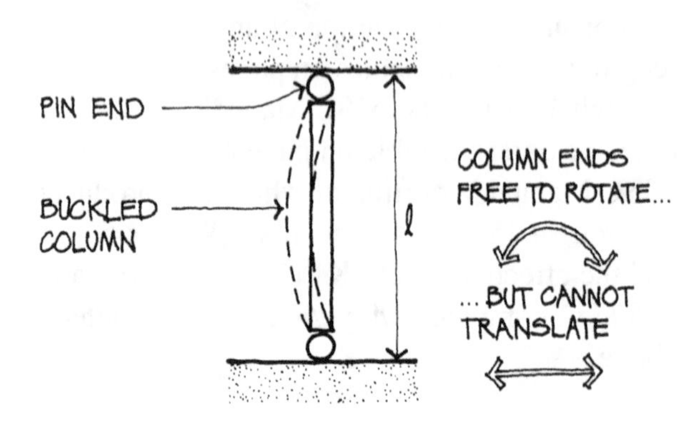

Figure 4.24

Example #7

A steel column 10 feet long is fixed at the base, as shown in Figure 4.26. The top of the column is braced against movement in the direction perpendicular to the paper, but is free to translate in the direction shown. If $r_{x\text{-}x} = 4.66$ and $r_{y\text{-}y} = 2.68$, what are the values of Kl/r with respect to both the x-x and y-y axes, and which value governs the design of the column?

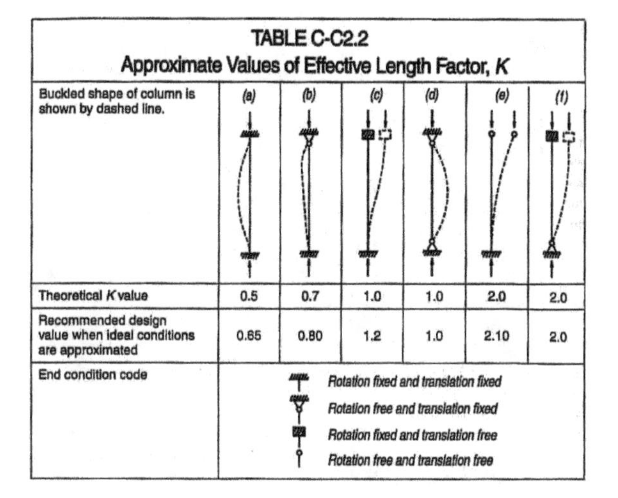

TABLE C-C2.2
Approximate Values of Effective Length Factor, *K*

Buckled shape of column is shown by dashed line.	(a)	(b)	(c)	(d)	(e)	(f)
Theoretical *K* value	0.5	0.7	1.0	1.0	2.0	2.0
Recommended design value when ideal conditions are approximated	0.65	0.80	1.2	1.0	2.10	2.0
End condition code	Rotation fixed and translation fixed Rotation free and translation fixed Rotation fixed and translation free Rotation free and translation free					

Figure 4.25

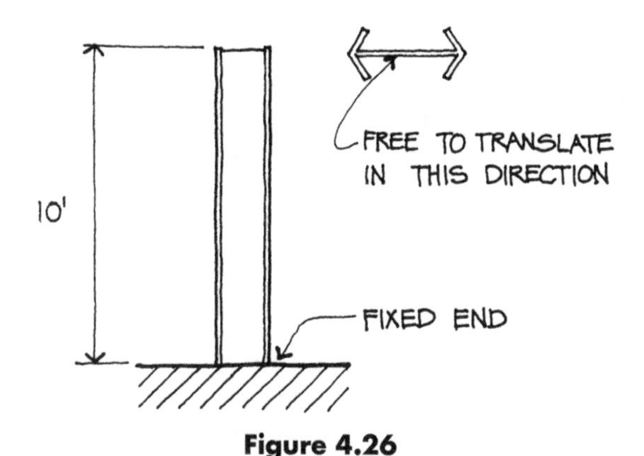

Figure 4.26

Solution:

Refer to Figure 4.25. About the x-x axis, the condition most closely approximates either (c) or (e). Since there is no indication that the top of the column is fixed, we conclude that (e) is the condition for the x-x axis, for which the recommended design value of K is 2.10.

About the y-y axis, the condition is closest to (a) or (b). Again, since nothing indicates that the top of the column is fixed, we conclude that (b) is closest to our condition; the recommended design value of K is 0.80.

$l = 120$ inches
$(Kl/r)_{x\text{-}x} = 2.10 \times 120/4.66 = \mathit{54.1}$
$(Kl/r)_{y\text{-}y} = 0.80 \times 120/2.68 = \mathit{35.8}$

As always, we use the *larger* value of Kl/r to design the column. *The governing value of Kl/r is therefore 54.1.*

Combining Bending and Axial Compression

Steel columns are sometimes subject to bending moment and axial compression at the same time. The bending moment may be caused by an eccentric load, by loads acting perpendicular to the length of column, or by the column acting as part of a moment-resisting rigid frame.

For combined bending and axial compression, the AISC Specification uses interaction formulas that are quite complex and therefore beyond the scope of this course and not likely to appear on the examination.

As with wood columns, the stress under combined bending and axial compression is P/A ± Mc/I (see page 39).

Column Base Details

Steel columns rest on and are generally welded to steel base plates, which must be designed to transfer all loads at the base of the column—axial, flexural, and shear—to the supporting foundation (Figure 4.27). Anchor bolts are used to connect the base plate to the foundation and are designed to transmit any shear or bending forces at the base of the column.

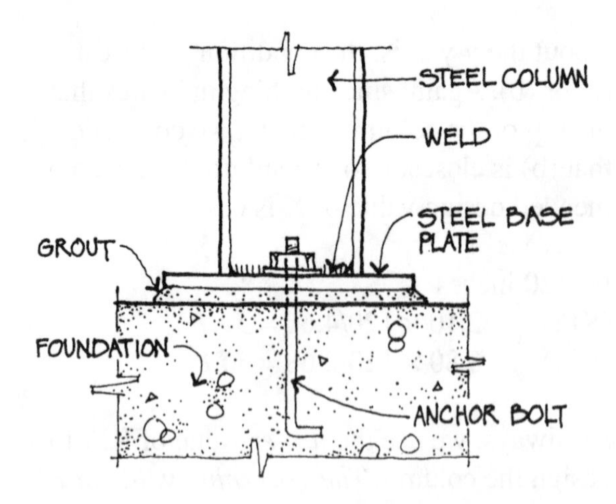

Figure 4.27

Example #8

A steel column supports an axial load of 200 kips and bears on a 12" × 16" base plate. What is the bearing pressure under the base plate?

Solution:

The bearing pressure is equal to the axial load divided by the area of the base plate = 200,000/12 × 16 = *1,042 psi*

LESSON 4 QUIZ

1. Select the CORRECT statements about shear stress in a steel beam.

   I. The unit shear stress f_v is equal to V/dt.

   II. The shear stress should be checked for all beams.

   III. The shear stress should be checked for beams with a short span and a heavy load.

   IV. The shear stress should be checked for beams with a large concentrated load near the support.

   V. If the beam is adequate to resist flexural stress, shear stress is not critical.

   A. I and II **C.** I and V
   B. I, III, and IV **D.** III and IV

2. In designing steel columns, the larger slenderness ratio is used because it

   A. results in a smaller allowable flexural stress.
   B. results in a larger buckling load.
   C. results in a smaller allowable axial load.
   D. is based on the larger value of radius of gyration.

3. What is the allowable axial load for a W8 × 40 column of ASTM A992 steel if Kl is 20 feet? For that section, $r_{x\text{-}x}$ = 3.53 in., $r_{y\text{-}y}$ = 2.04 in., A = 11.7 sq.in.

   A. 126 kips **C.** 195 kips
   B. 155 kips **D.** 207 kips

4. The carbon content of steel affects its

   I. modulus of elasticity.

   II. strength.

   III. ductility.

   IV. corrosion resistance.

   A. III and IV **C.** I, II, and IV
   B. II and III **D.** I, II, III, and IV

5. Select the INCORRECT statement. The modulus of elasticity of steel

   A. has a constant value.
   B. is higher than that of any other commonly used structural material.
   C. is a measure of the stiffness of steel.
   D. is a measure of the strength of steel.

6. A steel shape that is symmetrical about both axes and that is a very efficient structural member is called ________________.

7. A beam of ASTM A992 steel supports a uniformly distributed load of 1,250 pounds per foot on a 30-foot span. The beam is laterally supported at the third points of the span. What is the lightest section that may be used?

   A. W18 × 46 **C.** W12 × 53
   B. W16 × 40 **D.** W21 × 57

8. For the beam of Question 7, what is the lightest section that could be used if the beam were laterally supported?

   A. W21 × 44 **C.** W16 × 50
   B. W18 × 35 **D.** W12 × 53

9. If the calculated deflection of a steel beam is excessive, it is necessary to select a different beam section having a greater

A. area.

B. section modulus.

C. moment of inertia.

D. modulus of elasticity.

10. Select the INCORRECT statement about composite design.

A. A composite beam consists of a steel beam and a concrete slab that act together to resist bending stresses.

B. Because of the stiffness of a composite beam, it is unnecessary to calculate deflection.

C. The shear stress between the steel beam and the concrete slab is usually resisted by welded studs.

D. In composite design, the concrete slab effectively becomes part of the beam's top flange.

LESSON FIVE

REINFORCED CONCRETE CONSTRUCTION

INTRODUCTION

In the two previous lessons, we discussed framing members made of wood and structural steel. We come now to the third major structural building material—reinforced concrete. You will recall from Lesson Three that wood has about the same strength in both compression and tension. This is also true of steel, as discussed in Lesson Four. Therefore, wood and steel beams require the same amount of material above the neutral axis to resist compression as they have below the neutral axis to resist tension (Figure 5.1). Consequently, wood and steel beams are symmetrical and homogeneous.

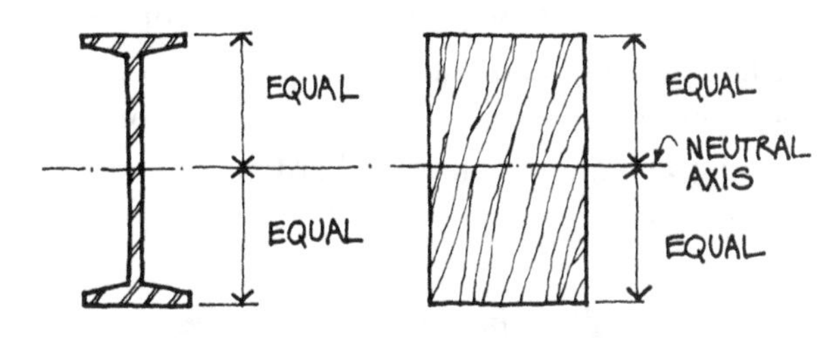

Figure 5.1

Reinforced concrete differs significantly in that concrete is strong in compression, but weak in tension. Hence, steel reinforcing bars are used to resist tensile stresses, while the concrete resists compressive stresses. Reinforced concrete beams are therefore unsymmetrical and non-homogeneous (they are composed of two materials, not one; see Figure 5.2).

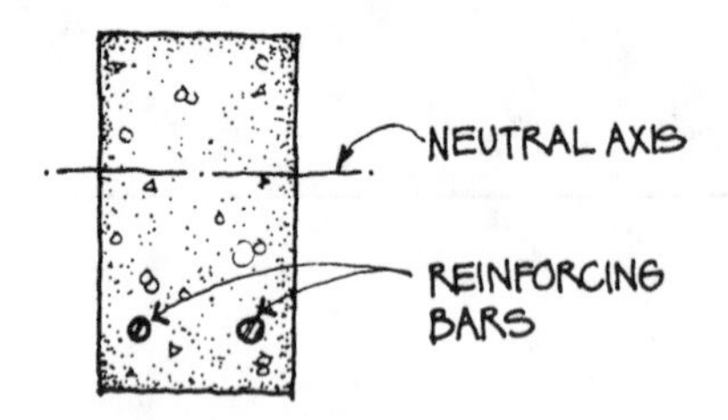

REINFORCED CONCRETE BEAM

- UNSYMMETRICAL
- NON-HOMOGENEOUS

Figure 5.2

This lack of symmetry and homogeneity invariably makes the theory and design of reinforced concrete more complex than that of wood or structural steel.

PROPERTIES OF CONCRETE

Concrete is a mixture of *aggregates* (sand and gravel) held together by a paste made from portland cement and water. The paste fills the spaces between the aggregates, and after being placed, the concrete cures or hardens to form a solid material.

The sand is called *fine aggregate*, while the gravel or crushed stone is known as *coarse aggregate*. Where reduction of weight is important, lightweight aggregates, made from expanded shale or clay, are often used instead of natural aggregates.

Portland cement is a very finely ground gray material manufactured mainly from limestones and clays or shales. It is available in bulk or in 94-pound sacks containing one cubic foot.

The ASTM provides for five types of portland cement:

Type I is suitable for all normal uses.

Type II is used where precaution against moderate sulfate attack is important. It usually generates less heat at a slower rate than Type I, especially important where the concrete is placed in warm weather.

Type III provides high early strength, usually in a week or less.

Type IV is intended for use in massive concrete structures, where the rate and amount of heat generated must be minimized. It develops strength at a slower rate than Type I.

Type V is used only in concrete exposed to severe sulfate action.

Almost any water that is drinkable and has no apparent taste or smell is satisfactory as mixing water for making concrete.

The most important factor that determines the strength of concrete is the *water-cement ratio*, usually expressed as the number of gallons of water per 94-pound sack of cement. The higher the water-cement ratio, the lower the concrete strength.

Sometimes, materials other than portland cement, water, and aggregates are added to concrete mixtures for a variety of purposes. These materials are called *admixtures*, and they may be used to improve durability, reduce the amount of mixing water required, retard or accelerate the setting time, improve workability, or reduce permeability.

The *workability* of concrete—the ease with which it can be placed and consolidated in the forms and around the reinforcing bars—is an important quality that is usually measured in the field by the slump test or the Kelly ball test.

In the *slump test*, a standard metal mold is filled with fresh concrete; after being filled, the mold is removed and the decrease in height of the concrete is measured. Stiff mixes have low slump, and conversely, more fluid mixes have greater slump. The slump should be as low as possible consistent with the necessary workability.

The other method of measuring workability is the *Kelly ball test*. In this test, a 30-pound, 6-inch-diameter hemisphere is dropped from a standard height onto the surface of the fresh concrete, and its penetration into the concrete is measured. When properly calibrated, the results of the Kelly ball test can be directly related to slump.

Concrete gains strength very rapidly during the first week or two after being placed. In order to realize its full strength, the concrete must be maintained at the proper moisture and temperature during this early period, a procedure known as *curing*.

Curing methods include keeping exposed concrete surfaces wet by ponding, sprinkling, or using wet coverings such as burlap; sealing surfaces by means of waterproof paper, membranes, or sealing compounds; and keeping wood forms moist by sprinkling.

One of the desirable qualities of concrete is its great compressive strength. Just how strong is concrete? Depending on the mix and the curing, it is possible to obtain concrete with a compressive strength of 10,000 psi or greater. However, the usual range is 3,000 to 6,000 psi, with values around 4,000 psi being the most common. When we refer to concrete having a certain specified strength, we always mean its strength when it is 28 days old. Unlike structural steel, which is factory-produced under strict quality control, concrete is usually made near, or even at, the construction site. How can we be certain that the field-produced concrete is as strong as specified? The principal way to measure the compressive strength of concrete is the *cylinder test*, in which standard cylinders 6" × 12" are filled with concrete during the placing of concrete for the project, moist-cured for 28 days, and then tested in a laboratory at a specified rate of loading. The compressive strength thus determined must be at least equal to that specified. Sometimes, cylinders are tested after seven days, at which time their strength is about 60 to 70 percent of the 28-day strength.

Sometimes, it is necessary to determine the strength of hardened concrete in place. The methods used for this purpose may either be nondestructive or destructive.

The most common nondestructive test is the *impact hammer test*, in which the rebound of a spring-loaded plunger is measured after it strikes a smooth concrete surface. The rebound reading gives an approximate indication of the concrete strength. This test is simple and nondestructive, but it is not a substitute for standard compression tests.

The most common test of hardened concrete is the *core test*, in which cores 2" to 4" in diameter are cut from the structure in areas where the concrete is suspected of being deficient. The cores are then tested in a laboratory to determine their compressive strength.

You will recall from the previous lesson that steel, regardless of its strength, has a constant value of modulus of elasticity E, which is about 29,000,000 psi. In contrast, the value of the modulus of elasticity of concrete (E_c) is not constant or exact, but varies with the 28-day compressive strength (f_c') and the unit weight of the concrete. For concrete of normal weight (about 145 pounds per cubic foot), the value of E_c is approximately $57{,}000\sqrt{f_c'}$. When f_c' is equal to 3,000 psi, the computed E_c value is 3,122,000 psi. For lightweight concrete (about 90 to 110 pounds per cubic foot), the value of E_c is much lower.

When concrete is loaded, it creeps—that is, it continues to deform with time, even if the load is not increased. The deformation finally stops after two to five years; in reinforced concrete, the final deflection is about 2.5 to 3.0 times the initial deflection.

One other factor to consider in the design and detailing of reinforced concrete construction is *shrinkage*. Concrete shrinks in volume as it dries, which can result in cracking and internal stresses. The amount of shrinkage is determined by the water and cement content of the concrete (water content being the more important factor) and the curing conditions. To minimize shrinkage, therefore, the water content of the concrete should be kept as low as possible; however it cannot be so low as to render the concrete unworkable. Reduction of the cement content and proper curing also help to reduce shrinkage. The freshly placed concrete must not be allowed to dry out rapidly. Hence, humid, overcast weather tends to minimize shrinkage; while wind, low humidity, and exposure to the sun increase the tendency for shrinkage.

REINFORCED CONCRETE THEORY

Background

Throughout history, and in almost every part of the world, concrete has been used as a building material. The ready availability of its constituent materials, the ease with which the plastic concrete can be poured into forms of virtually any practical shape, and the great compressive strength and durability of the hardened concrete have all contributed to its widespread use. Of course, the concrete of old was not the high quality material we use today—but being comprised of water, aggregate, and a cementing agent, it was concrete nonetheless.

Up until about 100 years ago, the use of concrete was generally limited to members subject only to compression, such as columns and arches, because its inherent weakness in tension precluded its economical use for beams.

The latter part of the 19th century saw the first practical use of reinforced concrete—the combination of concrete, with its great compressive strength, and steel reinforcing bars embedded in the concrete to provide the needed tensile strength. The union of the two materials has worked out very well, combining many of the advantages of each: low material cost; resistance to decay, insects, fire, and moisture; durability; and the ability to be formed into almost any shape. The concrete and reinforcing steel are compatible materials: they have almost the same coefficient of thermal expansion so that temperature changes do not introduce significant stresses, and the concrete is sufficiently impervious and fire-resistant to protect the reinforcing steel from both corrosion and fire.

There are, of course, some disadvantages associated with reinforced concrete. Forming, stripping, and finishing are time-consuming, expensive operations. Reinforced concrete

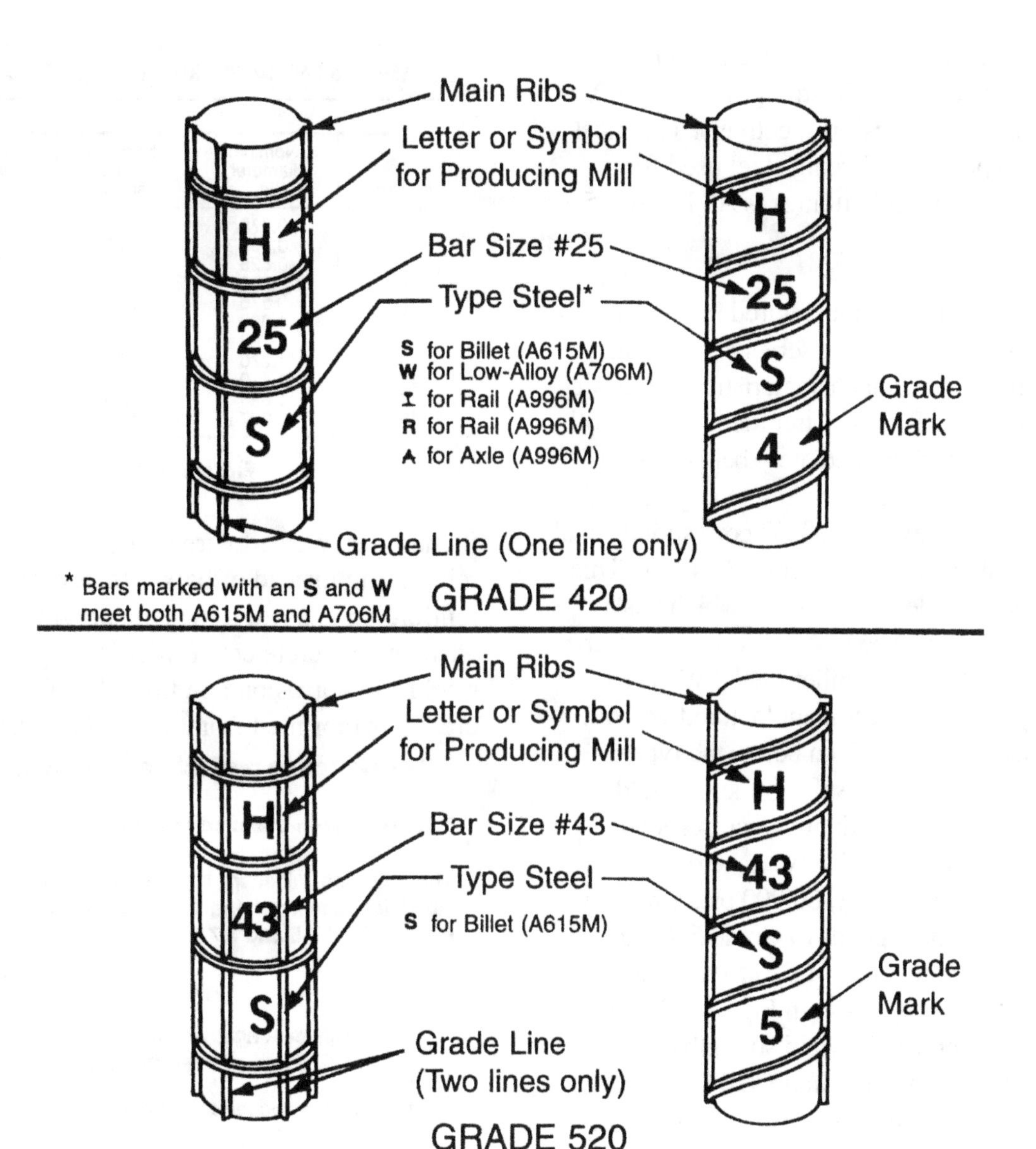

Figure 5.3

is subject to unsightly and possibly harmful cracking even when meticulous care is taken in design and construction. Finally, reinforced concrete members are usually heavy and large.

Reinforcing Steel

The reinforcing used for conventional reinforced concrete construction consists of round steel bars designated by numbers representing the bar diameter in eighths of an inch. The standard sizes are #3 (3/8" diameter) through #11 (1-3/8" diameter), #14 (1-3/4" diameter), and #18 (2-1/2" diameter). In order to assure that no slip occurs between the steel bars and surrounding concrete, reinforcing bars are made with surface deformations which interlock with the concrete.

Reinforcing steel is specified by the ASTM designation, such as ASTM A615 (billet steel), ASTM A706 (low-alloy steel), or ASTM A996 (axle and rail steel). In addition, a grade designation is used: grades 40, 60, and 75 have minimum specified yield strengths of 40,000, 60,000, and 75,000 psi, respectively. The most common type of reinforcing steel used for flexural members in buildings is ASTM A615 Grade 60.

All reinforcing bars manufactured in the United States are furnished with a series of markings that identify important characteristics of the bar (see Figure 5.3). The top letter identifies the rolling mill. The mark immediately below the top one gives the bar size according to a system of metric sizes. For example, "25" corresponds to a #25 bar, which has a diameter of 25.4 mm. This bar size is equivalent to a #8 bar, which has a diameter of 1.0 in. The next mark designates the type of steel: "S" is for billet steel, "W" is for low-alloy steel. For bars manufactured without grade lines, a mark is given below the type of steel; this mark identifies the grade (or yield stress) of the bar. A mark of "4" corresponds to Grade 420 bars, which have a yield stress of 420 MPa or, equivalently, 60,000 psi. Similarly, a mark of "5" corresponds to Grade 520 bars, which have a yield stress of 520 MPa (75,000 psi). Grade lines, as shown in Figure 5.3, are also used to identify the yield strength of the bar. One line designates Grade 420 bars and two lines designate Grade 520 bars.

The properties of reinforcing bars are tabulated in various handbooks. The table in Figure 5.4 is from the ACI Building Code Requirements (ACI 318-05).

ASTM STANDARD REINFORCING BARS

Bar size	Nominal diameter, in.	Nominal area, sq in.	Nominal weight, lb per ft
#3	0.375	0.11	0.376
4	0.500	0.20	0.668
5	0.625	0.31	1.043
6	0.750	0.44	1.502
7	0.875	0.60	2.044
8	1.000	0.79	2.670
9	1.128	1.00	3.400
10	1.270	1.27	4.303
11	1.410	1.56	5.313
14	1.693	2.25	7.650
18	2.257	4.00	13.600

Figure 5.4

Sometimes, welded wire fabric is used as reinforcing instead of bars. Whatever type of reinforcing is used, it must be provided with adequate concrete cover in order to protect the steel from corrosion and fire. The table in Figure 5.5 is reprinted from the ACI 318-05.

7.7 — Concrete protection for reinforcement

7.7.1 — Cast-in-place concrete (nonprestressed)

The following minimum concrete cover shall be provided for reinforcement, but shall not be less than required by 7.7.5 and 7.7.7:

Minimum cover, in.

(a) Concrete cast against and permanently exposed to earth 3

(b) Concrete exposed to earth or weather:

No. 6 through No. 18 bars............................ 2
No. 5 bar, W31 or D31 wire, and smaller 1-1/2

(c) Concrete not exposed to weather or in contact with ground:

Slabs, walls, joists:
No. 14 and No. 18 bars 1-1/2
No. 11 bar and smaller......................... 3/4

Figure 5.5

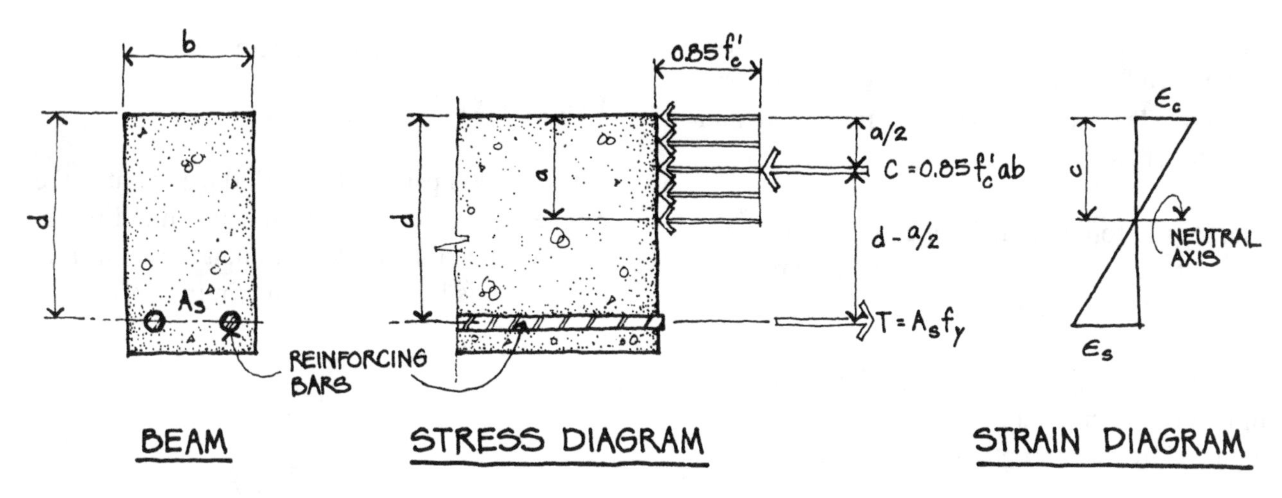

Figure 5.6

Strength Design

The method generally used for the design of reinforced concrete members is called *ultimate strength design*, or simply *strength design.* In this method, the internal stresses and strains in a reinforced concrete beam that is about to fail are used to determine the ultimate moment capacity of the member. This capacity is made larger than the greatest moment expected during the lifetime of the member, thus assuring structural safety.

As previously discussed, concrete is strong in compression, but weak in tension. Reinforcing bars are therefore embedded in the tension side of a beam and are assumed to resist all the tensile stresses, while the concrete is assumed to resist all the compressive stresses. The bond between the concrete and the reinforcing steel permits the two materials to act in this manner.

The stresses in a reinforced concrete beam at failure are shown in Figure 5.6, simplified for purposes of calculation.

Basic Notation

b = beam width in inches

d = distance from extreme compression fiber to centroid of tension reinforcement in inches

A_s = cross-sectional area of tensile reinforcement in square inches

a = depth of rectangular stress block = $\beta_1 c$ ($\beta_1 = 0.85$ up to $f'_c = 4{,}000$)

c = distance from extreme compression fiber to neutral axis

f_s = unit tensile stress in reinforcing steel in psi

f'_c = specified compressive strength of concrete in psi

f_y = specified yield strength of reinforcement

Ø = strength reduction factor

ρ = A_s/bd = ratio of tension reinforcement to beam area

A reinforced concrete beam can fail by (1) crushing of the concrete, which is assumed to take place when the concrete strain $\in_c$ reaches 0.003; or (2) yielding of the steel, which starts when the steel stress (f_s) reaches the yield strength (f_y). If (1) and (2) take place simultaneously, it is called a *balanced design.* In order to assure that the yielding of the steel takes place *before* failure of the concrete, the ACI Code requires that the net strain in the reinforcing bars that are closest to the tension face of the member must be greater than or equal to 0.004. This essentially limits the reinforcement ratio (= A_s/bd) to about 0.75 of that ratio that would produce a balanced design.

Because of possible variations in quality control, the ACI Code uses the strength reduction factor Ø. The value of Ø is 0.90 for flexure and 0.75 for shear.

The tensile force in the steel T is equal to $A_s f_y$, while the compressive force in the concrete C is equal to $0.85\ f'_c\ ab$.

The moment capacity M_u is equal to T multiplied by the moment arm (d – a/2) and then adjusted by the strength reduction factor Ø.

Thus, $M_u = ØA_s f_y(d - a/2)$

The moment arm (d – a/2) is sometimes referred to as $j_u d$, where j_u is approximately 0.9.

Thus, $M_u = ØA_s f_y j_u d$

Therefore, to increase the moment capacity of a reinforced concrete beam, we can

1. use more reinforcing steel A_s, or
2. use steel with a greater yield strength f_y, or
3. increase the effective depth d of the beam.

The moment capacity M_u may be computed, using the formulas above. This capacity must be at least equal to $1.2\ M_d + 1.6\ M_l$, where M_d = dead load moment and M_l = live load moment. 1.2 and 1.6 are called the *ultimate load factors* (U).

DESIGN OF REINFORCED CONCRETE BEAMS

Throughout this lesson, we refer to reinforced concrete beams, since the word *beam* is the most inclusive term we can use. We mean any flexural member made of reinforced concrete, whether it is called a beam, joist, girder, or slab.

Rectangular Beam Design

Example #1

If f_y = 60,000 psi and f'_c = 4,000 psi, what is the moment capacity of the beam shown in Figure 5.7? Assume that a = 4", and that the reinforcement ratio is less than the maximum permitted by the code.

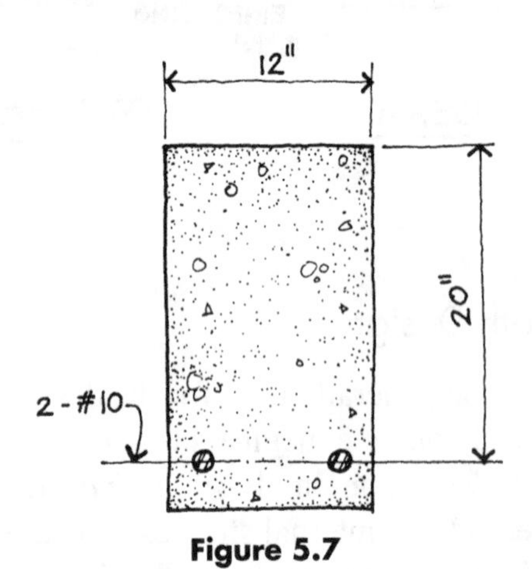

Figure 5.7

Solution:

The area of tensile reinforcement $A_s = 1.27 \times 2$ = 2.54 sq. in. (see in Figure 5.4).

$M_u = ØA_s f_y(d - a/2)$
$= 0.90(2.54)(60{,}000)(20 - 4/2)$
$= 137{,}160\ (20 - 2) = 2{,}468{,}880$ in.-lbs.

The moment may be converted from in.-lbs. to ft.-kips by dividing by 12,000.

M_u = 2,468,880/12,000 = *205.74 ft.-kips*

Example #2

Determine the safe moment M that may be applied to the beam of Example #1, if half of the total applied moment is dead load and half is live load.

Solution:

$$M_u = 1.2\ M_d + 1.6\ M_l$$
$$= 1.2(0.50\ M) + 1.6(0.50\ M)$$
$$= 0.60\ M + 0.80\ M = 1.40\ M$$
$$M = M_u/1.40 = 205.74/1.40 = \textit{146.96 ft.-kips}$$

Example #3

A reinforced concrete beam must resist a dead load moment of 80 ft.-kips and a live load moment of 115 ft.-kips. $f'_c = 3{,}000$, $f_y = 60{,}000$, b = 12", d = 32". Select the reinforcing steel.

Solution:

$$M_u = 1.2\ M_d + 1.6\ M_l$$
$$= 1.2(80) + 1.6(115)$$
$$= 96 + 184 = 280 \text{ ft.-kips}$$
$$A_s = \frac{M_u}{Øf_y j_u d}$$
$$= \frac{280{,}000 \times 12}{0.90 \times 60{,}000 \times 0.9 \times 32} = 2.16 \text{ sq. in.}$$

Use 2-#10, $A_s = 2 \times 1.27 = 2.54$ sq. in., from Figure 5.4.

Minimum Reinforcement

The ACI Code requires that wherever positive reinforcement is required, the ratio ρ shall not be less than

$$\rho_{min} = \frac{200}{f_y} \text{ and not less than } \frac{3\sqrt{f'_c}}{f_y}$$

For steel having a yield point of 40,000 psi, with $f'_c = 3{,}000$ psi, ρ_{min} is equal to

$$\frac{200}{40{,}000} = 0.005, \text{ or } \frac{3\sqrt{3{,}000}}{40{,}000} = 0.0041$$

The value of 0.005 governs.

Where the yield point is 60,000 psi, with $f'_c =$ 3,000,

$$\rho_{min} = \frac{200}{60{,}000} = 0.0033, \text{ or } \frac{3\sqrt{3{,}000}}{60{,}000} = 0.0027$$

The value of 0.0033 governs.

Example #4

Check the design of Example #3 for minimum reinforcing.

Solution:

From Example #3, b = 12", d = 32", $f_y =$ 60,000.

ρ_{min} for f_y of 60,000 psi = 0.0033

Min. $A_s = \rho_{min}\ bd = 0.0033 \times 12 \times 32 = 1.27$ sq. in.

A_s used = 2-#10 = 2.54 sq. in.

The reinforcement used exceeds the minimum code requirement.

T-Beam Design

Reinforced concrete beams are usually constructed integrally with the supported slab, and a portion of the slab may be considered to act with the beam to form a T-shaped beam, as shown in Figure 5.8.

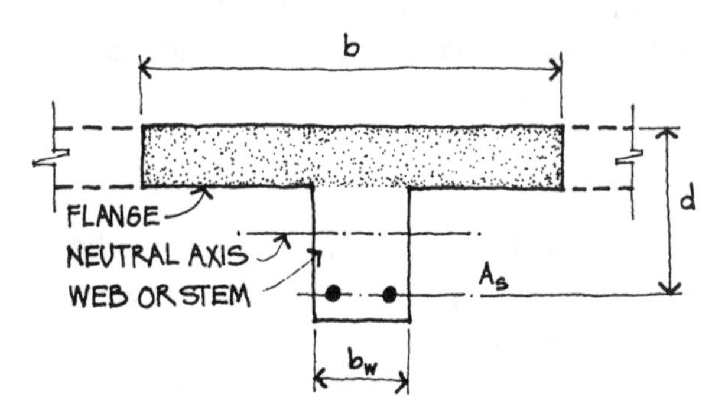

Figure 5.8

The ACI Code specifies that the assumed width b may not exceed 1/4 of the beam span, and the overhanging width on either side of the web may not exceed eight times the slab thickness nor 1/2 of the clear distance to the next beam.

The stress block (area ab) of T-beams usually falls within the flange, in which case the methods for rectangular beams may be used.

As with rectangular beams, the net strain in the reinforcement must be greater than or equal to 0.004.

Compressive Reinforcement

Sometimes the cross-sectional area of a beam is too small for the concrete to develop the compressive force necessary to resist the applied moment. In that case, reinforcing steel may be embedded in the compression side of the beam, to increase the beam's moment capacity (see Figure 5.9). The use of compressive reinforcement reduces the creep and hence the long-term deflection of the beam and also reduces the possibility of any sudden failure.

Compressive reinforcement in a beam tends to buckle outward when the beam is loaded. For that reason, such bars must be anchored by lateral ties, in the same manner as compressive bars in columns. The ties must be at least #3 in size, and the spacing of the ties may not exceed 16 times the diameter of the longitudinal bars or 48 times the diameter of the tie bars, as shown in Figure 5.9.

Shear

If the shear stress in a beam is excessive, the beam will develop cracks in the pattern shown in Figure 5.10.

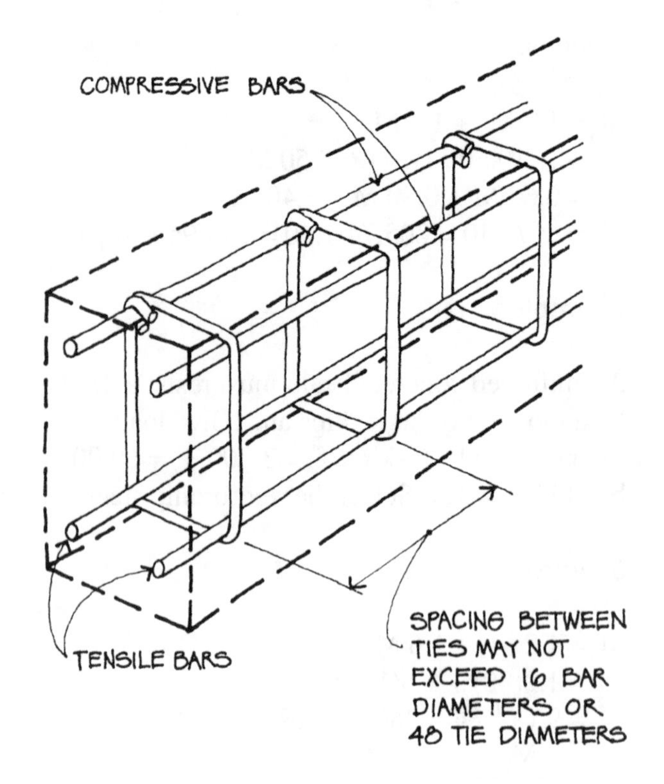

Figure 5.9

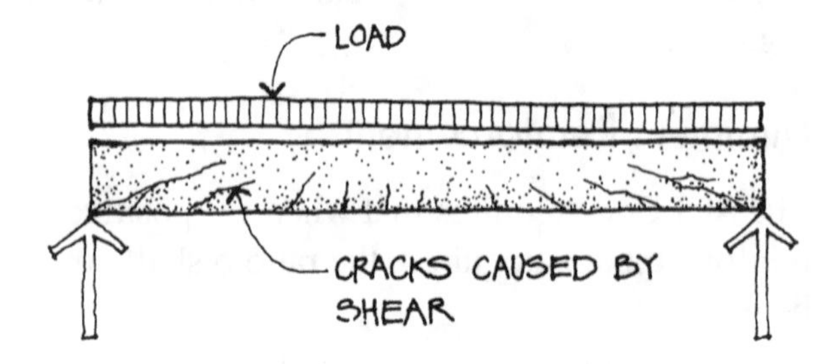

Figure 5.10

Therefore, in designing reinforced concrete beams, we generally verify that the shear strength of the concrete (V_c) multiplied by the strength reduction factor Ø (0.75 for shear) is at least equal to the ultimate shear force on the beam (V_u). If the shear strength of the concrete is inadequate, we add web reinforcement called stirrups (Figure 5.11). The stirrups are spaced close together where the shear force is highest, generally near the supports, and further apart where the shear force is less, with a maximum spacing of d/2.

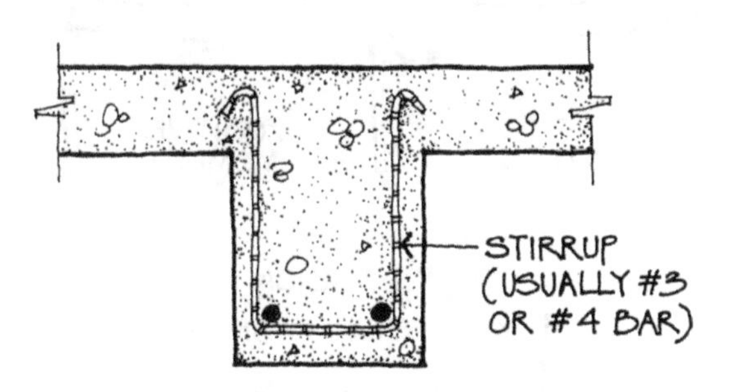

BEAM WITH SHEAR (WEB) REINFORCEMENT

Figure 5.11

Development Length

One of the basic requirements in all reinforced concrete is that the reinforcing steel and its surrounding concrete must act together, so that the bars cannot slip or pull out of the concrete. It is vital that the bars be anchored to the concrete sufficiently so that the tensile force can be *developed*, that is, transferred into and out of the reinforcing steel. Accordingly, a specified minimum length or extension of reinforcement beyond all points of peak stress in the bars is required. This *development length*, l_d, is a function of the bar size, the yield strength of the steel, and the compressive strength of the concrete.

Deflection

Although reinforced concrete construction is usually quite stiff—that is, resistant to deformation—this is not always the case. Reinforced concrete beams deflect, just as structural steel and wood beams do, and the amount of this deflection must be limited in order to prevent cracking, bouncy floors, unsightly sag, ponding of water on roofs, or possible structural damage.

It is not always necessary to compute the deflection of reinforced concrete beams. For most average loading and span conditions, maintaining minimum beam depths (thicknesses) shown in Table 9.5(a) from ACI 318-05 will keep the deflections within tolerable limits. Table 9.5(a) from ACI 318-05, reproduced in the appendix, can be used to obtain beam depths so that deflections need not be computed.

Sometimes, however, it is advisable to compute the deflection of reinforced concrete members. The amount of this deflection is less predictable than that of structural steel beams, and depends on a number of factors: the properties of the concrete and reinforcing steel, the amount of creep (increased deformation with time without any increase in load), and the amount of shrinkage. Therefore, the deflections computed for reinforced concrete members are always approximate.

Where the deflections are computed, the immediate deflection under load is calculated by the usual methods and formulas, using the value $57{,}000\sqrt{f'_c}$ as the modulus of elasticity E_c of normal weight concrete.

The moment of inertia used to calculate deflection is not the actual moment of inertia of the concrete section, but a modified value that is generally lower.

Shrinkage and creep due to sustained loads cause additional deflection over and above the immediate deflection that occurs when load is first placed on the structure. This additional deflection varies from one to two times the immediate deflection, depending on the duration of load.

The maximum deflections permitted by the IBC are shown in Table 1604.3 in the appendix.

Summary of Reinforced Concrete Beam Design

The design of reinforced concrete beams involves a number of procedures, which can be summarized as follows:

1. Select the specified concrete compressive strength f'_c and reinforcing steel yield strength f_y. Generally, these values are constant for all the beams in the building.
2. For the given beam, determine the maximum bending moment M_u and vertical shear V_u, using the load factors 1.2 for the dead load and 1.6 for the live load.
3. Select the width b and the effective depth d of the beam; these must be adequate to resist the maximum bending moment M_u. In general, the value of d should be about 1-1/2 times b. If the size of the beam is predetermined, verify that it is adequate. Use T-beam action when possible.
4. Compute the required area of tensile reinforcement A_s and select the bars. Verify that the steel area is at least the minimum required by the code.
5. If the depth of the beam is insufficient to resist the maximum moment, provide compressive reinforcement.
6. Check the shear capacity and provide stirrups where necessary.
7. Verify that all reinforcing bars are extended at least their development length beyond the point of peak stress.
8. Be sure that the beam is adequately sized for deflection.

CONTINUITY IN REINFORCED CONCRETE

By their very nature, most reinforced concrete beams are continuous, while most wood and steel beams are simply supported.

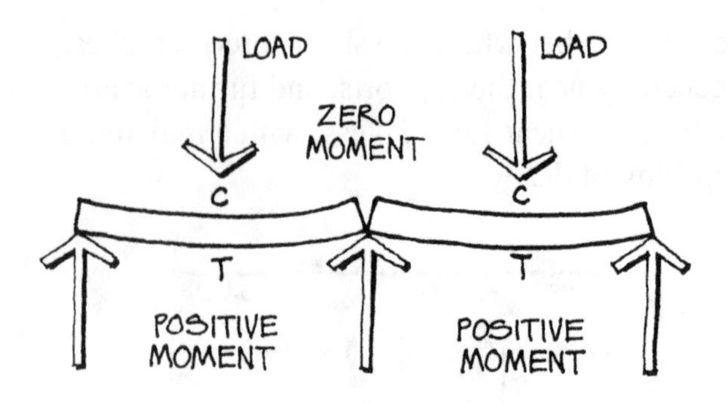
LOAD
LOAD
ZERO MOMENT
C
C
T
T
POSITIVE MOMENT
POSITIVE MOMENT
C = COMPRESSION FACE
T = TENSION FACE
TWO SEPARATE STEEL OR WOOD BEAMS DEVELOP NO MOMENT AT THE CENTER SUPPORT SINCE THEY ARE DISCONTINUOUS WITH EACH OTHER.
2 SIMPLE SPANS WITHOUT CONTINUITY

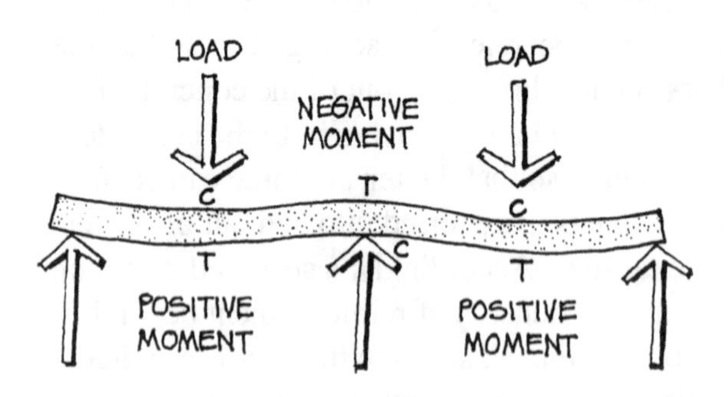
LOAD
LOAD
NEGATIVE MOMENT
C
T
C
T
C
T
POSITIVE MOMENT
POSITIVE MOMENT
C = COMPRESSION FACE
T = TENSION FACE
REINFORCED CONCRETE BEAM DEVELOPS NEGATIVE MOMENT AT THE CENTER SUPPORT BECAUSE BOTH SPANS ARE CONTINUOUS AND INTEGRAL WITH EACH OTHER.
CONTINUOUS 2-SPAN BEAM

Figure 5.12

The upper diagram in Figure 5.12 shows two adjacent simple spans without continuity, which is typical of most structural steel and wood construction.

If you have forgotten what is meant by simple and continuous beams and positive and negative moments, this would be a good time to review those concepts in Lesson Two.

The bottom of Figure 5.12 shows a continuous beam, which is typical of reinforced concrete construction. Of course, the number of spans is immaterial—we are merely discussing continuity over more than one span.

A continuous beam is statically indeterminate; that is, it cannot be solved from the equations of equilibrium alone. However, there are methods available, such as moment distribution, for solving continuous beams; but such methods are beyond the scope of this course. Within specified limits, moment coefficients that approximate the actual moments under uniformly distributed loads may be used in lieu of a more exact analysis. Figure 5.13 shows the moment coefficients permitted by ACI 318-05. Positive moment occurs between the supports (tension at the bottom), and negative moment occurs over the supports (tension at the top). For ease of fabrication and construction, use only straight top and straight bottom bars instead of bent up bars, which were popular in the past.

When a T-beam functions as a continuous beam, the flange is in compression only in the region of positive moment (between supports; see the top of Figure 5.14). In the region of negative moment (over the supports), the flange is in tension and the stem is in compression. (see the bottom of Figure 5.14.)

The concrete stress therefore is more critical over the supports, since the stem area available to resist compression is smaller than the flange area.

When continuous beams are solved by moment distribution or any other method instead of using moment coefficients, the live load arrangement should be varied to produce the maximum possible moments.

The full live load on alternate spans produces maximum positive moment between supports, while the full live load on two adjacent spans produces maximum negative moment over the support, as shown in Figure 5.15. Of course, the dead load acts simultaneously on all spans.

PRESTRESSED CONCRETE

Prestressed concrete is permanently loaded so as to cause stresses opposite in direction from those caused by dead and live loads.

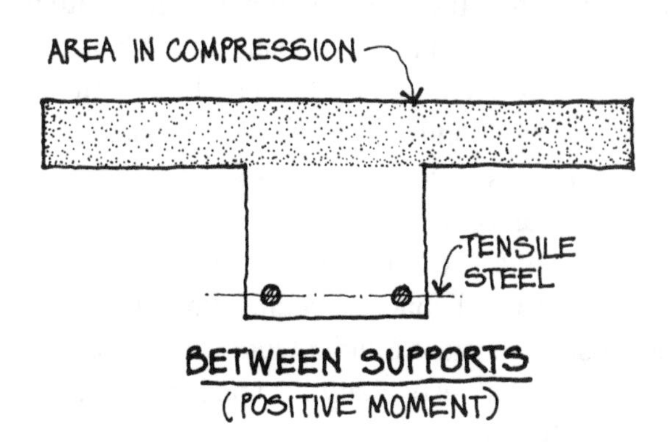

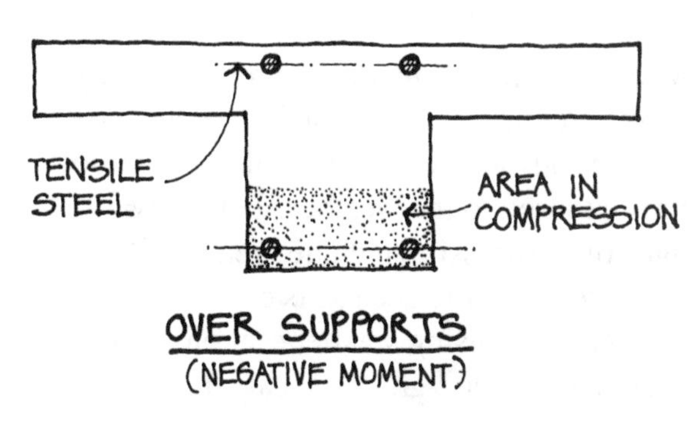

T-BEAM ACTION

Figure 5.13

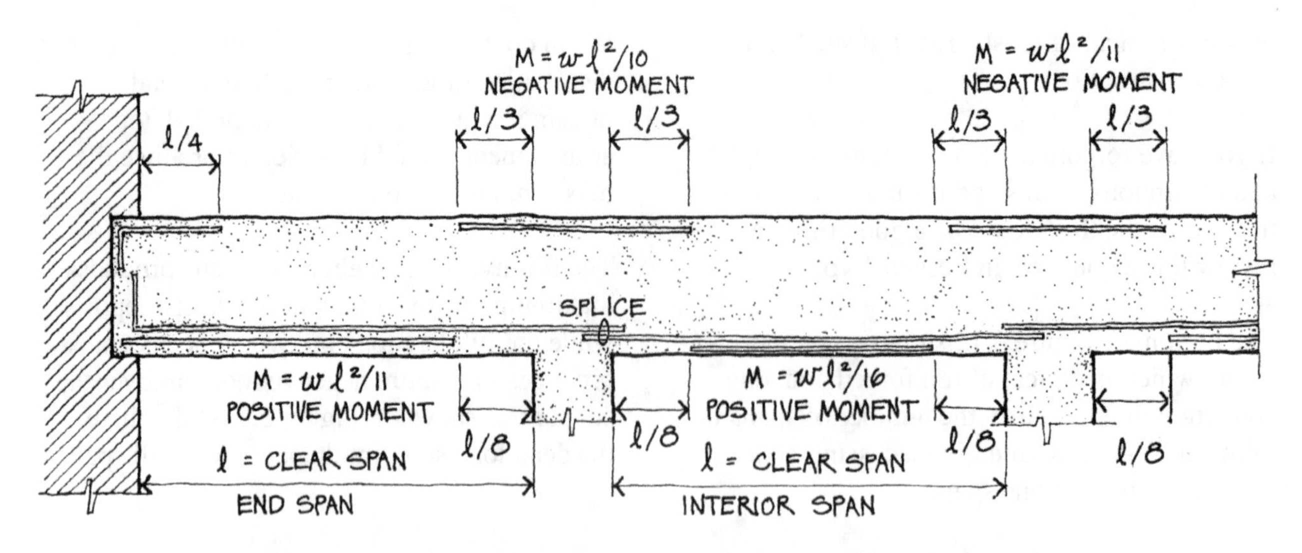

Figure 5.14

As shown in Figure 5.16, the prestressing wires placed eccentrically in the beam cause axial compression in the concrete along with negative moment. The dead and live loads on the beam produce positive moment. The combined effect of the prestress and the dead and live loads usually results in compression over the entire cross-section of the beam, in contrast to the relatively small compression area of conventional reinforced concrete beams. Prestressing thus results in more efficient and economical use of material, especially in repetitive long-span applications.

Where prestressed members are completely in compression, tension cracks are prevented, which is advantageous, particularly in structures exposed to the weather. Other advantages of prestressing include greater stiffness (because the entire section is effective), the practicality of using smaller sections (for the same reason), and greater shear strength.

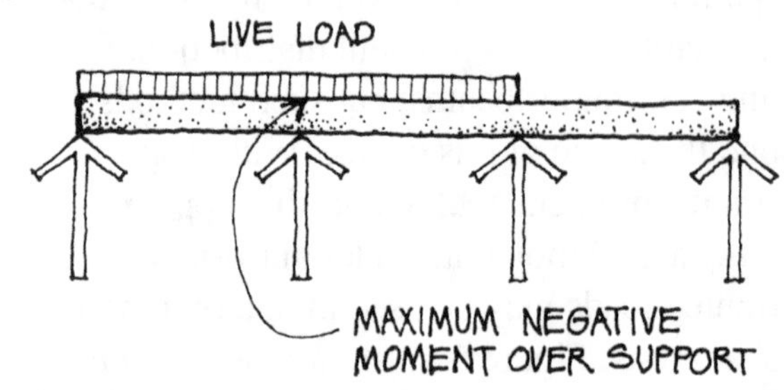

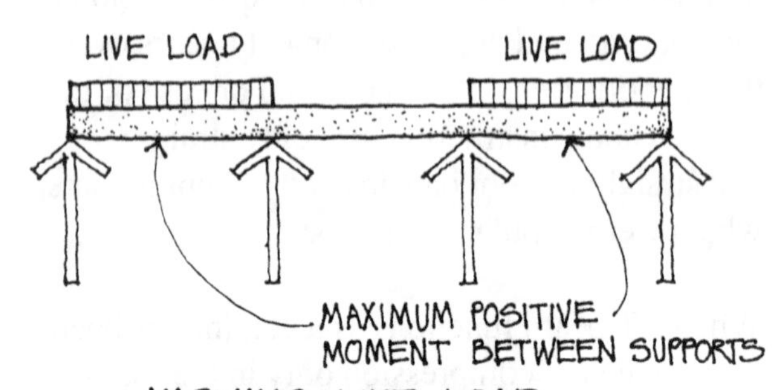

VARYING LIVE LOAD ARRANGEMENT

Figure 5.15

However, offsetting these advantages are greater material and labor costs and the need for closer quality control than with conventional reinforced concrete.

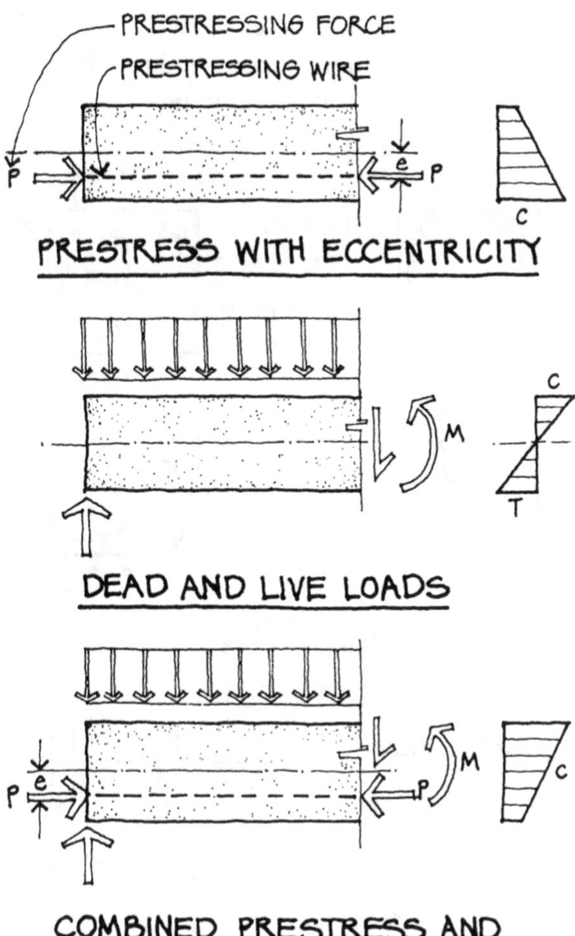

Figure 5.16

There are two procedures used to apply prestress: *pretensioning* and *posttensioning*. In pretensioning, a tensile force is put into high strength wires by stretching them between anchorages, usually at a casting yard away from the building site. The concrete is placed, the prestress wires bond to the concrete, and the concrete is cured. Then the prestress wires are cut. The tensile force in the wires is applied through bond as a compressive force to the concrete. Some loss of prestress occurs because of creep and shrinkage, slip, and friction.

In a posttensioned beam, the concrete is cast on the site with a hollow duct to encase the prestressing steel. The concrete is cured and the steel tendons stressed by jacking against anchorages at each end of the member.

The tendons are permanently locked under stress by special end anchors and grouted. The losses after tensioning, caused by friction, elastic shortening, and shrinkage, are usually less than with pretensioning.

REINFORCED CONCRETE COLUMNS

The design of reinforced concrete columns is far more complex than that of wood or steel columns; therefore, we will limit our discussion to general concepts only, rather than detailed design procedures.

Spiral Columns

Two types of reinforced concrete columns are commonly used: spiral columns and tied columns.

Spiral columns are square or round in shape and have longitudinal reinforcing bars arranged in a circle close to the face of the concrete and enclosed by a closely spaced continuous steel spiral (see Figure 5.17). The longitudinal bars help to carry the compressive load, while the spiral braces the longitudinal bars against buckling and confines the concrete. At least six longitudinal bars must be used, and the area of longitudinal reinforcement must be at least 0.01, but not more than 0.08, times the gross cross-sectional area of the column.

The spiral reinforcement consists of evenly spaced continuous spirals held firmly in place and true to line by vertical spacers. The minimum size of spiral reinforcement is 3/8 inch diameter, and the clear spacing between spirals must be at least one inch and not more than three inches.

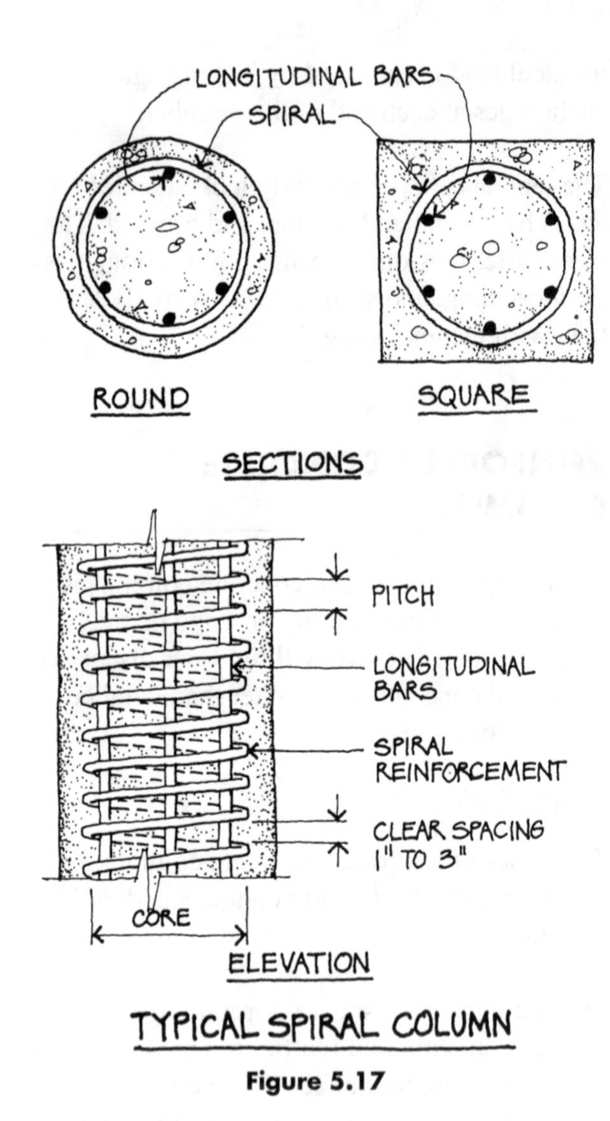

Figure 5.17

Tied Columns

Tied columns are usually square or rectangular in shape and have longitudinal reinforcing bars placed close to the face of the concrete, with separate lateral ties (see Figure 5.18). As with spiral columns, the longitudinal bars help to carry the compressive load, while the ties serve to hold the longitudinal bars in position, prevent the longitudinal bars from buckling outward, and also somewhat confine the concrete. The area of longitudinal reinforcement must be at least 0.01, but not more than 0.08, times the gross cross-sectional area of the column. Usually an even number of longitudinal bars is used, with the code requiring at least four for bars within rectangular ties. The longitudinal bars may be grouped together in bundles of two, three, or four bars each, which must be tied or wired together to ensure remaining in position.

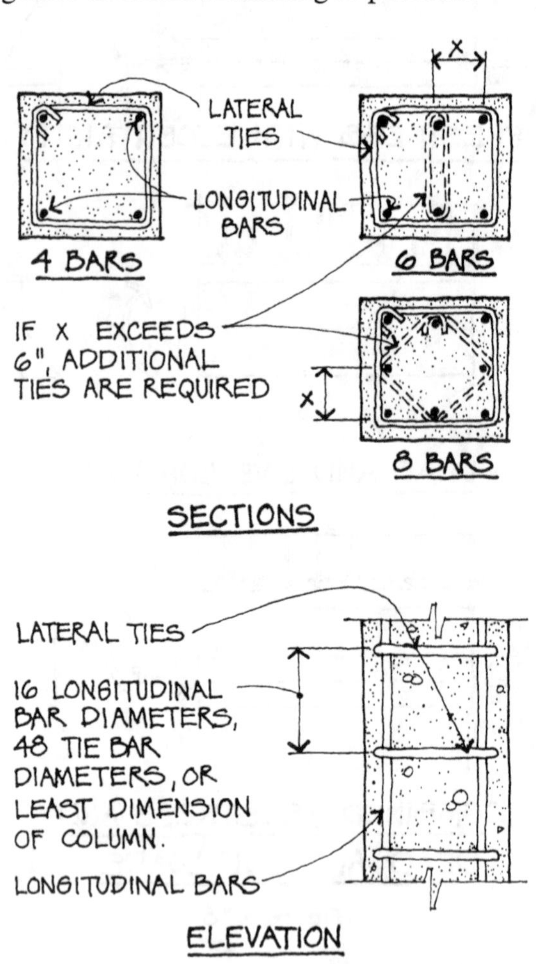

Figure 5.18

The lateral ties must be at least #3 in size for longitudinal bars #10 or smaller, and at least #4 in size for #11, #14, #18, and bundled longitudinal bars. The spacing of the ties must not exceed any of the following dimensions:

1. 16 longitudinal bar diameters, or
2. 48 tie bar diameters, or
3. the least dimension of the column.

In buildings assigned to Seismic Design Category C and higher, hoops are required, and their spacing is more stringent than that for typical column ties. Every corner and alternate longitudinal bar must be tied in both directions, and no bar may be further than 6" from such a laterally supported bar.

Some typical arrangements for tied columns are shown in Figure 5.18.

Required Strength

The load factors used to design reinforced concrete columns are the same as those for beams. For gravity loads, the required strength U = 1.2 D + 1.6L, where D is the dead load and L is the live load.

The strength reduction factor Ø is equal to 0.70 for spiral columns and 0.65 for tied columns, indicating that spiral columns have greater ductility or toughness.

The code also recognizes that pure axial load on a column seldom exists. Even if the load is theoretically axial, slight imperfections in construction and slight end restraints cause unintentional bending moments. Consequently, the code limits the axial load supported by a column to 80 or 85 percent of its nominal axial load strength.

Slenderness, which is such an important factor in the design of wood and steel columns, is less significant for reinforced concrete columns, since such columns are usually relatively stocky compared to their length. Therefore, the code allows the effects of slenderness to be neglected when the slenderness is below certain limits; above those limits, however, the slenderness must be considered.

The monolithic nature of reinforced concrete construction makes most reinforced concrete columns subject to combined bending moment and axial compression. As we've mentioned, the design procedures are relatively complex and beyond the scope of this course. Interested candidates are referred to the reinforced concrete texts and handbooks listed in the bibliography.

Column Splices

Where reinforced concrete columns are spliced at floor lines or at the foundation, the longitudinal bars are lapped, as shown in Figure 5.19, and must meet the criteria for development length as described on page 95.

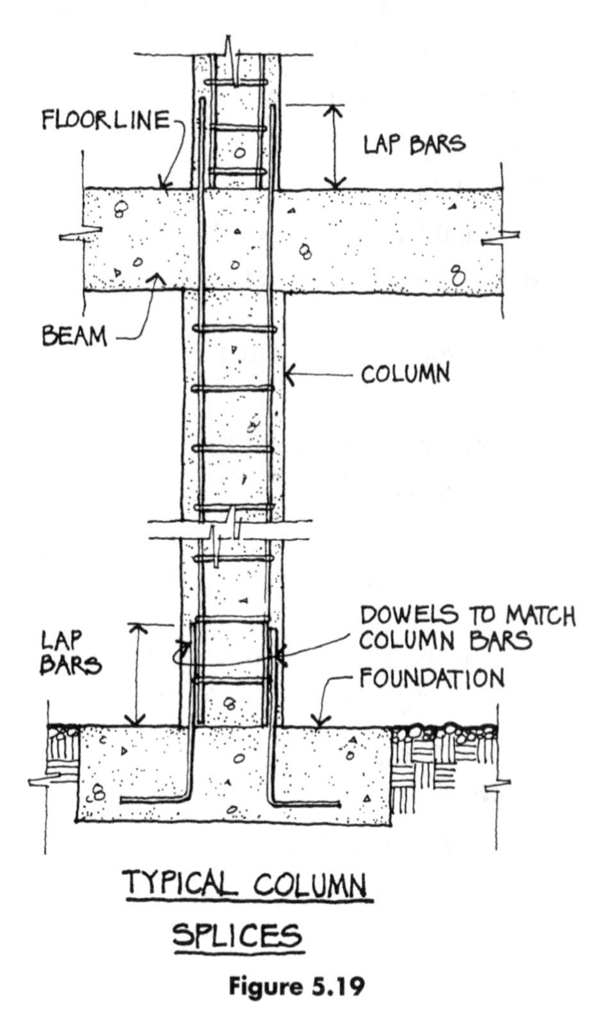

Figure 5.19

CONCLUSION

Reinforced concrete design is inherently more complex and time-consuming than structural steel or wood design.

Candidates for architectural licensing are not expected to be experts in reinforced concrete, but they should understand its basic concepts and be able to apply them in the solution of simple problems, such as those presented in this lesson.

LESSON 5 QUIZ

1. Which of the following tests are used to determine the workability of concrete?

I. Core test
II. Cylinder test
III. Impact hammer test
IV. Kelly ball test
V. Slump test

A. II and III
B. I, III, and V
C. IV and V
D. I and V

2. What property is measured by the cylinder test?

A. Compressive strength
B. Hardness
C. Tensile strength
D. Workability

3. What is the MOST important factor that determines the strength of concrete?

A. Size of aggregate
B. Water-cement ratio
C. Type of admixture
D. Type of cement

4. Select the CORRECT statements about prestressed concrete.

I. Smaller sizes may be used since the entire cross-section is effective in resisting the applied loads.
II. In pretensioned members, the tendons are grouted after being stressed.
III. In posttensioned members, the losses of prestress are usually less than in pretensioned members.
IV. Prestressed members are economical because they have lower material and labor costs than non-prestressed members.
V. Prestressing is not economical for long repetitive members.

A. III and IV
B. I, II, and IV
C. II and V
D. I and III

5. A rectangular concrete beam with b = 12" and d = 24" is to support a floor load. Assuming that f'_c = 3,000 psi and f_y = 60,000 psi, what is the minimum flexural reinforcement that may be used?

A. 2-#6 bars
B. 3-#7 bars
C. 3-#4 bars
D. 2-#7 bars

6. Select the CORRECT statement in regard to reinforced concrete columns.

   A. Tied columns generally have a circular cross-section.

   B. Spiral columns can support more load than tied columns, if all other factors are the same.

   C. Because of their stockiness, reinforced concrete columns are not generally subject to bending moment.

   D. In reinforced concrete columns, the longitudinal reinforcing bars carry all the vertical load, while the concrete serves to protect and confine the bars.

7. What is the ultimate moment capacity (M_u) of a reinforced concrete beam with b = 20 inches, d = 46 inches, 4-#8 bottom bars, f'_c = 4,000 psi, f_y = 60,000 psi? Assume that a = 9.2".

   A. 588.7 ft.-kips **C.** 603.1 ft.-kips

   B. 704.8 ft.-kips **D.** 670.1 ft.-kips

8. Which reinforcing steel arrangement is CORRECT for the two-span reinforced concrete beam shown below, when supporting a uniformly distributed load across both spans?

A.

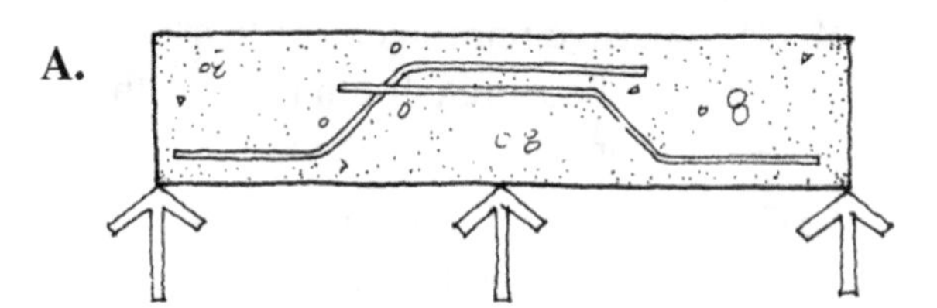

B.

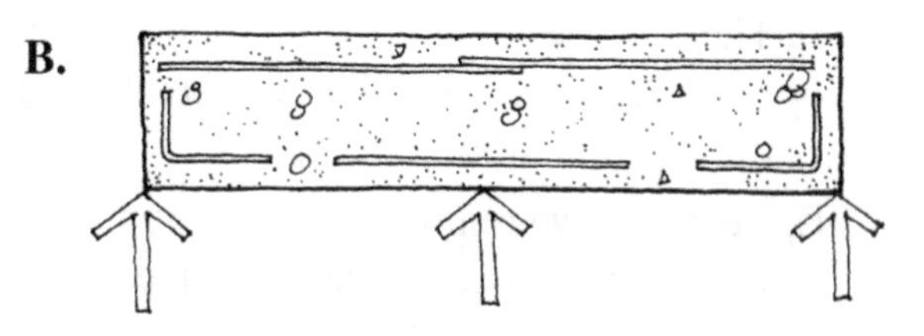

C.

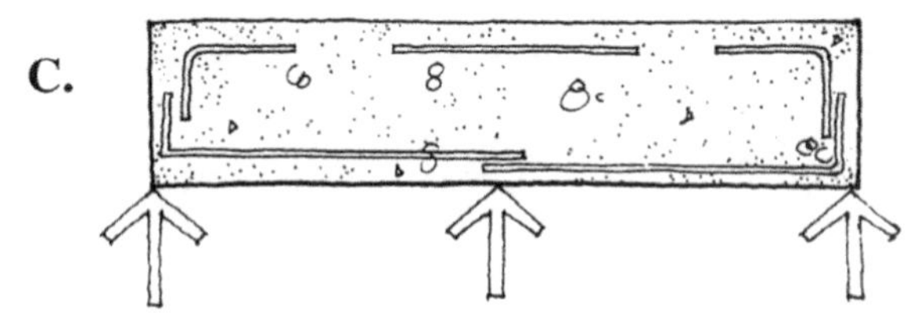

D.

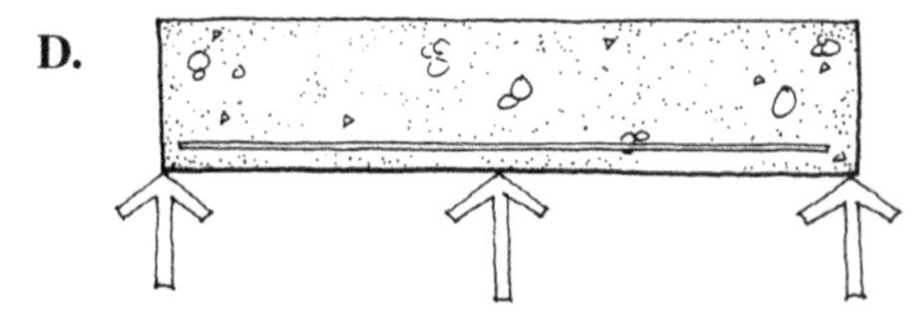

9. Where is T-beam action in a two-span reinforced concrete beam MOST effective?

   A. Over the middle support

   B. Where the moment is maximum

   C. At the points of inflection

   D. Between the supports

10. Select the CORRECT statements about concrete reinforcement.

A. Compressive reinforcement is used when the concrete section is inadequate to resist the moment.

B. Compressive reinforcement in a reinforced concrete beam reduces the beam's long-term deflection.

C. Compressive reinforcement reduces the possibility of sudden failure.

D. The cross-sectional area of compressive reinforcement is usually less than that of the tensile reinforcement.

E. Compressive reinforcement is used where shear stresses are excessive.

WALLS

INTRODUCTION

Walls are used to enclose space, provide support, and/or resist lateral loads from wind, earthquake, or retained earth. Walls may be classified as follows:

A *bearing wall* is one that supports vertical load in addition to its own weight.

A *nonbearing wall* is one that supports no vertical load other than its own weight.

A *shear wall* is one designed to resist lateral forces parallel to itself.

A *retaining wall* is one that resists the lateral displacement of soil or other materials.

Walls may be constructed from a number of different materials, principally wood, masonry, and concrete.

STUD WALLS

A type of wall construction frequently used for dwellings and other light frame structures consists of small closely-spaced wood members (studs), sheathed on both faces with a wall material such as wallboard, plaster, or wood siding. The studs are placed with their wide face perpendicular to the wall, and have a continuous bottom sill or plate and two continuous top plates, which are lapped and nailed at all joints. For most conventional construction, 2 × 4 studs spaced 16" on center are used for laterally unsupported heights up to 10 feet for bearing walls and 14 feet for nonbearing walls. Wood studs are actually small wood columns, which are laterally braced in the weak direction by the wall covering.

Wood stud walls may be either bearing or nonbearing, and may also be used as shear walls if they are adequately sheathed. In general, however, wood stud walls are not used as retaining walls, mainly because of their susceptibility to decay and insect infestation.

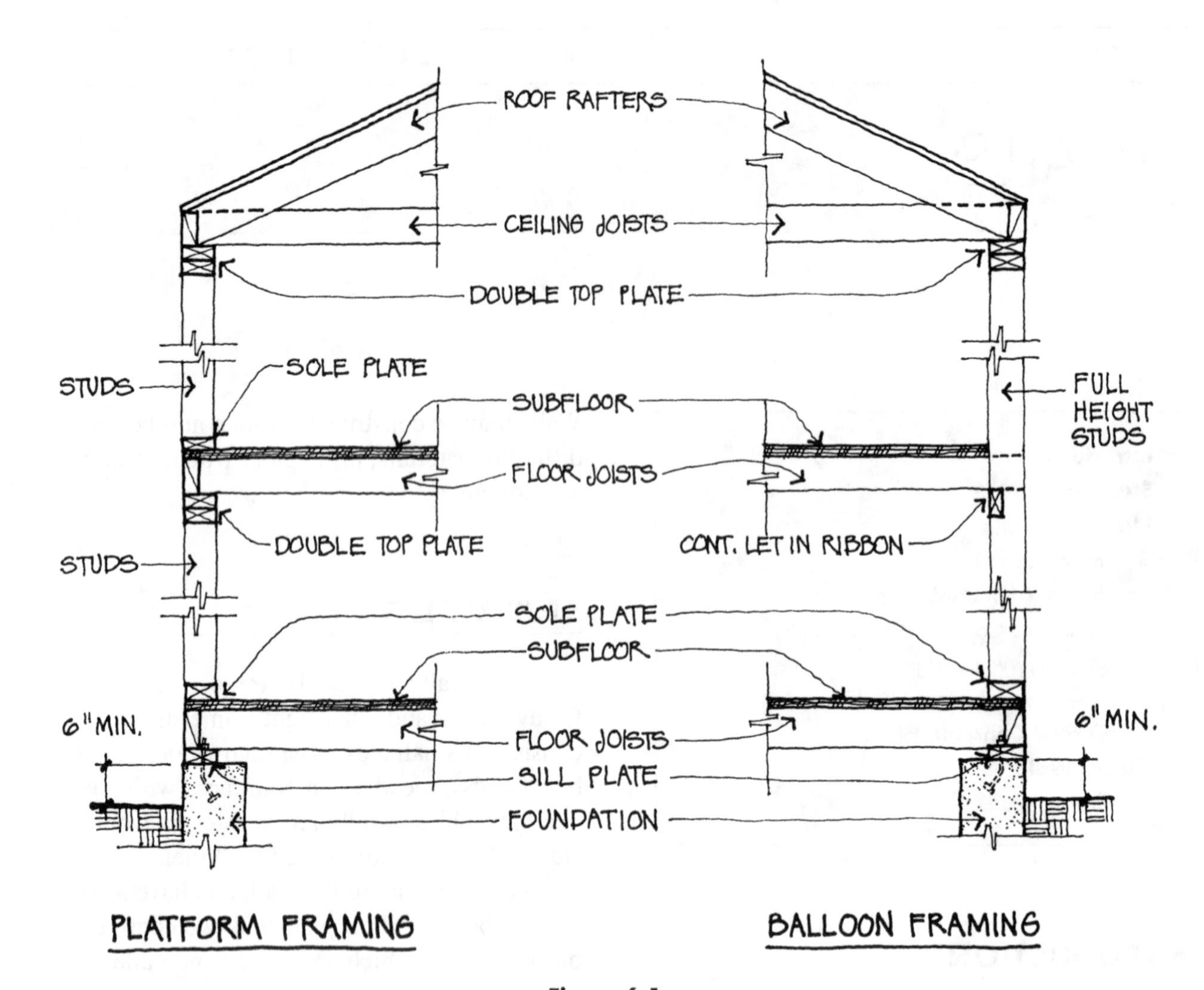

Figure 6.1

There are two ways to frame wood stud walls: *platform framing* and *balloon framing* (Figure 6.1). In platform framing, the studs are one story in height, and the floor joists bear on the top plates of the wall below. In balloon framing, the studs are continuous for the full height of the building (usually two stories), with the second floor joists bearing on a ribbon let into the studs. Platform framing is generally simpler to build, and is intended to equalize, but not minimize, shrinkage. Vertical shrinkage is held to a minimum in the balloon type of construction, making this system preferable where exterior walls are faced with brick veneer or stucco.

Where a wall is required to be incombustible, a *steel stud wall* is often used. The construction is similar to that of a wood stud wall; channel-shaped steel studs 2-1/2" to 6" in depth are spaced at 16 inches on center and welded or screwed to a continuous channel-shaped steel track at the top and bottom.

Stud walls are economical, simple to build, easily adapted to field conditions, lightweight, and allow for simple installation of plumbing and electrical wiring. However, stud walls are not as rigid or durable as masonry or concrete walls.

MASONRY WALLS

Introduction

The history of masonry structures traces back to the earliest human constructions. The classical temples of Greece and Rome and the cathedrals of medieval Europe were all built of masonry. Until relatively recent times, masonry structures were built without any steel reinforcing and could therefore resist only compressive loads. During the past 50 years, however, masonry with steel reinforcement has come into widespread use, principally for wall construction.

Reinforced masonry is analogous to reinforced concrete; masonry, like concrete, is strong in compression, but weak in tension. Steel bars, on the other hand, are subject to buckling under compression but are excellent for resisting tensile forces. The combination of the two materials, masonry for compression and steel for tension, produces a structure capable of resisting bending moments, whether caused by eccentric loading, lateral forces from wind or earthquake, or the pressure of retained earth.

Although there are numerous types of masonry walls, we will confine our discussion to the two kinds of reinforced masonry most often used for structural purposes: reinforced brick masonry and reinforced concrete block masonry.

Reinforced Brick Masonry

Reinforced brick masonry, also called reinforced grouted multi-wythe masonry, consists of brick units, usually two tiers or wythes, with a solidly grouted space between about 2" wide, in which #4 or #5 vertical and horizontal reinforcing bars are placed (Figure 6.2). The bar spacing is about 18" to 36" and is unrelated to the size of the brick units.

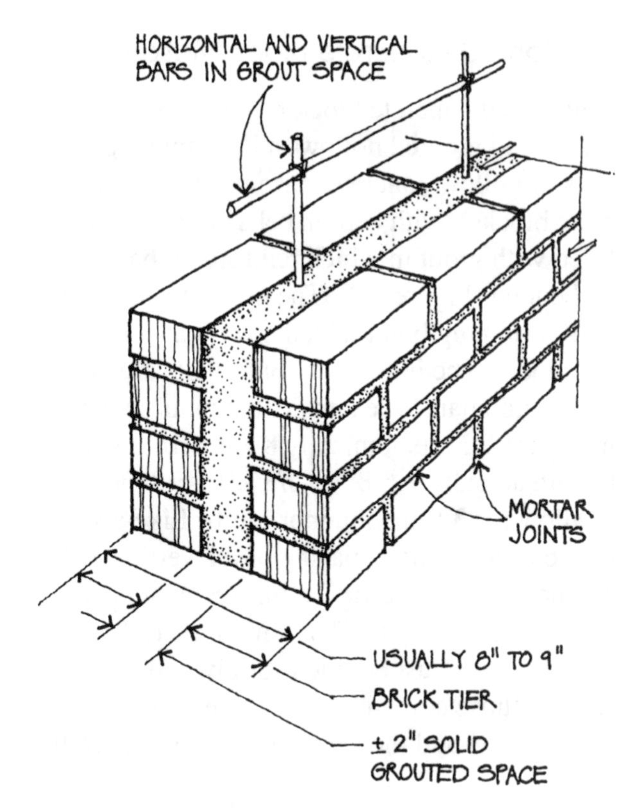

Figure 6.2

The brick units are laid in a leveling bed of mortar that bonds the units together and makes the wall weathertight.

The grout is a soupy concrete that can be poured without segregation. For narrow grout spaces, *fine grout* is used, consisting of 1 part portland cement, 2-1/4 to 3 parts sand, and not more than 1/10 part hydrated lime or lime putty. For wider grout spaces, *coarse grout* is used, composed of 1 part portland cement, 2-1/4 to 3 parts sand, and 1 to 2 parts pea gravel, to which may be added not more than 1/10 part hydrated lime or lime putty. All grout must have a minimum compressive strength of 2,000 psi at 28 days.

Reinforced Concrete Block Masonry

Reinforced concrete block masonry, also called reinforced grouted hollow-unit masonry, consists of hollow concrete masonry units (concrete blocks), with certain cells continuously filled with grout in which reinforcing bars are embedded (Figure 6.3). The units are generally laid in running bond with the vertical cells in alignment. The hollow masonry units are generally modular; for example, the units shown in Figure 6.3 are nominally 8 × 8 × 16 inches, but are actually 7-5/8 × 7-5/8 × 15-5/8 inches to allow for 3/8-inch mortar joints. Likewise, the center-to-center spacing of the vertical cells and hence of the vertical reinforcement is also modular, in this case 8" or a multiple of 8". Certain cells must be filled solidly with grout, such as those containing reinforcement or anchor bolts and those in contact with the earth.

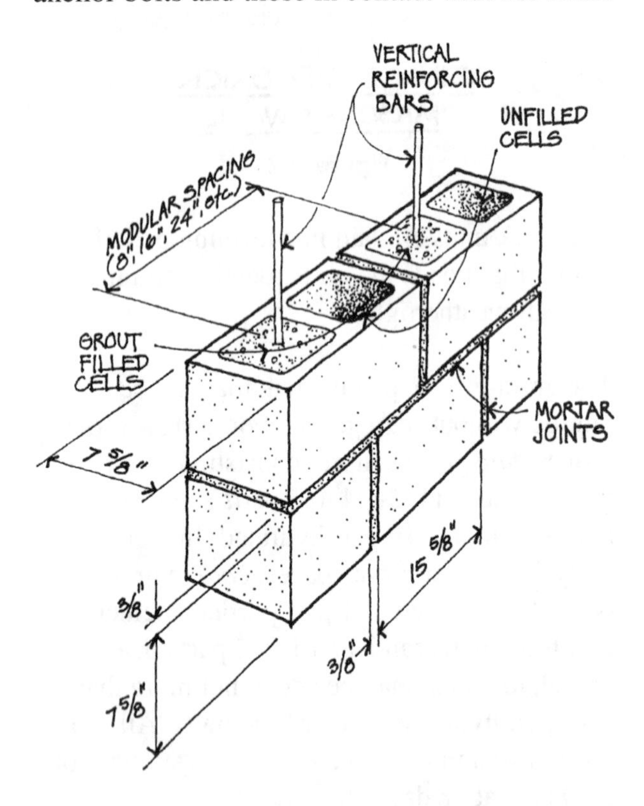

REINFORCED CONCRETE BLOCK MASONRY WALL

Figure 6.3

The other cells may be left unfilled, or in some cases filled with insulation. Horizontal reinforcing bars are placed in special grooved blocks, called bond beam blocks or lintel blocks, and are usually concentrated at roof and floor levels, at the tops of footings and parapet walls, and elsewhere if required. The spacing of horizontal reinforcement is also modular if possible, that is, at a multiple of the height of an individual unit.

The concrete block units are laid in a mortar leveling bed, and the grout that is poured into the vertical cells may be fine grout or coarse grout, as previously defined, depending on the cell dimensions and the height of the grout pour.

Masonry Wall Design

The structural design of reinforced masonry walls is somewhat similar to that of reinforced concrete, although the strength of masonry is less than that of concrete. Generally, the allowable stresses for reinforced brick masonry are higher than those for reinforced concrete block masonry. The IBC requires special inspections for all types of masonry construction. The special inspection is done by a qualified masonry inspector, who continuously observes the masonry construction.

The IBC requires that masonry walls be anchored to floors, roofs, and other structural elements that provide lateral support for the wall. The anchorage must be capable of resisting the horizontal forces specified by the code, with a minimum of 280 pounds per lineal foot of wall.

Reinforced masonry walls have many advantages: good fire resistance; durability; rigidity; the strength to resist vertical loads as well as horizontal forces from wind, seismic, or retained earth; an attractive appearance; and low maintenance. Some of the disadvantages include slow construction time, since the units are laid individually; the possibility of cracks and leakage; and sometimes, relatively high cost.

Openings in Masonry Walls

In reinforced masonry walls, openings for doors and windows are usually framed with reinforced masonry lintels, which are somewhat similar to reinforced concrete beams: reinforcing bars resist all the flexural tension stresses, while the masonry resists the flexural compression stresses. In unreinforced masonry, openings are framed either with reinforced masonry lintels or with steel angle lintels.

REINFORCED CONCRETE WALLS

Although reinforced concrete is stronger than reinforced masonry, reinforced concrete walls have many characteristics similar to reinforced masonry walls: fire resistance, durability, rigidity, strength, attractive appearance, low maintenance, slow construction time, possible cracks and leakage, and relatively high cost.

Reinforced concrete walls must be designed for any bending moment to which they are subjected, whether caused by eccentric loading, lateral forces from wind or earthquake, or the pressure of retained earth.

Like masonry walls, the IBC requires reinforced concrete walls to be anchored to structural elements that provide lateral support for the wall, and the anchorage must be capable of resisting the horizontal forces specified by the code, with a minimum of 280 pounds per lineal foot of wall.

TILT-UP WALLS

Tilt-up walls are reinforced concrete walls that are precast at the job site, usually in a flat position. When they have attained sufficient strength to withstand erection stresses, they are tilted up to a vertical position and set into place. The walls are then temporarily braced until the permanent attachment to the columns, floor slab, and roof structure are made.

In general, the requirements for cast-in-place concrete walls also apply to tilt-up and other precast concrete walls. There are some differences, however. Unlike cast-in-place walls, tilt-up walls are subject to very high stresses during construction, and therefore special care must be given to erection methods and stresses. While the construction methods are generally outside the scope of the architect's services, the architect should nevertheless be aware of them. The joinery of tilt-up walls to the surrounding members—columns, floor slab, roof structure, foundation, or other walls—is also extremely important.

Exterior columns may be cast-in-place concrete, precast concrete, or structural steel. Cast-in-place columns are formed after the tilt-up walls have been erected, and the wall reinforcing generally extends into the column.

The connection of tilt-up walls to precast columns may be done in a variety of ways. Steel members are often cast into the columns and walls and joined by field welding.

The joinery of tilt-up walls to structural steel columns is also done by field welding.

Some typical connections of tilt-up walls to columns are shown in Figure 6.4.

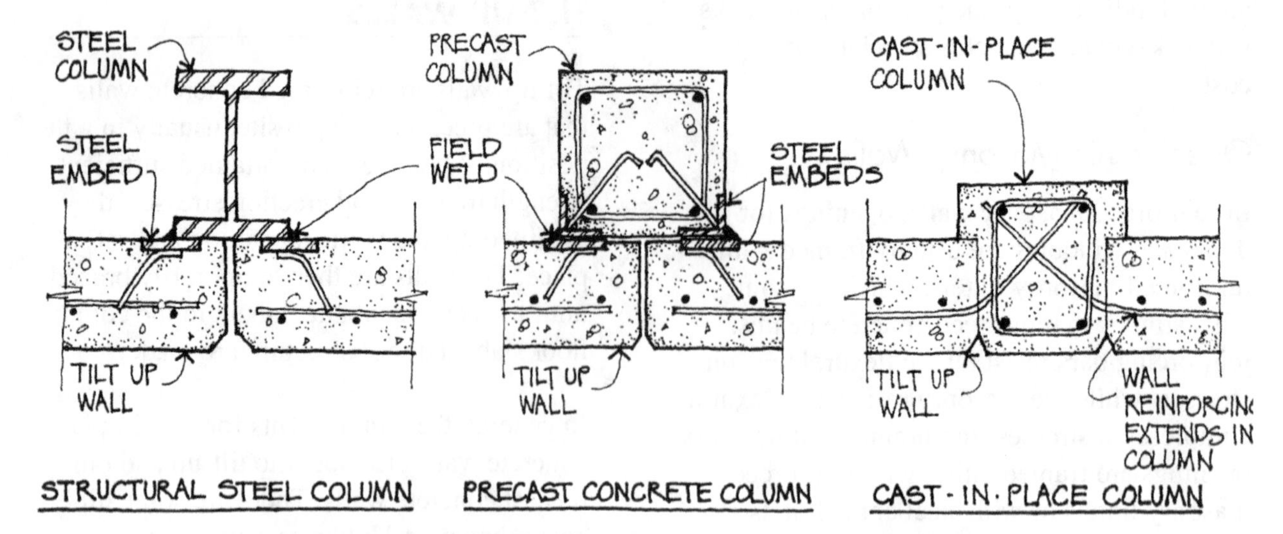

Figure 6.4

Tilt-up walls frequently function as deep beams and span between column footings. In such cases, a continuous footing is usually unnecessary, and if certain conditions are met, the height-to-thickness ratio is not limited.

Tilt-up walls are often used in simple, rectangular industrial buildings. They are constructed quickly, door and window openings and frames are easily installed, and they provide good fire resistance, great strength, low maintenance, and economy. However, tilt-up walls are not usually advantageous for buildings with a complex shape of nonrepetitive bays.

RETAINING WALLS

There are a number of ways by which ground elevation can be changed. The simplest way, of course, is to slope the ground surface; as long as the slope does not exceed the natural angle of repose of the soil, it will remain stable (Figure 6.5).

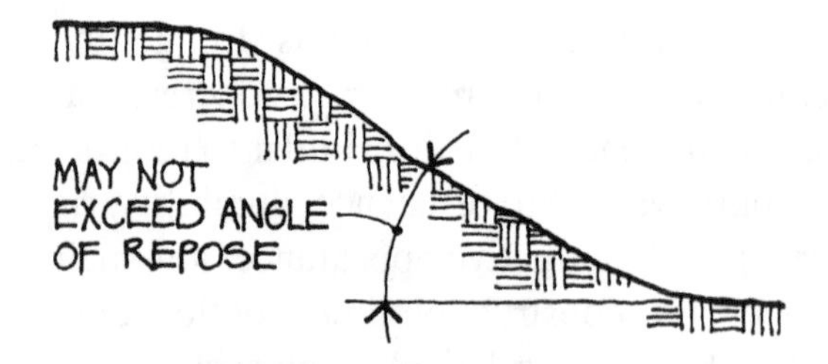

Figure 6.5

Where an abrupt change of ground elevation occurs, a retaining wall is necessary to hold the soil back. In the diagram shown in Figure 6.6, the retained earth behind (to the left of) the wall tends to push the wall to the right, and the wall and its foundation must be designed to resist this push.

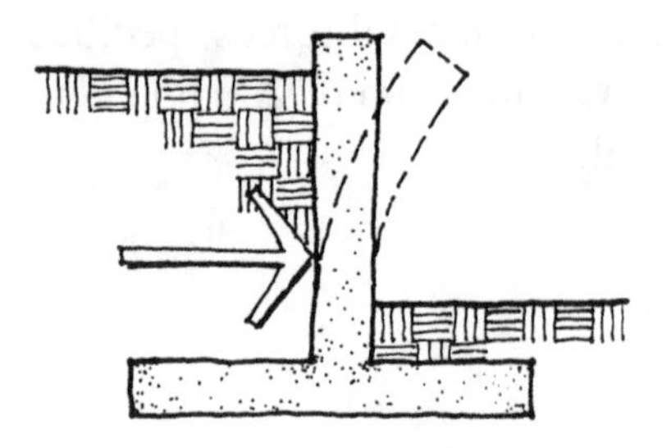

Figure 6.6

Low retaining walls are often constructed of reinforced concrete block masonry, while the taller walls are usually reinforced concrete.

There are three general types of retaining walls: gravity, cantilever, and counterfort, as shown in Figure 6.7.

The *gravity wall* depends entirely on its own weight to resist the pressure of the retained earth and provide stability. It is frequently constructed of plain (unreinforced) concrete.

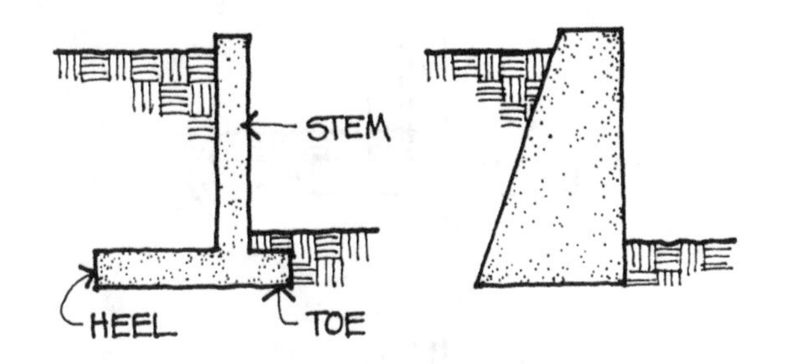

CANTILEVER WALL GRAVITY WALL

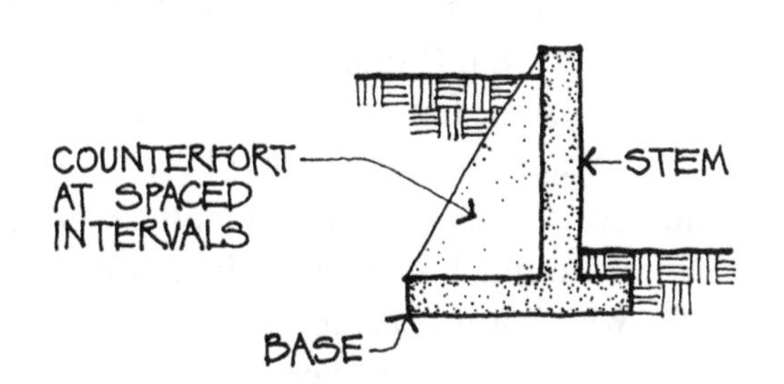

COUNTERFORT WALL

Figure 6.7

The *cantilever wall* is the most common type of retaining wall, in which the stem, heel, and toe act as cantilever slabs.

In the *counterfort wall*, the stem and base are connected at intervals by transverse walls called counterforts, which act as tension ties to support the stem. This type of wall is more economical than the cantilever wall for heights over about 25 feet.

The pressure exerted by the retained earth against the stem is known as *active pressure*, and depends on the type of soil, whether there is any vertical load on the ground surface behind the wall, and whether the ground surface behind the wall is level or sloping. Vertical load behind the wall or a sloping ground surface increases the earth pressure against the wall, and both are known as *surcharge*. For average conditions, with no surcharge, the active earth pressure may be assumed to be equivalent to that exerted by a fluid weighing 30 pounds per cubic foot. This results in a triangular distribution of earth pressure against the wall, varying from zero at the top to a maximum at the bottom equal to 30 times the height of the wall. Since the pressure diagram is a triangle, the resultant acts at the centroid of the triangle, which is located 1/3 of the height above the base.

Example #1

The stem of a cantilever retaining wall is 12 feet high. Assuming 30-pound equivalent fluid pressure, what is the magnitude of the earth pressure at the base, the total earth pressure, and the bending moment at the base caused by the earth pressure?

Solution:

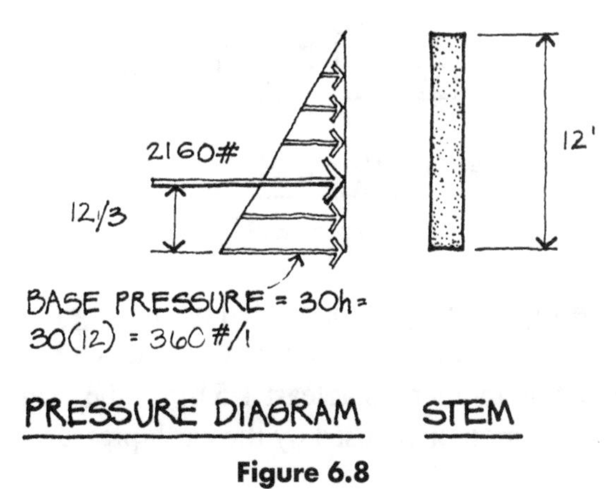

Figure 6.8

The soil pressure diagram is a triangle, varying from zero at the top to 30 × height at the bottom = 30 × 12 = *360 pounds per foot,* as shown in Figure 6.8.

The total earth pressure on the stem (for a one-foot length of wall) is equal to the area of the pressure diagram = base pressure × height/2 = 360 × 12/2 = *2,160#.*

The bending moment at the base is equal to the total earth pressure multiplied by the distance from the centroid of the triangle to the base = 2,160 × 12/3 = *8,640 foot-pounds.*

The moment causes tension in the face of the wall that is closest to the earth, and the reinforcing steel should therefore be placed in this face, as shown in Figure 6.15.

There are several ways by which a retaining wall may fail. In designing a wall, all of these must be carefully considered.

A wall can fail by *overturning* (Figure 6.9). This occurs when the base is too narrow.

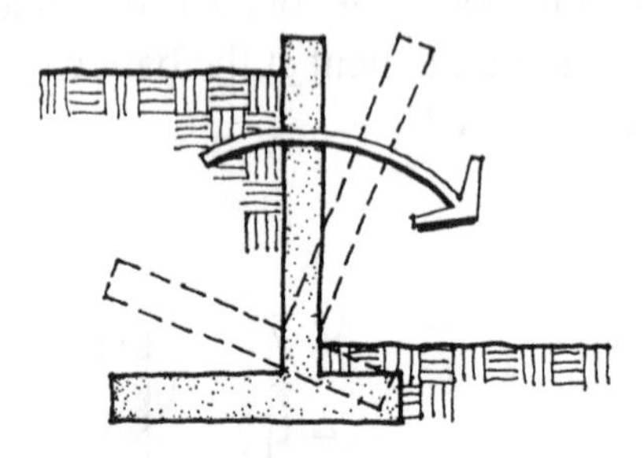

Figure 6.9

Overturning may be prevented by making the base sufficiently wide so that the dead load resisting moment is at least 1.5 times the overturning moment caused by the earth pressure.

The resistance of a retaining wall to *sliding* is provided by friction between the base and the underlying soil and by earth pressure in front of the toe (Figure 6.10). This sliding resistance must be at least 1.5 times the total earth pressure against the wall. If not, additional sliding resistance may be obtained by constructing the base with an integral key, or by deepening the base.

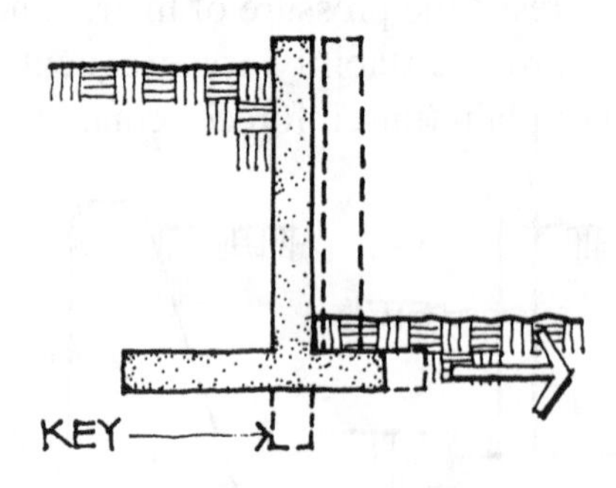

Figure 6.10

The least common type of failure is *structural failure* of the wall (Figure 6.11). This can occur because of faulty design or construction, or because the earth behind the wall is not adequately drained. In that event, the wall may be subject to much greater earth pressure than that for which it was designed. It is obviously important to provide proper drainage by means of weep holes through the wall or drains behind the wall, and by using gravel or other granular fill behind the wall.

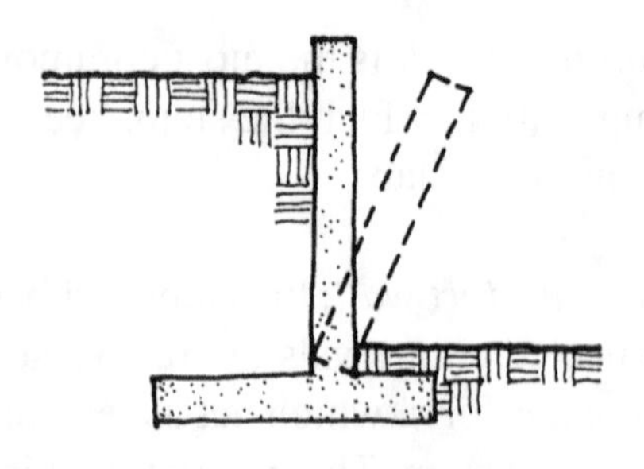

Figure 6.11

The pressure of the earth against a retaining wall causes a nonuniform soil pressure distribution under the base, as shown in Figure 6.14. The maximum soil pressure under the toe must be less than the allowable soil bearing pressure, and generally the resultant force on the base should fall within the middle third of the base.

Basement walls also resist the pressure of retained earth (Figure 6.12). However, unlike retaining walls, they are usually supported both at the bottom (by the foundation and the basement floor slab), and at the top (by the first floor construction). A basement wall acts like a slab spanning vertically between its upper and lower supports.

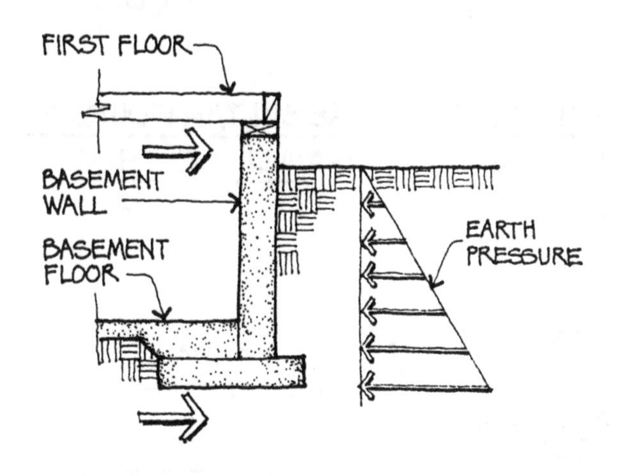

Figure 6.12

Example #2

A basement wall is 12 feet high. Assuming 30-pound equivalent fluid pressure, what is the magnitude of the earth pressure at the base, the total earth pressure, and the maximum bending moment in the wall?

Solution:

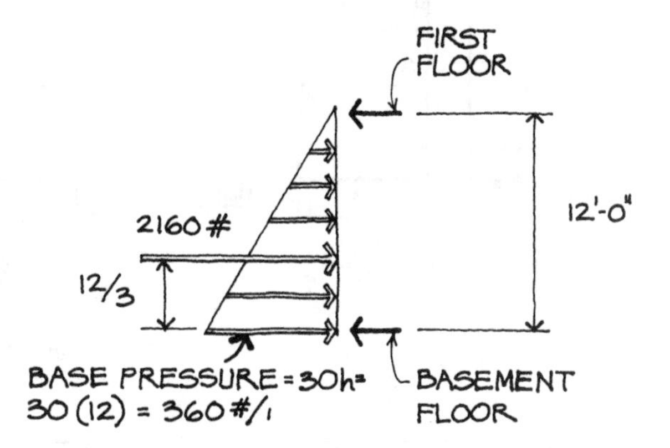

Figure 6.13

As in Example #1, the earth pressure at the base = 30 × 12 = *360 pounds per foot* (see Figure 6.13).

The total earth pressure on the wall (for a one-foot length of wall) = 360 × 12/2 = *2,160#*.

Unlike the wall of Example #1, which cantilevers off the base, the basement wall spans vertically between the basement floor and the first floor. From the beam diagram in Figure 6.16, reproduced from the AISC Manual, the maximum moment in a simple beam supporting a triangular load is .1283WL = .1283(2160)(12) = 3,326 ft.-lbs.

Thus, the basement wall is subject to a much smaller bending moment than the cantilever wall of Example #1.

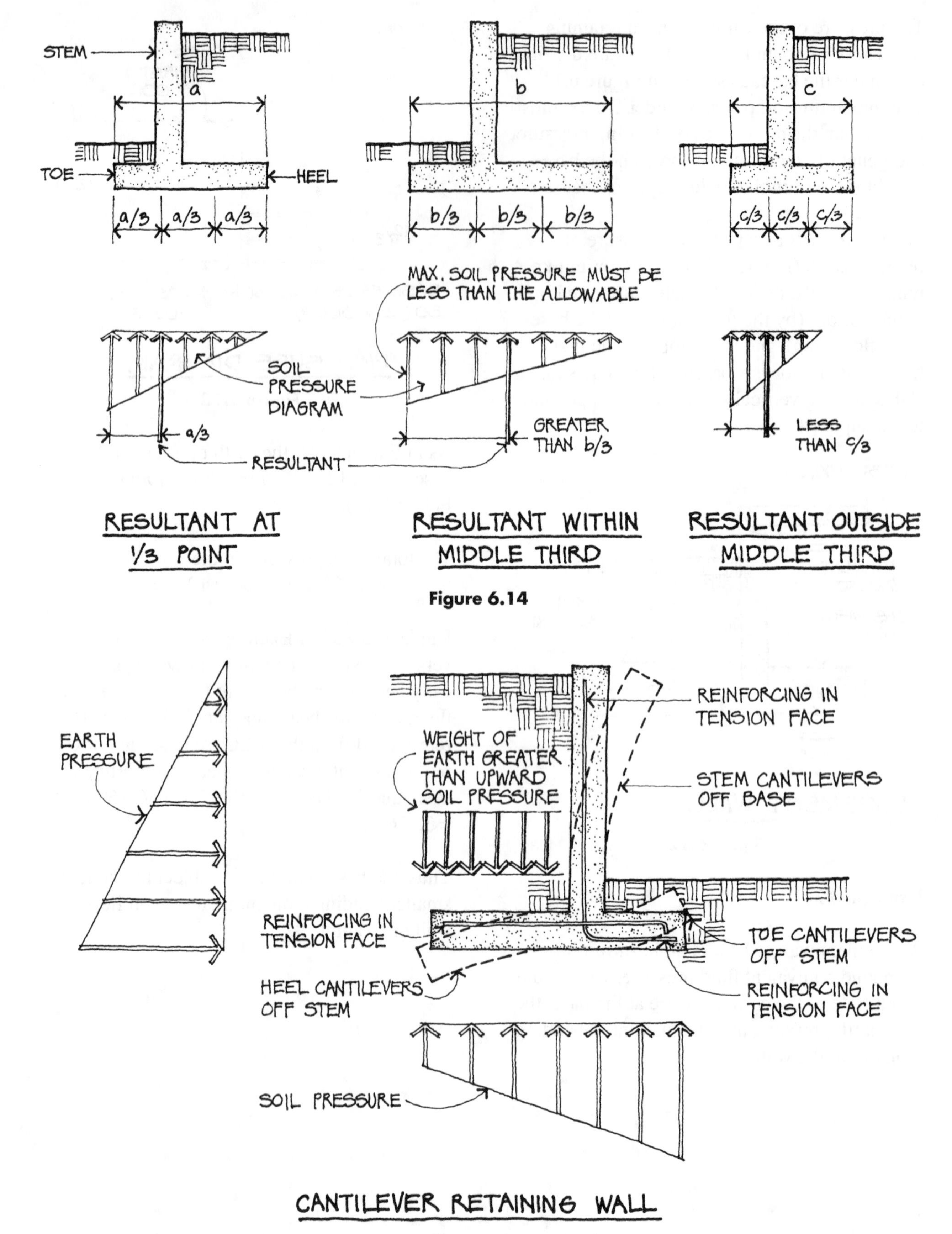

STEM
TOE
HEEL
a
b
c
a/3
b/3
c/3
MAX. SOIL PRESSURE MUST BE LESS THAN THE ALLOWABLE
SOIL PRESSURE DIAGRAM
GREATER THAN b/3
LESS THAN c/3
RESULTANT
RESULTANT AT 1/3 POINT
RESULTANT WITHIN MIDDLE THIRD
RESULTANT OUTSIDE MIDDLE THIRD

Figure 6.14

EARTH PRESSURE
WEIGHT OF EARTH GREATER THAN UPWARD SOIL PRESSURE
REINFORCING IN TENSION FACE
STEM CANTILEVERS OFF BASE
REINFORCING IN TENSION FACE
TOE CANTILEVERS OFF STEM
HEEL CANTILEVERS OFF STEM
REINFORCING IN TENSION FACE
SOIL PRESSURE
CANTILEVER RETAINING WALL

Figure 6.15

The reinforcing steel should be placed in the *inside face* of the basement wall, that is, the face furthest from the earth, which is the tension face (see Figure 6.15).

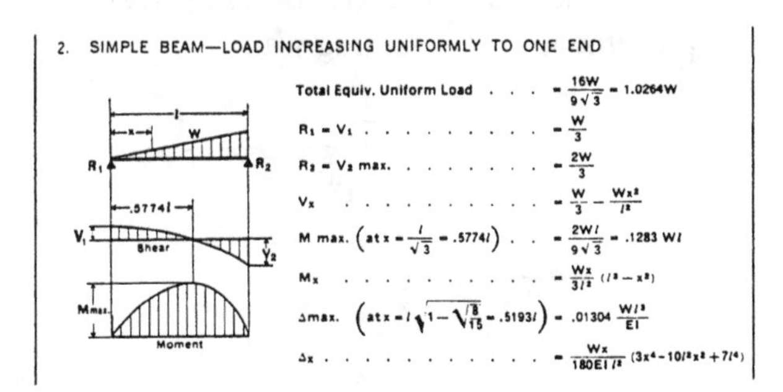

Figure 6.16

While the detailed design of retaining walls is generally beyond the scope of the licensing examination and hence of this course, candidates should at least have an understanding of the basic principles involved.

LESSON 6 QUIZ

1. Select the CORRECT statement about basement walls.

A. Drainage is usually provided by weep holes through the wall.

B. The wall should be designed to resist earth pressure unless a granular backfill is used.

C. The wall is usually designed to cantilever from its footing, since the floor construction is not permitted to laterally support the wall.

D. The reinforcing steel should be placed near the inside face of the wall.

2. Select the CORRECT statements about retaining walls.

I. Counterfort walls are not economical for walls 25 feet or higher.

II. Gravity walls are the most common type of retaining wall and consist of stem, heel, and toe.

III. A retaining wall is necessary whenever the ground elevation changes abruptly.

IV. The total earth pressure on the stem, for a one-foot length of wall, is equal to $30h^2/2$.

V. Low retaining walls are most often constructed of reinforced brick masonry.

A. I, III, IV, and V **C.** II and V

B. IV and V **D.** III and IV

3. Select the CORRECT reinforcing steel placement for a retaining wall.

A.

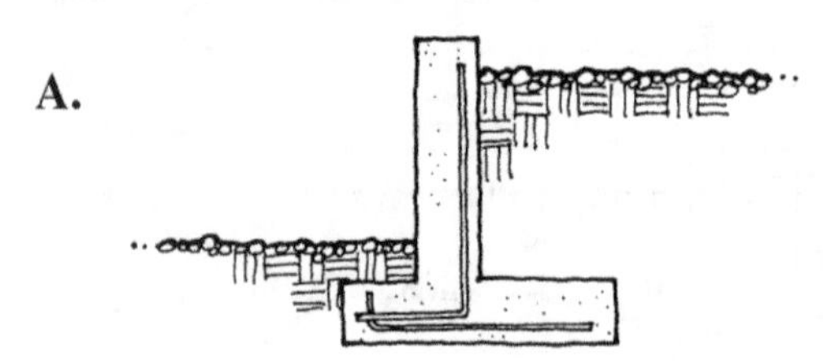

B.

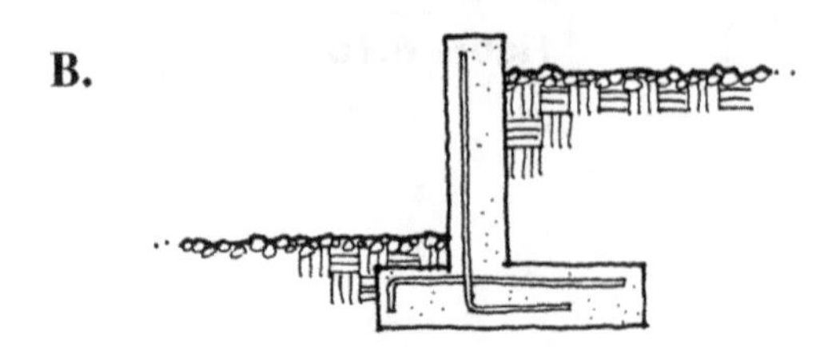

C.

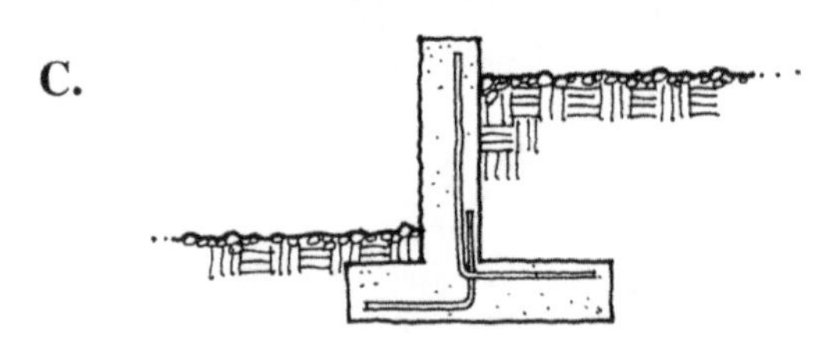

D.

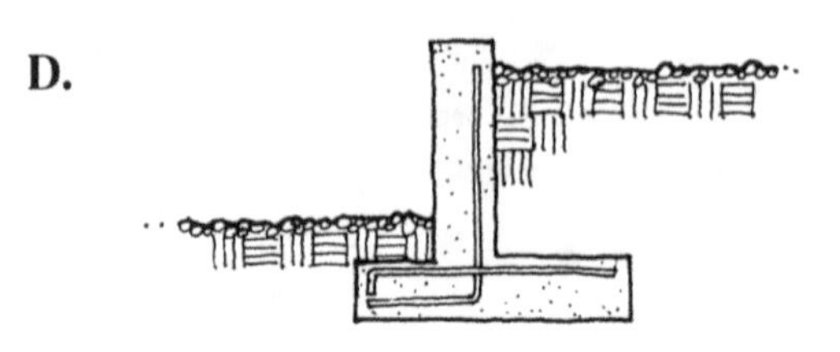

4. A type of concrete wall that is site-cast and often functions as a deep beam spanning between footing pads is a ____________ wall.

5. Which of the following statements about wood stud walls is CORRECT?

- **I.** Platform framing tends to minimize total vertical shrinkage.
- **II.** Balloon framing is preferable where exterior walls are faced with brick veneer.
- **III.** If properly detailed, wood stud walls may function as bearing walls, shear walls, or retaining walls.
- **IV.** In balloon framing, the studs are continuous and full height, with the second floor joists bearing on a continuous member that is let into the studs.
- **V.** Wood stud walls are not required to have a continuous concrete foundation, except in areas subject to ground freezing and thawing.

A. I, IV, and V **C.** II and IV

B. I, II, and V **D.** II, III, and IV

6. Select the INCORRECT statement about reinforced masonry construction.

A. Reinforced brick masonry is generally stronger than reinforced concrete block masonry.

B. In reinforced brick masonry, horizontal reinforcing bars are placed in the mortar joints between brick units.

C. The grout used in reinforced masonry construction must have a minimum compressive strength of 2,000 psi.

D. In reinforced concrete block construction, cells in contact with the earth should be filled solidly with grout.

7. Select the CORRECT statement.

A. Reinforced concrete walls are able to resist axial compressive loads, but not bending moment.

B. Reinforced concrete walls must be anchored to all floors and roofs that provide lateral support for the wall.

C. Openings in reinforced concrete walls are usually framed with steel angle lintels.

D. Reinforced concrete walls are often prestressed to prevent tension cracks.

8. The calculations for the design of a retaining wall show that the resultant force on the base falls outside the middle third of the base. How should the design be changed?

A. Not at all, the design is adequate.

B. The footing should be deepened.

C. The amount of reinforcing steel should be increased.

D. The footing should be made wider.

9. The stem of a cantilever retaining wall is 10 feet high. Assuming 30-pound equivalent fluid pressure, what is the bending moment at the base caused by the earth pressure?

A. 300 ft.-lbs. **C.** 5,000 ft.-lbs.

B. 1,500 ft.-lbs. **D.** 7,500 ft.-lbs.

10. Select the CORRECT statements about reinforced masonry construction.

I. Reinforced masonry walls can resist moments caused by earthquake forces or retained earth.

II. The reinforcing bars stabilize reinforced masonry walls against buckling.

III. Although grout contains more lime, it may be used interchangeably with mortar.

IV. Openings in reinforced masonry walls are generally framed with reinforced masonry lintels.

A. I and IV

B. III and IV

C. II and III

D. I, III, and IV

LESSON SEVEN 7

CONNECTIONS

INTRODUCTION

The strength and stability of any structure are largely dependent on the fastenings that hold its elements together.

In this lesson, we will consider wood-to-wood, wood-to-steel, and steel-to-steel connections. Concrete-to-concrete connections will not be considered here; proper lapping of the reinforcing bars and roughening or keying of the concrete joints usually provide an adequate connection without the need for any mechanical fasteners.

WOOD CONNECTIONS

Wood-to-wood and wood-to-steel connections utilize many types of fasteners. The most common of these are listed below, approximately in order of decreasing strength:

1. Timber connectors
2. Bolts
3. Lag screws
4. Wood screws
5. Nails

Usually these fasteners are required to resist loads in shear (single or double), rather than in tension (see Figure 7.1).

In general, the number of fasteners required to transfer a given load is determined by dividing the load by the design value of one fastener.

The factors to be considered in determining the design values for mechanical fasteners include (1) lumber species, (2) critical section, (3) angle of load to grain, (4) spacing of fasteners, (5) edge and end distances, and (6) conditions of loading.

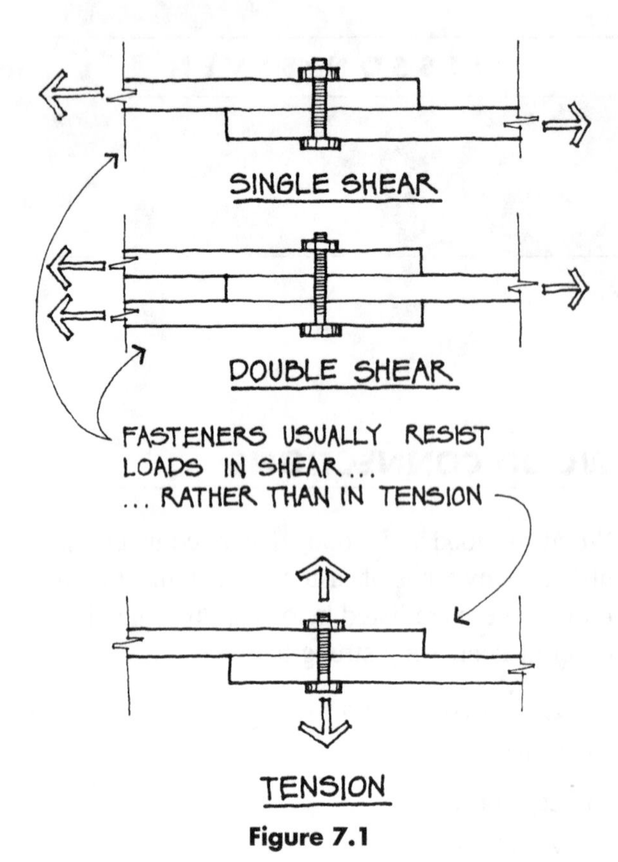

Figure 7.1

1. Lumber species

The design values for mechanical fasteners vary with the lumber species; bolts and other connectors in the denser woods, such as ash and beech, can resist greater loads than those in the lighter weight woods, such as aspen and Engelmann spruce.

2. Critical section

Holes for bolts or other connectors reduce the area of the wood available to resist the applied load. The critical section is the section where the net area is the least, and consequently, the stress in the member is maximum.

3. Angle of load to grain

Wood is stronger in bearing parallel to the grain than perpendicular to the grain. Therefore, the design value for certain fasteners depends on the angle of load to grain. The angle of load to grain is defined as the angle between the direction of the load acting on the member and the longitudinal axis of the member.

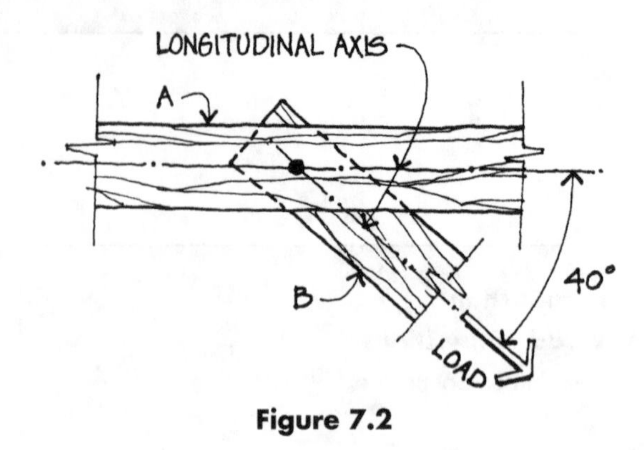

Figure 7.2

For member A depicted in Figure 7.2, the angle of load to grain is 40°, while the angle of load to grain for member B is 0°.

4. Spacing of fasteners

The spacing of fasteners is the distance between centers of the fasteners measured on a straight line joining their centers.

5. Edge and end distances

Edge distance is the distance from the edge of a member to the center of the fastener closest to that edge, measured perpendicular to the edge (Figure 7.3). The *loaded edge* is the edge toward which the fastener load acts. The *unloaded edge* is the edge away from which the fastener load acts. *End distance* is the distance, measured parallel to grain, from the center of the fastener to the end of the member.

6. Conditions of loading

The design values for fasteners may be increased for short-time loading the same as for wood members, as follows:

15% for 2 months' duration, as for snow

25% for 7 days' duration, as for roof live load or construction loads

60% for wind or earthquake

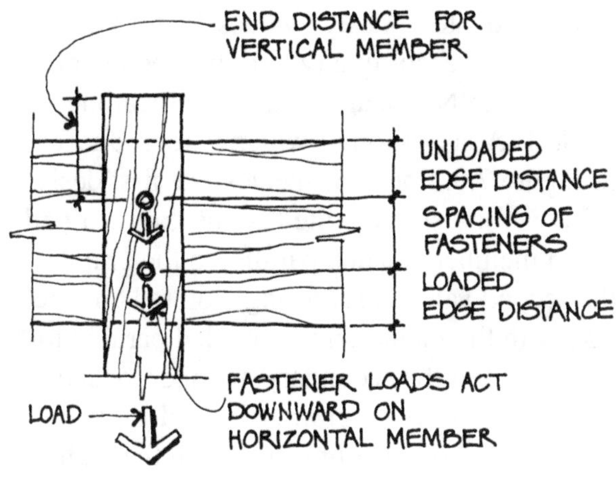

Figure 7.3

The impact load duraction factor does not apply to connections (see Section 10.3.2 of the Natural Design Specifications for Wood Construction).

Timber Connectors

There are three types of timber connector joints: (1) one split ring with its bolt in single shear for a wood-to-wood joint, (2) two shear plates used back-to-back in the contact faces of a wood-to-wood joint with their bolt in single shear, and (3) one shear plate with its bolt in single shear used in a wood-to-steel joint.

Split rings are available in 2-1/2" and 4" diameters and are primarily used in the assembly of trussed rafters and trusses (Figure 7.4). Circular grooves are precut by a special tool into the contacting faces of the wood members to be joined, so that half of the ring fits into each member. A bolt, concentric with the ring, is used to hold the two members together. The size and shape of the ring and its conforming groove provide for easy insertion, while also ensuring a tight fit when the ring is fully seated.

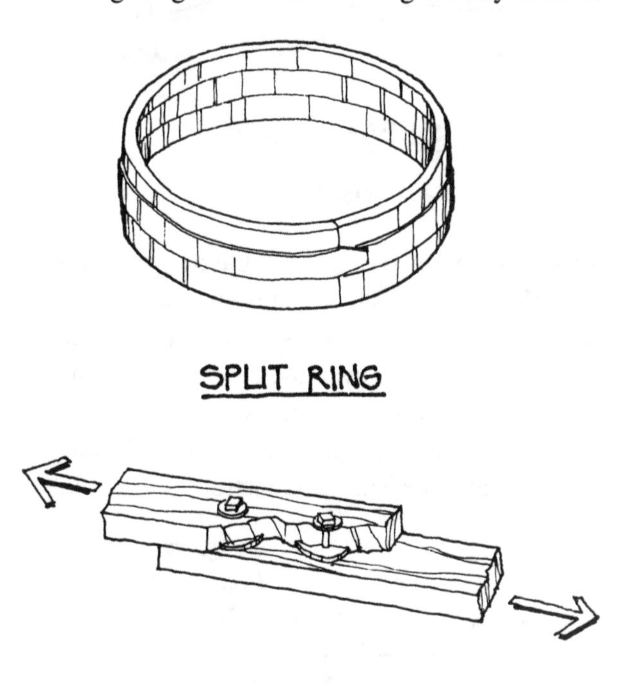

Figure 7.4

Shear plates are made of pressed steel or malleable iron and are available in two diameters, 2-5/8" and 4" (Figure 7.5). They are intended for wood-to-steel connections (using one shear plate), or for wood-to-wood connections (using two shear plates), where demountability is desired (Figure 7.6). They are placed in precut grooves and are completely embedded in the wood member, with the connector flush with the surface of the wood. Their advantage, as far as easy assembly and disassembly are concerned, lies in the fact that they may be installed immediately after fabrication and temporarily held in place by nails.

Since there are no projections, the members may easily be slid into position. A tight fitting bolt is then inserted into the central hole to hold the assembly together.

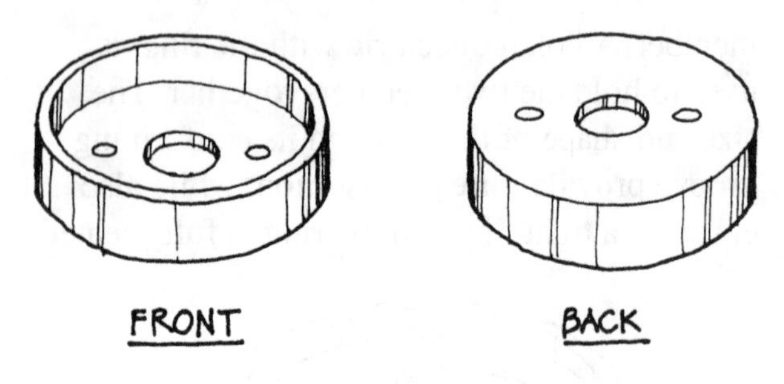

Figure 7.5

Tables or charts to determine the design values for split rings and shear plates and their required spacings and end distances are beyond the scope of this course. However, they are available in a number of reference sources, including the National Design Specification for Wood Construction and the Timber Construction Manual.

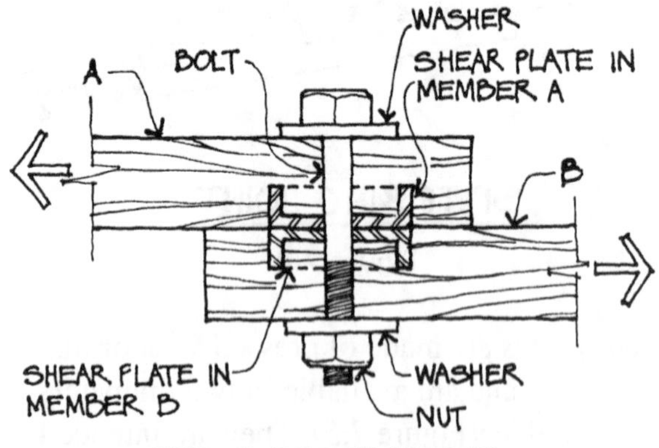

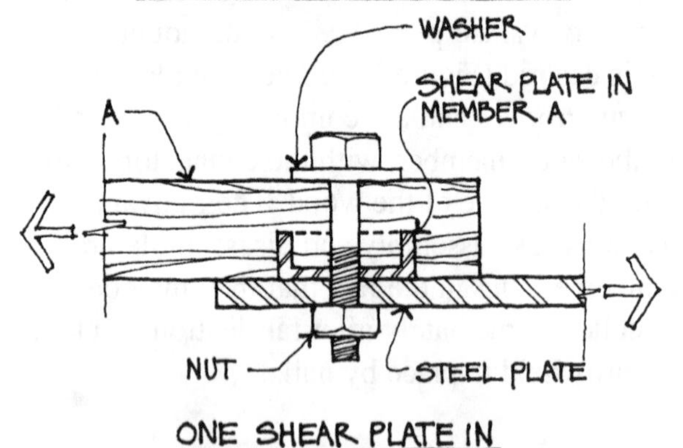

SHEAR PLATE JOINTS

Figure 7.6

Bolts

Bolts used for wood-to-wood or wood-to-steel connections are common (not high-strength) bolts. Bolt holes should be 1/32" to 1/16" larger than the bolt, and standard cut, plate, or malleable iron washers should be used between the bolt head and the wood and between the nut and the wood. An example of a single shear bolted joint is depicted in Figure 7.7.

Reference lateral design values, Z, for bolts in single shear joining two timber members are given in National Design Specification Table 11A, a portion of which is reproduced in Figure 7.8. For each species, the first design value is for a single bolted connection with all wood members loaded parallel to the grain; the second value is for a single bolted connection with the main member loaded parallel to the grain and the secondary member loaded perpendicular to the grain; the third value is for a single bolted connection with the main member loaded perpendicular to the grain and the secondary member loaded parallel to the grain; and the fourth value is for a single bolted connection with all wood members loaded perpendicular to the grain. A portion of Table 11F from the National Design Specification is also reproduced in Figure 7.9. This table gives design values for bolts in double shear (three-member connections).

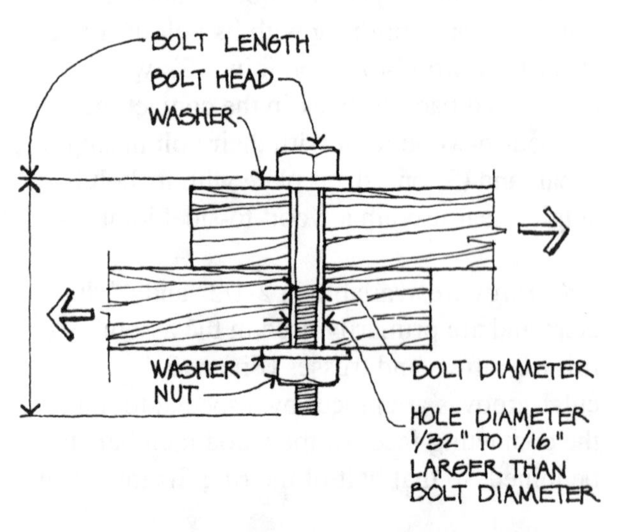

SINGLE SHEAR BOLTED JOINT

Figure 7.7

Table 11A BOLTS: Reference Lateral Design Values (Z) for Single Shear (two member) Connections[1,2]

for sawn lumber or SCL with both members of identical specific gravity

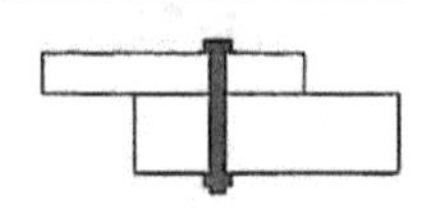

Thickness: Main Member t_m in.	Thickness: Side Member t_s in.	Bolt Diameter D in.	G=0.67 Red Oak $Z_{\parallel}$ lbs.	$Z_{s\perp}$ lbs.	$Z_{m\perp}$ lbs.	$Z_{\perp}$ lbs.	G=0.55 Mixed Maple Southern Pine $Z_{\parallel}$ lbs.	$Z_{s\perp}$ lbs.	$Z_{m\perp}$ lbs.	$Z_{\perp}$ lbs.	G=0.50 Douglas Fir-Larch $Z_{\parallel}$ lbs.	$Z_{s\perp}$ lbs.	$Z_{m\perp}$ lbs.	$Z_{\perp}$ lbs.	G=0.49 Douglas Fir-Larch(N) $Z_{\parallel}$ lbs.	$Z_{s\perp}$ lbs.	$Z_{m\perp}$ lbs.	$Z_{\perp}$ lbs.	G=0.46 Douglas Fir(S) Hem-Fir(N) $Z_{\parallel}$ lbs.	$Z_{s\perp}$ lbs.	$Z_{m\perp}$ lbs.	$Z_{\perp}$ lbs.
1-1/2	1-1/2	1/2	650	420	420	330	530	330	330	250	480	300	300	220	470	290	290	210	440	270	270	190
		5/8	810	500	500	370	660	400	400	280	600	360	360	240	590	350	350	240	560	320	320	220
		3/4	970	580	580	410	800	460	460	310	720	420	420	270	710	400	400	260	670	380	380	240
		7/8	1130	660	660	440	930	520	520	330	850	470	470	290	830	460	460	280	780	420	420	250
		1	1290	740	740	470	1060	580	580	350	970	510	510	310	950	510	510	300	890	480	480	280
1-3/4	1-3/4	1/2	760	490	490	390	620	390	390	290	560	350	350	250	550	340	340	250	520	320	320	230
		5/8	940	590	590	430	770	470	470	330	700	410	410	280	690	410	410	280	650	380	380	250
		3/4	1130	680	680	480	930	540	540	360	850	480	480	310	830	470	470	300	780	440	440	280
		7/8	1320	770	770	510	1080	610	610	390	990	550	550	340	970	530	530	320	910	500	500	300
		1	1510	860	860	550	1240	680	680	410	1130	610	610	360	1110	600	600	350	1040	560	560	320
2-1/2	1-1/2	1/2	770	480	540	440	660	400	420	350	610	370	370	310	610	360	360	300	580	340	330	270
		5/8	1070	660	630	520	930	560	490	390	850	520	430	340	830	520	420	330	780	470	390	300
		3/4	1360	890	720	570	1120	660	560	430	1020	590	500	380	1000	560	480	360	940	520	450	330
		7/8	1590	960	800	620	1300	720	620	470	1190	630	550	410	1170	600	540	390	1090	550	500	360
		1	1820	1020	870	660	1490	770	680	490	1360	690	600	440	1330	650	590	420	1250	600	550	390
3-1/2	1-1/2	1/2	770	480	560	440	660	400	470	360	610	370	430	330	610	360	420	320	580	340	400	310
		5/8	1070	660	760	590	940	560	620	500	880	520	540	460	870	520	530	450	830	470	490	410
		3/4	1450	890	900	770	1270	660	690	580	1200	590	610	510	1190	560	590	490	1140	520	550	450
		7/8	1890	960	990	830	1680	720	770	630	1590	630	680	550	1570	600	650	530	1470	550	600	480
		1	2410	1020	1080	890	2010	770	830	670	1830	680	720	590	1790	650	710	560	1680	600	660	520
3-1/2	1-3/4	1/2	830	510	590	480	720	420	510	390	670	380	470	350	660	380	460	340	620	360	440	320
		5/8	1160	680	820	620	1000	580	640	520	930	530	560	460	920	530	550	450	880	500	510	410
		3/4	1530	900	940	780	1330	770	720	580	1250	680	640	520	1240	660	620	500	1190	600	580	460
		7/8	1970	1120	1040	840	1730	840	810	640	1620	740	710	550	1590	700	690	530	1490	640	640	490
		1	2480	1190	1130	900	2030	890	880	670	1850	790	760	590	1820	750	760	570	1700	700	700	530
3-1/2	3-1/2	1/2	830	590	590	530	750	520	520	460	720	490	490	430	710	480	480	420	690	460	460	410
		5/8	1290	880	880	780	1170	780	780	650	1120	700	700	560	1110	690	690	550	1070	650	650	500
		3/4	1860	1190	1190	950	1690	960	960	710	1610	870	870	630	1600	850	850	600	1540	800	800	560
		7/8	2540	1410	1410	1030	2170	1160	1160	730	1930	1060	1060	680	1940	1040	1040	650	1810	980	980	590
		1	3020	1670	1670	1100	2480	1360	1360	820	2260	1230	1230	720	2210	1190	1190	690	2070	1110	1110	640
5-1/4	1-1/2	5/8	1070	660	760	590	940	560	640	500	880	520	590	460	870	520	590	450	830	470	560	430
		3/4	1450	890	990	780	1270	660	850	660	1200	590	790	590	1190	560	780	560	1140	520	740	520
		7/8	1890	960	1260	960	1680	720	1060	720	1590	630	940	630	1570	600	900	600	1520	550	830	550
		1	2410	1020	1500	1020	2150	770	1140	770	2050	680	1010	680	2030	650	970	650	1930	600	910	600
5-1/4	1-3/4	5/8	1160	680	820	620	1000	580	690	520	930	530	630	470	920	530	630	470	880	500	590	440
		3/4	1530	900	1050	800	1330	770	890	680	1250	680	830	630	1240	660	810	620	1190	600	780	590
		7/8	1970	1120	1320	1020	1730	840	1090	840	1640	740	960	740	1620	700	920	700	1550	640	850	640
		1	2480	1190	1530	1190	2200	890	1170	890	2080	790	1040	790	2060	750	1000	750	1990	700	930	700
5-1/4	3-1/2	5/8	1290	880	880	780	1170	780	780	680	1120	700	730	630	1110	690	720	620	1070	650	690	580
		3/4	1860	1190	1240	1080	1690	960	1090	850	1610	870	1030	780	1600	850	1010	750	1540	800	970	710
		7/8	2540	1410	1640	1260	2300	1160	1380	1000	2190	1060	1230	870	2170	1040	1190	840	2060	980	1100	770
		1	3310	1670	1940	1420	2870	1390	1520	1060	2660	1290	1360	940	2630	1260	1320	900	2500	1210	1230	830
5-1/2	1-1/2	5/8	1070	660	760	590	940	560	640	500	880	520	590	460	870	520	590	450	830	470	560	430
		3/4	1450	890	990	780	1270	660	850	660	1200	590	790	590	1190	560	780	560	1140	520	740	520
		7/8	1890	960	1260	960	1680	720	1090	720	1590	630	980	630	1570	600	940	600	1520	550	860	550
		1	2410	1020	1560	1020	2150	770	1190	770	2050	680	1060	680	2030	650	1010	650	1930	600	940	600
5-1/2	3-1/2	5/8	1290	880	880	780	1170	780	780	680	1120	700	730	630	1110	690	720	620	1070	650	690	580
		3/4	1860	1190	1240	1080	1690	960	1090	850	1610	870	1030	780	1600	850	1010	750	1540	800	970	710
		7/8	2540	1410	1640	1260	2300	1160	1410	1020	2190	1060	1260	910	2170	1040	1220	870	2060	980	1130	790
		1	3310	1670	1980	1470	2870	1390	1550	1100	2660	1290	1390	970	2630	1260	1340	930	2500	1210	1250	860
7-1/2	1-1/2	5/8	1070	660	760	590	940	560	640	500	880	520	590	460	870	520	590	450	830	470	560	430
		3/4	1450	890	990	780	1270	660	850	660	1200	590	790	590	1190	560	780	560	1140	520	740	520
		7/8	1890	960	1260	960	1680	720	1090	720	1590	630	1010	630	1570	600	990	600	1520	550	950	550
		1	2410	1020	1560	1020	2150	770	1350	770	2050	680	1270	680	2030	650	1240	650	1930	600	1190	600
7-1/2	3-1/2	5/8	1290	880	880	780	1170	780	780	680	1120	700	730	630	1110	690	720	620	1070	650	690	580
		3/4	1860	1190	1240	1080	1690	960	1090	850	1610	870	1030	780	1600	850	1010	750	1540	800	970	710
		7/8	2540	1410	1640	1260	2300	1160	1450	1020	2190	1060	1360	930	2170	1040	1340	900	2060	980	1280	850
		1	3310	1670	2090	1470	2870	1390	1830	1210	2660	1290	1630	1110	2630	1260	1570	1080	2500	1210	1470	1030

1. Tabulated lateral design values (Z) for bolted connections shall be multiplied by all applicable adjustment factors (see Table 10.3.1).
2. Tabulated lateral design values (Z) are for "full diameter" bolts (see Appendix L) with bending yield strength (F_{yb}) of 45,000 psi.

Figure 7.8

Table 11F BOLTS: Reference Lateral Design Values (Z) for Double Shear (three member) Connections[1,2]

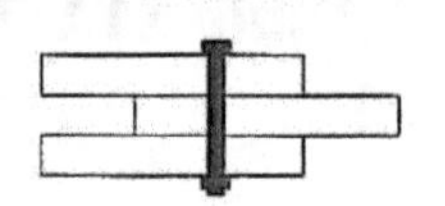

for sawn lumber or SCL with all members of identical specific gravity

Thickness: Main Member	Thickness: Side Member	Bolt Diameter	G=0.67 Red Oak			G=0.55 Mixed Maple Southern Pine			G=0.50 Douglas Fir-Larch			G=0.49 Douglas Fir-Larch(N)			G=0.46 Douglas Fir(S) Hem-Fir(N)		
t_m in.	t_s in.	D in.	$Z_{\parallel}$ lbs.	$Z_{s\perp}$ lbs.	$Z_{m\perp}$ lbs.	$Z_{\parallel}$ lbs.	$Z_{s\perp}$ lbs.	$Z_{m\perp}$ lbs.	$Z_{\parallel}$ lbs.	$Z_{s\perp}$ lbs.	$Z_{m\perp}$ lbs.	$Z_{\parallel}$ lbs.	$Z_{s\perp}$ lbs.	$Z_{m\perp}$ lbs.	$Z_{\parallel}$ lbs.	$Z_{s\perp}$ lbs.	$Z_{m\perp}$ lbs.
1-1/2	1-1/2	1/2	1410	960	730	1150	800	550	1050	730	470	1030	720	460	970	680	420
		5/8	1760	1310	810	1440	1130	610	1310	1040	530	1290	1030	520	1210	940	470
		3/4	2110	1690	890	1730	1330	660	1580	1170	590	1550	1130	560	1450	1040	520
		7/8	2460	1920	960	2020	1440	720	1840	1260	630	1800	1210	600	1690	1100	550
		1	2810	2040	1020	2310	1530	770	2100	1350	680	2060	1290	650	1930	1200	600
1-3/4	1-3/4	1/2	1640	1030	850	1350	850	640	1230	770	550	1200	750	530	1130	710	490
		5/8	2050	1370	940	1680	1160	710	1530	1070	610	1500	1060	600	1410	1000	550
		3/4	2460	1810	1040	2020	1550	770	1840	1370	680	1800	1310	660	1690	1210	600
		7/8	2870	2240	1120	2350	1680	840	2140	1470	740	2110	1410	700	1970	1290	640
		1	3280	2380	1190	2690	1790	890	2450	1580	790	2410	1510	750	2250	1400	700
2-1/2	1-1/2	1/2	1530	960	1120	1320	800	910	1230	730	790	1210	720	760	1160	680	700
		5/8	2150	1310	1340	1870	1130	1020	1760	1040	880	1740	1030	860	1660	940	780
		3/4	2890	1770	1480	2550	1330	1110	2400	1170	980	2380	1130	940	2280	1040	860
		7/8	3780	1920	1600	3360	1440	1200	3060	1260	1050	3010	1210	1010	2820	1100	920
		1	4690	2040	1700	3840	1530	1280	3500	1350	1130	3440	1290	1080	3220	1200	1000
3-1/2	1-1/2	1/2	1530	960	1120	1320	800	940	1230	730	860	1210	720	850	1160	680	810
		5/8	2150	1310	1510	1870	1130	1290	1760	1040	1190	1740	1030	1170	1660	940	1090
		3/4	2890	1770	1980	2550	1330	1550	2400	1170	1370	2380	1130	1310	2280	1040	1210
		7/8	3780	1920	2240	3360	1440	1680	3180	1260	1470	3150	1210	1410	3030	1100	1290
		1	4820	2040	2380	4310	1530	1790	4090	1350	1580	4050	1290	1510	3860	1200	1400
	1-3/4	1/2	1660	1030	1180	1430	850	1030	1330	770	940	1310	750	920	1250	710	870
		5/8	2310	1370	1630	1990	1160	1380	1860	1070	1230	1840	1060	1200	1760	1000	1090
		3/4	3060	1810	2070	2670	1550	1550	2510	1370	1370	2480	1310	1310	2370	1210	1210
		7/8	3940	2240	2240	3470	1680	1680	3270	1470	1470	3240	1410	1410	3110	1290	1290
		1	4960	2380	2380	4400	1790	1790	4170	1580	1580	4120	1510	1510	3970	1400	1400
	3-1/2	1/2	1660	1180	1180	1500	1040	1040	1430	970	970	1420	960	960	1370	920	920
		5/8	2590	1770	1770	2340	1560	1420	2240	1410	1230	2220	1390	1200	2150	1290	1090
		3/4	3730	2380	2070	3380	1910	1550	3220	1750	1370	3190	1700	1310	3090	1610	1210
		7/8	5080	2820	2240	4600	2330	1680	4290	2130	1470	4210	2070	1410	3940	1960	1290
		1	6560	3340	2380	5380	2780	1790	4900	2580	1580	4810	2520	1510	4510	2410	1400
5-1/4	1-1/2	5/8	2150	1310	1510	1870	1130	1290	1760	1040	1190	1740	1030	1170	1660	940	1110
		3/4	2890	1770	1980	2550	1330	1690	2400	1170	1580	2380	1130	1550	2280	1040	1480
		7/8	3780	1920	2520	3360	1440	2170	3180	1260	2030	3150	1210	1990	3030	1100	1900
		1	4820	2040	3120	4310	1530	2680	4090	1350	2360	4050	1290	2260	3860	1200	2100
	1-3/4	5/8	2310	1370	1630	1990	1160	1380	1860	1070	1270	1840	1060	1250	1760	1000	1180
		3/4	3060	1810	2110	2670	1550	1790	2510	1370	1660	2480	1310	1630	2370	1210	1550
		7/8	3940	2240	2640	3470	1680	2260	3270	1470	2100	3240	1410	2060	3110	1290	1930
		1	4960	2380	3240	4400	1790	2680	4170	1580	2360	4120	1510	2260	3970	1400	2100
	3-1/2	5/8	2590	1770	1770	2340	1560	1560	2240	1410	1460	2220	1390	1450	2150	1290	1390
		3/4	3730	2380	2480	3380	1910	2180	3220	1750	2050	3190	1700	1970	3090	1610	1810
		7/8	5080	2820	3290	4600	2330	2530	4390	2130	2210	4350	2070	2110	4130	1960	1930
		1	6630	3340	3570	5740	2780	2680	5330	2580	2360	5250	2520	2260	4990	2410	2100
5-1/2	1-1/2	5/8	2150	1310	1510	1870	1130	1290	1760	1040	1190	1740	1030	1170	1660	940	1110
		3/4	2890	1770	1980	2550	1330	1690	2400	1170	1580	2380	1130	1550	2280	1040	1480
		7/8	3780	1920	2520	3360	1440	2170	3180	1260	2030	3150	1210	1990	3030	1100	1900
		1	4820	2040	3120	4310	1530	2700	4090	1350	2480	4050	1290	2370	3860	1200	2200
	3-1/2	5/8	2590	1770	1770	2340	1560	1560	2240	1410	1460	2220	1390	1450	2150	1290	1390
		3/4	3730	2380	2480	3380	1910	2180	3220	1750	2050	3190	1700	2020	3090	1610	1900
		7/8	5080	2820	3290	4600	2330	2650	4390	2130	2310	4350	2070	2210	4130	1960	2020
		1	6630	3340	3740	5740	2780	2810	5330	2580	2480	5250	2520	2370	4990	2410	2200
7-1/2	1-1/2	5/8	2150	1310	1510	1870	1130	1290	1760	1040	1190	1740	1030	1170	1660	940	1110
		3/4	2890	1770	1980	2550	1330	1690	2400	1170	1580	2380	1130	1550	2280	1040	1480
		7/8	3780	1920	2520	3360	1440	2170	3180	1260	2030	3150	1210	1990	3030	1100	1900
		1	4820	2040	3120	4310	1530	2700	4090	1350	2530	4050	1290	2480	3860	1200	2390
	3-1/2	5/8	2590	1770	1770	2340	1560	1560	2240	1410	1460	2220	1390	1450	2150	1290	1390
		3/4	3730	2380	2480	3380	1910	2180	3220	1750	2050	3190	1700	2020	3090	1610	1940
		7/8	5080	2820	3290	4600	2330	2890	4390	2130	2720	4350	2070	2670	4130	1960	2560
		1	6630	3340	4190	5740	2780	3680	5330	2580	3380	5250	2520	3230	4990	2410	3000

1. Tabulated lateral design values (Z) for bolted connections shall be multiplied by all applicable adjustment factors (see Table 10.3.1).
2. Tabulated lateral design values (Z) are for "full diameter" bolts (see Appendix L) with bending yield strength (F_{yb}) of 45,000 psi.

Figure 7.9

Where steel side plates are used instead of wood side members, the design values can be found in Table 11B of the National Design Specification.

When the load is at an angle of load to grain between 0° and 90°, the design values are obtained from the Hankinson formula

$$N = \frac{PQ}{P\sin^2\theta + Q\cos^2\theta}$$

where

N = design value at angle θ to grain

P = design value parallel to grain

Q = design value perpendicular to grain

θ = angle between direction of load and direction of grain in degrees

The Hankinson formula may be solved analytically, or graphically using nomographs.

Other detailed requirements for bolts, such as spacings, end distances, and edge distances may be found in the National Design Specification for Wood Construction, and the Timber Construction Manual.

Example #1

Two 4" nominal Douglas fir-larch members are connected by 7/8" bolts as shown in Figure 7.10. What is the safe load P which can be resisted? Assume that the spacing, distances, and wood members are all adequate.

Solution:

This is a single shear joint.

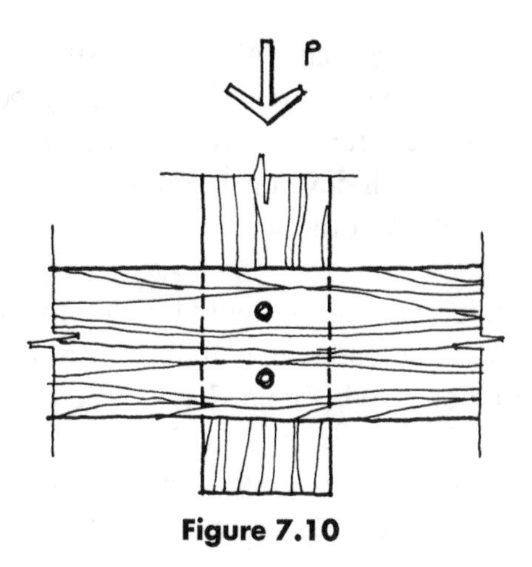

Figure 7.10

Using Table 11A in Figure 7.8, for main member and side member thickness of 3-1/2" (nominal 4" members) and 7/8" bolt, the design value for main members loaded parallel to the grain and side members loaded perpendicular to the grain is 1,060 pounds. The safe load P is therefore 1,060 pounds per bolt times 2 bolts = *2,120 pounds.*

Lag Screws

A lag screw, also called a lag bolt, is a wood screw with a head similar to that of a bolt and without a nut (Figure 7.11). Lag screws require prebored lead holes of the proper diameter. They are inserted in the lead hole by turning with a wrench, not by driving with a hammer. Soap or other lubricant may be used to facilitate insertion. As with bolts, washers should be installed between the wood member and the head of the lag screw.

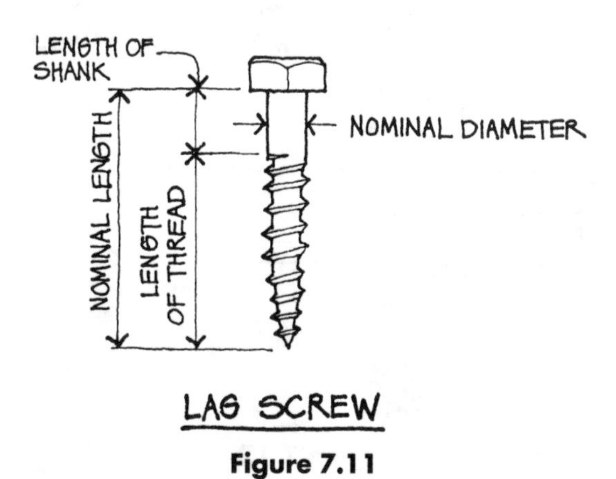

Figure 7.11

Lag screws have high withdrawal resistance, especially from side grain (Figure 7.12); penetration of the threaded portion of 7 to 11 times the diameter will develop approximately the tensile strength of the screw.

If possible, the design should be such that lag screws are not loaded in withdrawal from the end grain of wood (Figure 7.13). Where this condition is unavoidable, the design value in withdrawal from end grain should be no more than three-fourths of that for withdrawal from side grain.

The lateral resistance of lag screws, that is, the shear load they can safely resist, is generally lower than that of bolts of the same diameter (Figure 7.14). The lateral resistance of lag screws in side grain with a metal side plate is greater than the lateral resistance with a wood side member (Figure 7.15).

Design values for both withdrawal and lateral load are tabulated in the National Design Specification and the Timber Construction Manual. Lag screws are made in a variety of diameters and lengths, from 1/4" diameter and 1" long to 1-1/4" diameter and 12" long.

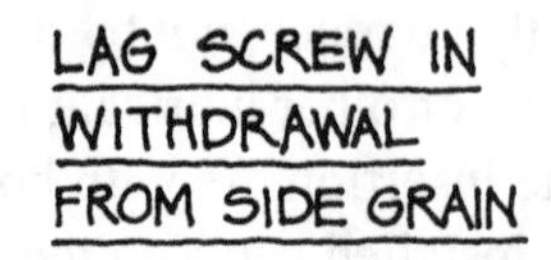

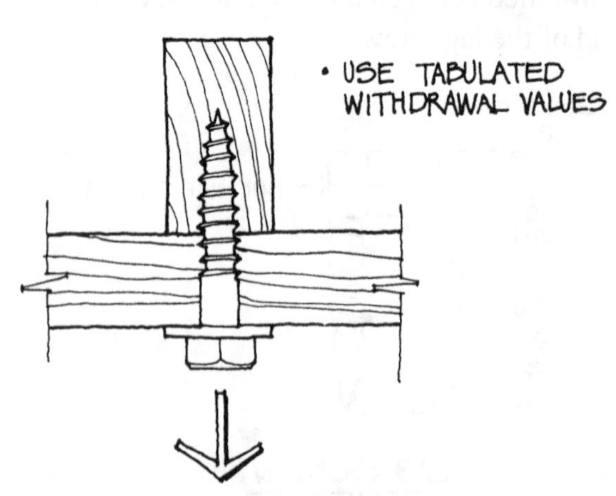

Figure 7.12

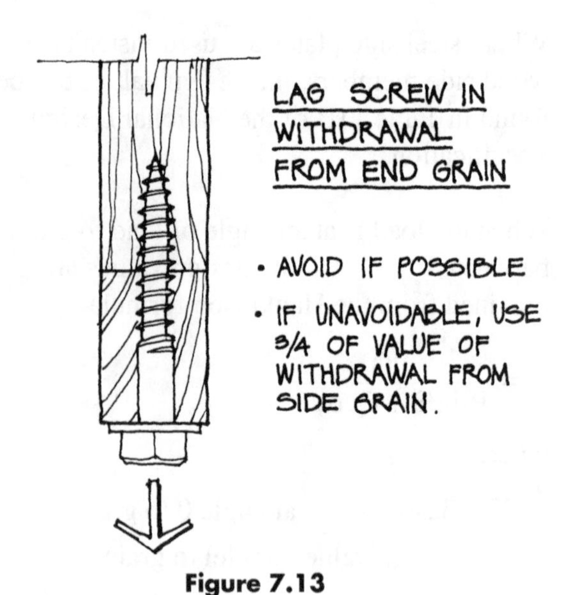

LAG SCREW IN WITHDRAWAL FROM END GRAIN
• AVOID IF POSSIBLE
• IF UNAVOIDABLE, USE 3/4 OF VALUE OF WITHDRAWAL FROM SIDE GRAIN.

Figure 7.13

Wood Screws

Wood screws, because they are threaded, have greater holding power than nails of the same diameter. The most common types of wood screws are flat head and round head (Figure 7.16). Wood screws are specified by diameter number, from #6 (0.138" diameter) to #24 (0.372" diameter), and by length. About two-thirds of the length of a standard wood screw is threaded. Wood screws require prebored lead holes of the proper diameter. They are inserted in the lead hole by turning with a screwdriver, not by driving with a hammer. Soap or other lubricant may be used to facilitate insertion.

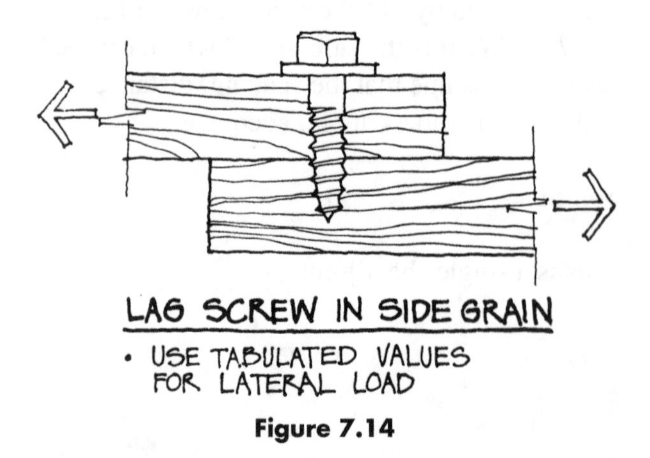

LAG SCREW IN SIDE GRAIN
• USE TABULATED VALUES FOR LATERAL LOAD

Figure 7.14

Although wood screws have tabulated withdrawal design values, it is preferable not to subject wood screws to withdrawal loads; loading wood screws in withdrawal from the end grain of wood is not permitted at all (see Figure 7.17).

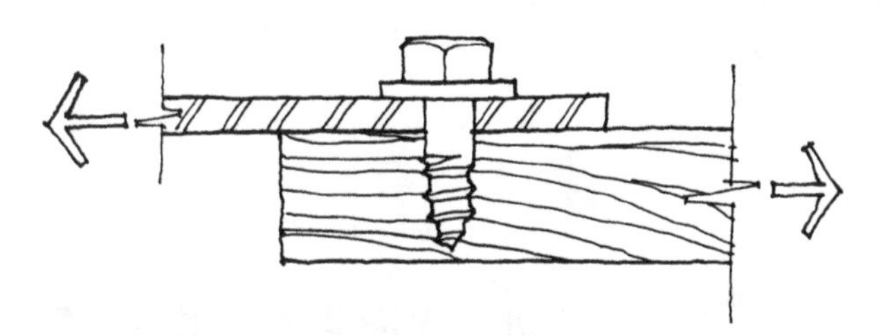

LAG SCREW IN SIDE GRAIN
WITH METAL SIDE PLATE
• TABULATED VALUES ARE HIGHER THAN WITH WOOD SIDE MEMBER

Figure 7.15

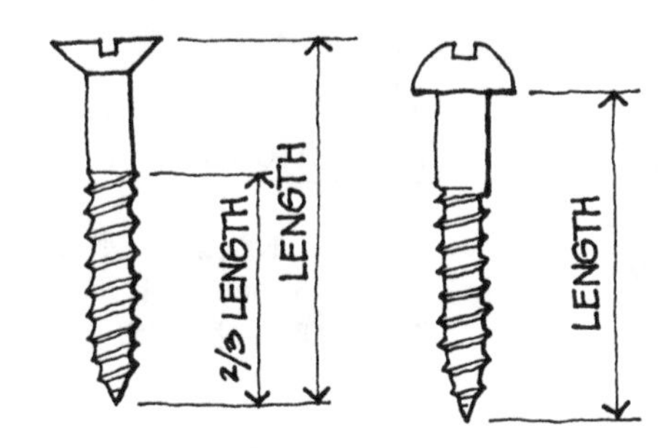

Figure 7.16

The design values for lateral resistance (shear) for wood screws are the same for any angle of load to grain, and are greater where metal side plates are used, rather than wood side members.

Nails

Nails are the most common means of connecting wood members, particularly for light frame construction. While there are many types and forms of nails, those most often used for structural purposes are common steel wire nails. These are available in sizes from 6d (6 penny), which is 2" long and has a diameter of 0.113", to 60d (60 penny), which is 6" in length and 0.263" in diameter.

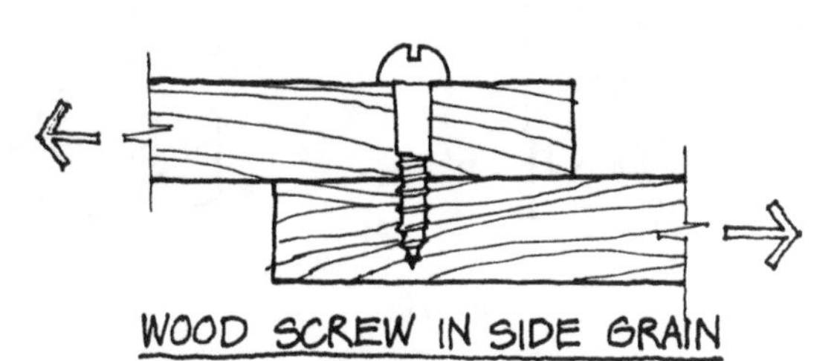

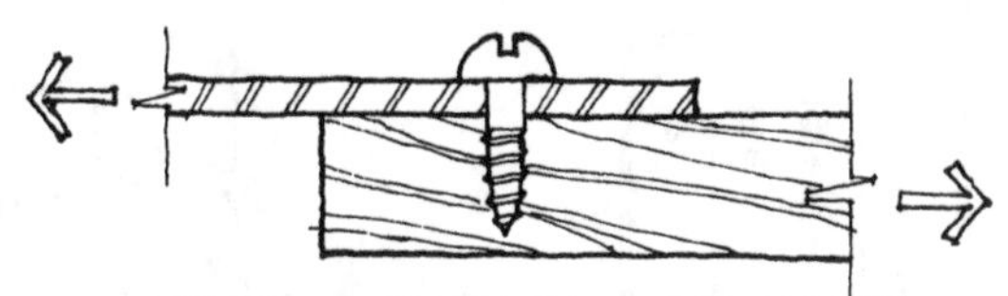

WOOD SCREW IN SIDE GRAIN
WITH METAL SIDE PLATE
• TABULATED VALUES ARE HIGHER THAN WITH WOOD SIDE MEMBER

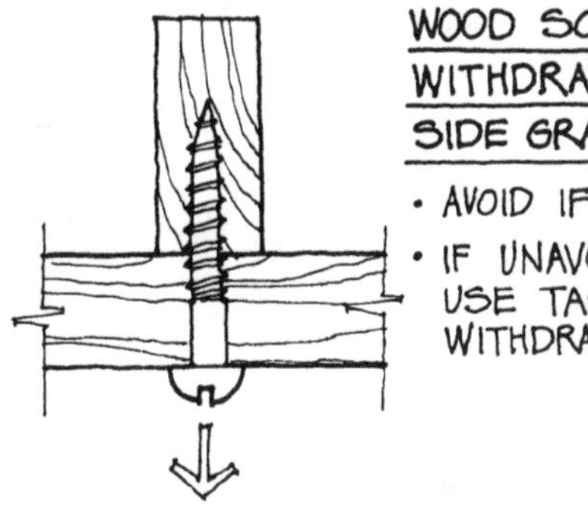

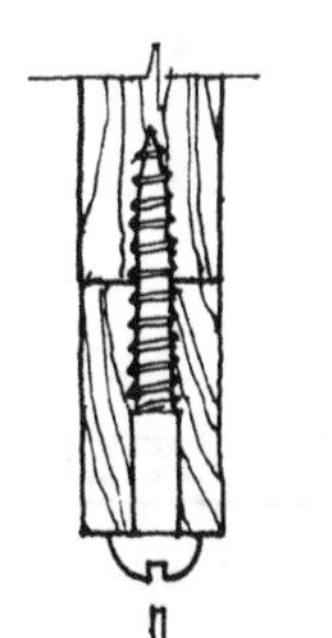

Figure 7.17

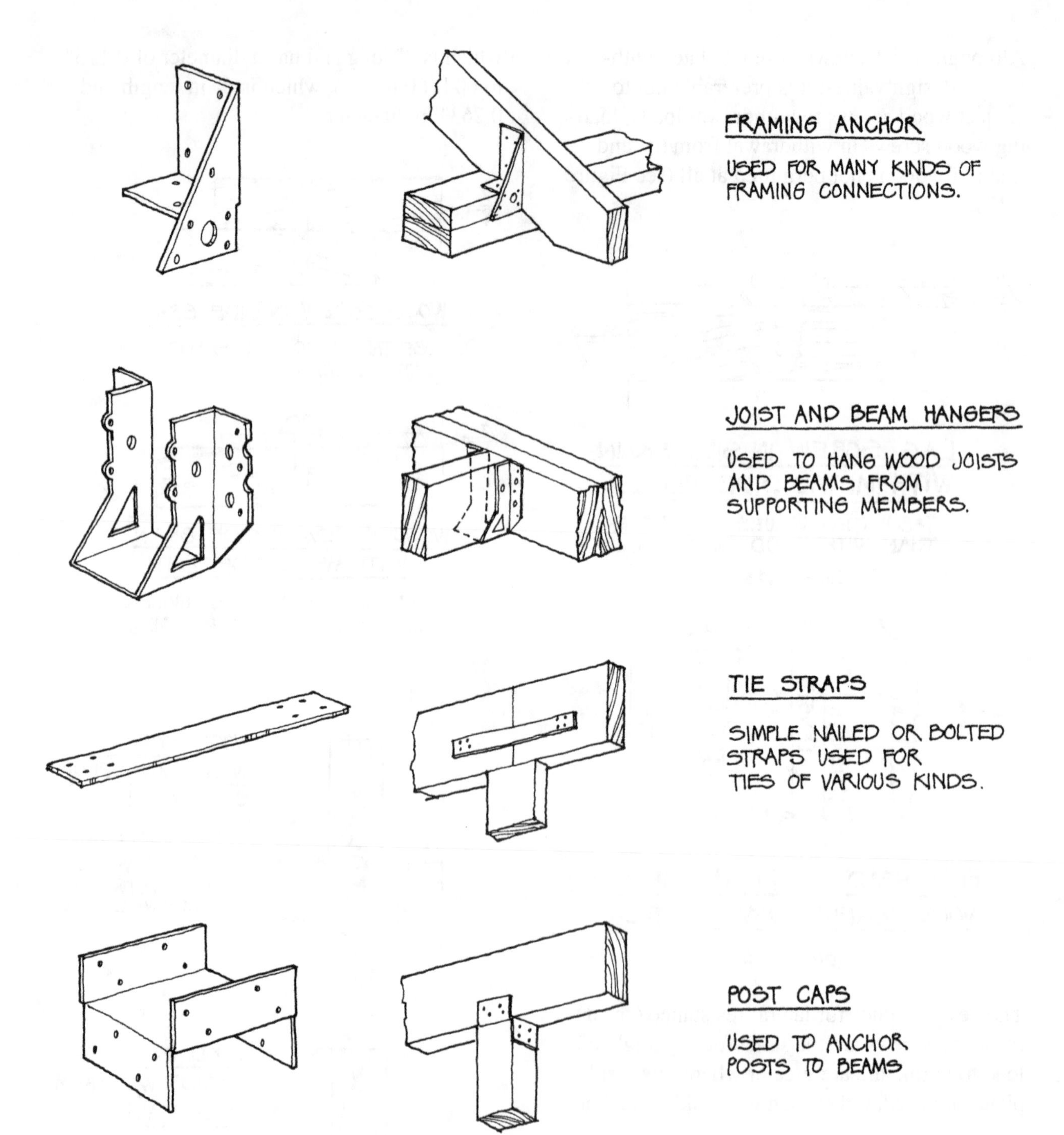
FRAMING ANCHOR
USED FOR MANY KINDS OF FRAMING CONNECTIONS.
JOIST AND BEAM HANGERS
USED TO HANG WOOD JOISTS AND BEAMS FROM SUPPORTING MEMBERS.
TIE STRAPS
SIMPLE NAILED OR BOLTED STRAPS USED FOR TIES OF VARIOUS KINDS.
POST CAPS
USED TO ANCHOR POSTS TO BEAMS
MISCELLANEOUS WOOD TO WOOD CONNECTORS

Figure 7.18

Nails have greater value when driven into the side grain (perpendicular to the wood fibers) than into the end grain of wood. Also, it is preferable that nails be used so that their lateral resistance (shear) is utilized, rather than withdrawal resistance. Nails are not permitted to be loaded in withdrawal from end grain of wood.

Nails have the same design values in shear (lateral resistance), regardless of angle of load to grain. Where metal side plates are used, rather than wood side members, the design values are greater. Since the requirements for nailed joints are similar to those using wood screws, the wood screw diagrams shown in Figures 7.12, 7.13, 7.14, 7.15, and 7.17 also apply to nails.

Miscellaneous Hardware

In addition to the fasteners described in this lesson, there are many types of hardware used for wood connections. These include framing anchors, joist and beam hangers, tie straps, post caps, and post bases. They are available in a variety of types and configurations, some of which are shown in Figure 7.18.

STEEL CONNECTIONS

Rivets and Bolts

Steel members are commonly connected in building construction by means of rivets, bolts, or welds. Rivets, which were the most commonly used fasteners, have been almost entirely replaced by high-strength bolts. Bolting is quieter and simpler than riveting and can be performed quickly by smaller labor crews, resulting in relatively low labor costs.

Connections made with mechanical fasteners are categorized as either *bearing-type* or *slip-critical*. This latter term replaces the previous designation of *friction-type*. Connections subject to significant stress reversal, fatigue loading, or any other condition where slip would be detrimental, are slip-critical. All other connections may be bearing-type.

In bearing-type connections, made with either rivets or bolts, the members transfer load to the connector by bearing, while the connector resists the load by shear. The strength of the joint is therefore determined by the shear strength of the connector or the bearing strength of the member against the connector, whichever is lower.

Slip-critical connections can only be made with high-strength bolts which conform to ASTM A325 or A490. In such connections, the bolts are tightened to a tension equal to 70 percent of their specified minimum tensile strength, thus tightly clamping the connected parts together. The resulting friction between the clamped parts resists the applied load and there is no actual shear or bearing stress. While slip-critical connections do not depend on bearing, bearing stress should be checked to maintain an adequate factor of safety in case the bolts slip.

The three methods of controlling bolt tension for slip-critical joints are the *calibrated wrench*, the *direct-tension indicator*, and the *turn-of-nut method*, which requires specified additional nut rotation after the bolts are *snug tight*. Special hardened washers are often required for slip-critical connections.

Joints, whether bearing-type or slip-critical, may be either *single shear* or *double shear* (Figure 7.19). In the single shear joint, there is one plane through the connector that resists shear; the allowable shear strength is therefore the cross-sectional area of the connector multiplied by the nominal shear stress of the connector. In the double shear joint, there are two planes through the connector that resist shear;

the allowable shear strength is therefore twice that of the single shear joint.

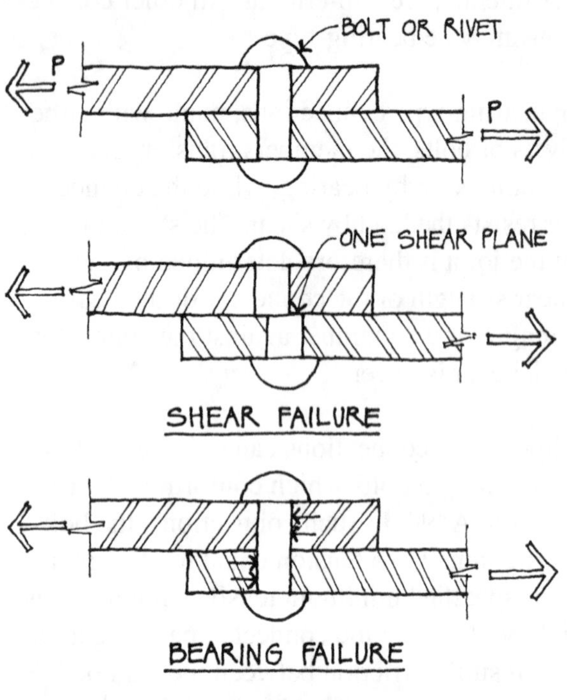

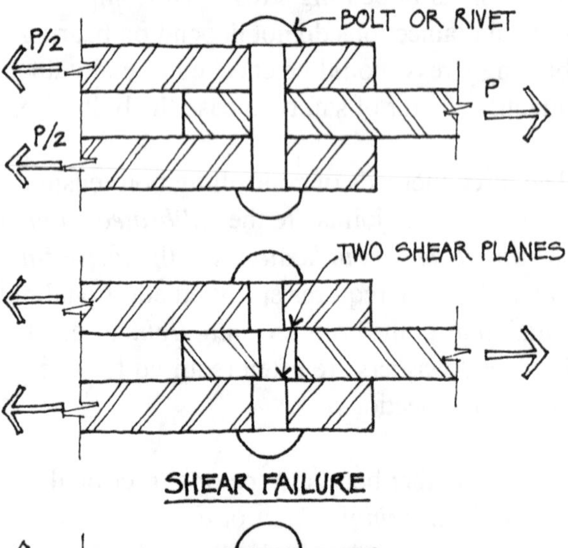

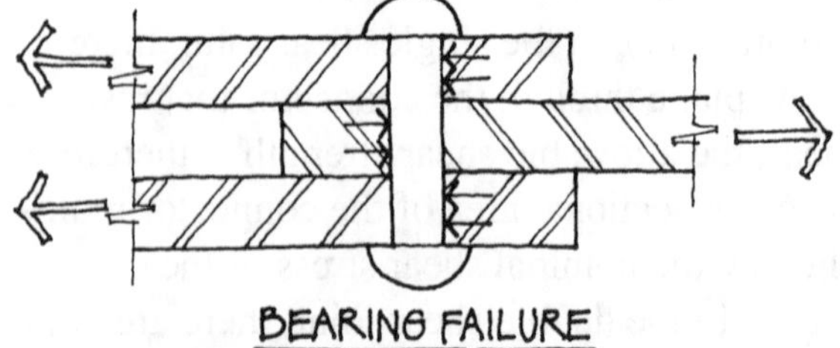

Figure 7.19

Single shear and double shear joints may also fail in bearing—crushing of the connected part against the shank of the connector. The area assumed to resist bearing is a rectangular area equal to the connector diameter multiplied by the thickness of the connected part.

The available shear strength of bolts is given in Table 7-1 of the AISC Manual (page 7-22) for bearing-type connections and in Table 7-3 for slip-critical connections (pages 7-24 through 7-25). Table 7-1 and a portion of Table 7-3 are reproduced in Figures 7.20 and 7.21. Available bolt shear strength is tabulated for various grades and sizes of bolts.

Note that in Table 7-1 two values are tabulated for bearing-type connections: N (threads included in shear plane) and X (threads excluded from shear plane). The drawing in Figure 7.22 shows these two conditions.

The maximum permitted bearing strength is 1.2 F_u, where F_u is the minimum tensile strength of the connected part. For ASTM A36 steel, F_u is equal to 58 ksi. Tables 7-5 and 7-6 in the AISC Manual (pages 7-28 through 7-31) give available bearing strengths at bolt holes based on bolt spacing and on edge distance, respectively. Portions of these two tables are reproduced in Figures 7.24 and 7.25.

Example #2

Two 1/2" plates are connected to a 3/4" plate using 2-1" bolts conforming to ASTM A325 in a bearing-type connection, with threads excluded from the shear plane and standard round holes, as shown in Figure 7.23. The plates are ASTM A36 steel with a minimum tensile strength of 58 ksi. Assume a 1.25" edge distance and a bolt spacing of 3". What is the allowable load P?

Table 7-1
Available Shear Strength of Bolts, kips

ASTM Desig.	Thread Cond.	F_{nv}/Ω (ksi) ASD	ϕF_{nv} (ksi) LRFD	Loading	5/8 (0.307 in.²) r_n/Ω_v ASD	5/8 $\phi_v r_n$ LRFD	3/4 (0.442 in.²) r_n/Ω_v ASD	3/4 $\phi_v r_n$ LRFD	7/8 (0.601 in.²) r_n/Ω_v ASD	7/8 $\phi_v r_n$ LRFD	1 (0.785 in.²) r_n/Ω_v ASD	1 $\phi_v r_n$ LRFD
A325 F1852	N	24.0	36.0	S	7.36	11.0	10.6	15.9	14.4	21.6	18.8	28.3
				D	14.7	22.1	21.2	31.8	28.9	43.3	37.7	56.5
	X	30.0	45.0	S	9.20	13.8	13.3	19.9	18.0	27.1	23.6	35.3
				D	18.4	27.6	26.5	39.8	36.1	54.1	47.1	70.7
A490	N	30.0	45.0	S	9.20	13.8	13.3	19.9	18.0	27.1	23.6	35.3
				D	18.4	27.6	26.5	39.8	36.1	54.1	47.1	70.7
	X	37.5	56.3	S	11.5	17.3	16.6	24.9	22.5	33.8	29.5	44.2
				D	23.0	34.5	33.1	49.7	45.1	67.6	58.9	88.4
A307	–	12.0	18.0	S	3.68	5.52	5.30	7.95	7.22	10.8	9.42	14.1
				D	7.36	11.0	10.6	15.9	14.4	21.6	18.8	28.3

ASTM Desig.	Thread Cond.	F_{nv}/Ω (ksi) ASD	ϕF_{nv} (ksi) LRFD	Loading	1 1/8 (0.994 in.²) r_n/Ω_v ASD	1 1/8 $\phi_v r_n$ LRFD	1 1/4 (1.23 in.²) r_n/Ω_v ASD	1 1/4 $\phi_v r_n$ LRFD	1 3/8 (1.48 in.²) r_n/Ω_v ASD	1 3/8 $\phi_v r_n$ LRFD	1 1/2 (1.77 in.²) r_n/Ω_v ASD	1 1/2 $\phi_v r_n$ LRFD
A325 F1852	N	24.0	36.0	S	23.9	35.8	29.5	44.2	35.6	53.5	42.4	63.6
				D	47.7	71.6	58.9	88.4	71.3	107	84.8	127
	X	30.0	45.0	S	29.8	44.7	36.8	55.2	44.5	66.8	53.0	79.5
				D	59.6	89.5	73.6	110	89.1	134	106	159
A490	N	30.0	45.0	S	29.8	44.7	36.8	55.2	44.5	66.8	53.0	79.5
				D	59.6	89.5	73.6	110	89.1	134	106	159
	X	37.5	56.3	S	37.3	55.9	46.0	69.0	55.7	83.5	66.3	99.4
				D	74.6	112	92.0	138	111	167	133	199
A307	–	12.0	18.0	S	11.9	17.9	14.7	22.1	17.8	26.7	21.2	31.8
				D	23.9	35.8	29.5	44.2	35.6	53.5	42.4	63.6

ASD	LRFD
$\Omega_v = 2.00$	$\phi_v = 0.75$

Figure 7.20

A325

Table 7-3
Slip-Critical Connections
Available Shear Strength, kips, when
Slip is a Serviceability Limit-State
(Class A Faying Surface, $\mu = 0.35$)

ASTM A325/F1852 Bolts

Hole Type	Loading	Nominal Bolt Diameter d, in.							
		5/8		3/4		7/8		1	
		Minimum ASTM A325/F1852 Bolt Pretension, kips							
		19		28		39		51	
		r_n/Ω_v	$\phi_v r_n$	r_n/Ω_v	$\phi_v r_n$	r_n/Ω_v	$\phi_v r_n$	r_n/Ω_v	$\phi_v r_n$
		ASD	LRFD	ASD	LRFD	ASD	LRFD	ASD	LRFD
STD	S	5.01	7.51	7.38	11.1	10.3	15.4	[illegible]	20.2
	D	10.0	15.0	14.8	22.1	[illegible]	[illegible]	26.9	40.3
SSLT	S	4.26	6.39	6.28	9.41	8.74	13.1	11.4	17.1
	D	8.52	12.8	12.6	18.8	17.5	26.2	22.9	34.3
LSLT	S	3.51	5.26	5.17	7.75	7.20	10.8	9.41	14.1
	D	7.01	10.5	10.3	15.5	14.4	21.6	18.8	28.2

Hole Type	Loading	Nominal Bolt Diameter d, in.							
		1 1/8		1 1/4		1 3/8		1 1/2	
		Minimum ASTM A325/F1852 Bolt Pretension, kips							
		56		71		85		103	
		r_n/Ω_v	$\phi_v r_n$	r_n/Ω_v	$\phi_v r_n$	r_n/Ω_v	$\phi_v r_n$	r_n/Ω_v	$\phi_v r_n$
		ASD	LRFD	ASD	LRFD	ASD	LRFD	ASD	LRFD
STD	S	14.8	22.1	18.7	28.1	22.4	33.6	27.2	40.7
	D	29.5	44.3	37.4	56.2	44.8	67.2	54.3	81.5
SSLT	S	12.6	18.8	15.9	23.9	19.0	28.6	23.1	34.6
	D	25.1	37.7	31.8	47.7	38.1	57.1	46.2	69.3
LSLT	S	10.3	15.5	13.1	19.7	15.7	23.5	19.0	28.5
	D	20.7	31.0	26.2	39.3	31.4	47.1	38.0	57.0

STD = Standard Hole
SSLT = Short-Slotted Hole transverse to the line of force
LSLT = Long-Slotted Hole transverse to line of force
S = Single Shear
D = Double Shear

ASD	LRFD	
$\Omega_v = 1.50$	$\phi_v = 1.00$	Note: For available slip resistance when slip is a strength limit state, see Table 7-4. For Class B faying surfaces ($\mu = 0.50$), multiply the tabulated available strength by 1.43. The required strength is determined using LRFD load combinations for LRFD design and ASD load combinations for ASD design.

Figure 7.21

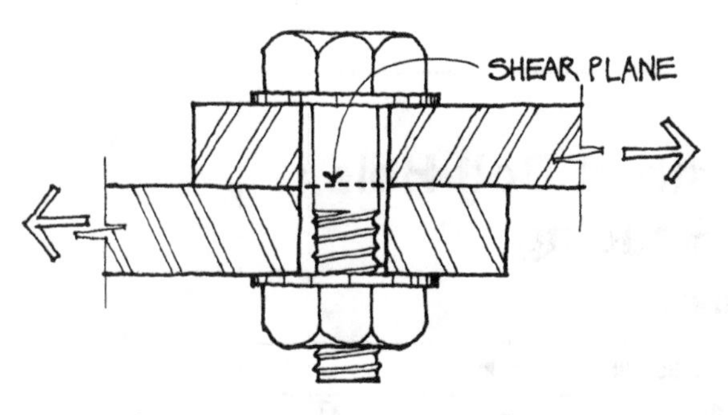

THREADS EXCLUDED FROM SHEAR PLANE

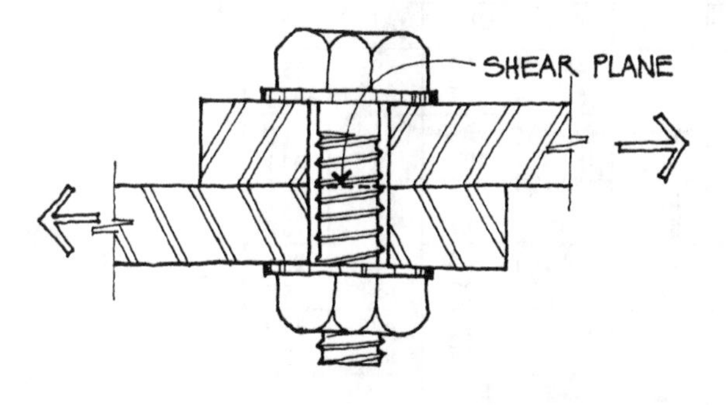

THREADS INCLUDED IN SHEAR PLANE

Figure 7.22

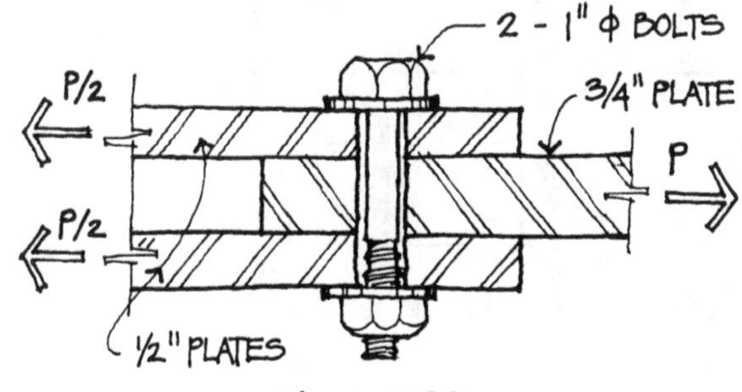

Figure 7.23

Solution:

This is a double shear joint; from Table 7-1 in Figure 7.20, the available sheer strength for each 1" diameter A325-X STD bolt is 47.1 kips. The total capacity of the two bolts in shear is 2 × 47.1 = 94.2 kips.

We must also check the bearing capacity. Each bolt bears on two 1/2" plates in one direction and one 3/4" plate in the other. The 3/4" plate governs, since it is thinner than the two 1/2" plates.

From Table 7-5 in Figure 7.24, based on a bolt spacing of 3", each 1" diameter bolt can transfer 67.4 kips per inch thickness of plate. Thus, for the 3/4" thick plate, each bolt can transfer 0.75 × 67.4 = 50.6 kips, or two bolts can transfer 2 × 50.6 = 101.2 kips. From Table 7-6 in Figure 7.25, based on an edge distance of 1.25", each 1" diameter bolt can transfer 25.0 kips per inch thickness of plate, or 0.75 × 25 = 18.8 kips. Two bolts can transfer 2 × 18.8 = 37.6 kips. Therefore, the available bearing strength is based on edge distance, and is equal to 37.6 kips.

Since the available strength of the bolts in bearing is less than in shear, the bearing strength governs, and the allowable load P is *37.6 kips.*

Example #3

For the connection in Example #2, what is the allowable load P if 1" bolts conforming to ASTM A307 (common bolts) are used?

Solution:

From Table 7-1 in Figure 7.20, the available shear strength for each 1" diameter A307 bolt in double shear is 18.8 kips. The total capacity of the two bolts in shear is 2 × 18.8 = 37.6 kips. The available bearing strength is 37.6 kips, as in Example #2. In this example, both the shear and bearing capacities are equal, and the allowable P is *37.6 kips.*

Example #4

Again, using the connection in Example #2, what is the allowable load P if 1" bolts conforming to ASTM A325 in a slip-critical connection are used?

Table 7-5
Available Bearing Strength at Bolt Holes Based on Bolt Spacing
kips/in. thickness

Hole Type	Bolt Spacing, s, in.	F_u, ksi	Nominal Bolt Diameter d_b, in.							
			$^5/_8$		$^3/_4$		$^7/_8$		1	
			r_n/Ω_v	$\phi_v r_n$	r_n/Ω_v	$\phi_v r_n$	r_n/Ω_v	$\phi_v r_n$	r_n/Ω_v	$\phi_v r_n$
			ASD	LRFD	ASD	LRFD	ASD	LRFD	ASD	LRFD
STD SSLT	$2^2/_3\, d_b$	58	34.1	51.1	41.3	62.0	48.6	72.9	55.8	83.7
		65	38.2	57.3	46.3	69.5	54.4	81.7	62.6	93.8
	3 in.	58	43.5	65.3	52.2	78.3	60.9	91.4	67.4	101
		65	48.8	73.1	58.5	87.8	68.3	102	75.6	113
SSLP	$2^2/_3\, d_b$	58	27.6	41.3	34.8	52.2	42.1	63.1	47.1	70.7
		65	30.9	46.3	39.0	58.5	47.1	70.7	52.8	79.2
	3 in.	58	43.5	65.3	52.2	78.3	60.9	91.4	58.7	88.1
		65	48.8	73.1	58.5	87.8	68.3	102	65.8	98.7
OVS	$2^2/_3\, d_b$	58	29.7	44.6	37.0	55.5	44.2	66.3	49.3	74.0
		65	33.3	50.0	41.4	62.2	49.6	74.3	55.3	82.9
	3 in.	58	43.5	65.3	52.2	78.3	60.9	91.4	60.9	91.4
		65	48.8	73.1	58.5	87.8	68.3	102	68.3	102
LSLP	$2^2/_3\, d_b$	58	3.62	5.44	4.35	6.53	5.08	7.61	5.80	8.70
		65	4.06	6.09	4.88	7.31	5.69	8.53	6.50	9.75
	3 in.	58	43.5	65.3	39.2	58.7	28.3	42.4	17.4	26.1
		65	48.8	73.1	43.9	65.8	31.7	47.5	19.5	29.3
LSLT	$2^2/_3\, d_b$	58	28.4	42.6	34.4	51.7	40.5	60.7	46.5	69.8
		65	31.8	47.7	38.6	57.9	45.4	68.0	52.1	78.2
	3 in.	58	36.3	54.4	43.5	65.3	50.8	76.1	56.2	84.3
		65	40.6	60.9	48.8	73.1	56.9	85.3	63.0	94.5
STD, SSLT, SSLP, OVS, LSLP	$s \geq s_{full}$	58	43.5	65.3	52.2	78.3	60.9	91.4	69.6	104
		65	48.8	73.1	58.5	87.8	68.3	102	78.0	117
LSLT	$s \geq s_{full}$	58	36.3	54.4	43.5	65.3	50.8	76.1	58.0	87.0
		65	40.6	60.9	48.8	73.1	56.9	85.3	65.0	97.5
Spacing for full bearing strength s_{full}[a], in.		STD, SSLT, LSLT	$1^{15}/_{16}$		$2^5/_{16}$		$2^{11}/_{16}$		$3^1/_{16}$	
		OVS	$2^1/_{16}$		$2^7/_{16}$		$2^{13}/_{16}$		$3^1/_4$	
		SSLP	$2^1/_8$		$2^1/_2$		$2^7/_8$		$3^5/_{16}$	
		LSLP	$2^{13}/_{16}$		$3^3/_8$		$3^{15}/_{16}$		$4^1/_2$	
Minimum Spacing[a] $= 2^2/_3\, d_b$, in.			$1^{11}/_{16}$		2		$2^5/_{16}$		$2^{11}/_{16}$	

STD = Standard Hole
SSLT = Short-Slotted Hole oriented transverse to the line of force
SSLP = Short-Slotted Hole oriented parallel to the line of force
OVS = Oversized Hole
LSLP = Long-Slotted Hole oriented parallel to the line of force
LSLT = Long-Slotted Hole oriented transverse to the line of force

ASD	LRFD
$\Omega_v = 2.00$	$\phi_v = 0.75$

Note: Spacing indicated is from the center of the hole or slot to the center of the adjacent hole or slot in the line of force. Hole deformation is considered. When hole deformation is not considered, see AISC Specification Section J3.10.

[a] Decimal value has been rounded to the nearest sixteenth of an inch.

Figure 7.24

Table 7-6
Available Bearing Strength at Bolt Holes Based on Edge Distance
kips/in. thickness

Hole Type	Edge Distance L_e, in.	F_u, ksi	Nominal Bolt Diameter d_b, in.							
			5/8		3/4		7/8		1	
			r_n/Ω_v	$\phi_v r_n$	r_n/Ω_v	$\phi_v r_n$	r_n/Ω_v	$\phi_v r_n$	r_n/Ω_v	$\phi_v r_n$
			ASD	LRFD	ASD	LRFD	ASD	LRFD	ASD	LRFD
STD SSLT	1 1/4	58	31.[illegible]	47.[illegible]	29.[illegible]	[illegible]	27.2	[illegible]	25.0	37.5
		65	35.3	53.0	32.9	49.4	30.5	45.7	28.0	42.0
	2	58	43.5	65.3	52.2	78.3	53.3	79.9	51.1	76.7
		65	48.8	73.1	58.5	87.8	59.7	89.6	57.3	85.9
SSLP	1 1/4	58	28.3	42.4	26.1	39.2	23.9	35.9	20.7	31.0
		65	31.7	47.5	29.3	43.9	26.8	40.2	23.2	34.7
	2	58	43.5	65.3	52.2	78.3	50.0	75.0	46.8	70.1
		65	48.8	73.1	58.5	87.8	56.1	84.1	52.4	78.6
OVS	1 1/4	58	29.4	44.0	27.2	40.8	25.0	37.5	21.8	32.6
		65	32.9	49.4	30.5	45.7	28.0	42.0	24.4	36.6
	2	58	43.5	65.3	52.2	78.3	51.1	76.7	47.9	71.8
		65	48.8	73.1	58.5	87.8	57.3	85.9	53.6	80.4
LSLP	1 1/4	58	16.3	24.5	10.9	16.3	5.44	8.16	—	—
		65	18.3	27.4	12.2	18.3	6.09	9.14	—	—
	2	58	42.4	63.6	37.0	55.5	31.5	47.3	26.1	39.2
		65	47.5	71.3	41.4	62.2	35.3	53.0	29.3	43.9
LSLT	1 1/4	58	26.3	39.4	24.5	36.7	22.7	34.0	20.8	31.3
		65	29.5	44.2	27.4	41.1	25.4	38.1	23.4	35.0
	2	58	36.3	54.4	43.5	65.3	44.4	66.6	42.6	63.9
		65	40.6	60.9	48.8	73.1	49.8	74.6	47.7	71.6
STD, SSLT, SSLP, OVS, LSLP	$L_e \geq L_{e\,full}$	58	43.5	65.3	52.2	78.3	60.9	91.4	69.6	104
		65	48.8	73.1	58.5	87.8	68.3	102	78.0	117
LSLT	$L_e \geq L_{e\,full}$	58	36.3	54.4	43.5	65.3	50.8	76.1	58.0	87.0
		65	40.6	60.9	48.8	73.1	56.9	85.3	65.0	97.5

Edge distance for full bearing strength $L_e \geq L_{e\,full}$[a], in.		5/8	3/4	7/8	1
	STD, SSLT, LSLT	1 5/8	1 15/16	2 1/4	2 9/16
	OVS	1 11/16	2	2 5/16	2 5/8
	SSLP	1 11/16	2	2 5/16	2 11/16
	LSLP	2 1/16	2 7/16	2 7/8	3 1/4

STD = Standard Hole
SSLT = Short-Slotted Hole oriented transverse to the line of force
SSLP = Short-Slotted Hole oriented parallel to the line of force
OVS = Oversized Hole
LSLP = Long-Slotted Hole oriented parallel to the line of force
LSLT = Long-Slotted Hole oriented transverse to the line of force

ASD	LRFD	
$\Omega_v = 2.00$	$\phi_v = 0.75$	— Indicates spacing less than minimum spacing required per AISC Specification Section J3.3. Note: Spacing indicated is from the center of the hole or slot to the center of the adjacent hole or slot in the line of force. Hole deformation is considered. When hole deformation is not considered, see AISC Specification Section J3.10. [a] Decimal value has been rounded to the nearest sixteenth of an inch.

Figure 7.25

Solution:

Referring to Table 7-3 in Figure 7.21, the available shear strength for each 1" diameter A325-SC bolt in double shear is 26.9 kips, or 2 × 26.9 = 53.8 kips for two bolts. Even though this is a slip-critical joint, we check the available strength in bearing, which is 37.6 kips, as in Example #2. The lower value governs, and the allowable load P is *37.6 kips.*

Welds

Welding is a method of transferring load between connected parts without the use of any mechanical fasteners. Since members can be attached together without using connecting plates or angles, the welded connection is usually simpler, more compact, and hence lighter in weight. Holes for rivets or bolts are avoided, and therefore the gross section rather than the net section is used to determine the cross-sectional area of members in tension.

The most usual welding process used in building construction is *electric arc welding.* In this process, one terminal of a power source is connected to the members to be joined and the other to the electrode or welding rod, which the welder holds using an insulated holder. When the electrode is brought close to the members, an electric arc is formed. The intense heat at the tip of the electrode melts a small part of each member and the filler metal of the electrode. When these areas cool and solidify, they are fused into one homogeneous piece. The depth of the base metal to the point where fusion stops is called *penetration.*

In recent years, great advances have been made in automatic and semi-automatic welding processes, so that manual welding today is generally limited to short welds or those of minor importance.

Welding may be done either in a fabricating shop (shop welding) or on the job site (field welding). As much welding as possible should be done in the shop, where equipment and techniques can be more efficiently employed for greater economy.

A number of nondestructive tests are available for examining welded joints. Among these are *radiographic inspection* and *ultrasonic testing.*

Radiography utilizes x-rays and gamma rays to penetrate welded joints and obtain a permanent record on film, similar to medical x-rays. Although radiographic inspection is an excellent means of detecting weld defects, it is sometimes a difficult or even impossible procedure, because of the configuration of certain welded joints.

Ultrasonic testing uses high frequency sound waves to locate and measure defects in welded joints. Although this testing method has limitations, it is often preferred over radiography, since it only requires access on one side of a section.

Most welds commonly seen in structural connections are *fillet welds* or *groove welds* (Figure 7.26). Fillet welds are placed in the angle formed by lapping or intersecting plates and are generally subject to shear stress. Groove welds usually act in direct tension or compression and are placed between the two butting pieces of metal to be joined. Groove welds are classified according to the way the members are prepared for welding, such as square, single-bevel, double-vee, etc. The three types of joint configurations commonly used are *lap, tee*, and *butt.*

Several welded joints are shown in Figure 7.28 along with their standardized symbols. You will note that the weld symbol approximates the shape of the weld. If the weld symbol is below the reference line, it means that the arrow side of the joint is welded. If the symbol is above the line, the side of the joint away from the arrow is welded. Of course, if the symbol is both above and below the line, it means that both sides of the joint are welded.

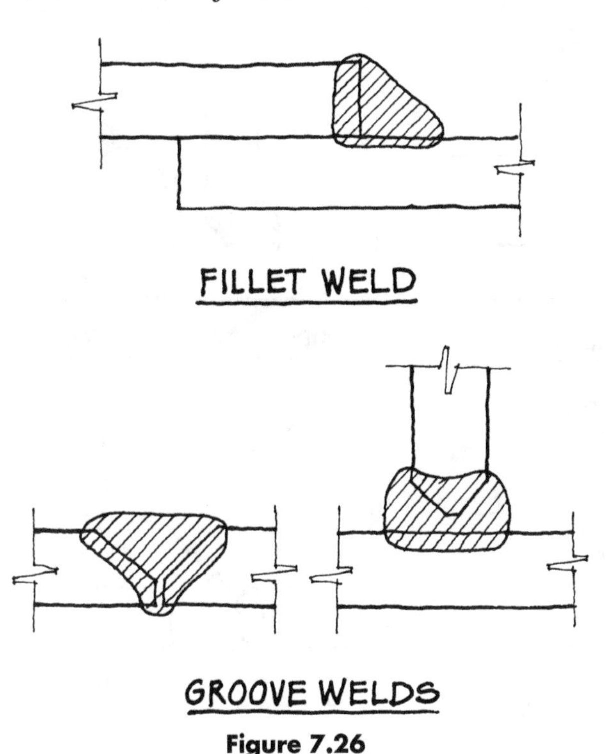

Figure 7.26

Other information that can be conveyed by the weld symbol includes field weld, weld all around, length of weld, spacing of welds, and size of weld.

The strength of a welded joint depends on the dimension of the throat of the weld.

In fillet welds, the stress is always considered to be shear stress on the minimum throat area regardless of the direction of the applied load, where the throat area is 0.707 times the weld size (Figure 7.27). The allowable shear strength is 18.0 ksi for welds made with E60 electrodes and 21.0 ksi for welds made with E70 electrodes when the base metal is ASTM A36. In general, the allowable stress for groove welds is the same as for the base metal.

Example #5

What is the maximum load per lineal inch on a 1/4" fillet weld made with E70 electrodes if the connected parts are ASTM A36 steel?

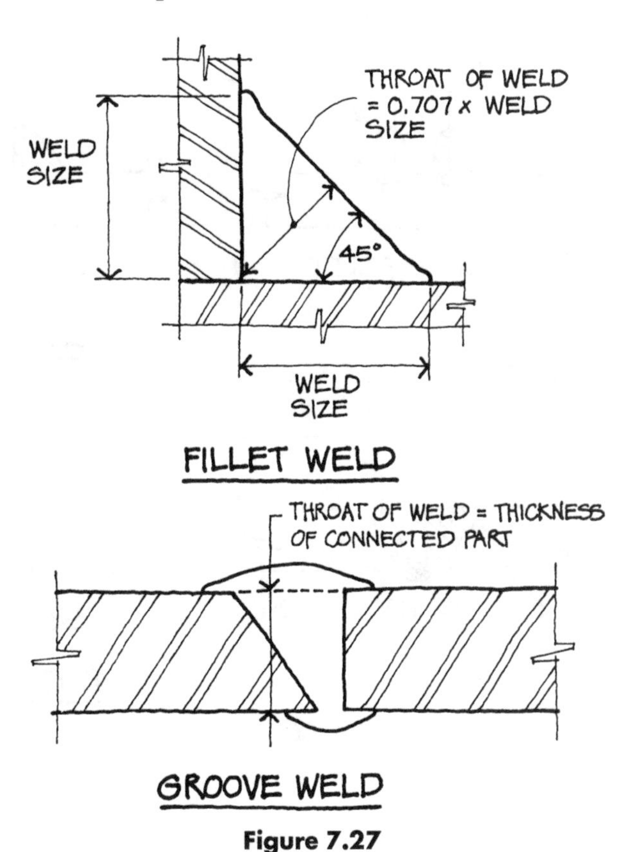

Figure 7.27

Solution:

The stress in a fillet weld is always considered to be shear on the throat area. Throat area = 0.707 × 0.25 = 0.17675 square inches, per inch of weld.

Allowable shear strength = 21.0 ksi

Maximum load per inch of weld = 21.0 × 0.17675 = *3.71 kips per lineal inch.*

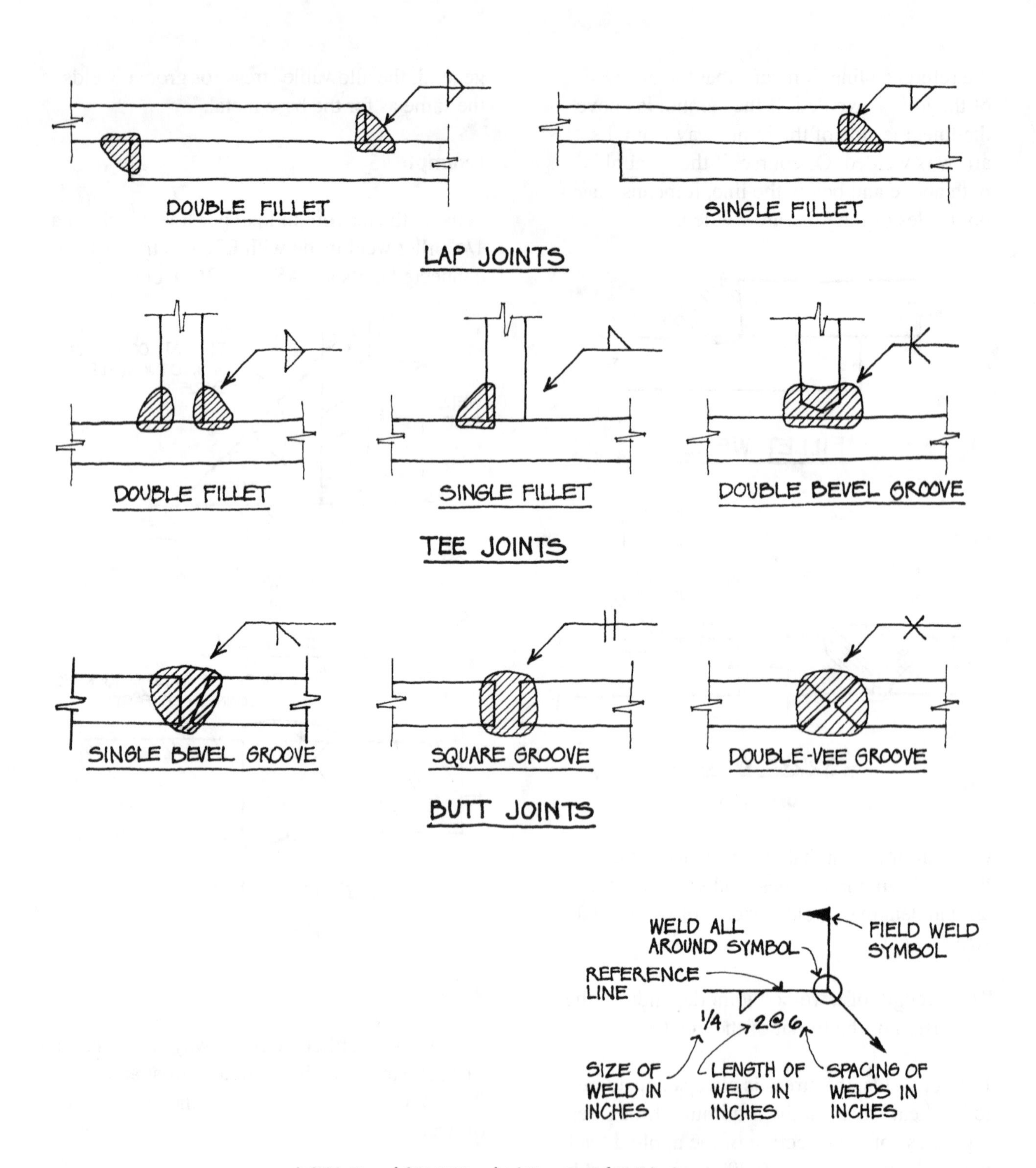
DOUBLE FILLET
SINGLE FILLET
LAP JOINTS
DOUBLE FILLET
SINGLE FILLET
DOUBLE BEVEL GROOVE
TEE JOINTS
SINGLE BEVEL GROOVE
SQUARE GROOVE
DOUBLE-VEE GROOVE
BUTT JOINTS
WELD ALL AROUND SYMBOL
FIELD WELD SYMBOL
REFERENCE LINE
1/4
2@6
SIZE OF WELD IN INCHES
LENGTH OF WELD IN INCHES
SPACING OF WELDS IN INCHES
WELD JOINTS AND SYMBOLS

Figure 7.28

Example #6

The 4" × 1/2" plate of ASTM A36 steel shown in Figure 7.29 is stressed to its maximum allowable tension. Determine the size and length of weld required to connect it to the larger plate, using E60 electrodes.

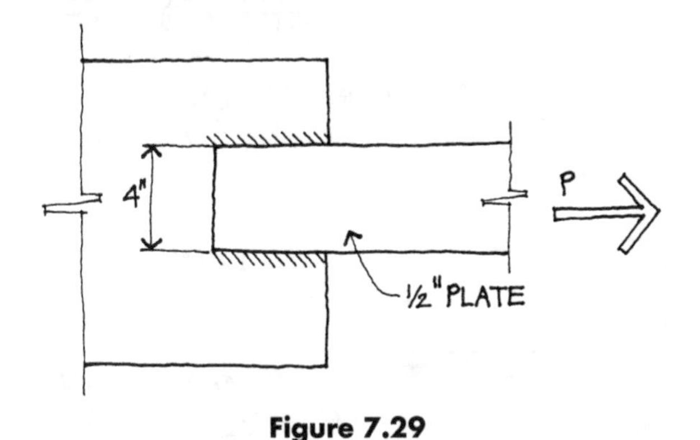

Figure 7.29

Solution:

We first calculate the allowable tension in the plate.

The allowable tensile stress = F_y/Ω = 36/1.67 = 21.6 ksi

Total allowable tension = 21.6 × 4 × 1/2 = 43.2 kips

The AISC Specification (Section J2.2b) specifies a maximum size of fillet weld of 1/16 inch less than the thickness of the material. Therefore, the largest fillet weld which may be applied to the 1/2 inch plate is 7/16 inch. The allowable load per inch of weld = 18.0 × 0.707 × 7/16 = 5.57 kips per inch.

Total required fillet weld length = 43.2/5.57 = 7.76"

We use *4" of 7/16" fillet weld along each edge of the plate*, making a total of 8". This also satisfies the requirement in the AISC Specification that the length must be at least equal to the perpendicular distance between welds, in this case four inches.

Simple Connections

There are numerous arrangements used for connecting steel framing members. Some of these are shown in Figures 7.30 and 7.31.

Example #7

A W21 × 62 beam is connected to a W27 × 94 girder with a framed beam connection consisting of 2 angles 4 × 3-1/2 × 3/8 with 5-3/4" diameter bolts in each leg. The beams are A992 steel and all connection parts are ASTM A36 steel. What load can be safely resisted by the connection using (1) ASTM A325 bolts slip-critical connection, (2) ASTM A325 bolts bearing-type connection, shear plane through threads, (3) ASTM A325 bolts bearing-type connection, threads excluded from shear plane.

Solution:

From page 1-18 of the AISC Manual, we read 0.400 inch as the web thickness of the W21 × 62 beam. The connection is shown on page 10-20 of the AISC Manual, which is reproduced in Figure 7.33.

1. ASTM A325 bolts slip-critical connection. From Figure 7.33, read value directly from table = 73.8 kips. The available strength of the web of the W21 × 62 must also be checked. This can be done by using the lower portion of Table 10-1 in Figure 7.33. The top flange of the W21 × 62 must be coped when the top flanges of the W21 × 62 and W27 × 94 are at the same elevation (see framed connection in Figure 7.30). Assuming a 1.25" by 1.75" cope, the available beam web strength is 216 kips per inch. For the 0.40-inch thick web, the available strength is 216 × 0.40 = 86.4 kips > 73.8 kips. Thus, safe load capacity = *73.8 kips*.

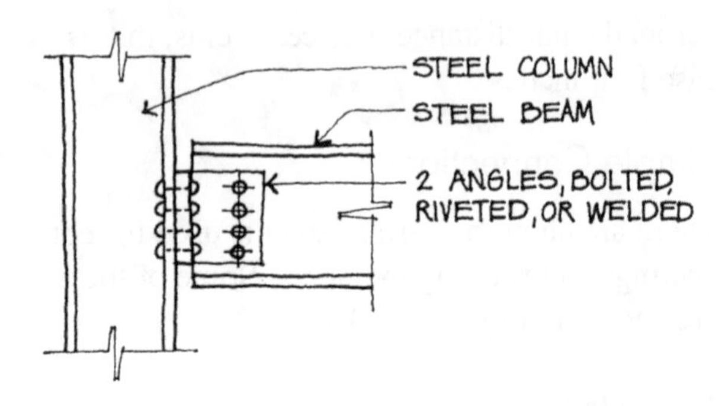

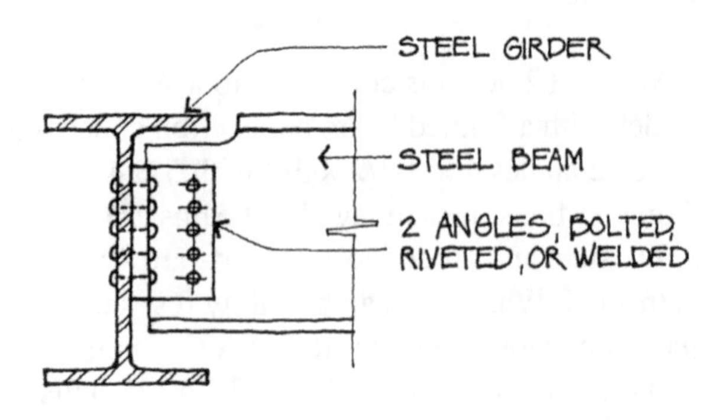

FRAMED CONNECTIONS
(SEE P. 10-13 THROUGH 10-45 OF AISC MANUAL)

Figure 7.30

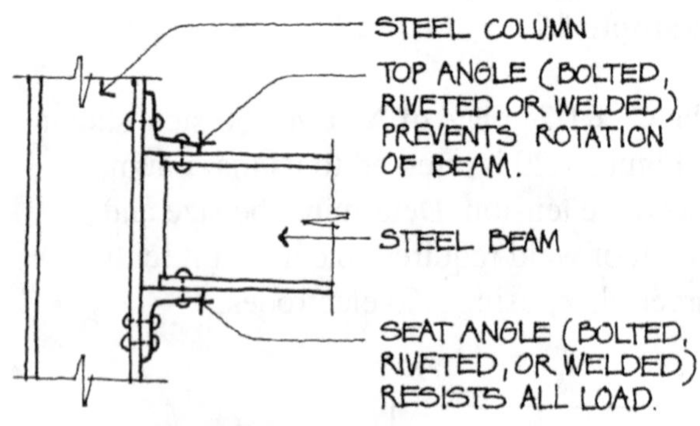

SEATED CONNECTION
(SEE P. 10-88 THROUGH 10-96 OF AISC MANUAL)

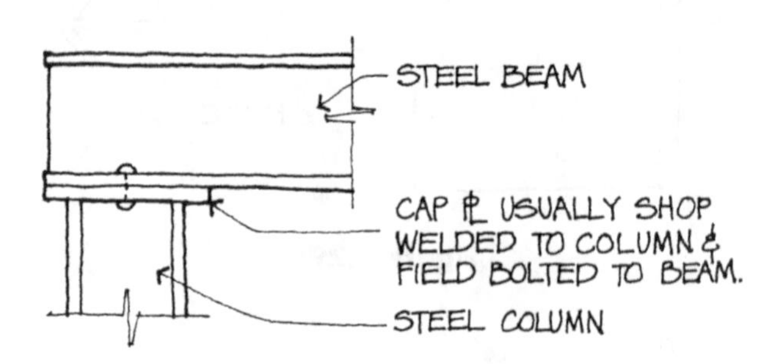

BEAM BEARING ON COLUMN

Figure 7.31

2. ASTM A325 bolts bearing-type connection, shear plane through threads. From Figure 7.33, for ASTM A325-N bolts, read value directly from table = 106 kips. Web strength is the same as case (1) = 86.4 kips < 106 kips. Therefore, safe load capacity = *86.4 kips.*
3. ASTM A325 bolts bearing-type connection, threads excluded. From Figure 7.33, for ASTM A325-X bolts, read value directly from table = 125 kips. Web strength is, again, 86.4 kips < 125 kips. Therefore, safe load capacity is equal to *86.4 kips*.

Moment Connections

The connections previously discussed are appropriate for the support of simple beams. However, we frequently use moment-resistant connections in steel construction, particularly for rigid frames and continuous beams. Moment-resistant connections are capable of resisting the required negative moment (tension at the top, compression at the bottom) using rivets, bolts, or welds. Two examples of moment-resistant connections are shown in Figure 7.32.

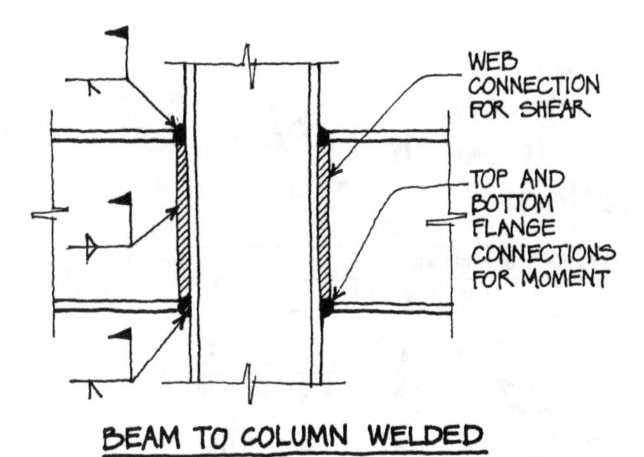
WEB
CONNECTION
FOR SHEAR
TOP AND
BOTTOM
FLANGE
CONNECTIONS
FOR MOMENT
BEAM TO COLUMN WELDED

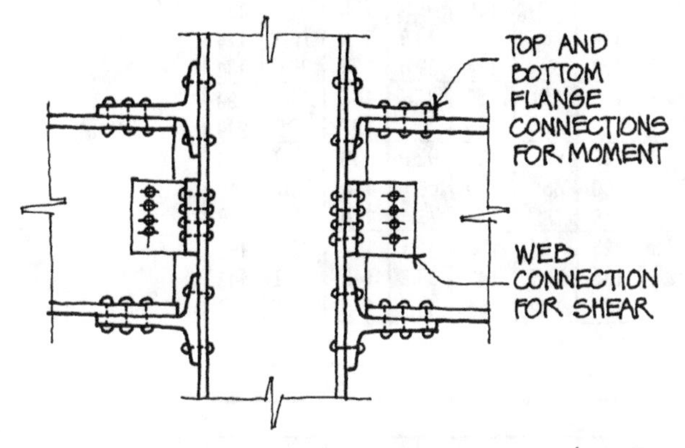
TOP AND
BOTTOM
FLANGE
CONNECTIONS
FOR MOMENT
WEB
CONNECTION
FOR SHEAR
BEAM TO COLUMN BOLTED OR RIVETED

Figure 7.32

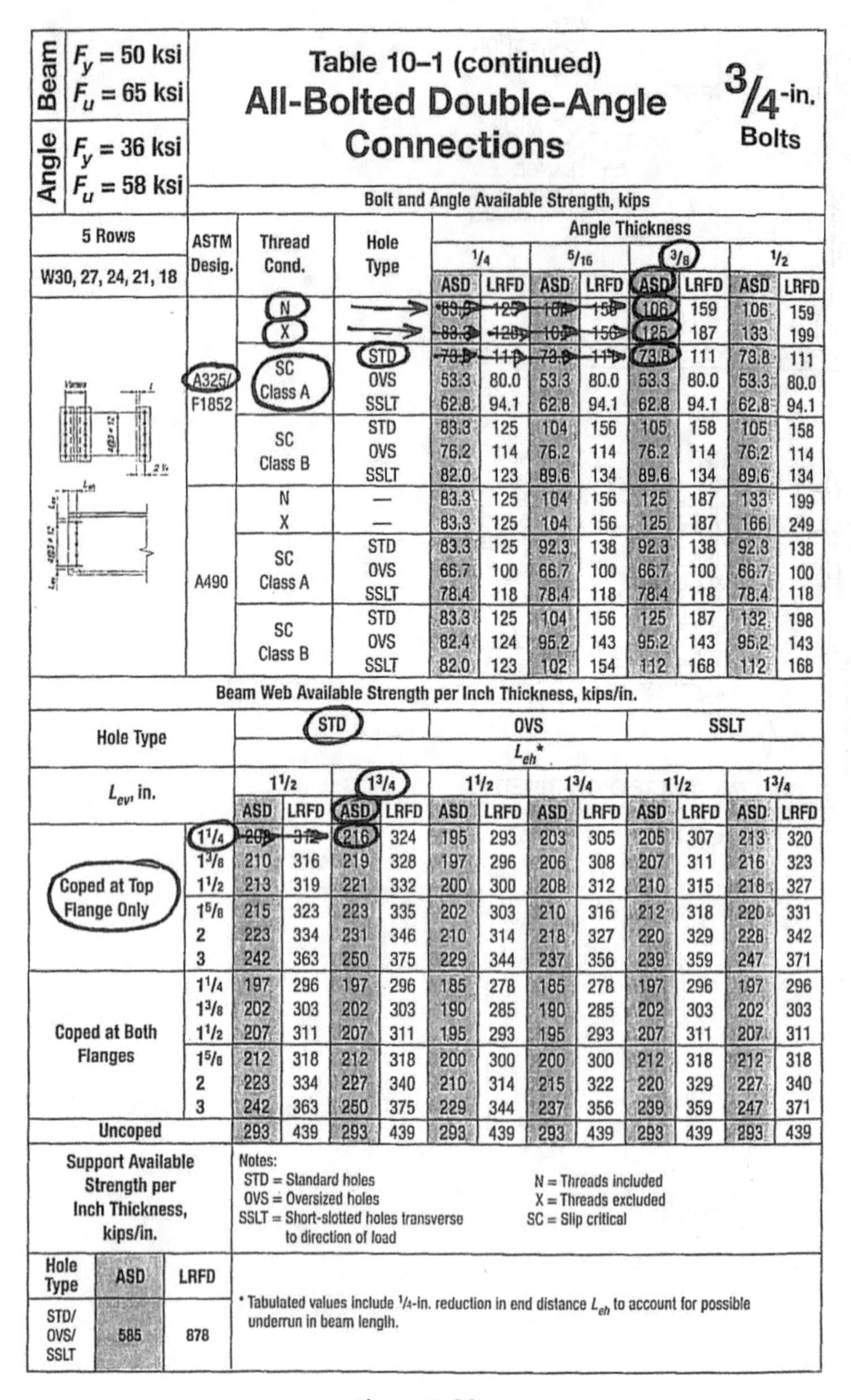

Beam F_y = 50 ksi, F_u = 65 ksi
Angle F_y = 36 ksi, F_u = 58 ksi

Table 10–1 (continued)
All-Bolted Double-Angle Connections
3/4-in. Bolts

Bolt and Angle Available Strength, kips

5 Rows W30, 27, 24, 21, 18	ASTM Desig.	Thread Cond.	Hole Type	1/4 ASD	1/4 LRFD	5/16 ASD	5/16 LRFD	3/8 ASD	3/8 LRFD	1/2 ASD	1/2 LRFD
	A325/ F1852	N	—	83.3	125	104	156	106	159	106	159
		X	—	83.3	125	104	156	125	187	133	199
		SC Class A	STD	73.8	111	73.8	111	73.8	111	73.8	111
			OVS	53.3	80.0	53.3	80.0	53.3	80.0	53.3	80.0
			SSLT	62.8	94.1	62.8	94.1	62.8	94.1	62.8	94.1
		SC Class B	STD	83.3	125	104	156	105	158	105	158
			OVS	76.2	114	76.2	114	76.2	114	76.2	114
			SSLT	82.0	123	89.6	134	89.6	134	89.6	134
	A490	N	—	83.3	125	104	156	125	187	133	199
		X	—	83.3	125	104	156	125	187	166	249
		SC Class A	STD	83.3	125	92.3	138	92.3	138	92.3	138
			OVS	66.7	100	66.7	100	66.7	100	66.7	100
			SSLT	78.4	118	78.4	118	78.4	118	78.4	118
		SC Class B	STD	83.3	125	104	156	125	187	132	198
			OVS	82.4	124	95.2	143	95.2	143	95.2	143
			SSLT	82.0	123	102	154	112	168	112	168

Beam Web Available Strength per Inch Thickness, kips/in.

Hole Type / L_{eh}*	L_{ev}, in.	STD 1½ ASD	STD 1½ LRFD	STD 1¾ ASD	STD 1¾ LRFD	OVS 1½ ASD	OVS 1½ LRFD	OVS 1¾ ASD	OVS 1¾ LRFD	SSLT 1½ ASD	SSLT 1½ LRFD	SSLT 1¾ ASD	SSLT 1¾ LRFD
Coped at Top Flange Only	1¼	208	312	216	324	195	293	203	305	205	307	213	320
	1⅜	210	316	219	328	197	296	206	308	207	311	216	323
	1½	213	319	221	332	200	300	208	312	210	315	218	327
	1⅝	215	323	223	335	202	303	210	316	212	318	220	331
	2	223	334	231	346	210	314	218	327	220	329	228	342
	3	242	363	250	375	229	344	237	356	239	359	247	371
Coped at Both Flanges	1¼	197	296	197	296	185	278	185	278	197	296	197	296
	1⅜	202	303	202	303	190	285	190	285	202	303	202	303
	1½	207	311	207	311	195	293	195	293	207	311	207	311
	1⅝	212	318	212	318	200	300	200	300	212	318	212	318
	2	223	334	227	340	210	314	215	322	220	329	227	340
	3	242	363	250	375	229	344	237	356	239	359	247	371
Uncoped		293	439	293	439	293	439	293	439	293	439	293	439

Support Available Strength per Inch Thickness, kips/in.

Hole Type	ASD	LRFD
STD/ OVS/ SSLT	585	878

Notes:
STD = Standard holes
OVS = Oversized holes
SSLT = Short-slotted holes transverse to direction of load
N = Threads included
X = Threads excluded
SC = Slip critical

* Tabulated values include 1/4-in. reduction in end distance L_{eh} to account for possible underrun in beam length.

Figure 7.33

LESSON 7 QUIZ

1. What are the factors to be considered in determining the allowable load for wood-to-wood mechanical fasteners?

   A. Lumber species

   B. Fastener spacing and edge and end distances

   C. Angle of load to grain and condition of loading

   D. All of the above

2. Two 3/4" steel plates are connected to a 1" plate using 3/4" bolts conforming to ASTM A325 in a slip-critical connection, as shown below. Assume bolt spacing and edge distance for full bearing strength. The plates are ASTM A36 steel. What is the allowable load P?

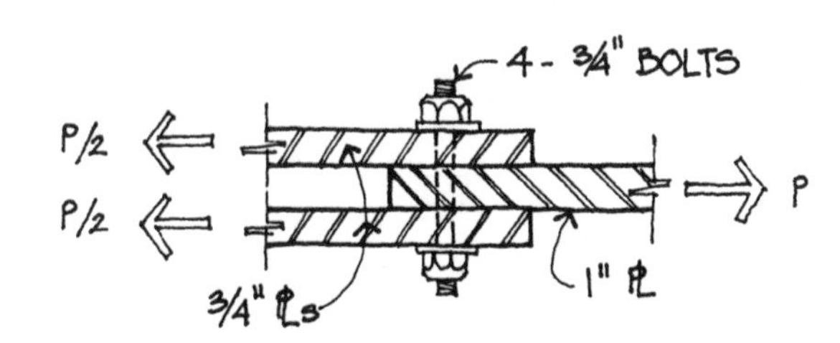

   A. 59.2 kips   **C.** 15.5 kips

   B. 30.8 kips   **D.** 65.3 kips

3. Which of the following welding symbols corresponds to the weld shown?

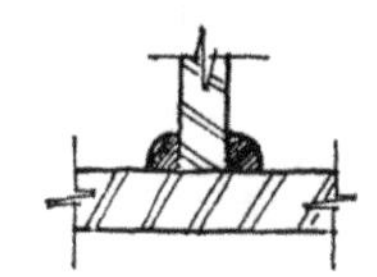

   A.   **C.**

   B.   **D.**

4. A connection is subject to vibration. Which of the following fasteners are appropriate?

   I. ASTM A325 bolts, slip-critical

   II. ASTM A325 bolts, bearing-type

   III. ASTM A307 bolts

   IV. ASTM A490 bolts, slip-critical

   V. ASTM A490 bolts, bearing-type

   A. I and IV   **C.** I, III, and IV

   B. II and V   **D.** I only

5. Which of the following fasteners has the greatest strength for wood-to-wood joints?

   I. Common bolt

   II. High-strength bolt

   III. Shear plates

   IV. Split ring

   V. Lag screw

   A. I and V   **C.** I, III, and IV

   B. II and IV   **D.** II and III

6. What is the MOST common fastener for steel-to-steel joints?

A. Common bolt

B. High-strength bolt

C. Rivet

D. Interference body bolt

7.

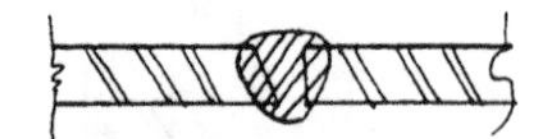

The type of welded joint shown above is a ___________________ joint.

8. A 4 × 10 is connected to 2-2 × 10 by 2-7/8" bolts as shown below. What is the safe load P? Use Table 11-F in Figure 7.9. All members are Douglas fir-larch, and assume that all spacings, distances, and members are adequate.

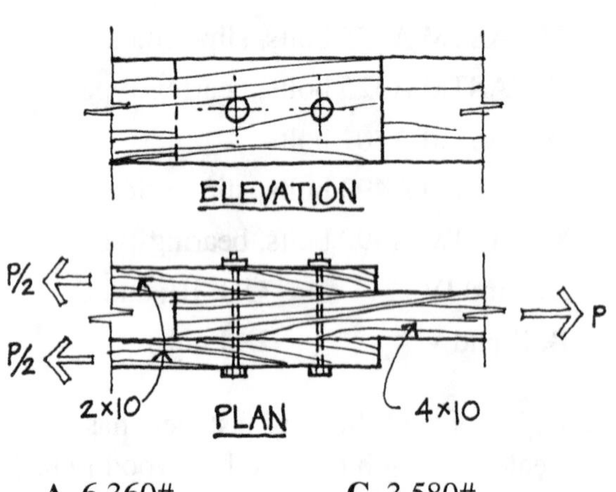

A. 6,360#

B. 3,320#

C. 3,580#

D. 2,780#

9. Two angles 3 × 3 × 3/8 of ASTM A36 steel are connected with 1/4" fillet welds as shown below, using E60 electrodes. What is the allowable load P?

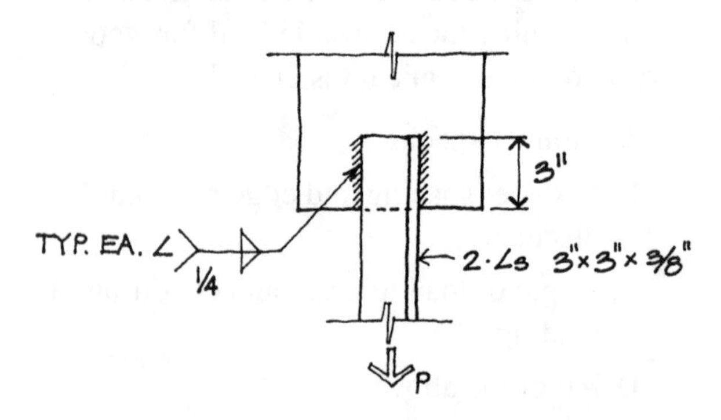

A. 22.26 kips

B. 44.52 kips

C. 19.08 kips

D. 38.16 kips

10. At what angle of load to grain do wood fasteners have the greatest strength?

A. 0°

B. 30°

C. 60°

D. 90°

FOUNDATIONS

INTRODUCTION

The foundation is that part of a building's structure that transmits the building's load to the underlying soil. The resulting compression of the soil causes settlement of the building. The goal of foundation design is to limit the settlement to some small amount that is tolerable. To accomplish this, the foundation must be placed within a soil stratum of adequate strength, and the load must be spread over a sufficiently large area so as not to exceed the soil's bearing capacity.

We also try to minimize *differential settlement* of the various parts of the building, and one way to achieve this is to design all the footings in a building for the same soil bearing pressure under dead load. In that way, every footing will settle about the same amount.

BEARING CAPACITY

How is the bearing capacity of a soil determined? There are a number of ways; for some buildings, the values given in the governing building code may be used. An example is IBC Table 1806.2 (see appendix).

For sizable structures, or where the nature of the underlying soil is not sufficiently known, a subsurface exploration of the site is usually made. This may be accomplished by digging *test pits*, or more commonly, by drilling *test borings*, which can be extended to a greater depth. Using either method, samples of the subsurface soils are obtained and examined and tested in the laboratory. The number of observations to be made (test pits or borings) is determined by the size and nature of the building, the size of the site, and the type of soil. A minimum of three or four observations is usually required, preferably located near the corners of the building. The distance between observations may be 25 to 150 feet for multi-story buildings, and 50 to 200 feet for one- or two-story buildings.

The depth of test pits is limited by the equipment available and usually does not exceed 10 to 15 feet.

The IBC requires that the average properties of the soil in the top 100 feet be determined in order to establish a site class definition at a building site. This site class, in turn, is used in determining seismic ground motion values. Thus, with some exceptions that are noted in the IBC, the depth of borings must be at least 100 feet at any site.

If a boring is unable to penetrate the subsoil, this may indicate that bedrock has been encountered. The bedrock profile may be roughly determined by a seismic exploration, which allows a large area to be investigated quickly. Elastic waves in the subsoil travel at different velocities in different materials, slower in soft soils and faster in rock. The seismic method uses a small explosive charge to produce an elastic wave, and sensing devices record the arrival time of the wave at a series of detectors. From this data, the depth of bedrock can be determined.

The location of each boring is shown on a plot plan, and a detailed log of each boring is made. A typical *soil boring log* is shown in Figure 8.1. You will notice that both the soil boring log and IBC Table 1806.2 use standard classification symbols (SM, SP, etc.). These symbols conform to the Unified Soil Classification System, which is widely used in this country and described in Figure 8.2.

Undisturbed samples from the borings are examined in the laboratory, and a number of different tests may be performed. These tests enable the soils engineer to determine various properties of the soil, including density, moisture content, and in the case of silts and clays, shear strength, bearing value, void ratio, liquid and plastic limits, and plasticity index. Based on the field exploration and laboratory tests, the soils engineer recommends the type of foundation to be used and the allowable soil bearing pressure. This pressure is the maximum unit load that can be placed on the underlying soil without causing excessive settlement or shear failure of the soil.

Sometimes load tests performed on the site are required to determine a soil's bearing capacity. Two such tests are the *plate load test*, used to determine the bearing capacity of a soil for static load on spread footings, and the *pile load test*, which is performed to check the design loads on piles.

In general, rock is the best bearing material, while sands and gravels are good materials, especially if they are compact and well-graded (containing a mixture of grains of various sizes). The fine-grained soils, such as silts and clays, may be satisfactory as foundation materials, but require careful investigation. Finally, organic soils, unconsolidated fills, and very soft or loose soils are generally unacceptable for the support of buildings.

If loose or soft soil is encountered at a site, it is often economical to excavate the deposit and replace it with properly compacted fill. Sometimes the on-site soils are acceptable as fill material; if not, soil from another location may be brought to the site and compacted in place. Sheepsfoot rollers are often used to compact soil in layers, and vibratory equipment is sometimes used to compact sandy soils. Where compaction is anticipated, typical soil samples should be obtained and subjected to a laboratory compaction test, known as the *Proctor*

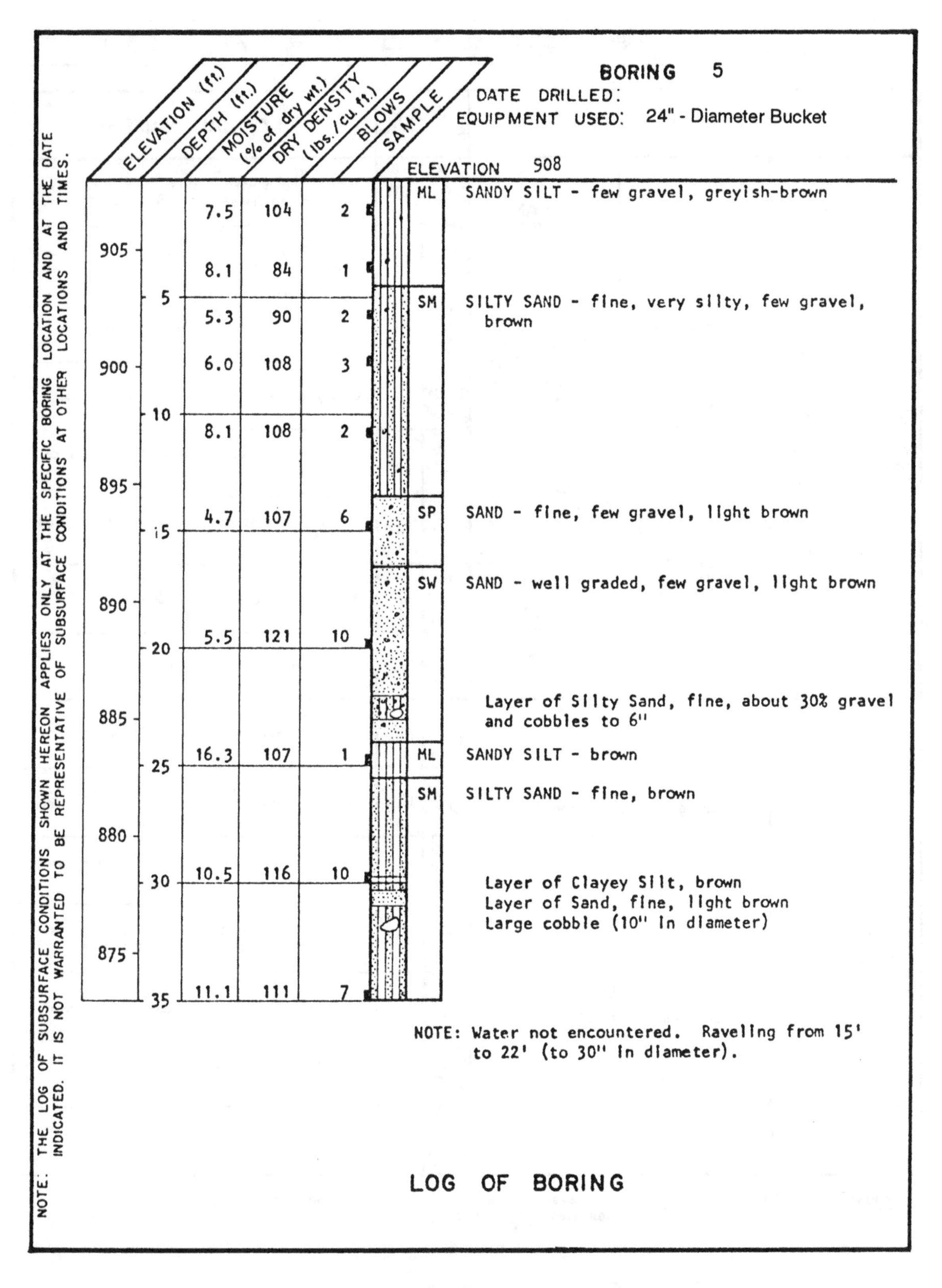
BORING 5
DATE DRILLED:
EQUIPMENT USED: 24" - Diameter Bucket
ELEVATION 908
ELEVATION (ft.)
DEPTH (ft.)
MOISTURE (% of dry wt.)
DRY DENSITY (lbs./cu.ft.)
BLOWS
SAMPLE
905
900
895
890
885
880
875
5
10
15
20
25
30
35
7.5 104 2
8.1 84 1
5.3 90 2
6.0 108 3
8.1 108 2
4.7 107 6
5.5 121 10
16.3 107 1
10.5 116 10
11.1 111 7
ML SANDY SILT - few gravel, greyish-brown
SM SILTY SAND - fine, very silty, few gravel, brown
SP SAND - fine, few gravel, light brown
SW SAND - well graded, few gravel, light brown
Layer of Silty Sand, fine, about 30% gravel and cobbles to 6"
ML SANDY SILT - brown
SM SILTY SAND - fine, brown
Layer of Clayey Silt, brown
Layer of Sand, fine, light brown
Large cobble (10" in diameter)
NOTE: Water not encountered. Raveling from 15' to 22' (to 30" in diameter).
NOTE: THE LOG OF SUBSURFACE CONDITIONS SHOWN HEREON APPLIES ONLY AT THE SPECIFIC BORING LOCATION AND AT THE DATE INDICATED. IT IS NOT WARRANTED TO BE REPRESENTATIVE OF SUBSURFACE CONDITIONS AT OTHER LOCATIONS AND TIMES.
LOG OF BORING

Figure 8.1

UNIFIED SOIL CLASSIFICATION SYSTEM

MAJOR DIVISIONS			GROUP SYMBOLS	TYPICAL NAMES
COARSE GRAINED SOILS (More than 50% of material is LARGER than No. 200 sieve size)	GRAVELS (More than 50% of coarse fraction is LARGER than the No. 4 sieve size)	CLEAN GRAVELS (Little or no fines)	GW	Well graded gravels, gravel-sand mixtures, little or no fines.
			GP	Poorly graded gravels or gravel-sand mixtures, little or no fines.
		GRAVELS WITH FINES (Appreciable amt. of fines)	GM	Silty gravels, gravel-sand-silt mixtures.
			GC	Clayey gravels, gravel-sand-clay mixtures.
	SANDS (More than 50% of coarse fraction is SMALLER than the No. 4 sieve size)	CLEAN SANDS (Little or no fines)	SW	Well graded sands, gravelly sands, little or no fines.
			SP	Poorly graded sands or gravelly sands, little or no fines.
		SANDS WITH FINES (Appreciable amt. of fines)	SM	Silty sands, sand-silt mixtures.
			SC	Clayey sands, sand-clay mixtures.
FINE GRAINED SOILS (More than 50% of material is SMALLER than No. 200 sieve size)	SILTS AND CLAYS (Liquid limit LESS than 50)		ML	Inorganic silts and very fine sands, rock flour, silty or clayey fine sands or clayey silts with slight plasticity.
			CL	Inorganic clays of low to medium plasticity, gravelly clays, sandy clays, silty clays, lean clays.
			OL	Organic silts and organic silty clays of low plasticity.
	SILTS AND CLAYS (Liquid limit GREATER than 50)		MH	Inorganic silts, micaceous or diatomaceous fine sandy or silty soils, elastic silts.
			CH	Inorganic clays of high plasticity, fat clays.
			OH	Organic clays of medium to high plasticity, organic silts.
HIGHLY ORGANIC SOILS			Pt	Peat and other highly organic soils.

BOUNDARY CLASSIFICATIONS: Soils possessing characteristics of two groups are designated by combinations of group symbols.

Figure 8.2

test, to determine the soil's optimum moisture content and density.

Specifications for compacted fill generally require that the fill be compacted to a density of between 90 and 100 percent of the optimum laboratory density, and that the moisture content at the time of compaction be within about 2 percent of the optimum laboratory moisture content. Close inspection and testing during the placing of the compacted fill is required to assure compliance with the specifications. Properly compacted fill is often suitable for the support of building footings, floor slabs, walks, and pavements.

CLIMATIC FACTORS

A number of factors other than bearing capacity may have to be considered in foundation design. Most of these factors are climatic in nature.

In areas of the country subject to frost, foundations may heave when moisture in the soil freezes. To minimize this possibility, it is usually advisable to place the foundation below the *frost line*, which is the deepest frost penetration expected in the area. The depth of the frost line varies, from zero in southern California and Florida to as much as 100" in northern Minnesota. Where frost is not a problem, the foundation is generally placed at least one or two feet below grade, to assure that the foundation is below the zone of disturbed soil, to increase the shear strength of the subsoil, and to prevent moisture infiltration under the building.

Another climatic effect occurs in some clay soils which swell when wet and shrink when dried. This *expansive soil* problem may be solved by placing the foundation below the depth of seasonal moisture change, so that the subsoil's moisture content remains relatively constant. Another solution is to use foundation piers that extend below the line of seasonal moisture change and that are insulated from the surrounding expansive soil.

The presence of *groundwater* may cause problems. The soil's bearing capacity may be reduced, water may leak into the building, and walls or slabs below the groundwater line may be subject to *hydrostatic pressure*—lateral and upward pressure caused by the water in the soil (see Figure 8.3). It may be possible to divert the water using a drainage system; if not, adequate waterproofing provisions should be made, and the structure should be designed to resist the hydrostatic pressure.

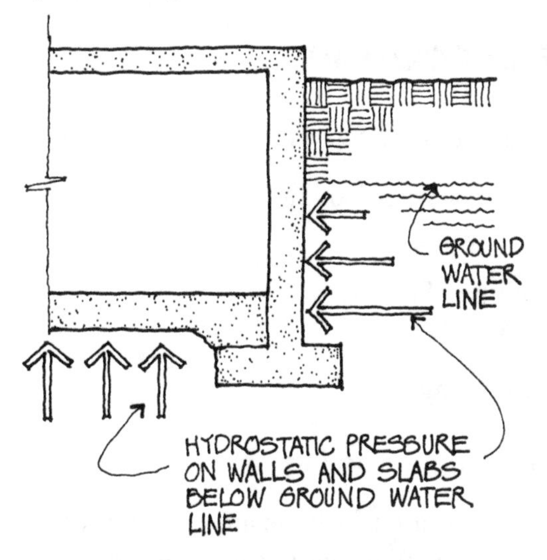

Figure 8.3

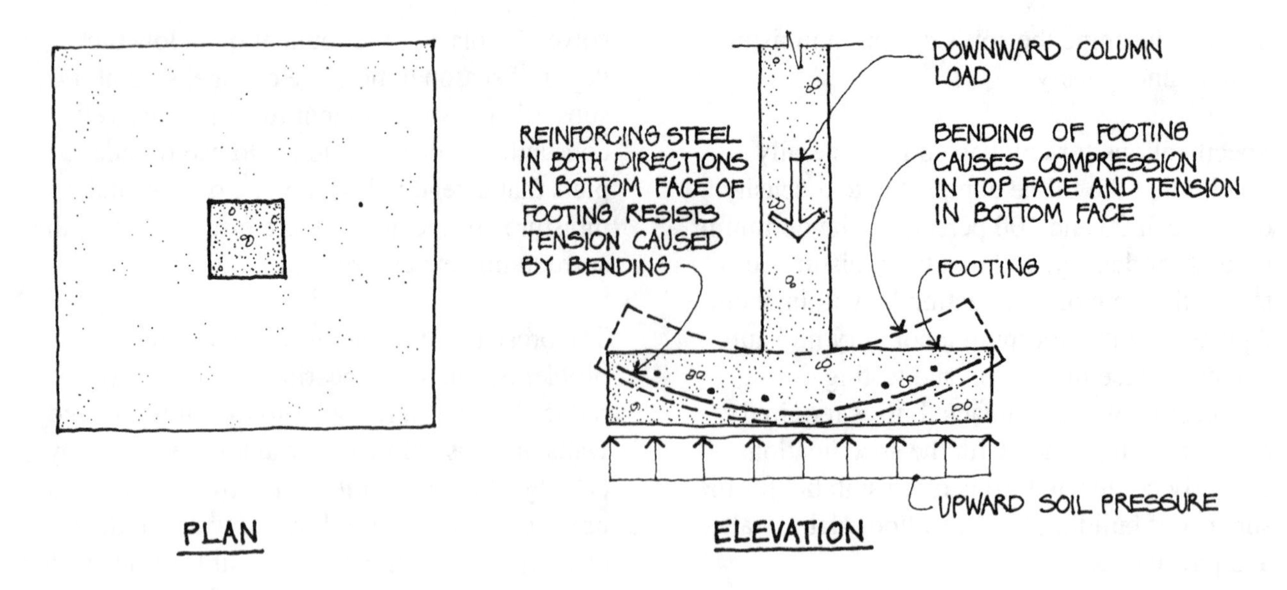

Figure 8.4

SPREAD FOOTINGS

There are five basic types of spread footings:

1. Single Column Footing
2. Wall Footing
3. Combined Footing
4. Cantilever Footing
5. Mat Footing

Single Column Footing

Most single column footings are square in plan, although rectangular footings are occasionally used (Figure 8.4). Usually, the column, or the base plate in the case of a steel column, bears directly on the concrete footing block. The required area of the footing is found by dividing the total load, including the estimated weight of the footing, by the bearing capacity of the soil.

In a single column footing, the downward load of the column is resisted by the soil pressure acting upward on the footing, which is usually assumed to be uniform over the entire footing area. The upward soil pressure causes the footing to cantilever in both directions from the face of the column, as shown in Figure 8.4. The resulting bending moment causes compression in the top face of the footing and tension in the bottom face. The footing is designed as a reinforced concrete member, with reinforcing steel in both directions near the bottom face resisting all the tension.

As with other reinforced concrete members, the footing must also be designed for shear. Since shear reinforcement is rarely used in footings, the effective depth d of the footing must be great enough so that the shear stress can be resisted by the concrete only. Two types of shear stress are investigated in single column footings: two-way shear as a slab and one-way shear as a beam. The critical section for two-way shear is located at a distance d/2 from the face of the column, while for one-way shear it is d from the column face (see Figure 8.5).

Example #1

A column is subject to a dead load of 200 kips and a live load of 150 kips. The allowable soil bearing pressure is 3,500 psf. What size footing should be used, assuming that the footing weighs 20 kips?

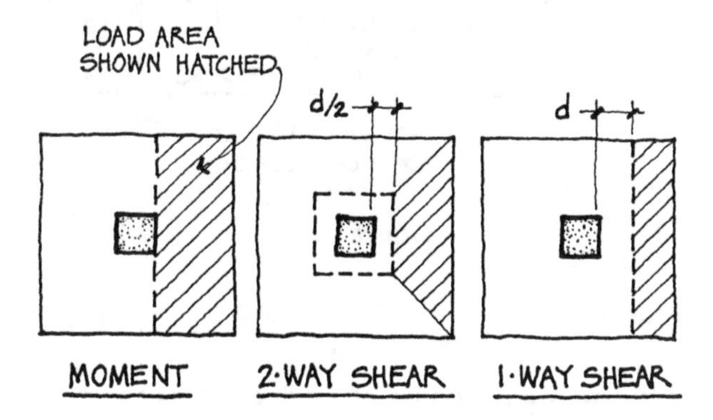

Figure 8.5

Solution:

Total load on footing = 200 + 150 + 20 = 370 kips

Required area of footing = 370 kips/3.5 ksf = 105.7 square feet

Use 10'-6" × 10'-6" footing (Area = 10.5 × 10.5 = 110.25 sq.ft.)

Wall Footing

A wall footing is a continuous spread footing supporting the vertical load on the wall, the weight of the wall itself, and the weight of the footing (Figure 8.6). The design is based on one lineal foot of the wall. The maximum bending moment is taken at the face of the wall, and the shear stress is investigated at a distance d from the face of the wall.

Example #2

A 16" concrete wall supports a dead load of 12,000# per lineal foot of wall and a live load of 6,000# per lineal foot. The allowable soil bearing pressure is 3,000 lbs. per square foot, and the footing is assumed to weigh 1,000# per lineal foot. What width footing should be used?

Solution:

Total load on footing = 12,000# + 6,000# + 1,000# = 19,000# per lin. ft.

Required width of footing = 19,000#/3,000 psf = 6.33 ft.

Use 6'-4" footing

Wall footings supporting light loads have low shears and moments, which can often be resisted by plain (unreinforced) concrete. Even so, it is good practice to provide some longitudinal reinforcing bars to help resist any longitudinal bending caused by non-uniform loading or non-uniform soil conditions. Such bars are usually one or two #4 or #5, located near the top of the foundation wall and near the bottom of the footing, as shown in Figure 8.6

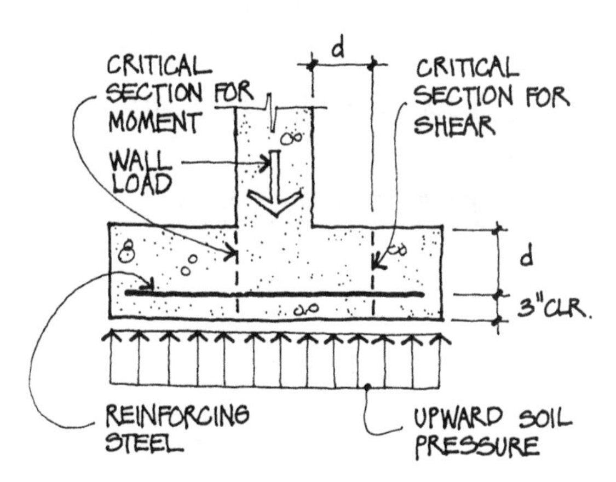

Figure 8.6

Combined Footing

A combined footing may be rectangular or trapezoidal in plan (Figure 8.7). It is used to support two or more columns that are so close together that single column footings for each would overlap, or where one column is too close to the property line to have a symmetrical footing. The resultant of the column loads should coincide with the centroid of the combined footing.

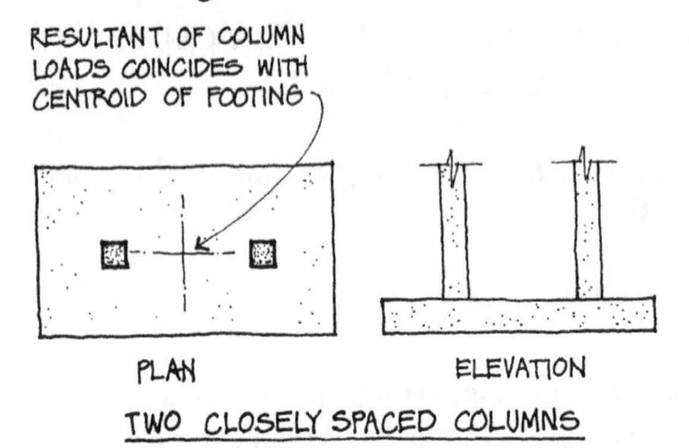

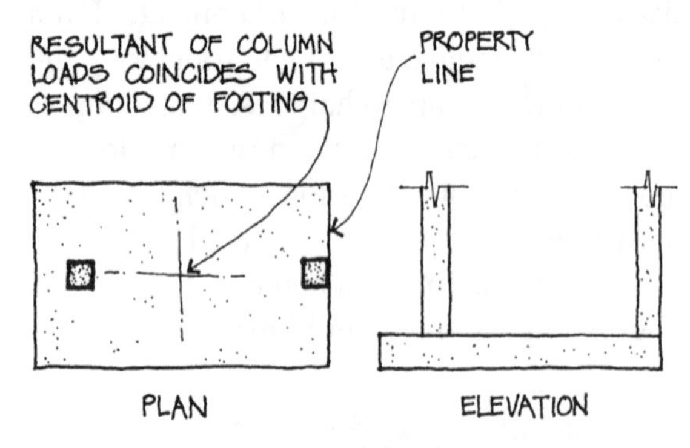

COMBINED FOOTINGS

Figure 8.7

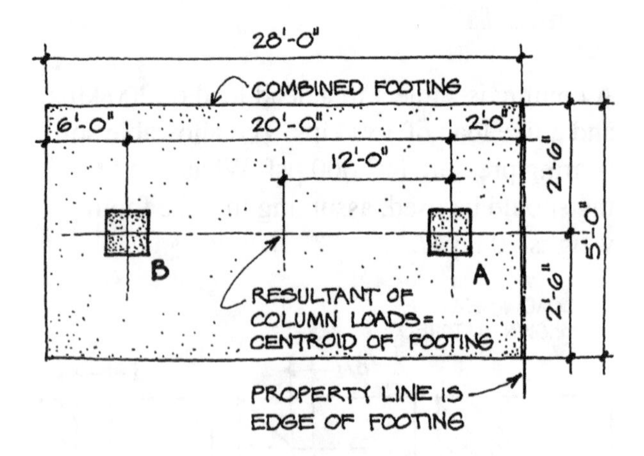

Figure 8.8

Example #3

Two 24" square columns are located as shown in Figure 8.8. Column A supports 120 kips dead load and 80 kips live load, and column B supports 180 kips dead load and 120 kips live load. The allowable soil bearing pressure is 4,000 psf. What size combined footing should be used? Assume a footing weight of 40 kips.

Solution:

Locate the resultant of the column loads by taking moments about A.

Distance of the resultant from column A

$$= \frac{P_B \times 20 \text{ ft.}}{P_B + P_A}$$

$$= \frac{(180 \text{ kips} + 120 \text{ kips})(20 \text{ ft.})}{(180 \text{ kips} + 120 \text{ kips}) + (120 \text{ kips} + 80 \text{ kips})}$$

= 12.0 feet from A

The centroid of the footing should coincide with the resultant of the column loads. If the footing extends to the property line, the length of the footing should be 2(12 ft. + 2 ft.) = 28 feet.

Required area of footing

$$= \frac{300 \text{ kips} + 200 \text{ kips} + 40 \text{ kips}}{4 \text{ ksf}}$$

= 135 sq.ft.

Required width of footing = 135/28 = 4.82 ft.

Use footing 28'-0" × 5'-0"

Cantilever Footing

Also known as a *strap footing*, a cantilever footing consists of an exterior column footing joined by a concrete beam to an interior column footing (Figure 8.9). It is used instead of a combined footing where the distance between exterior and interior columns is so great that a combined footing would have very high bending moments and therefore would be uneconomical.

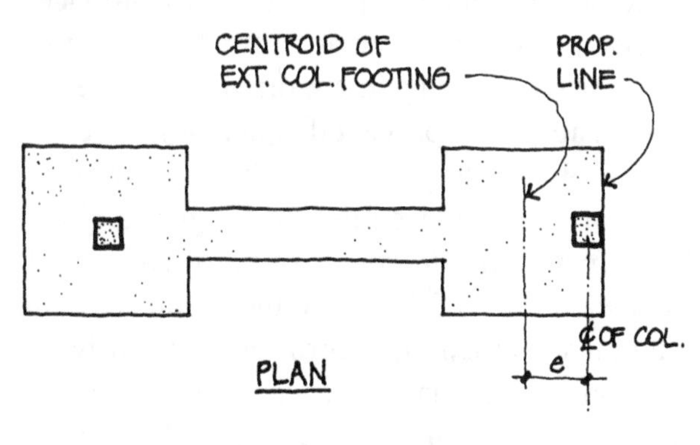

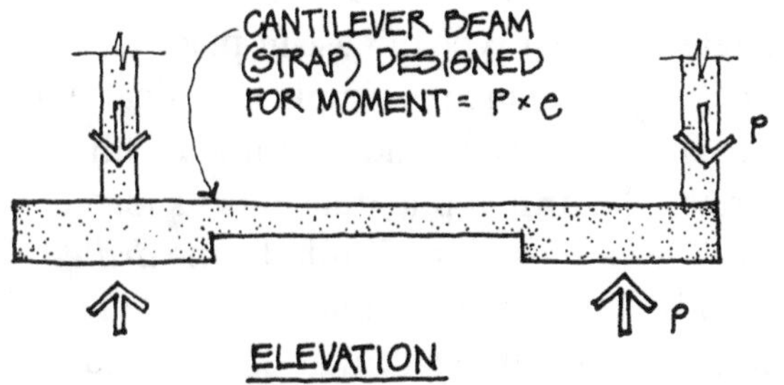

CANTILEVER FOOTING

Figure 8.9

Mat Foundation

The mat foundation is essentially one large footing under the entire building, which distributes the load over the entire building area (Figure 8.10). Also called a *raft foundation*, it is used when soil conditions are poor. The mat foundation is an inverted reinforced concrete structure, with the downward column loads applied at the top of the foundation resisted by the uniformly distributed soil pressure acting upward on the bottom of the foundation. The framing systems used are the same as those used for floor framing, such as flat slab, beam-and-girder, and flat plate.

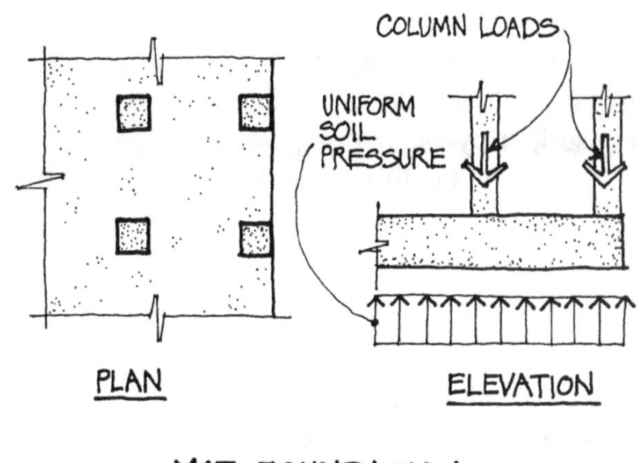

MAT FOUNDATION

Figure 8.10

PILE FOUNDATIONS

Introduction

If the upper soils have insufficient bearing capacity to support spread footings, the building loads are often transmitted to deeper, firmer soils by piles.

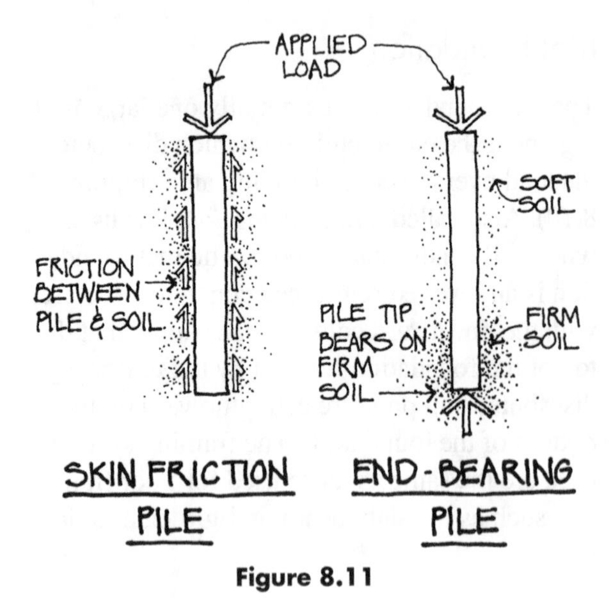

Figure 8.11

Piles may transfer the load to the soil by *skin friction* between the pile and the surrounding soil, or by *end bearing*, where the load is supported by the firm subsoil under the pile tip (Figure 8.11).

Pile Types

Piles may be made of timber, concrete, steel, or a composite of two materials. Timber piles must be treated with creosote or a similar preservative, unless they are permanently below the water table. They are usually round in cross-section, 20 to 40 feet long, with a capacity of about 30 to 50 tons, which they support by skin friction.

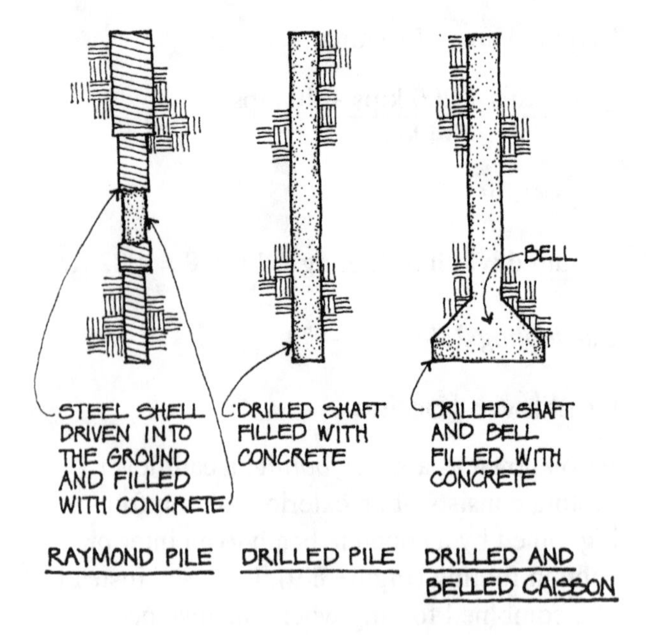

Figure 8.12

Concrete piles may be precast or cast-in-place. Precast piles may be square, round, or octagonal in cross-section and are often prestressed (pretensioned) to prevent damage or failure caused by handling stresses. There are several types of cast-in-place piles (see Figure 8.12). One type of pile (*Raymond*) employs a tapered steel shell that is driven into the ground with an internal steel mandrel (core). When the driving is complete, the mandrel is removed and the steel shell is left in the ground and filled with concrete. Another type of cast-in-place pile is the *drilled pile*, which is simply a vertical shaft drilled into the ground and then filled with concrete. The drilled pile transmits its load to the soil by skin friction. Similarly, the *drilled caisson* is an end-bearing pile constructed by pouring concrete into a drilled shaft. Frequently the bottom of the shaft is enlarged, or *belled*, to form a larger bearing area. Drilled piles and caissons sometimes require a steel casing, which is removed as the concrete is poured. The advantages of concrete over timber piles are greater durability and capacity.

Example #4

A column load of 200 kips is to be supported on four drilled piles 18 inches in diameter. If the frictional resistance of each pile is 500 pounds per square foot, how deep should the piles extend?

Solution:

Each pile supports a load of 200 kips/4 = 50 kips. Required frictional area = $\frac{50 \text{ kips}}{0.500 \text{ ksf}}$ = 100 sq.ft.

Circumference of the pile = $\pi d = \pi(1.5 \text{ ft.})$ = 4.71 ft.

Required pile length = 100 sq.ft./4.71 ft. = 21.23 ft., which we round up to *22'-0"*.

Example #5

The column of Example #4 is to be supported on a drilled and belled caisson. The allowable soil bearing value is 4,000 pounds per square foot. What should the bell diameter be?

Solution:

Required bearing area = 200 kips/4.0 ksf = 50.0 sq.ft.

Area of bell = 50.0 = $\pi d^2/4$

$$d^2 = \frac{4(50.0)}{\pi} = 63.66 \text{ ft.}^2$$

$$d = \sqrt{63.66} = 7.98 \text{ ft.}$$

Use 8'0" diameter bell

Steel piles are usually H-shaped rolled sections, which may be encased in concrete, or pipe, which is filled with concrete. The advantages of steel piles include durability, availability of long lengths, ease of splicing, and high capacity. Steel piles may be end-bearing, friction, or a combination of both.

All of the pile types described above, except for the drilled pile and drilled caisson, are driven into place with pile-driving hammers. Drop hammers are relatively slow and are frequently used on smaller jobs, while large projects often employ hammers operated by steam, compressed air, or diesel engines.

Bearing Capacity of Piles

The bearing capacity of driven piles may be determined by pile-driving formulas, static formulas, or pile load tests.

Although there are several pile-driving formulas, the *Engineering News formula* is probably used more often than any other. For a drop hammer, this formula is P = 12 WH/s + 1. Using a factor of safety of 6, the formula becomes $P = \frac{2WH}{s + 1}$

where

P = bearing capacity of pile, in pounds

W= weight of hammer, in pounds

H = drop of hammer, in feet

s = pile penetration under the final blow, in inches

Experience has shown that pile-driving formulas are not always accurate, particularly for piles driven into clay. The formulas must therefore be used with caution and judgment.

Static pile formulas establish the bearing capacity of piles from the properties of the soil and the pile dimensions. Here, too, the formulas do not always give reliable results.

The most dependable method of determining the bearing capacity of piles is the *pile load test*. Although pile load tests are often made to *verify* the design loads on piles, if made early enough they can *determine* the allowable pile loads to be used in the design. The load on the test pile is applied in increments of about 25 percent of the design load up to twice the design load and then reduced in similar increments to zero. Each load is maintained constant and the settlement measured at regular intervals until the rate of settlement becomes negligible. The test results are then plotted and used to determine the safe capacity of the pile.

Piles are frequently driven in groups. The capacity of a group of piles is usually less than the sum of the capacities of the individual piles.

Foundations on Piles

Individual building columns are generally supported by a group, or cluster, of piles. Between the column base and the piles, a thick reinforced concrete *pile cap* is placed, which distributes the column load to all the piles in the cluster. The piles are usually spaced about three feet apart. A typical arrangement for a column supported on four piles is shown in Figure 8.13.

The downward vertical load of the column is resisted by the upward reactions of the piles, and the pile cap is designed to resist the resulting shears and moments.

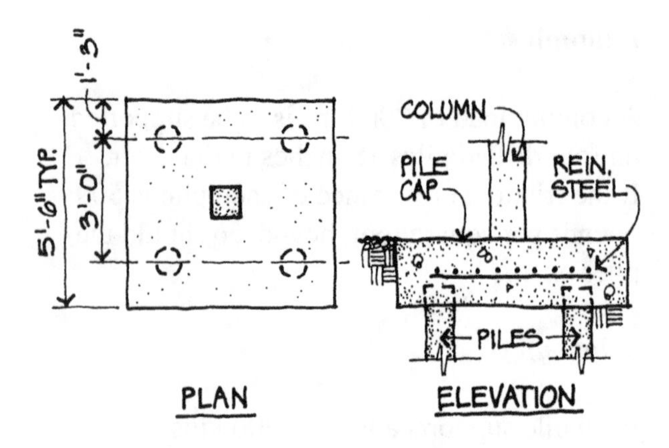

Figure 8.13

Piles supporting walls are generally spaced at least three feet apart and embedded in a reinforced concrete *grade beam* as shown in Figure 8.14. The grade beam is designed to support the load of the wall above and span between the piles.

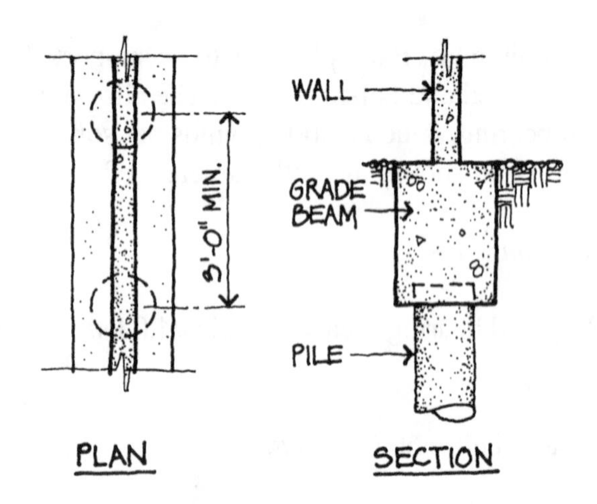

Figure 8.14

LESSON 8 QUIZ

1. Which of the following soil types are unacceptable for the support of a building?

I. Compacted fill
II. Fine-grained soil
III. Organic soil
IV. Unconsolidated fill

A. I and IV
B. II, III, and IV
C. III only
D. III and IV

2. What type of foundation is generally square in plan, with the reinforcing steel near its bottom face?

A. Strap footing
B. Single column footing
C. Mat foundation
D. Cantilever footing

3. What do we attempt to minimize in designing building foundations?

A. Bending moment
B. Differential settlement
C. Shear
D. The number of test borings

4. What test or procedure often forms the basis for the determination of a soil's bearing capacity?

A. Compaction test
B. Proctor test
C. Seismic exploration
D. Test boring

5. Select the CORRECT statements about pile foundations.

A. Timber piles are not treated with preservatives unless they are permanently below the water table.
B. Drilled piles transmit their load to the soil by skin friction.
C. Belled caissons transmit their load to the soil by end bearing.
D. Wood piles have a capacity of 30 to 50 tons and are usually end-bearing.
E. Through extensive use and field testing, pile driving formulas have proven to be very accurate.

6. Single column footings

A. are stressed in tension in the top face and compression in the bottom face.
B. are designed as simple beams.
C. are never governed by shear capacity.
D. have flexural reinforcing in two directions in the bottom face.

7. Two 18" square columns are located as shown below. Column A supports 200 kips and column B supports 245 kips. Allowable soil bearing pressure is 4,000 psf. Neglecting the weight of the columns, determine the size of a combined rectangular footing assuming a footing weight of 40 kips.

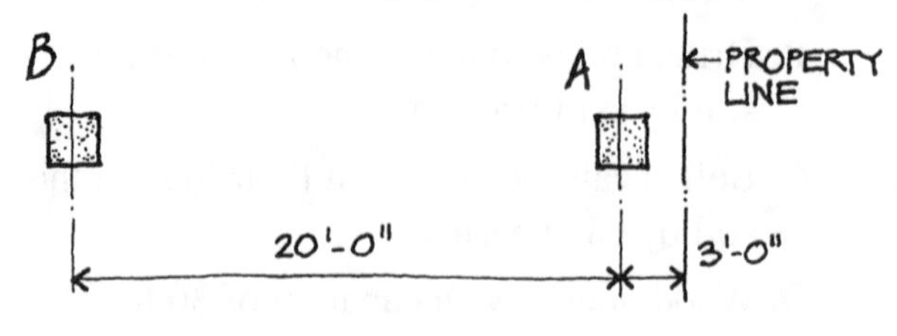

A. 28'-0" × 4'-4"
B. 22'-0" × 5'-6"
C. 26'-0" × 4'-4"
D. 30'-0" × 4'-0"

8. The calculations for a square column footing show that the footing is overstressed in shear. What is the BEST way to correct this situation?

A. Provide shear reinforcing.

B. Increase the thickness of the footing.

C. Provide additional flexural reinforcement.

D. Decrease the column load.

9.

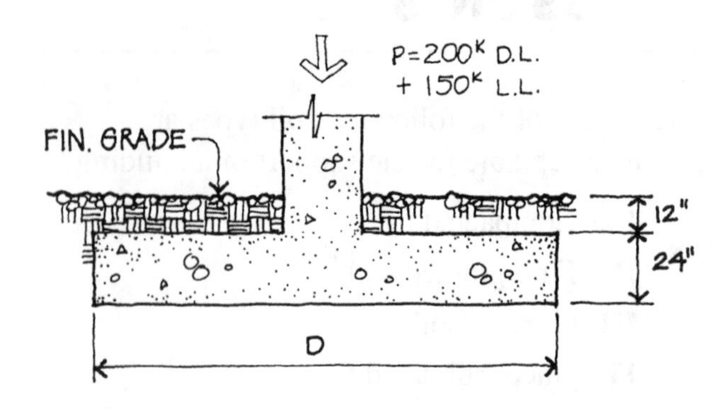

What should be dimension D of the square footing shown above? The soil weight is 100 pounds per cubic foot, the concrete weight is 150 pounds per cubic foot, and the soil bearing value is 5,000 pounds per square foot.

A. 8'-9"
B. 8'-5"
C. 7'-8"
D. 9'-4"

10. Soil borings indicate that a site has four to five feet of loose fill. Select the MOST CORRECT statement.

A. If the fill is removed and recompacted, it may be used to support floor slabs, walks, and pavements, but not building footings.

B. The fill may be used to support a building if its surface is properly compacted with a heavy sheepsfoot roller.

C. If the fill is removed and recompacted, it may be used to support building footings.

D. The foundation must penetrate the fill, whether or not it is compacted, and bear on the firmer underlying soil.

LESSON NINE

9

CONVENTIONAL STRUCTURAL SYSTEMS

BUILDING LOADS

The structure of every building must be designed to safely resist all the forces that are anticipated to act on the structure during its lifetime. These forces include dead and live loads, which act vertically; wind and earthquake loads, which are primarily horizontal but may act in other directions as well; and miscellaneous loads, including those caused by earth pressure, water pressure, and temperature change.

Dead Load

Dead load is the vertical load due to the weight of all permanent structural and non-structural components of a building, such as walls, floors, roofs, and fixed service equipment. Since dead load is caused by gravity, it always acts vertically. Dead load is constant, since a building's weight does not vary, and its magnitude can be determined quite accurately. A candidate should know the approximate weight of some of the common building materials, including the following:

Composition and gravel roofing	6.0 psf
1" nominal wood sheathing	2.5 psf
2" nominal wood sheathing	5.0 psf
Plywood (per in. of thickness)	3.0 psf

Lumber	35 pcf
Lath and plaster ceiling	8.0 psf
Wood stud wall, plastered two sides	16.0 psf
Concrete	150 pcf
Concrete (per in. of thickness)	12.5 psf
Brick	120 pcf
Brick (per in. of thickness)	10.0 psf
Structural Steel	See AISC Manual

For other materials, one should refer to reference books such as the AISC Manual.

Live Load

Live load is the load superimposed by the use and occupancy of the building, not including the wind load, earthquake load, or dead load. It includes the weight of people, furniture, supplies, etc., that are supported by the building's structure.

The live load varies as people move around, furniture is rearranged, and file cabinets are filled or emptied. The live load also has certain dynamic characteristics: in a light frame building, for example, you can sometimes feel the footsteps of a person walking. Nevertheless, the live load used in the design of a structure is usually assumed to be a static vertical load, whose magnitude is determined from the building code.

The minimum live loads in pounds per square foot required by the building code are reproduced in Figure 9.1. Table 1607.1 applies to floor and roof live loads.

Floors in buildings where partition locations are subject to change must be designed to support a load of 15 pounds per square foot, in addition to all other loads. The IBC permits floor live loads to be reduced by either a general or alternate method. In the alternate floor live load reduction method, the design live loads on any member supporting more than 150 square feet can be reduced in accordance with the formula $R = 0.08(A - 150)$

where

R = reduction in percent

A = area of floor supported by the member in square feet

The live load reduction must not exceed 40 percent for horizontal members, 60 percent for vertical members, or R as determined from the formula

$$R = 23.1(1 + D/L_o)$$

where

D = dead load per square foot supported by the member

L_o = unreduced design live per square foot supported by the member

It is important to note that a reduction in floor live loads is not permitted (1) in places of public assembly, (2) where the live load exceeds 100 psf (except that the live loads for members supporting two or more floors may be reduced by a maximum of 20 percent), and (3) in passenger vehicle parking garages (except that the live loads for members supporting two or more floors may be reduced by a maximum of 20 percent).

Example #1

What live load reduction is permitted for a floor member supporting 300 square feet, if the dead load is 80 pounds per square foot and the live load is 100 pounds per square foot?

TABLE 1607.1
MINIMUM UNIFORMLY DISTRIBUTED LIVE LOADS AND MINIMUM CONCENTRATED LIVE LOADS[g]

OCCUPANCY OR USE	UNIFORM (psf)	CONCENTRATED (lbs.)
1. Apartments (see residential)	—	—
2. Access floor systems Office use Computer use	 50 100	 2,000 2,000
3. Armories and drill rooms	150	—
4. Assembly areas and theaters Fixed seats (fastened to floor) Follow spot, projections and control rooms Lobbies Movable seats Stages and platforms	 60 50 100 100 125	—
5. Balconies On one- and two-family residences only, and not exceeding 100 sq ft	100 60	—
6. Bowling alleys	75	—
7. Catwalks	40	300
8. Dance halls and ballrooms	100	—
9. Decks	Same as occupancy served[h]	—
10. Dining rooms and restaurants	100	—
11. Dwellings (see residential)	—	—
12. Cornices	60	—
13. Corridors, except as otherwise indicated	100	—
14. Elevator machine room grating (on area of 4 in²)	—	300
15. Finish light floor plate construction (on area of 1 in²)	— —	200
16. Fire escapes On single-family dwellings only	100 40	—
17. Garages (passenger vehicles only) Trucks and buses	40 See Section 1607.6	Note a
18. Grandstands (see stadium and arena bleachers)	—	—
19. Gymnasiums, main floors and balconies	100	—
20. Handrails, guards and grab bars	See Section 1607.7	
21. Hospitals Corridors above first floor Operating rooms, laboratories Patient rooms	 80 60 40	 1,000 1,000 1,000
22. Hotels (see residential)	—	—
23. Libraries Corridors above first floor Reading rooms Stack rooms	 80 60 150[b]	 1,000 1,000 1,000
24. Manufacturing Heavy Light	 250 125	 3,000 2,000
25. Marquees	75	—
26. Office buildings Corridors above first floor File and computer rooms shall be designed for heavier loads based on anticipated occupancy Lobbies and first-floor corridors Offices	 80 — 100 50	 2,000 — 2,000 2,000
27. Penal institutions Cell blocks Corridors	 40 100	—
28. Residential One- and two-family dwellings Uninhabitable attics without storage[i] Uninhabitable attics with limited storage[i, j, k] Habitable attics and sleeping areas All other areas except balconies and decks Hotels and multiple-family dwellings Private rooms and corridors serving them Public rooms and corridors serving them	 10 20 30 40 40 100	—
29. Reviewing stands, grandstands and bleachers	Note c	
30. Roofs All roof surfaces subject to maintenance workers Awnings and canopies Fabric construction supported by a lightweight rigid skeleton structure All other construction Ordinary flat, pitched, and curved roofs Primary roof members, exposed to a work floor Single panel point of lower chord of roof trusses or any point along primary structural members supporting roofs: Over manufacturing, storage warehouses, and repair garages All other occupancies Roofs used for other special purposes Roofs used for promenade purposes Roofs used for roof gardens or assembly purposes	 5 nonreduceable 20 20 Note l 60 100	 300 2,000 300 Note l

Figure 9.1

TABLE 1607.1—continued
MINIMUM UNIFORMLY DISTRIBUTED LIVE LOADS AND MINIMUM CONCENTRATED LIVE LOADS[g]

OCCUPANCY OR USE	UNIFORM (psf)	CONCENTRATED (lbs.)
31. Schools Classrooms Corridors above first floor First-floor corridors	 40 80 100	 1,000 1,000 1,000
32. Scuttles, skylight ribs and accessible ceilings	—	200
33. Sidewalks, vehicular driveways and yards, subject to trucking	250[d]	8,000[e]
34. Skating rinks	100	—
35. Stadiums and arenas Bleachers Fixed seats (fastened to floor)	 100[c] 60[c]	—
36. Stairs and exits One- and two-family dwellings All other	 40 100	Note f
37. Storage warehouses (shall be designed for heavier loads if required for anticipated storage) Heavy Light	 250 125	
38. Stores Retail First floor Upper floors Wholesale, all floors	 100 75 125	 1,000 1,000 1,000
39. Vehicle barriers	See Section 1607.7.3	
40. Walkways and elevated platforms (other than exitways)	60	—
41. Yards and terraces, pedestrians	100	—

For SI: 1 inch = 25.4 mm, 1 square inch = 645.16 mm²,
1 square foot = 0.0929 m²,
1 pound per square foot = 0.0479 kN/m², 1 pound = 0.004448 kN,
1 pound per cubic foot = 16 kg/m³

a. Floors in garages or portions of buildings used for the storage of motor vehicles shall be designed for the uniformly distributed live loads of Table 1607.1 or the following concentrated loads: (1) for garages restricted to vehicles accommodating not more than nine passengers, 3,000 pounds acting on an area of 4.5 inches by 4.5 inches; (2) for mechanical parking structures without slab or deck which are used for storing passenger vehicles only, 2,250 pounds per wheel.

b. The loading applies to stack room floors that support nonmobile, double-faced library bookstacks, subject to the following limitations:
1. The nominal bookstack unit height shall not exceed 90 inches;
2. The nominal shelf depth shall not exceed 12 inches for each face; and
3. Parallel rows of double-faced bookstacks shall be separated by aisles not less than 36 inches wide.

c. Design in accordance with the ICC *Standard on Bleachers, Folding and Telescopic Seating and Grandstands.*

d. Other uniform loads in accordance with an approved method which contains provisions for truck loadings shall also be considered where appropriate.

e. The concentrated wheel load shall be applied on an area of 20 square inches.

f. Minimum concentrated load on stair treads (on area of 4 square inches) is 300 pounds.

g. Where snow loads occur that are in excess of the design conditions, the structure shall be designed to support the loads due to the increased loads caused by drift buildup or a greater snow design determined by the building official (see Section 1608). For special-purpose roofs, see Section 1607.11.2.2.

h. See Section 1604.8.3 for decks attached to exterior walls.

i. Attics without storage are those where the maximum clear height between the joist and rafter is less than 42 inches, or where there are not two or more adjacent trusses with the same web configuration capable of containing a rectangle 42 inches high by 2 feet wide, or greater, located within the plane of the truss. For attics without storage, this live load need not be assumed to act concurrently with any other live load requirements.

j. For attics with limited storage and constructed with trusses, this live load need only be applied to those portions of the bottom chord where there are two or more adjacent trusses with the same web configuration capable of containing a rectangle 42 inches high by 2 feet wide or greater, located within the plane of the truss. The rectangle shall fit between the top of the bottom chord and the bottom chord of any other truss member, provided that each of the following criteria is met:
i. The attic area is accessible by a pull-down stairway or framed opening in accordance with Section 1209.2, and
ii. The truss shall have a bottom chord pitch less than 2:12.
iii. Bottom chords of trusses shall be designed for the greater of actual imposed dead load or 10 psf, uniformly distributed over the entire span.

k. Attic spaces served by a fixed stair shall be designed to support the minimum live load specified for habitable attics and sleeping rooms.

l. Roofs used for other special purposes shall be designed for appropriate loads as approved by the building official.

Figure 9.1 (continued)

Solution:

The area supported exceeds 150 square feet and the live load does not exceed 100 psf; therefore a live load reduction is permitted.

1. The live load reduction R = 0.08 (A – 150) = 0.08(300 – 150) = 12%
2. The maximum allowable live load reduction R = 40.0% for horizontal members, or
3. $R = 23.1\ (1 + D/L_o)$
   $= 23.1\ (1 + 80/100)$
   $= 41.58\%$

The live load reduction permitted is therefore 12.0 *percent*. Thus, the design live load may be reduced from 100 psf to 100 – 0.12(100) = 88 psf.

The IBC permits the uniformly distributed roof live loads L_o in Table 1607.1 to be reduced according to the formula $L_r = L_o R_1 R_2$ where L_r = reduced roof live load per square foot of horizontal projection in pounds per square foot and R_1 and R_2 are reduction factors that depend on the tributary area supported by the roof member (A_t) and the roof type (flat, pitched, and curved). In no case shall L_r be taken less than 12 psf after both reduction factors have been applied.

The reduction factor R_1 is equal to 1.0 for tributary areas A_t less than or equal to 200 square feet. For A_t greater than 200 square feet but less than or equal to 600 square feet, R_1 is determined by the formula $R_1 = 1.2 - 0.001A_t$. Finally, for A_t greater than 600 square feet, R_1 is equal to 0.6.

The reduction factor R_2 is equal to 1.0 for F less than or equal to 4, where F is the number of inches of rise per foot for a sloped roof or is the rise-to-span ratio multiplied by 32 for an arch or domed roof. Where F is greater than 4 but less than or equal to 12, R_2 is determined by the formula $R_2 = 1.2 - 0.05F$. Finally, R_2 is equal to 0.6 for F greater than 12.

The IBC also gives live load provisions for special-purpose roofs (roof gardens, etc.), landscaped roofs, and awnings and canopies.

Example #2

What live load reduction is permitted for an ordinary roof member supporting 350 square feet, if the dead load is 20 pounds per square foot and the roof pitch is 5 in 12?

Solution:

From Table 1607.1 in Figure 9.1, the roof live load L_o for an ordinary flat, pitched, or curved roof is equal to 20 psf.

Since the tributary area A_t is equal to 350 square feet, the reduction factor R_1 is determined by the following formula:

$$R_1 = 1.2 - 0.001A_t = 1.2 - 0.001(350) = 0.85$$

With a roof pitch of 5 in 12, F = 5. Thus, reduction factor $R_2 = 1.2 - 0.05F = 1.2 - 0.05(5) = 0.95$.

Therefore, the live load reduction = 1 – (0.85 × 0.95) = 1 – 0.81 = 0.19, or 19 *percent.* The design roof live load may be reduced from 20 psf to 0.81 × 20 = 16.2 psf, which is greater than the minimum permitted reduced live load of 12 psf.

In general, roof members must be designed for the maximum effects due to dead load, roof live load, wind, and where applicable, snow and earthquake loads. The 2009 IBC references the 2005 edition of Minimum Design Loads for Buildings and Other Structures (ASCE 7-05) to determine snow loads. The determination of design snow loads can be complicated, as they depend on many factors, including local snow depths, exposure, thermal factors, and roof pitch to name a few. Once the design snow load is determined, it should be compared to the design roof live load and the larger of the two is used to design the roof members.

Wind and Earthquake Loads

Building structures must be designed to resist wind load or earthquake load, but not both acting simultaneously. Although wind forces on buildings are dynamic in nature, for design purposes they are generally assumed to be static forces acting horizontally on the gross area of the vertical projection of the building. The magnitude of the wind force varies with the height and location of the building.

Forces exerted on buildings from earthquake ground shaking are also complex and dynamic. However, for most buildings, the IBC permits a static design approach to be used. Using this approach, the horizontal earthquake forces are determined and distributed to the different levels of the structure, and then a structural system is designed to resist them. A more detailed discussion of these topics can be found in Lesson Twelve.

Miscellaneous Loads

In addition to the dead, live, wind, and earthquake loads, portions of buildings may have to be designed to resist certain special loads, among which are the following.

Retained Earth

Walls retaining drained earth are usually designed to resist a pressure equivalent to that exerted by a fluid weighing 30 pounds per cubic foot, in addition to any surcharge.

Example #3

What is the total pressure exerted on the retaining wall 12 feet high shown in Figure 9.2?

Solution:

The pressure at the bottom of the 12-ft.-high wall is equal to the height multiplied by the unit weight of the equivalent fluid = 12 × 30 = 360 lbs. per square foot. The total pressure is equal to the average pressure multiplied by the wall height = (360/2) × 12 = *2,160 lbs. per lineal foot of wall.*

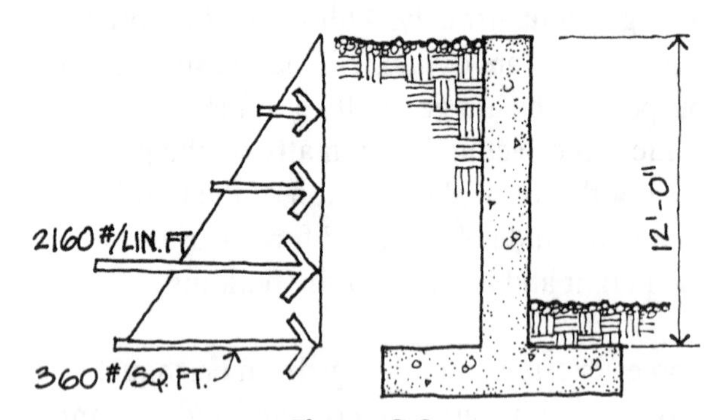

Figure 9.2

Hydrostatic Pressure

Structures containing water or any other fluid, such as pools, storage tanks, etc., must be designed to resist the pressure exerted by the liquid, which is called *hydrostatic pressure.* The magnitude of this pressure is equal to the unit weight of the liquid multiplied by the depth. For water, the unit weight is 62.4 pounds per cubic foot. At the bottom of a ten-foot-deep water storage tank, for example, the pressure is equal to 62.4 × 10 = 624 psf, and it acts perpendicular to all surface that it contacts.

Underground structures, such as basement walls, are usually designed to resist the pressure of the outside earth. However, if the ground water level is higher than the basement floor, the basement walls and floor are usually designed to resist the hydrostatic pressure exerted by the ground water.

Forces Caused by Temperature Change

An increase in temperature causes materials to expand, while a decrease in temperature causes them to contract. As long as the materials in a building are free to expand and contract, no forces are developed. However, any restraint on the free expansion or contraction will cause forces to be exerted on the structure.

Impact Loads

The sudden application of load will cause stresses much larger than those caused by a static load. Therefore, the stresses caused by moving loads, such as cranes and elevators, are generally computed by using an equivalent static load equal to the actual load multiplied by an impact factor varying between 1.20 and 2.00, which is equivalent to increasing the load 20 to 100 percent.

Railings
According to the IBC, balcony and other railings are to resist the effects from (1) a load of 50 pounds per lineal foot applied in any direction to the top of the railing, and (2) a concentrated load of 200 pounds applied in any direction at any point along the top of the railing. These two loads need not be applied to the railing at the same time. In one- and two-family dwellings, only the 200-pound concentrated load needs to be considered. In storage, high-hazard, factory and industrial, and certain institutional occupancies where there are fewer than 50 people (jails, detention centers, etc.), a 20 pound per lineal foot load can be used to design the railings in areas that are not accessible to the general public.

Vibration
Vibration of a building's structure may be caused by moving machinery or vehicles, or may be induced by people walking or dancing. Most buildings contain many nonstructural elements, such as partitions and ceilings, which tend to damp (decrease) the vibrations. Some buildings, however, have large open spaces that are especially vulnerable to vibration, particularly if their construction is light and flexible.

Vibrations are objectionable because they are annoying to the building's users, may be detrimental to delicate equipment, and may cause actual physical damage. Methods of reducing the effects of vibration include the use of springs and other shock absorbing devices to isolate vibrating machinery, the use of massive bases under vibrating equipment to absorb energy while transmitting a minimum amount to the structure, and avoiding the use of light wood or steel frame construction in buildings or portions of buildings subject to vibration, especially where damping from nonstructural elements may be inadequate.

Blast
Some buildings or portions of buildings must be designed to resist the effects of possible explosion. Such structures include bank vaults, some defense installations, and facilities using nuclear material. Federal buildings, courthouses, and other public facilities must typically also be designed to resist such effects.

BASIC CONCEPTS

Flexural Design Criteria

Most structural members resist load primarily by bending action, or *flexure.* The three basic criteria used in the design of flexural members are *shear stress, flexural stress*, and *deflection* (Figure 9.3). Shear stress is often an important design consideration in wood and reinforced concrete framing, particularly for short, heavily-loaded spans; however, shear is seldom critical in the design of structural steel framing. Flexural stress, which is tensile in one face of a member and compressive in the other, is always an important consideration in design. Deflection is sometimes critical, especially for members which are shallow or which span a great distance.

Strength and Stiffness

The two most important determinants of flexural strength and stiffness, or resistance to deflection, are the depth of a member and its material. The stiffness of a material is determined by its modulus of elasticity E: *the greater the value of E, the greater the stiffness.* For example, the E value of steel is about 29,000,000 psi, which is about 20 times greater than that of wood and about 10 times greater than that of concrete. The flexural strength of structural steel is roughly 10 to 15 times that of wood or reinforced concrete.

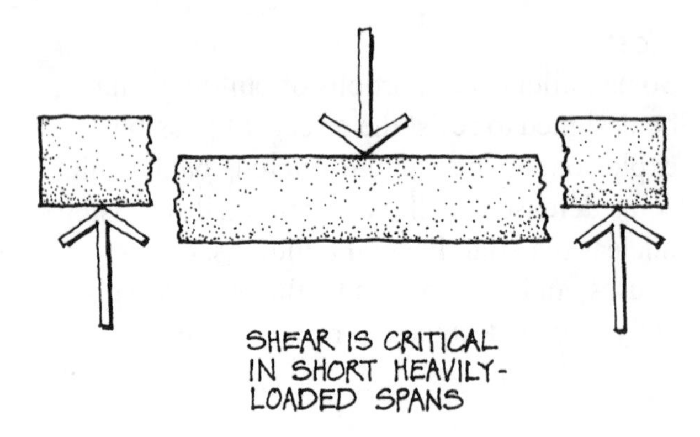
SHEAR IS CRITICAL
IN SHORT HEAVILY-
LOADED SPANS

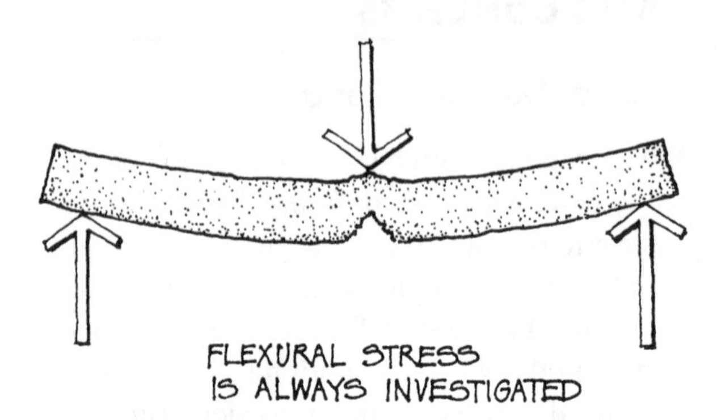
FLEXURAL STRESS
IS ALWAYS INVESTIGATED

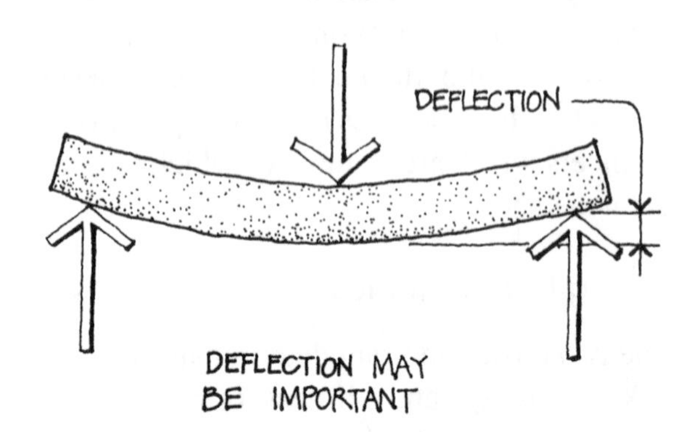
DEFLECTION
DEFLECTION MAY
BE IMPORTANT

CRITERIA FOR DESIGN OF FLEXURAL MEMBERS

Figure 9.3

The depth of a member is much more important than its width as far as strength and stiffness are concerned; for a rectangular member, the flexural strength varies as the square of the depth, but only the first power of the width (since the section modulus is equal to $bd^2/6$). Put another way, doubling a member's depth increases its flexural strength fourfold, while doubling its width only increase its flexural strength twofold.

A member's stiffness is proportional to the third power of its depth and the first power of its width. Thus, if a member's depth doubles, its moment of inertia (which is a measure of its stiffness) increases eightfold ($2 \times 2 \times 2$). But if its width doubles, its stiffness increases only twofold. These relationships are summarized in Figure 9.4.

Ideally, then, flexural members should be narrow and deep. There are other considerations, however: deep members increase the total depth of construction, which may be uneconomical. And very narrow members tend to be unstable and may require some type of bracing for stability. At the other extreme, members that are very wide and shallow will tend to deflect too much and may also be uneconomical. In structural design, all of these factors are considered in order to arrive at an optimum solution.

Efficient Flexural Members

The further away a member's material is from the neutral axis, the more efficiently it can resist bending. Examples include structural steel wide flange beams, steel and wood trusses, and prestressed concrete I-shaped girders. In such members, the material concentrated near the outer edges comprises the flanges or chords and resists flexure, while the material between (the web) mainly resists the shear (Figure 9.5).

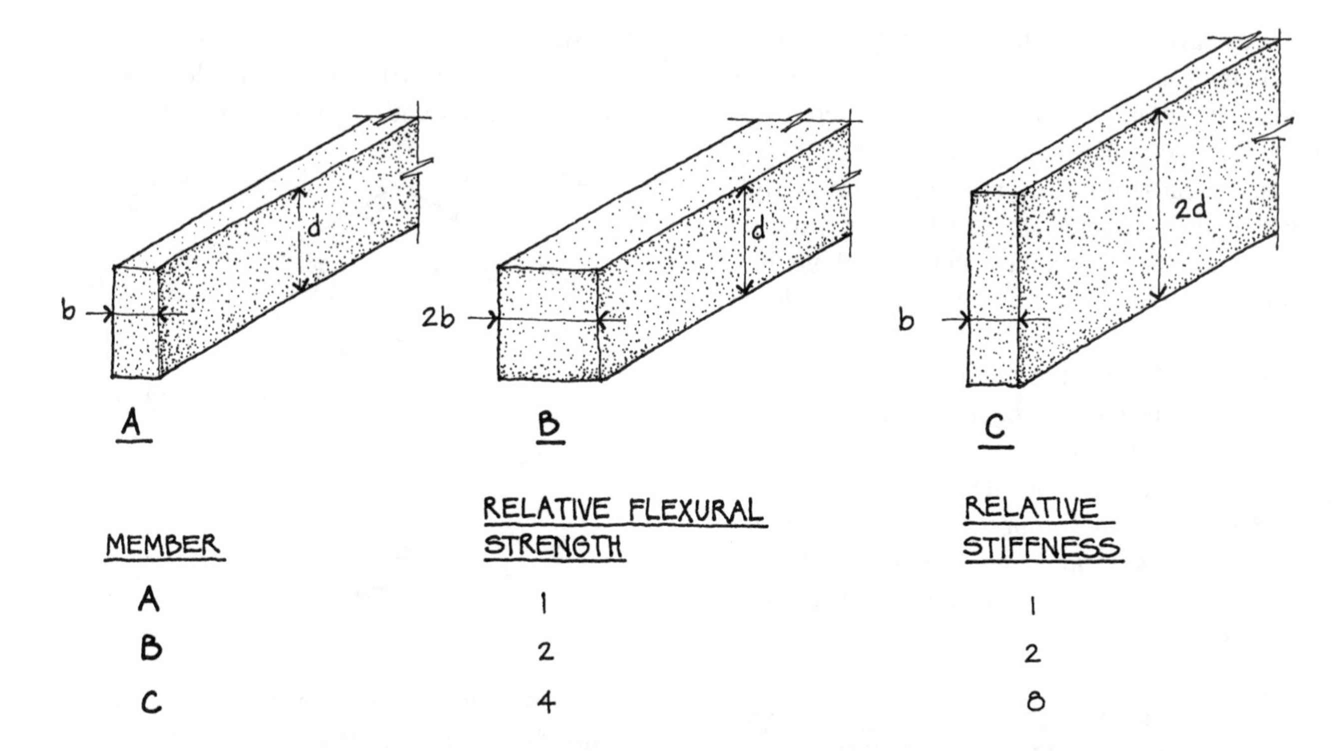

MEMBER	RELATIVE FLEXURAL STRENGTH	RELATIVE STIFFNESS
A	1	1
B	2	2
C	4	8

PROPERTIES AFFECTED BY WIDTH AND DEPTH

Figure 9.4

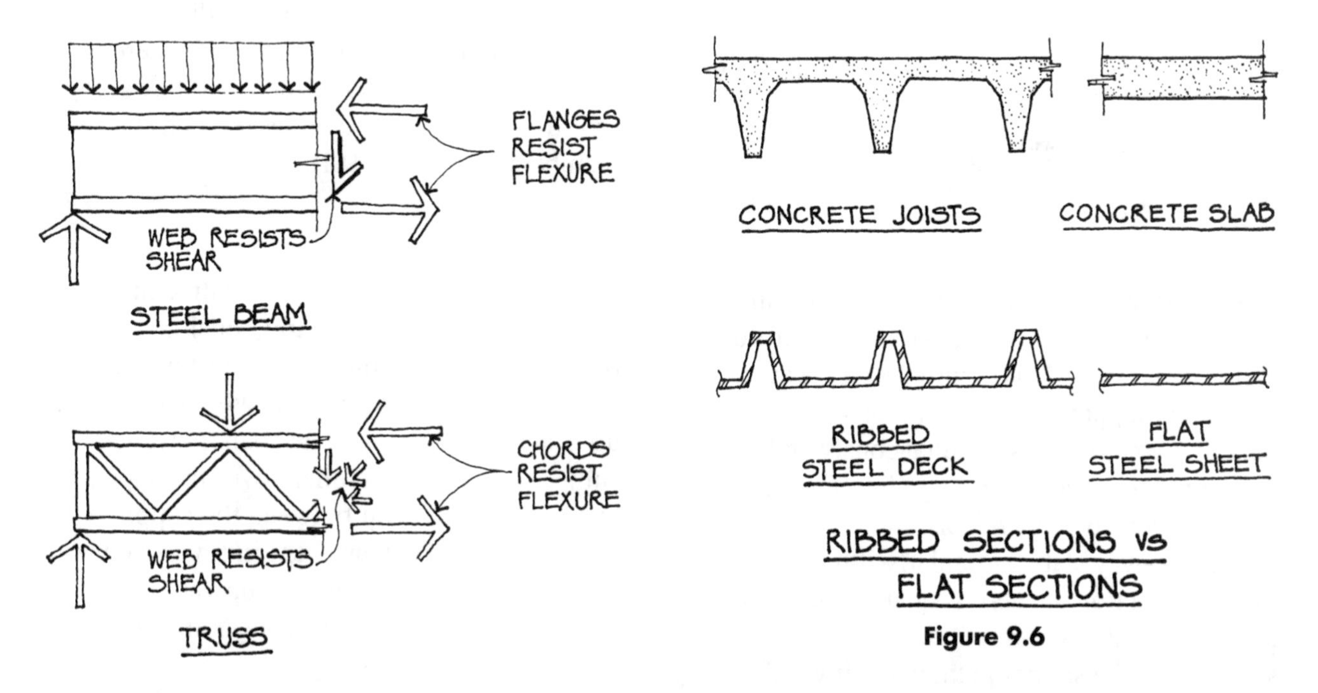

RIBBED SECTIONS vs FLAT SECTIONS

Figure 9.6

FLEXURAL ACTION

Figure 9.5

Ribbed sections are more efficient than flat sections in resisting bending. Wide, flat beams, such as concrete slabs and wood planking, are usually limited to fairly short spans. Two examples are shown in Figure 9.6. The concrete joist system may contain no more concrete than the slab, yet it can span a much greater distance because it is deep and ribbed. Similarly, a ribbed or corrugated steel deck is much stronger and stiffer than a flat sheet of steel, yet it does not necessarily contain any more material.

If the flat element acts together with the deep element to resist bending, the efficiency is improved and the deflection reduced. Two examples are a reinforced concrete *T-beam* and a concrete slab connected to a steel beam to form a *composite beam* (Figure 9.7). This type of action is seldom achieved with wood framing.

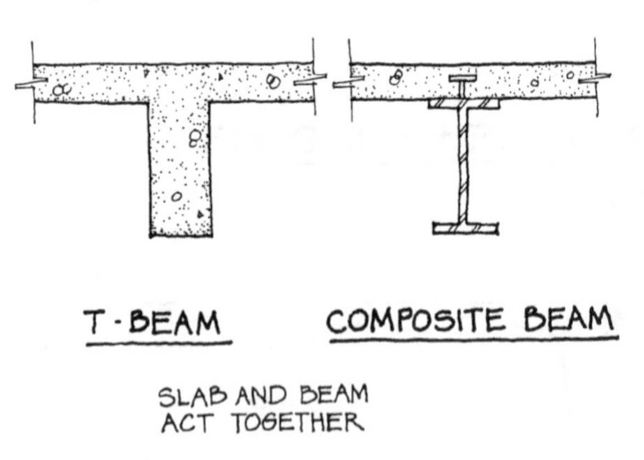

Figure 9.7

If a slab is supported along two sides, it can only bend in one direction, and it is therefore called a *one-way slab* (Figure 9.8). However, if the slab is supported along all four sides, it can bend in two directions simultaneously and is therefore known as a *two-way slab* (Figure 9.9). A two-way slab supports load more efficiently than a one-way slab, particularly if the spans in each direction are about equal. If the spans are unequal, the forces will tend to travel to the supports by the shorter path; in other words, more load will be carried in the short direction than in the long direction. When the long side of a slab becomes much longer than the short side, most of the two-way action is lost, and the slab becomes essentially a one-way slab.

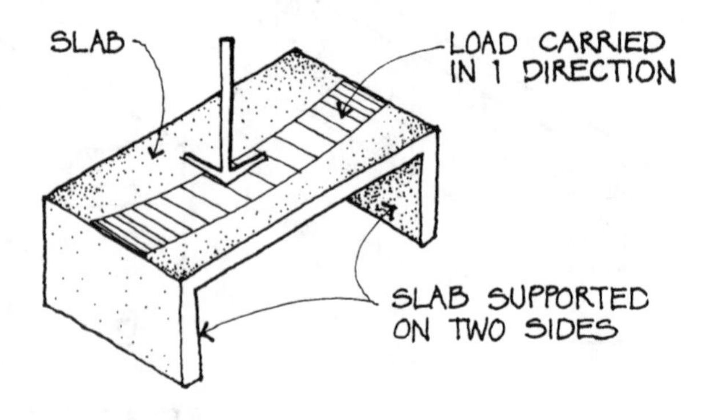

Figure 9.8

Two-way construction often occurs in reinforced concrete construction because of its monolithic nature, but seldom occurs in wood or structural steel construction.

Members may be constructed with a profile which approximates the moment diagram, so that maximum strength is provided where it is needed. Tapered roof girders and bowstring roof trusses are two examples of such shaped members (see Figure 9.10).

Continuity

Simple beams are subject to positive moment only (compression at the top, tension at the bottom) and bend concave upward when loaded (Figure 9.11). Beams which are continuous across more than one span, however, are subject to negative moment over the supports and positive moment between the supports (Figure 9.12). They are bent in reverse curvature, changing from concave upward where the moment is positive to convex upward where the moment is negative.

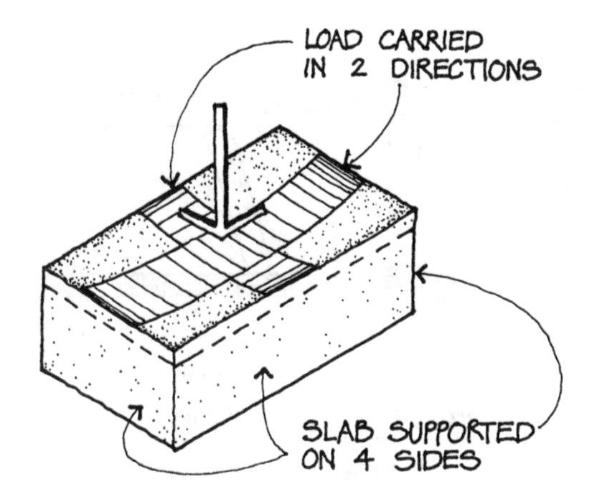

TWO-WAY SLAB

Figure 9.9

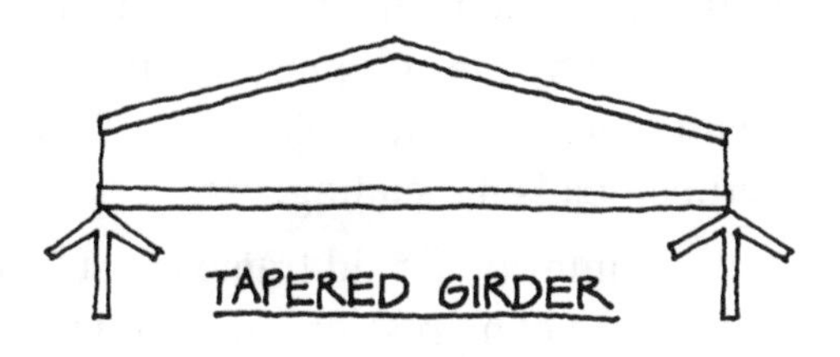

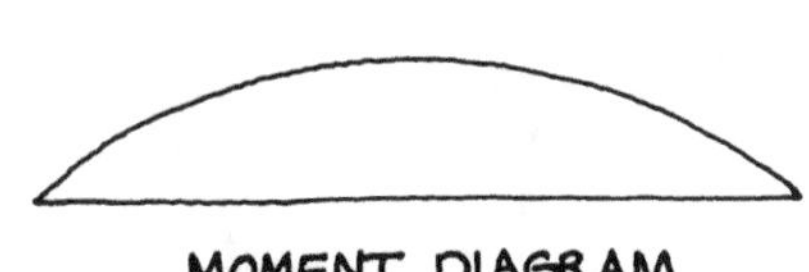

MOMENT DIAGRAM
FOR UNIFORM LOAD

MEMBERS WITH VARYING PROFILE

Figure 9.10

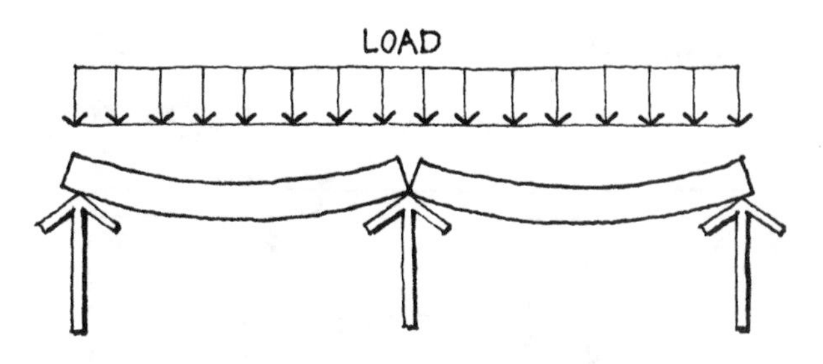

SIMPLE BEAMS

- POSITIVE MOMENT ONLY
- CURVATURE IS CONCAVE UP

Figure 9.11

Continuous beams are statically indeterminate; that is, the equations of equilibrium alone are not sufficient to compute their reactions and moments. Additional methods must therefore be used to analyze such beams.

Continuous beams generally have less deflection than simple beams having the same load, span, and cross-section. Also, the maximum bending moments in continuous beams are lower than in simple beams subject to the same load on the same span.

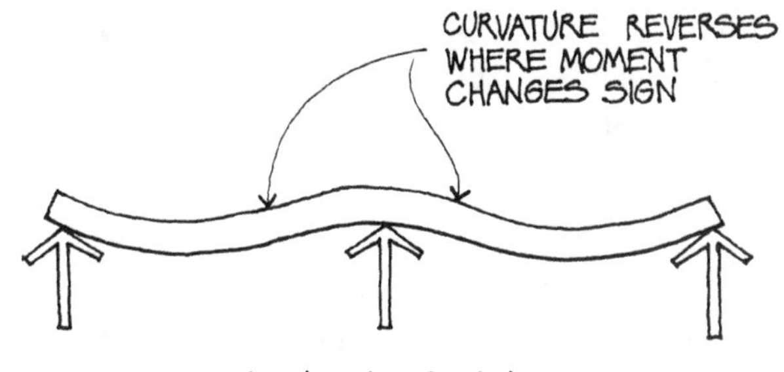

CONTINUOUS BEAM

- POSITIVE AND NEGATIVE MOMENTS
- REVERSED CURVATURE
- LESS DEFLECTION THAN SIMPLE BEAM
- LOWER MAX. MOMENT THAN SIMPLE BEAM

Figure 9.12

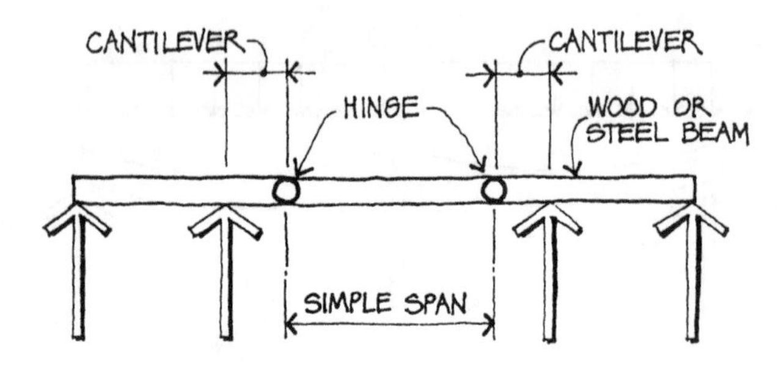

Figure 9.13

The advantages of continuity, smaller moments and deflections, are realized most often in cast-in-place reinforced concrete construction, because of its inherently monolithic nature. With special detailing, structural steel beams can also be designed and constructed to take advantage of continuity. In precast concrete and wood construction, continuity is more difficult to achieve. Simulated continuity is sometimes provided with wood or steel beams, as shown in Figure 9.13. The center beam is hung from the outer cantilever beams, thereby reducing the positive moment in the end spans. In addition, the effective span of the center beam is not the distance between supports, but the much shorter distance between hinges.

Rigid Frames

The simplest structural system is the *post-and-beam* (Figure 9.14). A horizontal beam subject to pure bending rests on posts, which are stressed in pure compression. This system is satisfactory for supporting vertical loads, but is unable to resist horizontal loads in its own plane.

Now if the connections between the beam and posts are made rigid, so that the beam and posts act together, we have a *rigid frame*, which is stronger for both vertical and horizontal loads (Figure 9.15).

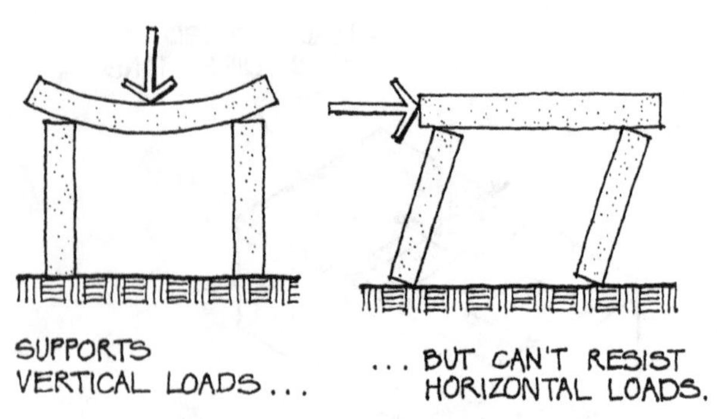

Figure 9.14

The behavior of a rigid frame differs in several respects from that of a post-and-beam system:

1. In the rigid frame, the beam ends are restrained by the columns, making the beam more rigid and able to resist greater vertical loads in bending.
2. The columns of a rigid frame are subject not only to compression, but also to bending moment because of their rigid connections to the beam.
3. The base of a rigid frame requires horizontal forces, as well as vertical forces, to hold the frame in equilibrium.
4. The rigid frame is able to resist horizontal loads in its own plane.

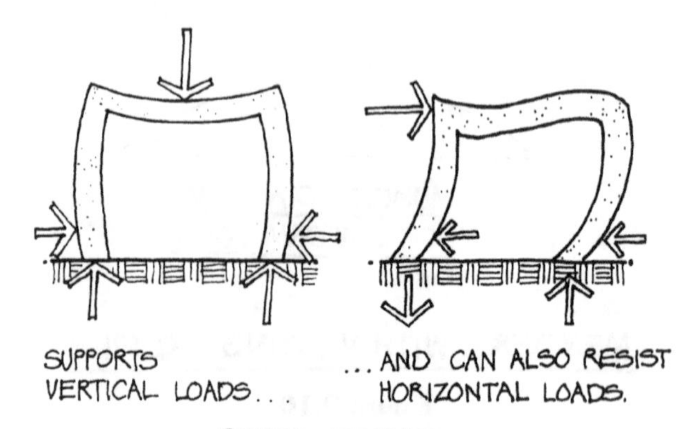

Figure 9.15

Although we have been discussing one-story one-bay rigid frames, the same principles apply to multi-story multi-bay rigid frames, which are used in many of today's steel and concrete buildings.

Reinforced concrete rigid frames rely on the monolithic nature of cast-in-place concrete construction and the proper lapping of reinforcing bars to achieve rigidity, while structural steel rigid frames require special detailing at the beam-column joints.

Welding is particularly efficient for rigid joints, but bolting is also frequently employed. Rigid frames of wood are seldom used, except for special structures such as gabled frames.

CONVENTIONAL STRUCTURAL SYSTEMS

General

A structural system is an arrangement of structural components that resists a building's vertical and horizontal loads. There are numerous factors that influence the choice of structural system. Some of these include fire resistive requirements, availability of various materials and elements, economy, building code requirements, column spacing, functional requirements, time available for construction, the building's size and shape, and adaptability to potential changes.

Systems to resist lateral forces from wind or earthquake consist primarily of horizontal diaphragms that distribute the lateral loads to vertical resisting elements, which may be shear walls, braced frames, rigid frames, or a combination of such elements. Many of the elements of the lateral force resisting system may also be used to support vertical loads, particularly diaphragms, shear walls, and rigid frames.

Lateral force resisting systems were discussed in greater detail in Lessons Four and Five.

There is a seemingly endless variety of structural systems, and seldom is any one system ideal. What the architect and structural engineer try to do before starting the final design is to narrow the field to a few acceptable systems, and then select the optimum system from those.

Most conventional one-way structural systems consist essentially of beams or trusses that span distances by flexure, or bending action (Figure 9.16). The main members are spaced relatively far apart, and lighter secondary members are used to span between the main members. Although flexural systems are not always efficient in the use of material, this is generally more than offset by their simple straight line arrangements, which are easy and economical to construct. Also, being horizontal, they are particularly appropriate for framing floors.

A short summary of some of the conventional roof and floor framing systems in common use is shown in Figures 9.18 through 9.20.

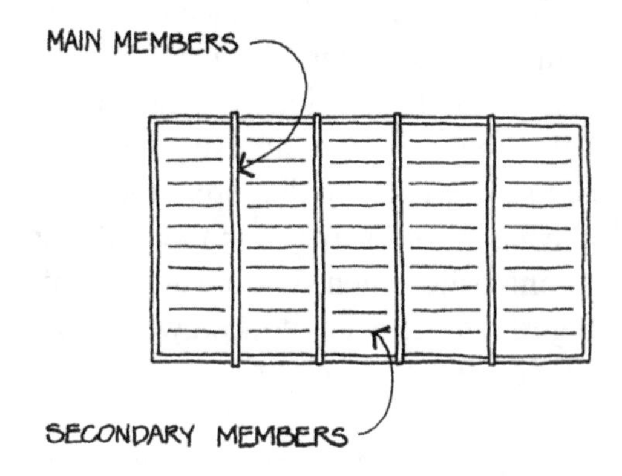

ONE-WAY FLEXURAL SYSTEM

Figure 9.16

We show three wood systems: joists, plank-and-beam, and trussed rafters, of which the joist system is the most common. We also show three steel systems: open web joists, beam-and-girder, and stub girder, all of which must be fire-protected. Open web steel joists are described in more detail in Lesson Ten. This system utilizes a thin concrete slab over closely-spaced joists, while the beam-and-girder system uses a relatively thick slab over widely-spaced beams. The beam-and-girder system is often designed for composite action between the beams and slab, and the structural floor may be a solid slab placed directly over the beams or steel decking with a concrete fill.

The open web joist system is frequently more economical than the beam-and-girder system. However, joist-supported floors tend to be more springy than beam-and-girder floors, especially in large spaces that have few interior partitions to damp vibrations. In addition, providing fire protection and under-floor electrical ducts is often easier with the beam-and-girder system.

The stub girder system is a variation of the beam-and-girder system, in which the beams sit on top of a girder, instead of framing into a girder. Since the beams are clear of the girder, they may be designed for simulated continuity, as shown in Figure 9.13. The space between floor beams provides room for the mechanical and electrical distribution systems. Short lengths of stub girders the same depth as the floor beams are welded to the top of the girder to provide a connection between the girder and slab for composite action. The advantages of the stub girder system are reduced weight of steel and reduced story height.

Because concrete construction is monolithic, concrete framing systems can be either one-way or two-way. For bays that are approximately square, two-way systems such as flat plate, flat slab, and waffle slab are often used. Where bays are rectangular, one-way systems such as joist slab and one-way beam-and-slab are favored. If the structure is exposed to view, flat plate, banded slab, and waffle slab are preferred. For light to moderate loads, any of the systems shown can be used. For heavy loads, such as library stacks or heavy storage, flat slab and waffle slab are most appropriate.

There are many possible variations of the systems shown on the following pages. For example, *lift slabs* are flat plate floors that are cast one over the other at grade level with embedded steel collars that fit around steel columns. After curing, the slabs are lifted to their proper elevations by hydraulic jacks, and the embedded collars are welded to the columns. The design procedures for lift slabs are essentially the same as those used for the flat plate system. Two additional factors that must be considered are (1) possible unequal jack movements, which would induce stresses in the slab, and (2) the connections of slab to collar and collar to column. Lift slabs are most appropriate for multi-story buildings with short spans and relatively light loads, as in dormitories and apartment buildings.

Reinforced concrete columns, beams, slabs, or walls may be precast instead of cast-in-place. Such elements may be also pretensioned for longer spans.

Where columns are widely spaced in both directions, beam-and-slab systems may have intermediate beams to reduce the slab span.

Multi-story Frames

Multi-story frames are usually constructed of reinforced concrete, either cast-in-place or precast, or structural steel.

In cast-in-place reinforced concrete frames, continuity at the beam-column joints is provided by the monolithic nature of the material and proper lapping of the reinforcing bars. This continuity enables the frame joints to transfer bending moments, thus allowing the frame to resist horizontal loads.

Multi-story frames constructed of precast concrete may also have the ability to resist horizontal forces, provided that the connections are designed and constructed to transfer bending moments. If not, the frames can only support vertical loads, and horizontal loads must be resisted by other elements, such as diagonal bracing placed within permanent partitions, or shear walls.

Structural steel multi-story frames may be designed as rigid frames with full continuity at the joints, or as semi-rigid frames with moment connections at the beam-column joints sufficient to resist lateral forces only. If the beam-column connections have no rigidity, horizontal loads must be resisted by shear walls or diagonal bracing.

Stressed Skin

Stressed skin panels are made of spaced members solidly sheathed on one or both sides. They behave like a series of built-up I-beams, with the sheathing (the stressed skin) forming the flanges and resisting flexural tension and compression, while the spaced members comprise the webs and resist shear (Figure 9.17).

The joint between the sheathing and the spaced members must transfer the horizontal shear stress without any slip. The most typical application of this system uses plywood sheathing and wood stringers to form a panel which can support roof loads over spans of 12 to 32 feet.

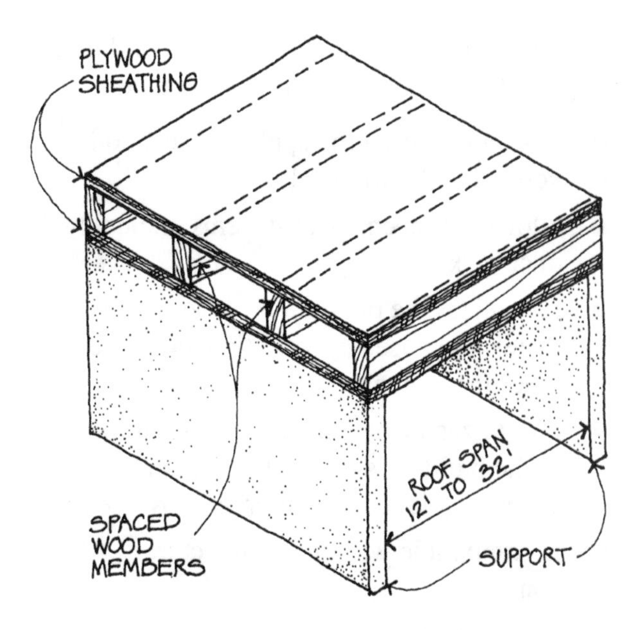

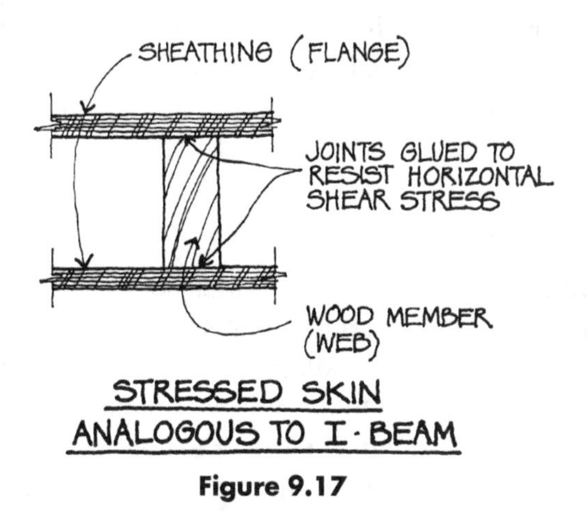

Figure 9.17

SELECTING A SYSTEM

General

Knowledge of structural concepts and systems is useful only if it leads to the optimum solution for a particular building. The goal in selecting the structural system for a building is to provide the most efficient system that

- will safely resist the applied loads, both vertical and horizontal;
- allows for mechanical systems, including ductwork;
- minimizes the possibility of damage to materials caused by vibration, deflection, or other movement;
- minimizes inconvenience or discomfort to the building's occupants caused by vibration, deflection, or other movement;
- is compatible with the architectural concept;
- meets fire-resistive requirements; and
- can be constructed within the available time.

It is usually economical to select one structural system and use it throughout the project. Using different systems is justified only on larger projects or those that involve more than one use.

Fire Resistance

Building codes require various degrees of fire resistance for a building's components, depending on the building's size and use. Fire resistance refers to the ability of a material or assembly to confine a fire, or to continue to perform its structural function during a fire, or both. The fire-resistive period of a material or assembly (the time during which it retains its fire resistance) is determined by standard test procedures, and periods of one to four hours are commonly required for the various elements of a building.

Concrete construction can provide any amount of fire resistance required. Steel construction, on the other hand, has no fire resistance by itself, but must be protected from fire by other materials, such as plaster or concrete.

Construction Time

Construction time is frequently of the essence, and the architect must then select structural systems and materials that can be constructed quickly. While this varies with time and place, it is generally true that prefabricated systems can be erected quickly, while those that are entirely built at the site require more time to construct. Thus, structural steel, precast concrete, or any pre-assembled system requires relatively little time; while cast-in-place concrete or conventional masonry construction requires more time. However, local costs, practices, and availability must always be considered before a specific material or system is selected.

Construction Cost

Cost is always a consideration in the selection of structural materials or systems. There is so much variation in cost, depending on site conditions, the contractor's experience, local practices, project size, local availability of materials and labor, etc., that rules of thumb regarding cost can only be general in nature. For example, on one project a structural steel system may be the most economical, while for a similar project in a different locality, a reinforced concrete system may be the least expensive. Therefore, the final selection of materials or systems should only be based on cost estimates of several competitive schemes.

Nevertheless, we can make some general comments regarding cost. In this country, the cost of labor is relatively high compared to the cost of materials, and field labor is usually more expensive than shop labor. Therefore, prefabricated or pre-assembled systems or assemblies should be utilized whenever feasible in order to minimize field labor.

Structural elements should perform more than one function if possible. For example, exposed concrete frames or walls can resist both vertical and horizontal loads, while at the same time providing an attractive architectural feature.

For one-way systems, the framing should be oriented in the same direction throughout the project if possible.

For smaller projects and where permitted by the building code, wood construction is usually less expensive than structural steel or reinforced concrete.

The cost of cast-in-place concrete construction is made up of three elements: formwork, which often represents over 50 percent of the total, followed by reinforcing steel and concrete. Therefore the key to economy in concrete construction is simplicity and maximum reuse of forms. Some suggestions in this regard are as follows:

1. Bay sizes and floor heights should be repetitive.
2. Columns should be the same size and walls should have the same thickness throughout the building, if possible.
3. Beams should also be the same size throughout, and beams and joists should have the same depth.
4. Joists or waffle slabs should be formed with standard forms of one size.

Complex building shapes are generally less economical than those that are simple; of course, aesthetics or other considerations may outweigh economy in any given decision.

What percentage of the total cost of construction is represented by the structure? Again, this varies a great deal. For conventional office buildings, the structural cost may represent roughly 25 to 30 percent of the total construction cost, with the remainder divided between architectural items (45 to 60 percent) and mechanical and electrical costs (20 to 30 percent). For projects with extensive mechanical and electrical requirements, such as hospitals and laboratories, the structural percentage might drop to around 15 to 20 percent, the mechanical/electrical portion might increase to as much as 50 percent, with the balance being architectural. For bare-bones industrial buildings, such as warehouses, the reverse is true: the structural percentage would increase to 50 percent or more, while the percentages for architectural and mechanical/electrical work would drop.

Structural costs also vary with a building's height and spans. For example, if an office building has 50-foot floor spans, its structural cost is roughly 25 percent greater than if its floor spans are 25 feet.

A low- to medium-rise building is designed primarily for vertical loads, and the weight of steel framing in such a building is about 10 to 20 pounds per square foot of floor area. For a tall building, however, the primary structural consideration is providing resistance to lateral loads from wind or earthquake. Tall buildings therefore require roughly 25 to 35 pounds of steel per square foot, resulting in structural costs that might be double those of lower buildings.

Although we always try to select structural systems which are economical, remember that the structure represents only about 25 percent of the total construction cost. Suppose the budgeted construction cost for a building is $10,000,000, of which 25 percent, or $2,500,000, is the cost of structure. If we use a very efficient structural system, we might be able to reduce the structural cost by 20 percent, from $2,500,000 to $2,000,000. This saving of $500,000, although significant, is only 5 percent of the total cost of construction.

CONCLUSION

In this lesson, the subject matter has been presented conceptually, rather than mathematically. Careful study will help the candidate gain a better understanding of the structural concepts and systems used in building construction. It is not our goal in this lesson to teach a candidate how to prepare the design calculations, but rather to help him or her understand intuitively the nature of structural behavior.

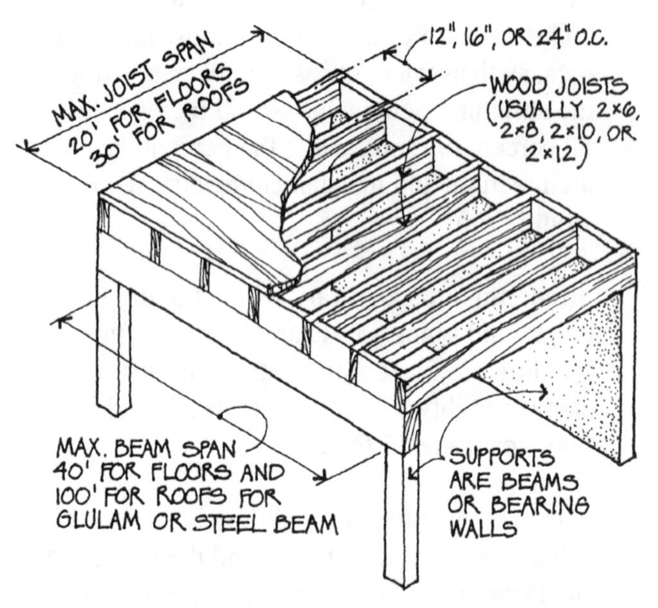

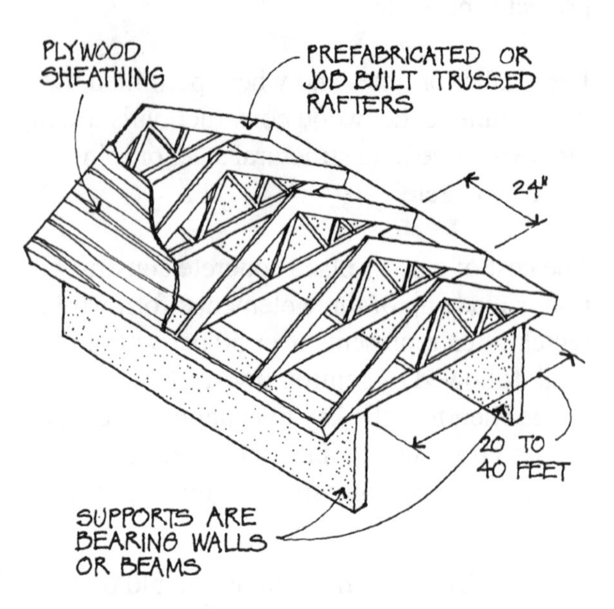

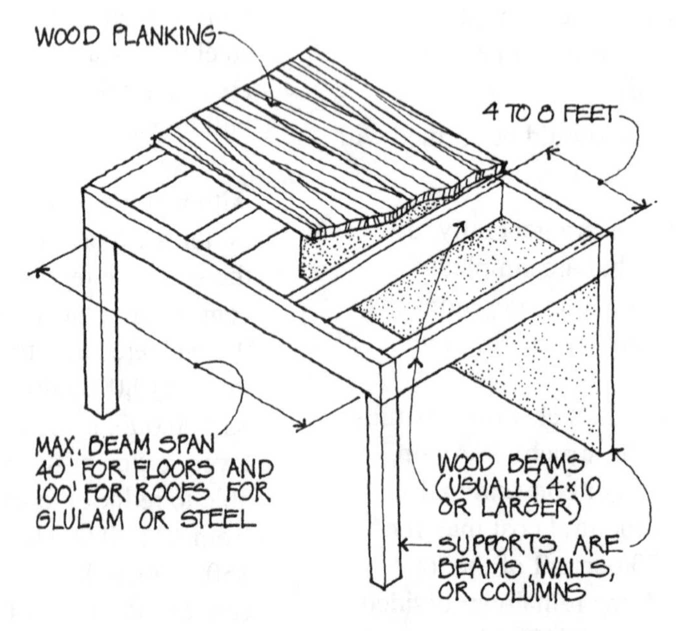

Figure 9.18

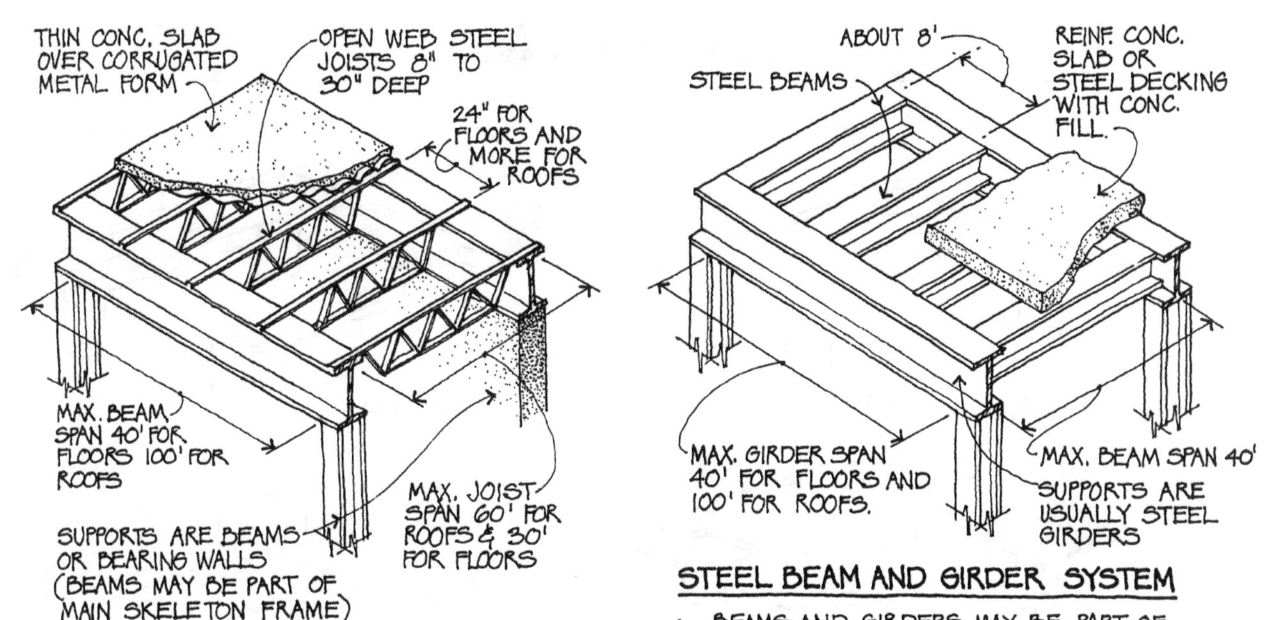

STEEL JOIST SYSTEM

STEEL BEAM AND GIRDER SYSTEM

- BEAMS AND GIRDERS MAY BE PART OF MAIN SKELETON FRAME
- COMPOSITE ACTION BETWEEN BEAM AND SLAB POSSIBLE
- ECONOMICAL FOR MOST BUILDING LOADS.

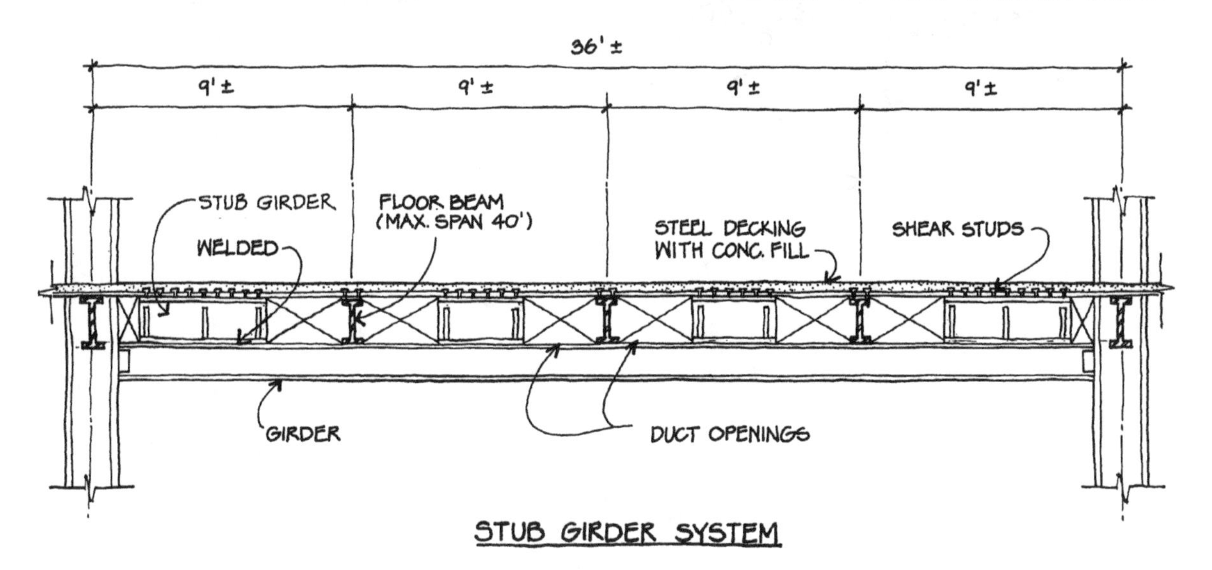

STUB GIRDER SYSTEM

- SIMULATED CONTINUITY FOR FLOOR BEAMS
- COMPOSITE ACTION FOR BEAMS AND GIRDERS

Figure 9.19

CONCRETE FRAMING SCHEMES

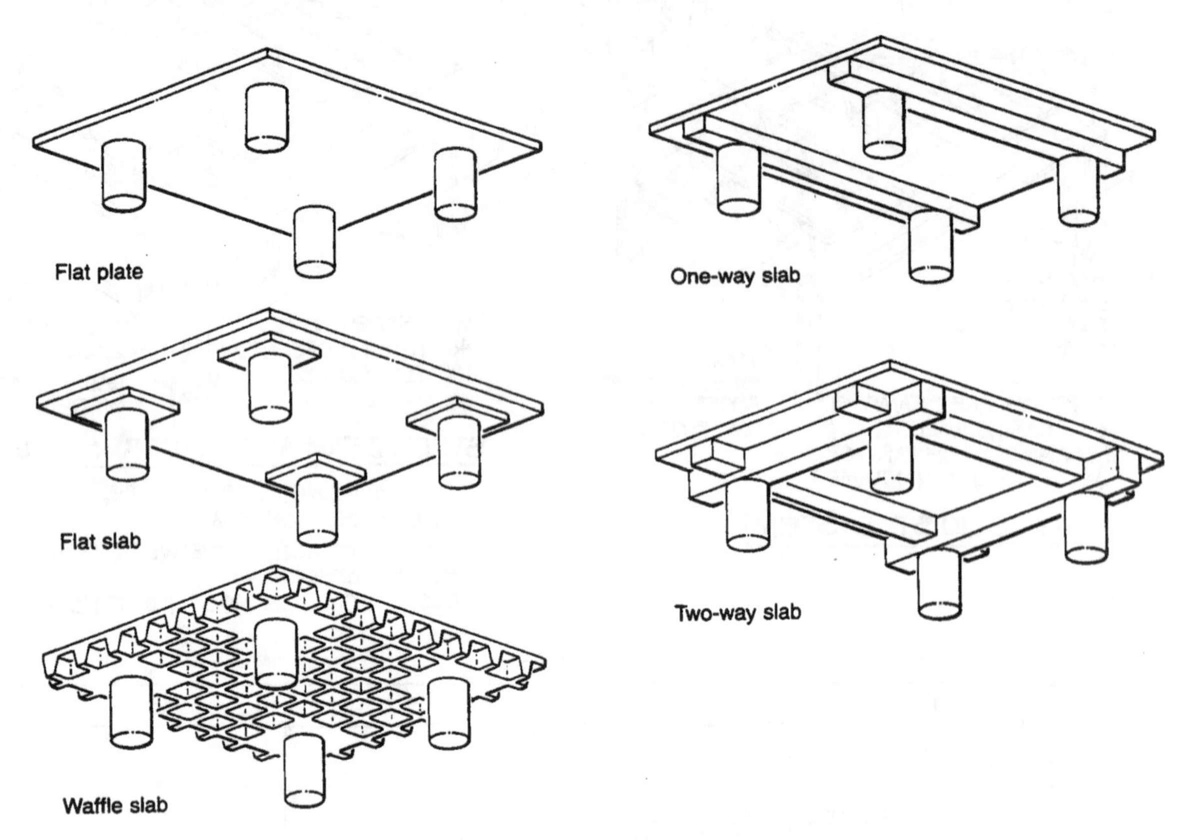

Figure 9.20

LESSON 9 QUIZ

1. Which of the following structural systems consist of a flat element acting together with a deep element to resist bending? Check all that apply.

   A. Composite beam

   B. Flat plate

   C. Flat slab

   D. T-beam

2. Of the structural systems listed below, which does NOT resist load by two-way flexural action?

   A. Flat plate

   B. Rigid frame

   C. Flat slab

   D. Waffle slab

3. What structural system has a profile which approximates its moment diagram?

   A. Tapered girder

   B. T-beam

   C. Truss

   D. Composite beam

4. A reinforced concrete slab is 15 feet by 15 feet in plan and has beams on all four sides. Select the CORRECT statement.

   A. The slab acts as a one-way slab.

   B. An intermediate beam is required to reduce the slab span.

   C. The slab acts as a two-way slab.

   D. Most of the load is carried diagonally.

5. Select the INCORRECT statement about flexural members.

   A. Shear stress is often critical for short, heavily-loaded spans.

   B. The two most important determinants of flexural stiffness are the depth of a member and its material.

   C. The further away a member's material is from its neutral axis, the more efficiently it can resist bending.

   D. Stiffness and strength are structural terms that may be used interchangeably.

6. Select the CORRECT statement.

   A. Rigid frames are unable to resist any horizontal load in their own plane.

   B. In rigid frames, both beams and columns must resist bending and axial load.

   C. There is always bending moment at the base of a rigid frame.

   D. Under symmetrical vertical loading, the base of a rigid frame requires no horizontal reaction.

7. Select the CORRECT statement.

   A. Doubling the floor spans in an office building increases the cost of construction about 25 percent.

   B. Tall buildings have unit structural costs about 10 percent greater than short buildings, because of the need for heavier columns.

   C. The most expensive part of cast-in-place concrete construction is reinforcing steel.

   D. For conventional office buildings, the structural cost is roughly 25 percent of the total construction cost.

8. Compared to a simple beam, a continuous beam has

A. greater positive moment.

B. less shear.

C. more deflection.

D. greater negative moment.

9. Select the CORRECT statements about floor live load reduction, in accordance with the International Building Code.

I. A live load reduction is permitted only when the live load exceeds 100 psf.

II. A live load reduction is not permitted in a place of public assembly.

III. A live load reduction is permitted only when the member supports more than 150 square feet.

IV. For a member supporting 250 square feet of floor area, the allowable live load reduction is 40 percent.

A. All of the above

B. I and III

C. II and III

D. II and IV

10. Which of the following is LEAST likely to be an efficient structural system for a building whose column spacing is 20 feet in one direction and 40 feet in the other direction?

A. Concrete slab over steel beams and girders

B. Concrete slab over open web steel joists and steel girders

C. Concrete joist system

D. Waffle slab system

LESSON TEN

10

LONG SPAN STRUCTURAL SYSTEMS

INTRODUCTION

When the Architect Registration Examination was first introduced, it included three structural tests, one of which covered only long span structures. That test was merged with the general structures test in 1988, but the subject of long span structures continues to be an important part of the ARE. In this lesson, we will discuss the background of long span structures and describe some of the structural systems used to span long distances.

Historical Background

Long span roofs are not new; more than 2,000 years ago, the Romans were using arches and vaults to span long distances for large public buildings, bridges, and aqueducts. Roman arches were built of stone masonry or brick and were stressed principally in compression (Figure 10.1). The vertical pull of gravity on each stone was held in equilibrium by the compressive forces of the stones above and below it. This compressive action continued around the arch to the supports, where the horizontal thrust of the arch was resisted either by another arch or by heavy walls.

The arch form also engendered the vault and the dome. The vault can be thought of as a

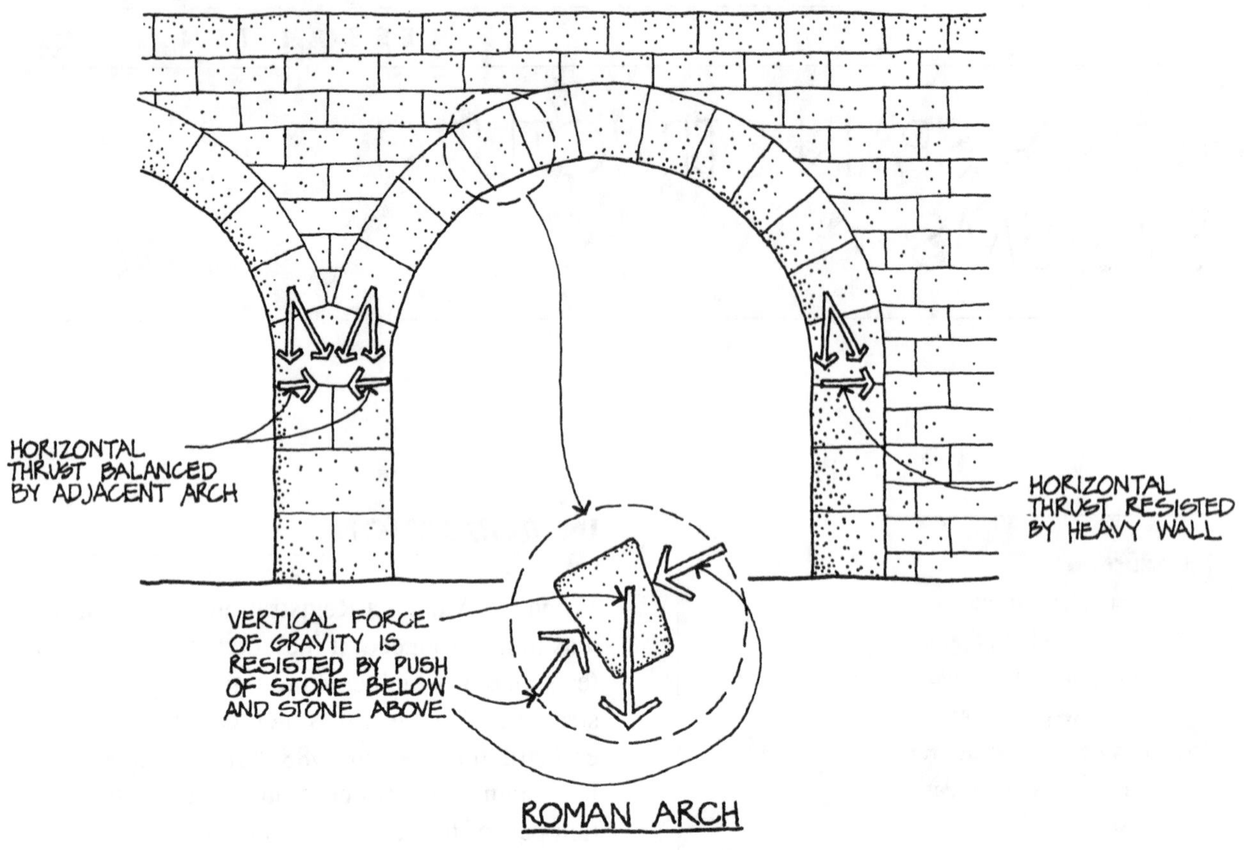
HORIZONTAL THRUST BALANCED BY ADJACENT ARCH
HORIZONTAL THRUST RESISTED BY HEAVY WALL
VERTICAL FORCE OF GRAVITY IS RESISTED BY PUSH OF STONE BELOW AND STONE ABOVE
ROMAN ARCH

Figure 10.1

series of arches placed side by side to form a continuous structure, while the dome is essentially an arch rotated about its vertical axis to form a curved surface.

The arch became the basic structural form in early European architecture, and it continues to be widely used today. Because the structural action of an arch is essentially compressive, the materials used for arch construction have naturally been those that are strong in compression, such as masonry and concrete, and later wood and steel.

In this country and elsewhere, long span roofs built during the 19th century and the early part of this century were used to cover railroad stations, factories, and armories. The structural systems used were generally one-way systems that had been extensively employed for railroad bridges, such as arches and simple span trusses. During this period, a number of roofs spanning up to 300 feet and even greater were successfully designed and built.

Because of technological changes, most of these structures have been modernized, replaced, or destroyed. Fortunately, however, as the historic preservation movement in this country has gained strength, increasing numbers of historic architectural and engineering structures are being adapted to modern functions. A street car barn becomes a shopping center, a tanning factory is converted to an apartment building, a railroad bridge is turned into a restaurant. But such cases are still the exceptions; it is far more common for a historic structure to be destroyed than preserved.

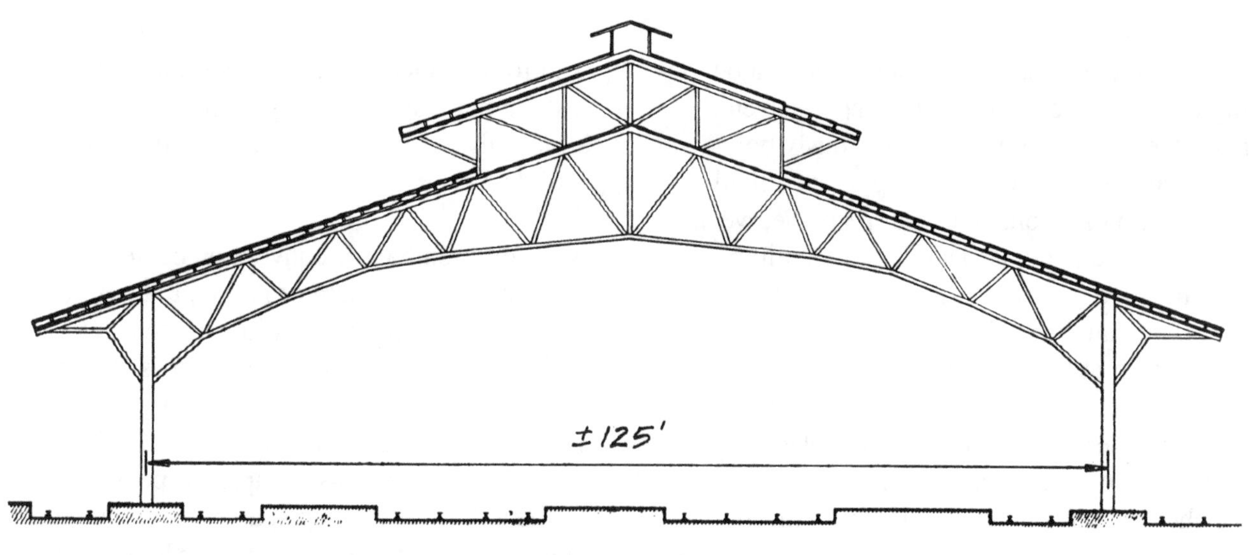

Figure 10.2

One structure which has survived is the Union Station Train Shed in Richmond, Virginia. This trussed-roof structure, built in 1900, is representative of the long span buildings of that era. A section through the building is shown in Figure 10.2.

After World War II, a strong demand developed in the United States and elsewhere for large facilities with column-free spaces, such as auditoriums, sports arenas and stadiums, fieldhouses, and airplane hangars. Ways were sought, and found, to provide these large column-free areas economically. Many of these designs would not have been possible without the help of the computer, which made the analysis of highly complex long span structures practicable.

In recent years, there have been several collapses of long span structures, causing a number of deaths and millions of dollars in damage. As a result, nationwide attention has been focused on structures of this type.

Definition of Long Span

What is a long span structure? There is no specific span dimension above which the structure is considered to be long span. According to the AIA Long-Span Building Panel, *60 feet for an office building is a long span. Sixty feet for a gymnasium is not a long span. Structural material is also a factor. Sixty feet may be a long span for a sawn timber beam, but a short span for a steel beam.*

Does the Steel Joist Institute have a better definition for long span? Not exactly, but it is interesting to note that standard open web steel joists are tabulated up to a span of sixty feet, longspans up to 96 feet, and deep longspan joists up to a span of 144 feet.

So, for our purposes, let's define *long span* as any span in excess of 60 feet.

Long Span Building Types

Long span structural systems are used for a variety of buildings, including arenas, theaters, exhibition halls, stadiums, and airplane hangars. Let's discuss some of these building types.

In the English language, there are often many names for the same thing. For example, a building in which spectators witness sports events and other entertainment may variously be called an arena, a coliseum, a garden, a fieldhouse, or a pavilion. Whatever its name, we are referring to an enclosed structure containing a rectangular playing surface large enough to accommodate basketball or hockey and surrounded on at least two sides by elevated seating for spectators. In addition to basketball and hockey, other events that may take place in an arena include rock concerts, circuses, rodeos, ice shows, tennis matches, and track meets.

If the arena floor is designed to be large enough for hockey, with an ice surface 85 feet by 200 feet, it will be adequate for most other indoor sports events and entertainment activities.

Spectator seating capacity ranges from a few thousand in smaller cities or schools to around 15,000 for major university arenas and up to about 20,000 for major arenas serving professional hockey and basketball teams, such as Madison Square Garden in New York and the Great Western Forum in Los Angeles.

The size and shape of the main arena roof are determined by the size of the central floor space and the configuration and capacity of the spectator seating area. The shape of the roof may be square, rectangular, circular, or oval, and the span is usually between 150 and 300 feet. Ceiling heights are usually at least 50 feet, and almost all modern arenas are column-free for unobstructed vision.

A theater is a building smaller than an arena and used for dramatic or musical presentations, motion pictures, or concerts. Again, as with arenas, a theater may be called by many names: auditorium, concert hall, opera house, to name a few. Older theaters did not always have clear span design, and seats behind columns were sold for a reduced price or were not sold at all. But the main space in large modern theaters is almost always column-free, requiring long span roof design.

An exhibition hall or convention center includes a large square or rectangular space used for the display of exhibits in connection with conventions or consumer shows. The size may vary from 25,000 square feet to 300,000 square feet or more. For maximum flexibility, columns are spaced as far apart as possible. While a 30-foot column spacing is not uncommon, some of the large modern exhibition halls have column spacings of about 150 feet.

A stadium is a structure, either open or enclosed, used for baseball or football. In the old days, stadiums were always open, used most often for major league baseball, and boasting of few spectator amenities. But all of that has changed. Professional football is now a major spectator sport, and major league baseball has expanded to more and more cities. With the proliferation of sports events on television, greater comfort and convenience must be offered to spectators to entice them out of their homes and into the ballparks. But the most dramatic change has been the development of the enclosed stadium—the enormous domed roof covering nine or ten acres, sheltering 50,000 to 80,000 spectators and spanning 700 feet or more. Although only a few of these super-mega-domes have been built, there are many who believe that the topless stadium will go the way of the horse and buggy, and that all major stadiums built in the future will be enclosed. In any event, no discussion of long span roof construction would be complete without mentioning the enclosed stadium.

Although roofs enclosing structures used for public entertainment and sports events provide the most visible and dramatic examples of long span construction, there are long span structures for many other needs: warehouses, industrial plants, airplane hangars, airport terminal buildings, retail stores, and they come in a variety of sizes and shapes.

Special Requirements

Before we get into specific structural systems, let's discuss some requirements that are typical of all long span structures. It may sound a bit simplistic, but all long span structures are *long*. Hence, they undergo great changes of length when subject to temperature changes, particularly in the case of open air structures. Special detailing may be required to allow for these movements.

Shipping long members may be a problem; although members up to 120 feet in length have been shipped, the usual maximum length is about 60 feet for truck shipment and about 80 feet for shipment by railroad flat car. What this means is that longer members must be partially assembled in the field, where labor costs are high.

Long span members are often deep, and this too may cause a problem in shipping. Generally, members up to 100 inches in depth can easily be shipped. Deeper members require special permits or special escorts, and the absolute limit for shipment is about 14 feet. Deeper members require field assembly.

Since some field assembly is often required for long span structures, what kind of connections should be used? In general, bolts are the most economical connections for field use, and it is usually more economical to assemble structures on the ground and then lift them into place than to assemble them in the air.

Although the building codes in different parts of the country vary, they generally permit the fire protection for roof structures to be omitted if the roof is high enough above the floor. In particular, the International Building Code states in footnote c of Table 601: *Except in Group F-1, H, M and S-1 occupancies [Moderate-Hazard Factory Industrial; High-Hazard; Mercantile; and Moderate-Hazard Storage], fire protection of structural members shall not be required, including protection of roof framing and decking where every part of the roof construction is 20 feet or more above any floor immediately below.*

Thus, many long span roofs over arenas and stadiums are built of exposed structural steel, providing a structure that satisfies both economic and aesthetic considerations (see IBC Table 601).

ONE-WAY FLEXURAL SYSTEMS

One-way flexural systems, as described in Lesson Nine, are also widely used for long span applications. Let us examine some of these systems.

Steel Beams and Girders

The main structural steel framing members for floors and roofs are usually rolled beam sections, most often wide flange (W) shapes.

Rolled beams are available up to 44 inches in depth, and their economical span limit is generally around 80 feet.

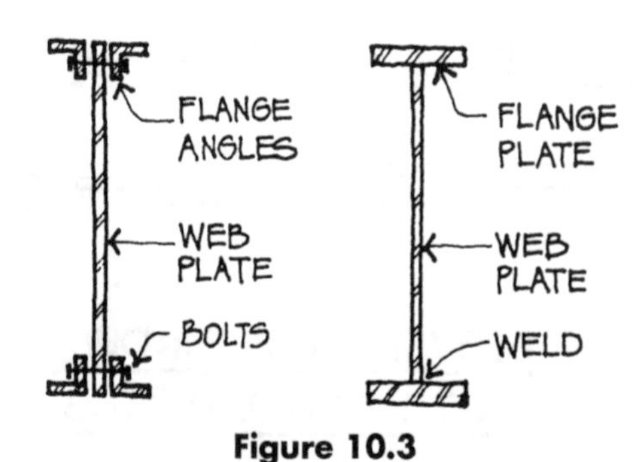

Figure 10.3

Where longer spans or heavy loads occur, rolled beams of the required size are often unavailable or uneconomical. In such cases, built-up members known as *plate girders* may be used instead. These girders comprise an assembly of plates, or plates and angles, which are welded or bolted together to form an integral member. Two plate girder assemblies are shown in Figure 10.4.

The detailed design of plate girders is a lengthy procedure beyond the scope of this course. However, it is advisable for candidates to have some conceptual knowledge about plate girders.

As with rolled beams, bending causes one flange of a plate girder to be stressed in compression while the other flange is in tension. All shear stresses are resisted by the web. The connections between each flange and the web, whether bolted or welded, must be adequate to transfer horizontal shear stress. The web of a plate girder usually needs to be reinforced against buckling, by means of vertical stiffener plates or angles located at the supports, at concentrated loads, and at intermediate locations.

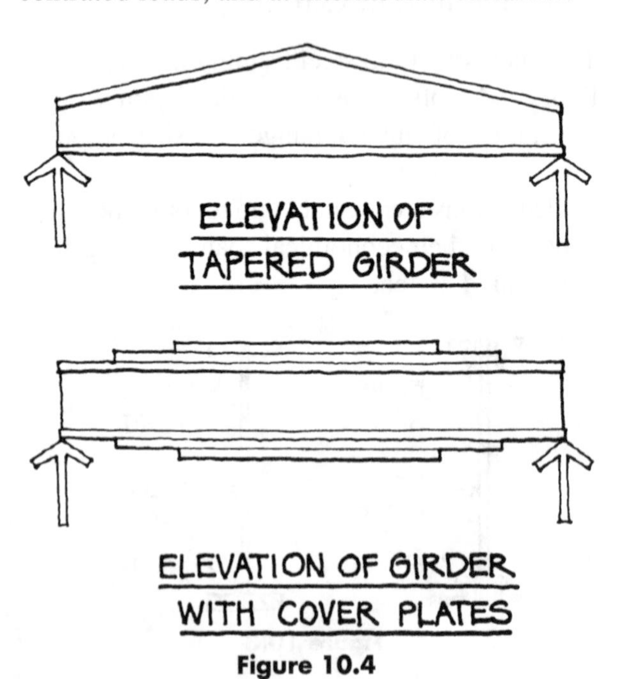

Figure 10.4

Since a plate girder is built up from component parts, it is relatively easy to vary its cross section to provide the maximum section modulus where the bending moment is maximum (generally at midspan), and conversely, the minimum section modulus where the bending moment is minimum, which is usually at the supports. Two ways to vary a plate girder's cross section are (1) tapering the girder and (2) using cover plates, as shown in Figure 10.4.

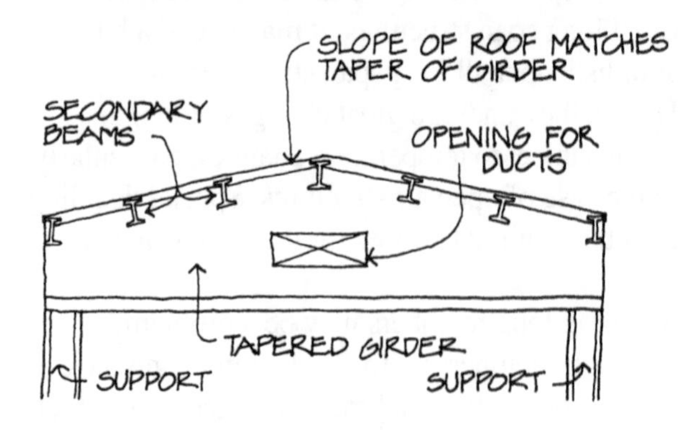

Figure 10.5

Two additional advantages of tapered girders are (1) for roof construction, the slope of the tapered top flange may be made the same as the roof slope and thus simplify the construction, and (2) openings through the girder web may easily be provided for the passage of air conditioning ducts. These are illustrated in Figure 10.5.

Trusses

For long spans and/or heavy loads, trusses are often more economical than beams and girders. They are used most often in roof construction, although they are sometimes used in long span floors as well. Occasionally they are used to support the load of upper columns which do not extend down to the floor below (Figure 10.6).

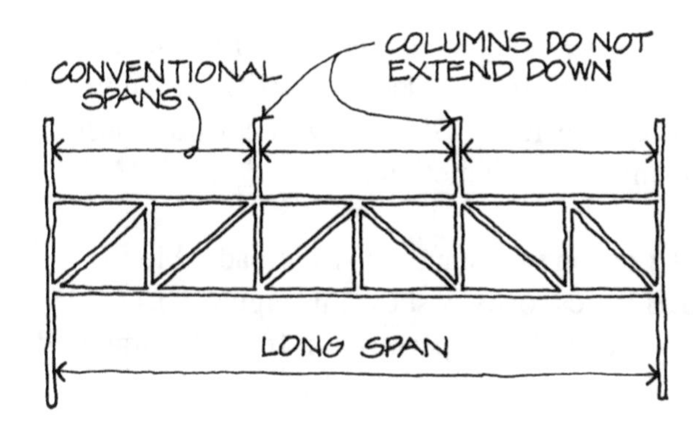

Figure 10.6

Trusses are discussed in greater detail in the following lesson.

Vierendeel Trusses

A Vierendeel truss, named after its Belgian inventor, is one without diagonals (Figure 10.7). Strictly speaking, therefore, it is not a truss at all, but a kind of rigid frame. As in other trusses, the bending moment is resisted by a couple, consisting of a compressive force in the top chord and a tensile force in the bottom chord (Figure 10.8). However, since there are no diagonal members, the chords must also resist shear, which is perpendicular to the axis of the chords. This produces local moments in the chord members, in addition to the axial tension or compression. The vertical members are also subject to bending moment caused by shear. Vierendeel trusses also tend to have high deflection. Therefore, to provide members adequate to resist both bending moment and direct stress, and to keep the deflection within acceptable limits, a Vierendeel truss requires much more material than a triangulated truss. Vierendeel trusses made of reinforced concrete or structural steel are used where diagonal members would conflict with door or windows, or where the Vierendeel configuration might be more desirable aesthetically than that of a truss with diagonals. Typical applications include bridges and deep transfer members supporting discontinuous columns above.

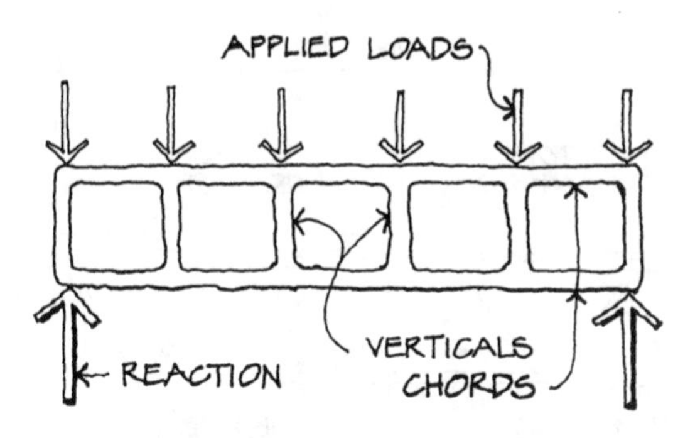

Figure 10.7

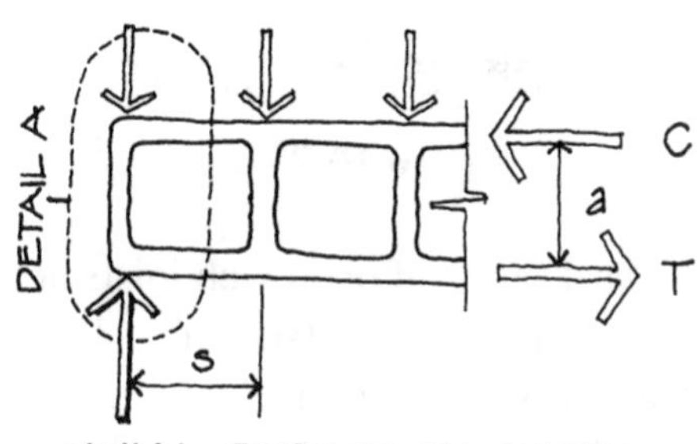

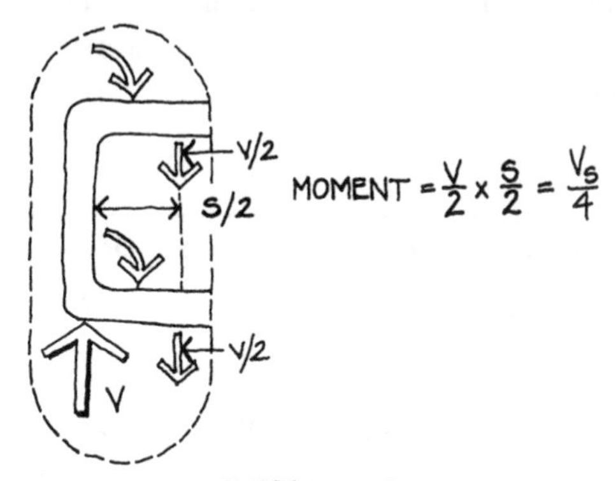

Figure 10.8

Open Web Steel Joists

Open web steel joists are closely-spaced, shop fabricated, standardized lightweight steel trusses that span between main girders, trusses, or bearing walls and support roof or floor loads (Figure 10.9). Floor or roof decking made of wood, steel, or concrete is placed over the top chords of the joists.

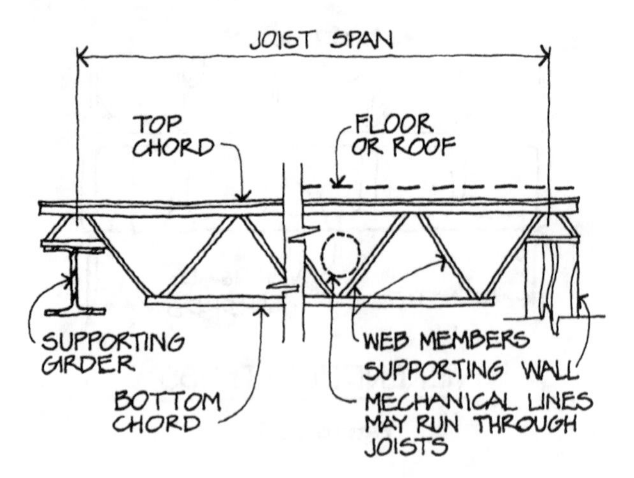

Figure 10.9

Three standard series of open web joists are available: K, LH, and DLH (see Table 10.1). The K series vary in depth in 2" increments from 8" to 30", the LH series from 18" to 48" in increments of 2" to 4", and the DLH series from 52" to 72" in 4" increments. Joists are generally made of steel with a yield point F_y of 50 ksi.

Series	Description	Depth	Usual Span
K	Standard	8"–30"	8'–60'
LH	Longspan	18"–48"	25'–96'
DLH	Deep Longspan	52"–72"	89'–144'

Table 10.1

Joists are designated by the joist depth, the series, and a number indicating the chord section. Thus, a 16K7 refers to a 16" deep joist of the K series with number 7 chord sections.

If you are given the uniform load and the span and asked to select the most economical joist, you use load tables in the Standard Specifications of the Steel Joist Institute, which is listed in the Bibliography.

Joists are designed using either Load and Resistance Factor Design (LRFD) or Allowable Stress Design (ASD). We will consider design using the ASD method.

A portion of a standard ASD load table for LH series joists is shown in Figure 10.13, reproduced with permission from the Standard Specifications of the Steel Joist Institute.

Open web joists are usually spaced two to three feet on center for floor construction and three to six feet on center for roofs. In all cases, the spacing should not be greater than the safe span of the deck material over the joists.

Joists are usually economical for supporting light loads on relatively long spans, and the ability to penetrate the joist webs with mechanical ducts is a great advantage. The economy of steel joists results from standardized shop fabrication and the ease with which they are erected because of their standardization and light weight. The depth-to-span ratio of joists is limited to 1/24, so a joist spanning 60 feet must have a depth of at least $60 \times 12/24 = 30$ inches.

Steel joists are deep and slender and therefore require bridging for stability and load distribution. The allowable spacing for bridging can be found in SJI tables, which are reproduced in Figures 10.10 and 10.11 by permission of the Steel Joist Institute.

The top chords of joists, which are in compression, are designed on the basis of being laterally supported in one direction by the web members and in the other direction by the decking. It is important, therefore, that the decking be well anchored to the joists, and the joists to their supports. Some typical joist details are shown in Figure 10.12.

TABLE 5.4-1

U. S. UNITS NUMBER OF ROWS OF TOP CHORD BRIDGING** Refer to the K-Series Load Table and Specification Section 6 for required bolted diagonal bridging. Distances are Joist Span lengths in feet - See "Definition of Span" preceding Load Table.					
*Section Number	One Row	Two Rows	Three Rows	Four Rows	Five Rows
#1	Up thru 16	Over 16 thru 24	Over 24 thru 28		
#2	Up thru 17	Over 17 thru 25	Over 25 thru		
#3	Up thru 18	Over 18 thru 28	Over 28 thru 38	Over 38 thru 48	
#4	Up thru 19	Over 19 thru 28	Over 28 thru 38	Over 38 thru 48	
#5	Up thru 19	Over 19 thru 29	Over 29 thru 39	Over 39 thru 50	Over 50 thru 52
#6	Up thru 19	Over 19 thru 29	Over 29 thru 39	Over 39 thru 51	Over 51 thru 56
#7	Up thru 20	Over 20 thru 33	Over 33 thru 45	Over 45 thru 58	Over 58 thru 60
#8	Up thru 20	Over 20 thru 33	Over 33 thru 45	Over 45 thru 58	Over 58 thru 60
#9	Up thru 20	Over 20 thru 33	Over 33 thru 46	Over 46 thru 59	Over 59 thru 60
#10	Up thru 20	Over 20 thru 37	Over 37 thru 51	Over 51 thru 60	
#11	Up thru 20	Over 20 thru 38	Over 38 thru 53	Over 53 thru 60	
#12	Up thru 20	Over 20 thru 39	Over 39 thru 53	Over 53 thru 60	

* Last digit(s) of joist designation shown in Load Table
** See Section 5.11 for additional bridging required for uplift design.

Figure 10.10

Table 104.5-1			
LH-DLH SECTION* NUMBER	MAX. SPACING OF LINES OF TOP CHORD BRIDGING	NOMINAL** HORIZONTAL BRACING FORCE	
		lbs	(N)
02,03,04	11'-0" (3352 mm)	400	(1779)
05,06	12'-0" (3657 mm)	500	(2224)
07,08	13'-0" (3962 mm)	650	(2891)
09,10	14'-0" (4267 mm)	800	(3558)
11,12	16'-0" (4876 mm)	1000	(4448)
13,14	16'-0" (4876 mm)	1200	(5337)
15,16	21'-0" (6400 mm)	1600	(7117)
17	21'-0" (6400 mm)	1800	(8006)
18,19	26'-0" (7924 mm)	2000	(8896)

Number of lines of bridging is based on joist clear span dimensions.
* Last two digits of joist designation shown in load table.
** Nominal bracing force is unfactored.

Figure 10.11

Example #1

Long span steel joists spanning 72 feet between girders are used to support a roof. The joists are spaced 6 feet on center, the dead load is 20 pounds per square foot (including the joist weight), the live load is 20 pounds per square foot, and the live load deflection is limited to 1/360 of the span. What is the lightest joist that can support the load? Use the table in Figure 10.13.

Solution:

First we convert the load in pounds per square foot to pounds per lineal foot of span, by multiplying the square foot loading by the joist spacing. Therefore, the total load supported by each joist is equal to $(20 + 20)6 = 240$ pounds per lineal foot. The live load per joist is equal to 20 × 6 = 120 pounds per lineal foot. For the joist designation 36LH09 we read the numbers 267 and 127 for a span of 72 feet, and similarly, for the joist designation 40LH09, we read the numbers 276 and 147. This means that the 36LH09 can safely support a total load of 267 pounds per lineal foot, and that a live load of 127 pounds per lineal foot will produce a deflection of 1/360 of the span. Since the actual total load is 240 pounds per lineal foot, which is less than 267 pounds per lineal foot, and the actual live load is 120 pounds per lineal foot, which is less than 127 pounds per lineal foot, the *36LH09* joist is satisfactory. Similarly, we can see that the *40LH09* joist is also satisfactory. Since both joists have the same weight of 21 pounds per lineal foot, either one is the lightest joist that can safely support the load. Note that joists ending with 08 or lower are inadequate to support the load, and those ending with 10 or higher are not economical, since they weigh more than the 36LH09 or 40LH09.

Joist Girders

A joist girder is a shop fabricated steel truss that supports evenly spaced steel joists along its top chord and usually spans up to about 60 feet. Knowing the span of the joist girder, the span of the steel joists, the load per square foot, and the joist spacing, the joist girder may be selected as shown in the following example.

Example #2

A joist girder spans 60 feet and supports joists spaced at 6'-0" on center which span 40 feet, as shown in Figure 10.14. The total dead and live load is 40 pounds per square foot including the estimated joist girder weight. What is the correct designation for the joist girder?

Solution:

1. Compute the number of joist spaces (N). Girder span/joist spacing = 60/6 = 10.

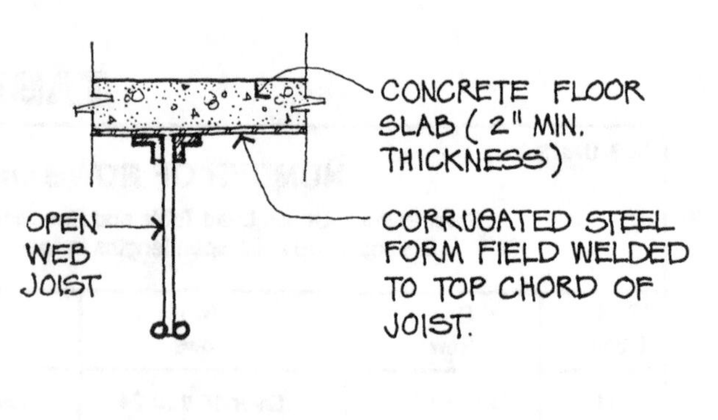

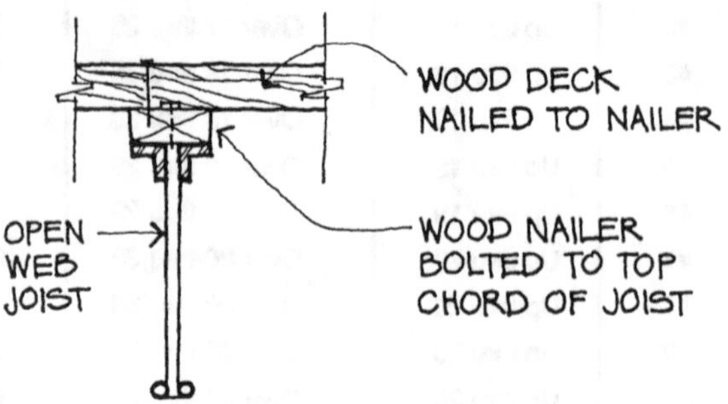

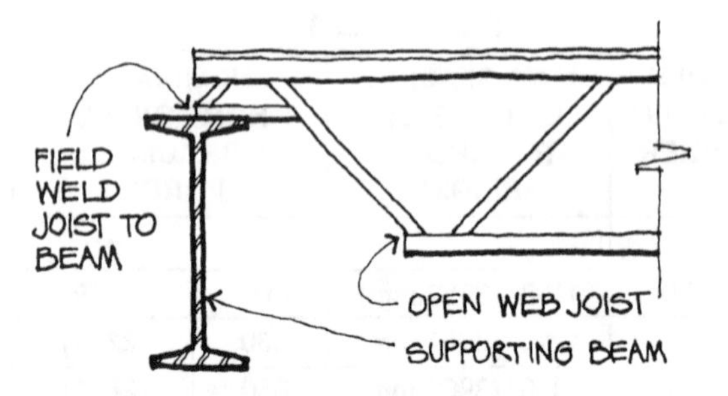

Figure 10.12

2. Compute the joist load per lineal foot = joist spacing × load per square foot = 6 × 40 = 240 pounds per lineal foot.
3. Compute the concentrated load at the top chord panel points = load per lineal foot × joist span = 240 × 40 = 9,600 pounds = 9.6 kips.

STANDARD LOAD TABLE/LONGSPAN STEEL JOISTS, LH-SERIES
Based on a Maximum Allowable Tensile Stress of 30 ksi

Joist Designation	Approx. Wt. in Lbs. per Linear Ft. (Joists Only)	Depth in Inches	SAFE LOAD* in Lbs. Between		CLEAR SPAN IN FEET															
			28–32		33	34	35	36	37	38	39	40	41	42	43	44	45	46	47	48
24LH03	11	24	11500		342 / 235	339 / 226	336 / 218	323 / 204	307 / 188	293 / 175	279 / 162	267 / 152	255 / 141	244 / 132	234 / 124	224 / 116	215 / 109	207 / 102	199 / 96	191 / 90
24LH04	12	24	14100		419 / 288	398 / 265	379 / 246	360 / 227	343 / 210	327 / 195	312 / 182	298 / 169	285 / 158	273 / 148	262 / 138	251 / 130	241 / 122	231 / 114	222 / 107	214 / 101
24LH05	13	24	15100		449 / 308	446 / 297	440 / 285	419 / 264	399 / 244	380 / 226	363 / 210	347 / 198	331 / 182	317 / 171	304 / 160	291 / 150	280 / 141	269 / 132	258 / 124	248 / 117
24LH06	16	24	20300		604 / 411	579 / 382	555 / 356	530 / 331	504 / 306	480 / 284	457 / 263	437 / 245	417 / 228	399 / 211	381 / 197	364 / 184	348 / 172	334 / 161	320 / 152	307 / 142
24LH07	17	24	22300		665 / 452	638 / 421	613 / 393	588 / 367	565 / 343	541 / 320	516 / 297	491 / 276	468 / 257	446 / 239	426 / 223	407 / 208	389 / 195	373 / 182	357 / 171	343 / 161
24LH08	18	24	23800		707 / 480	677 / 447	649 / 416	622 / 388	597 / 362	572 / 338	545 / 314	520 / 292	497 / 272	475 / 254	455 / 238	435 / 222	417 / 208	400 / 196	384 / 184	369 / 173
24LH09	21	24	28000		832 / 562	808 / 530	785 / 501	764 / 460	731 / 424	696 / 393	663 / 363	632 / 337	602 / 313	574 / 292	548 / 272	524 / 254	501 / 238	480 / 223	460 / 209	441 / 196
24LH10	23	24	29600		882 / 596	856 / 559	832 / 528	809 / 500	788 / 474	768 / 439	737 / 406	702 / 378	668 / 351	637 / 326	608 / 304	582 / 285	556 / 266	533 / 249	511 / 234	490 / 220
24LH11	25	24	31200		927 / 624	900 / 588	875 / 555	851 / 525	829 / 498	807 / 472	787 / 449	768 / 418	734 / 388	701 / 361	671 / 337	642 / 315	616 / 294	590 / 276	567 / 259	544 / 243
			33–40		41	42	43	44	45	46	47	48	49	50	51	52	53	54	55	56
28LH05	13	28	14000		337 / 219	323 / 205	310 / 192	297 / 180	286 / 169	275 / 159	265 / 150	255 / 142	245 / 133	237 / 126	228 / 119	220 / 113	213 / 107	206 / 102	199 / 97	193 / 92
28LH06	16	28	18600		448 / 289	429 / 270	412 / 253	395 / 238	379 / 223	364 / 209	350 / 197	337 / 186	324 / 175	313 / 166	301 / 156	291 / 148	281 / 140	271 / 133	262 / 126	253 / 120
28LH07	17	28	21000		505 / 326	484 / 305	464 / 285	445 / 267	427 / 251	410 / 236	394 / 222	379 / 209	365 / 197	352 / 186	339 / 176	327 / 166	316 / 158	305 / 150	295 / 142	285 / 135
28LH08	18	28	22500		540 / 348	517 / 325	496 / 305	475 / 285	456 / 268	438 / 252	420 / 236	403 / 222	387 / 209	371 / 196	357 / 185	344 / 175	331 / 165	319 / 156	308 / 148	297 / 140
28LH09	21	28	27700		667 / 428	639 / 400	612 / 375	586 / 351	563 / 329	540 / 309	519 / 291	499 / 274	481 / 258	463 / 243	446 / 228	430 / 216	415 / 204	401 / 193	387 / 183	374 / 173
28LH10	23	28	30300		729 / 466	704 / 439	679 / 414	651 / 388	625 / 364	600 / 342	576 / 322	554 / 303	533 / 285	513 / 269	495 / 255	477 / 241	460 / 228	444 / 215	429 / 204	415 / 193
28LH11	25	28	32500		780 / 498	762 / 475	736 / 448	711 / 423	682 / 397	655 / 373	629 / 351	605 / 331	582 / 312	561 / 294	540 / 278	521 / 263	502 / 249	485 / 236	468 / 223	453 / 212
28LH12	27	28	35700		857 / 545	837 / 520	818 / 496	800 / 476	782 / 454	766 / 435	737 / 408	709 / 383	682 / 361	656 / 340	632 / 321	609 / 303	587 / 285	566 / 270	546 / 256	527 / 243
28LH13	30	28	37200		895 / 569	874 / 543	854 / 518	835 / 495	816 / 472	799 / 452	782 / 433	766 / 415	751 / 396	722 / 373	694 / 352	668 / 332	643 / 314	620 / 297	598 / 281	577 / 266
			38-46	47-48	49	50	51	52	53	54	55	56	57	58	59	60	61	62	63	64
32LH06	14	32	16700	16700	338 / 211	326 / 199	315 / 189	304 / 179	294 / 169	284 / 161	275 / 153	266 / 145	257 / 138	249 / 131	242 / 125	234 / 119	227 / 114	220 / 108	214 / 104	208 / 99
32LH07	16	32	18800	18800	379 / 235	366 / 223	353 / 211	341 / 200	329 / 189	318 / 179	308 / 170	298 / 162	288 / 154	279 / 146	271 / 140	262 / 133	254 / 127	247 / 121	240 / 116	233 / 111
32LH08	17	32	20400	20400	411 / 255	397 / 242	383 / 229	369 / 216	357 / 205	345 / 194	333 / 184	322 / 175	312 / 167	302 / 159	293 / 151	284 / 144	275 / 137	267 / 131	259 / 125	252 / 120
32LH09	21	32	25600	25600	516 / 319	498 / 302	480 / 285	463 / 270	447 / 256	432 / 243	418 / 230	404 / 219	391 / 208	379 / 198	367 / 189	356 / 180	345 / 172	335 / 164	325 / 157	315 / 149
32LH10	21	32	28300	28300	571 / 352	550 / 332	531 / 315	512 / 297	495 / 282	478 / 267	462 / 254	445 / 240	430 / 228	416 / 217	402 / 206	389 / 196	376 / 186	364 / 178	353 / 169	342 / 162
32LH11	24	32	31000	31000	625 / 385	602 / 363	580 / 343	560 / 325	541 / 308	522 / 292	505 / 277	488 / 263	473 / 251	458 / 239	443 / 227	429 / 216	416 / 206	403 / 196	390 / 187	378 / 179
32LH12	27	32	36400	36400	734 / 450	712 / 428	688 / 406	664 / 384	641 / 364	619 / 345	598 / 327	578 / 311	559 / 295	541 / 281	524 / 267	508 / 255	492 / 243	477 / 232	463 / 221	449 / 211
32LH13	30	32	40600	40600	817 / 500	801 / 480	785 / 461	771 / 444	742 / 420	715 / 397	690 / 376	666 / 354	643 / 336	621 / 319	600 / 304	581 / 288	562 / 275	544 / 262	527 / 249	511 / 238
32LH14	33	32	41800	41800	843 / 515	826 / 495	810 / 476	795 / 458	780 / 440	766 / 417	738 / 395	713 / 374	688 / 355	665 / 337	643 / 321	622 / 304	602 / 290	583 / 276	564 / 264	547 / 251
32LH15	35	32	43200	43200	870 / 532	853 / 511	837 / 492	821 / 473	805 / 454	791 / 438	776 / 422	763 / 407	750 / 393	725 / 374	701 / 355	678 / 338	656 / 322	635 / 306	616 / 292	597 / 279
			42-46	47-56	57	58	59	60	61	62	63	64	65	66	67	68	69	70	71	72
36LH07	16	36	16800	16800	292 / 177	283 / 168	274 / 160	266 / 153	258 / 146	251 / 140	244 / 134	237 / 128	230 / 122	224 / 117	218 / 112	212 / 107	207 / 103	201 / 99	196 / 95	191 / 91
36LH08	18	36	18500	18500	321 / 194	311 / 185	302 / 176	293 / 168	284 / 160	276 / 153	268 / 146	260 / 140	253 / 134	246 / 128	239 / 123	233 / 118	227 / 113	221 / 109	215 / 104	209 / 100
36LH09	21	36	23700	23700	411 / 247	398 / 235	386 / 224	374 / 214	363 / 204	352 / 195	342 / 186	333 / 179	323 / 171	314 / 163	306 / 157	297 / 150	289 / 144	282 / 138	275 / 133	267 / 127
36LH10	21	36	26100	26100	454 / 273	440 / 260	426 / 248	413 / 236	401 / 225	389 / 215	378 / 206	367 / 197	357 / 188	347 / 180	338 / 173	328 / 165	320 / 159	311 / 152	303 / 146	295 / 140
36LH11	23	36	28500	28500	495 / 297	480 / 283	465 / 269	451 / 257	438 / 246	425 / 234	412 / 224	401 / 214	389 / 205	378 / 196	368 / 188	358 / 180	348 / 173	339 / 166	330 / 159	322 / 153
36LH12	25	36	34100	34100	593 / 354	575 / 338	557 / 322	540 / 307	523 / 292	508 / 279	493 / 267	478 / 255	464 / 243	450 / 232	437 / 222	424 / 213	412 / 204	400 / 195	389 / 187	378 / 179
36LH13	30	36	40100	40100	697 / 415	675 / 395	654 / 376	634 / 359	615 / 342	596 / 327	579 / 312	562 / 298	546 / 285	531 / 273	516 / 262	502 / 251	488 / 240	475 / 231	463 / 222	451 / 213
36LH14	36	36	44200	44200	768 / 456	755 / 434	729 / 412	706 / 392	683 / 373	661 / 356	641 / 339	621 / 323	602 / 309	584 / 295	567 / 283	551 / 270	535 / 259	520 / 247	505 / 237	492 / 228
36LH15	36	36	46600	46600	809 / 480	795 / 464	781 / 448	769 / 434	744 / 413	721 / 394	698 / 375	677 / 358	656 / 342	637 / 327	618 / 312	600 / 299	583 / 286	567 / 274	551 / 263	536 / 252

Figure 10.13

STANDARD LOAD TABLE/LONG SPAN STEEL JOISTS, LH-SERIES
Based on a Maximum Allowable Tensile Stress of 30 ksi

Joist Designation	Approx. Wt in Lbs. Per Linear Ft. (Joists Only)	Depth in inches	SAFELOAD* in Lbs. Between		CLEAR SPAN IN FEET															
			47-59	60-64	65	66	67	68	69	70	71	72	73	74	75	76	77	78	79	80
40LH08	16	40	16600	16600	254	247	241	234	228	222	217	211	206	201	196	192	187	183	178	174
					150	144	138	132	127	122	117	112	108	104	100	97	93	90	86	83
40LH09	21	40	21800	21800	332	323	315	306	298	291	283	276	269	263	256	250	244	239	233	228
					196	188	180	173	166	160	153	147	141	136	131	126	122	118	113	109
40LH10	21	40	24000	24000	367	357	347	338	329	321	313	305	297	290	283	276	269	262	255	249
					216	207	198	190	183	176	169	162	156	150	144	139	134	129	124	119
40LH11	22	40	26200	26200	399	388	378	368	358	349	340	332	323	315	308	300	293	286	279	273
					234	224	215	207	198	190	183	176	169	163	157	151	145	140	135	130
40LH12	25	40	31900	31900	486	472	459	447	435	424	413	402	392	382	373	364	355	346	338	330
					285	273	261	251	241	231	222	213	205	197	189	182	176	169	163	157
40LH13	30	40	37600	37600	573	557	542	528	514	500	487	475	463	451	440	429	419	409	399	390
					334	320	307	295	283	271	260	250	241	231	223	214	207	199	192	185
40LH14	35	40	43000	43000	656	638	620	603	587	571	556	542	528	515	502	490	478	466	455	444
					383	367	351	336	323	309	297	285	273	263	252	243	233	225	216	209
40LH15	36	40	48100	48100	734	712	691	671	652	633	616	599	583	567	552	538	524	511	498	486
					427	408	390	373	357	342	328	315	302	290	279	268	258	248	239	230
40LH16	42	40	53000	53000	808	796	784	772	761	751	730	710	691	673	655	638	622	606	591	576
					469	455	441	428	416	404	387	371	356	342	329	316	304	292	282	271
			52-59	60-72	73	74	75	76	77	78	79	80	81	82	83	84	85	86	87	88
44LH09	19	44	20000	20000	272	265	259	253	247	242	236	231	226	221	216	211	207	202	198	194
					158	152	146	141	136	131	127	122	118	114	110	106	103	99	96	93
44LH10	21	44	22100	22100	300	293	286	279	272	266	260	254	249	243	238	233	228	223	218	214
					174	168	162	155	150	144	139	134	130	125	121	117	113	110	106	103
44LH11	22	44	23900	23900	325	317	310	302	295	289	282	276	269	264	258	252	247	242	236	232
					188	181	175	168	162	157	151	146	140	136	131	127	123	119	115	111
44LH12	25	44	29600	29600	402	393	383	374	365	356	347	339	331	323	315	308	300	293	287	280
					232	224	215	207	200	192	185	179	172	166	160	155	149	144	139	134
44LH13	30	44	35100	35100	477	466	454	444	433	423	413	404	395	386	377	369	361	353	346	338
					275	265	254	246	236	228	220	212	205	198	191	185	179	173	167	161
44LH14	31	44	40400	40400	549	534	520	506	493	481	469	457	446	436	425	415	406	396	387	379
					315	302	291	279	268	259	249	240	231	223	215	207	200	193	187	181
44LH15	36	44	47000	47000	639	623	608	593	579	565	551	537	524	512	500	488	476	466	455	445
					366	352	339	326	314	303	292	281	271	261	252	243	234	227	219	211
44LH16	42	44	54200	54200	737	719	701	684	668	652	637	622	608	594	580	568	555	543	531	520
					421	405	390	375	362	348	336	324	313	302	291	282	272	263	255	246
44LH17	47	44	58200	58200	790	780	769	759	750	732	715	699	683	667	652	638	624	610	597	584
					450	438	426	415	405	390	376	363	351	338	327	316	305	295	285	276
			56-59	60-80	81	82	83	84	85	86	87	88	89	90	91	92	93	94	95	96
48LH10	21	48	20000	20000	246	241	236	231	226	221	217	212	208	204	200	196	192	188	185	181
					141	136	132	127	123	119	116	112	108	105	102	99	96	93	90	87
48LH11	22	48	21700	21700	266	260	255	249	244	239	234	229	225	220	216	212	208	204	200	196
					152	147	142	137	133	129	125	120	117	113	110	106	103	100	97	94
48LH12	25	48	27400	27400	336	329	322	315	308	301	295	289	283	277	272	266	261	256	251	246
					191	185	179	173	167	161	156	151	147	142	138	133	129	126	122	118
48LH13	29	48	32800	32800	402	393	384	376	368	360	353	345	338	332	325	318	312	306	300	294
					228	221	213	206	199	193	187	180	175	170	164	159	154	150	145	141
48LH14	32	48	38700	38700	475	464	454	444	434	425	416	407	399	390	383	375	367	360	353	346
					269	260	251	243	234	227	220	212	206	199	193	187	181	176	171	165
48LH15	36	48	44500	44500	545	533	521	510	499	488	478	468	458	448	439	430	422	413	405	397
					308	298	287	278	269	260	252	244	236	228	221	214	208	201	195	189
48LH16	42	48	51300	51300	629	615	601	588	576	563	551	540	528	518	507	497	487	477	468	459
					355	343	331	320	310	299	289	280	271	263	255	247	239	232	225	218
48LH17	47	48	57600	57600	706	690	675	660	646	632	619	606	593	581	569	558	547	536	525	515
					397	383	371	358	346	335	324	314	304	294	285	276	268	260	252	245

* The safe uniform load for the clear spans shown in the Safe Load Column is equal to (Safe Load)/(Clear span + 0.67). (The added 0.67 feet (8 inches) is required to obtain the proper length on which the Load Tables were developed).

In no case shall the safe uniform load, for clear spans less than the minimum clear span shown in the Safe Load Column, exceed the uniform load calculated for the minimum clear span listed in the Safe Load Column.

To solve for live loads for clear spans shown in the Safe Load Column (or lesser clear spans), multiply the live load of the shortest clear span shown in the Load Table by the (the shortest clear span shown in the Load Table + 0.67 feet)^2 and divide by (the actual clear span + 0.67 feet)^2. The live load shall not exceed the safe uniform load.

Figure 10.13 (continued)

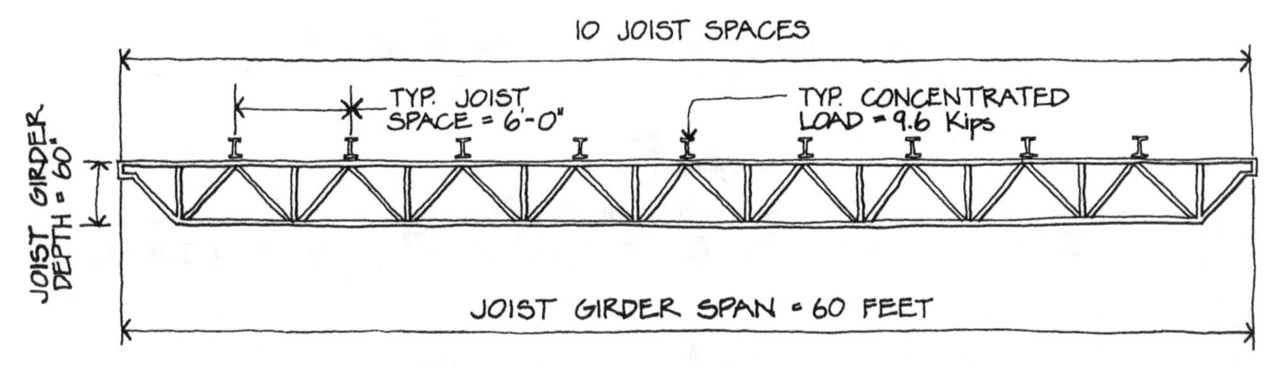

Figure 10.14

4. Select the joist girder depth. Assume one inch of depth for each foot of span (depth to span ratio = 1/12). Therefore select a joist girder depth of 60 inches.
5. The joist girder is designated as follows:

60G	10N	9.6K
Depth in inches	No of joist spaces	Kip load on each panel point

6. Enter the joist girder table in the Standard Specifications of the Steel Joist Institute, a portion of which is reproduced in Figure 10.15 by permission of the Steel Joist Institute. The table shows the weight for a 60G10N10K as 57 pounds per lineal foot. The designer should verify that this is not greater than the weight assumed in estimating the dead load.
7. Check the deflection under live load and limit it to 1/360 of the span for plastered ceilings; however, this requirement rarely governs. For this purpose, the moment of inertia of the joist girder may be approximated using an empirical formula.

Note that the exact configuration of the joist girder will vary, depending on the manufacturer.

Glued Laminated Beams

The only wood beams used for long span construction are glued laminated beams (glulams), which combine the natural beauty of wood with permanence, fire resistance, strength, and economy. Glued laminated beams are often used in schools, churches, and one-story industrial buildings, and maximum spans are about 40 feet for floors and 100 feet for roofs. A glued laminated beam is an assembly of wood laminations, 1" or 2" in nominal thickness, in which the grain of all laminations is approximately parallel longitudinally, and in which the laminations are bonded with adhesives. Some of the characteristics of glued laminated beams are the following:

1. The wood species is most often Douglas Fir or Southern Pine.
2. The individual laminations are properly seasoned and joined, and the assembly of laminations is glued together in a shop under strict quality control.
3. Using the proper adhesives, the joints are durable and at least as strong as the wood itself.
4. Because the individual laminations are thoroughly dried before assembly, finished members are relatively free from seasoning action and therefore tend to be dimensionally stable.

DESIGN GUIDE WEIGHT TABLE FOR JOIST GIRDERS

ASD

GIRDER SPAN (ft.)	JOIST SPACES (ft.)	GIRDER DEPTH (in.)	JOIST GIRDER WEIGHT – POUNDS PER LINEAR FOOT																	
			LOAD ON EACH PANEL POINT – KIPS																	
			4	5	6	7	8	9	10	11	12	13	14	16	18	20	22	24	26	28
55	5N@ 11.00	44	21	21	24	25	29	32	35	38	41	43	47	53	59	63	71	82	83	86
		48	21	21	23	24	28	30	[illegible]	35	38	41	43	49	56	60	64	71	73	83
		52	20	22	23	25	27	29	32	33	36	39	42	44	52	57	65	66	74	74
		56	20	21	24	24	26	28	31	33	36	37	39	44	51	53	58	66	66	74
		60	23	24	24	24	27	27	31	33	35	38	38	45	47	52	60	61	67	68
		66	24	24	24	25	26	28	28	33	34	37	37	42	47	48	55	56	62	69
	6N@ 9.17	44	19	22	26	29	33	36	38	43	45	51	52	59	66	75	86	86	98	101
		48	20	22	24	28	31	33	[illegible]	40	44	46	50	56	64	68	75	87	89	98
		52	20	22	24	26	29	33	35	37	41	59	59	66	74	86	93	99	109	110
		56	18	21	24	25	28	31	35	36	39	42	47	52	55	63	70	71	78	91
		60	20	21	24	25	29	30	33	35	38	39	43	48	55	60	64	71	75	80
		66	19	20	22	24	28	30	31	33	36	39	40	47	50	56	62	65	73	73
	7N@ 7.86	44	21	24	28	33	36	39	44	50	53	59	59	70	75	87	97	102	111	120
		48	21	24	27	31	34	38	[illegible]	45	51	54	56	65	72	76	89	98	103	110
		52	21	23	26	29	33	36	39	44	46	52	55	62	69	74	86	91	100	105
		56	20	22	25	28	31	35	38	40	46	48	53	55	64	70	79	87	92	101
		60	21	22	24	27	30	33	36	39	41	47	49	56	64	68	72	81	93	94
		66	22	22	24	26	30	32	35	37	40	43	48	52	58	65	70	74	83	84
	9N@ 6.11	44	24	29	34	39	46	52	55	60	67	74	74	87	98	105	116	135	137	
		48	24	28	32	38	40	47	[illegible]	57	61	68	69	81	97	103	107	118	129	139
		52	25	30	33	39	43	47	52	57	65	65	73	77	90	104	105	114	125	133
		56	24	29	32	38	43	46	51	53	59	66	67	75	87	92	105	107	117	128
		60	24	27	32	36	40	45	47	52	56	60	67	71	80	93	95	108	109	118
		66	24	27	31	35	39	42	46	49	54	58	61	71	78	83	91	97	111	113
	11N@ 5.00	44	30	36	43	49	55	63	67	74	87	88	97	106	126	137				
		48	28	33	39	45	54	61	[illegible]	69	76	87	89	103	112	128	139			
		52	27	34	37	44	52	55	62	66	73	77	88	99	105	115	131	142		
		56	27	33	39	42	48	54	60	64	68	77	80	93	102	107	118	134	146	
		60	26	31	37	40	47	49	58	64	67	72	77	82	95	108	110	121	137	148
		66	26	31	36	39	45	50	54	60	65	68	74	82	97	98	113	117	126	141
60	5N@ 12.00	48	21	23	27	29	33	35	[illegible]	43	44	49	51	57	63	69	76	87	89	94
		52	21	22	27	28	31	33	36	40	44	45	47	52	60	65	69	77	85	90
		56	22	23	24	28	30	31	34	36	41	44	45	52	59	63	69	74	78	87
		60	22	23	24	28	29	32	34	35	40	42	45	49	53	60	66	70	75	80
		66	24	24	24	26	30	30	33	35	36	38	42	47	51	56	61	67	72	73
		72	25	25	25	25	27	30	31	35	36	37	39	45	48	56	56	63	69	70
	6N@ 10.00	48	20	24	29	32	36	38	[illegible]	47	49	56	60	67	72	80	93	93	112	113
		52	20	23	28	30	33	37	39	46	48	50	57	62	69	78	80	94	94	113
		56	19	24	25	30	33	38	39	42	48	49	51	58	66	69	79	83	95	96
		60	19	23	24	29	32	34	39	40	43	49	50	57	63	70	75	83	83	96
		66	19	23	24	27	32	32	34	40	42	44	50	52	61	65	69	77	84	85
		72	22	22	24	27	28	33	34	36	41	43	44	52	54	63	68	71	75	87
	8N@ 7.50	48	24	29	34	39	43	49	[illegible]	57	64	72	72	80	93	112	123	125	136	148
		52	23	29	31	37	40	48	50	57	58	66	72	81	94	103	114	125	127	139
		56	23	26	31	36	38	44	49	51	58	60	66	75	83	96	104	116	127	129
		60	23	26	32	33	39	42	47	50	53	59	61	69	77	85	98	106	118	129
		66	28	30	33	34	41	43	46	48	53	57	62	70	78	82	90	100	108	120
		72	29	30	31	34	36	41	46	47	52	58	59	66	73	80	90	92	104	110
	10N@ 6.00	48	26	32	37	44	49	55	60	67	74	79	87	97	105	118	137	138		
		52	28	34	38	44	50	56	64	65	71	75	88	97	103	113	130	138		
		56	27	33	37	43	46	51	[illegible]	66	65	72	76	90	104	105	123	131	143	
		60	25	31	37	30	35	51	57	60	66	70	73	86	93	104	111	126	134	
		66	27	32	37	42	49	51	56	62	65	72	74	85	95	102	120	122	134	145
		72	26	32	33	38	42	47	50	55	59	66	69	74	83	96	98	111	111	121
	12N@ 5.00	48	33	39	46	53	59	68	75	86	87	97	102	111	135					
		52	31	37	45	51	57	65	69	76	88	89	98	104	118	139				
		56	29	36	41	48	55	62	66	72	77	89	91	104	113	129	140			
		60	30	35	39	47	54	56	64	73	74	79	91	102	106	116	133	145		
		66	32	35	41	48	53	61	62	70	77	80	87	100	110	122	134	147	164	
		72	29	33	38	42	50	52	60	61	69	72	77	86	100	110	114	127	142	151
	15N@ 4.00	48	40	49	64	72	80	93	102	113	124	126	136							
		52	39	48	57	66	74	81	94	103	114	126	127	150						
		56	38	46	53	67	71	80	83	96	104	116	127	140	153					
		60	38	42	51	60	68	76	83	89	98	106	118	132	144					
		66	35	41	49	55	62	70	81	87	87	103	110	123	136	153	167			
		72	35	44	46	55	64	66	77	85	90	93	106	125	139	142	160	171		

Figure 10.15

5. Glued laminated members are stronger than conventional sawn timbers, for two main reasons: (1) the individual laminations, being only 1" or 2" in nominal thickness, are easily seasoned before fabrication, and (2) the laminations are selected free of defects normally found in larger timbers and any defects are small and dispersed.
6. Two types of glue are used: casein glue for interior applications and waterproof glue (phenol, resorcinol, or melamine) when exposed to weather or excessive interior moisture.
7. Glulam beams resist destruction by fire and are therefore considered to be *heavy timber* construction by the International Building Code, if they are at least 6 inches wide and 10 inches deep. This is essentially the equivalent of one-hour fire-resistive construction.
8. The basic methods of design are the same as those for sawn timbers, with modifications made to the design values for a number of factors, including curvature factor, volume factor, beam stability factor, and short-time loading.
9. The standard finished widths of glued laminated members are less than the nominal widths. For example, a glulam with a nominal width of 4" has a net finished width of 3-1/8".
10. The standard depths are based on an exact number of laminations; the net depth of a 2" nominal lamination is 1-1/2", and the net depth of a 1" nominal lamination is 3/4".

In long span roof construction, glulam beams are generally used as primary members, with lighter secondary joists or purlins spanning between them. Glued laminated beams are usually simple beams, but for multi-span construction, it is sometimes economical to design beams for simulated continuity, as shown in Figure 9.13.

Prestressed Concrete Beams

Although conventional reinforced concrete members can be used to span long distances, they tend to be excessively bulky. More often, prestressed concrete members are used, which are usually smaller in cross section, lighter in weight, and deflect and crack less than conventional reinforced concrete.

In prestressed concrete, internal stresses are introduced that counteract the stresses caused by external dead and live loads. Consequently, tensile stresses are reduced or eliminated, cracking is minimized or eliminated altogether, deflections are limited, and member sizes are made smaller.

It is interesting to contrast the development of prestressed concrete in Europe with that in the United States. Historically, material costs in this country have been relatively low, while labor costs have been relatively high. In Europe and other parts of the world, the reverse has been true. Consequently, prestressed concrete designs in Europe have tended to be sophisticated, in order to reduce concrete and steel material to a minimum, despite the great amount of skilled labor required to execute such designs. In the United States, on the other hand, most prestressed concrete is in the form of relatively small, mass-produced precast members. Of course, all of this is changing, now that labor costs in Europe have caught up with and even surpassed those in this country.

Prestress force can be applied to a concrete beam in two ways: *pretensioning* and *posttensioning*. In pretensioning, high strength steel wires are tensioned between abutments in a casting yard before placing the concrete in the beam forms (Figure 10.16). The concrete is then placed, and after the concrete has attained adequate strength, the prestress wires are cut. This transfers the tensile force in the wires as a compressive force to the concrete by bond and friction along the steel strands. Some loss of prestress occurs because of creep and shrinkage, slip, and friction. Pretensioning lends itself to mass production, since the casting beds can be hundreds of feet long, the entire length cast at one time, and individual beams cut to the required lengths.

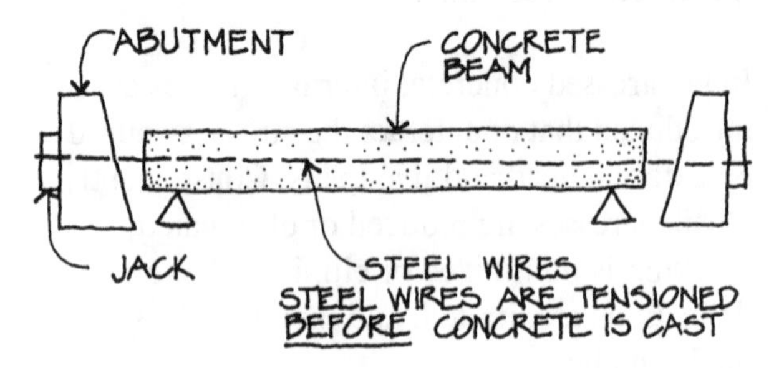

Figure 10.16

In posttensioned construction, the concrete is cast with a hollow conduit or sleeve through which the prestressing wires are passed (Figure 10.17). The concrete is cured and after it has acquired sufficient strength, the steel tendons are stressed by jacking against anchorages at each end of the member. Frequently, one end of the prestressing tendon is anchored, and the force is applied at the other end. After the desired amount of prestress force is reached, the wires are permanently locked against the concrete, and the jacking equipment is removed. The prestress losses caused by friction, elastic shortening, and shrinkage are generally less than with pretensioning. A number of proprietary systems, jacks, and fittings have been developed, and the contractor is generally free to select any system that meets the designer's requirements.

The two most common shapes for precast, pretensioned concrete members are the single T and the double T, which are illustrated in Figure 10.18. The single T is used for spans up to 100 feet or more. The double T is probably the most popular cross section used in this country. A flat surface is provided, usually four feet in width, and spans usually do not exceed 60 feet.

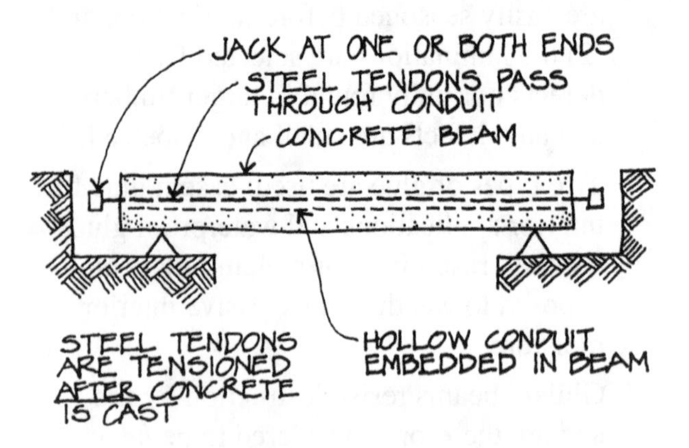

Figure 10.17

The maximum span-depth ratio for single and double Ts is about 36 to 44, about double that of a conventional reinforced concrete beam. Thus, for a given span, a prestressed T will be about half as deep as a conventional reinforced concrete beam.

Other standard shapes that are used include I sections, channel shaped slabs, and box girders, which are shown in Figure 10.19.

Special shapes may also be used, but they are only economical when there is enough repetition to justify the cost of special forms.

T, I, and box sections with thin webs and flanges are usually more efficient than members with thick parts. However, there are a number of limiting factors. Overhanging elements in compression become unstable if they are too thin, thin parts are vulnerable to breakage in shipping and handling, concrete is difficult to place in very thin sections, and sufficient space must be provided for the steel prestressing wires.

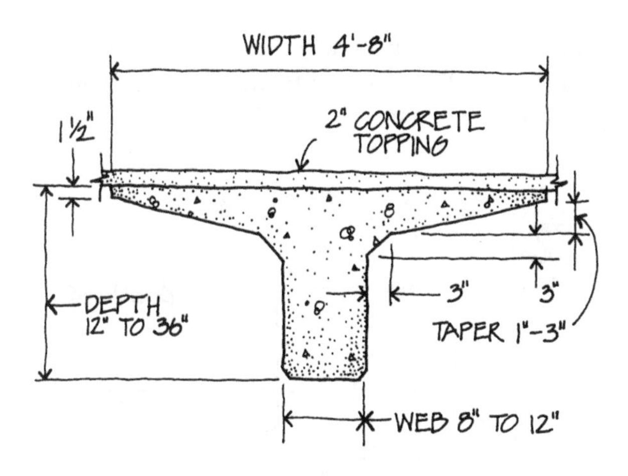

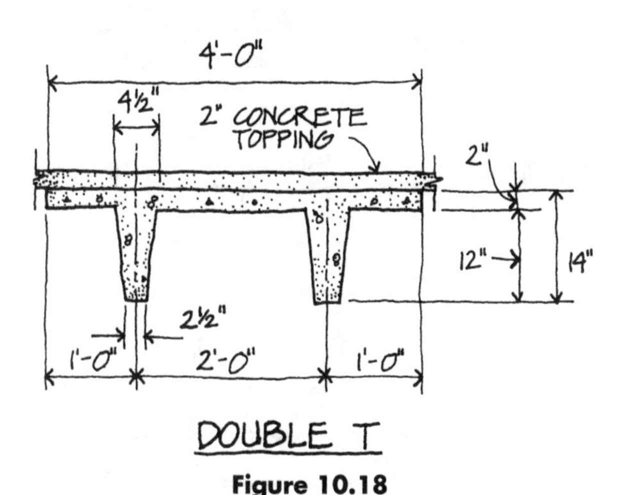

Figure 10.18

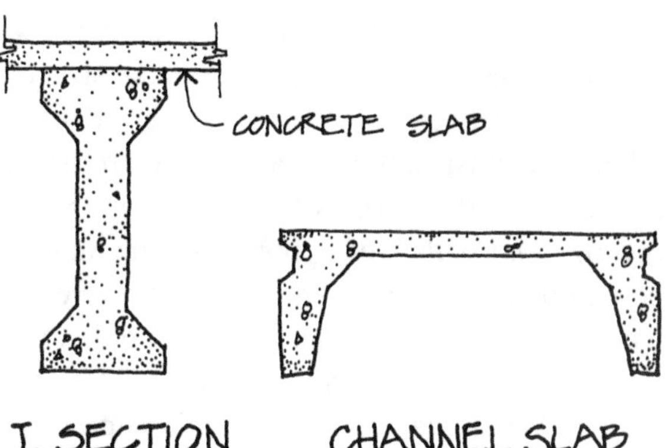

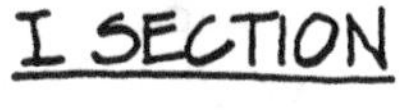

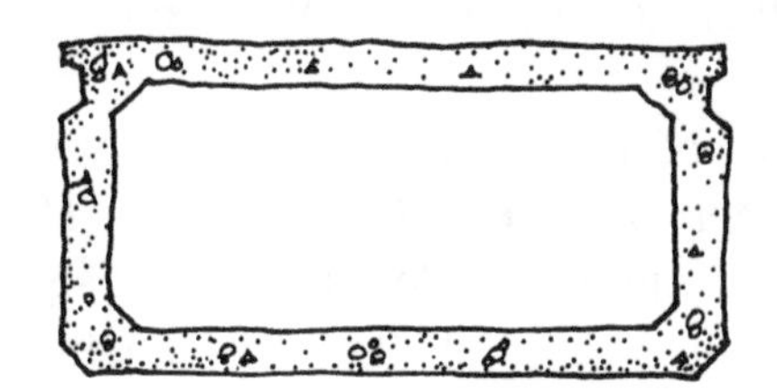

BOX GIRDER

Figure 10.19

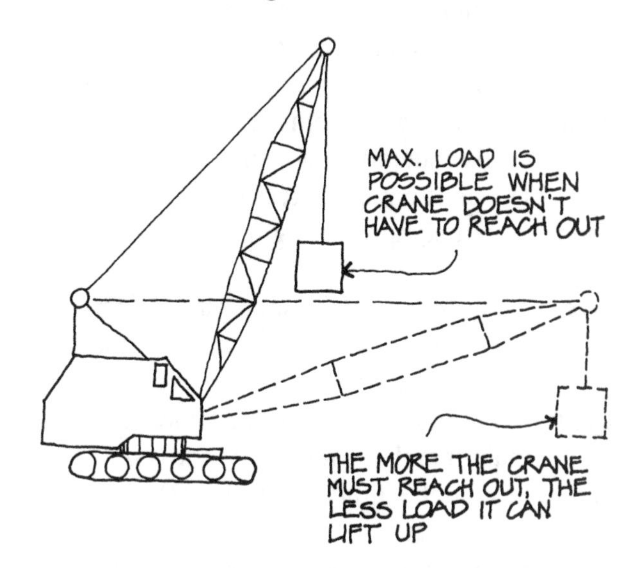

Figure 10.20

The architect must also consider the handling, shipping, erection, and connection problems involved with precast, prestressed members. It is generally advisable to use the largest size member that can be shipped and erected, in order to minimize the number of field connections. See page 185 for maximum shipping lengths and widths. The maximum weight of each member to be erected depends on the crane capacity and how far the crane must reach out in order to place the member into position (Figure 10.20). The less the crane has to reach out, the heavier the piece can be. Usually the weight of an individual member should not exceed ten tons.

TWO-WAY FLEXURAL SYSTEMS

Up to now, we have discussed one-way systems, in which the main members span in one direction and secondary members run perpendicular to them. A two-way system consists of a series of trusses or girders that intersect each other in a consistent grid pattern and are rigidly connected at their points of intersection (Figure 10.21). Loads are supported by the members in both directions, and the entire system works as a unit, in a manner similar to that of a reinforced concrete two-way slab or flat plate system. When the spans in both directions are approximately equal, a two-way system is often more efficient than a one-way system.

The depth of a two-way system usually varies between 1/12 and 1/20 of the span. Thus, a two-way system to enclose a 200-foot square space might be around 10 feet deep, while a comparable one-way system might be as much as 20 feet in depth. The shallower depth of the two-way system results in less wall area, reduced building volume, and consequently, lower construction and operating costs. However, two-way systems are more complex to build than one-way systems. Two-way systems are statically indeterminate, and their structural analysis is complex, requiring the use of sophisticated computer programs.

Two-way systems may be supported by perimeter walls or by piers or columns. They may involve only simple spans or cantilevers. Cantilevers reduce the positive moment resisted by the main members, resulting in less material and reduced deflection (Figure 10.22).

Two-way systems form a grid pattern, consisting of 30 foot to 60 foot square modules (Figure 10.23). How is this square area between main members infilled structurally? Being square, it lends itself ideally to a secondary two-way system, such as a two-way concrete slab, a hyperbolic paraboloid, a pyramidal structure, or a small two-way framed system. However, a one-way system may also be employed, such as beams or joists which span in alternate directions to form a checkerboard pattern.

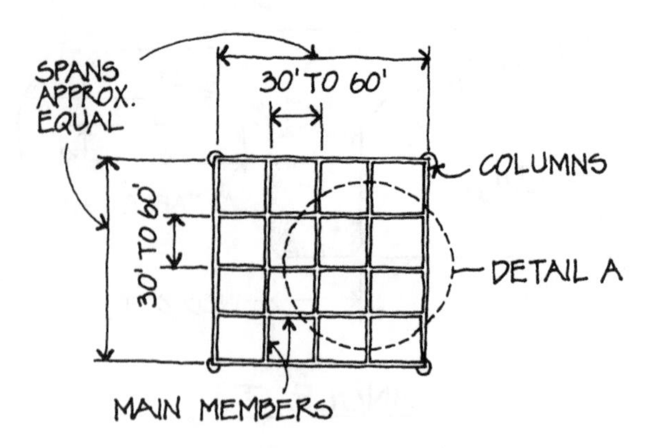

Figure 10.21

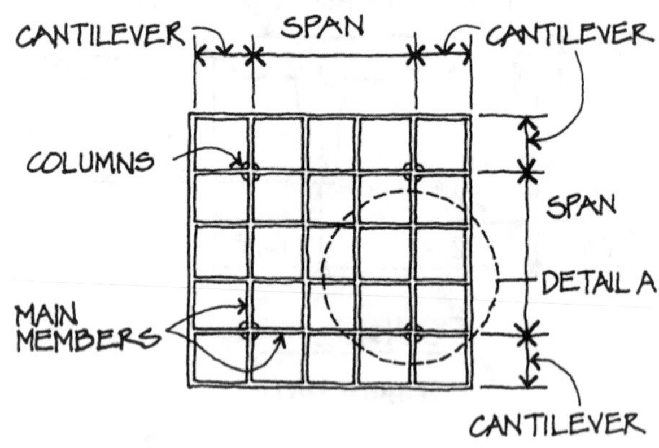

Figure 10.22

The main members in a two-way system may be plate girders or trusses. Plate girders are relatively simple to fabricate and erect, but they generally weigh more than trusses. The simplest two-way truss system consists of a series of trusses intersecting at right angles to form a grid, as shown in Figure 10.24.

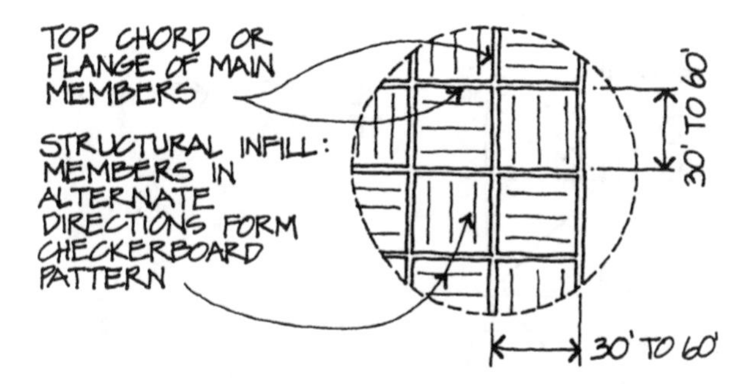

Figure 10.23

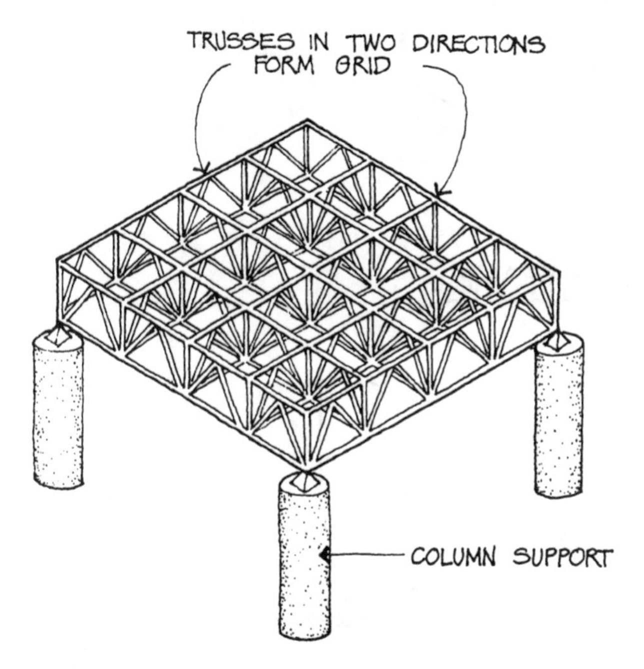

Figure 10.24

The two-way systems described above comprise trusses or girders that are in a vertical plane. If we use equal depth trusses and offset the top and bottom chords in both directions by half a module, we create a two-way layout called a *space truss* or *space frame.* The trusses are not in a vertical plane, but are inclined. Each chord member serves two adjacent trusses, and each web member is a diagonal of a truss in each direction. Thus, in the plan in Figure 10.25, web member 1 is a part of both east-west truss A and north-south truss B.

As in conventional trusses, the truss members may be tubular sections, tees, single or double angles, or wide flange sections. Connections may be welded or bolted. Welded connections are often preferred because they eliminate the need for connection material and devices and they have the neatest appearance. For field assembly, however, bolted connections are usually preferable.

Space frames may be supported by continuous perimeter walls or by isolated columns. Where columns are used, they are sometimes tree-like, with four or more branches at the top, each extending out to support a joint of the space frame.

Space frames are often economical for enclosing large column-free spaces with spans of 80 to 250 feet, as in arenas and exhibition halls, and the key to their economy is the use of repetitive members and connections that can be mass produced and easily assembled. Other advantages include reduced material weight, greater stiffness, and reduced depth of construction. In addition to their economy, space frames also have great visual appeal. The enclosing of vast spaces with a structure of apparent lightness and interesting geometry provides an expression that is both practical and aesthetically satisfying.

There are a number of interesting variations of the two-way flexural system. For example, a three-way system may be used, which has a triangular grid in plan and forms a series of tetrahedrons, which are internally rigid, unlike the rectangular pyramidal modules formed by two-way space frames.

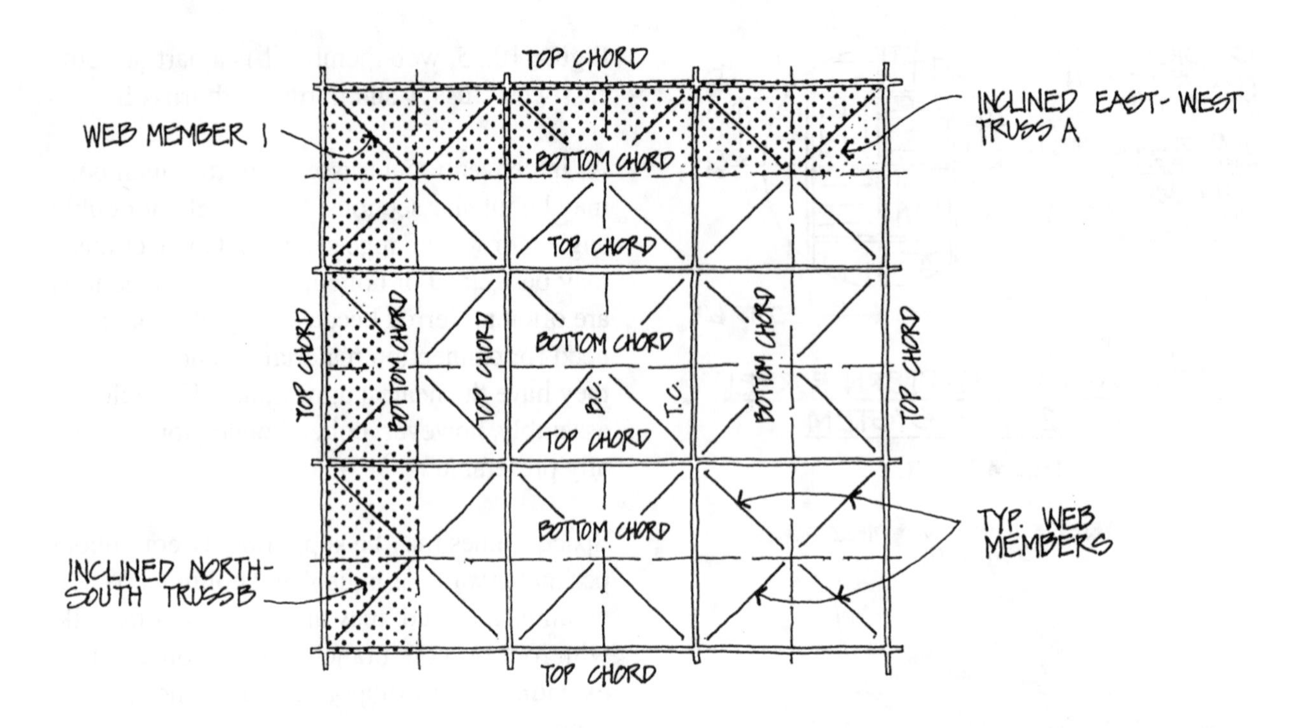
TOP CHORD
WEB MEMBER 1
INCLINED EAST-WEST TRUSS A
BOTTOM CHORD
TOP CHORD
TOP CHORD
BOTTOM CHORD
TOP CHORD
BOTTOM CHORD
B.C.
T.C.
BOTTOM CHORD
TOP CHORD
TOP CHORD
BOTTOM CHORD
TYP. WEB MEMBERS
INCLINED NORTH-SOUTH TRUSS B
TOP CHORD
PLAN OF TYPICAL SPACE FRAME

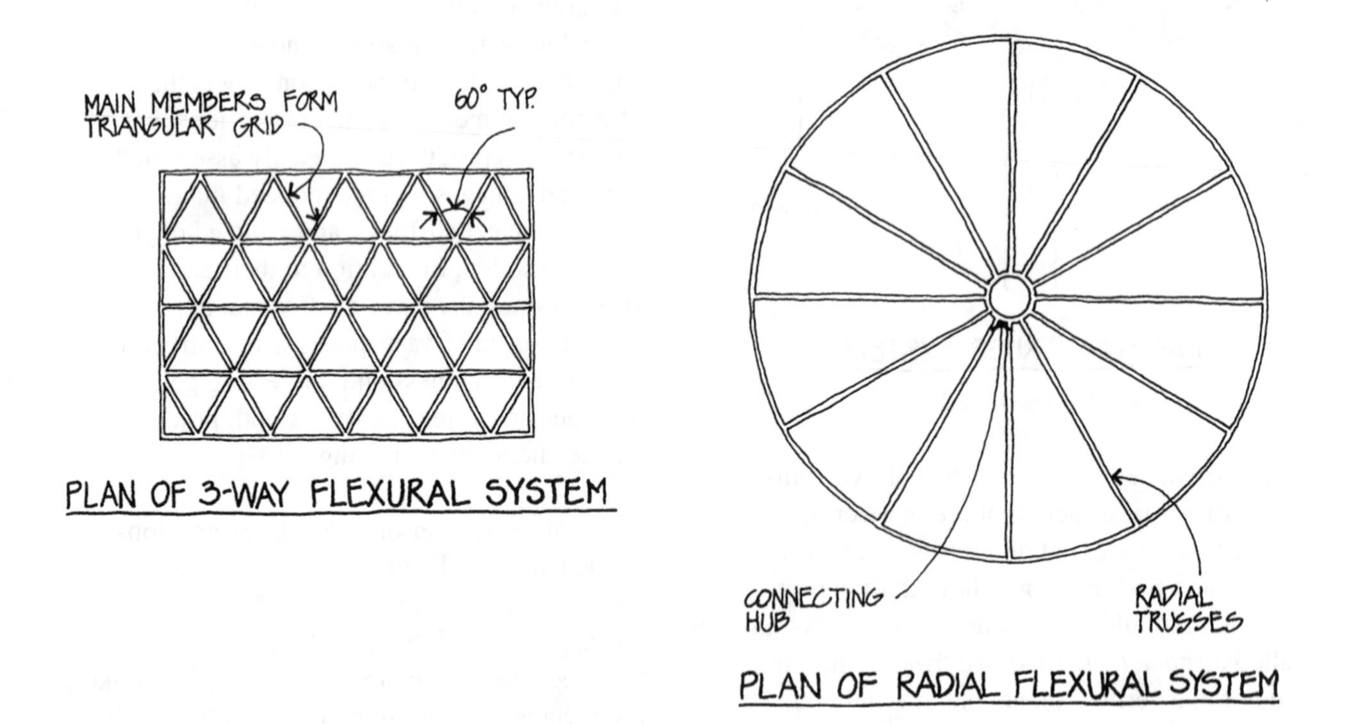
MAIN MEMBERS FORM TRIANGULAR GRID
60° TYP.
PLAN OF 3-WAY FLEXURAL SYSTEM
CONNECTING HUB
RADIAL TRUSSES
PLAN OF RADIAL FLEXURAL SYSTEM

Figure 10.25

For circular or oval buildings with flat roofs, radial trusses may be used, which intersect at a central hub.

AXIAL SYSTEMS

As we have seen, long span structures can be built using flexural structural elements, such as beams, girders, or trusses, spanning in one or both directions. However, it is often more economical to use a system made up of curved members, such as arches or cables, in which the stresses are axial, that is acting parallel to the length of the member, rather than flexural. The internal stresses in an arch are primarily compression, while those in a cable are purely tension.

Arches

As previously discussed, arches have been used for thousands of years to span long distances. This is because an arch is essentially a compressive structure, and natural materials strong in compression, such as stone and bricks made from mud, have always been readily available. The Romans, the greatest builders of antiquity, were masters of arch construction. Arch bridges were used profusely along their road network, their aqueducts carried water over the tops of arches, and their buildings and monuments used arches in a variety of ways. Today, we still use arches to span long distances, but the stone masonry and mud brick construction of the ancients have given way to reinforced concrete, structural steel, and glued laminated wood.

Although arches are often efficient and visually appealing, their use is mostly limited to long span roofs, such as hangars, industrial buildings, and gymnasiums, where their curvature can be accommodated more readily than in floor construction.

Arch supports may be hinged or fixed. Hinged supports permit the arch base to rotate, while fixed supports do not. Hinged arches are flexible, and do not develop high bending moments caused by temperature changes or foundation settlement; fixed arches, on the other hand, are more rigid and develop high bending stresses under temperature variations or soil settlement. Some hinged arches have an additional hinge at the high point, or crown of the arch. Such arches, called three-hinged arches, are statically determinate, and find their widest application in glued laminated arches. All arches other than three-hinged arches are statically indeterminate, and require methods additional to the laws of statics for their analysis.

For an arch with hinged bases supporting a uniformly distributed load across the entire span, the shape of the pure arch is a parabola. By *pure* arch, we mean one that is stressed only in compression, without any bending when supporting a load. Most arches are not pure arches: although they are stressed primarily in compression, they are also subject to some bending moment. There are two main reasons for this:

1. The arch may have a shape other than parabolic. This creates some bending moment, whose magnitude depends on how much the arch's profile deviates from a parabola.
2. An arch is pure for only one specific loading condition. Thus, if the arch is stressed purely in compression under dead load only, it will subject to some bending moment when it supports any other load, such as a non-uniform live load or a horizontal load. Bending moments may also result from foundation settlement or temperature.

Let's examine the basic action of a three-hinged arch (refer to Figure 10.26).

By symmetry, the vertical reaction at each end of the arch is equal to one-half of the total load supported, or wL/2. The horizontal reaction can easily be computed by taking moments about the crown of the arch.

$\Sigma M = 0$
$VL/2 - Hh - wL/2 \times L/4 = 0$
$Hh = wL/2 \times L/2 - wL^2/8 = wL^2/4 - wL^2/8 = wL^2/8$
$H = wL^2/8h$

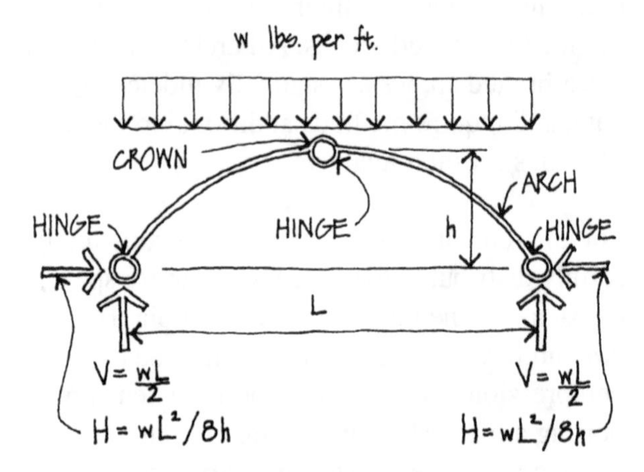

Figure 10.26

This analysis tells us a couple of things:

1. When an arch is loaded, it requires a horizontal reaction at its base. Without this horizontal force, called the arch *thrust*, the arch would spread outward and collapse. The thrust is usually resisted either by a tension tie rod connecting the bases or by an abutment (see Figure 10.27).
2. The horizontal thrust is equal to $wL^2/8h$. Therefore, *the greater the rise of the arch, the smaller the horizontal thrust.* In general, the rise of an arch is about 1/8 to 1/5 the span (Figure 10.28).

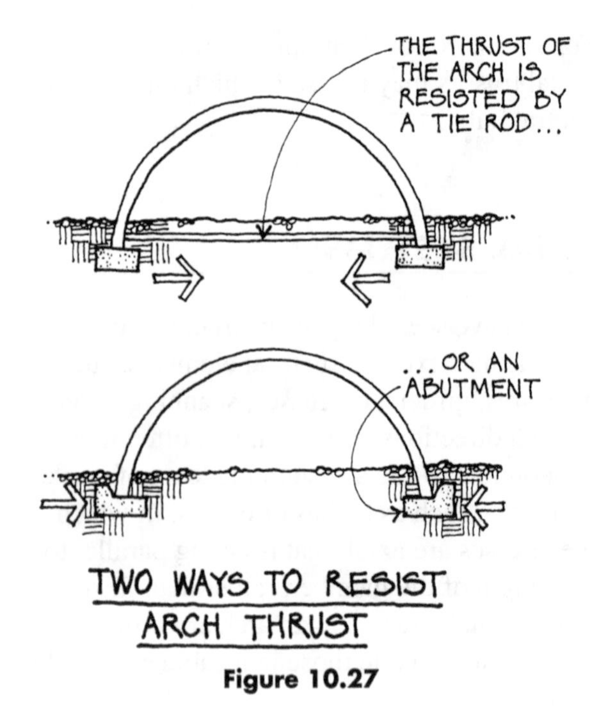

Figure 10.27

Example #3

A three-hinged arch with a rise of 20 feet spans 100 feet and supports a uniformly distributed load of 1,500 pounds per foot. What is the horizontal thrust at each end of the arch?

Solution:

From the previous analysis, the horizontal thrust $H = wL^2/8h$

$= 1{,}500\ (100)^2/8 \times 20$
$= 93{,}750\#$

Now, what would the thrust be if the rise were 30 feet instead of 20 feet?

$H = wL^2/8h$
$= 1{,}500(100)^2/8 \times 30$
$= 62{,}500\#$

So we can see, once again, that the greater the rise of an arch, the smaller the thrust.

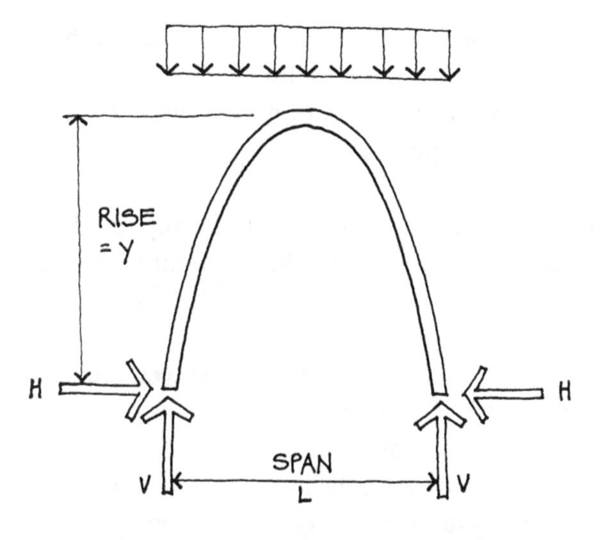

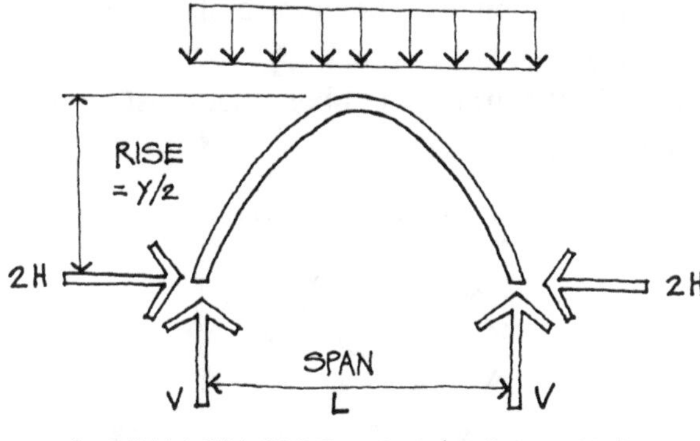

RELATIONSHIP BETWEEN RISE AND THRUST OF AN ARCH

Figure 10.28

The analysis above is strictly correct for a three-hinged arch. For a two-hinged arch, one with hinged bases only and no hinge at the crown, the structural behavior is similar, but the horizontal reactions are somewhat different, causing some local bending in the arch in addition to its basic axial compression.

The action of a fixed arch, one whose bases are fixed against rotation, is similar to that of a three-hinged arch, except that a fixed arch develops bending moment at its bases and elsewhere, in addition to the primary compressive force.

So we see that an arch, although stressed primarily in compression, often must resist some bending moment as well. However, because of the arch's curvature, this moment is much smaller than the moment in a simple beam spanning the same distance and supporting the same load.

Arch Construction

There are two basic ways in which arches are used in building construction: ribs and vaults (Figure 10.29). Arch ribs are individual arches connected by elements that span between them, such as curved concrete slabs. A vault is essentially a series of arches placed side by side to form a continuous structure, as shown in Figure 10.29.

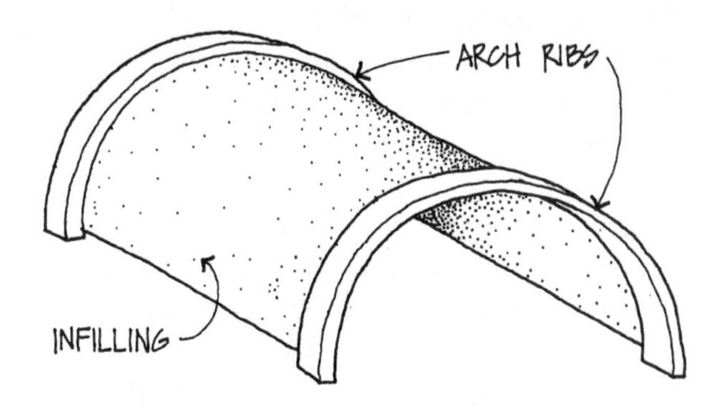

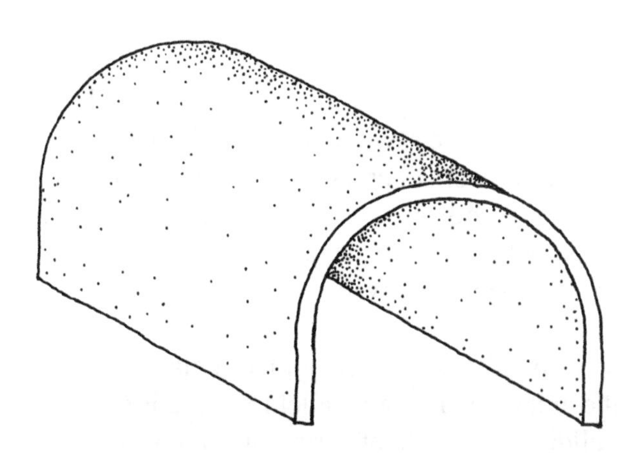

Figure 10.29

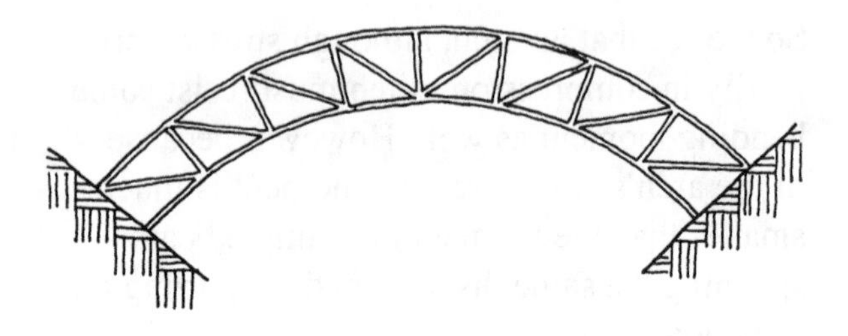

Figure 10.30

Both ribs and vaults may be constructed of reinforced concrete, structural steel, or wood. Cast-in-place concrete is a material naturally suited for arch construction because of its inherent compressive strength and the ease with which it can be formed into virtually any shape. Ribs and transverse infilling are usually cast together to form a monolithic structure. Individual arch ribs, on the other hand, are often made of precast concrete.

Structural steel arch ribs are often stiffened vertically with trusses (Figure 10.30). Is such a structure a truss or an arch? If the primary stress is compression, with some secondary bending, it is an arch. If it spans essentially by flexural action, it is a truss. Infilling material may be concrete, wood, or steel decking, with or without purlins depending on the distance between ribs.

Wood arch ribs are usually glued laminated members, since sawn timbers are limited to relatively small cross sections and short lengths. Another interesting application of arch construction in wood is the use of plywood vaults to span relatively short distances.

Most buildings are rectangular in plan, with the arches spanning parallel to one axis of the building. A couple of variations to this direct approach are the *diagonal arch* and the *lamella* roof. With the diagonal arch, the arches are placed along the diagonals, rather than parallel to the sides of the building. This design approach was used by Yamasaki in the Lambert St. Louis Airport.

Pioneered by the brilliant Italian engineer Pier Luigi Nervi, the lamella roof structure consists of a series of parallel arches that are skewed with respect to the axes of the building, and that intersect another series of skewed arches (Figure 10.31). The arches, which may be made of steel, concrete, or wood, interact with each other in much the same way as a two-way flexural system. They form a diamond pattern, which is infilled with either concrete or wood, with or without secondary beams. The lamella roof makes efficient use of material, but erection costs can be high because of the many short framing members that must be field assembled.

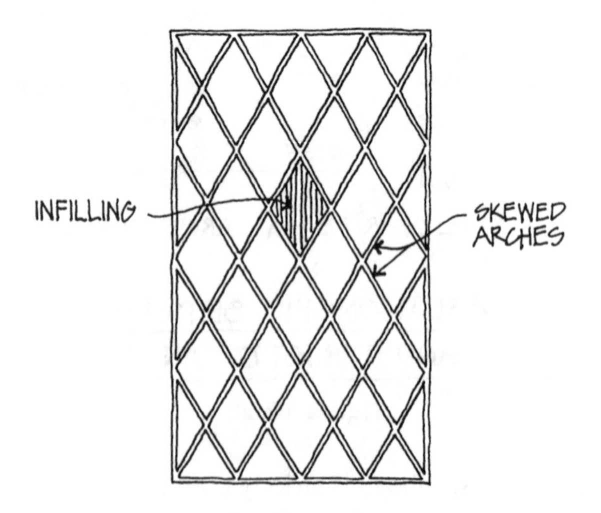

Figure 10.31

Suspension Structures

All modern suspension structures rely on the steel cable, which is able to resist very high tension forces, but cannot resist compression or bending. Greatly simplified, a cable may be considered as the reverse of an arch. Where the arch is stressed in compression, the cable is stressed in tension. For gravity loads, the arch is curved downward, while the cable is curved upward. The arch tends to spread outward at its base, resulting in inward-acting horizontal thrusts, as shown in Figure 10.32. The cable tends to pull its support inward, resulting in outward-acting horizontal thrusts (Figure 10.33). The rise of an arch may be compared to the sag, or drape of a cable. And just as steep arches have small thrusts, cables with a deep sag have small thrusts.

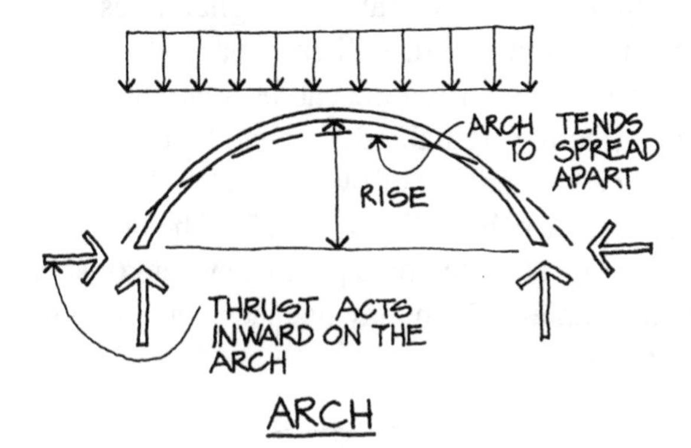

Figure 10.32

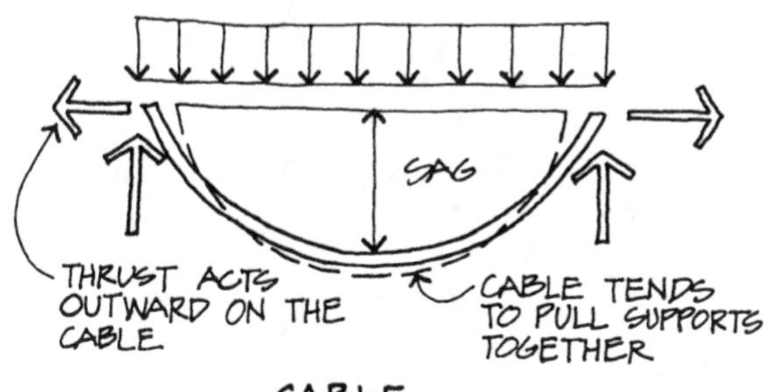

Figure 10.33

Thus, an arch and a cable are analogous structures. But there is one very significant difference, and that is the way they respond to changing loads. When an arch is subject to changing loads, it develops bending moments; but a cable cannot resist any bending moment since it can only resist tension. A cable therefore responds to changing loads by *changing its shape.* For example, under the action of a single concentrated load, a cable forms two straight lines meeting at the point of application of the load; when two concentrated loads act on the cable, it forms three straight lines, and so on (Figure 10.34). If the loads are uniformly distributed horizontally across the entire span, the cable assumes the shape of a parabola, the same shape as the pure arch resisting the same load. If the loads are distributed uniformly along the length of the cable, rather than horizontally, the cable becomes a catenary, a curve very similar to a parabola (Figure 10.35). If we suspend a cable of constant cross section between two supports, so that the only load acting on it is its own weight, the natural shape it assumes is a catenary.

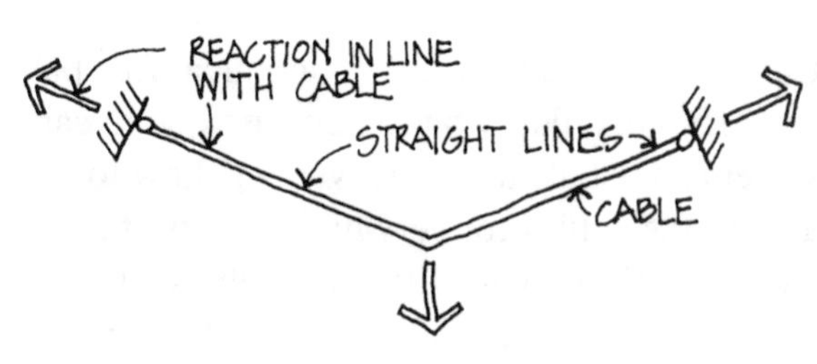

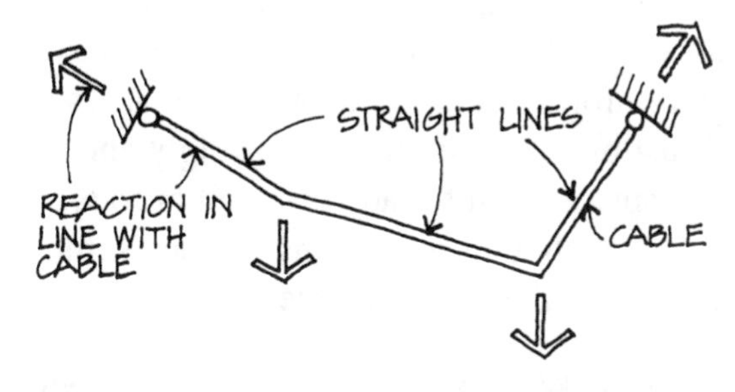

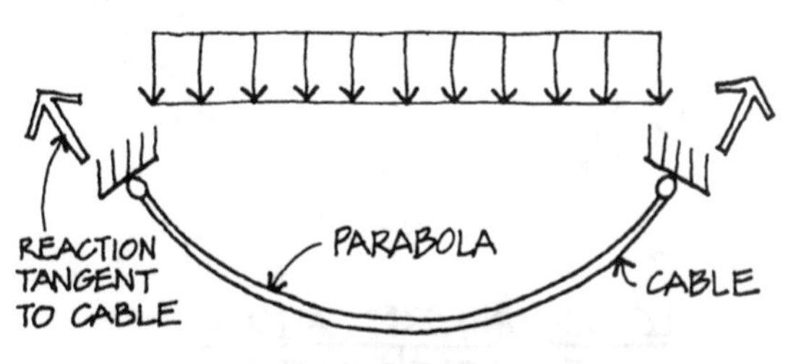

Figure 10.34

The tension in a cable is related to its sag: the greater the sag, the lower the tension. A large sag increases the length of the cable, but decreases its tension and hence its required cross-sectional area. What is the most economical amount of sag? It is that sag that will result in the least volume of cable under a given load. For a single concentrated load at midspan, the optimum sag is one-half the span. For a parabolic cable, the most economical sag is three-tenths of the span, while for a catenary, it is one-third. However, sags of this magnitude are generally not practicable in building design (a 300 foot span would require a sag of 100 feet). More often, the sag is about 1/15 of the span.

Cable structures are often economical for spans over 150 feet, but two problems unique to cable systems must always be solved: (1) how to anchor the cables for very high thrusts at their ends, and (2) how to stiffen the cables against movement or flutter caused by wind, earthquake, or changing loads.

The cables may be anchored much as they are in suspension bridges, by fastening them into heavy reinforced concrete abutments that resist the thrust by means of their own weight and/or the pressure against the adjacent soil or rock. Although abutments are expensive, they sometimes provide the best available solution.

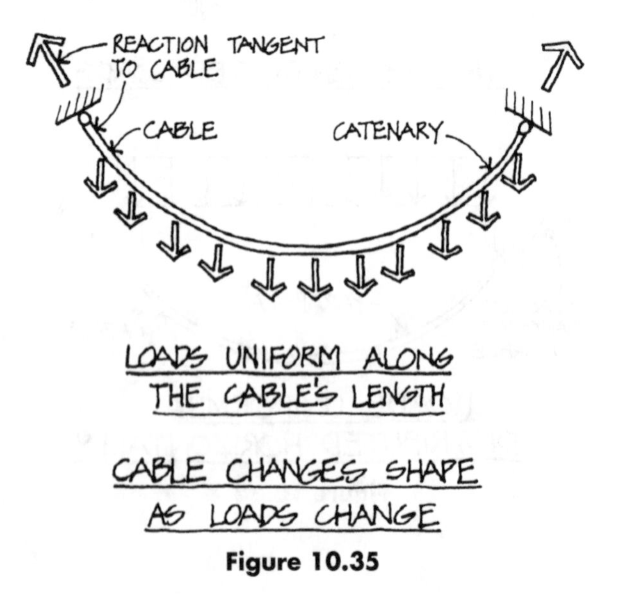

Figure 10.35

For a circular building, radial cables can be anchored into closed rings. In the example shown in Figure 10.36, the tension in the cables causes the outer ring to be in compression and the inner ring in tension.

Stiffening or stabilizing the cables may be done in a variety of ways. One method is to place a heavy roof material, such as concrete, over the cables. Another is to string guy wires from the cable roof structure to the foundation. Sometimes, the primary cables are stabilized by another series of cables at right angles to them to form a grid. Still another method is to use trusses to stiffen the cables. One interesting system, originally developed by engineer Lev Zetlin, consists of two connected layers of prestressed cables (Figure 10.37). The primary cables are stiffened by a parallel system of stabilizing cables, which may be placed either above or below them.

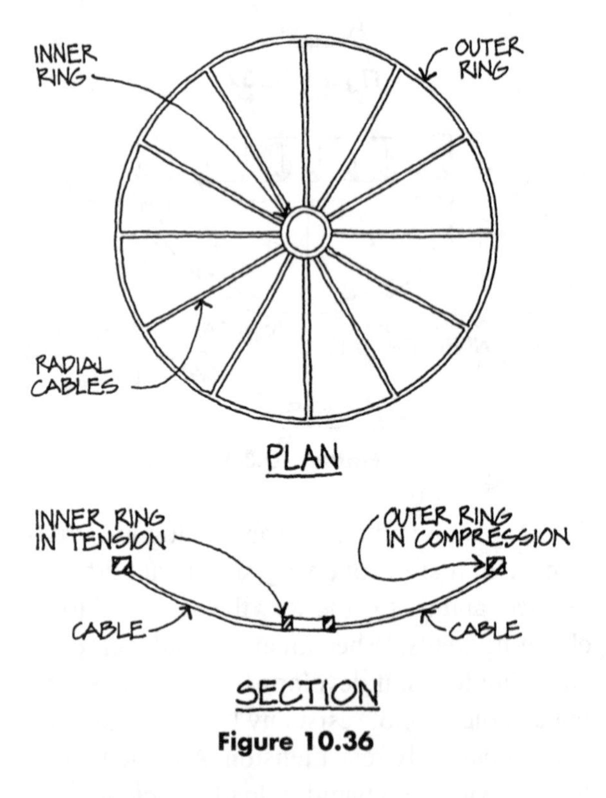

Figure 10.36

Cable roofs provide the greatest potential for ingenious, creative design solutions for large roof areas, and the use of suspension structures is likely to increase in the years ahead.

Rigid Frames

We have classified structural systems for long span construction according to their primary action: one-way or two-way flexural systems, where there is bending only and no axial stress; and axial systems, where there is axial stress and little or no bending. A rigid frame combines elements of both: it has both bending stress and axial stress. The relative magnitude of these stresses depends on how much the profile of the rigid frame deviates from the pure arch form. Thus, a rectangular frame acts primarily in flexure, while a gabled frame has less bending but more compression (see Figure 10.38).

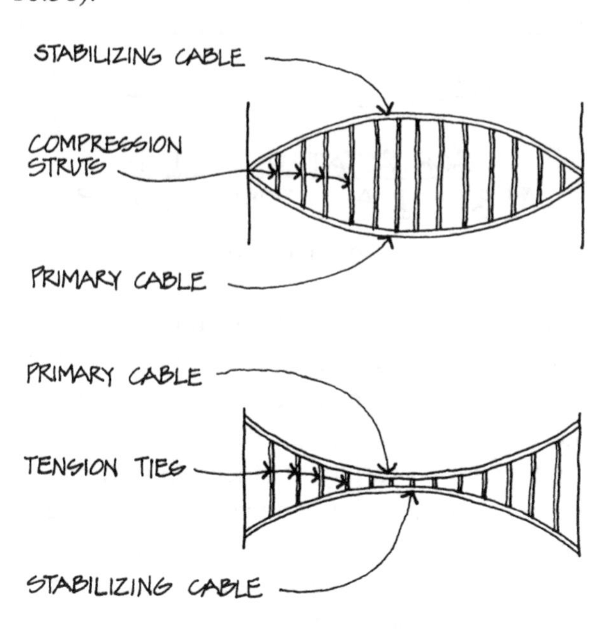

Figure 10.37

The characteristics of a rigid frame are described in Lesson Nine. The advantage of a rigid frame over a simple beam supported by columns (post and beam) is the frame's greater ability to resist both vertical and horizontal loads. The advantage of a rigid frame over an arch is that the frame, comprising only straight members, utilizes interior space better than a curved arch.

Rigid frames may be made of wood, structural steel, or reinforced concrete. The most common frame in wood construction is the three-hinged gabled frame, also called the three-hinged arch, which is statically determinate, and relatively easy to design, fabricate, and erect. Such frames are widely used where appearance is of utmost concern, in churches, schools, libraries, gymnasiums, and auditoriums. Spans are usually under 100 feet, although greater spans are possible. Purlins are generally used to span between the frames, with wood decking applied directly over the purlins. As with curved arches, a horizontal tension tie or abutment is required to resist the horizontal thrust at each base. A three-hinged glued laminated arch is depicted in Figure 10.39.

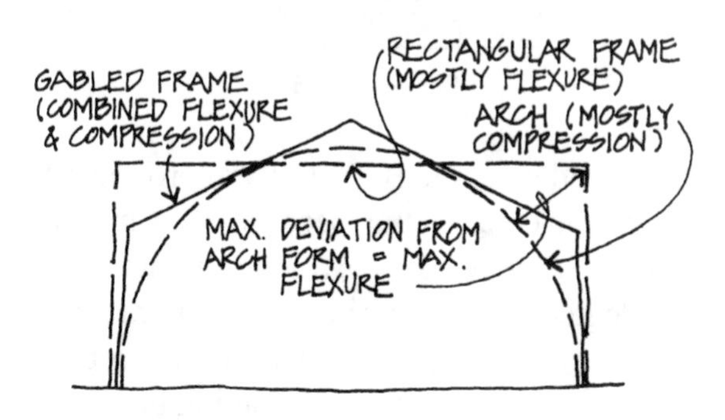

Figure 10.38

Structural steel is often used for long span rigid frames, both rectangular and gabled. Rectangular frames are most often two-hinged, while gabled frames may be two- or three-hinged. The members may be built up from plates, or they may be standard rolled sections, such as wide flange beams. For longer spans, trussed members are often more economical than solid members. The joints must be carefully considered. Welding provides a high degree of continuity, while bolting allows simpler field assembly. Both methods are commonly employed. Spans are generally under 150 feet, but longer spans are possible. Purlins are commonly used to span between frames, and a means of resisting the horizontal thrust must be provided, either by abutments or a horizontal tie. A number of manufacturers provide standard steel gabled frames that are pre-engineered and prefabricated for easy and economical shipment, assembly, and erection.

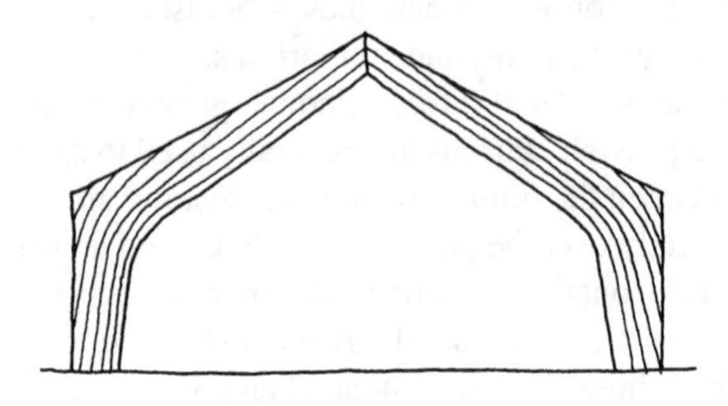

3-HINGED GLUED LAMINATED ARCH

Figure 10.39

Reinforced concrete is naturally appropriate for rigid frame construction because of its monolithic nature. Even at construction joints, a high degree of resistance to shear and moment is provided by the use of keys and the proper extension and lapping of reinforcing bars. Where precast concrete frames are employed the joints must be carefully designed and detailed to preserve, as much as possible, the monolithic character of the concrete.

Domes

The dome is the most dramatic and spectacular of all roof structures, be it St. Peter's or St. Paul's, or the Houston Astrodome. For hundreds of years, domes built of stone or brick were synonymous with religious architecture, evoking images of majesty and grandeur.

Because the dome symbolically represents power and authority, many government buildings built in this country during the nineteenth century were topped with central domes, starting with the cast iron dome of the U.S. Capitol in 1865, and followed by many state capitol buildings which were modeled after the U.S. Capitol.

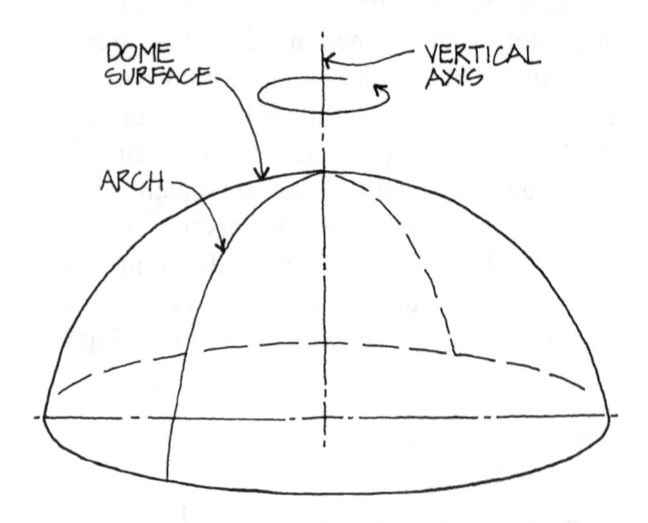

ARCH ROTATED ABOUT ITS VERTICAL AXIS FORMS A DOME

Figure 10.40

In more recent times, we have come to recognize that the dome is a structure of remarkable efficiency. Although still used for religious and government buildings, domes are now employed for a variety of other applications where a large circular or square area must be covered with a high roof without any interior columns. These include water and oil storage tanks, observatories, and—most spectacularly—public assembly buildings such as stadiums.

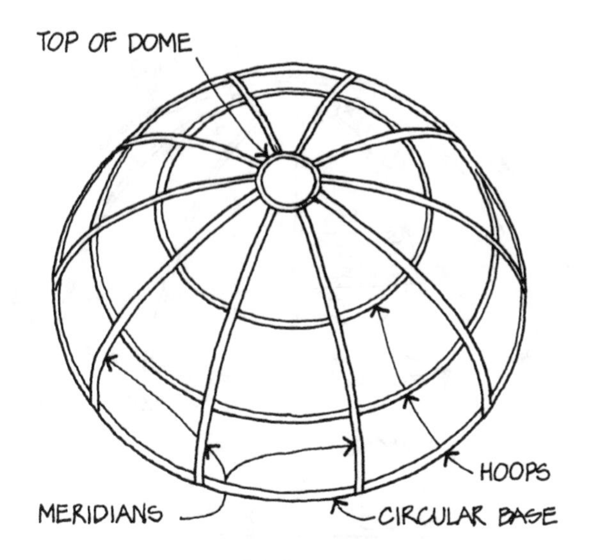

Figure 10.41

A dome can be thought of as an arch rotated about its vertical axis to form a curved surface (Figure 10.40). Although this surface can have a variety of shapes—spherical, elliptical, parabolic, conical—it is most often spherical.

The dead and live loads supported by a dome are transmitted to the ground primarily by compression along the curved meridians, which act like a series of circular arches extending from the circular base to the top of the dome (Figure 10.41). But if we provide horizontal circumferential hoops, or rings, something else happens. Typically in domes, the upper part of the meridians tends to move down, while the bottom tends to move out (Figure 10.42). These movements are prevented by the action of the horizontal hoops, which are stressed in compression in the upper hoops and tension in the lower ones. This combined arch and hoop action makes the dome much stiffer than an arch and very much stiffer than a flexural member.

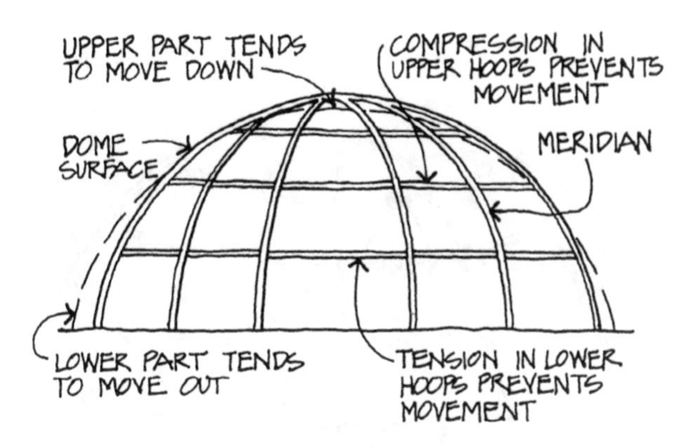

Figure 10.42

Thus, a shell dome can be made very thin (see Figure 10.47), a framed dome can be built of relatively small members, and all domes have very small deflections. Also, no external members are required to resist the dome's thrust at the top and base because the hoops, which are an integral part of the dome's structure, resist these stresses.

Domes can be built in a number of different ways. Among steel-framed domes, the most common is the *Schwedler dome*, named after its German inventor (Figure 10.43). In this dome, the individual arch ribs (meridians) extend from the base at A to a compression ring at B. Additional hoops are placed at C, D, and E. Purlins usually span between the hoops, with steel decking over the purlins, and frequently, diagonal bracing in the plane of the roof surface is used. The meridian ribs may be short segments (such as A-C, C-D, etc.), but more often they are continuous from A to B, which eliminates the need for most construction formwork.

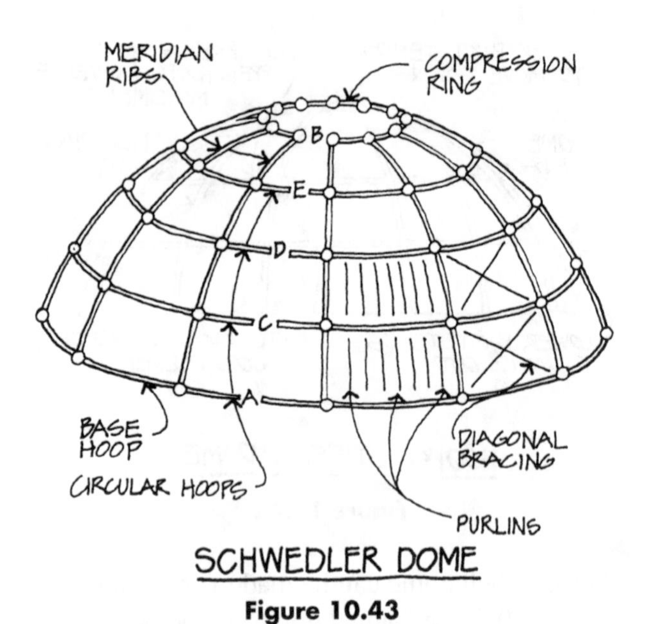

Figure 10.43

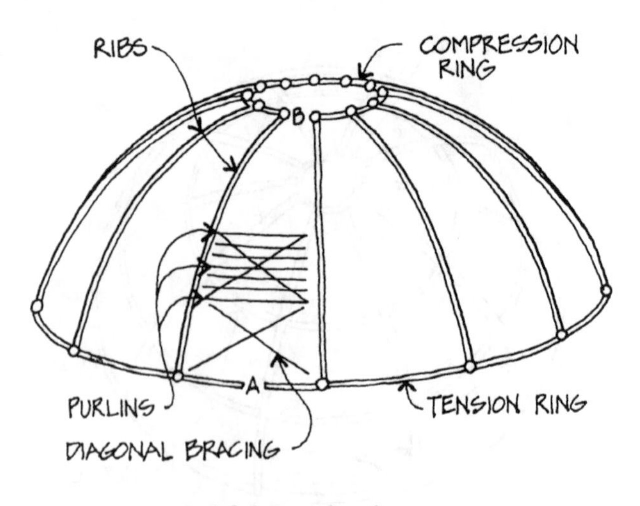

Figure 10.44

To build a spherical dome in this manner would require that all the members be curved, which would not be economical. The more usual approach is to have the panel points, the points where the meridians and hoops intersect, lie on the true spherical surface, connected by straight members. The dome is then polygonal, rather than spherical, but by using a sufficient number of ribs and hoops, the surface approaches that of a sphere.

If we omit hoops C, D, and E, we have a *ribbed dome*, in which the individual ribs behave like a series of three-hinged arches, each supporting a triangular or trapezoidal load, which is minimum at the top and maximum at the bottom (Figure 10.44). The ribs are supported by a top compression ring at B and a bottom tension ring at A, and because the ribs are subject to some bending, they are often trussed. A ribbed dome does not provide the stiffness or economy of a dome with hoops, such as the Schwedler dome, but it is simple to fabricate and erect and visually attractive.

Recently, other types of steel-framed domes have become popular. These include the *lattice dome*, which comprises a two-way grid of arches; the *lamella dome*, consisting of a skewed grid of arches; and of course, Buckminster Fuller's *geodesic dome* (Figure 10.45).

We have been discussing domes in general and steel-framed domes in particular. But domes can also be framed in wood or concrete, using the same principles and design methods. The main members in wood domes are usually glued laminated sections, and spans up to 350 feet or even greater can be economically built. Concrete domes may comprise either precast or cast-in-place ribs. If there are no ribs, the dome becomes a thin shell, which we discuss later.

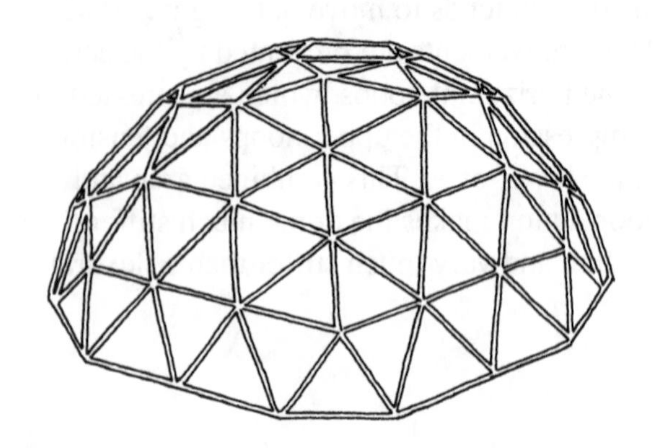

GEODESIC DOME

Figure 10.45

FORM-RESISTANT STRUCTURES

For most buildings, flexural systems such as beams, girders, or trusses provide the most economical structural solution.

The simple, flat, straight-line arrangement of members is easy to fabricate and erect and lends itself ideally to floor construction. These factors usually outweigh the uneconomical use of material inherent in such systems.

When longer spans are required, however, flexural systems become less economical, and curved systems, such as arches and cable structures, offer attractive and economical solutions.

Continuing with this categorization of structural systems, some of the most visually exciting long span roofs are form-resistant structures. A form-resistant structure is a surface structure whose strength is derived from its shape, and which resists load by developing stresses in its own plane. The two broad categories of such structures are *shells* and *membranes.*

Shells

A shell, or thin shell, is a structure with a curved surface that supports load by compression, shear, or tension in its own plane, but which is too thin to resist any bending stresses (Figure 10.46).

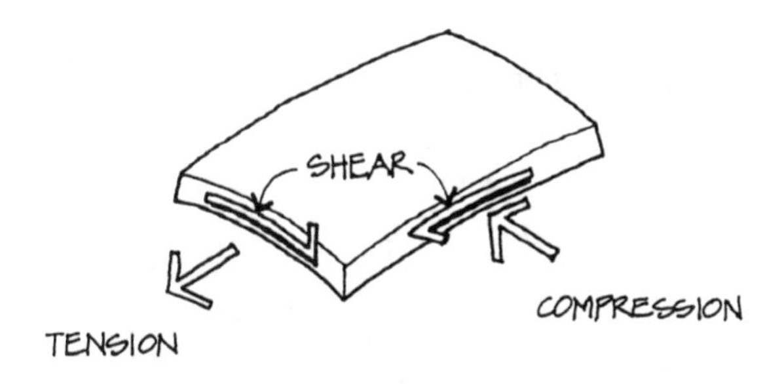

Figure 10.46

The best example of a shell in nature is the egg, which, although thin, can resist a great deal of uniformly distributed load because of its shape.

Shells are very efficient in the use of material, since they resist load by their shape, rather than by bending action. They are particularly strong in resisting uniformly applied loads, but cannot resist any substantial concentrated loads or edge restraints, which tend to induce bending. One other limitation in the use of shells is their curved shape, which makes them suitable only for roof structures and not for floors.

The variety of shapes that can be used in thin shell construction is vast. In fact, a shell can theoretically have any shape, whether or not it is mathematically definable. However, in practice, if a particular shape is not structurally efficient, or is difficult (and hence expensive) to analyze or construct, it is often abandoned.

Although shells have been constructed of wood and structural steel, they are most suited to reinforced concrete, because of its ability to be formed into virtually any shape. However, the high cost of formwork and scaffolding often makes shells uneconomical. Among the more popular thin shell shapes are the dome, the cylindrical or barrel shell, the vault, and the hyperbolic paraboloid.

Domes

We have already discussed domes in general, and particularly framed domes. The action of a thin shell dome is similar, except that in the framed dome, the meridians and hoops are separate, connected structural members. But in the shell dome, each element of the shell can be considered to be part of a meridian and also part of a hoop. The shell therefore resists meridional stresses, which are compressive, and hoop stresses, which may be compressive or tensile, and both stresses lie within the plane of the dome.

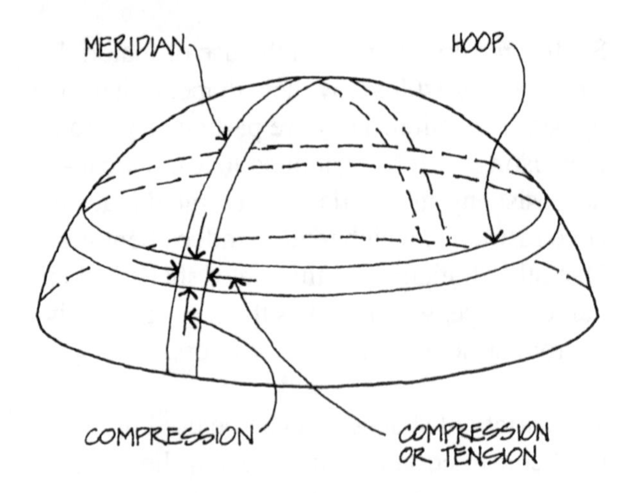

Figure 10.47

The thin shell dome, as shown in Figure 10.47, is a very stiff structure: a reinforced concrete dome only three inches thick can span 100 feet! A dome is stable under any load, whether symmetrical or unsymmetrical; that is, it does not change its shape. When there is an unbalanced load, the shell develops shear stresses in its own plane, in addition to the basic compressive and tensile stresses.

Barrel Shells

Cylindrical or barrel shell roofs made of reinforced concrete may provide an economical roof structure for rectangular buildings because their formwork is simple and can be reused many times (Figure 10.48). The shell behaves like an arch in the transverse direction. If the barrel is long (length much greater than the radius) and supported at its curved ends, the entire barrel acts as a beam with a curved cross section that spans end to end longitudinally, developing compressive stresses above the neutral axis and tensile stresses below (Figure 10.49).

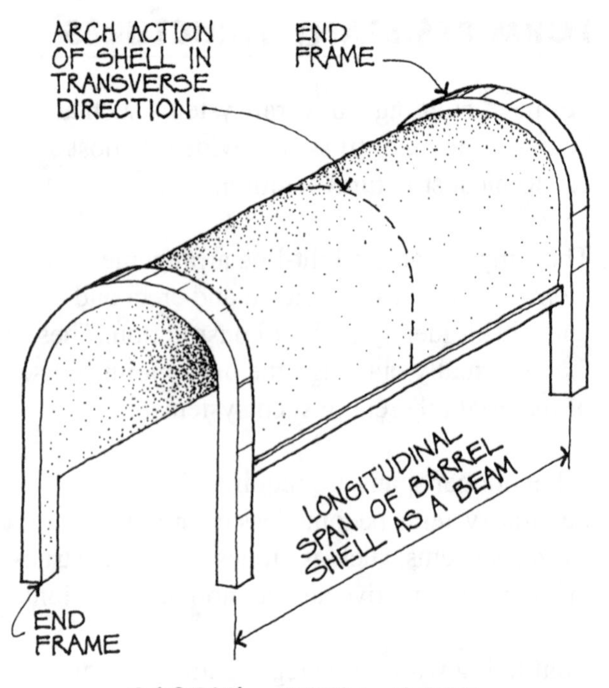

Figure 10.48

Long barrels are often used for warehouses and other industrial applications, where the longitudinal span is about 100 feet and the transverse span is 20 to 40 feet.

If the barrel is short (length less than the radius), it behaves like an arch between the longitudinal edges, which act as deep beams spanning between supports. Short barrels are used in structures such as auditoriums and airplane hangars, where the transverse span is 100 to 300 feet. Barrel shells are not as stiff as domes.

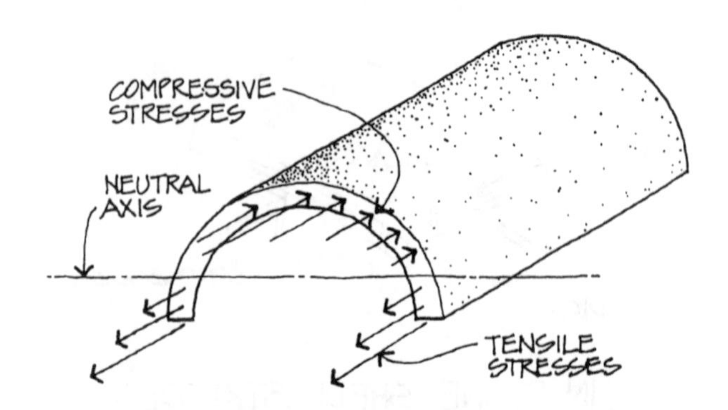

Figure 10.49

If the barrel is supported along its longitudinal edges, it becomes a *vault*, which can be thought of as a series of arches placed next to each other to form a continuous structure (Figure 10.50). As with other arches, horizontal thrusts are developed at the base of the vault, which must be resisted by tie rods or abutments (Figure 10.51).

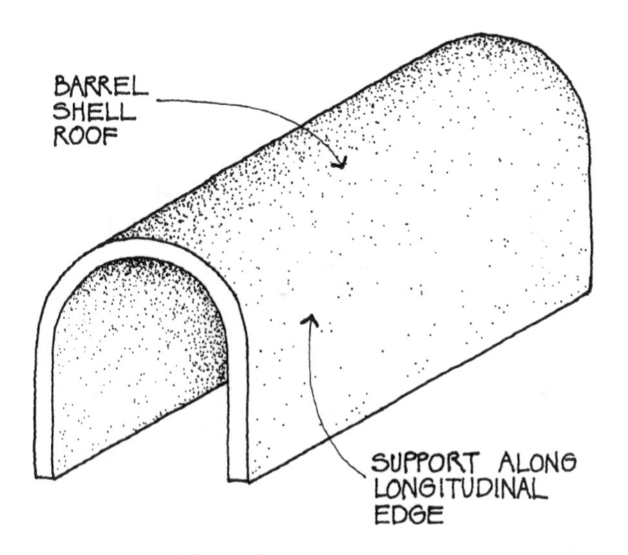

Figure 10.50

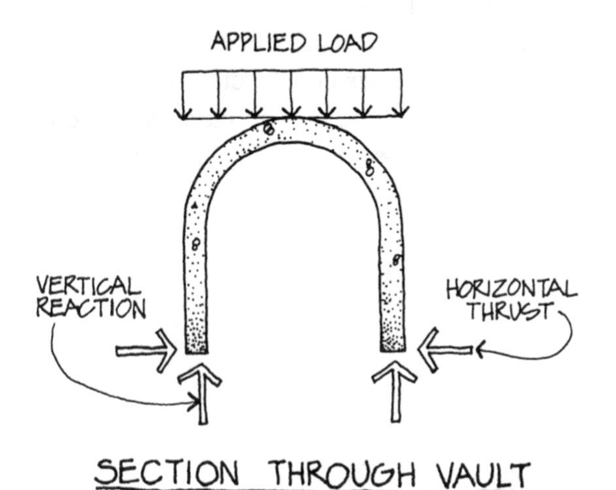

Figure 10.51

Hyperbolic Paraboloids

The hyperbolic paraboloid (HP) is popular for roof construction because of its economy in the use of materials and its striking appearance. The hyperbolic paraboloid is a saddle-shaped surface formed by moving a vertical parabola with downward curvature (y-y) along and perpendicular to another parabola with upward curvature (x-x) (Figure 10.52).

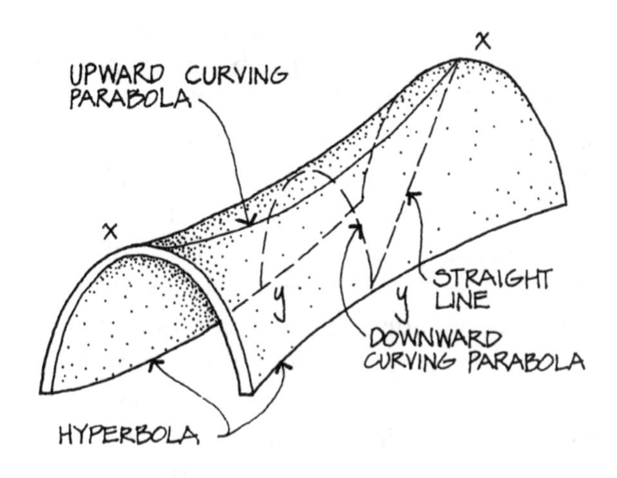

Figure 10.52

Thus, if we cut a vertical section along axis x-x, we obtain a parabola curved upward, and similarly, if we cut a vertical section along axis y-y, we get a downward-curving parabola. The intersection of a horizontal plane with the hyperbolic paraboloid consists of two curves that form a hyperbola. It is interesting to note that we can draw lines along the curved surface that are straight lines. In fact, if a straight line is moved between two other straight lines that are not parallel, a hyperbolic paraboloid surface is formed.

Referring to Figure 10.53, if we start with plane A-B-C-D and lift point D to D', the figure A-B-C-D' is a warped surface. If we connect line A-B to line C-D' with a series of straight lines or strings, the surface produced is a hyperbolic paraboloid. Section A-D' is an upward-curving parabola and section B-C is a parabola that is curved downward. So, although the surface of a hyperbolic paraboloid is doubly curved, it is relatively easy to construct because it can be formed by a series of straight timber boards.

The downward-curved sections behave essentially like compressive arches, while the sections perpendicular to these, which curve upward, act as tensile cables. This two-way action makes hyperbolic paraboloids very stiff; a concrete HP only a few inches thick can span or cantilever tremendous distances, all because of the shape of its surface. HP roofs can be constructed in a number of different ways; for example, four paraboloids can form a corner-supported roof, or four paraboloids can form a centrally-supported umbrella roof, or a single paraboloid can be supported at two points to form a *butterfly* roof. Hyperbolic paraboloids provide roofs which are dramatic in appearance and economical in the use of material.

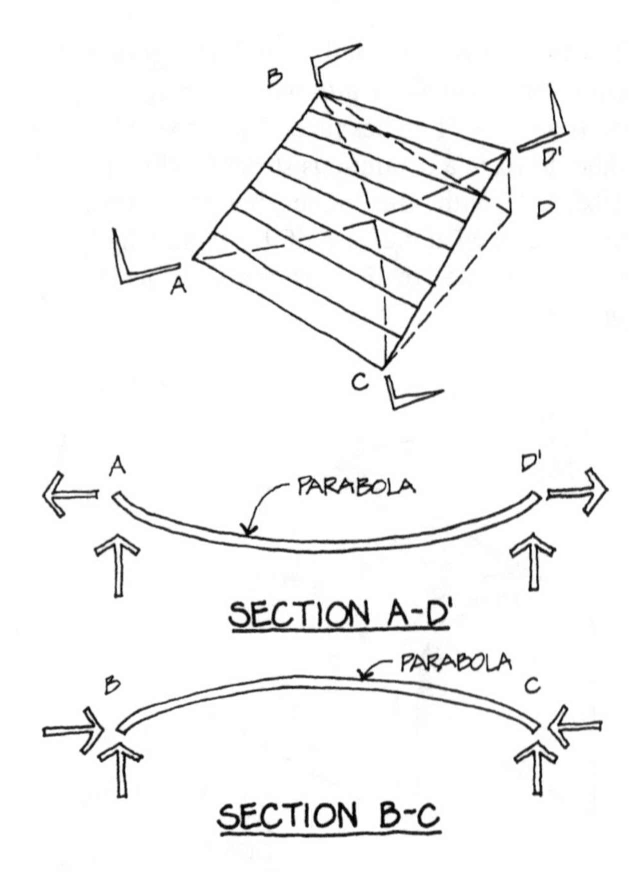

Figure 10.53

Folded Plates

A folded plate basically consists of two inclined plates, 1 and 2, as shown in Figure 10.54. A folded plate is not a shell, strictly speaking, because in the transverse direction, each plate spans like a beam or slab between the eave and the ridge. At the eave, vertical support is provided by a beam or wall. However, along the ridge, which is called a *fold line*, vertical support is provided by the action of the two plates leaning against each other. This results in forces in the plane of each plate. Each plate behaves like a beam and resists the forces in its own plane by spanning between end frames. This beam action develops compressive stresses above the neutral axis and tensile stresses below, somewhat similar to a barrel shell. At the end frames, vertical and horizontal support must be provided.

Folded plates are not as stiff as barrel shells, and their spans are usually under 100 feet. Although sometimes constructed of wood or steel, they are particularly economical in reinforced concrete because of the simple, reusable formwork.

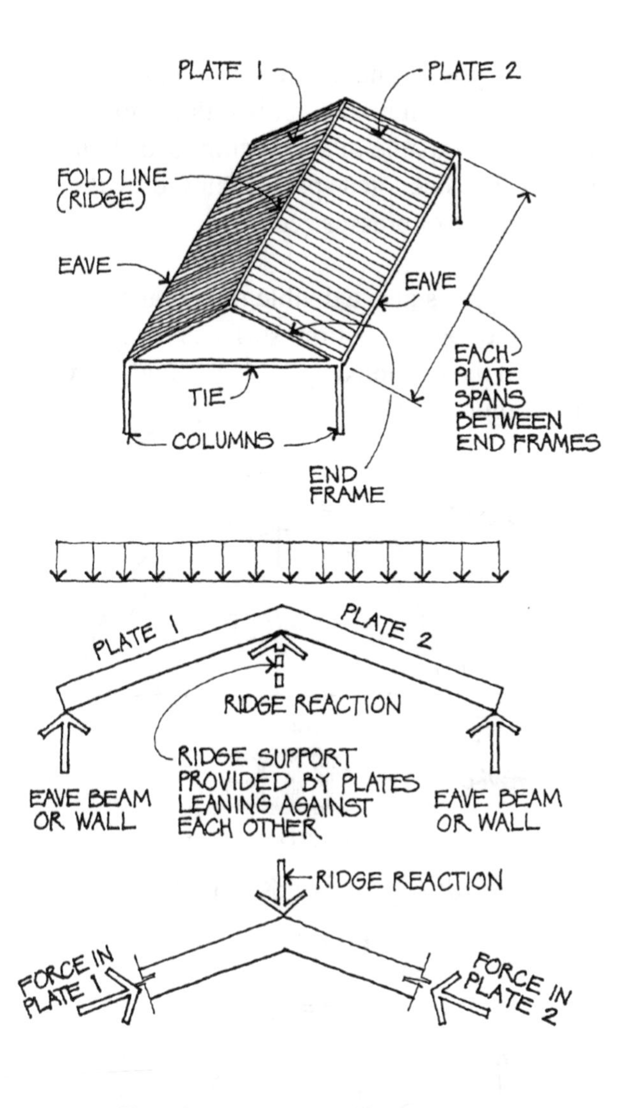

Figure 10.54

We have briefly described a few of the more popular thin shell forms. Many other shapes are also possible, limited only by the architect's imagination. Of course, not every shape is structurally efficient, aesthetically pleasing, or capable of mathematical analysis. Although most thin shells are very efficient structures and are among the most sophisticated expressions of modern design, they are sometimes difficult and expensive to form, placing and finishing the concrete may be difficult, and they often present weather-proofing, thermal, or acoustical problems. As we gain more experience in thin shell construction, solutions to these problems will undoubtedly be found.

Membranes

This review of systems used for long span structures concludes with the method which is both the most sophisticated and the most primitive—the tent (Figure 10.55). Whether used 10,000 years ago or in the late 20th century, the major problem faced by tentmakers (Omar or Frei Otto) is wind. Fabric tends to flutter and tear apart under the action of the wind, and the way to solve this problem is to pull the fabric taut—to pretension it. Therefore, the technology of membranes is concerned with developing methods for pretensioning the fabric and developing high-strength durable fabric materials. The modern tent or membrane structure has been successfully used for some very large roofs with spans of 200 to 1,000 feet.

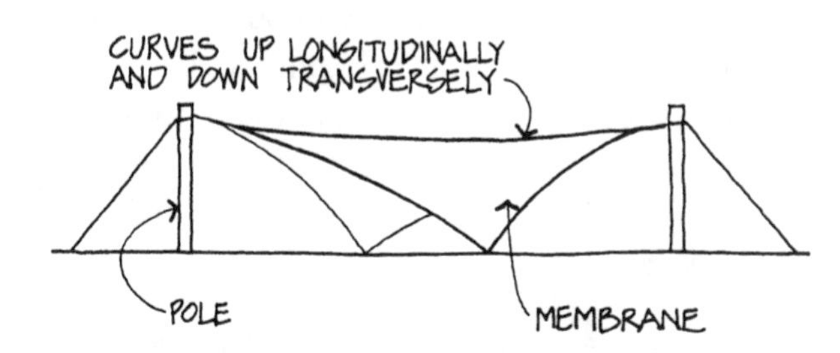

SADDLE-SHAPED TENT

Figure 10.55

What is a membrane? It is a thin sheet that cannot resist compression, bending, or shear, but only tension. Its action is similar to that of a two-way grid of cables, and like cables, membranes are unstable. When stabilized by pretensioning, membranes are able to support load, as for example, the trampoline. Thus, a roof surface may consist of a membrane stretched tightly over a steel frame; the greater the tension in the membrane, the stiffer it is. The need for materials with high tensile strength has led to the development of synthetic fabrics that are strong, fire-resistant, durable, and translucent.

One very successful and visible membrane structure is Frei Otto's design for the Munich Olympic Stadium, built in 1972. The largest area covered by tents up to this time was the Haj Terminal of the New Jeddah Airport, in Saudi Arabia, which comprises 210 tent units 150 feet square, covering a total of 105 acres. While evoking the tents that have sheltered Muslim pilgrims for centuries, the design, by SOM and Geiger Berger Associates, is very sophisticated and required a complex computer program to solve.

Air-Supported Structures

A membrane can be inflated by air pressure to completely enclose an occupied space, and several notable sports arenas with inflatable roofs have been built. A pressure differential of only five pounds per square foot between inside and outside is sufficient to preload the structure so it can resist the applied load without wrinkling or buckling the membrane (Figure 10.56). Such structures require a mechanical system to provide a continuous flow of air. If the mechanical system should fail, the roof will deflate and possibly fail.

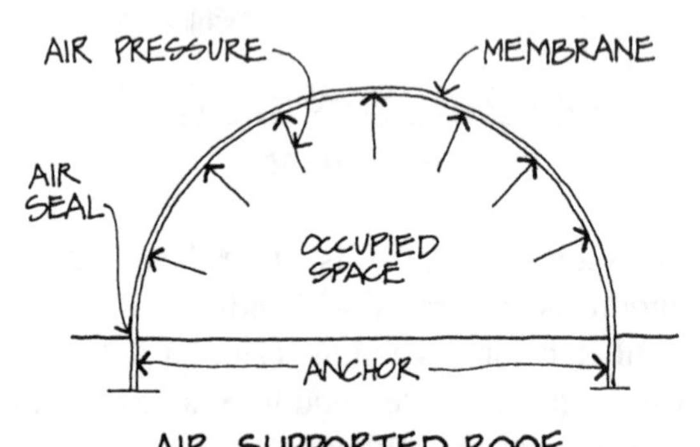

Figure 10.56

Some air-supported roof structures with a low profile have been built, in which the fabric membrane spans a relatively short distance between steel cables, which in turn span a long distance.

Another type of pneumatic structure consists of a double membrane that is inflated to form a structure capable of supporting load (Figure 10.57). The occupied space enclosed by the structure is not pressurized.

A third type is also a double membrane, in which the top and bottom surfaces form a pillow, and the occupied space is not under pressure (Figure 10.58).

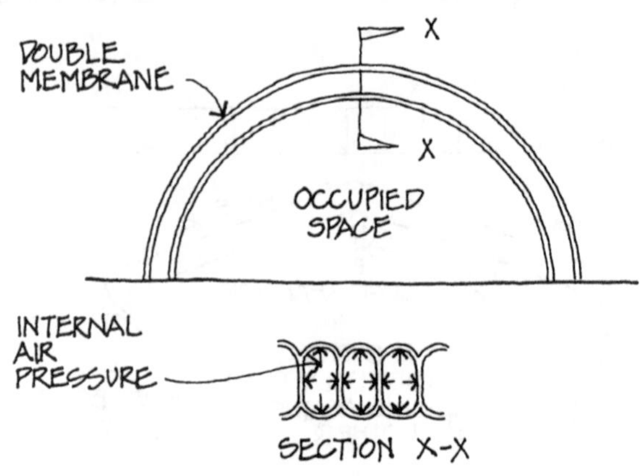

Figure 10.57

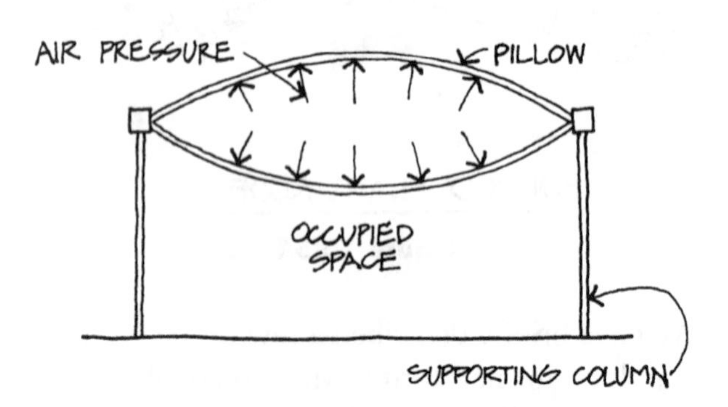

Figure 10.58

Inflatable structures tend to vibrate or flutter, and therefore, most such structures built to date have been for temporary, rather than permanent, use. However, their low initial cost, coupled with the improved durability of modern fabrics, should increase the popularity of these structures, especially for large sports facilities.

The use of membranes runs contrary to our structural intuition, knowledge, and prejudice. Instead of rigidity, membranes provide flexibility. Instead of stability, they provide change and movement. And instead of being supported by columns in compression, they must be held down so they don't float away. The possibilities for membrane structures are exciting and unlimited, and they offer a whole new way of perceiving structure.

CONCLUSION

The enclosing of large column-free areas, whether for public entertainment or assembly, or for a multitude of other commercial or industrial applications, will continue to challenge the ingenuity and creativity of architects. But problems and factors unique to long span buildings exist, and these must be thoroughly understood and considered by all those involved in the design and construction of such structures. Many of these considerations have come to light only as a result of the disastrous failures of long span structures that have occurred in the recent past. With proper care and understanding, however, it is hoped that such failures will be avoided in the future.

Long span structures are not merely short span structures with the scale changed. As we have seen, the structural systems used and the way they behave are frequently very different from the more conventional systems that we are accustomed to.

Factors that can safely be neglected in short span design often become significant and even dominant in long span structures. Deflection, for example, is one such factor. If a structure spanning 20 feet deflects one inch, this can usually be accommodated. But what if a 200 foot span structure deflects 10 inches? Will that interfere with proper roof drainage? Will it be disturbing visually? And how do you accommodate such large deflections in the architectural details? These and other questions must be faced and answered early in the design process.

Then there is the matter of the loads to be supported, particularly snow, water, and wind. In short span structures, these loads are considered, but they do not usually pose much of a problem, since so many columns and walls are available to resist the loads. But this is not true for long span structures; not only are there few supporting members, but the nature of the structure itself may make these loads more significant. Sometimes the shape of the roof will create local concentrations and abrupt variations in load; for example, snow and ice tend to accumulate in the valleys of folded plates and other undulating roofs. Similarly, design for uplift due to wind is not critical for short span structures, but it may be in long span construction, especially in roofs having low inherent stiffness, such as cable or membrane structures. Wind gusts or turbulence, which pose no problem for conventional, rigid structures, may cause flutter or vibration in light, tension type roof structures.

Secondary effects may become important in long span design. The end rotation of a short span beam, for example, causes small movements which are neglected in design. But that same amount of rotation in a 300 foot span space frame will create a significant amount of deformation. Movements and stresses caused by temperature change, creep, and shrinkage must be understood and provided for in design and detailing.

As we have pointed out, some long span structures are sensitive to concentrated loads or partial loads and this may severely restrict the equipment that can be suspended from or supported by the structure.

Long span structures have little redundancy; their safety depends on a few critical elements. If local overload or failure should occur, there is often no way for the system to redistribute loads to other parts of the structure, and so there is greater potential for sudden failure of the entire structure.

Many failures can be attributed, at least in part, to mistakes in the details. Accurate analysis is important, but so are well-designed joints or connections. This is especially true when the details are designed by someone other than the architect or engineer, such as a fabricator. Therefore, the use of higher factors of safety at critical connections might be advisable.

What all of this suggests is that greater care must be taken during design, as well as during the construction phase, by all design professionals.

The usual building department plan check, which may be perfectly satisfactory for conventional short span structures, is inadequate for long span buildings. Building departments often lack the technical sophistication or personnel to adequately review unusual or advanced structural systems. Perhaps review by one's peers should be considered for critical designs, even though this procedure may be objectionable to some architects and engineers.

The review of shop drawings, which is often treated as a perfunctory task for conventional structures, should be very carefully done by competent professionals familiar with the design of the long span structure in question. What may appear to be minor changes in detailing may significantly affect the behavior of the structure.

The architect is not usually responsible for construction methods, sequences, or procedures. But in long span construction, a specific sequence or erection is sometimes necessary to achieve a desired geometric and structural result. Hence, it may be advisable for the design architect and engineer to exercise some degree of control over the erection sequence. But how does the architect exercise that control without exposing himself or herself to significantly greater potential liability?

On-site observation of long span structures by the design architect and engineer is more critical than for short span structures, and testing and inspection firms that may be qualified for conventional structures may be inexperienced, and hence unqualified, when it comes to long span structures.

What this means is that the architect must be involved during the construction phase to a much greater extent than with conventional systems.

In short, long span structures are inherently more hazardous than short span structures. But this is no doomsday scenario. Failures don't have to happen. They are the result of many factors, some of which we've touched on here. And with awareness and a prudent sense of responsibility, they are avoidable.

LESSON 10 QUIZ

1. The oldest long span structures were

   A. arches, domes, and vaults.

   B. one-way systems, such as simple span trusses.

   C. generally built of wood.

   D. railroad bridges.

2. Exposed structural steel is often used for long span roofs used to enclose arenas and stadiums. This is possible because

   A. steel is incombustible.

   B. sprinkler systems are generally used in such structures.

   C. building codes permit unprotected steel construction when the roof is high above the floor.

   D. such buildings are only used intermittently.

3. Steel girders are efficient bending members because most of their material is in the

   ________________________________.

4. Select the INCORRECT statement. Secondary stresses in trusses are

   A. often ignored.

   B. bending and shear stresses caused by joint restraint.

   C. stresses in truss members other than the chords.

   D. necessary to consider when designing primary trusses spanning long distances.

5. The joist designation 18LH06 indicates which of the following?

   I. It is a long span steel joist.

   II. It is made of steel with a yield strength of 36,000 psi.

   III. Its depth is 18 inches.

   IV. Its top chord is section number 06.

   V. It is underslung and has parallel chords.

   A. I, II, III, IV, and V

   B. I, III, and IV

   C. I and III

   D. II and IV

6. Select the INCORRECT statement about glued laminated beams.

   A. The wood species is generally Douglas Fir or Southern Pine.

   B. They are more dimensionally stable than sawn timber beams.

   C. They are stronger than conventional sawn timbers.

   D. Individual laminations may be either shop or field glued.

7. A two-way truss system

   A. is efficient when the spans in both directions are about equal.

   B. is deeper than a comparable one-way system.

   C. does not lend itself to cantilever action.

   D. is simpler to design and build than a one-way system.

8. The bases of an arch

A. are subject to a horizontal thrust.

B. are generally hinged.

C. must be tied together.

D. must resist bending moment.

9. How does a cable respond to changing loads?

A. It develops bending moments.

B. It assumes the shape of a catenary.

C. It develops tension.

D. It changes its shape.

10. Select the CORRECT statements. A hyperbolic paraboloid

I. can be formed by a series of straight lines.

II. is a very stiff structure.

III. is economical in the use of material.

IV. is curved downward in one direction and upward in the other.

A. I and III

B. II and IV

C. I, III, and IV

D. I, II, III, and IV

LESSON ELEVEN

11 TRUSSES

INTRODUCTION

As discussed in the previous lessons, a beam is the most common and efficient type of framing member used for short to medium spans. Wide flange steel beams are particularly efficient, since most of their material is concentrated in the flanges, at the greatest possible distance from the neutral axis.

As the spans and loads increase, however, the dead load of a beam increases disproportionately, making the beam a less efficient bending member. Why not separate the flanges a greater distance and connect them with diagonal and vertical members to form, in effect, a built-up beam with a discontinuous web. Such a member is called a *truss*.

Trusses are used primarily as main carrying members for long span roofs. They are usually constructed of structural steel or wood; while concrete trusses are theoretically possible, they are seldom used because they are very heavy and difficult to handle.

The usual spacing between trusses is around 20 to 40 feet, depending on the purlins that span between them. Typical truss spans range from 40 to 200 feet, although for spans greater than 120 feet, the truss may exceed 12 feet in depth and consequently be difficult to ship. The optimum depth to span ratio of a truss is about 1/10, so a truss spanning 100 feet should be around 10 feet deep. Because trusses are deep and slender, they must be provided with adequate bracing.

The word *truss* denotes a structure designed to support vertical or horizontal loads and composed of straight members that form a number of triangles. Why triangles and not some other shape? Simply because the triangle is stable; that is, its shape cannot be changed without changing the length of one or more of its sides. A truss, therefore, must be composed of triangles in order to be stable.

The perimeter members of a truss are called *chords* (see Figure 11.1 for truss nomenclature). The interior members connecting the chords are called *web members*, or more specifically, *diagonals* or *verticals*, depending on their direction. Any truss member not necessary for stability is called a *redundant member*.

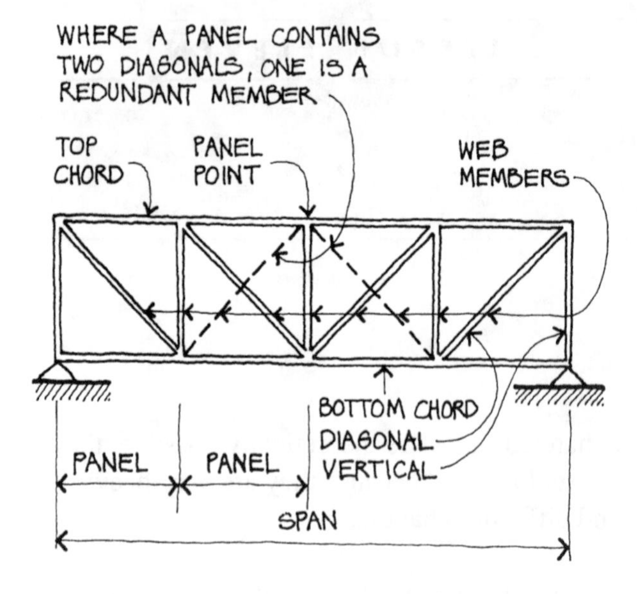

Figure 11.1

A truss is analogous to a steel beam: the truss chords are like the flanges of a beam, while the web members of a truss are similar to the web of a beam (see Figure 11.2). For a simple span truss, the top chord is in compression and the bottom chord in tension, just as in a simple beam, the top flange is in compression and the bottom flange in tension.

In a parallel chord truss, the forces in the chords increase toward the center of the span, just as in a beam, the flange forces increase toward the center. However, in a bowstring truss, whose depth varies from a minimum at the supports to a maximum at midspan, the chord forces remain relatively constant across the entire span.

The shear in a beam is resisted by shear stresses in the web, while the shear in a truss is resisted by the tensile or compressive forces in the web members, depending on their direction. And just as the web shear in a beam decreases toward the center of the span, similarly the forces in the web members of a truss decrease toward the center.

Trusses can have a variety of configurations, some of which are shown in Figure 11.3.

Trusses are usually designed as if their joints are pinned or hinged to allow unrestrained rotation, in which case the stresses in the truss members would be pure tension or compression. In fact, the joints of some of the early truss bridges were actually provided with hinges. Nowadays, pins or hinges are rarely used; it is more economical to connect the members by welding or bolting. This restrains rotation and induces small bending and shear stresses into the members, in addition to the primary axial stresses of tension or compression. These so-called *secondary stresses* are generally ignored, but they must be considered when designing major trusses spanning great distances.

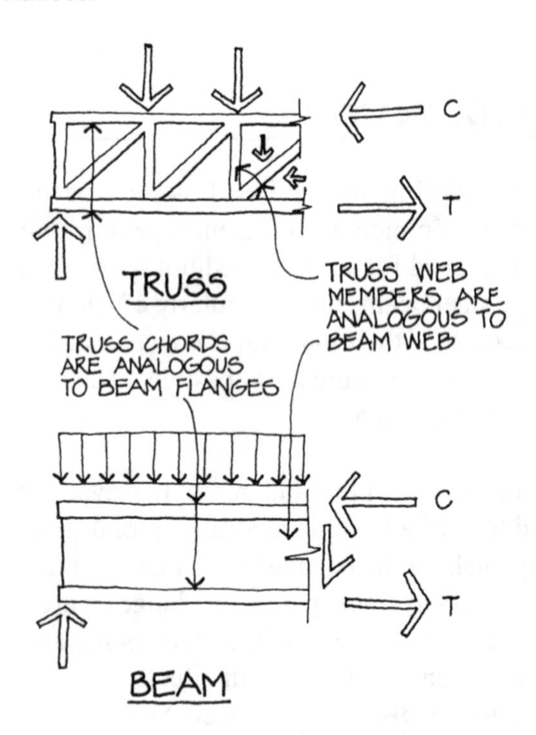

Figure 11.2

The axial force in any truss member may be determined analytically (by the method of joints or the method of sections) or graphically.

In the method of joints and the method of sections, free body diagrams are drawn, and the following rules apply:

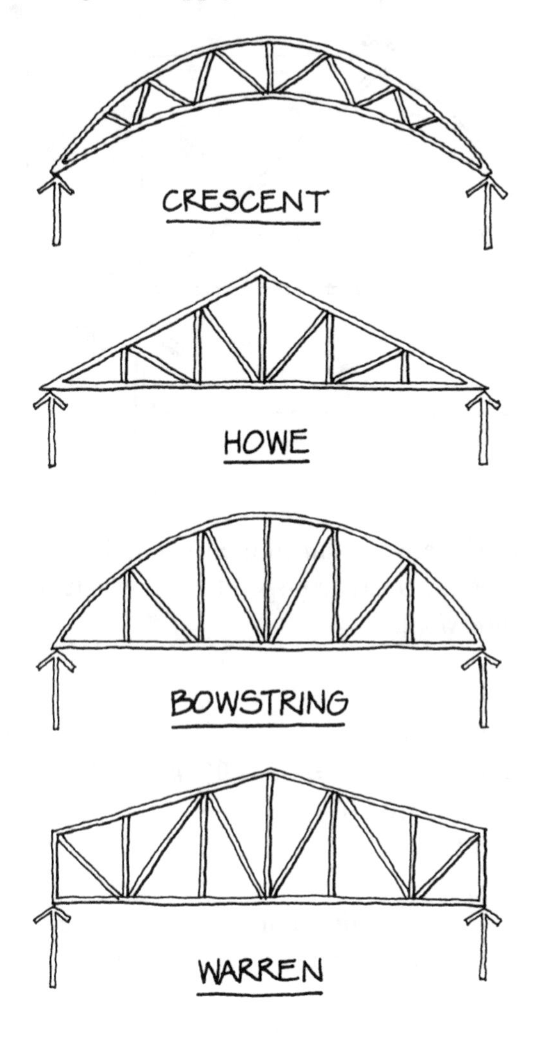

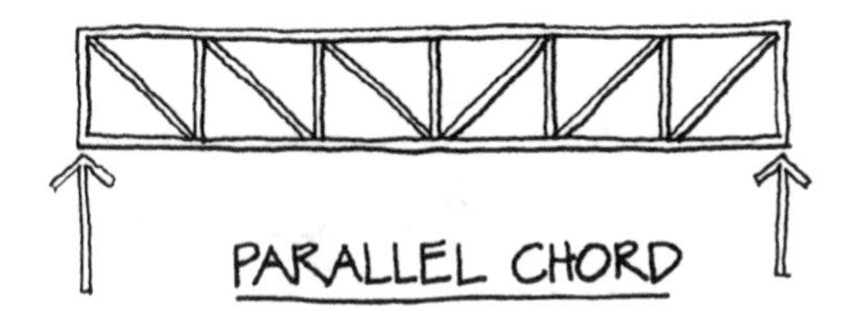

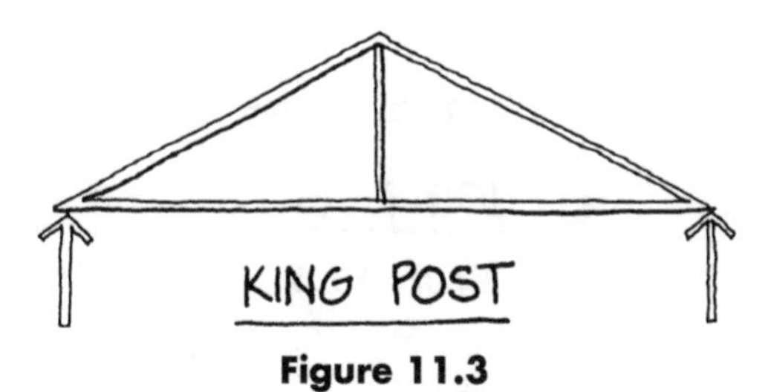

Figure 11.3

1. External loads and reactions are shown acting in their known directions.
2. Internal axial forces (tension or compression), once computed or known, are shown acting in their known directions.
3. All unknown axial forces are assumed to be tension.
4. Internal axial forces in members are tension if acting *away from* the cut section or *away from* the joint, and compression if acting *toward* the section or joint.

METHOD OF JOINTS

In this method, we isolate each joint, draw a free body diagram, and solve for the unknowns using the equations of equilibrium.

Example #1

Find the forces in the members of the truss shown in Figure 11.4.

Solution:

First solve for the reactions:

$\Sigma M_A = 0$
$-R_E(40\text{ ft.}) + 3{,}000\#(10\text{ ft.}) + 5{,}000\#(10 + 20)\text{ft.} = 0$
$R_E = (30{,}000 + 150{,}000)/40 = \mathit{4{,}500\#}$
$\Sigma M_E = 0$
$R_A(40\text{ ft.}) - 5{,}000\#(10\text{ ft.}) - 3{,}000\#(10 + 20)\text{ft.} = 0$
$R_A = (50{,}000 + 90{,}000)/40 = \mathit{3{,}500\#}$
Check: $\Sigma F_y = 0$
$3{,}500 + 4{,}500 - 3{,}000 - 5{,}000 = 0 \qquad 0 = 0\ \surd$

Next consider a free body at joint A, which is shown in Figure 11.5. By the laws of equilibrium we can find the forces in members AB and AC (F_{AB} and F_{AC}).

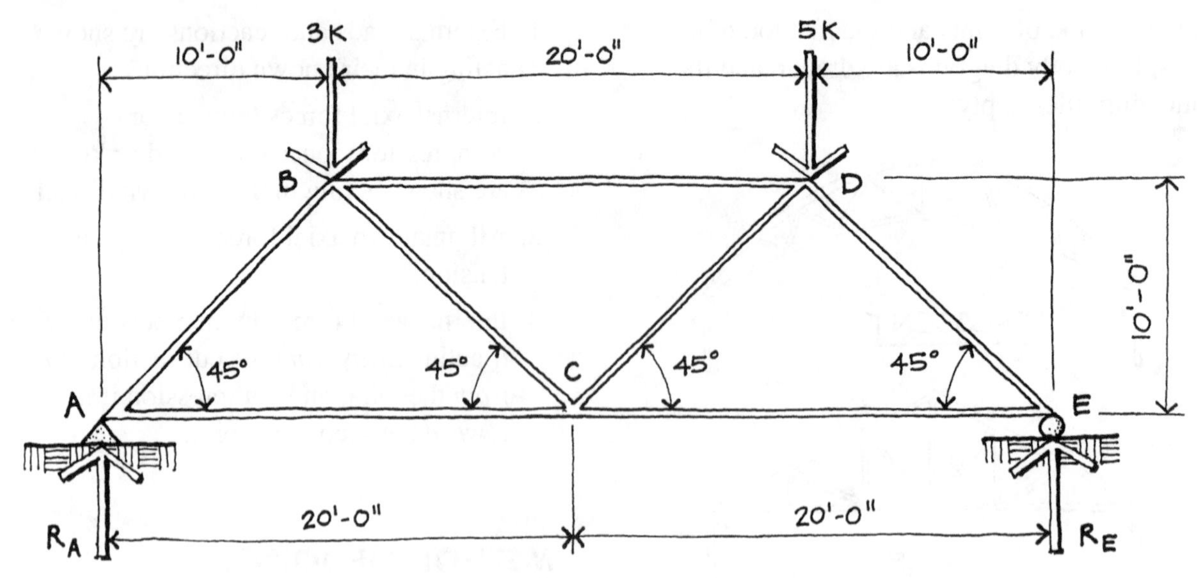

Figure 11.4

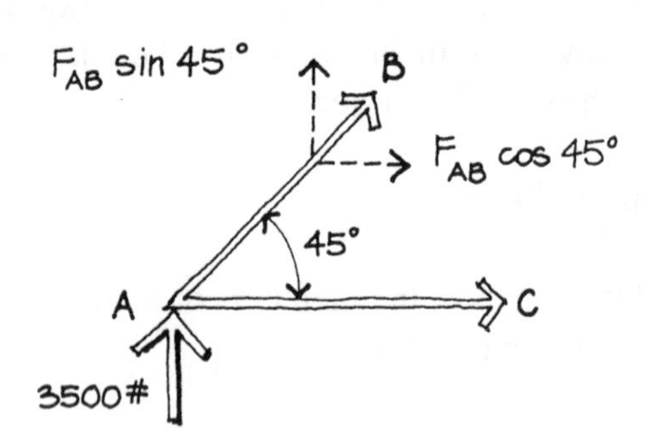

Figure 11.5

The only known force is the reaction (R_A), which we have determined as 3,500#.

We resolve F_{AB} into its vertical component ($F_{AB} \times \sin 45°$) and its horizontal component ($F_{AB} \times \cos 45°$).

$\Sigma F_y = 0$
$3{,}500\# + F_{AB} \sin 45° = 0$
$F_{AB} = -3{,}500/\sin 45° = -3{,}500/0.707$
$= -4{,}950\#$
$= 4950\#$ *compression*

The minus sign means that the force in AB acts in the opposite direction from that assumed.

Thus, since we assumed the force to be tension (acting away from joint A), it is actually *compression.*

$\Sigma F_x = 0$
$F_{AC} + F_{AB} \cos 45° = 0$
$F_{AC} = -F_{AB} \cos 45° = -(-4{,}950\#) \cos 45°$
$= 4{,}950\#(0.707)$
$F_{AC} = 3{,}500\#$

Since F_{AC} is positive, our original assumed direction is correct and F_{AC} is *tension.*

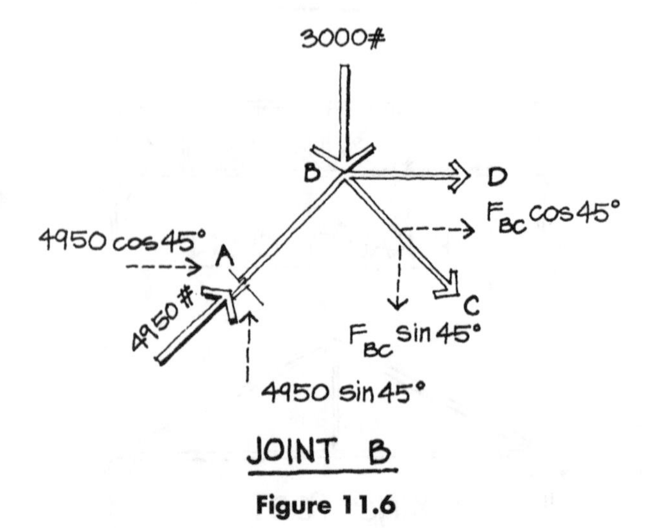

Figure 11.6

We next consider joint B (see Figure 11.6)

$\Sigma F_y = 0$
$-3{,}000\# + F_{AB} \sin 45° - F_{BC} \sin 45° = 0$
$-3{,}000\# + (4{,}950\#) \sin 45° - F_{BC} \sin 45° = 0$
$F_{BC} = \dfrac{-3000\# + 3{,}500}{\sin 45°} = \mathit{707\#\ tension}$
$\Sigma F_x = 0$
$F_{BD} + F_{AB} \cos 45° + F_{BC} \cos 45° = 0$
$F_{BD} = (-4{,}950\#) \cos 45° - (707\#) \cos 45°$
$= \mathit{-4{,}000\#}$
$= \mathit{4{,}000\#\ compression}$

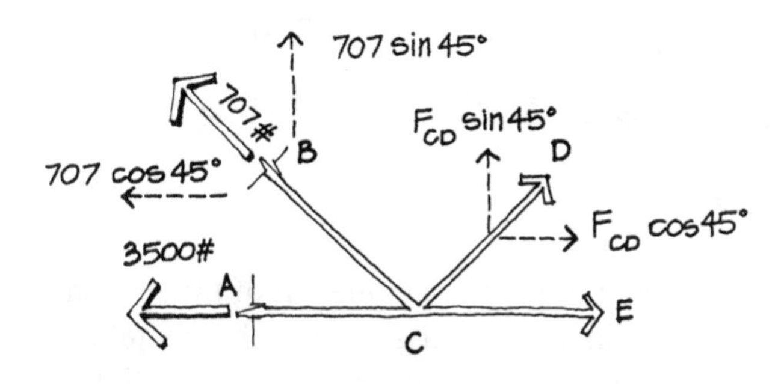

Figure 11.7

At joint C (see Figure 11.7),

$\Sigma F_y = 0$
$F_{BC} \sin 45° + F_{CD} \sin 45° = 0$
$F_{CD} = -F_{BC} = \mathit{-707\#}$
$= 707\#$ compression
$\Sigma F_x = 0$
$-F_{AC} - F_{BC} \cos 45° + F_{CD} \cos 45° + F_{CE} = 0$
$-3{,}500\# - 500\# + (-500\#) + F_{CE} = 0$
$F_{CE} = \mathit{4500\#\ tension}$

We finally come to joint E (see Figure 11.8)

$\Sigma F_x = 0$
$-4{,}500\# - F_{DE} \cos 45° = 0$
$F_{DE} = -4{,}500\#/\cos 45°$
$= \mathit{-6{,}365\#}$
$= \mathit{6{,}365\#\ compression}$

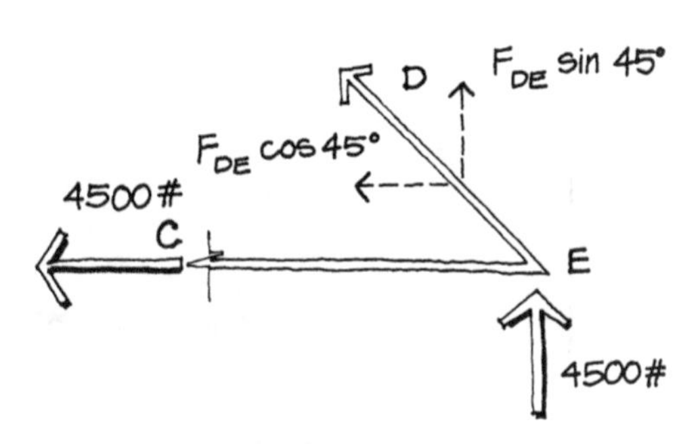

Figure 11.8

Check: $\Sigma F_y = 0$
$4{,}500\# + F_{DE} \sin 45° = 0$
$4{,}500\# + (-6{,}365\#)\ 0.707 = 0$
$0 = 0 \checkmark$

This example problem is more complex and time-consuming than any exam questions are likely to be. However, if you were able to follow along, you should be able to solve any truss problem on the exam which involves the method of joints.

METHOD OF SECTIONS

In this method, we cut an imaginary section through the truss which divides it into two parts. The member whose internal force is to be found must be cut by the section. We then draw a free body diagram of one of the two parts of the truss and apply the equations of equilibrium. The following example illustrates the use of this method.

Example #2

Find the forces in members 2-3 and 3-4 of the truss shown in Figure 11.9.

Like Example #1, this problem is probably more difficult and time-consuming than any you will encounter on the actual exam. However, if you

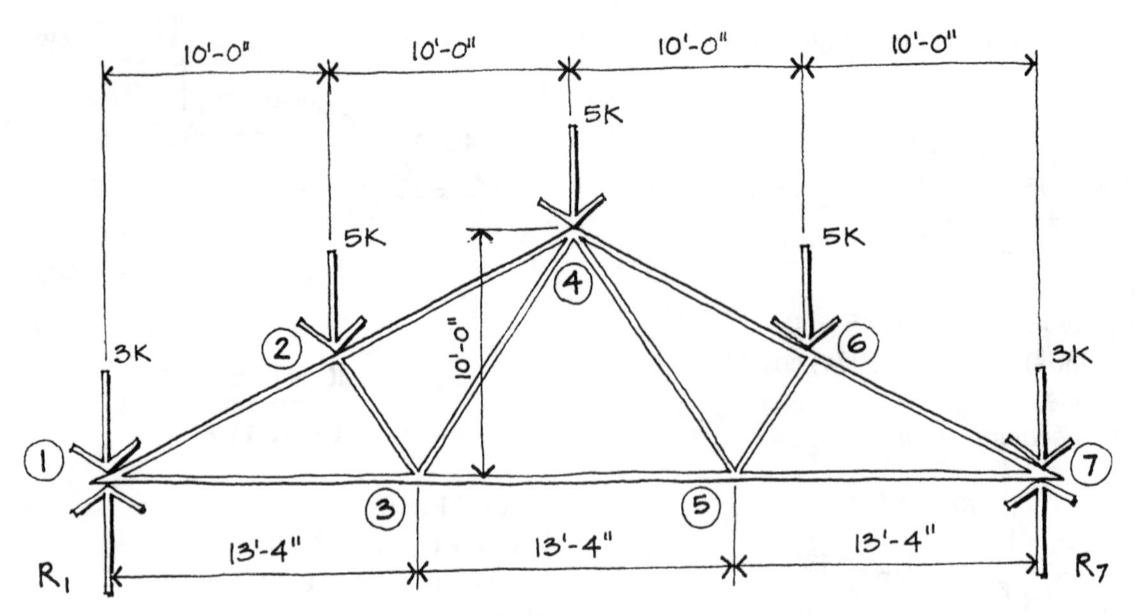

Figure 11.9

understand the procedure, you should be able to solve any truss problem on the exam that involves the method of sections.

Solution:

First solve for the reactions.

$\Sigma M_1 = 0$

$(5 \text{ kips} \times 10 \text{ ft.}) + (5 \text{ kips} \times 20 \text{ ft.})$
$+ (5 \text{ kips} \times 30 \text{ ft.}) + (3 \text{ kips} \times 40 \text{ ft.})$
$- R_7 \times 40 \text{ ft.}$
$= 0$
$R_7 = 10.5 \text{ kips}$

$\Sigma M_7 = 0$

$R_1 \times 40 \text{ ft.} - (5 \text{ kips} \times 10 \text{ ft.}) - (5 \text{ kips} \times 20 \text{ ft.}) - (5 \text{ kips} \times 30 \text{ ft.}) - (3 \text{ kips} \times 40 \text{ ft.})$
$= 0$

$R_1 = 10.5 \text{ kips}$

Check: $\Sigma F_y = 0$
$3 + 5 + 5 + 5 + 3 - 10.5 - 10.5 = 0$
$0 = 0$ √

We cut section A-A through the truss as shown in Figure 11.10 and draw a free body diagram of the part of the truss to the left of the section. In addition to member 2-3, two other members whose forces are unknown are cut by section A-A: members 2-4 and 1-3. We take moments about point 1, where the lines of action of these two members intersect.

We translate force 2-3 along its line of action to point 3. The horizontal component of force $F_{2\text{-}3}$ ($F_{x2\text{-}3}$) also passes through point 1; therefore,

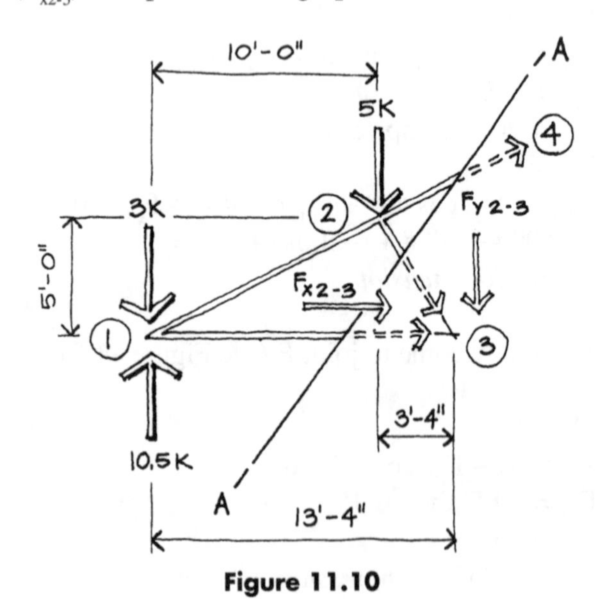

Figure 11.10

$\Sigma M_1 = 0$
$(5 \text{ kips} \times 10 \text{ ft.}) + F_{y2-3}\,(13.33 \text{ ft.}) = 0$
$-50 = F_{y2-3}(13.33)$
$F_{y2-3} = -50/13.33 = -3.75 \textit{ kips}$
$= 3.75 \text{ k compression}$

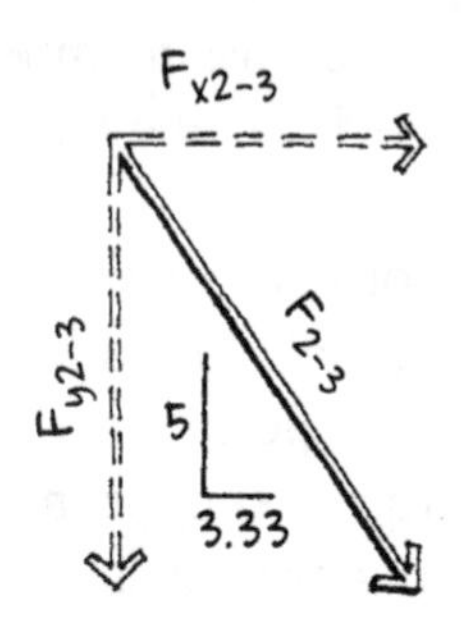

COMPONENTS OF FORCE IN DIAGONAL 2-3

Figure 11.11

Since $\dfrac{F_{x2-3}}{F_{y2-3}} = \dfrac{3.33}{5}$ (see Figure 11.11):

$F_{x2-3} = \dfrac{3.33}{5}(F_{y2-3})$

$= \dfrac{3.33}{5}(-3.75) = -2.5 \text{ kips}$

and $(F_{2-3}) = \sqrt{3.75^2 + 2.5^2}$
$= \textit{4.5 k compression}$

To find the force in diagonal 3-4, we cut section B-B through the truss, which cuts member 3-4, and draw another free body diagram (see Figure 11.12).

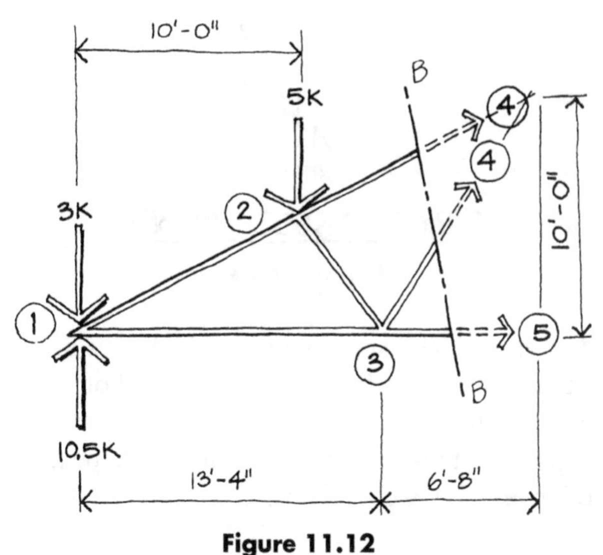

Figure 11.12

Using the same techniques as before, we again take moments about point 1.

$\Sigma M_1 = 0$
$(5 \text{ kips} \times 10 \text{ ft.}) - F_{y3-4}(13.33 \text{ ft.}) = 0$
$F_{y3-4} = 50/13.33 = \textit{3.75 kips (tension)}$

Since $\dfrac{F_{x3-4}}{F_{y3-4}} = \dfrac{6.67}{10}$ (see Figure 11.13):

$F_{x3-4} = \dfrac{6.67}{10} \times 3.75 = 2.50 \text{ kips}$

and $F_{3-4} = \sqrt{3.75^2 + 2.5^2} = 4.50 \textit{ kips (tension)}$

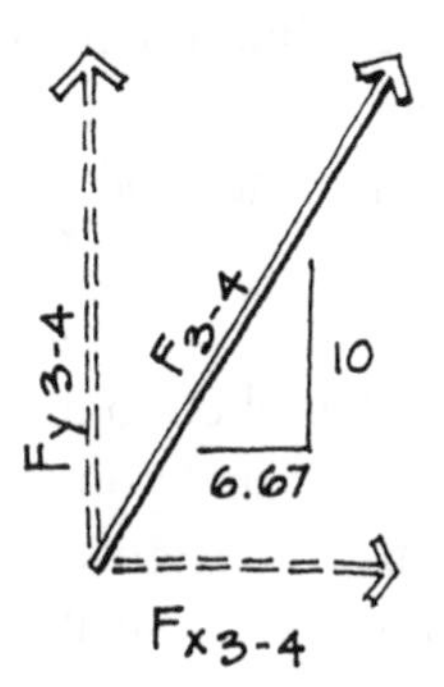

COMPONENTS OF FORCE IN DIAGONAL 3-4

Figure 11.13

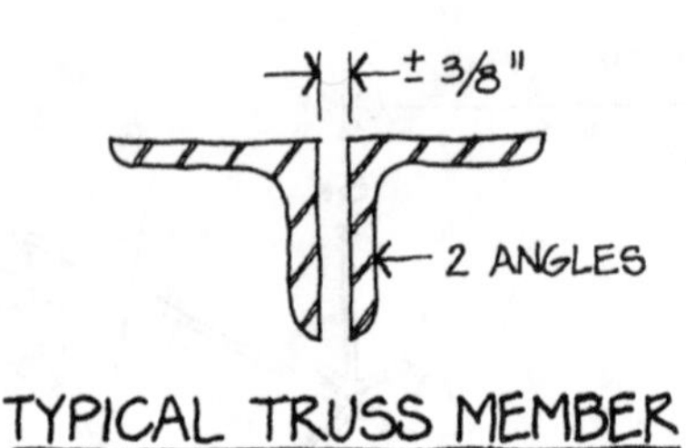

Figure 11.14

The analytical determination of the forces in truss members, using either the method of joints or the method of sections, is especially applicable to trusses having a simple shape and relatively few web members.

TRUSS DESIGN

After the forces in all the truss members have been determined, either graphically or analytically, the truss members are designed. Compression members are designed as columns, and tension members must have sufficient net area, after deducting the area of the fasteners, so that the allowable tensile unit stress is not exceeded. It should also be noted that although the length is an important factor in the design of compression members, because of the buckling tendency, it is less important for tension members. The slenderness ratio l/r of steel tension members, other than rods, preferably should not exceed 300.

Sometimes, loads are applied to the top or bottom chord *between* the panel points; in such cases, the chord must be designed to resist the resulting bending moment as well as the axial force caused by the truss action.

Steel Trusses

The chords and webs of steel trusses are usually made of double angles, separated by about 3/8" to allow for a gusset plate to connect the members together at the panel points (Figure 11.14). If the force in a truss member is very low, the member may be a single angle.

Other types of sections are sometimes used for steel truss members, including tubular sections where appearance is important and wide flange sections for main trusses supporting heavy loads and/or spanning great distances.

A typed joint in a steel truss is illustrated in Figure 11.16. The members that meet at a truss joint are connected to the gusset plate by bolts or welds. The centroidal axes of truss members should meet at a point, in order to avoid eccentricity, which would induce bending moment in the members. In steel trusses, the usual practice is to have the *gage lines* of the double angle members, rather than the centroidal axes, meet at a point. The gage line of an angle member is the longitudinal axis corresponding to the standard gage for that size angle. The standard gages for angles are shown in Figure 11.15, reproduced from the AISC Manual by permission of the American Institute of Steel Construction.

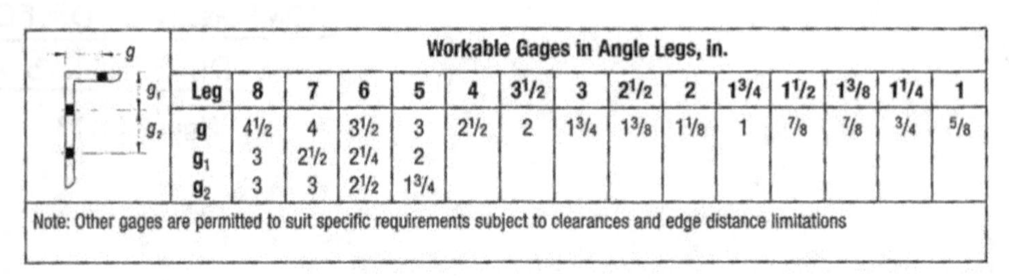

Workable Gages in Angle Legs, in.

Leg	8	7	6	5	4	3 1/2	3	2 1/2	2	1 3/4	1 1/2	1 3/8	1 1/4	1
g	4 1/2	4	3 1/2	3	2 1/2	2	1 3/4	1 3/8	1 1/8	1	7/8	7/8	3/4	5/8
g_1	3	2 1/2	2 1/4	2										
g_2	3	3	2 1/2	1 3/4										

Note: Other gages are permitted to suit specific requirements subject to clearances and edge distance limitations

Figure 11.15

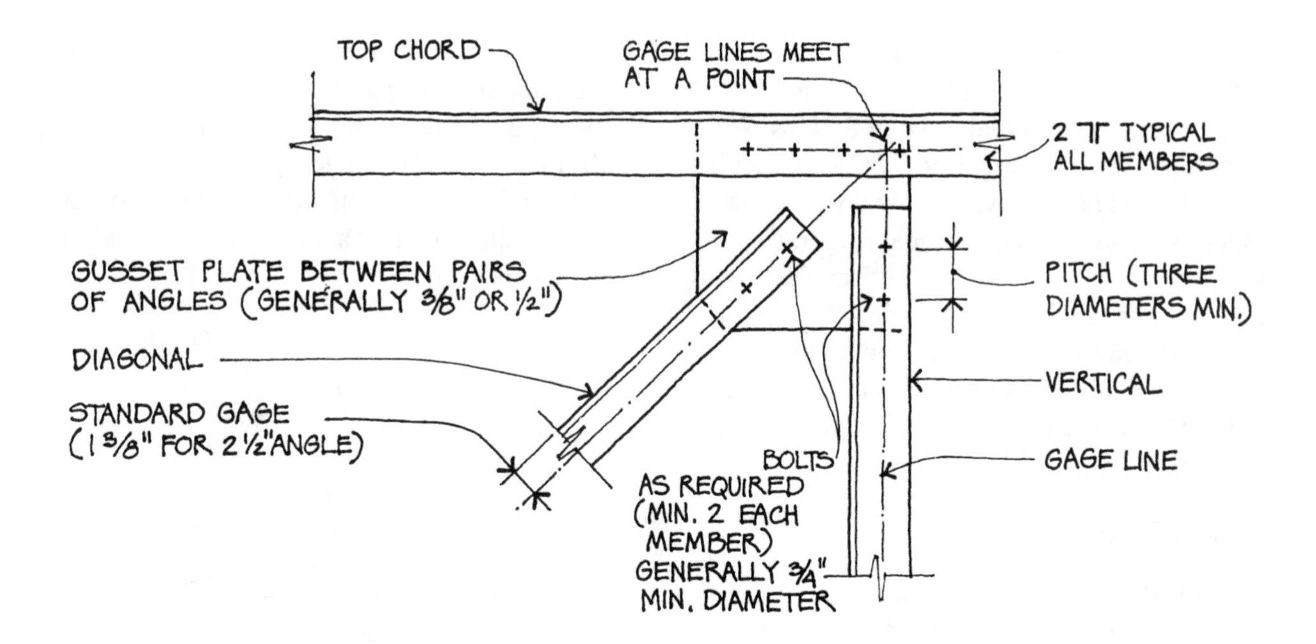

Figure 11.16

Angles for compression members usually have unequal legs, with the long legs placed back to back, in order that the value of the radius of gyration r be approximately equal in both directions. The properties of 2 angles are found on pages 1-100 through 1-107 of the AISC Manual, and the compressive load capacities of 2 angles are found on pages 4-118 through 4-156 of the AISC Manual.

As discussed previously, steel members may be designed using principles of Load and Resistance Factor Design (LRFD) or Allowable Stress Design (ASD). As before, we will consider ASD here.

Example #4

The maximum force in the top chord of a truss is 80 kips compression. Its unsupported length is 10 feet. Select a double angle member of ASTM A36 steel, assuming 3/8" space between the angle legs.

Solution:

The value of K for truss members is usually assumed to equal 1. Therefore, Kl = 1 × 10 = 10 feet. The simplest way to solve this kind of problem is to use the load capacity tables in the AISC Manual.

The table on page 4-133 of the AISC Manual is reproduced in Figure 11.18, by permission of the American Institute of Steel Construction.

Under the table headed 5 × 3-1/2 × 3/8 (the size of the 2 angles), read down to 10 feet and read the allowable load 96.4 kips with respect to the x-x axis and 81.7 kips with respect to the y-y axis. Since the lower value governs, the safe capacity of the 2 angles is 81.7 kips, which is greater than the actual force of 80 kips. We therefore select 2 *angles 5 × 3-1/2 × 3/8*, with the long legs back to back. How do we know that this is the most economical solution? From the table, we note that the 2 angles weigh 20.8 pounds per foot. By glancing through the tables, we can see that there is no double angle section weighing less which has a capacity of at least 80 kips with respect to both axes for a Kl value of 10 feet.

Wood Trusses

As with steel trusses, wood trusses are of three general types: parallel chord or flat, sloping chord or triangular, and bowstring (see Figure 11.3). Members may be sawn, glued laminated, or mechanically laminated. They may consist of a single piece of lumber or of two or more pieces. Webs and chords may be in the same plane and attached by means of gusset plates, or they may lap each other. Connections may be made with bolts or with other types of connectors. Some wood trusses use steel rods, rather than wood members, for web members subject only to tension.

A type of wood truss in widespread use for dwellings and other small structures is the *roof truss* (Figure 11.17). This is a prefabricated lightweight truss generally designed to span 20 to 40 feet and placed about 24 inches on center. Chords and webs are usually single members, such as 2 × 4 or 2 × 6, and connected by means of toothed steel gusset plates, which are factory pressed to the members at the panel points.

Some trusses may require greater depth than most other long span framing systems. However, trusses offer a number of advantages: they can be erected quickly, mechanical ducts can easily penetrate the relatively unobstructed truss webs, and they are often economical in the use of material.

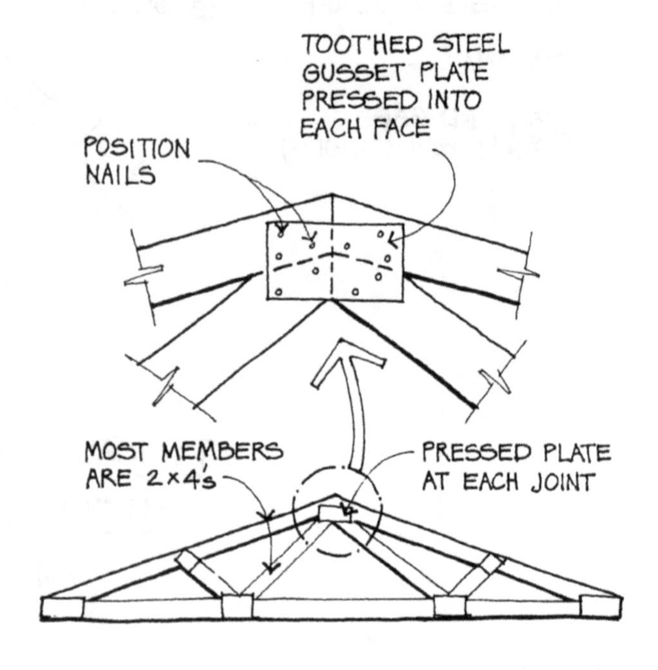

Figure 11.17

F_y = 36 ksi

Table 4-9 (continued)
Available Strength in Axial Compression, kips
Double Angles—LLBB

2L5 LLBB

Shape		2L5×3½× 3/4		5/8		1/2		3/8[c]		5/16[c]		No. of connectors[a]
Wt/ft		39.6		33.5		27.2		20.8		17.4		
Design		P_n/Ω_c	$\phi_c P_n$	P_n/Ω_c	$\phi_c P_n$	P_n/Ω_c	$\phi_c P_n$	P_n/Ω_c	$\phi_c P_n$	P_n/Ω_c	$\phi_c P_n$	
		ASD	LRFD	ASD	LRFD	ASD	LRFD	ASD	LRFD	ASD	LRFD	
Effective length KL (ft) with respect to indicated axis: X-X Axis	0	250	376	212	319	173	259	129	194	101	151	
	2	247	372	209	315	171	256	128	192	99.6	150	
	4	238	358	202	303	164	247	123	185	96.4	145	
	6	223	336	190	285	155	232	116	175	91.4	137	
	8	204	307	174	261	142	213	107	161	84.8	127	
	10	182	274	155	234	127	191	96.4	145	76.9	116	
	12	159	238	136	204	111	167	84.7	127	68.3	103	b
	14	134	202	115	173	95.0	143	72.7	109	59.4	89.3	
	16	111	167	95.6	144	79.1	119	61.0	91.7	50.6	76.0	
	18	89.4	134	77.2	116	64.2	96.4	49.9	75.0	42.1	63.3	
	20	72.4	109	62.6	94.0	52.0	78.1	40.4	60.8	34.3	51.6	
	22	59.8	89.9	51.7	77.7	43.0	64.6	33.4	50.2	28.4	42.6	
	24	50.3	75.6	43.4	65.3	36.1	54.2	28.1	42.2	23.8	35.8	
Y-Y Axis	0	250	376	212	319	173	259	129	194	101	151	
	2	239	359	199	299	157	235	108	162	79.1	119	
	4	229	344	190	286	150	226	106	159	77.8	117	
	6	213	320	177	267	140	210	101	152	75.1	113	
	8	193	290	160	241	127	190	92.7	139	70.1	105	
	10	170	255	141	212	111	167	81.7	123	62.8	94.4	
	12	145	218	120	180	94.9	143	69.4	104	54.3	81.6	2
	14	120	180	99.3	149	78.4	118	57.0	85.7	45.3	68.1	
	16	101	152	79.6	120	62.7	94.2	45.3	68.1	36.6	55.0	
	18	80.7	121	66.2	99.5	49.8	74.9	36.3	54.5	29.5	44.3	
	20	65.5	98.4	53.8	80.8	40.5	60.9	29.7	44.6	24.2	36.4	
	22	54.2	81.5	44.5	66.9	33.6	50.5	24.7	37.1	20.2	30.4	
	24	45.6	68.5	37.5	56.3	28.3	42.6	20.8	31.3	17.1	25.7	3
Properties of 2 angles—3/8 in. back to back												
A_g (in.²)		11.6		9.84		8.01		6.10		5.12		
r_x (in.)		1.55		1.56		1.58		1.59		1.60		
r_y (in.)		1.53		1.50		1.48		1.46		1.44		
Properties of single angle												
r_z (in.)		0.744		0.746		0.750		0.755		0.758		

ASD	LRFD
$\Omega_c = 1.67$	$\phi_c = 0.90$

[a] For Y-Y axis, welded or pretensioned bolted intermediate connectors must be used.
[b] For required number of intermediate connectors, see the discussion of Table 4-8.
[c] Shape is slender for compression with F_y = 36 ksi.
Note: Heavy line indicates Kl/r equal to or greater than 200.

Figure 11.18

LESSON 11 QUIZ

1. Choose the INCORRECT statement.

A. In a parallel chord truss, the forces in the chords increase toward the center of the span, while the forces in the web members decrease toward the center.

B. Although they require relatively complex fabrication and great depth, trusses are often economical framing members.

C. In a bowstring truss, where the depth varies from a minimum at the supports to a maximum at midspan, the chord forces are relatively constant across the entire span.

D. The optimum depth-to-span ratio of a truss is about 1/20.

2. Which statement is CORRECT for the truss shown below?

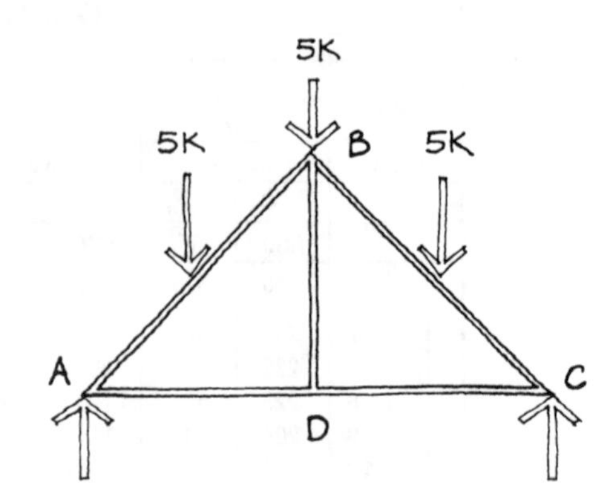

A. The force in member BD is zero, and members AB and BC should be designed for combined bending stress and axial stress.

B. The force in member BD is 5 kips, and members AB and BC should be designed for combined bending stress and axial stress.

C. The force in members AD and DC is zero, and members AB and BC are in compression.

D. Members AD, DC, and BD have equal internal forces of 2.5 kips each.

3. Select the CORRECT statement.

A. A truss may have various configurations, as long as it is comprised of triangles.

B. A truss cannot support any load between panel points.

C. A truss cannot resist any horizontal load.

D. A truss member is in tension if its internal axial force acts toward the joint to which it is connected.

4. A redundant member is a truss member that

A. can resist either tension or compression.

B. is not necessary for stability.

C. consists of two structural sections.

D. resists stress only if another member fails.

5. Which of the following is a type of lightweight prefabricated truss that is in widespread use?

A. Bowstring truss

B. Howe truss

C. Wood roof truss

D. Warren truss

6.

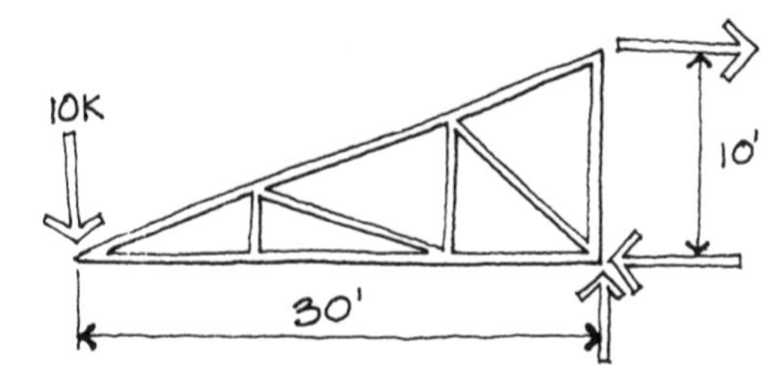

Select the CORRECT statements about the cantilever truss shown above.

A. All of the diagonals are in tension.

B. The top chord is in tension and the bottom chord is in compression.

C. All of the verticals are in compression.

D. All of the diagonals have zero stress.

E. All of the verticals have zero stress.

7. Select the CORRECT statement about double angle compression members in a steel truss.

A. The design is based on $r_{x\text{-}x}$.

B. The design is based on $r_{y\text{-}y}$.

C. The design is based on $r_{z\text{-}z}$.

D. The design may be based on either $r_{x\text{-}x}$ or $r_{y\text{-}y}$.

8.

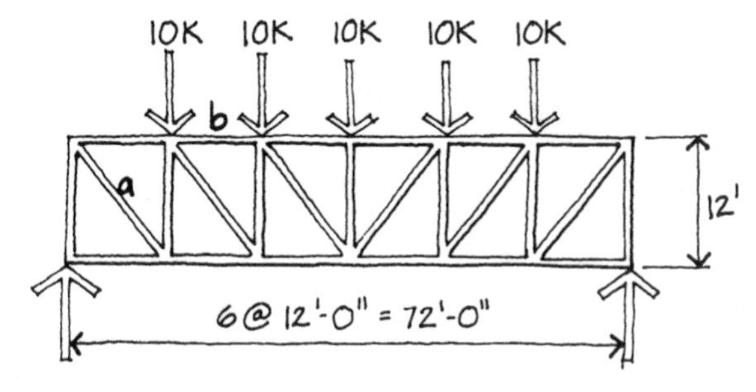

In the truss shown above, the force in member a is

A. 25 kips tension.

B. 25 kips compression.

C. 35.4 kips tension.

D. 35.4 kips compression.

9. In the truss of Question 8, what is the force in member b?

A. 40 kips compression

B. 40 kips tension

C. 10 kips compression

D. 20 kips compression

LESSON TWELVE

EARTHQUAKE DESIGN

12

INTRODUCTION

The Structural Systems division tests candidates for their knowledge of earthquake and wind design. Whether or not you live in an area prone to earthquakes, you must have sufficient understanding of earthquake design to pass this exam. This lesson will provide that understanding. The next lesson covers wind design.

NATURE OF EARTHQUAKES

Of all the natural disasters that can occur, earthquakes are probably the most feared and least predictable. With little or no warning, tremendous forces are unleashed that last only seconds, yet can cause widespread destruction of life and property.

Although advances have been made in the prediction of earthquakes, they remain capricious: reliable earthquake prediction is still a long way off. And we still have no way of controlling the vast amount of energy released by an earthquake. However, detailed study of many earthquakes in recent years by seismologists and engineers has brought about an increased understanding of the physical effects of earthquakes on buildings and other structures. As a result, our ability to design and construct buildings to resist earthquakes has significantly improved.

What causes earthquakes? Centuries ago, people believed that these inexplicable movements of the earth were caused by the restlessness of a monster that supported the world on its back. Depending on your tribe, the monster might be an immense frog, spider, or tortoise. Until comparatively recent times, seismology, the study of earthquakes, had not advanced much beyond those primitive beliefs.

Even now, the causes of earthquakes are not totally understood. However, the theory which best explains earthquake phenomena is called *plate tectonics*. According to this theory, the earth's crust is made up of great rock plates floating on a layer of molten rock. As these plates move, friction and other forces lock them together. As the strain increases over long periods of time, the frictional force between the plates is finally overcome and the plates violently slip past each other to a new unstrained position. This slip releases complex shock waves, which travel outward at great speeds through the rock and overlying soil. The ground through which these waves travel experiences shaking or vibration, and buildings founded in this shaking ground are pushed and pulled to-and-fro, side-to-side, and up-and-down. In a short time, perhaps 10 to 20 seconds, the worst part of the earthquake is usually over, often followed by a number of aftershocks of lesser severity. The intent of earthquake, or seismic, design is to provide buildings with sufficient resistance to the effects of this ground shaking so that they can survive even a major earthquake without collapsing.

About 80 percent of the world's earthquakes, including many of the strongest on record, occur around the rim of the Pacific Ocean, from the tip of South America up through Alaska and down along Japan and China to the South Pacific. In the United States, the most vulnerable areas are in California and Alaska. However, many locations east of the Rockies are also subject to high seismic risk: one of the most severe earthquakes in U.S. history occurred in the Mississippi River valley near the confluence of the Ohio and Mississippi Rivers in 1811.

The term *fault* refers to the boundary between adjacent tectonic plates, such as the San Andreas fault in California. It also denotes the numerous lesser breaks and fractures in the earth's crust. A major fault, such as the San Andreas, is frequently a system of faults comprising a fault zone of considerable width. A strong earthquake may result in lateral or vertical shifts along the fault of anywhere from a few inches to as much as 20 feet.

The location in the earth's crust where the rock slippage begins is called the *focus* or *hypocenter* of the earthquake. Its depth is usually five to ten miles in California, although foci several hundred miles deep have been located in other parts of the world. The projection of the focus on the ground surface is called the *epicenter*.

MEASURING EARTHQUAKES

Earthquakes are measured by two distinctly different scales: the *Richter scale* and the *Modified Mercalli scale*. The Richter scale is a measure of the earthquake's magnitude—the amount of energy it releases. The scale is logarithmic, with each number representing about 33 times the energy of the preceding number. Thus, an earthquake of magnitude 7 releases 33 times more energy than one of magnitude 6, or about 1,000 times (33 × 33) more energy than one of magnitude 5.

The largest earthquakes ever recorded had a magnitude of about 8.9. The 1906 San Francisco earthquake has been estimated at about 8.3, the Alaska earthquake of 1964 measured 8.4, the San Fernando earthquake of 1971 registered 6.4, the Northern California earthquake of 1989 had a magnitude of 7.1, and the 1994 Northridge earthquake measured 6.7.

The Richter scale does not indicate potential damage to buildings: an earthquake of moderate magnitude centered in a city would be more damaging than a greater quake located far from population centers.

The damage of great earthquakes may not be much greater than those of smaller quakes; the difference is that the greater quakes generally last longer and cause damage over a much greater area.

Earthquakes of magnitude 5 are potentially damaging, 6 to 7 are considered moderate and potentially destructive, 7 to 7.75 are termed *major*, while those above 7.75 are called *great earthquakes*.

The Modified Mercalli scale measures the intensity of an earthquake (its effects on people and buildings). It is expressed in Roman numerals I to XII, each number corresponding to a specific description of the earthquake's effects. For example, I is *not felt except under especially favorable circumstances*, while XII has *damage nearly total. Large rock masses displaced. Lines of sight and level distorted. Objects thrown into the air.*

There is no precise correlation between these scales. The Richter is a fairly accurate measure of physical energy, but it does not reveal the effects of the earthquake. The Modified Mercalli, on the other hand, gives us an indication of damage, but is subjective and imprecise.

BUILDING RESPONSE

An earthquake causes the ground to shake erratically, both vertically and horizontally. The vertical motions are usually neglected in design since they are generally smaller than the horizontal. Also, since buildings are basically designed for vertical gravity loads, they usually have considerable excess strength in the vertical direction.

The ground does not move at a uniform velocity. Quite the contrary: its velocity is constantly changing during an earthquake. The rate of change of velocity is called *acceleration*. It is convenient to measure this acceleration as a fraction or percentage of g, the acceleration of gravity. (The acceleration of gravity is the acceleration experienced by a freely falling body and is equal to 32 feet per second per second.) Thus, we may say that the ground acceleration at a given time is .10 g or 10 percent g, which is equal to .10 × 32 = 3.2 feet per second.

Information on ground acceleration during an earthquake is recorded on *strong-motion accelerographs*. These seismological instruments are normally inoperative; when subject to strong earth motion, however, they become activated, record the earth motion, and then shut off. As recorded by these instruments, the ground acceleration during an earthquake changes rapidly and irregularly, varying from zero to a maximum in each direction.

If these varying ground accelerations are applied to a series of idealized structures, each having a different natural period, the maximum acceleration of each of these structures can be determined, and the results plotted.

The resulting curve is called a *response spectrum*. Idealized design response spectra generated from ASCE 7 are shown in Figure 12.1 for buildings located in regions of low and high seismicity. In each region, response spectra are depicted for three different soil types. While the spectrum will be different for every accelerograph record and hence every earthquake, the spectra all have similar shapes, from which certain conclusions may be drawn.

What is meant by *period?* If you take a pendulum, consisting of a weight or mass on top of a rod with a certain stiffness, and you start the pendulum moving, it will take a certain amount of time for it to go through one complete back-and-forth motion (Figure 12.2). This amount of time is called its *natural period* or *fundamental period* and is independent of how hard you push on the pendulum.

The natural period is dependent on only two factors: the mass, or weight, of the pendulum and the stiffness, or rigidity, of the rod. Thus, making the mass heavier or the rod more flexible changes the natural period of the pendulum.

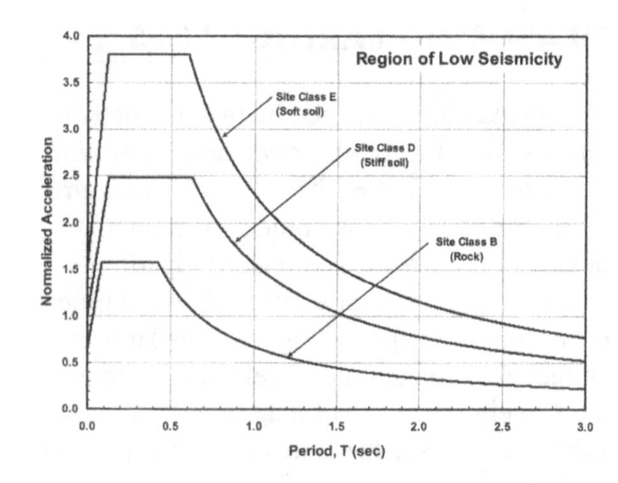

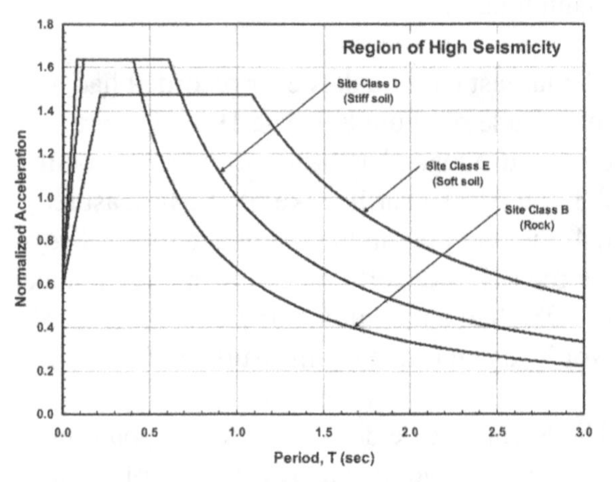

Figure 12.1

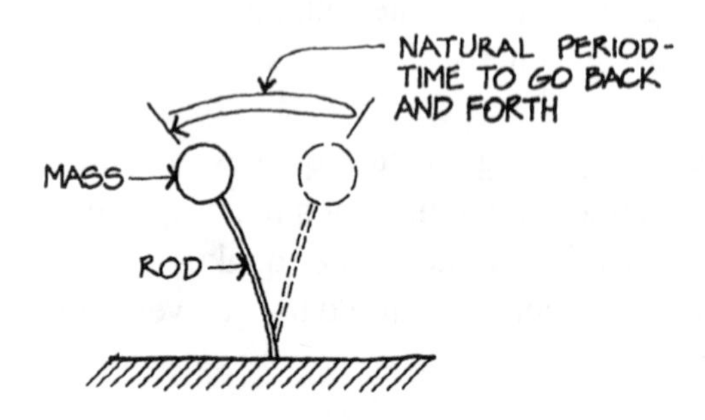

Figure 12.2

Getting back to the response spectra, what they tell us is that for a structure having a natural period of 0 seconds (an infinitely rigid structure), the maximum acceleration of the structure will be the same as that of the ground. As the period of the structure increases, its peak acceleration also increases, reaching a maximum when the period is about 0.5 seconds. This maximum acceleration is generally two to three times the maximum ground acceleration. However, even larger magnification of ground acceleration is possible. In the Mexico City earthquake of 1985, for example, accelerations in the upper floors of some buildings were six times the maximum ground acceleration.

As the period increases beyond 0.5 seconds or so, the acceleration decreases and soon becomes less than the ground acceleration. Thus, in general, we can say that a building that is less stiff, or *more flexible*, will have a *longer natural period* and hence a *lower maximum acceleration.*

In regions of low seismicity, the response spectra show that the building acceleration will be greater if the building is on soft soil (Site Class E) and less if it is on stiff soil (Site Class D) or rock (Site Class B). For buildings with periods less than about 0.5 second in regions of high seismicity, building accelerations tend to be smaller when the building is on soft soil and larger when the building is on stiff soil or rock. This can be seen from the response spectra: the accelerations for buildings on Site Class E are smaller than those on Site Class D or B up to a period of about 0.7 second. The accelerations become larger for buildings on Site Class E when the period is greater than about 0.7 second.

Imagine for a moment an earthquake in which the ground moves instantaneously in a given horizontal direction, as shown in Figure 12.3. A building sitting on the ground will momentarily have its base offset relative to the upper portion of the building. The physical property that causes the upper portion of the building to remain in its original position while the base is moved is called *inertia.*

The same distorted shape of the building could be obtained by applying simulated horizontal forces to the upper portion of the building. The internal forces created by these simulated external forces are equal in magnitude to the actual forces induced in the structure by the earthquake motion. In other words, the applied earthquake forces are not real externally applied forces, as are dead, live, and wind forces, but are simulated forces that produce the same effect on the building structure as the actual earthquake motion.

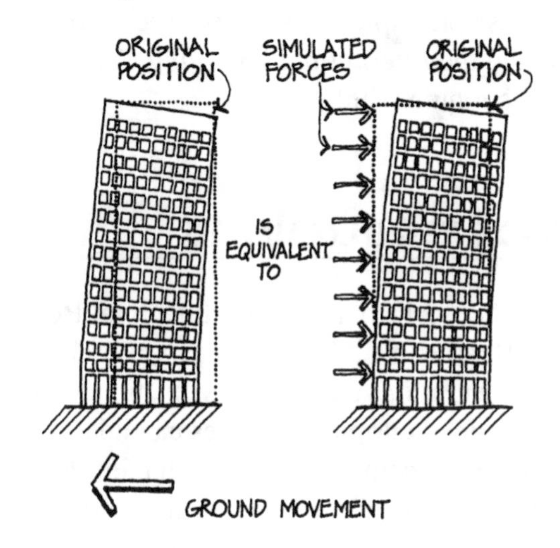

Figure 12.3

If we know the building's natural period, we can determine its acceleration for a given earthquake by using the response spectrum.

We can then convert acceleration to force by using Newton's second law of motion: F = Ma, or force is equal to mass times acceleration. Mass is weight divided by g, the acceleration of gravity (g = 32 feet per second per second). *Thus, the magnitude of the inertia force is equal to the mass of the structure multiplied by its acceleration.* Since long period, or flexible, buildings have less acceleration than short period, or stiff, buildings, they induce less force.

And since heavy buildings have more mass than light buildings, they have greater earthquake forces. It can be seen, therefore, that there is a significant difference between conventional structural design and seismic design. In the former, the magnitude and nature of the applied dead, live, and wind loads are *independent* of the stiffness of the structure. In seismic design, however, increasing the structure's stiffness increases the induced seismic force; the seismic force is therefore *dependent* on the stiffness of the structure. If we have some understanding of these principles, we will see the rationale behind the *International Building Code* earthquake requirements.

CODE REQUIREMENTS

The National Council of Architectural Registration Boards (NCARB) no longer references the *Uniform Building Code* in its exam guidelines. The council currently requires candidates to familiarize themselves only with the *International Building Code* (IBC) or the *National Building Code of Canada.*

The IBC has been adopted by many jurisdictions in the U.S. and other parts of the world. The 2009 edition adopts by reference the 2005 edition of the American Society of Civil Engineers' *Minimum Design Loads for Buildings and Other Structures,* commonly known as ASCE 7 or ASCE 7-05. The discussion that follows therefore relies on the IBC and, in turn, ASCE 7.

Introduction

Structures designed in accordance with IBC provisions should generally be able to

- resist minor earthquakes without damage,
- resist moderate earthquakes without structural damage, but possibly some nonstructural damage, and
- resist major earthquakes without collapse, but possibly some structural and nonstructural damage.

The code is intended to safeguard against major failures and loss of life; the protection of property per se is not its purpose (see Figure 12.4). While it is believed that the code provides reasonably for protection of life, even that cannot be completely assured.

The code provides for resistance to earthquake ground shaking only, but not for earth slides, subsidence, settlement, faulting in the immediate vicinity of the structure, or for soil liquefaction, which is the transformation of soil into a liquefied state similar to quicksand.

Collapsed apartment building, San Francisco, 1989

Figure 12.4

The code provides a number of methods, both static and dynamic, for determining earthquake forces. Not all are within the scope of this lesson, but in general the building's Seismic Design Category (SDC), structural system, dynamic properties, and regularity determine the preferred method.

The *Seismic Response History Procedures* are dynamic analysis procedures and are always acceptable. *Modal Response Spectrum Analysis* is essentially a static procedure and is also acceptable in any situation. *Equivalent Lateral Force Analysis*, a static procedure, is allowed only under certain conditions involving building regularity, occupancy, and period.

In the Seismic Response History Procedures, a mathematical model of the structure is developed and then subjected to appropriate ground motions. A software program determines the model's dynamic response to the input motions, from which the forces to be used in the design can be determined. The details of the dynamic procedure are quite technical and therefore beyond the scope of this course.

In Modal Response Spectrum Analysis, the distribution of seismic forces is based on the deformed shapes of the natural modes of vibration—which are determined from the distribution of mass and stiffness of the building. The details of the Seismic Response History Procedures and Modal Response Spectrum Analysis are quite technical and beyond the scope of this lesson.

The Equivalent Lateral Force Procedure can be used to design most buildings. In this method, static (non-varying) horizontal forces are applied to the building, producing internal forces similar to those induced by actual earthquakes. As discussed, an earthquake actually applies no external forces to the building; the forces used in static analysis are thus not real forces.

Before examining the Equivalent Lateral Force Procedure, we will discuss how to determine both seismic ground motion values at a site and a building's Seismic Design Category (SDC). As mentioned, the SDC determines the procedure for analyzing the effects of earthquake ground motion on a given building. The SDC also plays a key role in many other aspects of seismic design, such as detailing structural members to resist the effects of earthquakes.

Seismic Ground Motion Values

It is important to establish the amount of ground shaking (or equivalently, ground acceleration) expected to occur at a given site during an earthquake. As discussed, the magnitudes of the simulated horizontal forces applied to a building directly relate to the amount of ground shaking.

A *maximum considered earthquake* (MCE) measures the magnitude of ground shaking or acceleration expected at a site. For most of the U.S., the MCE is an earthquake that has a chance of occurring about once every 2500 years. In coastal California, where a significant amount of ground motion data has been acquired over the years, the MCE is the largest earthquake that can be generated by known seismic sources.

United States Geological Survey (USGS) seismic hazard maps provide MCE ground accelerations for the United States and its territories. These contour maps are given in IBC Figures 1613.5(1) through 1613.5(14). Mapped accelerations S_S are for short periods (period equal to 0.2 second) and S_1 are for a period equal to 1.0 second. Both accelerations are given for Site Class B (rock). Figure 1613.5(1), reproduced in Figure 12.5, gives S_S values for the 48 contiguous states.

While contour maps provide accelerations S_S and S_1, more accurate values of these accelerations can be obtained online. Enter the zip code or the latitude and longitude of a site into the ground motion parameter calculator available on the USGS Web site (*http://earthquake.usgs.gov/research/hazmaps/design*) for more detailed information than these maps provide.

As noted above, mapped values of S_S and S_1 are based on Site Class B (rock).

Modify these accelerations for other site classes (soils) using the following formulas:

$S_{MS} = F_a S_S$
$S_{M1} = F_v S_1$

where

S_{MS} = MCE acceleration at short periods adjusted for site class effects

F_a = site coefficient defined in IBC Table 1613.5.3(1) and ASCE 7 Table 11.4-1

S_{M1} = MCE acceleration at a period of 1 second adjusted for site class effects

F_v = site coefficient defined in IBC Table 1613.5.3(2) and ASCE 7 Table 11.4-2

IBC Tables 1613.5.3(1) and 1613.5.3(2), shown in Figures 12.6 and 12.7, reveal that values of F_a and F_v are higher in areas of low seismicity and lower in areas of high seismicity.

IBC Table 1613.5.2, shown in Figure 12.8, provides site class definitions. A geotechnical engineer usually determines site class. In cases where soil properties are unknown, Site Class D must be used, unless it has been determined that Site Class E or F soils are present at the site.

Finally, design accelerations S_{DS} and S_{D1} are determined by the following formulas:

$S_{DS} = 2S_{MS}/3$
$S_{D1} = 2S_{M1}/3$

These accelerations are used in determining SDC and the horizontal forces on the building caused by ground shaking effects.

Seismic Design Category

Every structure requires a Seismic Design Category. The SDC helps determine all of the following:

- Structural system
- Building height and irregularity limitations
- Components that must be designed for seismic resistance
- Horizontal force analysis method

The SDC is a function of building occupancy or use and the design accelerations at the site. IBC Table 1604.5, shown in Figure 12.9, defines the various occupancy categories.

Generally, the SDC is determined twice: first as a function of S_{DS} by IBC Table 1613.5.6(1) (Figure 12.10), and second as a function of S_{D1} by IBC Table 1613.5.6(2) (Figure 12.11). You should assign the building the more severe of the two results.

The SDC can be determined based on Table 1613.5.6(1) alone, *if* all of the following apply:

- In each direction, the approximate period of the building T_a is less than $0.8T_s = 0.8S_{D1}/S_{DS}$
- In each direction, the period of the building used to calculate drift is less than T_s (The period $T_s = S_{D1}/S_{DS}$ is the period at which the short-period part of the design spectrum transitions into the long-period part of the spectrum.)
- The seismic response coefficient C_s is determined from $S_{DS}/(R/I)$
- The diaphragms are rigid or, for diaphragms that are flexible, the distance between elements of the seismic-force resisting system is 40 feet or less

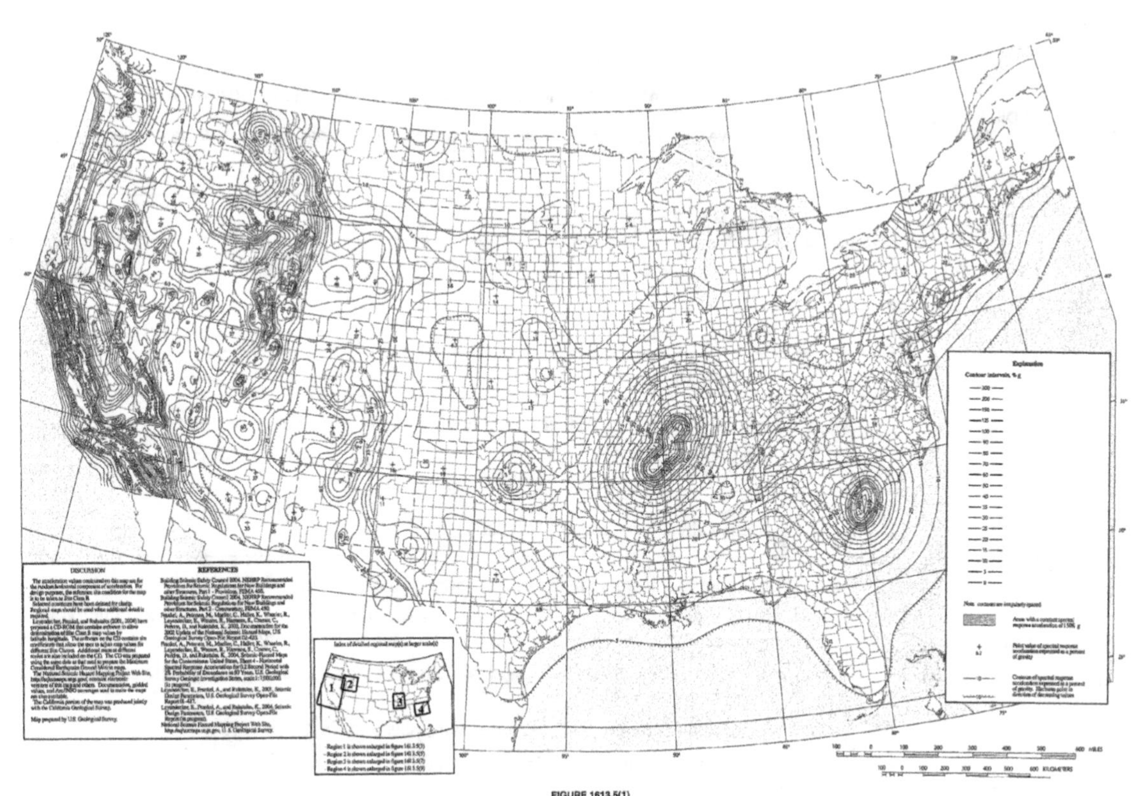

FIGURE 1613.5(1)
MAXIMUM CONSIDERED EARTHQUAKE GROUND MOTION FOR THE CONTERMINOUS UNITED STATES OF 0.2 SEC SPECTRAL RESPONSE ACCELERATION (5% OF CRITICAL DAMPING), SITE CLASS B

Figure 12.5

TABLE 1613.5.3(1)
VALUES OF SITE COEFFICIENT F_a [a]

SITE CLASS	MAPPED SPECTRAL RESPONSE ACCELERATION AT SHORT PERIOD				
	$S_s \le 0.25$	$S_s = 0.50$	$S_s = 0.75$	$S_s = 1.00$	$S_s \ge 1.25$
A	0.8	0.8	0.8	0.8	0.8
B	1.0	1.0	1.0	1.0	1.0
C	1.2	1.2	1.1	1.0	1.0
D	1.6	1.4	1.2	1.1	1.0
E	2.5	1.7	1.2	0.9	0.9
F	Note b	Note b	Note b	Note b	Note b

a. Use straight-line interpolation for intermediate values of mapped spectral response acceleration at short period, S_s.
b. Values shall be determined in accordance with Section 11.4.7 of ASCE 7.

Figure 12.6

TABLE 1613.5.3(2)
VALUES OF SITE COEFFICIENT F_V [a]

SITE CLASS	MAPPED SPECTRAL RESPONSE ACCELERATION AT 1-SECOND PERIOD				
	$S_1 \le 0.1$	$S_1 = 0.2$	$S_1 = 0.3$	$S_1 = 0.4$	$S_1 \ge 0.5$
A	0.8	0.8	0.8	0.8	0.8
B	1.0	1.0	1.0	1.0	1.0
C	1.7	1.6	1.5	1.4	1.3
D	2.4	2.0	1.8	1.6	1.5
E	3.5	3.2	2.8	2.4	2.4
F	Note b	Note b	Note b	Note b	Note b

a. Use straight-line interpolation for intermediate values of mapped spectral response acceleration at 1-second period, S_1.
b. Values shall be determined in accordance with Section 11.4.7 of ASCE 7.

Figure 12.7

TABLE 1613.5.2
SITE CLASS DEFINITIONS

SITE CLASS	SOIL PROFILE NAME	AVERAGE PROPERTIES IN TOP 100 feet, SEE SECTION 1613.5.5		
		Soil shear wave velocity, $\bar{v}_s$, (ft/s)	Standard penetration resistance, $\bar{N}$	Soil undrained shear strength, $\bar{s}_u$, (psf)
A	Hard rock	$\bar{v}_s > 5{,}000$	N/A	N/A
B	Rock	$2{,}500 < \bar{v}_s \le 5{,}000$	N/A	N/A
C	Very dense soil and soft rock	$1{,}200 < \bar{v}_s \le 2{,}500$	$\bar{N} > 50$	$\bar{s}_u \ge 2{,}000$
D	Stiff soil profile	$600 \le \bar{v}_s \le 1{,}200$	$15 \le \bar{N} \le 50$	$1{,}000 \le \bar{s}_u \le 2{,}000$
E	Soft soil profile	$\bar{v}_s < 600$	$\bar{N} < 15$	$\bar{s}_u < 1{,}000$
E	—	Any profile with more than 10 feet of soil having the following characteristics: 1. Plasticity index $PI > 20$, 2. Moisture content $w \ge 40\%$, and 3. Undrained shear strength $\bar{s}_u < 500$ psf		
F	—	Any profile containing soils having one or more of the following characteristics: 1. Soils vulnerable to potential failure or collapse under seismic loading such as liquefiable soils, quick and highly sensitive clays, collapsible weakly cemented soils. 2. Peats and/or highly organic clays ($H >$ 10 feet of peat and/or highly organic clay where H = thickness of soil) 3. Very high plasticity clays ($H >$25 feet with plasticity index $PI >$75) 4. Very thick soft/medium stiff clays ($H >$120 feet)		

For SI: 1 foot = 304.8 mm, 1 square foot = 0.0929 m^2, 1 pound per square foot = 0.0479 kPa. N/A = Not applicable

Figure 12.8

This exception is often advantageous: it is generally unnecessary and wasteful to determine the SDC of a short-period building by long-period ground motion.

Figure 12.12 describes the relationship between Seismic Design Categories and Occupancy Categories.

For buildings assigned to SDC A, static lateral forces are applied at each floor level to simulate the effects of the earthquake on the structure. The following formula determines the magnitude of these forces:

$$F_x = 0.01 w_x$$

where

F_x = lateral force applied at story x in the building

w_x = portion of the total dead load of the structure located or assigned to level x

Thus, the total shear at the base of the building is equal to 10 percent of its total dead load.

TABLE 1604.5
OCCUPANCY CATEGORY OF BUILDINGS AND OTHER STRUCTURES

OCCUPANCY CATEGORY	NATURE OF OCCUPANCY
I	Buildings and other structures that represent a low hazard to human life in the event of failure, including but not limited to: • Agricultural facilities. • Certain temporary facilities. • Minor storage facilities.
II	Buildings and other structures except those listed in Occupancy Categories I, III and IV
III	Buildings and other structures that represent a substantial hazard to human life in the event of failure, including but not limited to: • Covered structures whose primary occupancy is public assembly with an occupant load greater than 300. • Buildings and other structures with elementary school, secondary school or day care facilities with an occupant load greater than 250. • Buildings and other structures with an occupant load greater than 500 for colleges or adult education facilities. • Health care facilities with an occupant load of 50 or more resident patients, but not having surgery or emergency treatment facilities. • Jails and detention facilities. • Any other occupancy with an occupant load greater than 5,000. • Power-generating stations, water treatment for potable water, waste water treatment facilities and other public utility facilities not included in Occupancy Category IV. • Buildings and other structures not included in Occupancy Category IV containing sufficient quantities of toxic or explosive substances to be dangerous to the public if released.
IV	Buildings and other structures designated as essential facilities, including but not limited to: • Hospitals and other health care facilities having surgery or emergency treatment facilities. • Fire, rescue and police stations and emergency vehicle garages. • Designated earthquake, hurricane or other emergency shelters. • Designated emergency preparedness, communication, and operation centers and other facilities required for emergency response. • Power-generating stations and other public utility facilities required as emergency backup facilities for Occupancy Category IV structures. • Structures containing highly toxic materials as defined by Section 307 where the quantity of the material exceeds the maximum allowable quantities of Table 307.1.(2). • Aviation control towers, air traffic control centers and emergency aircraft hangars. • Buildings and other structures having critical national defense functions. • Water treatment facilities required to maintain water pressure for fire suppression.

Figure 12.9

TABLE 1613.5.6(1)
SEISMIC DESIGN CATEGORY BASED ON SHORT-PERIOD RESPONSE ACCELERATIONS

VALUE OF S_{DS}	OCCUPANCY CATEGORY		
	I or II	III	IV
$S_{DS} < 0.167g$	A	A	A
$0.167g \le S_{DS} < 0.33g$	B	B	C
$0.33g \le S_{DS} < 0.50g$	C	C	D
$0.50g \le S_{DS}$	D	D	D

Figure 12.10

TABLE 1613.5.6(2)
SEISMIC DESIGN CATEGORY BASED ON 1-SECOND PERIOD RESPONSE ACCELERATION

VALUE OF S_{D1}	OCCUPANCY CATEGORY		
	I or II	III	IV
$S_{D1} < 0.067g$	A	A	A
$0.067g \le S_{D1} < 0.133g$	B	B	C
$0.133g \le S_{D1} < 0.20g$	C	C	D
$0.20g \le S_{D1}$	D	D	D

Figure 12.11

SEISMIC DESIGN CATEGORY (SDC)	OCCUPANCY CATEGORY	DESCRIPTION
A	All	Buildings of any occupancy where anticipated ground shaking is minor.
B	I, II, III	Buildings in regions where moderately destructive ground shaking is anticipated.
C	IV	Buildings in regions where moderately destructive ground shaking may occur.
	I, II, III	Buildings in regions with somewhat more severe ground shaking potential.
D	All	Buildings in regions expected to experience destructive ground shaking, but not located close to major active faults.
E	I, II, III	Buildings in regions located close to major active faults.
F	IV	Buildings in regions located close to major active faults.

Figure 12.12

For buildings assigned to SDC B and higher, more rigorous analysis methods—such as the Equivalent Lateral Force Procedure, discussed later—are needed to determine the horizontal forces due to ground shaking.

Some structures are exempt from earthquake requirements. In particular, the following structures need not be designed to IBC requirements:

- Detached one- and two-family dwellings located where S_S is less than 0.4g or where the SDC is A, B, or C
- Detached one- and two-family wood-frame dwellings not included in the exception above and not more than 2 stories in height. These types of structures must be designed and detailed in accordance with the International Residential Code.
- Agricultural storage structures intended only for incidental human occupancy
- Structures such as bridges, electrical transmission towers, hydraulic structures, buried utility lines, and nuclear reactors, which require special consideration

Equivalent Lateral Force Procedure

The ASCE 7 Equivalent Lateral Force Procedure consists of determining the total lateral force, or shear at the base (V) from the following formula:

$$V = C_s W$$

where

C_s = seismic response coefficient = $S_{DS}/(R/I)$

W = effective seismic weight, including the total dead load of the building

This formula can be rewritten as:

$$V = S_{DS} WI/R$$

It might help to understand this formula if we go back to the basic relationship between force and acceleration, which is F = Ma. This is equivalent to $V = S_{DS}W$.

The importance factor I is introduced, in order to increase V for certain critical structures. Thus, $V = S_{DS}WI$.

Now we take that value of V and reduce it by the coefficient R, to account for the ability of the structural system to accommodate loads and absorb energy considerably in excess of the usual allowable stresses without collapsing, and thus,

$V = \frac{S_{DS}I}{R} W$, the basic seismic formula.

In the paragraphs that follow, we will explain the terms in this formula in more detail.

The seismic force V is generally evaluated in two horizontal directions parallel to the main axes of the building. For buildings assigned to SDC A and B, the structure must be able to resist effects caused by seismic forces in either direction, although not in both directions simultaneously. The same is true for buildings assigned to SDC C, except for those buildings possessing a nonparallel systems horizontal irregularity, where the vertical elements of the seismic-force resisting system are not parallel to or symmetric about the major orthogonal (right-angle) axes of the seismic-force-resisting system. Building irregularities are discussed in greater detail below. When such a horizontal discontinuity exists, structural members must be designed to resist the effects from the simultaneous application of 100 percent of the seismic forces in one direction and 30 percent of the seismic forces in the orthogonal direction. Buildings assigned to SDC D through F must conform to these same requirements.

Additionally, any column or wall that is part of two intersecting seismic-force-resisting systems and that is subjected to an axial load due to seismic forces greater than 20 percent of the axial design strength of the member must be designed for the most critical effects due to seismic forces acting in any direction. ASCE 7 gives procedures on how to satisfy the direction of loading requirements for SDC D through F.

Seismic Response Coefficient, C_s

The formulas for C_s given in the code are used to determine the base shear V. These formulas form the basis of the design spectrum of the Equivalent Lateral Force Procedure. The design spectrum is illustrated in Figure 12.13 for a typical building. Note the similarities between the shape of the design spectrum and the response spectra presented in Figure 12.13.

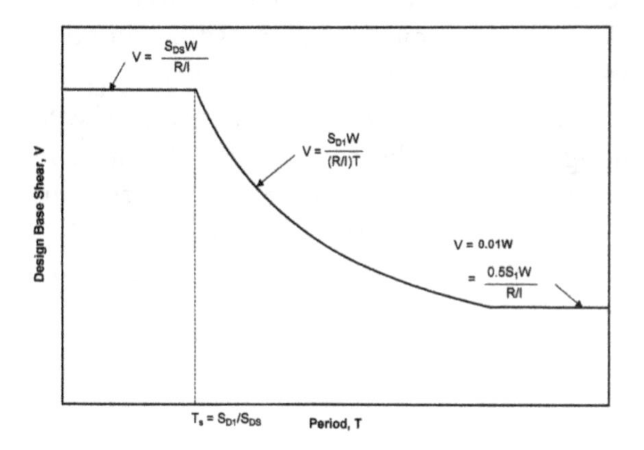

Figure 12.13

As noted previously, the seismic response coefficient C_s is determined by the following formula:

$$C_s = S_{DS}/(R/I)$$

The previous formula is independent of the period of the building and represents the horizontal line (short-period response) of the design spectrum. Note that in the Equivalent Lateral Force Procedure, the horizontal portion of the spectrum begins at a period equal to zero and extends to a period equal to $T_s = S_{D1}/S_{DS}$, which is slightly different than what is depicted in the previous figures of the response spectra. Ignoring the initial linear segment of the curve at low periods is mainly for simplicity, since that portion of the spectrum rarely has an effect on the design of a typical building.

The value of C_s need not exceed the following:

$C_s = S_{D1}/T(R/I)$ for T less than or equal to T_L
$C_s = S_{D1}T_L/T^2(R/I)$ for T greater than T_L

It is evident from these formulas that C_s is dependent on the fundamental period of the building T, which is discussed later.

The first of these formulas represents the descending portion of the design spectrum (long-period response) up to the period T_L, where T_L is the long-period transition period. ASCE 7 Figures 22-15, which contains T_L values for the conterminous U.S. is reproduced in Figure 12.14. The second formula represents the descending portion of the design spectrum for periods greater than T_L. For typical buildings, the fundamental period of the building T is usually much less than T_L.

The code also requires that C_s be no less than 0.01. This essentially sets a minimum base shear V equal to 1 percent of the effective seismic weight W, which is the same minimum horizontal force requirement for buildings assigned to SDC A. This minimum V may govern in cases where the period of the building is large. In addition, for buildings located in regions of high seismic risk where S_1 is greater than or equal to 0.6g, the minimum C_s is determined by the following formula:

$C_s = 0.5S_1/(R/I)$

A form of this lower-bound formula originally appeared in the 1997 Uniform Building Code, and was first adopted in response to the 1994 Northridge earthquake in Southern California.

The code sets an upper limit on the value of S_S when computing C_s by the above formulas. In particular, the value of S_S need not exceed 1.5g

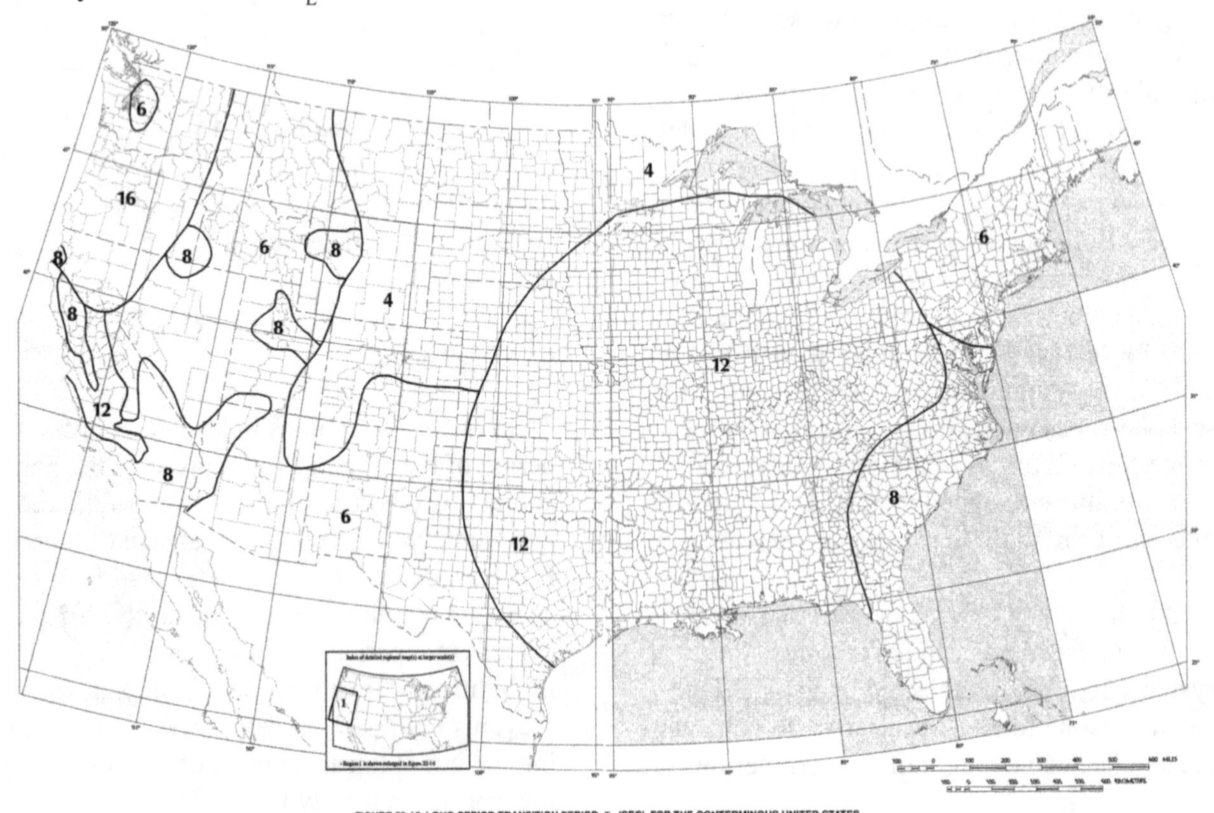

FIGURE 22-15 LONG-PERIOD TRANSITION PERIOD, T_L (SEC), FOR THE CONTERMINOUS UNITED STATES

Figure 12.14

in regular buildings that are five stories or less in height and that have a fundamental period T less than or equal to 0.5 second.

The value of the fundamental period of vibration of the building T may be approximated from the following formula:

$T = C_t(h_n)^x$

where C_t and x are parameters that depend on the structure type and h_n is the height of the building in feet. Values of C_t and x are given in Table 12.8-2 of ASCE 7, which is reproduced in Figure 12.15. Note that in the formula for T, h_n is raised to the x power where x varies from 0.75 to 0.9. These calculations must be done on a calculator that has a function of raising numbers to a power.

TABLE 12.8-2 VALUES OF APPROXIMATE PERIOD PARAMETERS C_t AND x

Structure Type	C_t	x
Moment-resisting frame systems in which the frames resist 100% of the required seismic force and are not enclosed or adjoined by components that are more rigid and will prevent the frames from deflecting where subjected to seismic forces:		
Steel moment-resisting frames	0.028 (0.0724)[a]	0.8
Concrete moment-resisting frames	0.016 (0.0466)[a]	0.9
Eccentrically braced steel frames	0.03 (0.0731)[a]	0.75
All other structural systems	0.02 (0.0488)[a]	0.75

[a]Metric equivalents are shown in parentheses.

Figure 12.15

Two other formulas are provided in the code that may be used to determine T when certain conditions are satisfied. The previous formula can be used in any situation.

I Factor

I is the importance factor as defined by ASCE 7 Table 11.5-1, which is reproduced in Figure 12.16.

TABLE 11.5-1 IMPORTANCE FACTORS

Occupancy Category	*I*
I or II	1.0
III	1.25
IV	1.5

Figure 12.16

You can see that the value of I for earthquake is either 1.0, 1.25, or 1.5, depending on the occupancy category of the building.

Hence, essential facilities such as hospitals and fire and police stations are designed for seismic forces 50 percent greater than normal (I = 1.5). In this way, such emergency facilities are expected to be safe and usable following an earthquake. Buildings containing certain highly toxic substances must also be designed to these higher standards.

Similarly, high-occupancy buildings that represent a substantial hazard to human life—schools, health care facilities, and stadiums—must be designed for seismic forces 25 percent greater than normal (I = 1.25).

The importance factor for wind will be discussed in Lesson Thirteen.

Response Modification Factor, R

When we design a structural system to resist dead, live, and wind loads, we attempt to keep deformations and stresses within acceptably low limits. However, it would be economically prohibitive to use those same limitations when designing for the maximum expected earthquake motion. Instead, the basic philosophy of seismic design is that the structure be able to accommodate the maximum expected earthquake *without collapse*. Although the structure is expected to ride out the earthquake, inelastic deformation is expected to occur, as well as structural and nonstructural damage.

The Response Modification Factor R, which is determined by the type of lateral load resisting system used, is a measure of the system's ability to accommodate earthquake loads and absorb energy without collapse. A stiff, brittle structure has a low value of R, while a resilient, ductile system has a high value of R. Later in this lesson we will discuss the R values of various lateral load resisting systems in more detail.

Effective Seismic Weight, W

The effective seismic weight W of a building includes all of the dead loads and applicable portions of other loads, as follows:

- In storage occupancies (including warehouses), 25 percent of the floor live load. Note that floor live load in public garages and open parking structures need not be included.
- Where a partition load is used in the floor design, the actual weight of the partitions, or a minimum weight of 10 psf
- The operating weight of any equipment permanently attached to the building
- Where the flat roof snow load exceeds 30 psf, 20 percent of the uniform design snow load, regardless of the roof slope

LATERAL LOAD RESISTING SYSTEMS

Introduction

When earthquake motion causes a building to move, energy is induced into the building structure. The function of the lateral load resisting system is to absorb this energy by moving, or deforming, without collapse. The building will probably be damaged during a major earthquake, but the damage is expected to be repairable.

The ability of structural systems and materials to deform and absorb energy, without failure or collapse, is termed *ductility*. Materials or systems that are able to absorb energy through movement are called *ductile*, whereas those which are less able to do so are termed *non-ductile* or *brittle*. An extreme example of a ductile, though nonstructural, material would be a rubber band, while an example of a non-ductile material would be unreinforced concrete.

All structural systems resist forces in three basic ways: by bending (flexure), shear, or axial tension and compression. In general, systems that resist forces by bending are more ductile than those that are stressed in shear or axial tension and compression.

There are three basic types of lateral load resisting systems: moment-resisting frames, shear walls, and braced frames. Generally, shear walls are the most rigid, that is, they deflect the least when subject to a given load. Braced frames are usually less rigid than shear walls, and moment-resisting frames are the least rigid.

Moment-Resisting Frames

A moment-resisting frame resists lateral forces by bending action, as shown in Figures 12.17 and 12.18.

Moment-resisting frames may be constructed of either structural steel or reinforced concrete.

There are three categories of moment-resisting frames:

1. A *special moment-resisting frame* (SMRF) is a moment-resisting frame made of structural steel or reinforced concrete that has the ability to absorb a large amount of energy in the inelastic range, that is, when the material is stressed above its yield point, without failure and without

deforming unacceptably. Concrete frames assigned to SDC D and above must be special moment-resisting frames.

2. An *intermediate moment-resisting frame* (IMRF) is a steel or concrete frame that has less stringent requirements than a SMRF, and in general may only be used in buildings assigned to SDC C and below.
3. An *ordinary moment-resisting frame* (OMRF) is a steel or concrete frame that does not meet the special detailing requirements for ductile behavior. For concrete, it is permitted only in buildings assigned to SDC A and B. For steel, it is permitted without any limitations in buildings assigned to SDC A, B, and C. In SDC D and higher, it is permitted only when certain conditions are met.

Special moment-resisting frames of concrete are designed and detailed to assure ductility, and the code requirements for such frames are quite specific, including the use of special transverse reinforcement, the use of top and bottom reinforcement throughout the length of beams, special attention to details such as reinforcement splices, and very close special inspection.

Such frames are usually cast-in-place; however, precast beam-column elements are permitted if it can be shown that they will have the same capabilities as cast-in-place concrete.

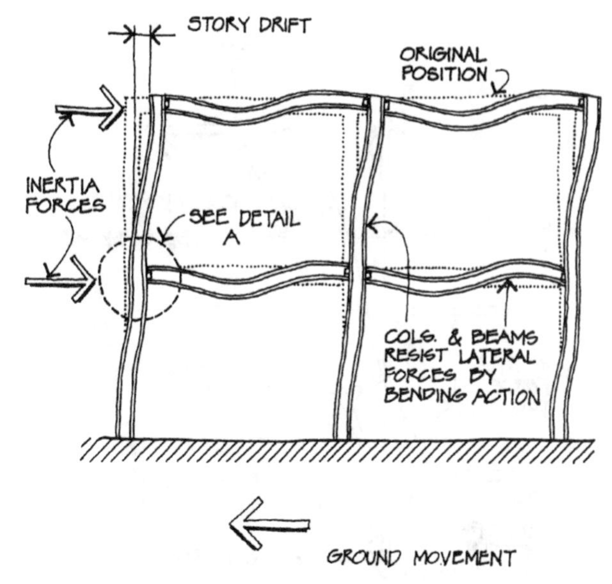

Figure 12.17

A moment-resisting frame made of structural steel does not automatically qualify as a special moment-resisting frame. It must comply with special criteria in the code, including requirements for the design of beam-to-column connections, requirements for nondestructive testing of welded connections, and limiting the material that may be used to certain specified grades of steel only.

Since a moment-resisting frame resists lateral loads by bending, it is the most ductile lateral load resisting system.

Shear Walls

A shear wall is one that resists lateral forces by developing shear in its own plane and cantilevering from its base, as shown in Figure 12.19.

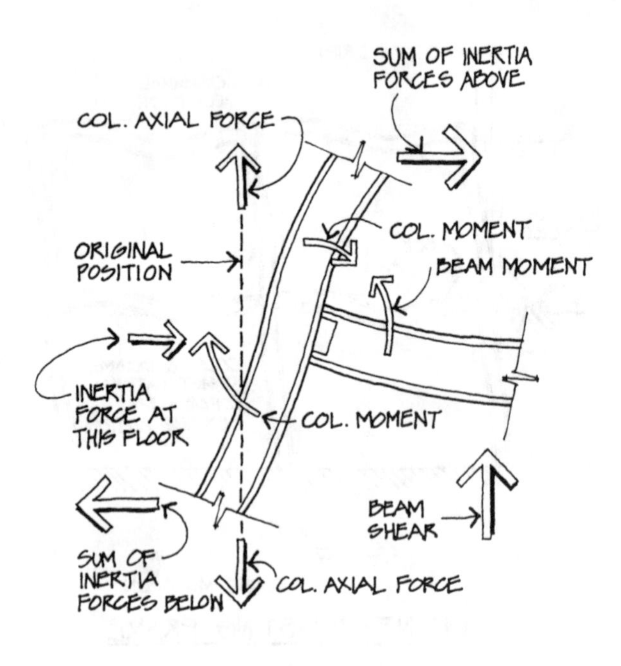

Figure 12.18

Thus, a shear wall is essentially a very deep cantilever beam that develops flexural stresses in addition to its basic shear stress. The wall tends to lift up at one end, and there must be sufficient dead load to prevent this, or if not, it must be adequately tied down to its foundation. Also, the connection of the wall to the foundation must be sufficient to prevent sliding.

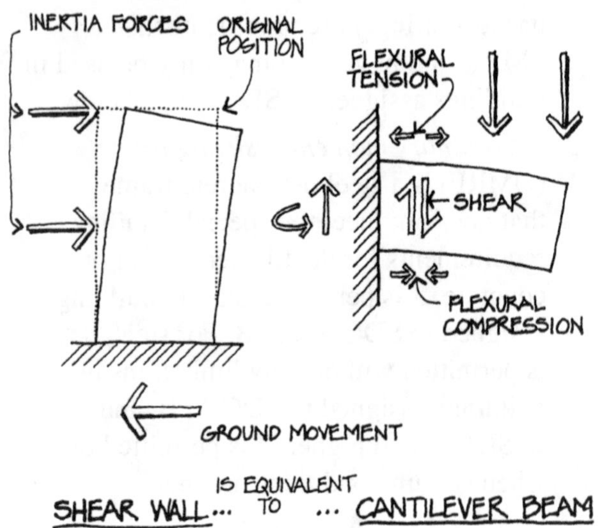

Figure 12.19

TABLE 2306.4.1
ALLOWABLE SHEAR (POUNDS PER FOOT) FOR WOOD STRUCTURAL PANEL SHEAR WALLS WITH FRAMING OF DOUGLAS FIR-LARCH OR SOUTHERN PINE[a] FOR WIND OR SEISMIC LOADING[b, h, i, j, l]

PANEL GRADE	MINIMUM NOMINAL PANEL THICKNESS (inch)	MINIMUM FASTENER PENETRATION IN FRAMING (inches)	PANELS APPLIED DIRECT TO FRAMING					PANELS APPLIED OVER $^1/_2$" OR $^5/_8$" GYPSUM SHEATHING				
			NAIL (common or galvanized box) or staple size[k]	Fastener spacing at panel edges (inches)				NAIL (common or galvanized box) or staple size[k]	Fastener spacing at panel edges (inches)			
				6	4	3	2[e]		6	4	3	2[e]
Structural I Sheathing	$^5/_{16}$	$1^1/_4$	6d (2 × 0.113" common, 2" × 0.099" galvanized box)	200	300	390	510	8d ($2^1/_2$" × 0.131" common, $2^1/_2$" × 0.113" galvanized box)	200	300	390	510
		1	$1^1/_2$ 16 Gage	165	245	325	415	2 16 Gage	125	185	245	315
	$^3/_8$	$1^3/_8$	8d ($2^1/_2$" × 0.131" common, $2^1/_2$" × 0.113" galvanized box)	230[d]	360[d]	460[d]	610[d]	10d (3" × 0.148" common, 3" × 0.128" galvanized box)	280	430	550[f]	730
		1	$1^1/_2$ 16 Gage	155	235	315	400	2 16 Gage	155	235	310	400
	$^7/_{16}$	$1^3/_8$	8d ($2^1/_2$" × 0.131" common, $2^1/_2$" × 0.113" galvanized box)	255[d]	395[d]	505[d]	670[d]	10d (3" × 0.148" common, 3" × 0.128" galvanized box)	280	430	550[f]	730
		1	$1^1/_2$ 16 Gage	170	260	345	440	2 16 Gage	155	235	310	400
	$^{15}/_{32}$	$1^3/_8$	8d ($2^1/_2$" × 0.131" common, $2^1/_2$" × 0.113" galvanized box)	280	430	550	730	10d (3" × 0.148" common, 3" × 0.1218" galvanized box)	280	430	550[f]	730
		1	$1^1/_2$ 16 Gage	185	280	375	475	2 16 Gage	155	235	300	400
		$1^1/_2$	10d (3" × 0.148" common, 3" × 0.128" galvanized box)	340	510	665[f]	870	10d (3" × 0.148" common, 3" × 0.128" galvanized box)	—	—	—	—
Sheathing, plywood siding[g] except Group 5 Species	$^5/_{16}$ or $^1/_4$[c]	$1^1/_4$	6d (2" × 0.113" common, 2" × 0.099" galvanized box)	180	270	350	450	8d ($2^1/_2$" × 0.131" common, $2^1/_2$" × 0.113" galvanized box)	180	270	350	450
		1	$1^1/_2$ 16 Gage	145	220	295	375	2 16 Gage	110	165	220	285
	$^3/_8$	$1^1/_4$	6d (2" × 0.113" common, 2" × 0.099" galvanized box)	200	300	390	510	8d ($2^1/_2$" × 0.131" common, $2^1/_2$" × 0.113" galvanized box)	200	300	390	510
		$1^3/_8$	8d ($2^1/_2$" × 0.131" common, $2^1/_2$" × 0.113" galvanized box)	220[d]	320[d]	410[d]	530[d]	10d (3" × 0.148" common, 3" × 0.128" galvanized box)	260	380	490[f]	640
		1	$1^1/_2$ 16 Gage	140	210	280	360	2 16 Gage	140	210	280	360
	$^7/_{16}$	$1^3/_8$	8d ($2^1/_2$" × 0.131" common, $2^1/_2$" × 0.113" galvanized box)	240[d]	350[d]	450[d]	585[d]	10d (3" × 0.148" common, 3" × 0.128" galvanized box)	260	380	490[f]	640
		1	$1^1/_2$ 16 Gage	155	230	310	395	2 16 Gage	140	210	280	360
	$^{15}/_{32}$	$1^3/_8$	8d ($2^1/_2$" × 0.131" common, $2^1/_2$" × 0.113" galvanized box)	260	380	490	640	10d (3" × 0.148" common, 3" × 0.128" galvanized box)	260	380	490[f]	640
		$1^1/_2$	10d (3" × 0.148" common, 3" × 0.128" galvanized box)	310	460	600[f]	770	—	—	—	—	—
		1	$1^1/_2$ 16 Gage	170	255	335	430	2 16 Gage	140	210	280	360
	$^{19}/_{32}$	$1^1/_2$	10d (3" × 0.148" common, 3" × 0.128" galvanized box)	340	510	665[f]	870	—	—	—	—	—
		1	$1^3/_4$ 16 Gage	185	280	375	475	—	—	—	—	—
			Nail Size (galvanized casing)					Nail Size (galvanized casing)				
	$^5/_{16}$[c]	$1^1/_4$	6d (2" × 0.099")	140	210	275	360	8d ($2^1/_2$" × 0.113")	140	210	275	360
	$^3/_8$	$1^3/_8$	8d ($2^1/_2$" × 0.113")	160	240	310	410	10d (3" × 0.128")	160	240	310[f]	410

Notes to Table 2306.4.1

For SI: 1 inch = 25.4 mm, 1 pound per foot = 14.5939 N/m.

a. For framing of other species: (1) Find specific gravity for species of lumber in AF&PA NDS. (2) For staples find shear value from table above for Structural I panels (regardless of actual grade) and multiply value by 0.82 for species with specific gravity of 0.42 or greater, or 0.65 for all other species. (3) For nails find shear value from table above for nail size for actual grade and multiply value by the following adjustment factor: Specific Gravity Adjustment Factor = [1-(0.5 - SG)], where SG = Specific Gravity of the framing lumber. This adjustment factor shall not be greater than 1.

b. Panel edges backed with 2-inch nominal or wider framing. Install panels either horizontally or vertically. Space fasteners maximum 6 inches on center along intermediate framing members for $^3/_8$-inch and $^7/_{16}$-inch panels installed on studs spaced 24 inches on center. For other conditions and panel thickness, space fasteners maximum 12 inches on center on intermediate supports.

c. $^3/_8$-inch panel thickness or siding with a span rating of 16 inches on center is the minimum recommended where applied direct to framing as exterior siding.

d. Allowable shear values are permitted to be increased to values shown for $^{15}/_{32}$-inch sheathing with same nailing provided (a) studs are spaced a maximum of 16 inches on center, or (b) panels are applied with long dimension across studs.

e. Framing at adjoining panel edges shall be 3 inches nominal or wider, and nails shall be staggered where nails are spaced 2 inches on center.

f. Framing at adjoining panel edges shall be 3 inches nominal or wider, and nails shall be staggered where both of the following conditions are met: (1) 10d (3" × 0.148") nails having penetration into framing of more than $1^1/_2$ inches and (2) nails are spaced 3 inches on center.

g. Values apply to all-veneer plywood. Thickness at point of fastening on panel edges governs shear values.

h. Where panels applied on both faces of a wall and nail spacing is less than 6 inches o.c. on either side, panel joints shall be offset to fall on different framing members, or framing shall be 3-inch nominal or thicker at adjoining panel edges and nails on each side shall be staggered.

i. In Seismic Design Category D, E or F, where shear design values exceed 350 pounds per linear foot, all framing members receiving edge nailing from abutting panels shall not be less than a single 3-inch nominal member, or two 2-inch nominal members fastened together in accordance with Section 2306.1 to transfer the design shear value between framing members. Wood structural panel joint and sill plate nailing shall be staggered in all cases. See Section 2305.3.11 for sill plate size and anchorage requirements.

j. Galvanized nails shall be hot dipped or tumbled.

k. Staples shall have a minimum crown width of $^7/_{16}$ inch and shall be installed with their crowns parallel to the long dimension of the framing members.

l. For shear loads of normal or permanent load duration as defined by the AF&PA NDS, the values in the table above shall be multiplied by 0.63 or 0.56, respectively.

Figure 12.20

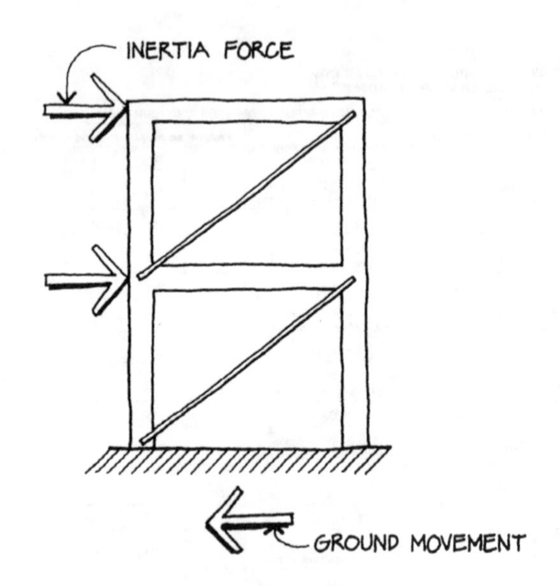
INERTIA FORCE
GROUND MOVEMENT
CONCENTRIC DIAGONAL BRACING

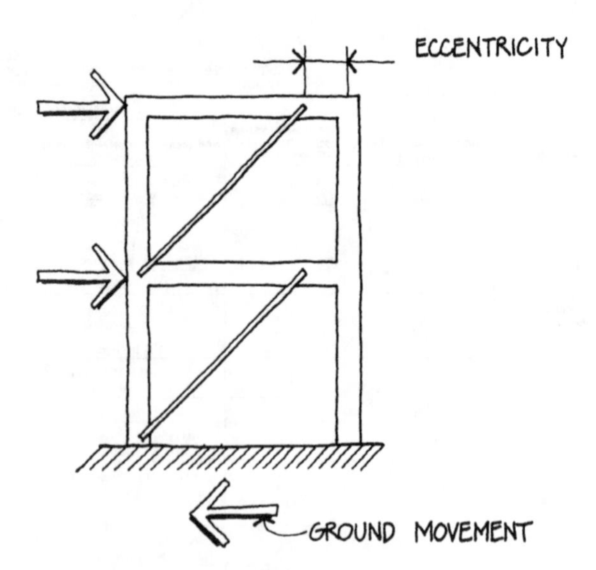
ECCENTRICITY
GROUND MOVEMENT
ECCENTRIC DIAGONAL BRACING

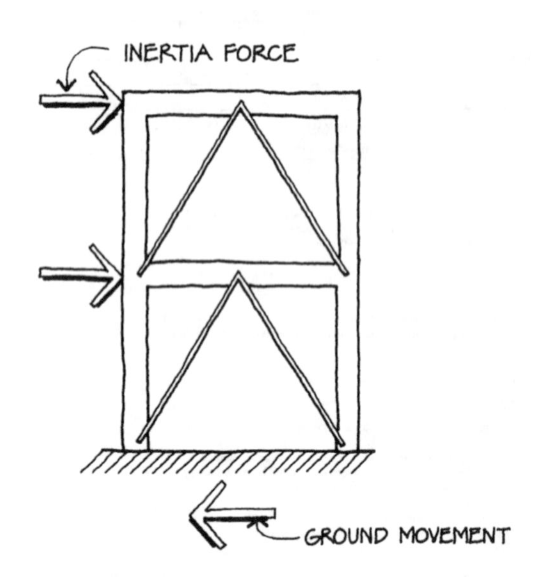
INERTIA FORCE
GROUND MOVEMENT
CONCENTRIC CHEVRON BRACING

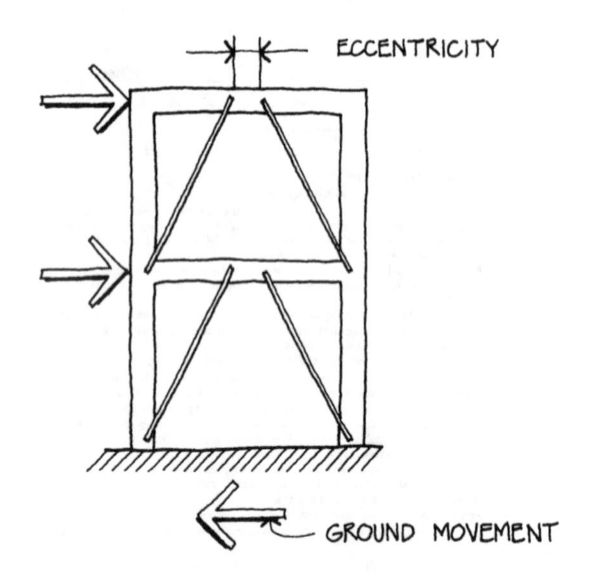
ECCENTRICITY
GROUND MOVEMENT
ECCENTRIC CHEVRON BRACING

BRACED FRAMES

Figure 12.21

TABLE 12.2-1 DESIGN COEFFICIENTS AND FACTORS FOR SEISMIC FORCE–RESISTING SYSTEMS

Seismic Force–Resisting System	ASCE 7 Section where Detailing Requirements are Specified	Response Modification Coefficient, R^a	System Overstrength Factor, $\Omega_0{}^g$	Deflection Amplification Factor, $C_d{}^b$	Structural System Limitations and Building Height (ft) Limit[c]				
					Seismic Design Category				
					B	C	D^d	E^d	F^e
A. BEARING WALL SYSTEMS									
1. Special reinforced concrete shear walls	14.2 and 14.2.3.6	5	2½	5	NL	NL	160	160	100
2. Ordinary reinforced concrete shear walls	14.2 and 14.2.3.4	4	2½	4	NL	NL	NP	NP	NP
3. Detailed plain concrete shear walls	14.2 and 14.2.3.2	2	2½	2	NL	NP	NP	NP	NP
4. Ordinary plain concrete shear walls	14.2 and 14.2.3.1	1½	2½	1½	NL	NP	NP	NP	NP
5. Intermediate precast shear walls	14.2 and 14.2.3.5	4	2½	4	NL	NL	40^k	40^k	40^k
6. Ordinary precast shear walls	14.2 and 14.2.3.3	3	2½	3	NL	NP	NP	NP	NP
7. Special reinforced masonry shear walls	14.4 and 14.4.3	5	2½	3½	NL	NL	160	160	100
8. Intermediate reinforced masonry shear walls	14.4 and 14.4.3	3½	2½	2¼	NL	NL	NP	NP	NP
9. Ordinary reinforced masonry shear walls	14.4	2	2½	1¾	NL	160	NP	NP	NP
10. Detailed plain masonry shear walls	14.4	2	2½	1¾	NL	NP	NP	NP	NP
11. Ordinary plain masonry shear walls	14.4	1½	2½	1¼	NL	NP	NP	NP	NP
12. Prestressed masonry shear walls	14.4	1½	2½	1¾	NL	NP	NP	NP	NP
13. Light-framed walls sheathed with wood structural panels rated for shear resistance or steel sheets	14.1, 14.1.4.2, and 14.5	6½	3	4	NL	NL	65	65	65
14. Light-framed walls with shear panels of all other materials	14.1, 14.1.4.2, and 14.5	2	2½	2	NL	NL	35	NP	NP
15. Light-framed wall systems using flat strap bracing	14.1, 14.1.4.2, and 14.5	4	2	3½	NL	NL	65	65	65
B. BUILDING FRAME SYSTEMS									
1. Steel eccentrically braced frames, moment resisting connections at columns away from links	14.1	8	2	4	NL	NL	160	160	100
2. Steel eccentrically braced frames, non-moment-resisting, connections at columns away from links	14.1	7	2	4	NL	NL	160	160	100
3. Special steel concentrically braced frames	14.1	6	2	5	NL	NL	160	160	100
4. Ordinary steel concentrically braced frames	14.1	3¼	2	3¼	NL	NL	35^j	35^j	NP^j
5. Special reinforced concrete shear walls	14.2 and 14.2.3.6	6	2½	5	NL	NL	160	160	100
6. Ordinary reinforced concrete shear walls	14.2 and 14.2.3.4	5	2½	4½	NL	NL	NP	NP	NP
7. Detailed plain concrete shear walls	14.2 and 14.2.3.2	2	2½	2	NL	NP	NP	NP	NP
8. Ordinary plain concrete shear walls	14.2 and 14.2.3.1	1½	2½	1½	NL	NP	NP	NP	NP
9. Intermediate precast shear walls	14.2 and 14.2.3.5	5	2½	4½	NL	NL	40^k	40^k	40^k
10. Ordinary precast shear walls	14.2 and 14.2.3.3	4	2½	4	NL	NP	NP	NP	NP
11. Composite steel and concrete eccentrically braced frames	14.3	8	2	4	NL	NL	160	160	100
12. Composite steel and concrete concentrically braced frames	14.3	5	2	4½	NL	NL	160	160	100
13. Ordinary composite steel and concrete braced frames	14.3	3	2	3	NL	NL	NP	NP	NP
14. Composite steel plate shear walls	14.3	6½	2½	5½	NL	NL	160	160	100
15. Special composite reinforced concrete shear walls with steel elements	14.3	6	2½	5	NL	NL	160	160	100
16. Ordinary composite reinforced concrete shear walls with steel elements	14.3	5	2½	4½	NL	NL	NP	NP	NP
17. Special reinforced masonry shear walls	14.4	5½	2½	4	NL	NL	160	160	100
18. Intermediate reinforced masonry shear walls	14.4	4	2½	4	NL	NL	NP	NP	NP
19. Ordinary reinforced masonry shear walls	14.4	2	2½	2	NL	160	NP	NP	NP
20. Detailed plain masonry shear walls	14.4	2	2½	2	NL	NP	NP	NP	NP
21. Ordinary plain masonry shear walls	14.4	1½	2½	1¼	NL	NP	NP	NP	NP

Figure 12.22

TABLE 12.2-1 DESIGN COEFFICIENTS AND FACTORS FOR SEISMIC FORCE–RESISTING SYSTEMS (continued)

Seismic Force–Resisting System	ASCE 7 Section where Detailing Requirements are Specified	Response Modification Coefficient, R^a	System Overstrength Factor, Ω_0^g	Deflection Amplification Factor, C_d^b	Structural System Limitations and Building Height (ft) Limit^c — Seismic Design Category				
					B	C	D^d	E^d	F^e
22. Prestressed masonry shear walls	14.4	1½	2½	1¾	NL	NP	NP	NP	NP
23. Light-framed walls sheathed with wood structural panels rated for shear resistance or steel sheets	14.1, 14.1.4.2, and 14.5	7	2½	4½	NL	NL	65	65	65
24. Light-framed walls with shear panels of all other materials	14.1, 14.1.4.2, and 14.5	2½	2½	2½	NL	NL	35	NP	NP
25. Buckling-restrained braced frames, non-moment-resisting beam-column connections	14.1	7	2	5½	NL	NL	160	160	100
26. Buckling-restrained braced frames, moment-resisting beam-column connections	14.1	8	2½	5	NL	NL	160	160	100
27. Special steel plate shear wall	14.1	7	2	6	NL	NL	160	160	100
C. MOMENT-RESISTING FRAME SYSTEMS									
1. Special steel moment frames	14.1 and 12.2.5.5	8	3	5½	NL	NL	NL	NL	NL
2. Special steel truss moment frames	14.1	7	3	5½	NL	NL	160	100	NP
3. Intermediate steel moment frames	12.2.5.6, 12.2.5.7, 12.2.5.8, 12.2.5.9, and 14.1	4.5	3	4	NL	NL	$35^{h,i}$	NP^h	NP^i
4. Ordinary steel moment frames	12.2.5.6, 12.2.5.7, 12.2.5.8, and 14.1	3.5	3	3	NL	NL	NP^h	NP^h	NP^i
5. Special reinforced concrete moment frames	12.2.5.5 and 14.2	8	3	5½	NL	NL	NL	NL	NL
6. Intermediate reinforced concrete moment frames	14.2	5	3	4½	NL	NL	NP	NP	NP
7. Ordinary reinforced concrete moment frames	14.2	3	3	2½	NL	NP	NP	NP	NP
8. Special composite steel and concrete moment frames	12.2.5.5 and 14.3	8	3	5½	NL	NL	NL	NL	NL
9. Intermediate composite moment frames	14.3	5	3	4½	NL	NL	NP	NP	NP
10. Composite partially restrained moment frames	14.3	6	3	5½	160	160	100	NP	NP
11. Ordinary composite moment frames	14.3	3	3	2½	NL	NP	NP	NP	NP
D. DUAL SYSTEMS WITH SPECIAL MOMENT FRAMES CAPABLE OF RESISTING AT LEAST 25% OF PRESCRIBED SEISMIC FORCES	12.2.5.1								
1. Steel eccentrically braced frames	14.1	8	2½	4	NL	NL	NL	NL	NL
2. Special steel concentrically braced frames	14.1	7	2½	5½	NL	NL	NL	NL	NL
3. Special reinforced concrete shear walls	14.2	7	2½	5½	NL	NL	NL	NL	NL
4. Ordinary reinforced concrete shear walls	14.2	6	2½	5	NL	NL	NP	NP	NP
5. Composite steel and concrete eccentrically braced frames	14.3	8	2½	4	NL	NL	NL	NL	NL
6. Composite steel and concrete concentrically braced frames	14.3	6	2½	5	NL	NL	NL	NL	NL
7. Composite steel plate shear walls	14.3	7½	2½	6	NL	NL	NL	NL	NL
8. Special composite reinforced concrete shear walls with steel elements	14.3	7	2½	6	NL	NL	NL	NL	NL
9. Ordinary composite reinforced concrete shear walls with steel elements	14.3	6	2½	5	NL	NL	NP	NP	NP
10. Special reinforced masonry shear walls	14.4	5½	3	5	NL	NL	NL	NL	NL
11. Intermediate reinforced masonry shear walls	14.4	4	3	3½	NL	NL	NP	NP	NP
12. Buckling-restrained braced frame	14.1	8	2½	5	NL	NL	NL	NL	NL
13. Special steel plate shear walls	14.1	8	2½	6½	NL	NL	NL	NL	NL

Figure 12.22 (continued)

TABLE 12.2-1 DESIGN COEFFICIENTS AND FACTORS FOR SEISMIC FORCE–RESISTING SYSTEMS (continued)

Seismic Force-Resisting System	ASCE 7 Section where Detailing Requirements are Specified	Response Modification Coefficient, R[a]	System Overstrength Factor, Ω_0[g]	Deflection Amplification Factor, C_d[b]	Structural System Limitations and Building Height (ft) Limit[c] — Seismic Design Category B	C	D[d]	E[d]	F[e]
E. DUAL SYSTEMS WITH INTERMEDIATE MOMENT FRAMES CAPABLE OF RESISTING AT LEAST 25% OF PRESCRIBED SEISMIC FORCES	12.2.5.1								
1. Special steel concentrically braced frames[f]	14.1	6	2½	5	NL	NL	35	NP	NP[h,k]
2. Special reinforced concrete shear walls	14.2	6½	2½	5	NL	NL	160	100	100
3. Ordinary reinforced masonry shear walls	14.4	3	3	2½	NL	160	NP	NP	NP
4. Intermediate reinforced masonry shear walls	14.4	3½	3	3	NL	NL	NP	NP	NP
5. Composite steel and concrete concentrically braced frames	14.3	5½	2½	4½	NL	NL	160	100	NP
6. Ordinary composite braced frames	14.3	3½	2½	3	NL	NL	NP	NP	NP
7. Ordinary composite reinforced concrete shear walls with steel elements	14.3	5	3	4½	NL	NL	NP	NP	NP
8. Ordinary reinforced concrete shear walls	14.2	5½	2½	4½	NL	NL	NP	NP	NP
F. SHEAR WALL-FRAME INTERACTIVE SYSTEM WITH ORDINARY REINFORCED CONCRETE MOMENT FRAMES AND ORDINARY REINFORCED CONCRETE SHEAR WALLS	12.2.5.10 and 14.2	4½	2½	4	NL	NP	NP	NP	NP
G. CANTILEVERED COLUMN SYSTEMS DETAILED TO CONFORM TO THE REQUIREMENTS FOR:	12.2.5.2								
1. Special steel moment frames	12.2.5.5 and 14.1	2½	1¼	2½	35	35	35	35	35
2. Intermediate steel moment frames	14.1	1½	1¼	1½	35	35	35[h]	NP[h,i]	NP[h,i]
3. Ordinary steel moment frames	14.1	1¼	1¼	1¼	35	35	NP	NP[h,i]	NP[h,i]
4. Special reinforced concrete moment frames	12.2.5.5 and 14.2	2½	1¼	2½	35	35	35	35	35
5. Intermediate concrete moment frames	14.2	1½	1¼	1½	35	35	NP	NP	NP
6. Ordinary concrete moment frames	14.2	1	1¼	1	35	NP	NP	NP	NP
7. Timber frames	14.5	1½	1½	1½	35	35	35	NP	NP
H. STEEL SYSTEMS NOT SPECIFICALLY DETAILED FOR SEISMIC RESISTANCE, EXCLUDING CANTILEVER COLUMN SYSTEMS	14.1	3	3	3	NL	NL	NP	NP	NP

[a] Response modification coefficient, R, for use throughout the standard. Note R reduces forces to a strength level, not an allowable stress level.
[b] Reflection amplification factor, C_d, for use in Sections 12.8.6, 12.8.7, and 12.9.2
[c] NL = Not Limited and NP = Not Permitted. For metric units use 30.5 m for 100 ft and use 48.8 m for 160 ft. Heights are measured from the base of the structure as defined in Section 11.2.
[d] See Section 12.2.5.4 for a description of building systems limited to buildings with a height of 240 ft (73.2 m) or less.
[e] See Section 12.2.5.4 for building systems limited to buildings with a height of 160 ft (48.8 m) or less.
[f] Ordinary moment frame is permitted to be used in lieu of intermediate moment frame for Seismic Design Categories B or C.
[g] The tabulated value of the overstrength factor, Ω_0, is permitted to be reduced by subtracting one-half for structures with flexible diaphragms, but shall not be taken as less than 2.0 for any structure.
[h] See Sections 12.2.5.6 and 12.2.5.7 for limitations for steel OMFs and IMFs in structures assigned to Seismic Design Category D or E.
[i] See Sections 12.2.5.8 and 12.2.5.9 for limitations for steel OMFs and IMFs in structures assigned to Seismic Design Category F.
[j] Steel ordinary concentrically braced frames are permitted in single-story buildings up to a height of 60 ft (18.3 m) where the dead load of the roof does not exceed 20 psf (0.96 kN/m^2) and in penthouse structures.
[k] Increase in height to 45 ft (13.7 m) is permitted for single story storage warehouse facilities.

Figure 12.22 (continued)

Shear walls may be made of reinforced concrete, reinforced masonry, or steel. They may also be constructed of a facing of plywood, particleboard, or fiberboard applied over wood studs. The allowable shear in a shear wall constructed of plywood sheathing over wood studs is found in IBC Table 2306.4.1, reproduced in Figure 12.20. In this table, the term *wood structural panel* refers to various wood-based structural panels, including plywood.

For example, the allowable shear in a plywood shear wall using 15/32" plywood of Structural I grade with 8d nails spaced at six inches at plywood panel edges is 280 pounds per foot.

Reinforced concrete and reinforced masonry shear walls generally have allowable shears much greater than those for plywood shear walls.

Braced Frames

A braced frame is a vertical truss that resists lateral forces by axial tension and compression in the truss members. Braced frames are most often constructed of structural steel, although reinforced concrete or timber braced frames are possible.

There are two types of braced frames: *concentric* and *eccentric.*

In the concentric braced frame, which has been widely used for many years, the center lines of all intersecting members meet at a point, and therefore the members are subjected primarily to axial forces.

The eccentric braced frame (EBF) is a steel braced frame in which at least one end of each brace is eccentric to the beam-column joint or the opposing brace. The intent is to make the braced frame more ductile and therefore able to absorb a significant amount of energy without buckling the braces. Examples of concentric and eccentric braced frames are shown in Figure 12.21.

Because shear walls and concentric braced frames resist lateral load by shear and axial forces, they cannot undergo as much motion without collapse, or absorb as much energy, as a moment-resisting frame. Hence, these systems are not as ductile as moment-resisting frames.

On the other hand, eccentric braced frames, if properly designed and detailed, can perform in a ductile manner similar to moment-resisting frames.

R Values

Now that we have described the different types of lateral load resisting systems, we can refer to ASCE Table 12.2-1, reproduced in Figure 12.22, to determine the R values for different systems. First we need to define a few terms.

A *bearing wall system* is one in which the vertical load is supported by bearing walls or bracing systems and the required lateral forces are resisted by shear walls. In other words, the shear walls support vertical, as well as lateral, loads. If these elements fail during an earthquake, their ability to support vertical loads may be eliminated and the structure may collapse. Consequently, bearing wall systems are assigned low values of R, varying from 1½ to 6½. Height limits in those systems permitted in SDC D and above vary from 35 feet to 160 feet.

A *building frame system* is one with an essentially complete frame that provides support for vertical loads. Lateral loads are resisted by shear walls or braced frames, which support minimal vertical load. The vertical load frame can provide some secondary resistance to lateral loads, and can prevent collapse if the shear walls or braced frames are damaged during an earthquake. Therefore, these systems have intermediate values of R, varying from 1½ to 8.

A *moment-resisting frame system* is one with an essentially complete frame that provides support for vertical loads. Lateral loads are resisted by moment-resisting frames. Because of the great ductility of special moment-resisting frames (SMRF), these systems are assigned the maximum R value of 8, and their height is unlimited in buildings assigned to SDC D and above.

Ordinary and intermediate moment-resisting frames (OMRF and IMRF) are much less ductile and therefore have lower R values, varying from 3 to 5. Of these systems, only OMRF of structural steel are permitted in buildings assigned to SDC D and above, when certain conditions are met.

A *dual system* is one with an essentially complete frame that provides support for vertical loads. Lateral loads are resisted by *both* moment-resisting frames (special and intermediate) and shear walls or braced frames, in proportion to their relative rigidities. The moment-resisting frame acting independently must be able to resist at least 25 percent of the total required lateral force. These systems have varying degrees of ductility, and their R values vary correspondingly, from 3 to 8. For those systems permitted in SDC D and above, the height limit varies from 35 feet to 160 feet.

Shear wall-frame interactive systems are permitted in SDC A and B, and consist of ordinary reinforced concrete moment frames and ordinary reinforced concrete shear walls. Like dual systems, the earthquake forces are proportioned to the moment frames and shear walls based on relative rigidity. However, the moment frame need not be designed independently for 25 percent of the earthquake load. This system has an R value of 4½.

Steel systems that are not specifically detailed for seismic resistance are assigned an R value of 3, and are not permitted in SDC D and higher. Any type of steel system can be used to resist the earthquake effects in this case.

The following examples show how the terms in the basic seismic formula are determined.

Example #1

A building is located on a site where it has been determined that the mapped accelerations S_S and S_1 are 0.276g and 0.117g, respectively. For Site Class E, determine the design accelerations S_{DS} and S_{D1}.

Solution:

To determine S_{DS} and S_{D1}, we must first calculate the soil-modified accelerations S_{M1} and S_{MS} for Site Class E from previously provided formulas.

$$S_{MS} = F_a S_S$$
$$S_{M1} = F_v S_1$$

The coefficients F_a and F_1 are obtained from IBC Tables 1613.5.3(1) and 1613.5.3(2), respectively.

For S_S = 0.276g and Site Class E, F_a is between 2.5 and 1.7—see Table 1613.5.3(1). The code permits you to use straight-line interpolation for immediate values of F_a, so for 0.276g, F_a is determined as follows:

$$F_a = 1.7 + [(2.5 - 1.7)(0.5 - 0.276)/(0.5 - 0.25)] = 2.42$$

Similarly, for $S_1 = 0.117g$ and Site Class E, $F_v = 3.45$ using straight-line interpolation on the values in Table 1613.5.3(2).

Thus,

$S_{MS} = 2.42 \times 0.276 = 0.67g$
$S_{M1} = 3.45 \times 0.117 = 0.40g$

The design accelerations are two-thirds of the soil-modified accelerations:

$S_{DS} = 2 \times 0.67/3 = 0.45g$
$S_{D1} = 2 \times 0.40/3 = 0.27g$

Example #2

Which is more ductile, a building frame system with special reinforced concrete shear walls, or a moment-resisting frame system with reinforced concrete special moment-resisting frames?

Solution:

In general, moment-resisting frames are more ductile than shear walls. Referring to Table 12.2-1 in Figure 12.22, the building frame system with special reinforced concrete shear walls is system B.5, which has an R value of 6. The moment-resisting frame system with reinforced concrete SMRF is system C.5, which has an R value of 8. The greater value of R indicates greater ductility, and therefore, as expected, *the moment-resisting frame system is more ductile*.

Example #3

An eight-story apartment building is 100 feet high and 120 feet long, and has special reinforced concrete shear walls (building frame system) to resist earthquake forces. Also, it has been determined that $T_L = 12$ seconds. Given the design accelerations S_{DS} and S_{D1} from Example 1, determine C_s.

Solution:

From Example 1, $S_{DS} = 0.45g$ and $S_{D1} = 0.27g$.

We first calculate the period of the building, T.

From page 249, $T = C_t(h_n)^x$

For a building with shear walls, $C_t = 0.02$ and $x = 0.75$ ("All other structural systems")

Therefore, $T = 0.02(100)^{0.75} = 0.63$ second

The importance factor I = 1.0, since the apartment building would fall under Occupancy Category II.

For a building frame system with special reinforced concrete shear walls, R = 6 (see system B.5 in ASCE 7 Table 12.2-1).

From page 247, $C_s = S_{DS}/(R/I) = 0.45/(6/1) = 0.075$

For T = 0.63 seconds, which is less than $T_L = 12$ seconds, C_s need not be greater than the following:

$C_s = S_{D1}I/TR = (0.27 \times 1.0)/(0.63 \times 6) = 0.071 < 0.075$

Note that 0.071 is also greater than the minimum $C_s = 0.01$.

Thus, $C_s = 0.071$.

Example #4

A three-story apartment building with a height of 40 feet in SDC D has wood-framed bearing walls with wood structural shear panels. Its soil conditions are not known. What is the base shear as a percentage of the dead load of the building, given $S_{DS} = 0.93g$ and $S_{D1} = 0.50g$?

Solution:

The base shear $V = C_sW$

where $C_s = S_{DS}/(R/I)$

I = 1.0 since this is not a high occupancy, essential, or hazardous facility.

Referring to ASCE 7 Table 12.2-1 in Figure 12.22, the structure is bearing wall system A.13. Buildings of this type have demonstrated excellent earthquake resistance when properly designed and detailed, and therefore have a relatively high R value of 6.5.

$C_s = 0.93/(6.5/1.0) = 0.14$

Also, $C_s = S_{D1}I/TR$

From page 249, $T = C_t(h_n)^x$

For a building with shear walls, $C_t = 0.02$ and $x = 0.75$ ("All other structural systems")

Therefore, $T = 0.02(40)^{0.75} = 0.32$ second

$C_s = (0.50 \times 1.0)/(0.32 \times 6.5) = 0.24 > 0.14$

Minimum $C_s = 0.01$.

Therefore, $C_s = 0.14$.

$V = C_sW = 0.14W$

or *14 percent of the dead load.*

Example #5

What is the minimum base shear that may be used in the seismic design of a building, expressed as a percentage of the building's dead load?

Solution:

We apply the basic seismic formula

$V = C_sW$

In cases where $S_1 < 0.6g$, $C_s = 0.01$ so that

$V = 0.01W$

or *1 percent of the dead load when $S_1 < 0.6g$.*

However, when S_1 is greater than or equal to 0.6g, minimum C_s is

$C_s = 0.5S_1I/R$

This formula shows that the minimum C_s, or equivalently the minimum V, depends on the occupancy, ground acceleration, and seismic-force-resisting system. We can make some assumptions to get a percentage range for a typical case.

Assuming a building is not high occupancy, essential, or hazardous, I = 1.0.

Let's also assume a range in R from between 4 and 8, which would be typical for buildings located in areas with such relatively high ground accelerations. Thus, for $S_1 = 0.6g$,

$C_s = 0.5 \times 0.6 \times 1.0/4 = 0.075$

and

$C_s = 0.5 \times 0.6 \times 1.0/8 = 0.038$

Therefore,

$V = 0.075W$

and

$V = 0.038W$

or *3.8 to 7.5 percent of the dead load when S_1 = 0.6g.*

Other percentages can be found for other ground accelerations.

DISTRIBUTION OF BASE SHEAR

Once the base shear or total horizontal force has been determined, we must distribute this force to the various levels of the structure.

The force applied to any level x =

$F_x = C_{vx}V$

and

$$C_{vx} = \frac{w_x(h_x)^k}{\sum_{i=1}^{n} w_i(h_i)^k}$$

where

C_{vx} = vertical distribution factor

w_i and w_x = portion of W located at or assigned to level i or x

h_i and h_x = height in feet above the base to level i or x

k = exponent related to the structure period as follows:

- For structures having T less than or equal to 0.5 second, k = 1
- For structures having T greater than or equal to 2.5 seconds, k = 2
- For structures having a period between 0.5 and 2.5 seconds, k = 2, or it can be determined by linear interpolation between 1 and 2

For relatively stiff, short buildings with equal weights on each floor level, the shear will be distributed in the form of a triangle (k = 1), as shown in Figure 12.23. For taller, less stiff buildings, the distribution is parabolic (k = 2). The larger the value of k, the higher the proportion of V distributed to the upper portions of a building. This accounts for a whiplash effect that can be significant in these taller buildings.

OVERTURNING

Every building must be designed to resist the overturning moments caused by earthquake forces. In computing uplift caused by overturning moment, however, only a percent of the dead load may be used to resist uplift forces.

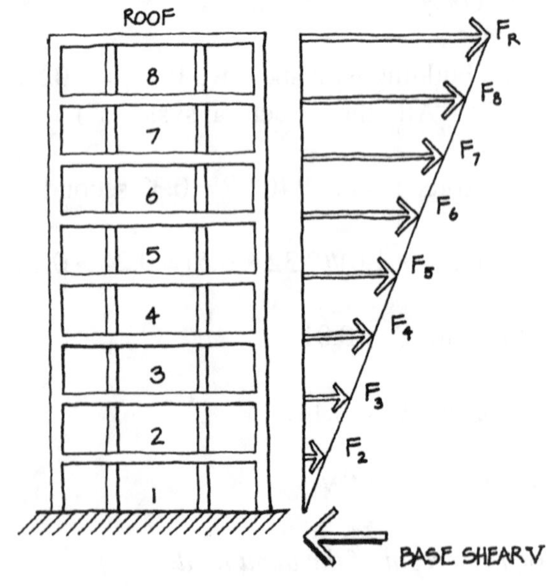

Figure 12.23

In particular, when using the basic combinations for strength design, the code requires that only $(0.9 - 0.2S_{DS})$ of the dead load be used to resist uplift forces, where S_{DS} is the short-period design acceleration. Decreasing the dead load by an additional amount of 20 percent of S_{DS} accounts for the vertical effects that an earthquake can impart on a building. Similarly, when using the basic combinations for allowable stress design, only $(0.6 - 0.14S_{DS})$ of the dead load can be used to resist overturning.

Example #6

A two-story braced frame is subject to the seismic loads and dead loads shown in Figure 12.24. What is the overturning moment at the base? For how much uplift should each column be designed? Assume $S_{DS} = 1.39g$ and use strength design load combinations.

Solution:

The overturning moment = 20 kips (10 + 10) ft. + 10 kips (10 ft.) = 400 + 100 = 500 *ft.-kips.*

The gross uplift at the left column =

$$\frac{500 \text{ ft.-kips}}{20 \text{ ft.}} = 25 \text{ kips}$$

This is reduced by $(0.9 - 0.2S_{DS})$ of the dead load = $[0.9 - (0.2 \times 1.39)](20) = 12.4$ kips

The net uplift is therefore 25 – 12.4 = *12.6 kips.*

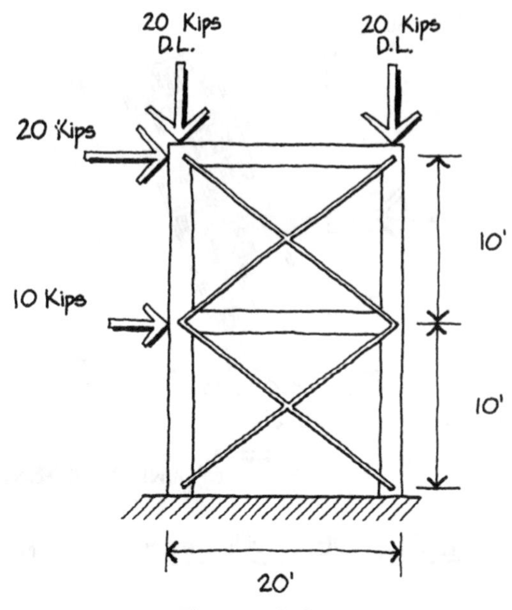

Figure 12.24

Since seismic loads can come from either direction, each column should be designed to resist this uplift.

Overturning effects are most critical for buildings with a high height-to-width ratio. In other words, a tall, slender building is more susceptible to overturning than a short, squat building (see the upper portion of Figure 12.25).

Top-heavy buildings tend to overturn more than pyramidal buildings (see the lower portion of Figure 12.25).

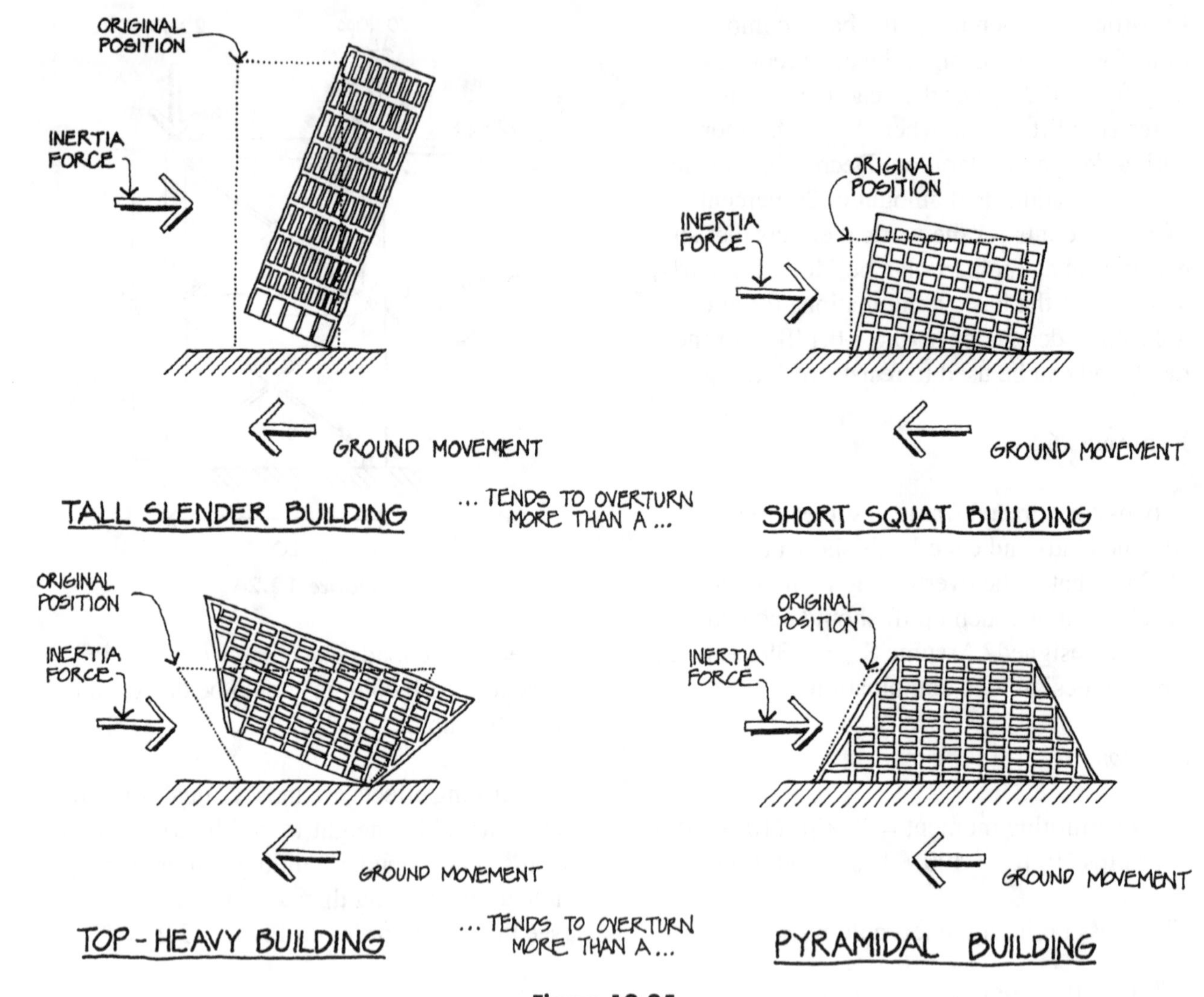

Figure 12.25

Heavy equipment, such as water tanks, should be located as low in the structure as possible in order to minimize the overturning effect on the building.

DEFLECTION AND DRIFT

When a building vibrates with the earthquake motion, the structure deflects and the stories move horizontally relative to each other. The story-to-story horizontal movement is called the *story drift*. Stiff systems, such as shear walls and braced frames, have relatively small drifts, while more flexible moment-resisting frames typically have larger drifts.

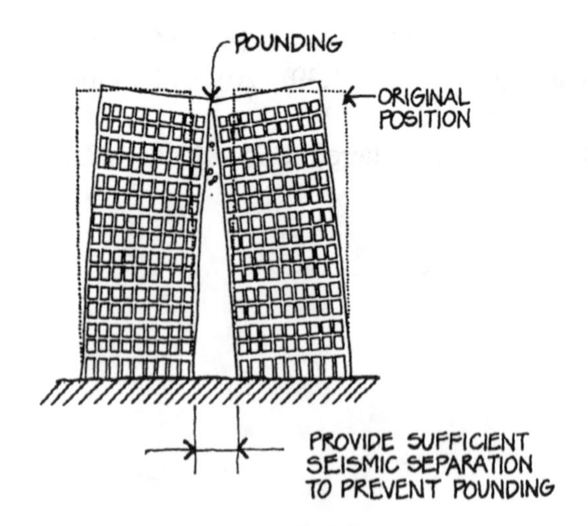

Figure 12.26

The code limits the amount of this movement in order to insure structural integrity, minimize discomfort to the building's occupants, and restrict damage to brittle nonstructural elements such as glass, plaster walls, etc.

Two buildings next to each other may move differently during an earthquake, and may therefore collide. This collision, called *pounding,* has occurred in many earthquakes and can be disastrous (Figure 12.26). To minimize the possibility of pounding, a *seismic separation* between the buildings should be provided at least equal to *the sum of the expected drifts* of the two buildings.

Another phenomenon caused by story drift is the *P-Delta effect.* If the story drift due to seismic forces is (delta), and the vertical load in a column is P, then the bending moments in the story are increased by an amount equal to P times Δ or P-Delta. This effect must be taken into account unless it is very low relative to the story bending moments.

DIAPHRAGMS

A diaphragm is the horizontal floor or roof system that distributes lateral forces to the vertical resisting elements, such as shear walls, braced frames, or moment-resisting frames. The action of a diaphragm is similar to that of a horizontal girder spanning between the vertical resisting elements (see Figure 12.27). In this analogy the floor itself (concrete slab, steel deck, plywood sheathing, etc.) may be compared to the girder web that resists shear forces, while the boundary members form the girder flanges or chords, and resist tensile and compressive forces. Although diaphragms are usually considered to be horizontal, they are sometimes inclined, as in a sloping roof.

The code gives the following formula to calculate the seismic force acting on floor and roof diaphragms at any given level x in the building:

$$F_{px} = \frac{\sum_{i=x}^{n} F_i}{\sum_{i=x}^{n} W_i} Wpx$$

where

F_i = design force applied to level i

w_i = weight tributary to level i

w_{px} = weight tributary to the diaphragm at level x

Note that F_{px} need not exceed $0.4S_{DS}Iw_{px}$. Also, the minimum F_{px} is equal to $0.2S_{DS}Iw_{px}$.

In general, the weight w_{px} that is tributary to the diaphragm at any level in the building depends on the direction in which the forces are being calculated. For example, assuming that seismic forces are being computed in the north-south direction of a building, the weight that is tributary to a diaphragm at any level equals the weight of the diaphragm at that level plus the weight of the north wall and the south wall tributary to that level. A similar situation would occur for seismic forces computed in the east-west direction.

The design force F_i at a floor level is determined from the formula presented on page 262 from the Equivalent Lateral Force Procedure. This force is a portion of the base shear V that has been distributed over the height of the building according to that formula.

Note that in a one-story building, the force F_i at the roof is equal to the base shear V. In such cases, the above formula for the diaphragm force reduces to $F_{px} = C_s w_{px}$, where w_{px} is the weight tributary to the roof diaphragm.

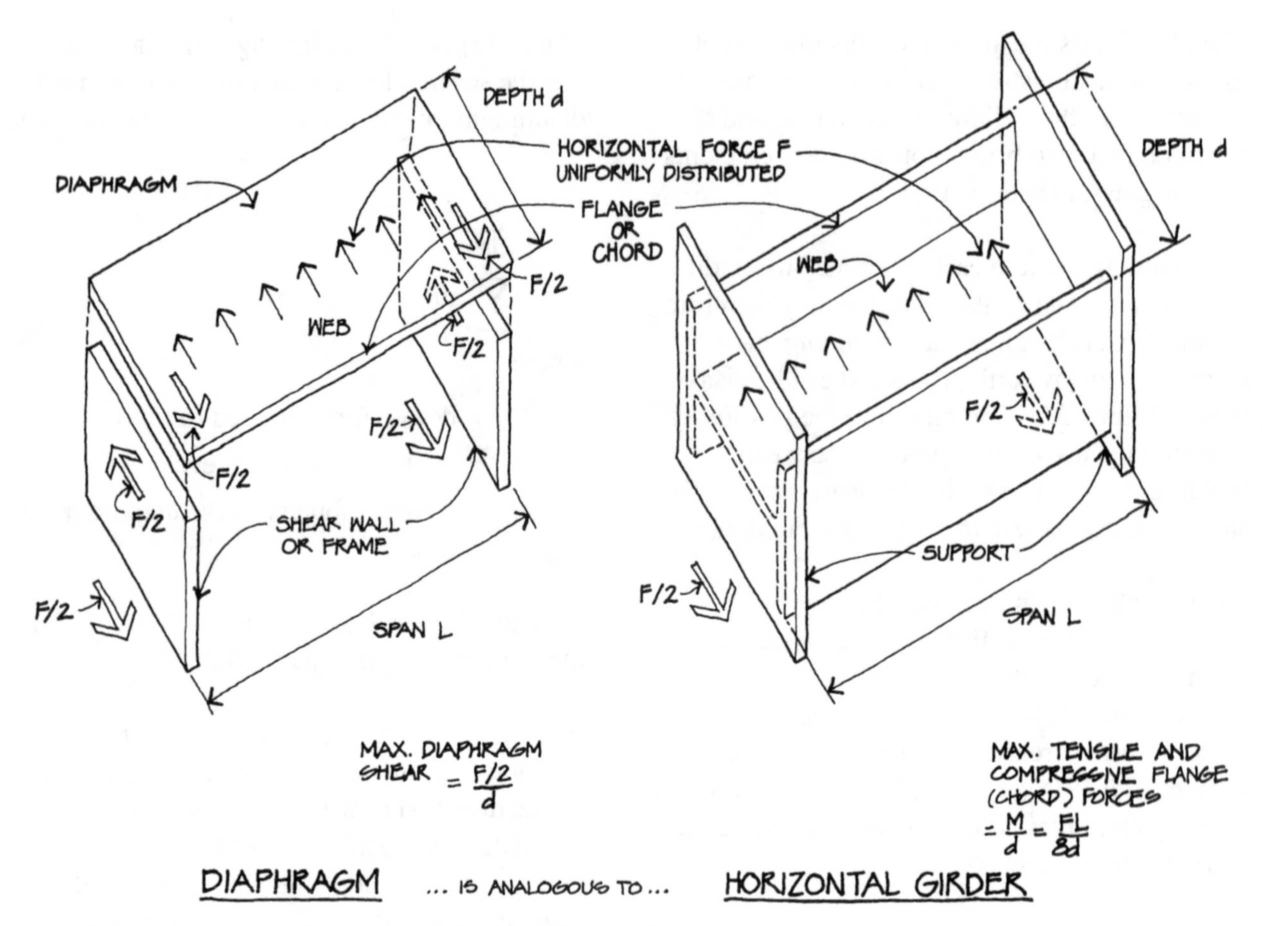

Figure 12.27

A *flexible diaphragm* acts as a simple beam in distributing the horizontal forces to the vertical resisting elements, which therefore resist the horizontal forces on a tributary area basis. Wood and some steel deck diaphragms are considered flexible.

In the flexible diaphragm shown in Figure 12.28, the rigidity values of the walls are not considered. Each end wall resists 1/4 of the lateral load, while the center wall resists 1/2 of the lateral load.

A *rigid diaphragm* distributes the horizontal forces to the vertical resisting elements in proportion to their relative rigidities, rigidity being defined as resistance to deformation. Concrete and some steel deck diaphragms are considered rigid.

In the rigid diaphragm shown in Figure 12.29, the horizontal forces are distributed to the three walls in proportion to their rigidities. $\Sigma R = 2 + 1 + 2 = 5$. Each end wall has 2/5 of the total rigidity and therefore resists 2/5 of the total lateral load. The middle wall has 1/5 of the total rigidity and therefore resists 1/5 of the total lateral load.

Note that the distribution of lateral loads to the shear walls is very different from the distribution with a flexible diaphragm.

The allowable shear values for plywood diaphragms can be found in IBC Table 2306.2.1,

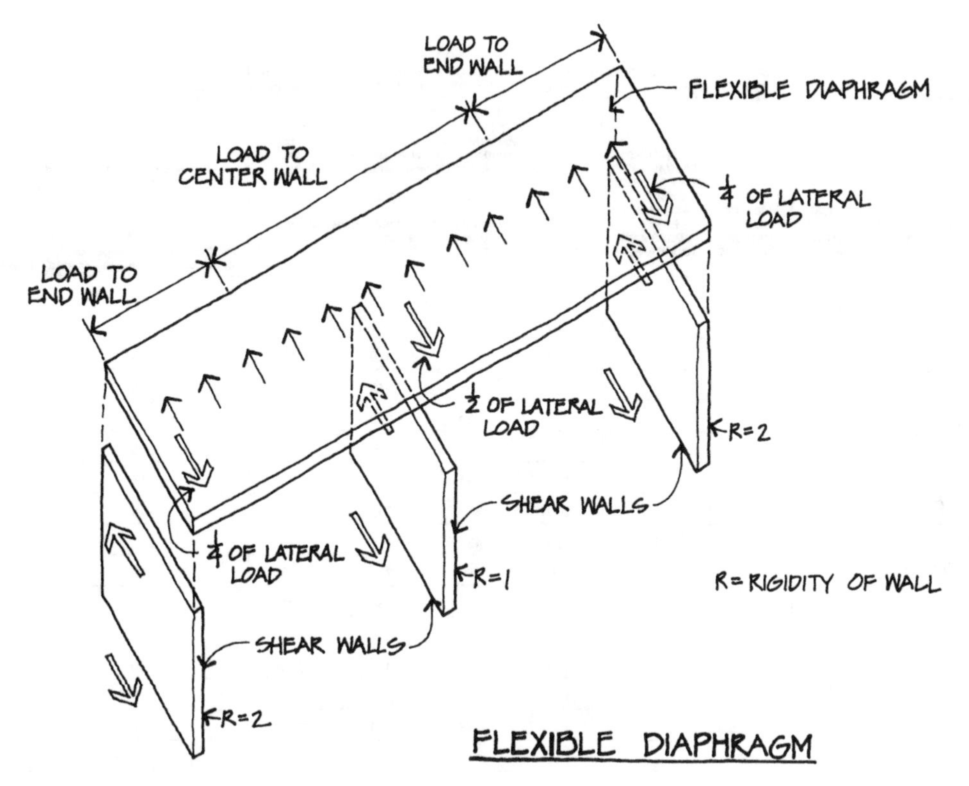

Figure 12.28

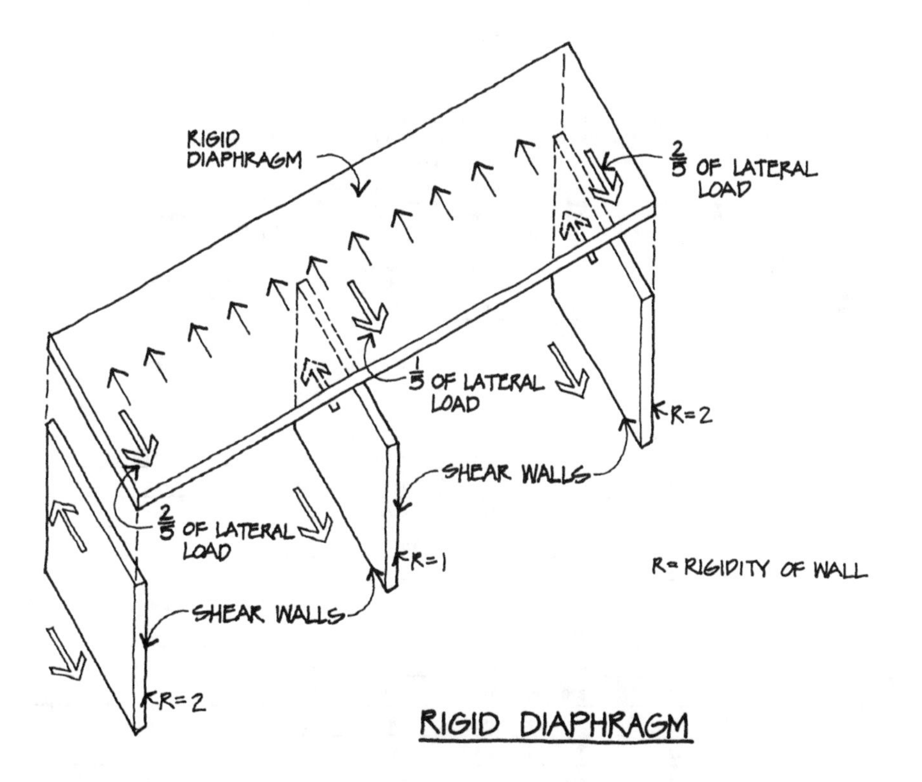

Figure 12.29

TABLE 2306.2.1
ALLOWABLE SHEAR (POUNDS PER FOOT) FOR WOOD STRUCTURAL PANEL DIAPHRAGMS WITH FRAMING OF DOUGLAS FIR-LARCH, OR SOUTHERN PINE[a] FOR WIND OR SEISMIC LOADING[h]

PANEL GRADE	COMMON NAIL SIZE OR STAPLE[f] LENGTH AND GAGE	MINIMUM FASTENER PENETRATION IN FRAMING (inches)	MINIMUM NOMINAL PANEL THICKNESS (inch)	MINIMUM NOMINAL WIDTH OF FRAMING MEMBERS AT ADJOINING PANEL EDGES AND BOUNDARIES[g] (inches)	BLOCKED DIAPHRAGMS: Fastener spacing (inches) at diaphragm boundaries (all cases) at continuous panel edges parallel to load (Cases 3, 4), and at all panel edges (Cases 5, 6)[b]: 6 / Fastener spacing (inches) at other panel edges (Cases 1, 2, 3 and 4)[b]: 6	4 / 6	2 1/2[c] / 4	2[c] / 3	UNBLOCKED DIAPHRAGMS: Fasteners spaced 6" max. at supported edges[b]: Case 1 (No unblocked edges or continuous joints parallel to load)	All other configurations (Cases 2, 3, 4, 5 and 6)
Structural I Grades	6d[e] (2" × 0.113")	1 1/4	5/16	2	185	250	375	420	165	125
				3	210	280	420	475	185	140
	1 1/2 16 Gage	1		2	155	205	310	350	135	105
				3	175	230	345	390	155	115
	8d (2 1/2" × 0.131)	1 3/8	3/8	2	270	360	530	600	240	180
				3	300	400	600	675	265	200
	1 1/2 16 Gage	1		2	175	235	350	400	155	115
				3	200	265	395	450	175	130
	10d[d] (3" × 0.148")	1 1/2	15/32	2	320	425	640	730	285	215
				3	360	480	720	820	320	240
	1 1/2 16 Gage	1		2	175	235	350	400	155	120
				3	200	265	395	450	175	130
Sheathing, single floor and other grades covered in DOC PS 1 and PS 2	6d[e] (2" × 0.113)	1 1/4	5/16	2	170	225	335	380	150	110
				3	190	250	380	430	170	125
	1 1/2 16 Gage	1		2	140	185	275	315	125	90
				3	155	205	310	350	140	105
	6d[e] (2" × 0.113)	1 1/4	3/8	2	185	250	375	420	165	125
				3	210	280	420	475	185	140
	8d (2 1/2" × 0.131)	1 3/8		2	240	320	480	545	215	160
				3	270	360	540	610	240	180

(continued)

TABLE 2306.2.1—continued
ALLOWABLE SHEAR (POUNDS PER FOOT) FOR WOOD STRUCTURAL PANEL DIAPHRAGMS WITH FRAMING OF DOUGLAS FIR-LARCH, OR SOUTHERN PINE[a] FOR WIND OR SEISMIC LOADING[h]

PANEL GRADE	COMMON NAIL SIZE OR STAPLE[f] LENGTH AND GAGE	MINIMUM FASTENER PENETRATION IN FRAMING (inches)	MINIMUM NOMINAL PANEL THICKNESS (inch)	MINIMUM NOMINAL WIDTH OF FRAMING MEMBERS AT ADJOINING PANEL EDGES AND BOUNDARIES[g] (inches)	BLOCKED DIAPHRAGMS: Fastener spacing (inches) at diaphragm boundaries (all cases) at continuous panel edges parallel to load (Cases 3, 4), and at all panel edges (Cases 5, 6)[b]: 6 / Fastener spacing (inches) at other panel edges (Cases 1, 2, 3 and 4)[b]: 6	4 / 6	2 1/2[c] / 4	2[c] / 3	UNBLOCKED DIAPHRAGMS: Fasteners spaced 6" max. at supported edges[b]: Case 1 (No unblocked edges or continuous joints parallel to load)	All other configurations (Cases 2, 3, 4, 5 and 6)
Sheathing, single floor and other grades covered in DOC PS 1 and PS 2 (continued)	1 1/2 16 Gage	1	3/8	2	160	210	315	360	140	105
				3	180	235	355	400	160	120
	8d (2 1/2"× 0.131")	1 3/8	7/16	2	255	340	505	575	230	170
				3	285	380	570	645	255	190
	1 1/2 16 Gage	1		2	165	225	335	380	150	110
				3	190	250	375	425	165	125
	8d (2 1/2" × 0.131")	1 3/8	15/32	2	270	360	530	600	240	180
				3	300	400	600	675	265	200
	10d[d] (3" × 0.148")	1 1/2		2	290	385	575	655	255	190
				3	325	430	650	735	290	215
	1 1/2 16 Gage	1		2	160	210	315	360	140	105
				3	180	235	355	405	160	120
	10d[d] (3" × 0.148")	1 1/2	19/32	2	320	425	640	730	285	215
				3	360	480	720	820	320	240
	1 3/4 16 Gage	1		2	175	235	350	400	155	115
				3	200	265	395	450	175	130

(continued)

Figure 12.30

TABLE 2306.2.1—**continued**
ALLOWABLE SHEAR (POUNDS PER FOOT) FOR WOOD STRUCTURAL PANEL DIAPHRAGMS WITH FRAMING OF DOUGLAS FIR-LARCH, OR SOUTHERN PINE[a] FOR WIND OR SEISMIC LOADING[h]

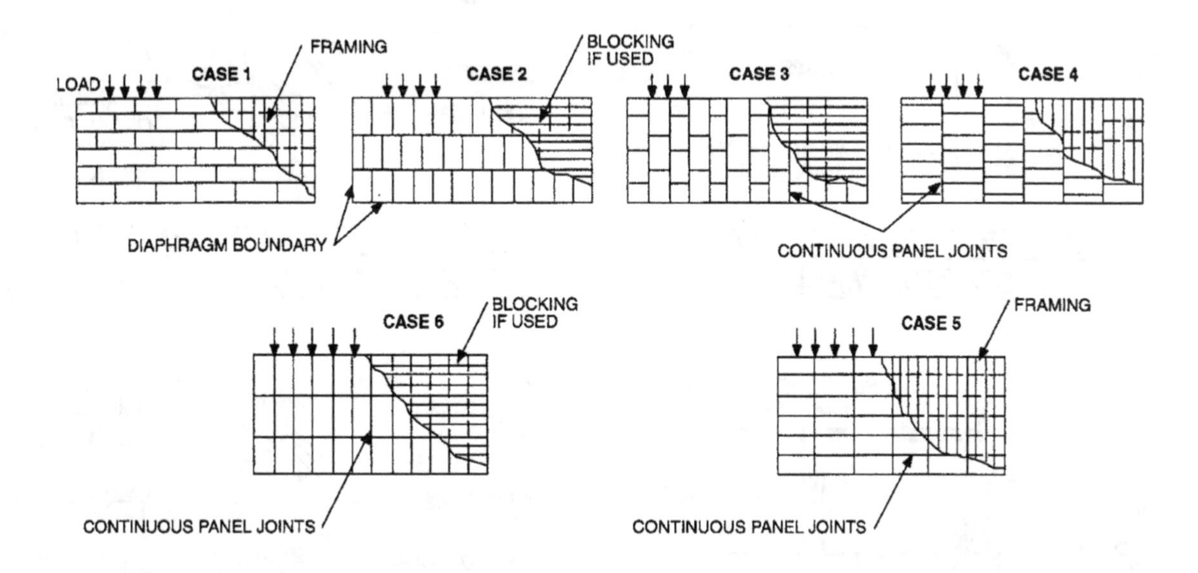

For SI: 1 inch = 25.4 mm, 1 pound per foot = 14.5939 N/m.

a. For framing of other species: (1) Find specific gravity for species of lumber in AF&PA NDS. (2) For staples find shear value from table above for Structural I panels (regardless of actual grade) and multiply value by 0.82 for species with specific gravity of 0.42 or greater, or 0.65 for all other species. (3) For nails find shear value from table above for nail size for actual grade and multiply value by the following adjustment factor: Specific Gravity Adjustment Factor = [1-(0.5 - SG)], where SG = Specific Gravity of the framing lumber. This adjustment factor shall not be greater than 1.
b. Space fasteners maximum 12 inches o.c. along intermediate framing members (6 inches o.c. where supports are spaced 48 inches o.c.).
c. Framing at adjoining panel edges shall be 3 inches nominal or wider, and nails shall be staggered where nails are spaced 2 inches o.c. or $2\,^{1}/_{2}$ inches o.c.
d. Framing at adjoining panel edges shall be 3 inches nominal or wider, and nails shall be staggered where both of the following conditions are met: (1) 10d nails having penetration into framing of more than $1\,^{1}/_{2}$ inches and (2) nails are spaced 3 inches o.c. or less.
e. 8d is recommended minimum for roofs due to negative pressures of high winds.
f. Staples shall have a minimum crown width of $^{7}/_{16}$ inch and shall be installed with their crowns parallel to the long dimension of the framing members.
g. The minimum nominal width of framing members not located at boundaries or adjoining panel edges shall be 2 inches.
h. For shear loads of normal or permanent load duration as defined by the AF&PA NDS, the values in the table above shall be multiplied by 0.63 or 0.56, respectively.

Figure 12.30 (continued)

shown in Figure 12.30. In this table, the term *wood structural panel* refers to various wood-based structural panels, including plywood. Plywood is widely used for diaphragms in small and medium-sized buildings. It is lightweight, economical, quickly and easily installed, and moderately strong. As shown in the table, its shear value depends on the plywood thickness, the size and spacing of the nails, and whether blocking is used at the edges of the plywood panels (a higher value is allowed where blocking is used). Adequate nailing must always be provided around the perimeter of the plywood diaphragm.

For example, the allowable shear in a 3/8" plywood blocked diaphragm of Structural I grade with 8d nails spaced at 6" at plywood panel edges and 3" framing members is 300 pounds per foot for seismic loading.

Diaphragms of steel deck have shear values varying from about 100 pounds per lineal foot to 2,600 pounds per lineal foot, depending on the steel deck thickness, the size and spacing of welds or other attachments between the deck

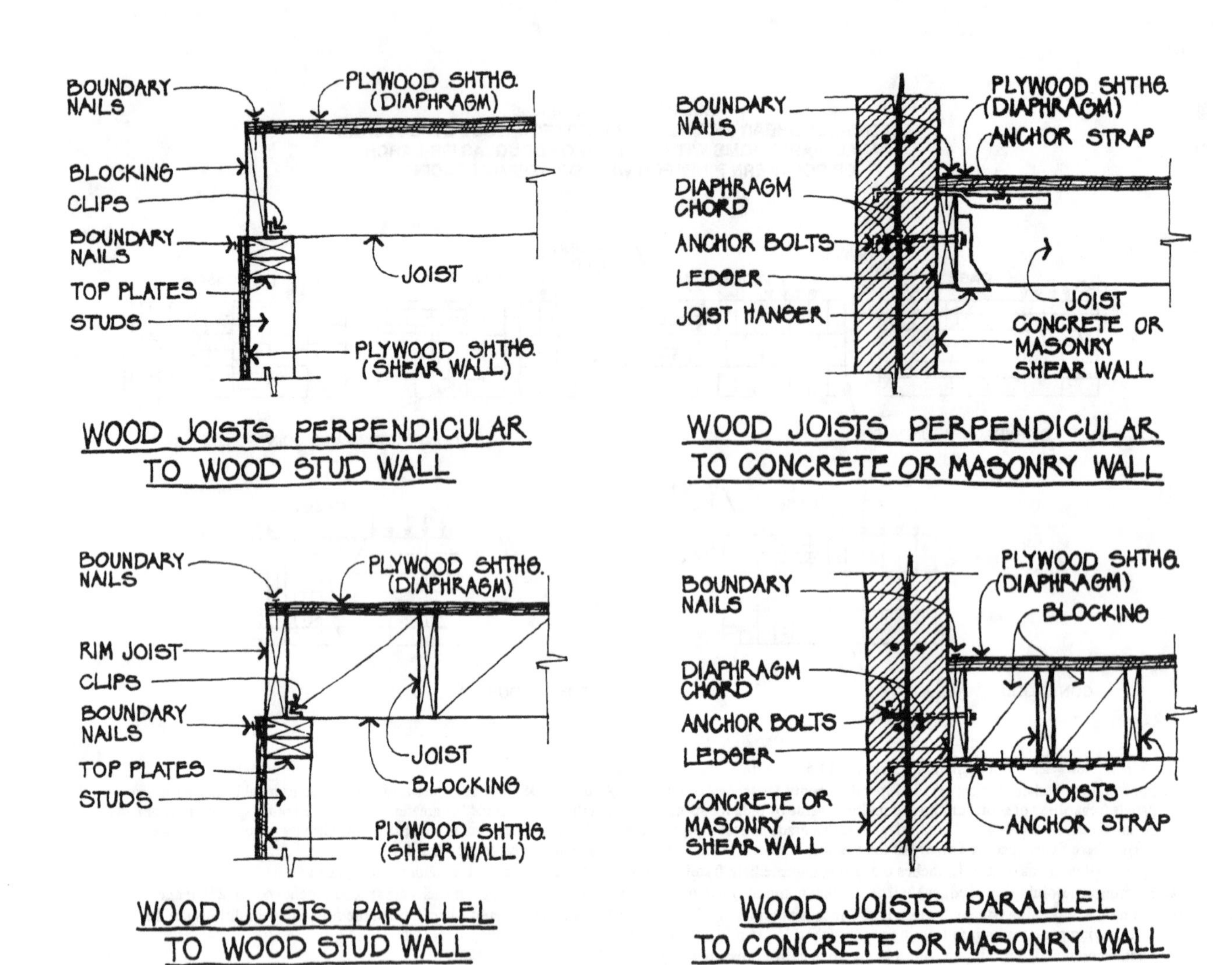

Figure 12.31

and the framing and between deck sheets, and the deck span. The shear value of steel deck can also be increased by placing concrete fill over the steel deck.

Concrete slabs can also be used as diaphragms, and a six-inch concrete slab, for example, has an allowable shear of about 10,000 pounds per lineal foot.

Although most structural decking materials, such as plywood, steel deck, and concrete, can function as diaphragms, some cannot. Among these are straight tongue-and-groove sheathing and light gauge corrugated metal.

In such cases, alternate methods, such as a horizontal bracing system, must be used to distribute the lateral forces to the vertical resisting elements.

Another factor that must be considered in diaphragm design is the horizontal deflection of the diaphragm. Referring again to the girder analogy, as the span of the diaphragm between vertical resisting elements increases, the deflection of the diaphragm increases at a greater rate. If the diaphragm deflection becomes too great, there can be excessive damage, and possibly even collapse, of elements laterally supported by the diaphragm. Accordingly, the code limits the span-to-depth ratio of plywood diaphragms to 4:1, and further requires that the diaphragm deflection not exceed that for which the diaphragm and any attached resisting elements will maintain their structural integrity without danger to the occupants of the structure.

Diaphragms must be adequately connected to the vertical resisting elements, such as shear walls. The connections must be designed to do the following:

1. Transfer the diaphragm shear force, which is parallel to the shear wall, into the wall.
2. Connect the diaphragm to the chord, or flange.
3. Connect the wall to the diaphragm for seismic forces perpendicular to the wall.

Examples of several diaphragm connection details are shown in Figure 12.31. Although there is an infinite variety of diaphragm connection details, if you understand the concepts of those shown, it will help you answer any questions on this subject which may appear on the exam.

In the two details in the left column of Figure 12.31, the diaphragm shear is transferred to the shear wall as follows: from the diaphragm (plywood sheathing) through the roof boundary nails to the rim joist or blocking, then through the clips to the top plates, then through the wall boundary nails to the shear wall (plywood wall sheathing).

The chord is either the rim joist or the top plates, connected to the diaphragm as described above.

Seismic or wind forces perpendicular to the wall are small and can be transferred from the top plates to the joists or blocking through toenails, and then through nails to the diaphragm.

In the two details in the right column of Figure 12.31, the diaphragm shear is transferred from the sheathing through the boundary nails to the ledger, and then from the ledger to the wall through the anchor bolts.

The diaphragm chord consists of a couple of reinforcing bars in the wall, and the connection of the diaphragm to the wall is as described previously.

Seismic forces perpendicular to the wall are transferred through the anchor straps to the joists or blocking, and then through nails to the sheathing.

The joist hangers support the joists for vertical load.

In Figure 12.32, the diaphragm shear is transferred from the steel deck through the welds to the continuous angle, then through the anchor bolts to the wall.

A couple of reinforcing bars in the wall form the diaphragm chord, and the connection of the diaphragm to the wall is as described above.

Seismic forces perpendicular to the wall are transferred through the dowels into the concrete fill.

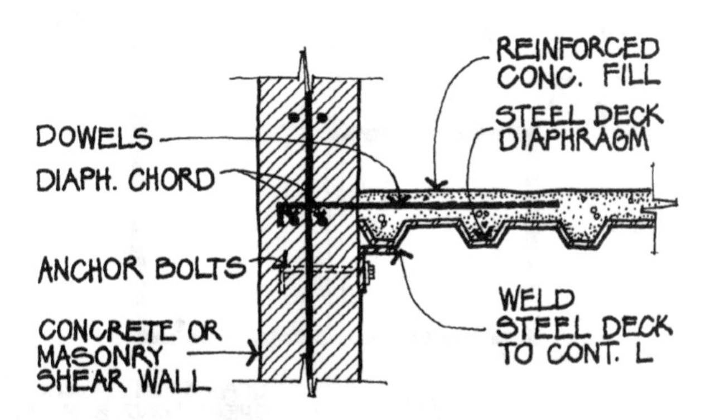

STEEL DECK CONNECTION TO CONCRETE OR MASONRY WALL

Figure 12.32

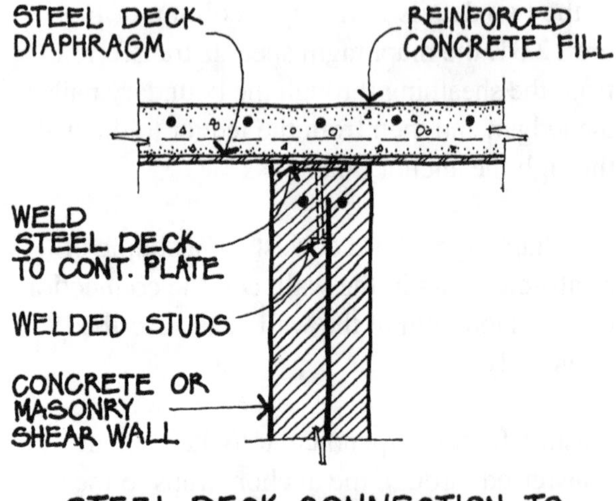

Figure 12.33

In Figure 12.33, the diaphragm shear is transferred from the steel deck through the welds to the continuous plate, then through the welded studs to the wall. Seismic forces perpendicular to the wall are transferred to the steel deck diaphragm in the same way. Since this type of detail would probably be used over an interior wall, not at the diaphragm boundary, a diaphragm chord is not required.

COLLECTORS

Where a shear resisting element is discontinuous, a collector member, often called a *strut* or *drag strut*, is used to collect seismic load from the diaphragm to which it is attached and deliver it to the shear resisting element. Two examples are shown in Figure 12.34.

TORSION

Torsion is the rotation caused in a diaphragm when the center of mass (where the resultant load is applied) does not coincide with the center of rigidity (where the resultant load is resisted), as shown in Figure 12.35. Torsion occurs only in rigid diaphragms.

The torsional moment causes forces in the shear resisting elements in direct proportion to (1) the magnitude of the torsional moment, (2) the distance of the shear resisting element from the center of rigidity, and (3) the rigidity of the shear resisting element.

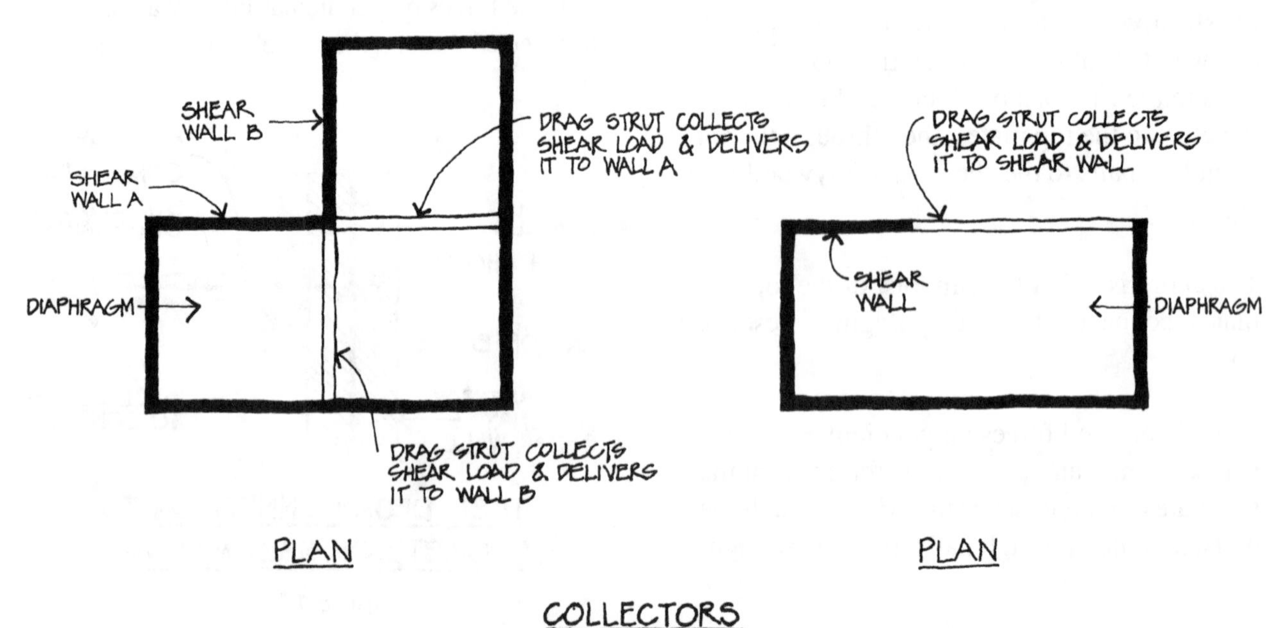

Figure 12.34

The code requires that when the force caused by torsional moment is in the same direction as the force caused by the applied seismic force, the forces must be added. However, when the force caused by torsional moment is in the opposite direction from that caused by the applied seismic force, it is not subtracted.

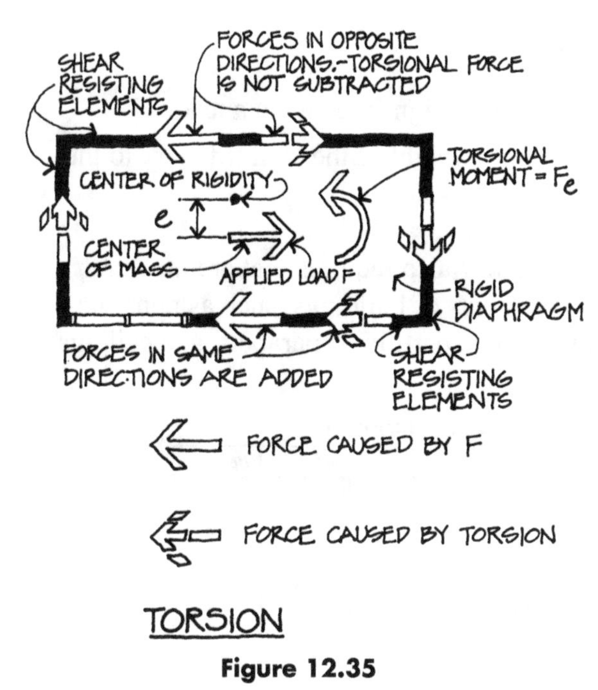

Figure 12.35

In addition, the code requires that an arbitrary amount of *accidental torsion* must be provided for in the design, even if the building is completely symmetrical. This is to account for conditions such as nonuniform distribution of vertical loads, unsymmetrical location of floor openings, eccentricity in rigidity caused by nonstructural elements, and seismic ground motion that may include a torsional component.

To provide for this accidental torsion, the mass at each level is assumed to be displaced a distance equal to 5 percent of the building dimension at that level perpendicular to the direction of the force, as shown in Figure 12.36.

Thus, in the north-south direction, the accidental torsional moment is equal to V times .05a, and in the east-west direction, it is equal to V times .05b.

Torsional effects are most significant in unsymmetrical buildings. These include buildings with an L configuration and structures whose shear resisting elements are in a core far from the building's perimeter, particularly where the core is not near the center of the building. While torsional effects cannot be entirely eliminated, they can be reduced by locating shear resisting elements at the perimeter of a structure, and by making the building and its system of shear resisting elements symmetrical. A more complete discussion of symmetry and regularity is found starting on page 276.

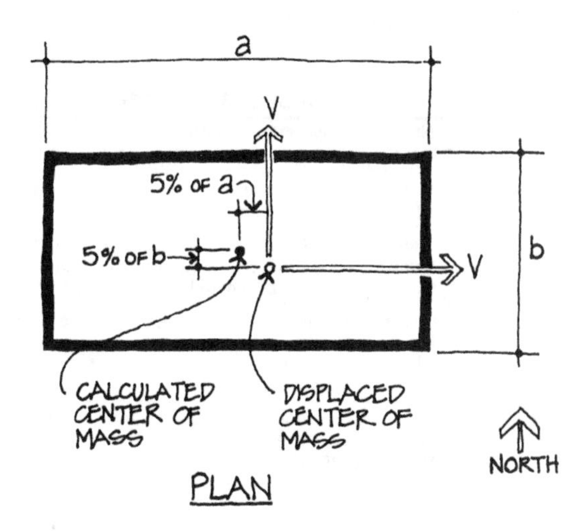

Figure 12.36

PARTS OF STRUCTURES

The code requires that structural walls and their anchorages be designed for seismic forces that are applied normal to the face of the wall in addition to forces in the plane of the wall.

In particular, bearing walls, shear walls, and their anchorages are to be designed to resist an out-of-plane force normal to the wall surface equal to the following:

$F_p = 0.4S_{DS}IW$

where W is the weight of the wall. Note that the minimum value of F_p is equal to 10 percent of the weight of the wall.

Concrete and masonry walls are to be anchored to other parts of the structure, and the anchors must be designed to resist the greater of the following:

- $0.4S_{DS}IW$ greater than or equal to 0.1W
- $400S_{DS}I$ pounds per linear foot of wall
- 280 pounds per linear foot

In addition to these requirements, when concrete or masonry walls are anchored to flexible diaphragms in buildings assigned to SDC C and higher, the anchors must be designed for a force no less than

$F_p = 0.8S_{DS}IW_p$

where

F_p = design force in an anchor
W_p = weight of the wall tributary to the anchor

The code 7 also requires that architectural components of buildings, such as nonstructural walls, partitions, parapets, and ceilings,

TABLE 13.5-1 COEFFICIENTS FOR ARCHITECTURAL COMPONENT

Architectural Component or Element	a_p[a]	R_p[b]
Interior Nonstructural Walls and Partitions[b]		
Plain (unreinforced) masonry walls	1.0	1.5
All other walls and partitions	1.0	2.5
Cantilever Elements (Unbraced or braced to structural frame below its center of mass)		
Parapets and cantilever interior nonstructural walls	2.5	2.5
Chimneys and stacks where laterally braced or supported by the structural frame	2.5	2.5
Cantilever Elements (Braced to structural frame above its center of mass)		
Parapets	1.0	2.5
Chimneys and Stacks	1.0	2.5
Exterior Nonstructural Walls[b]	1.0[b]	2.5
Exterior Nonstructural Wall Elements and Connections[b]		
Wall Element	1.0	2.5
Body of wall panel connections	1.0	2.5
Fasteners of the connecting system	1.25	1.0
Veneer		
Limited deformability elements and attachments	1.0	2.5
Low deformability elements and attachments	1.0	1.5
Penthouses (except where framed by an extension of the building frame)	2.5	3.5
Ceilings		
All	1.0	2.5
Cabinets		
Storage cabinets and laboratory equipment	1.0	2.5
Access Floors		
Special access floors (designed in accordance with Section 13.5.7.2)	1.0	2.5
All other	1.0	1.5
Appendages and Ornamentations	2.5	2.5
Signs and Billboards	2.5	2.5
Other Rigid Components		
High deformability elements and attachments	1.0	3.5
Limited deformability elements and attachments	1.0	2.5
Low deformability materials and attachments	1.0	1.5
Other Flexible Components		
High deformability elements and attachments	2.5	3.5
Limited deformability elements and attachments	2.5	2.5
Low deformability materials and attachments	2.5	1.5

[a] A lower value for a_p shall not be used unless justified by detailed dynamic analysis. The value for a_p shall not be less than 1.00. The value of $a_p = 1$ is for rigid components and rigidly attached components. The value of $a_p = 2.5$ is for flexible components and flexibly attached components. See Section 11.2 for definitions of rigid and flexible.
[b] Where flexible diaphragms provide lateral support for concrete or masonry walls and partitions, the design forces for anchorage to the diaphragm shall be as specified in Section 12.11.2.

Figure 12.37

be designed for seismic forces. The following formula is to be used to compute the seismic force, which is assumed to act at the center of gravity of the component:

$$F_p = \frac{0.4a_p S_{DS} W_p}{\left(\frac{R_p}{I_p}\right)}\left(1+2\frac{z}{h}\right)$$

where

- a_p = component amplification factor that varies from 1.00 to 2.50 (see ASCE 7 Table 13.5-1 in Figure 12.37)
- I_p = component importance factor that varies from 1.00 to 1.50 (components that are required for life-safety, contain hazardous materials, or are located in or attached to an Occupancy Category IV structure are assigned an I_p equal to 1.50; in all other cases, an I_p equal to 1.00 is assigned)
- W_p = component operating weight
- R_p = component response modification factor that varies from 1.00 to 3.50 (see ASCE 7 Table 13.5-1)
- z = height in structure of point of attachment of component with respect to the base of the building (value of z/h need not exceed 1.0)
- h = average roof height of the building with respect to the base

Note that F_p is not required to be taken as greater than

$$F_p = 1.6S_{DS}I_pW_p$$

and that F_p shall not be taken less than

$$F_p = 0.3S_{DS}I_pW_p$$

Example #7

The 8-inch thick reinforced concrete parapet wall depicted in Figure 12.38 is part of a residential building assigned to SDC D. The parapet extends 4 feet above the roof line. S_{DS} = 1.0g and S_{D1} = 0.61g. For what moment should the wall be designed?

Solution:

$$F_p = \frac{0.4a_p S_{DS} W_p}{\left(\frac{R_p}{I_p}\right)}\left(1+2\frac{z}{h}\right)$$

From ASCE 7 Table 13.5-1 (Figure 12.37), a_p = 2.5 and R_p = 2.5 for a cantilevered parapet wall that is braced to the structural frame below its center of mass. Also, I_p = 1.0. Since the parapet is at the roof of the building, z/h = 1.0. For normal weight concrete, W_p = 8/12 × 150 = 100 pounds per square foot.

Thus, $F_p = \frac{0.4\times2.5\times1.0\times100}{\left(\frac{2.5}{1.0}\right)}(1+2)$

= 120 pounds per square foot

Maximum $F_p = 1.6S_{DS}I_pW_p$ = 1.6 x 1.0 × 1.0 × 100 = 160 pounds per square foot

Minimum $F_p = 0.3S_{DS}I_pW_p$ = 0.3 × 1.0 × 1.0 × 100 = 30 pounds per square foot

Since F_p is equal to 120 pounds per square foot and the wall weighs 100 pounds per square foot, we can say that the parapet wall has a 120 percent seismic factor; in other words, it is designed for a horizontal force perpendicular to its surface equal to 120 percent of its own weight.

The moment = 120 × 4 × 4/2 = *960 foot-pounds*

COMBINED VERTICAL AND HORIZONTAL FORCES

All building components must be designed for load combinations using either (1) strength design or load and resistance factor design or (2) allowable stress design. The appropriate set of load combinations to use usually depends on the design requirements for the material specified for the building. For example, for reinforced concrete, load combinations based on strength design are used, while for steel, load combinations for strength design *or* allowable stress design are used.

Many load combinations are given for the various types of loads that can act on a building. The following combinations include dead loads (D), live loads (L), roof live loads (L_r), snow loads (S), and earthquake loads (E) only. A complete set of load combinations can be found in the IBC or ASCE 7.

Strength design load combinations:

1. $1.2D + 1.0E + f_1L + f_2S$
2. $0.9D + 1.0E$

where

f_1 = 1 for floors in places of public assembly, for live loads in excess of 100 pounds per square foot, and for parking garage live load

f_1 = 0.5 for other live loads

f_2 = 0.7 for roof configurations (such as saw tooth) that do not shed snow off the structure

f_2 = 0.2 for other roof configurations

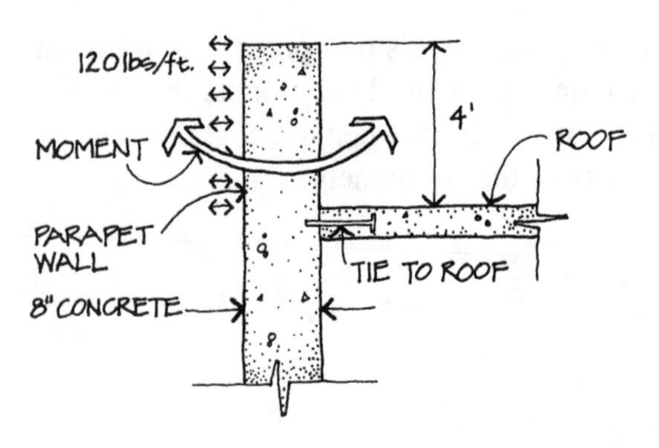

Figure 12.38

Allowable stress design load combinations:

1. $D + 0.7E$
2. $D + 0.75(0.7E) + 0.75L + 0.75(L_r \text{ or } S)$
3. $0.6D + 0.7E$

The code also gives another set of alternate basic load combinations for allowable stress design, which were previously given in the Uniform Building Code.

Load combinations reflect the probabilities of various loads acting on the structure at the same time.

REGULAR AND IRREGULAR STRUCTURES

According to the IBC, structures are designated as structurally regular or irregular. A regular structure has no significant discontinuities in plan, vertical configuration, or lateral force resisting systems. An irregular structure, on the other hand, has significant discontinuities such as those in ASCE 7 Tables 12.3-1 (horizontal irregularities) and 12.3-2 (vertical irregularities), reproduced in Figures 12.39 and 12.40.

TABLE 12.3-1 HORIZONTAL STRUCTURAL IRREGULARITIES

	Irregularity Type and Description	Reference Section	Seismic Design Category Application
1a.	**Torsional Irregularity** is defined to exist where the maximum story drift, computed including accidental torsion, at one end of the structure transverse to an axis is more than 1.2 times the average of the story drifts at the two ends of the structure. Torsional irregularity requirements in the reference sections apply only to structures in which the diaphragms are rigid or semirigid.	12.3.3.4 12.8.4.3 12.7.3 12.12.1 Table 12.6-1 Section 16.2.2	D, E, and F C, D, E, and F B, C, D, E, and F C, D, E, and F D, E, and F B, C, D, E, and F
1b.	**Extreme Torsional Irregularity** is defined to exist where the maximum story drift, computed including accidental torsion, at one end of the structure transverse to an axis is more than 1.4 times the average of the story drifts at the two ends of the structure. Extreme torsional irregularity requirements in the reference sections apply only to structures in which the diaphragms are rigid or semirigid.	12.3.3.1 12.3.3.4 12.7.3 12.8.4.3 12.12.1 Table 12.6-1 Section 16.2.2	E and F D B, C, and D C and D C and D D B, C, and D
2.	**Reentrant Corner Irregularity** is defined to exist where both plan projections of the structure beyond a reentrant corner are greater than 15% of the plan dimension of the structure in the given direction.	12.3.3.4 Table 12.6-1	D, E, and F D, E, and F
3.	**Diaphragm Discontinuity Irregularity** is defined to exist where there are diaphragms with abrupt discontinuities or variations in stiffness, including those having cutout or open areas greater than 50% of the gross enclosed diaphragm area, or changes in effective diaphragm stiffness of more than 50% from one story to the next.	12.3.3.4 Table 12.6-1	D, E, and F D, E, and F
4.	**Out-of-Plane Offsets Irregularity** is defined to exist where there are discontinuities in a lateral force-resistance path, such as out-of-plane offsets of the vertical elements.	12.3.3.4 12.3.3.3 12.7.3 Table 12.6-1 16.2.2	D, E, and F B, C, D, E, and F B, C, D, E, and F D, E, and F B, C, D, E, and F
5.	**Nonparallel Systems-Irregularity** is defined to exist where the vertical lateral force-resisting elements are not parallel to or symmetric about the major orthogonal axes of the seismic force–resisting system.	12.5.3 12.7.3 Table 12.6-1 Section 16.2.2	C, D, E, and F B, C, D, E, and F D, E, and F B, C, D, E, and F

Figure 12.39

TABLE 12.3-2 VERTICAL STRUCTURAL IRREGULARITIES

	Irregularity Type and Description	Reference Section	Seismic Design Category Application
1a.	**Stiffness-Soft Story Irregularity** is defined to exist where there is a story in which the lateral stiffness is less than 70% of that in the story above or less than 80% of the average stiffness of the three stories above.	Table 12.6-1	D, E, and F
1b.	**Stiffness-Extreme Soft Story Irregularity** is defined to exist where there is a story in which the lateral stiffness is less than 60% of that in the story above or less than 70% of the average stiffness of the three stories above.	12.3.3.1 Table 12.6-1	E and F D, E, and F
2.	**Weight (Mass) Irregularity** is defined to exist where the effective mass of any story is more than 150% of the effective mass of an adjacent story. A roof that is lighter than the floor below need not be considered.	Table 12.6-1	D, E, and F
3.	**Vertical Geometric Irregularity** is defined to exist where the horizontal dimension of the seismic force–resisting system in any story is more than 130% of that in an adjacent story.	Table 12.6-1	D, E, and F
4.	**In-Plane Discontinuity in Vertical Lateral Force-Resisting Element Irregularity** is defined to exist where an in-plane offset of the lateral force-resisting elements is greater than the length of those elements or there exists a reduction in stiffness of the resisting element in the story below.	12.3.3.3 12.3.3.4 Table 12.6-1	B, C, D, E, and F D, E, and F D, E, and F
5a.	**Discontinuity in Lateral Strength–Weak Story Irregularity** is defined to exist where the story lateral strength is less than 80% of that in the story above. The story lateral strength is the total lateral strength of all seismic-resisting elements sharing the story shear for the direction under consideration.	12.3.3.1 Table 12.6-1	E and F D, E, and F
5b.	**Discontinuity in Lateral Strength–Extreme Weak Story Irregularity** is defined to exist where the story lateral strength is less than 65% of that in the story above. The story strength is the total strength of all seismic-resisting elements sharing the story shear for the direction under consideration.	12.3.3.1 12.3.3.2 Table 12.6-1	D, E, and F B and C D, E, and F

Figure 12.40

Regular and symmetrical structures exhibit more favorable and predictable seismic response characteristics than irregular structures. Therefore, the use of irregular structures in earthquake-prone areas should be avoided if possible.

However, the IBC does not prohibit irregular structures. Instead, it contains specific design requirements for each type of irregularity. In some cases of irregularity, the static lateral force procedure as described in this lesson is not permitted, and a dynamic procedure is required.

The following is a summary of some recommended design practices concerning structural regularity.

1. Structures should be regular in stiffness and geometry, both in plan and elevation (Figure 12.41).
2. Abrupt changes of shape, stiffness, or resistance, such as a *soft* first story, should be avoided.
3. Portions of a building that are different in size, shape, or rigidity should have a seismic separation, as shown in Figure 12.43.

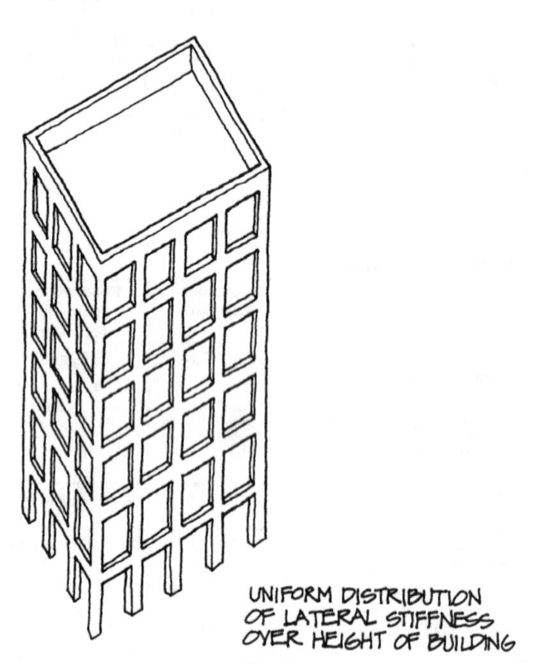

Figure 12.41

4. Reentrant corners should be avoided. Such corners occur in L-, T-, U-, and cross-shaped plans (Figure 12.42).
5. If a reentrant corner is unavoidable, it should be strengthened by using drag struts, or, preferably, a seismic separation should be provided, as shown in Figure 12.44.
6. Torsion should be minimized by making the building symmetrical and regular in geometry and stiffness, and by providing lateral load resisting elements at the building's perimeter as shown in Figure 12.46.
7. Open-front buildings have high torsional stresses because the rear wall, which generally has minimal openings, is very rigid, while the front is very flexible. The most practical solution is to make the front as rigid as possible, by providing a moment-resisting frame, as shown in Figure 12.45.
8. Complex or asymmetrical building shapes, which may introduce stress concentrations and/or torsion, should be avoided. If necessary, seismic separations should be provided.

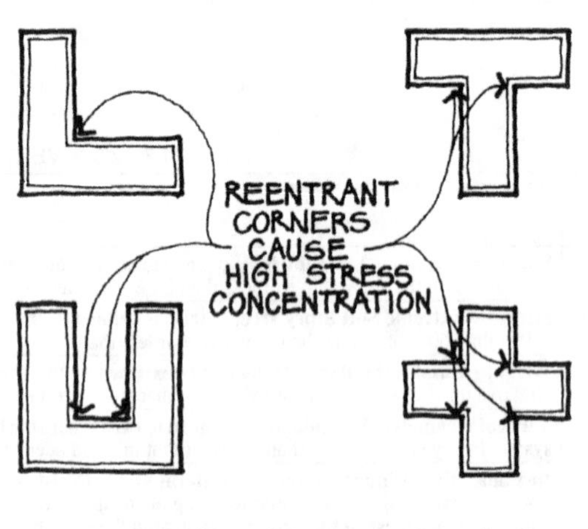

Figure 12.42

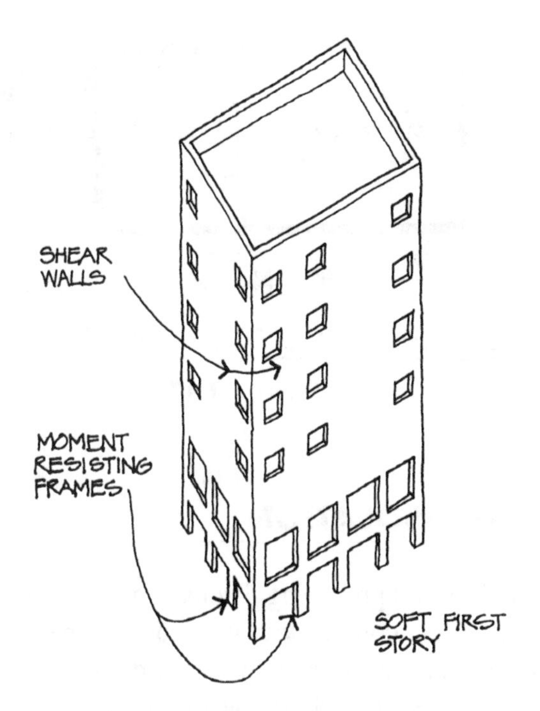

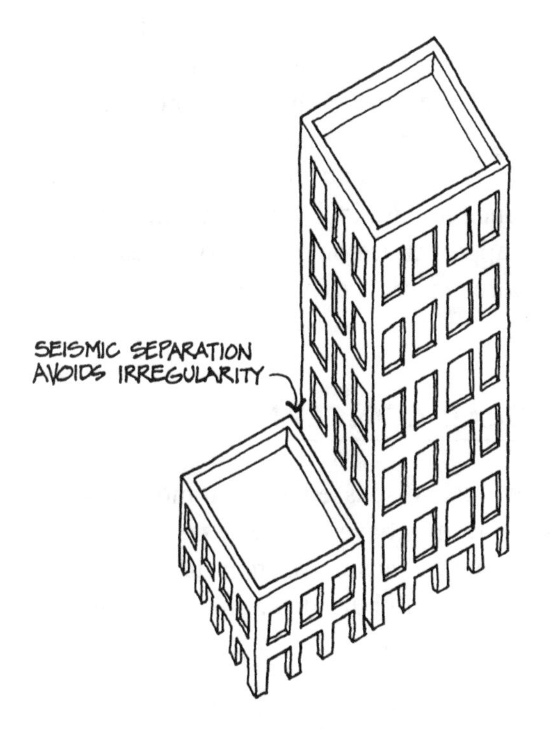

Figure 12.43

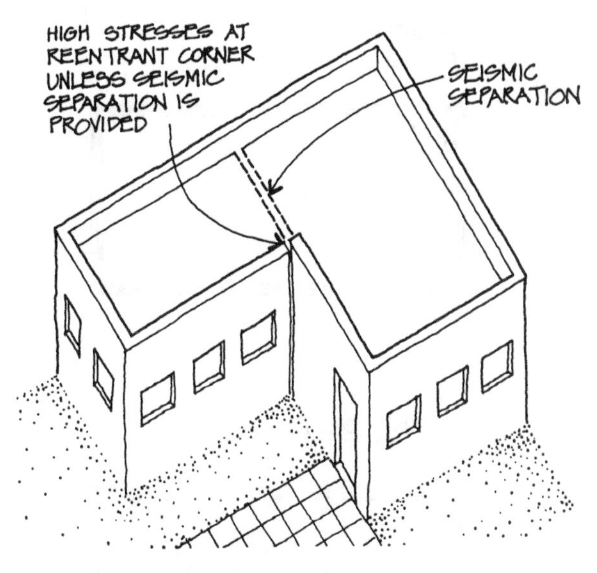

Figure 12.44

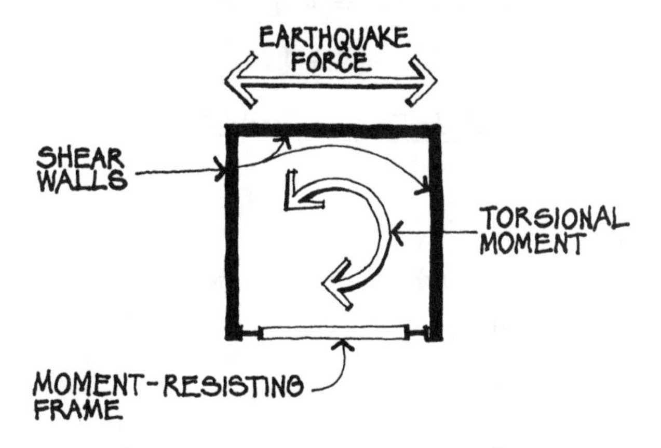

Figure 12.45

9. There should be a direct path for force transfer. Shear walls or other lateral load resisting elements should be continuous to the foundation. In-plane or out-of-plane offsets should be avoided (Figure 12.48).
10. Diaphragms with abrupt discontinuities, as shown on the top right of Figure 12.47, should be avoided.
11. Column stiffness variation, as shown in Figure 12.49, should be avoided, if possible.

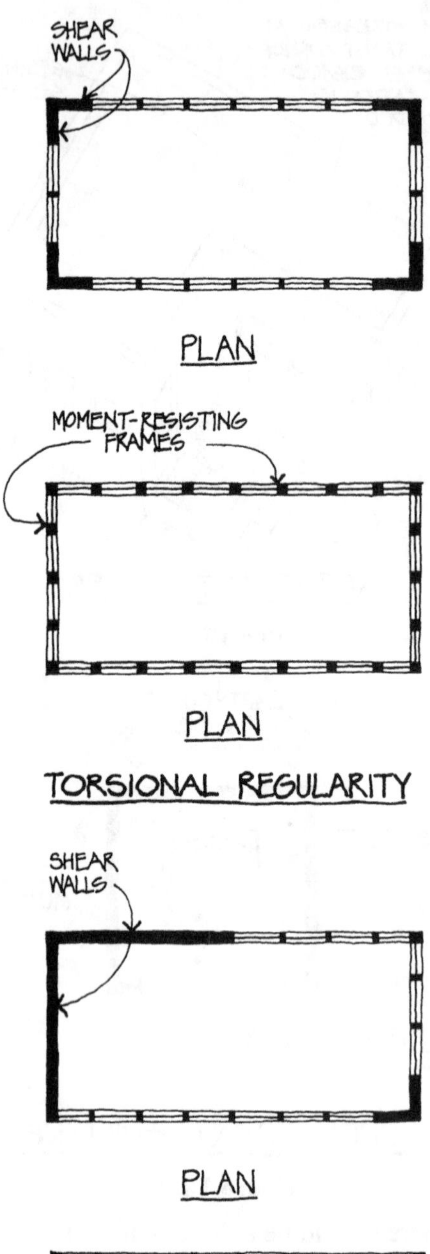
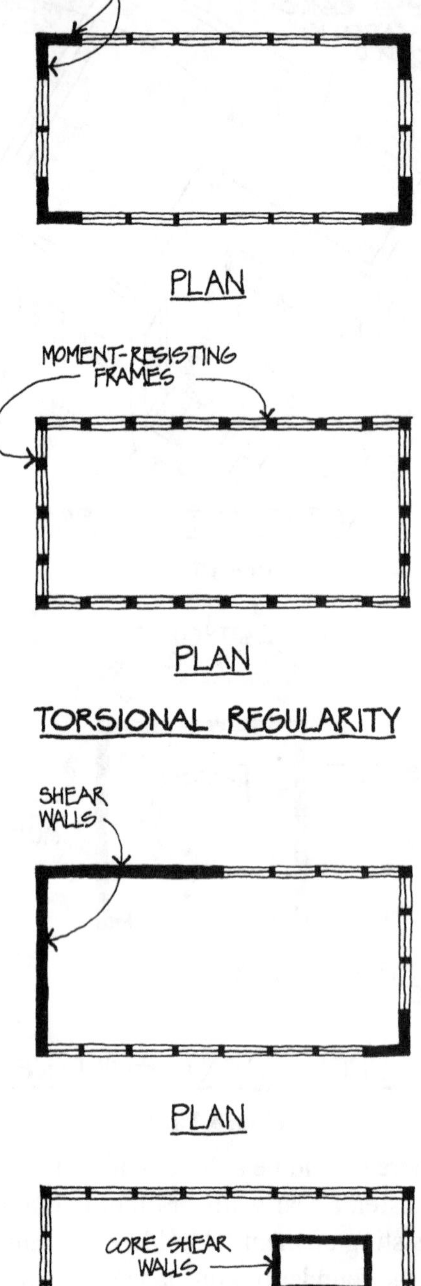

Figure 12.46

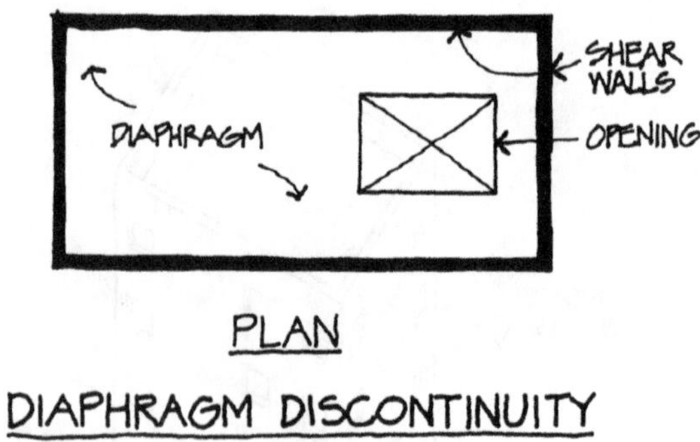

Figure 12.47

BASE ISOLATION

In all of the preceding discussion, the structure is in direct contact with the ground. Therefore, during an earthquake, the ground motion is transmitted directly to the structure.

A more recent approach, which has been used on numerous buildings, is base isolation. In this method, the structure is isolated from the ground by specially designed bearings and dampers that absorb earthquake forces, thus reducing the building's acceleration from earthquake ground motion (see Figure 12.50). The acceptance of base isolation as an alternative design approach is continuing to increase as we gain more experience in its use.

TUBULAR CONCEPT

As previously discussed, special moment-resisting frames have generally performed well in earthquakes because of their ductility. However, for very tall buildings, this system may not be economical.

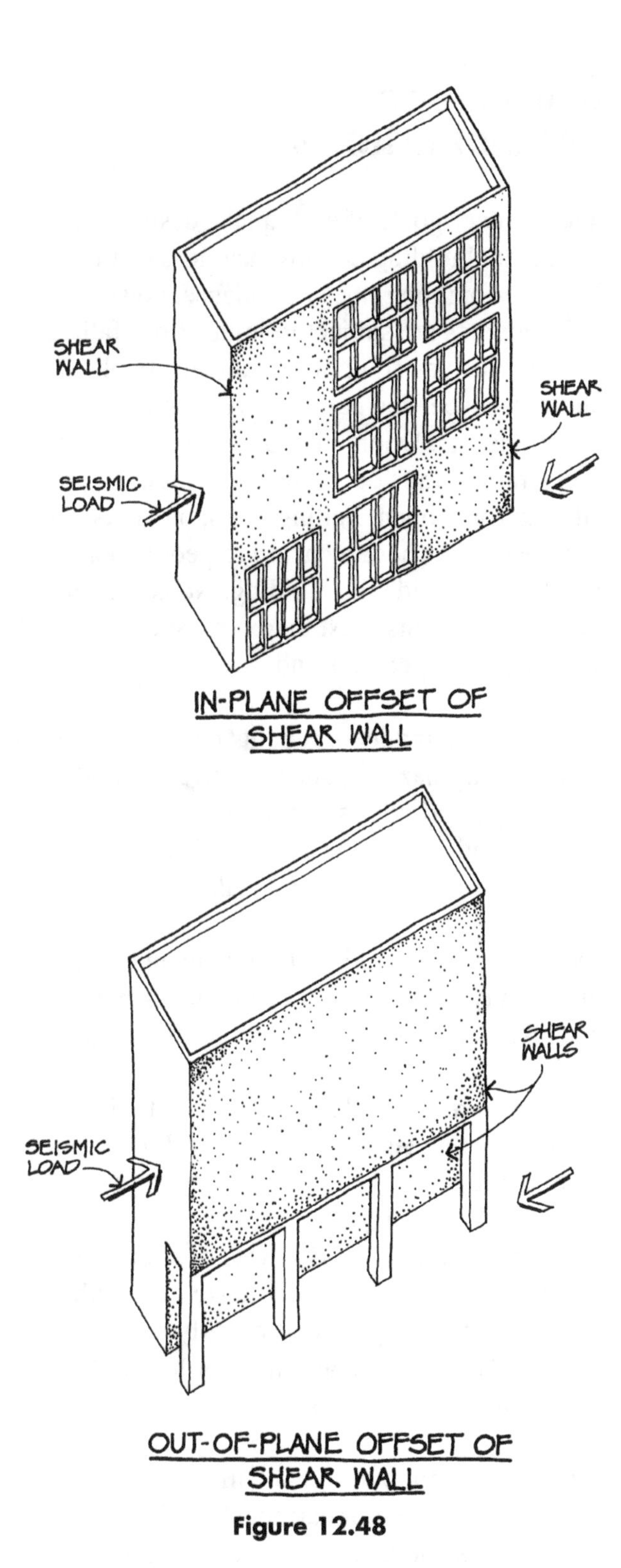

Figure 12.48

A relatively recent development, pioneered by the late Fazlur Khan of S.O.M., is the concept of tubular behavior for tall buildings subject to lateral loads from earthquake or wind. This consists of closely spaced columns at the perimeter of the building connected by deep spandrel beams at each floor to form, in effect, a perforated wall at each facade (see Figure 12.51). This system behaves like a tube, or box beam, that cantilevers from the ground when subject to lateral earthquake or wind forces.

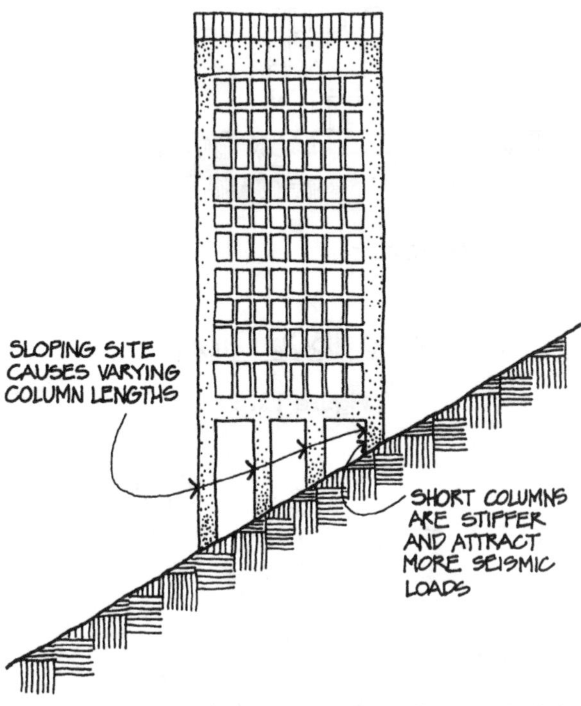

Figure 12.49

Tubular systems can provide strength and stiffness economically, particularly for tall buildings. Some of the tallest buildings constructed in the United States employ tubular systems, including the John Hancock Building and Willis Tower, both in Chicago, and the former Twin Towers of the World Trade Center in New York.

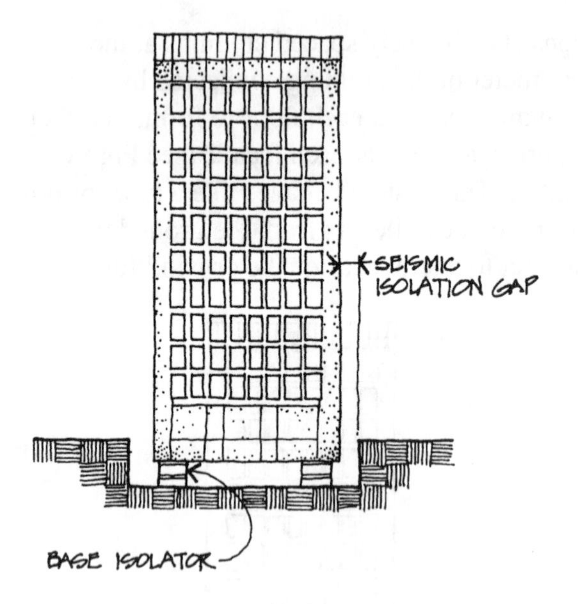

Figure 12.50

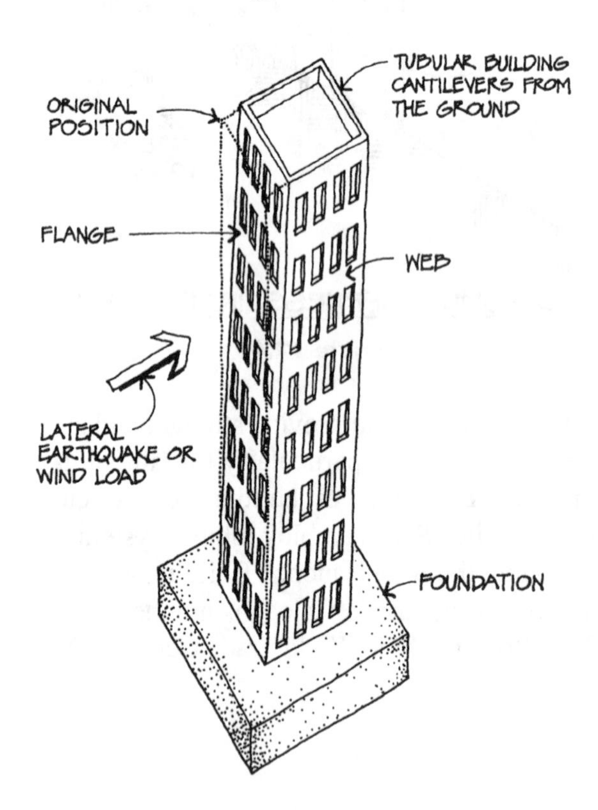

Figure 12.51

NONSTRUCTURAL CONSIDERATIONS

The primary goals of earthquake design are to prevent building collapse and loss of life. Even if a building doesn't collapse, however, its occupants are subject to injury from falling objects, such as equipment cabinets, bookshelves, suspended ceilings, lighting fixtures, partitions, etc. Doors may become jammed in their frames, thus becoming difficult or impossible to force open. Falling glass, parapets, and other elements pose a hazard to people both inside and outside the building. Nonstructural building elements must therefore be designed with these dangers in mind.

The building and its occupants may be subject to secondary hazards, such as fire and flooding. Alarm and sprinkler systems are vulnerable to damage and should therefore be designed to be operable following an earthquake.

Evacuation of a building following an earthquake may be difficult or even impossible because of debris, lack of light, or flooding.

Elevators are particularly vulnerable to damage. Therefore, the design should carefully consider all exit paths.

Mechanical and electrical systems should be designed to survive an earthquake, if possible, so that the building may be quickly returned to service. This is especially true for hospitals and other emergency centers.

It is not enough to design a building so that it can survive a major earthquake. All of its components, elements, and systems must be designed and braced so that damage during and after an earthquake is minimized. Sketches of a couple of typical details for nonstructural elements are shown in Figures 12.52 and 12.53.

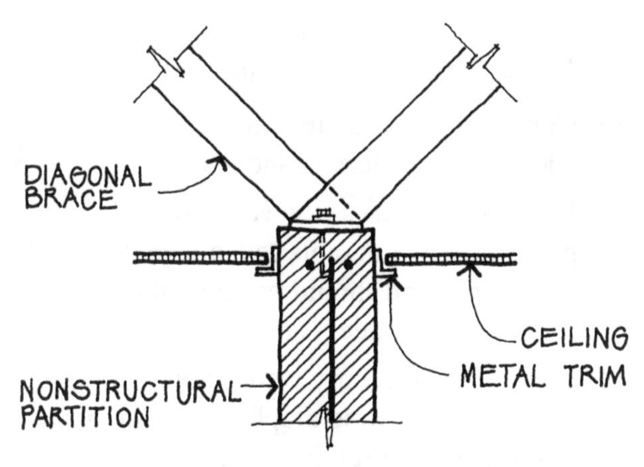

Figure 12.52

EXISTING BUILDINGS

Sometimes, an architect must evaluate the seismic resistance of an existing building. Whether a building is new or existing, essentially the same seismic concepts apply. Thus, regularity, continuity, redundancy, and other seismic design practices discussed in this lesson are pertinent to existing buildings.

Some general considerations in the seismic evaluation of existing buildings are as follows:

1. There should be a complete lateral force resisting system, and all of its elements including connections, anchorages, and diaphragms should be adequate to resist the seismic loads.
2. Significant corrosion, deterioration, or decay of the structural materials used in the vertical and lateral force resisting systems can impair their capacity and may require corrective action.
3. Chimneys, cornices, and parapets should be adequately reinforced and anchored to prevent them from breaking off and falling.
4. Exterior cladding, veneer, and wall panels should be adequately attached to the framing to prevent them from falling off.
5. Large openings in diaphragms should be reinforced so that the diaphragm has sufficient strength and stiffness to resist seismic loads.
6. Walls with garage doors or other large openings may be inadequate to resist

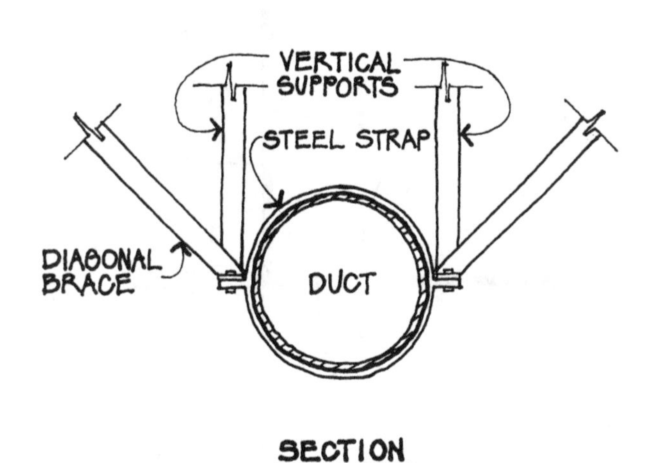

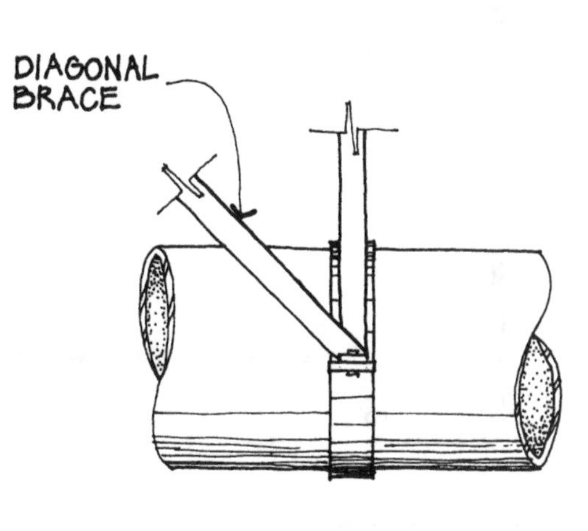

TYPICAL DUCT DETAIL

Figure 12.53

seismic loads and may undergo distress or collapse unless stiffened or reinforced.

7. Glazing should be sufficiently isolated from the structure to prevent it from shattering and falling out.
8. Wood frame buildings generally perform well in an earthquake and do not pose a significant threat to life. However, they may be damaged and their contents may be badly shaken.
9. Wood stud walls should be adequately bolted to the foundation, to prevent the structure from sliding off the foundation.
10. Wood stud cripple walls below the first floor should be braced, to prevent the structure from falling onto the foundation.
11. Wood framed roofs or floors with straight sheathing may not have adequate diaphragm strength and may require diagonal rod bracing or plywood sheathing.
12. Split level floors should be properly connected to avoid separation.
13. Concrete and masonry walls should be adequately reinforced and anchored.
14. Unreinforced masonry walls are particularly vulnerable to damage, falling, or collapse and should be strengthened and anchored, or if necessary, removed.
15. Concrete or masonry walls should be anchored to a wood diaphragm with metal ties, rather than relying on a wood ledger, which can fail by cross-grain bending and possibly cause collapse of the roof and/or falling out of the wall. Refer to Figure 12.31.
16. Diagonal cracks in masonry or concrete walls may have been caused by previous earthquakes and should be analyzed and repaired.

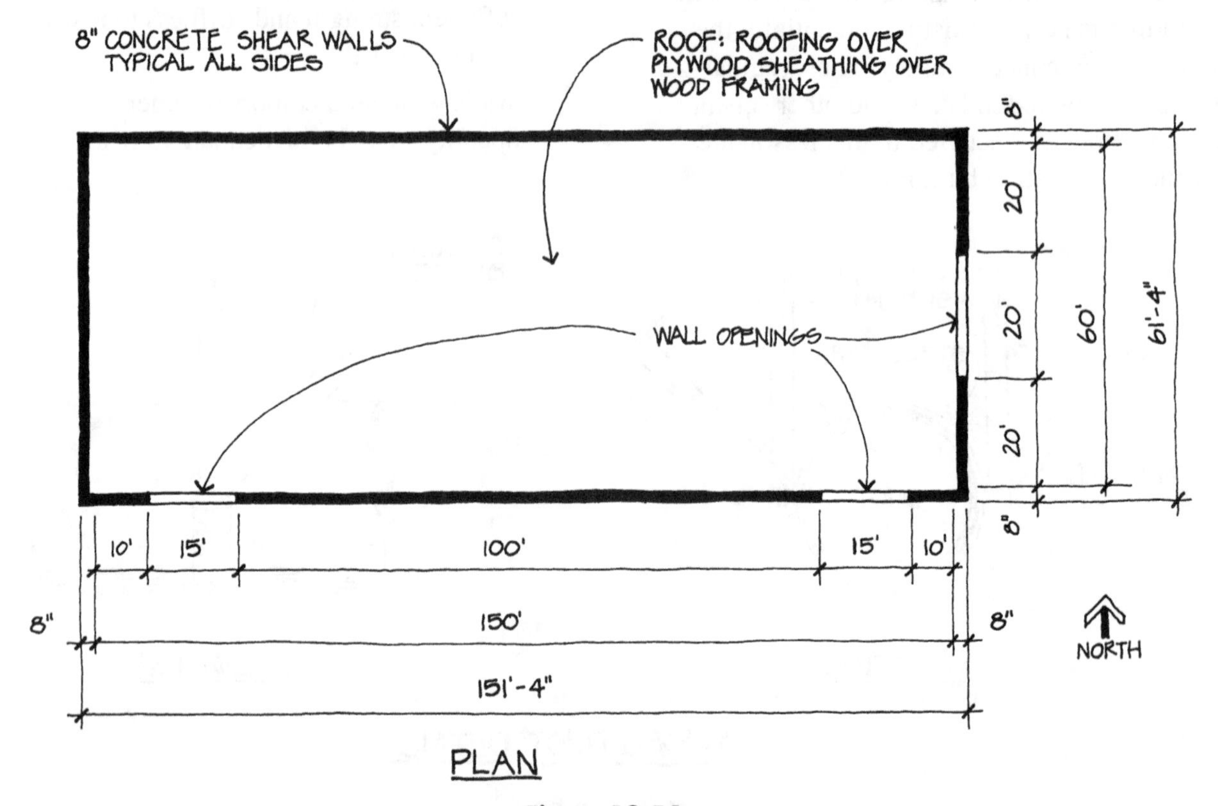

Figure 12.54

17. Metal deck floors and roofs without a reinforced concrete topping slab may not have adequate diaphragm strength and may require strengthening.
18. The restoration of historic buildings in seismic zones poses a special challenge. The preferred, but often difficult, solution involves the reinforcement of existing elements, rather than the addition of new nonhistoric elements such as visible walls or braces.

EARTHQUAKE DESIGN EXAMPLE

The following is a step-by-step design of a small commercial building assigned to SDC D, shown in Figures 12.54 and 12.55, to resist earthquake forces. The exam will not have a problem as lengthy or complex as this. However, if you study this design example, it should help you understand the basic principles of earthquake design.

Loads

The loads are as follows:

Roof live load	20.0 psf
Roof dead load	
Roofing	3.0
Plywood sheathing	1.5
Wood framing	2.5
Insulation	2.0
Ceiling	3.0
Mech. & elec.	2.0
Misc.	2.0
Total	16.0 psf

Dead load of 8" concrete walls = 8"/12" × 150#/ft^3 = 100.0 psf

Base Shear

What is the total base shear? Assume S_{DS} = 0.97g and S_{D1} = 0.49g.

Solution:

We determine the base shear from the following formula:

$V = C_s W$

where $C_s = S_{DS}/(R/I)$

Since we are not given any information to the contrary, we must assume that the building does not contain a complete frame providing support for vertical loads.

Therefore, its structural system is a bearing wall system. From ASCE 7 Table 12.2-1 in Figure 12.22, this is system A.1, which has an R value of 5.

The building is not high occupancy, essential, or hazardous, so from ASCE 7 Table 11.5-1 in Figure 12.16, I = 1.0.

Thus, $C_s = 0.97/(5/1.0) = 0.194$

Also, C_s need not exceed the following:

$C_s = S_{D1}/T(R/I)$

From page 249, $T = C_t(h_n)^x = 0.020(14)^{0.75}$ = 0.145 second. Note that h_n is the height to the diaphragm, not to the top of the parapet.

Thus, $C_s = (0.49 \times 1)/(0.145 \times 5) = 0.676$

Minimum $C_s = 0.01$.

Therefore, $C_s = 0.194$.

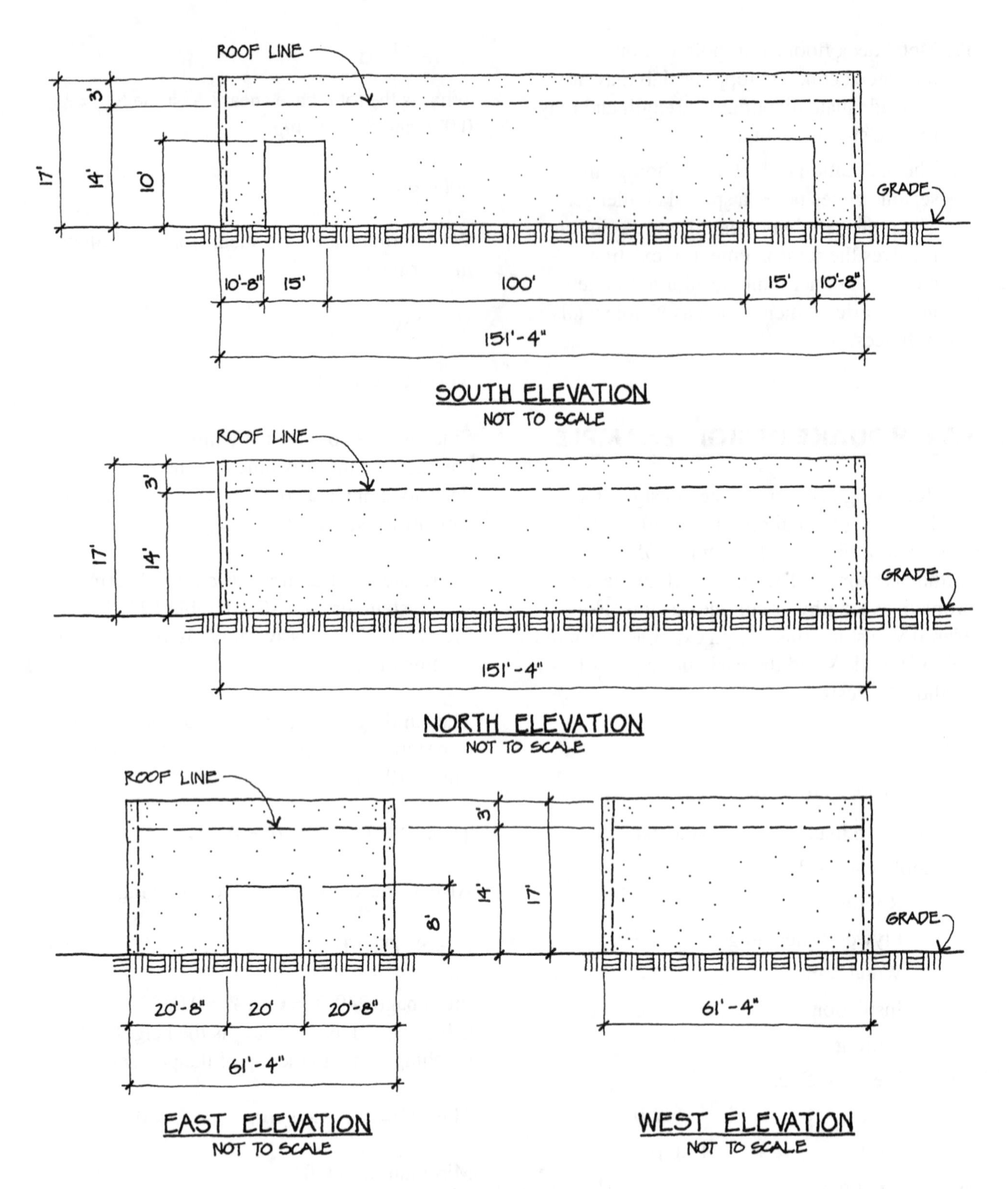
ROOF LINE
3'
14'
17'
10'
GRADE
10'-8"
15'
100'
15'
10'-8"
151'-4"
SOUTH ELEVATION
NOT TO SCALE
ROOF LINE
3'
17'
14'
GRADE
151'-4"
NORTH ELEVATION
NOT TO SCALE
ROOF LINE
3'
8'
14'
17'
GRADE
20'-8"
20'
20'-8"
61'-4"
61'-4"
EAST ELEVATION
NOT TO SCALE
WEST ELEVATION
NOT TO SCALE

Figure 12.55

We next determine W as follows:

Roof dead load = 16 × 60 × 150 = 144,000#

North and south wall dead loads = 100 × 17 × 150 × 2 = 510,000#

East and west wall dead loads = 100 × 17 × 61.33 × 2 = 208,522#

Total dead load = 862,522#

V = 0.194W = 0.194(862,522) = *167,329#*

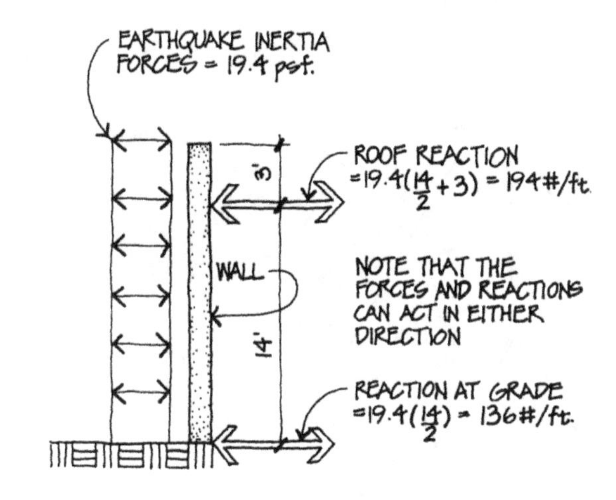

Figure 12.56

Diaphragm Design for North-South Direction

Design the roof diaphragm. Determine the diaphragm shear, select a plywood diaphragm, and calculate the maximum chord force in the diaphragm.

Solution:

For a one-story building, $F_{px} = C_s w_{px} = 0.194w_{px}$

Maximum $F_{px} = 0.4S_{DS}Iw_{px} = (0.4 \times 0.97 \times 1.0) w_{px} = 0.388w_{px}$

Minimum $F_{px} = 0.2S_{DS}Iw_{px} = (0.2 \times 0.97 \times 1.0) w_{px} = 0.194w_{px}$

Thus, $F_{px} = 0.194w_{px}$

The north and south walls span vertically between grade and the roof diaphragm when subject to earthquake inertia forces. These forces are equal to 0.194 × wall weight = 0.194 × 100 = 19.4 pounds per square foot as shown in Figure 12.56.

The bottom reaction of 136 pounds per foot is resisted by the footings and soil.

The top reaction of 194 pounds per foot is resisted by the roof diaphragm. In addition, the diaphragm resists the inertia forces caused by the roof dead load = 0.194 × 16 × 60 = 186 pounds per foot.

The roof diaphragm acts as a horizontal beam spanning between the east and west shear walls and resists a total load of 194 + 194 + 186 = 574 pounds per foot, as shown in Figure 12.57.

The plywood roof diaphragm is a flexible diaphragm, which acts as a simple beam.

The reaction to each shear wall is therefore 574#/ft. × 150 ft./2 = 574 × 75 = 43,050#.

The diaphragm shear is equal to this reaction divided by the diaphragm width = 43,050#/60 ft. = *718#/ft.*

Recall that the 718#/ft force is a strength level force. In order to determine an appropriate diaphragm from IBC Table 2306.2.1 (see Figure 12.30), this diaphragm shear force must be converted to an allowable force using the previously presented allowable load combination:

Allowable shear force = 0.7(strength level shear force) = 0.7 × 718 = 503#/ft.

We select a plywood diaphragm having an allowable shear of at least 503#/ft. from IBC Table 2306.2.1. Several diaphragms meet this requirement, and we select the following:

15/32" plywood, Structural I, with 10d nails at 2.5" o.c. at boundaries and continuous panel edges parallel to load, and at 4" o.c. at other panel edges, with blocking at all unsupported panel edges.

This diaphragm has an allowable shear of 640#/ft., which is greater than the 503#/ft. required.

The chord force is the tension or compression in the flanges, or chords, of the diaphragm. Since the diaphragm acts like a uniformly loaded simple beam, its maximum moment = $wL^2/8$.

The chord force is the maximum diaphragm moment divided by the diaphragm depth

$$\frac{574\#/\text{ft.} \times (150\text{ ft.})^2}{8 \times 60\text{ ft.}} = 26{,}906\#$$

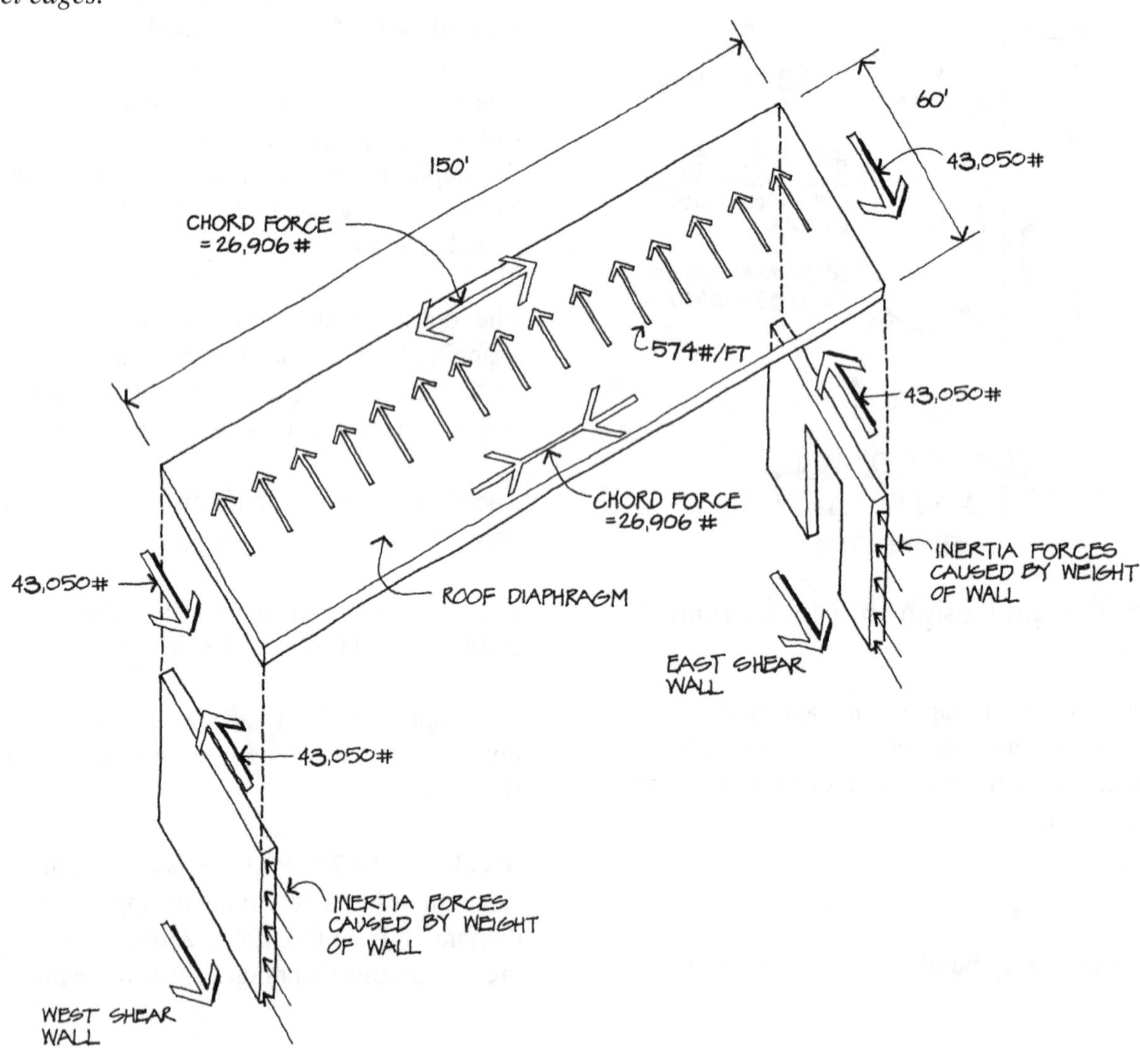

Figure 12.57

This chord force can be resisted by a continuous wood or steel member directly attached to the diaphragm, or by reinforcing steel in the north and south walls adjacent to the roof diaphragm.

Connection of Roof Diaphragm to East and West Shear Walls

The roof diaphragm is connected to the concrete walls with a wood ledger and anchor bolts, as shown in the details in Figure 12.31. What should be the maximum spacing of 3/4" anchor bolts to transfer the diaphragm shear into the walls? Neglect any vertical loads to be transferred by the anchor bolts, and assume an allowable load of 1,400# per bolt in the wood ledger and 2,940# per bolt in the concrete wall.

Solution:

The diaphragm shear travels through the plywood diaphragm and is transferred from the plywood to the ledger by the boundary nailing, and from the ledger to the concrete shear wall by the anchor bolts. The anchor bolts must be able to transfer the diaphragm shear of 503#/ft. Since the allowable bolt shear in the concrete wall is greater than its allowable shear in the wood ledger, the wood ledger value must be used in the design.

The required anchor bolt spacing is equal to

$$\frac{1{,}400\text{\#/bolt}}{503\text{\#/ft.}} = 2.78 \text{ ft./bolt} = 33.4 \text{ inches/bolt}$$

Use 3/4" anchor bolts at 2 feet 6 inches on center.

East and West Shear Walls

The east and west shear walls must resist the shear from the diaphragm, which is 43,050#, plus the inertia forces caused by the weight of the walls themselves. The walls must be adequate to resist this shear. Note that the unit shear stress is greater in the east wall because the 20-foot opening reduces the net area of the wall available to resist the shear stress.

The walls must also resist the overturning moment caused by the shear. If the resisting moment from the dead load of the wall and roof is less than the overturning moment, the wall must be tied to the footing to provide additional overturning resistance.

Because of the building's configuration, earthquake forces in the north-south direction cause greater stresses in the diaphragm and shear walls than east-west forces. Therefore, in this case, design for the east-west direction is not necessary.

SUMMARY

The way a building responds to an earthquake depends on many factors, including the nature of the earthquake itself, the soil type, the distance of the structure from the earthquake's epicenter, and the form, structural system, mass, structural materials, and construction quality of the building. Like a chain, a building is only as strong as its weakest element, and you can be sure that an earthquake will always discover what that is (see Figure 12.58).

It is therefore important for architects to exercise sound judgment when designing structures in earthquake-prone areas. In this regard, the following is a summary of recommended seismic design practices, some of which have been covered in detail in this lesson.

1. Buildings should be as regular and symmetrical as possible, to avoid torsion and/or stress concentrations.

Diagonal cracks in concrete piers between windows, Olive View Hospital, San Fernando earthquake, 1971 (N.O.A.A. photograph)

Figure 12.58

2. Drift limitations should be strictly adhered to, to prevent excessive movement. Adjacent buildings or portions of buildings should have adequate seismic separations to avoid pounding.
3. There should be a direct path for force transfer. Lateral load resisting systems should be continuous to the foundation.
4. Overturning effects should be minimized by keeping the building's height-to-width ratio low and by locating heavy elements as low in the building as possible.
5. All structural elements should be interconnected so that they move as a unit and thereby reduce the possibility of local failure.
6. All nonstructural elements should have adequate lateral bracing.
7. Redundancy should be provided by secondary systems that can resist part of the lateral force if the primary system fails or is damaged.
8. Infilling of a moment-resisting frame with a nonstructural wall tends to stiffen the frame and thereby attract more lateral load. The design and detailing of the frame and wall should therefore provide for this. Alternatively, the infilling wall should be structurally separated from the frame so that it can move independently, and yet be held in place.
9. Stairs should be designed and detailed to resist the lateral forces they may attract, or else detached from the building structure.
10. Openings in shear walls that may result in stress concentrations should be avoided, or if unavoidable, they should be adequately reinforced.

The subject of earthquake design is often mystifying or intimidating to architects, especially those from areas that seldom or never experience earthquakes. In this lesson, we hope we have made earthquake design clearer and more comprehensible, so that architectural candidates can approach the exam with confidence.

LESSON 12 QUIZ

1. The allowable shear in a blocked plywood diaphragm depends on all of the following EXCEPT

   A. plywood thickness.

   B. width of framing members.

   C. nail spacing.

   D. direction of framing members.

2. In a tall building, which of the following systems is usually MOST effective in resisting earthquakes?

   A. Shear walls at the perimeter

   B. Moment-resisting frames at the perimeter

   C. Braced frames at the perimeter

   D. Moment-resisting frames set back from the perimeter and rigidly anchored to the perimeter elements

3. Select the INCORRECT statement.

   A. A shear wall resists lateral forces in its own plane.

   B. A concrete diaphragm distributes horizontal forces in proportion to the rigidities of the vertical resisting elements.

   C. A plywood diaphragm distributes horizontal forces to the vertical resisting elements in proportion to the tributary area of each element.

   D. Torsion is the rotation that occurs in a diaphragm when the center of mass does not coincide with the centroid of the building.

4. Vertical ground motions are usually neglected in earthquake design because

   I. an earthquake causes only horizontal ground shaking.

   II. the vertical motions are generally smaller than the horizontal motions.

   III. most buildings have considerable excess strength in the vertical direction.

   IV. design live loads already include an earthquake factor.

   A. I only

   B. II, III, and IV

   C. II and IV

   D. II and III

5. A building assigned to SDC D has a steel frame that supports all the vertical load and concrete shear walls designed to resist all the lateral earthquake load. What is the R value of the building?

   A. 1½

   B. 2

   C. 5

   D. 6

6. A building in which core walls provide the only significant lateral force resisting elements may be inadequate to resist

   A. bending.

   B. torsion.

   C. vertical acceleration.

   D. diaphragm stresses.

7. The fundamental period of vibration of a structure

A. is unrelated to the height of the structure.

B. decreases as the height of the structure increases.

C. increases as the height of the structure increases.

D. may either increase or decrease as the height of the structure increases.

8. Special moment-resisting frames may be constructed of which of the following materials? Check all that apply.

A. Cast-in-place concrete

B. Precast concrete

C. Structural steel

D. Reinforced masonry

9. Buildings designed in accordance with the *International Building Code* earthquake provisions are believed to be

A. capable of resisting major earthquakes without damage.

B. capable of resisting shaking from major earthquakes without collapse.

C. capable of resisting ground movement, sliding, subsidence, or faulting due to an earthquake.

D. capable of resisting shaking from major earthquakes without collapse, if the lateral load resisting system is constructed of structural steel.

10. The Richter scale is a measure of an earthquake's

A. magnitude.

B. intensity.

C. acceleration.

D. force.

LESSON THIRTEEN

13

WIND DESIGN

NATURE OF WIND FORCES

Although windstorms are not as unpredictable as earthquakes, they are equally important in the design of buildings. In the United States, the loss of life and damage to property caused by windstorms exceeds that from earthquakes because of their more frequent occurrence and the large geographic areas affected. Also, winds are an important design factor to a greater or lesser degree in all parts of the country, while earthquakes are predominant only in specific geographic areas.

Wind is the movement of the earth's atmosphere caused by the unequal heating of the earth's surface and the rotation of the earth. These atmospheric movements are modified by local topographical conditions, often causing violent recurrent windstorms, such as the *chinooks* of the eastern slopes of the Rockies and the *Santa Anas* in Southern California.

Severe tropical storms are called *hurricanes* and occur mainly along the Gulf and South Atlantic coasts, although they have extended as far north as New England. Hurricane wind speeds usually vary between 30 and 120 miles per hour, with gusts up to 180 mph. Winds of 50 to 60 mph can strip small branches and leaves off trees. At 70 to 80 mph, shallow rooted trees and thin walls can be blown down, an entire roof sometimes lifted off, and large glass windows blown in. Above 80 mph, lifting of roofs and snapping of trees are common.

Tornadoes are the most violent of atmospheric storms and consist of rotating air masses of small diameter. They are much smaller than hurricanes but have greater wind speeds. The rotational speed of a tornado is estimated to be nearly 250 miles per hour, and may occasionally exceed 500 mph. The paths of tornadoes average several hundred yards in width and 16 miles in length, although large variations from these averages are not uncommon.

MEASUREMENT OF WINDS

Over the years, different methods have been used to measure wind speed. Wind speed has traditionally been measured by a device called a revolving cup anemometer. More recently, inertial navigation and global positioning systems have been employed to determine wind speed.

Until about 10 years ago, the *fastest-mile wind speed* was the standard wind measure in the United States. This speed was determined by the time it took a column of air one mile long to pass the anemometer (a mile was measured by a specific number of anemometer revolutions). The National Weather Service phased out the fastest-mile wind speed and replaced it with a *3-second gust wind speed.* The peak gust is associated with an averaging time of about 3 seconds. Within the United States, wind data are currently measured at approximately 750 stations (mostly airports and other open areas). The National Weather Service operates about 225 of these stations.

The wind speed also varies with the height above the ground, since the friction between the wind and the ground surface reduces the velocity of the wind. To make the wind station measurements comparable, all wind speed data have been standardized to a height of 10 meters, or 33 feet.

Basic wind speeds are shown in Figures 13.1 through 13.4 from IBC Figure 1609. When using the map, pay great attention to local wind records and conditions, which may indicate basic wind speeds higher than those on the map. Areas shown on the map as *special wind regions*, which include areas that are mostly mountainous terrain and near gorges, have experienced basic wind speeds so different from those in surrounding geographic areas that they are specifically excluded from the map. In such areas, wind speeds must be determined from local records.

The basic wind speeds for most of the United States have a chance of occurring once every 50 years. The lowest speed on the map is 85 miles per hour, which occurs on the Pacific Coast.

Hurricanes create high wind speeds along the Gulf and Atlantic coasts, as indicated on the map. Note that the wind speed map does *not* account for tornadoes.

RESPONSE OF BUILDINGS TO WIND LOADS

Buildings respond to wind forces in a variety of complex and dynamic ways. The most obvious of these is the direct pressure of the wind as it is stopped by the windward side of the building, that is, the side facing the direction from which the wind is blowing.

The wind creates a negative pressure, or *suction*, on the leeward side of the building, which is the side opposite the windward side, and the side walls parallel to the wind direction. Uplift pressure is created on horizontal or sloping surfaces, such as the roof.

The building structure also vibrates because of the gusting effects of the wind, similar to the way a building responds to earthquake motion. These vibrations are important not only because of their effect on the building structure, but also because such motions in taller, more flexible, buildings can be uncomfortable to the occupants.

As the wind passes over the building, friction causes drag forces to act in the direction of the wind.

The corners, edges, and eave overhangs of the building are subjected to especially complex forces as the wind passes these obstructions, causing much higher local suction forces than those on the building as a whole. In fact, most wind damage to buildings occurs in these areas.

These different responses of a building to wind loads are shown in Figure 13.5. Of course, any combination of these responses, or all of them, can occur at the same time, thus making the overall building response very complex.

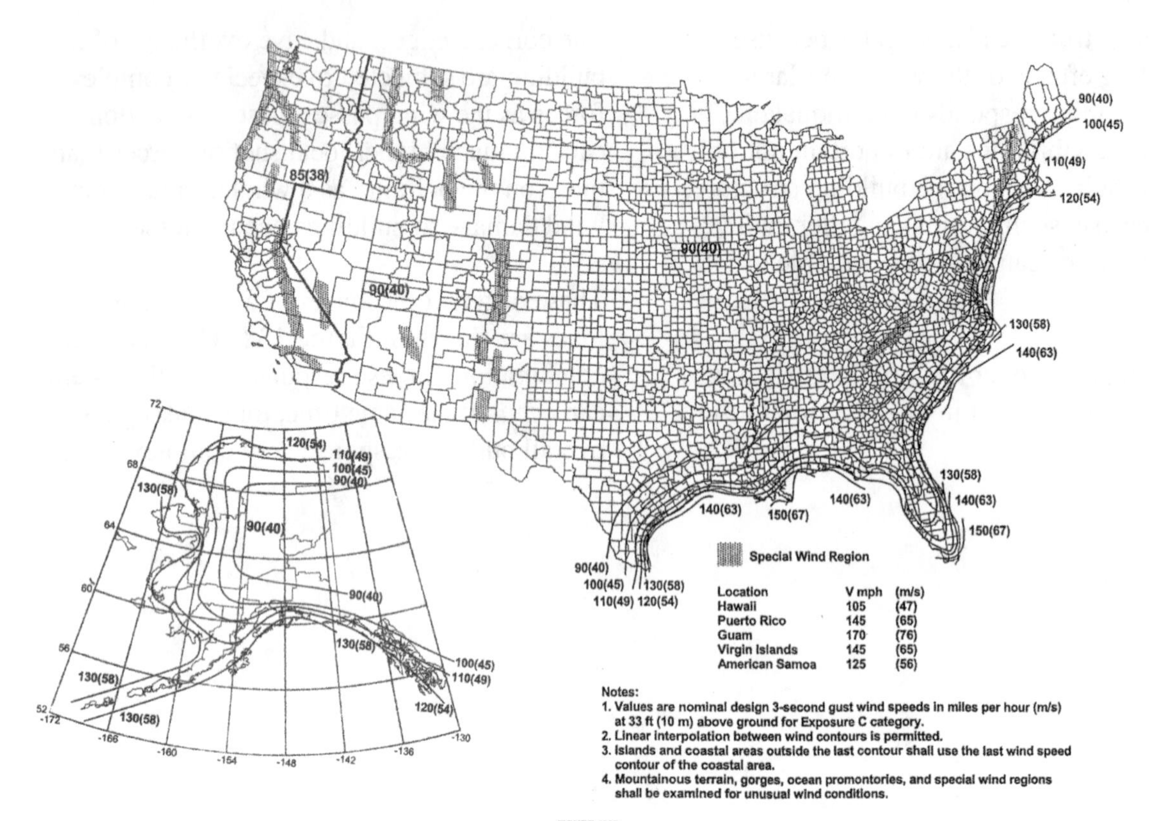

FIGURE 1609
BASIC WIND SPEED (3-SECOND GUST)

Figure 13.1

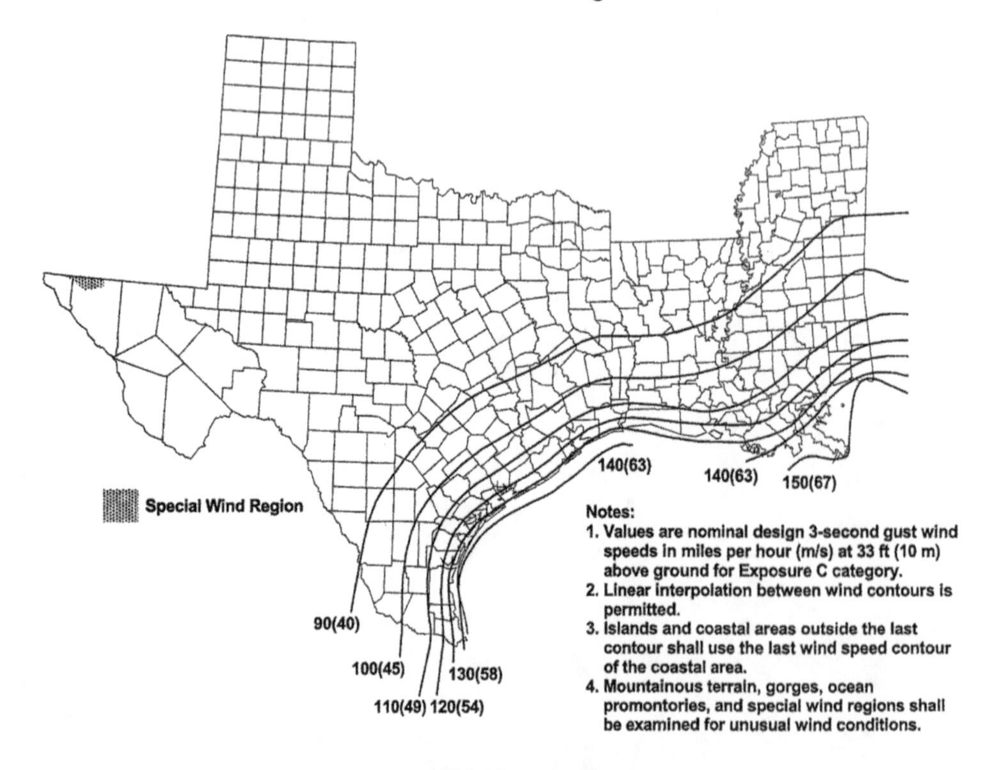

FIGURE 1609–continued
BASIC WIND SPEED (3-SECOND GUST)
WESTERN GULF OF MEXICO HURRICANE COASTLINE

Figure 13.2

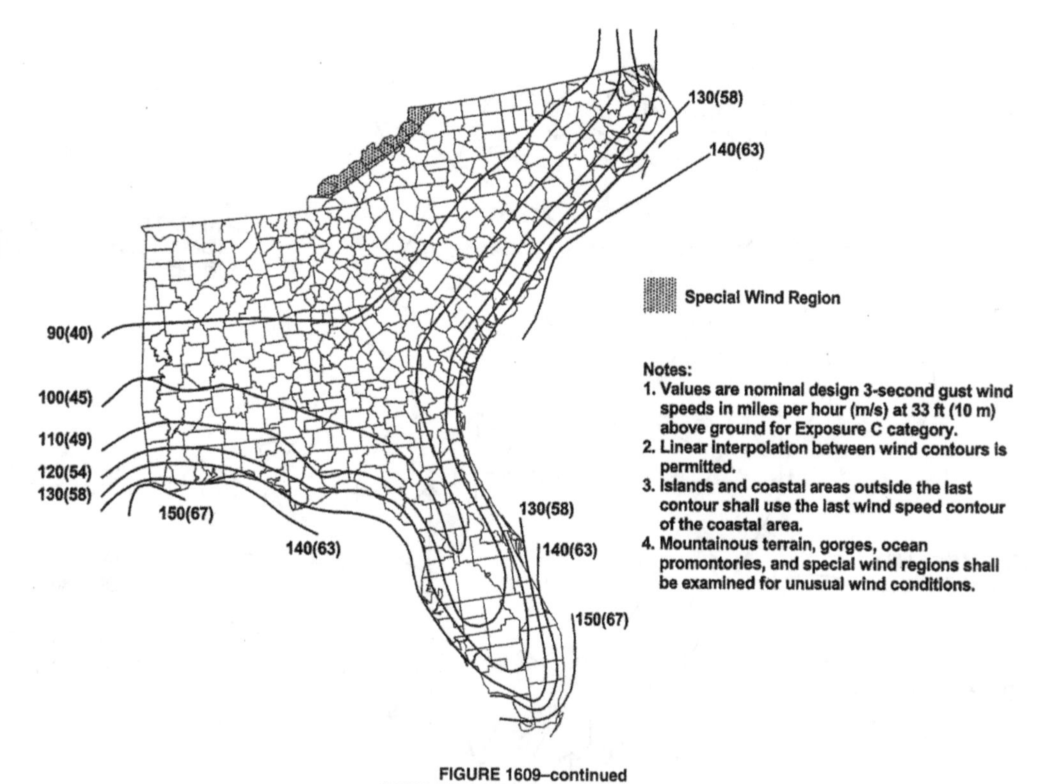
130(58)
140(63)
Special Wind Region
90(40)
100(45)
110(49)
120(54)
130(58)
150(67)
140(63)
130(58)
140(63)
150(67)
Notes:
1. Values are nominal design 3-second gust wind speeds in miles per hour (m/s) at 33 ft (10 m) above ground for Exposure C category.
2. Linear interpolation between wind contours is permitted.
3. Islands and coastal areas outside the last contour shall use the last wind speed contour of the coastal area.
4. Mountainous terrain, gorges, ocean promontories, and special wind regions shall be examined for unusual wind conditions.

FIGURE 1609–continued
BASIC WIND SPEED (3-SECOND GUST)
EASTERN GULF OF MEXICO AND SOUTHEASTERN U.S. HURRICANE COASTLINE

Figure 13.3

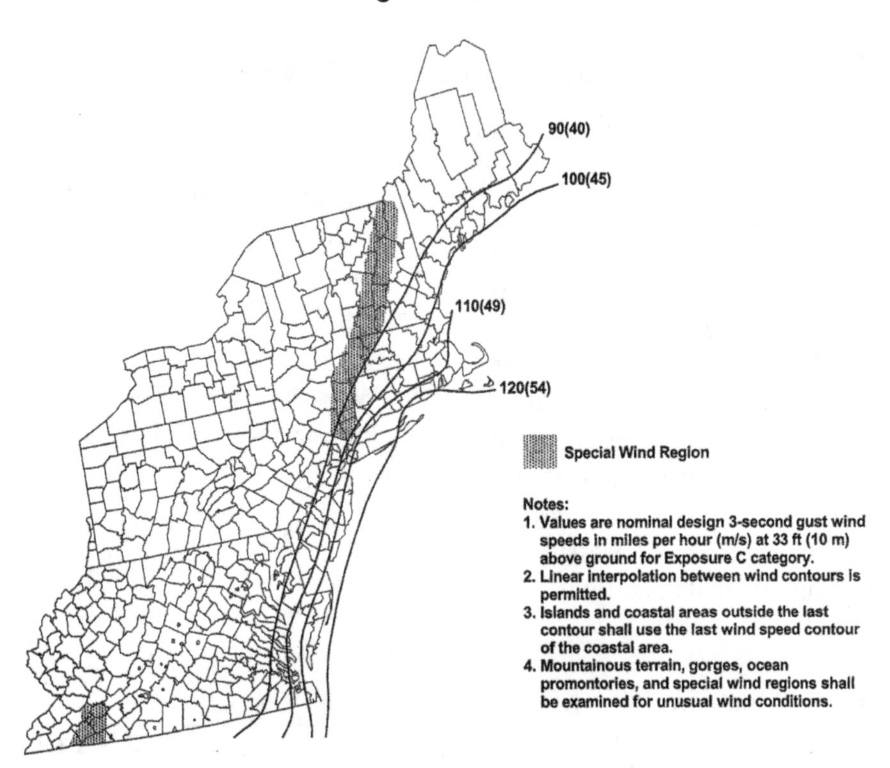
90(40)
100(45)
110(49)
120(54)
Special Wind Region
Notes:
1. Values are nominal design 3-second gust wind speeds in miles per hour (m/s) at 33 ft (10 m) above ground for Exposure C category.
2. Linear interpolation between wind contours is permitted.
3. Islands and coastal areas outside the last contour shall use the last wind speed contour of the coastal area.
4. Mountainous terrain, gorges, ocean promontories, and special wind regions shall be examined for unusual wind conditions.

FIGURE 1609–continued
BASIC WIND SPEED (3-SECOND GUST)
MID AND NORTHERN ATLANTIC HURRICANE COASTLINE

Figure 13.4

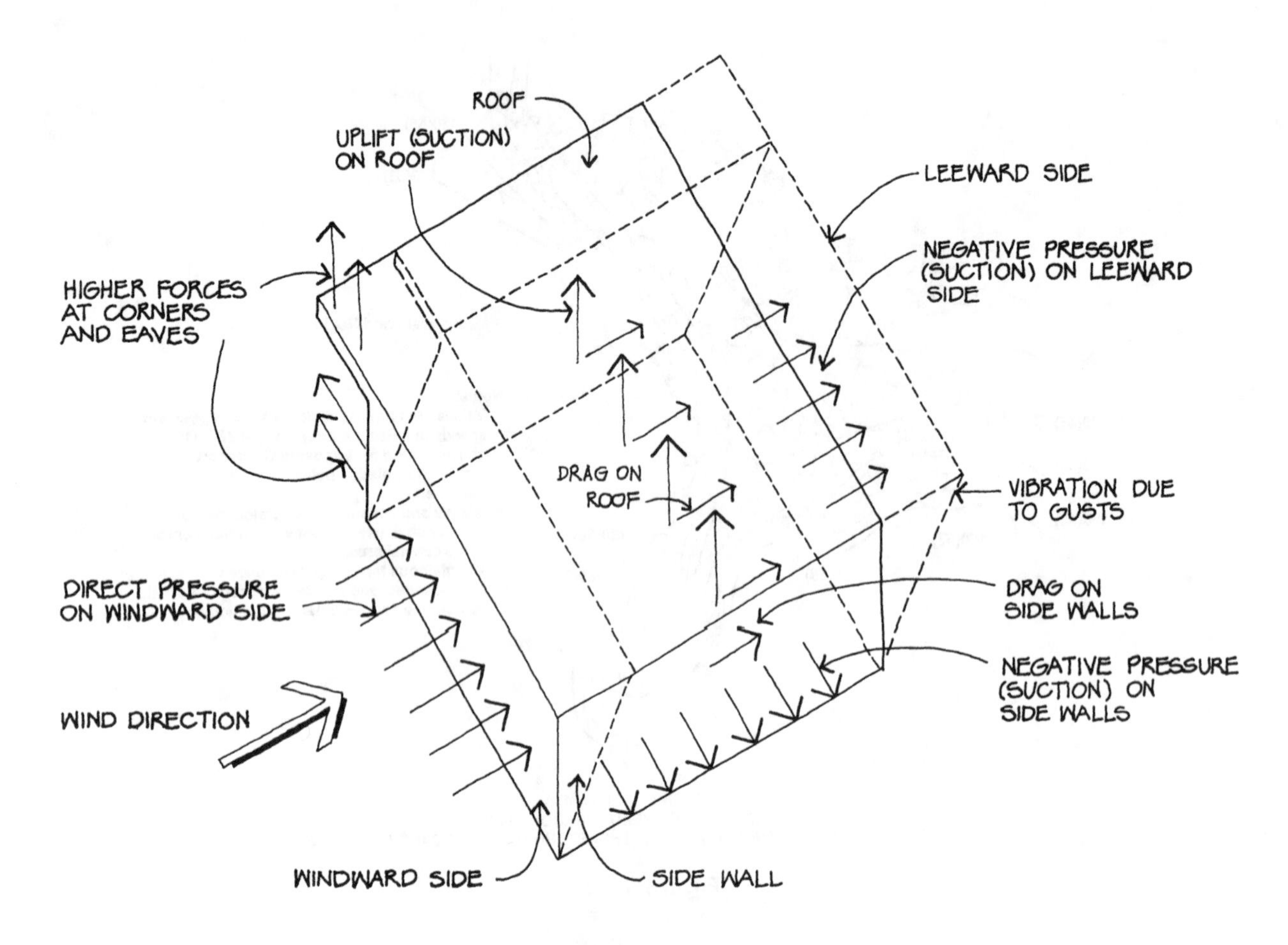

Figure 13.5

The direct wind pressure, also called the *stagnation pressure*, in pounds per square foot, on a vertical surface is related to the 3-second gust velocity in miles per hour, by the formula $p = 0.00256\ V^2$. Thus, if the 3-second gust velocity is 90 miles per hour, the wind pressure is $p = 0.00256 \times (90)^2 = 20.7$ psf. Since the pressure varies as the square of the velocity, *doubling the wind velocity increases the wind pressure four-fold*. The wind velocities and pressures for some common climatological conditions are shown in Table 13.1.

There are other factors that affect the wind pressure on a building. As previously discussed, the friction between the wind and the ground causes the wind velocity, and thus the pressure, to decrease close to the ground, and conversely, to increase as the height increases (Figure 13.6).

Wind Pressure lbs per sq. ft.	Wind Velocity miles per hour	Description
160	250	Tornado
58	150	Strong Hurricane
21	90	Basic U.S. wind speed
9	85	Basic West Coast wind speed

Table 13.1

Another important factor affecting wind velocity and pressure is the terrain surrounding the building. If it is built-up with buildings or other features, the resulting turbulence reduces the wind velocity and pressure. On the other hand, buildings on open sites, such as those on ridges or facing large bodies of water, must bear the full brunt of the wind and thus have greater pressures.

Also affecting the wind pressure are the size and shape of the building. Narrow buildings allow the wind to flow around them and therefore have lower pressures, while wide, tall buildings block more of the wind and thus have greater pressures.

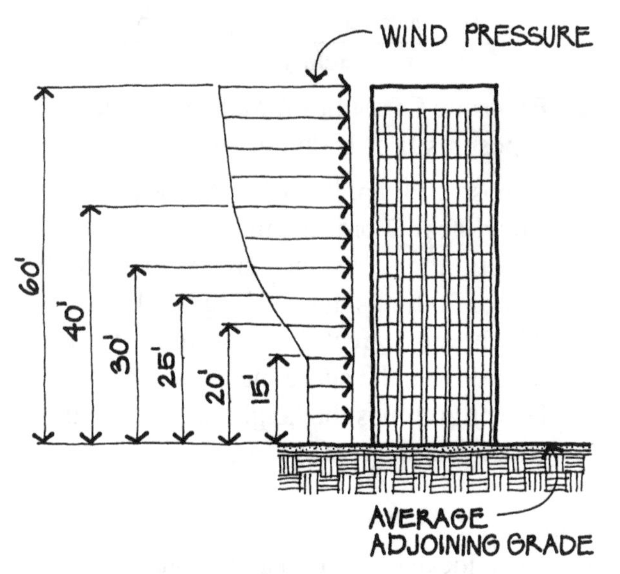

Figure 13.6

CODE REQUIREMENTS

As discussed in the previous lesson, the National Council of Architectural Registration Boards (NCARB) no longer references the *Uniform Building Code* in its exam guidelines. As in the previous lesson, the discussion that follows relies on the IBC and, in turn, ASCE 7.

The wind requirements discussed here apply only to regularly shaped buildings lacking characteristics that would cause them to respond to wind in any direction other than the direction of the wind itself. This book does not address buildings that are very tall and slender or that have unusual or irregular geometric shape. Such structures may require wind tunnel testing to determine applicable design loads.

The effects of tornadoes are also not considered in these code requirements and are beyond the scope of this course.

As discussed earlier, the actual wind forces are dynamic, constantly varying in magnitude and direction. The code, however, uses a static analysis method, similar to that used in earthquake design, in which static or non-varying external forces are applied to the building structure that simulate the actual varying wind forces.

Methods of Analysis

In general, buildings must be designed to resist the forces caused by wind blowing in any direction. ASCE 7 provides three methods to determine wind pressure.

Method 1: *Simplified Procedure.* Wind pressures can be selected from tables in the code with minimal calculation when the building meets all the requirements specified in the code. This procedure is used primarily to design regularly shaped low-rise buildings.

Method 2: *Analytical Procedure.* Wind pressures are determined using formulas, tables, and figures provided in the code. This method is used to design many types of structures.

Method 3: *Wind Tunnel Procedure*. This procedure provides general guidelines for conducting a wind tunnel test. Such tests may be performed on any building to determine wind pressures. Typically, wind tunnel tests are performed only for irregular or very tall, slender buildings—primarily due to costs.

ASCE 7 prescribes a minimum design wind pressure of 10 pounds per square foot, regardless of the method used to determine wind pressures. This minimum pressure is projected on a vertical plane of the building perpendicular to the direction of wind. This load case must be considered separately from the load cases set forth in the three methods described above.

Detailed descriptions of Methods 1 and 2 are given below, but *site exposure* and *importance factors,* common to both methods, must be discussed first.

Site Exposure

In order to determine site exposure, we must first know the surface roughness present at the site (since surface roughness impacts wind velocity). Table 13.2 summarizes the three surface roughness categories.

Once a roughness category has been established, each wind direction is assigned an exposure category based on the following definitions:

- Exposure B: applies when Surface Roughness B prevails in the upwind direction for a distance of at least 2,600 feet or 20 times the height of the building, whichever is greater. Note that the upwind distance may be reduced to 1,500 feet for buildings with mean roof heights 30 feet or less.

Surface Roughness Category	Description
B	Urban and suburban areas, wooded areas, or other terrain having numerous closely spaced obstructions the size of single-family dwellings or larger.
C	Open terrain with scattered obstructions generally shorter than 30 feet. This category includes flat open country, grasslands, and all water surfaces in hurricane-prone regions.
D	Flat, unobstructed areas and water surfaces outside hurricane-prone regions. This category includes smooth mud flats, salt flats, and unbroken ice.

Table 13.2

- Exposure C: applies for all cases where exposures B or D do not apply.
- Exposure D: applies when Surface Roughness D prevails in the upwind direction for a distance greater than 5,000 feet or 20 times the building height, whichever is greater. Exposure D extends inland from the shoreline a distance of 600 feet or 20 times the height of the building, whichever is greater.

Based on these definitions, it is possible for a site to be located in a transition zone between exposure categories. In such cases, the category resulting in the largest wind pressures is to be used, unless a more rational analysis is performed.

Exposure categories may be different in different wind directions depending on the terrain surrounding a building. However, a single exposure category is usually specified, based on the exposure that results in the highest wind pressures.

Importance Factor, I

The importance factor here is similar to that used in earthquake design. Referring to ASCE 7 Table 6-1 in Figure 13.7, you can see that I varies between 0.77 and 1.15, depending on occupancy category and on whether the building is located in a hurricane-prone region. Hurricane-prone regions in the United States and its territories are defined by the code as (1) Atlantic and Gulf coasts where the basic wind speed is greater than 90 miles per hour and (2) Hawaii, Puerto Rico, Guam, the Virgin Islands, and American Samoa.

Essential facilities such as hospitals and fire and police stations are designed for wind forces 15 percent greater than normal (I = 1.15). Such emergency facilities are expected to be safe and usable following a severe windstorm.

Category	Non-Hurricane Prone Regions and Hurricane Prone Regions with V = 85-100 mph and Alaska	Hurricane Prone Regions with V > 100 mph
I	0.87	0.77
II	1.00	1.00
III	1.15	1.15
IV	1.15	1.15

Figure 13.7

Using an I factor of 1.15 for certain occupancy categories in non-hurricane regions effectively designs such buildings for a windstorm with a mean recurrence interval of 100 years—for a 100-year windstorm. Importance factor values of 0.87 and 1.0 correspond to mean recurrence intervals of 25 and 50 years, respectively. In hurricane-prone regions, the mean recurrence interval will vary along the coast as I varies between 0.77 and 1.15. The risk levels associated with these importance factors are consistent with those in non-hurricane regions.

Method 2—Analytical Procedure

Perhaps counterintuitively, Method 2 forms the basis of Method 1. As such we will discuss Method 2 first. Some of the tables and figures referenced in this section are several pages long: for ease of reading, they are grouped at the end of this section, beginning on page 304.

The Analytical Procedure is used to determine wind pressures or forces for many types of buildings. Wind pressures are calculated on windward, leeward, side, and roof surfaces using formulas, tables, and figures given in ASCE 7.

Figure 6-6 in ASCE 7, shown in Figure 13.8, illustrates concepts fundamental to the Analytical Procedure. On the windward face, wind pressure acts *toward* the surface of the building (pressure) and varies with respect to the mean roof height. On the leeward face, pressure acts *away* from the surface (suction) and is assumed to be constant with respect to height. Constant pressure is also assumed to act on the side walls and roof. Suction occurs on the side walls and leeward side of the roof, while either pressure or suction occurs on the windward side of the roof, depending on its slope. We will now use the Analytical Procedure to determine wind pressure on a building's surfaces. We will focus on enclosed, rigid buildings—buildings that have a fundamental period of one second or less. Flexible buildings are beyond the scope of this lesson.

- Windward face

  The following formula gives wind pressure on the windward face of a rigid building (which varies with height):

  $$p_w = q_z GC_p - q_i(GC_{pi})$$

  where

  q_z = velocity pressure evaluated at height z above the ground
  = $0.00256 K_z K_{zt} K_d V^2 I$

  K_z = velocity pressure exposure coefficient (ASCE 7 Table 6-3, shown in Figure 13.9)

  K_{zt} = topographic factor (ASCE 7 Figure 6-4, shown in Figure 13.10)

  K_d = wind directionality factor
  = 0.85 for buildings (ASCE 7 Table 6-4, shown in Figure 13.11)

  V = basic wind speed (ASCE 7 Figure 6-1 or IBC Figure 1609, shown in Figure 13.1)

  I = importance factor (ASCE 7 Table 6-1, shown in Figure 13.7)

  G = gust effect factor
  = 0.85 for rigid structures

  C_p = external pressure coefficients (ASCE Figure 6-6, shown in Figure 13.8)

  q_i = positive internal pressure, which may be conservatively evaluated at height h ($q_i = q_h$)

  GC_{pi} = product of the internal pressure coefficient and gust-effect factor (ASCE 7 Table 6-1)

Obviously, many quantities contribute to the total windward pressure on a building; a brief summary of each follows. Figure 13.8 shows how wind pressure varies with height on a windward face.

Let us begin by discussing the velocity pressure q_z. As previously shown, basic wind pressure is equal to $0.00256V^2$. This basic pressure is multiplied by the velocity pressure exposure coefficient K_z, which accounts for the height of the building and the exposure, or roughness, of the terrain. The topographic factor K_{zt} accounts for the increase in wind speed at the tops of hills and escarpments. For relatively flat terrain, $K_{zt} = 1.0$. The wind directionality factor K_d accounts for the reduced probability of the maximum wind blowing in a direction that will produce the maximum pressure on a building. Finally, I accounts for relative importance.

The velocity pressure is multiplied by G and C_p. The gust effect factor G accounts for the loading effects due to wind gusts in the direction of the wind. ASCE 7 permits G to be taken as 0.85, or it may be calculated from a very complex set of formulas. It is much simpler to assume G = 0.85. The external pressure coefficients C_p in ASCE 7 Figure 6-6 reflect the type of loading (pressure or suction) on each surface of a building as a function of wind direction. In essence, these coefficients represent the ratios of the average pressure on a surface divided by the pressure at the stagnation point.

The pressure $q_z GC_p$ is the total external pressure acting on the windward face. Similarly, the pressure $q_i(GC_{pi})$ is the total pressure inside a building. This pressure can act either toward or away from the external walls and roof. Internal pressures play a key role when designing individual walls or the roof. However, since they cancel out in the horizontal direction, they do not play a role in the design of the main wind-force resisting system.

- Leeward face, side walls, and roof

  The wind pressure on the leeward face, side walls, and roof of a rigid building do not vary with height. These pressures are calculated as follows:

  $p_l = q_h GC_p - q_i(GC_{pi})$

  where

  q_h = velocity pressure evaluated at the mean roof height h above the ground
  $= 0.00256 K_h K_{zt} K_d V^2 I$

  K_h = velocity pressure exposure coefficient evaluated at the mean roof height h above the ground

  All other quantities are defined above for the windward face formula.

The total horizontal wind pressure at any point z along the height of the building is the summation of p_z and p_l. Wind forces due to these pressures are typically determined at each floor level. Such forces are calculated by multiplying the total wind pressure by the tributary area at that level, which is perpendicular to the direction of wind.

The main wind-force resisting system of any building designed in accordance with the Analytical Procedure must be designed for the load cases defined in ASCE 7 Figure 6-9, shown in Figure 13.12. Note that torsional effects and wind acting in two orthogonal directions at the same time must be considered.

The code permits the use of the following formula to determine wind pressure on low-rise buildings (mean roof height 60 feet or less, and mean roof height less than or equal to the least horizontal dimension) instead of the above formulas:

$p = q_h[(GC_{pf}) - (GC_{pi})]$

where

q_h = velocity pressure evaluated at the mean roof height h above the ground
$= 0.00256 K_h K_{zt} K_d V^2 I$

GC_{pf} = external pressure coefficient (ASCE 7 Figure 6-10, shown in Figure 13.13)

Pressure coefficients are provided for different surfaces of the building. In general, eight different load cases must be considered; however, for buildings with symmetric plans, the number of load cases can be reduced significantly.

This formula forms the basis of the Simplified Procedure, which is covered in the next lesson.

The code also provides a formula to determine the design wind pressure on parapets:

$p_p = q_p GC_{pn}$

where

q_p = velocity pressure evaluated at the top of the parapet

GC_{pn} = combined net pressure coefficient
= +1.5 for windward parapet
= −1.0 for leeward parapet

The pressure p_p is the net pressure on the parapet due to the pressures on the front and back surfaces of the parapet. This is illustrated in ASCE 7 Commentary Figure C6-12.

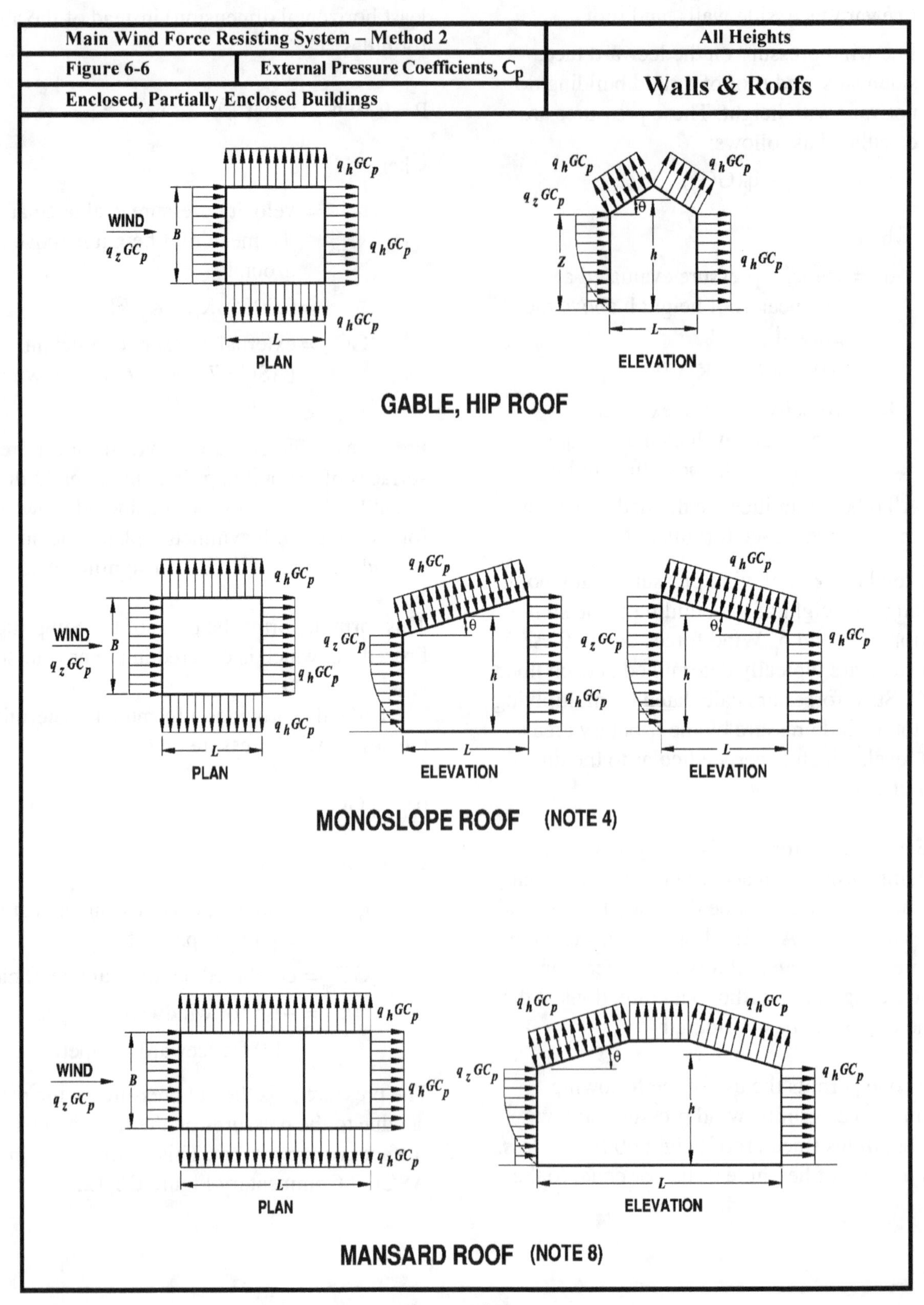
Main Wind Force Resisting System – Method 2
All Heights
Figure 6-6
External Pressure Coefficients, Cp
Walls & Roofs
Enclosed, Partially Enclosed Buildings
WIND
qzGCp
qhGCp
B
L
h
Z
θ
PLAN
ELEVATION
GABLE, HIP ROOF
MONOSLOPE ROOF (NOTE 4)
MANSARD ROOF (NOTE 8)

Figure 13.8

Main Wind Force Resisting System – Method 2		All Heights
Figure 6-6 (con't)	External Pressure Coefficients, C_p	Walls & Roofs
Enclosed, Partially Enclosed Buildings		

Wall Pressure Coefficients, C_p

Surface	L/B	C_p	Use With
Windward Wall	All values	0.8	q_z
Leeward Wall	0-1	-0.5	q_h
	2	-0.3	
	≥4	-0.2	
Side Wall	All values	-0.7	q_h

Roof Pressure Coefficients, C_p, for use with q_h

Wind Direction	Windward									Leeward		
	Angle, θ (degrees)									Angle, θ (degrees)		
	h/L	10	15	20	25	30	35	45	≥60#	10	15	≥20
Normal to ridge for θ ≥ 10°	≤0.25	-0.7 -0.18	-0.5 0.0*	-0.3 0.2	-0.2 0.3	-0.2 0.3	0.0* 0.4	0.4	0.01 θ	-0.3	-0.5	-0.6
	0.5	-0.9 -0.18	-0.7 -0.18	-0.4 0.0*	-0.3 0.2	-0.2 0.2	-0.2 0.3	0.0* 0.4	0.01 θ	-0.5	-0.5	-0.6
	≥1.0	-1.3** -0.18	-1.0 -0.18	-0.7 -0.18	-0.5 0.0*	-0.3 0.2	-0.2 0.2	0.0* 0.3	0.01 θ	-0.7	-0.6	-0.6

Wind Direction	h/L	Horiz distance from windward edge	C_p
Normal to ridge for θ < 10 and Parallel to ridge for all θ	≤ 0.5	0 to h/2	-0.9, -0.18
		h/2 to h	-0.9, -0.18
		h to 2 h	-0.5, -0.18
		> 2h	-0.3, -0.18
	≥ 1.0	0 to h/2	-1.3**, -0.18
		> h/2	-0.7, -0.18

*Value is provided for interpolation purposes.

**Value can be reduced linearly with area over which it is applicable as follows

Area (sq ft)	Reduction Factor
≤ 100 (9.3 sq m)	1.0
200 (23.2 sq m)	0.9
≥ 1000 (92.9 sq m)	0.8

Notes:

1. Plus and minus signs signify pressures acting toward and away from the surfaces, respectively.
2. Linear interpolation is permitted for values of *L/B, h/L* and θ other than shown. Interpolation shall only be carried out between values of the same sign. Where no value of the same sign is given, assume 0.0 for interpolation purposes.
3. Where two values of C_p are listed, this indicates that the windward roof slope is subjected to either positive or negative pressures and the roof structure shall be designed for both conditions. Interpolation for intermediate ratios of h/L in this case shall only be carried out between C_p values of like sign.
4. For monoslope roofs, entire roof surface is either a windward or leeward surface.
5. For flexible buildings use appropriate G_f as determined by Section 6.5.8.
6. Refer to Figure 6-7 for domes and Figure 6-8 for arched roofs.
7. Notation:
   B: Horizontal dimension of building, in feet (meter), measured normal to wind direction.
   L: Horizontal dimension of building, in feet (meter), measured parallel to wind direction.
   h: Mean roof height in feet (meters), except that eave height shall be used for θ ≤ 10 degrees.
   z: Height above ground, in feet (meters).
   G: Gust effect factor.
   q_z, q_h: Velocity pressure, in pounds per square foot (N/m^2), evaluated at respective height.
   θ: Angle of plane of roof from horizontal, in degrees.
8. For mansard roofs, the top horizontal surface and leeward inclined surface shall be treated as leeward surfaces from the table.
9. Except for MWFRS's at the roof consisting of moment resisting frames, the total horizontal shear shall not be less than that determined by neglecting wind forces on roof surfaces.

#For roof slopes greater than 80°, use C_p = 0.8

Figure 13.8 (continued)

Velocity Pressure Exposure Coefficients, K_h and K_z

Table 6-3

Height above ground level, z		Exposure (Note 1)			
		B		C	D
ft	(m)	Case 1	Case 2	Cases 1 & 2	Cases 1 & 2
0-15	(0-4.6)	0.70	0.57	0.85	1.03
20	(6.1)	0.70	0.62	0.90	1.08
25	(7.6)	0.70	0.66	0.94	1.12
30	(9.1)	0.70	0.70	0.98	1.16
40	(12.2)	0.76	0.76	1.04	1.22
50	(15.2)	0.81	0.81	1.09	1.27
60	(18)	0.85	0.85	1.13	1.31
70	(21.3)	0.89	0.89	1.17	1.34
80	(24.4)	0.93	0.93	1.21	1.38
90	(27.4)	0.96	0.96	1.24	1.40
100	(30.5)	0.99	0.99	1.26	1.43
120	(36.6)	1.04	1.04	1.31	1.48
140	(42.7)	1.09	1.09	1.36	1.52
160	(48.8)	1.13	1.13	1.39	1.55
180	(54.9)	1.17	1.17	1.43	1.58
200	(61.0)	1.20	1.20	1.46	1.61
250	(76.2)	1.28	1.28	1.53	1.68
300	(91.4)	1.35	1.35	1.59	1.73
350	(106.7)	1.41	1.41	1.64	1.78
400	(121.9)	1.47	1.47	1.69	1.82
450	(137.2)	1.52	1.52	1.73	1.86
500	(152.4)	1.56	1.56	1.77	1.89

Notes:

1. **Case 1:** a. All components and cladding.
   b. Main wind force resisting system in low-rise buildings designed using Figure 6-10.

   Case 2: a. All main wind force resisting systems in buildings except those in low-rise buildings designed using Figure 6-10.
   b. All main wind force resisting systems in other structures.

2. The velocity pressure exposure coefficient K_z may be determined from the following formula:

   For 15 ft. $\leq z \leq z_g$ — $K_z = 2.01\,(z/z_g)^{2/\alpha}$

   For $z < 15$ ft. — $K_z = 2.01\,(15/z_g)^{2/\alpha}$

   Note: z shall not be taken less than 30 feet for Case 1 in exposure B.

3. α and z_g are tabulated in Table 6-2.
4. Linear interpolation for intermediate values of height z is acceptable.
5. Exposure categories are defined in 6.5.6.

Figure 13.9

Topographic Factor, K_{zt} – Method 2

Figure 6-4

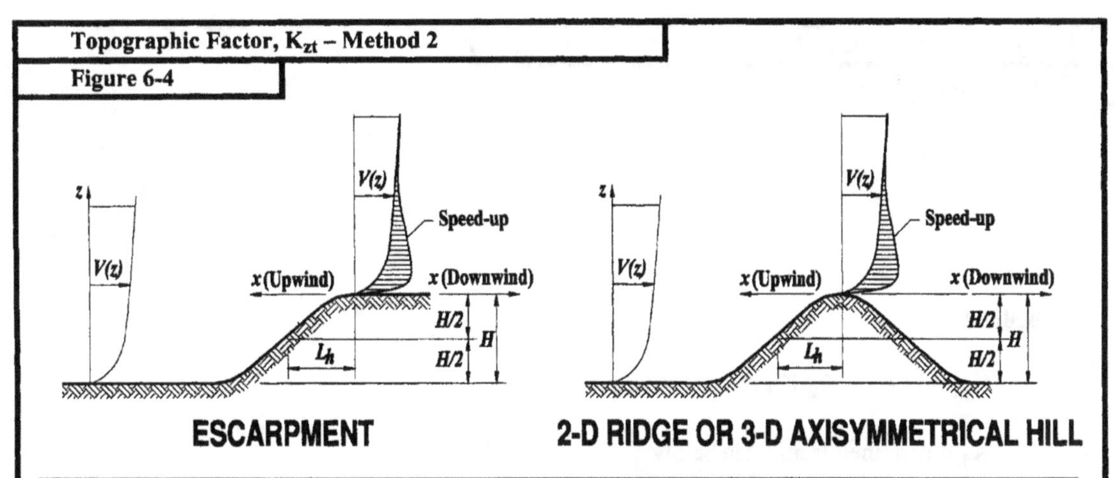

Topographic Multipliers for Exposure C

H/L_h	K_1 Multiplier 2-D Ridge	K_1 Multiplier 2-D Escarp.	K_1 Multiplier 3-D Axisym. Hill	x/L_h	K_2 Multiplier 2-D Escarp.	K_2 Multiplier All Other Cases	z/L_h	K_3 Multiplier 2-D Ridge	K_3 Multiplier 2-D Escarp.	K_3 Multiplier 3-D Axisym. Hill
0.20	0.29	0.17	0.21	0.00	1.00	1.00	0.00	1.00	1.00	1.00
0.25	0.36	0.21	0.26	0.50	0.88	0.67	0.10	0.74	0.78	0.67
0.30	0.43	0.26	0.32	1.00	0.75	0.33	0.20	0.55	0.61	0.45
0.35	0.51	0.30	0.37	1.50	0.63	0.00	0.30	0.41	0.47	0.30
0.40	0.58	0.34	0.42	2.00	0.50	0.00	0.40	0.30	0.37	0.20
0.45	0.65	0.38	0.47	2.50	0.38	0.00	0.50	0.22	0.29	0.14
0.50	0.72	0.43	0.53	3.00	0.25	0.00	0.60	0.17	0.22	0.09
				3.50	0.13	0.00	0.70	0.12	0.17	0.06
				4.00	0.00	0.00	0.80	0.09	0.14	0.04
							0.90	0.07	0.11	0.03
							1.00	0.05	0.08	0.02
							1.50	0.01	0.02	0.00
							2.00	0.00	0.00	0.00

Notes:

1. For values of H/L_h, x/L_h and z/L_h other than those shown, linear interpolation is permitted.
2. For $H/L_h > 0.5$, assume $H/L_h = 0.5$ for evaluating K_1 and substitute 2H for L_h for evaluating K_2 and K_3.
3. Multipliers are based on the assumption that wind approaches the hill or escarpment along the direction of maximum slope.
4. Notation:

H: Height of hill or escarpment relative to the upwind terrain, in feet (meters).

L_h: Distance upwind of crest to where the difference in ground elevation is half the height of hill or escarpment, in feet (meters).

K_1: Factor to account for shape of topographic feature and maximum speed-up effect.

K_2: Factor to account for reduction in speed-up with distance upwind or downwind of crest.

K_3: Factor to account for reduction in speed-up with height above local terrain.

x: Distance (upwind or downwind) from the crest to the building site, in feet (meters).

z: Height above local ground level, in feet (meters).

μ: Horizontal attenuation factor.

γ: Height attenuation factor.

Figure 13.10

Topographic Factor, K_{zt} – Method 2

Figure 6-4 (cont'd)

Equations:

$$K_{zt} = (1 + K_1 K_2 K_3)^2$$

K_1 determined from table below

$$K_2 = (1 - \frac{|x|}{\mu L_h})$$

$$K_3 = e^{-\gamma z / L_h}$$

Parameters for Speed-Up Over Hills and Escarpments						
Hill Shape	**$K_1/(H/L_h)$ Exposure**			**γ**	**μ**	
	B	**C**	**D**		**Upwind of Crest**	**Downwind of Crest**
2-dimensional ridges (or valleys with negative H in $K_1/(H/L_h)$)	1.30	1.45	1.55	3	1.5	1.5
2-dimensional escarpments	0.75	0.85	0.95	2.5	1.5	4
3-dimensional axisym. hill	0.95	1.05	1.15	4	1.5	1.5

Figure 13.10 (continued)

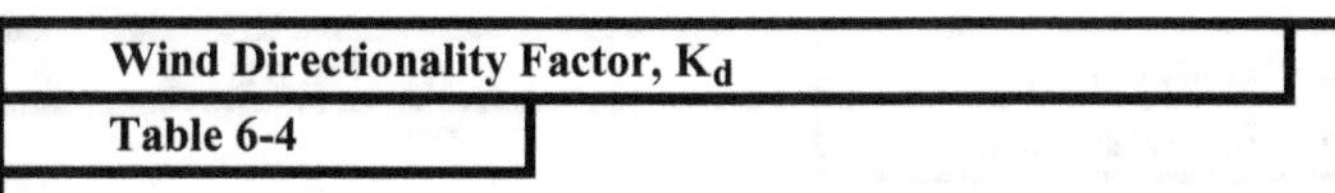
Wind Directionality Factor, K_d

Table 6-4

Structure Type	Directionality Factor K_d*
Buildings	
Main Wind Force Resisting System	0.85
Components and Cladding	0.85
Arched Roofs	0.85
Chimneys, Tanks, and Similar Structures	
Square	
Hexagonal	0.90
Round	0.95
	0.95
Solid Signs	0.85
Open Signs and Lattice Framework	0.85
Trussed Towers	
Triangular, square, rectangular	0.85
All other cross sections	0.95

*Directionality Factor K_d has been calibrated with combinations of loads specified in Section 2. This factor shall only be applied when used in conjunction with load combinations specified in 2.3 and 2.4.

Figure 13.11

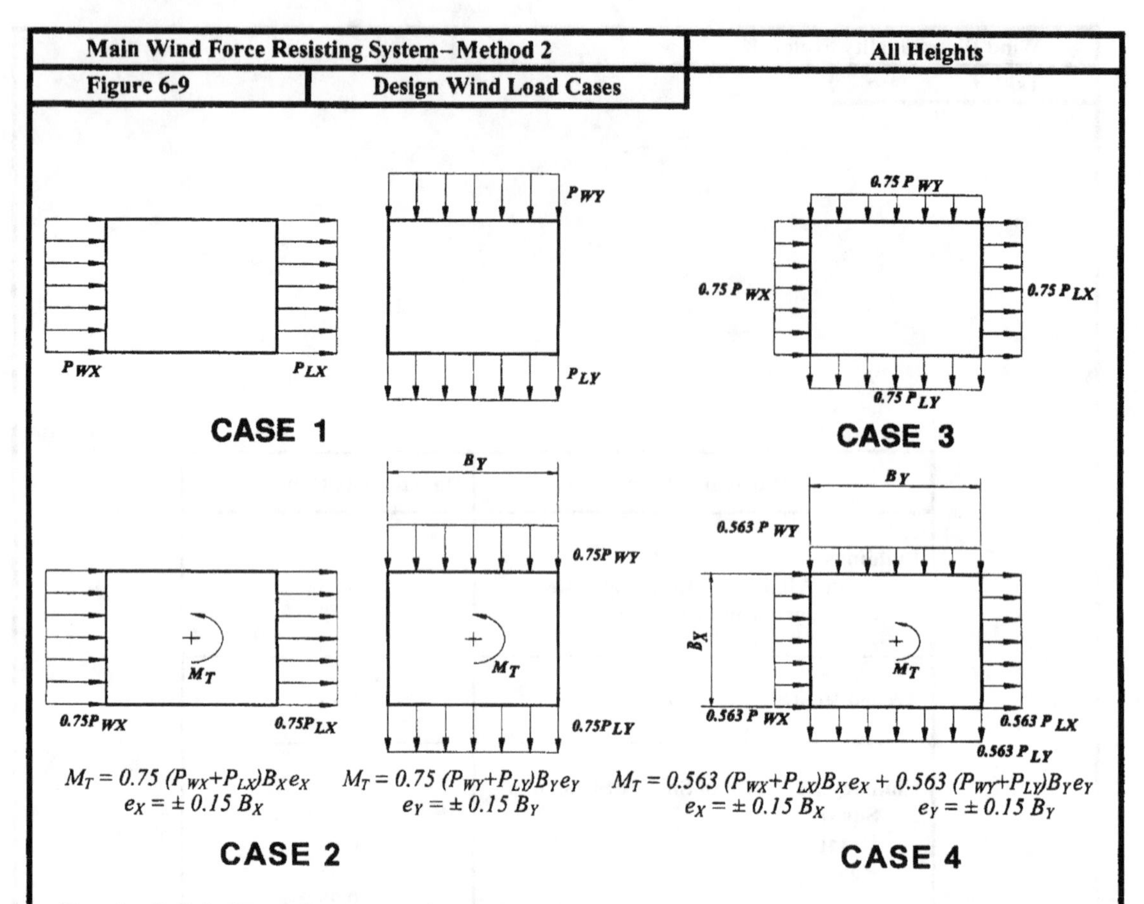

Case 1. Full design wind pressure acting on the projected area perpendicular to each principal axis of the structure, considered separately along each principal axis.

Case 2. Three quarters of the design wind pressure acting on the projected area perpendicular to each principal axis of the structure in conjunction with a torsional moment as shown, considered separately for each principal axis.

Case 3. Wind loading as defined in Case 1, but considered to act simultaneously at 75% of the specified value.

Case 4. Wind loading as defined in Case 2, but considered to act simultaneously at 75% of the specified value.

Notes:

1. Design wind pressures for windward and leeward faces shall be determined in accordance with the provisions of 6.5.12.2.1 and 6.5.12.2.3 as applicable for building of all heights.
2. Diagrams show plan views of building.
3. Notation:

   P_{WX}, P_{WY}: Windward face design pressure acting in the x, y principal axis, respectively.

   P_{LX}, P_{LY}: Leeward face design pressure acting in the x, y principal axis, respectively.

   $e\ (e_X, e_Y)$: Eccentricity for the x, y principal axis of the structure, respectively.

   M_T: Torsional moment per unit height acting about a vertical axis of the building.

Figure 13.12

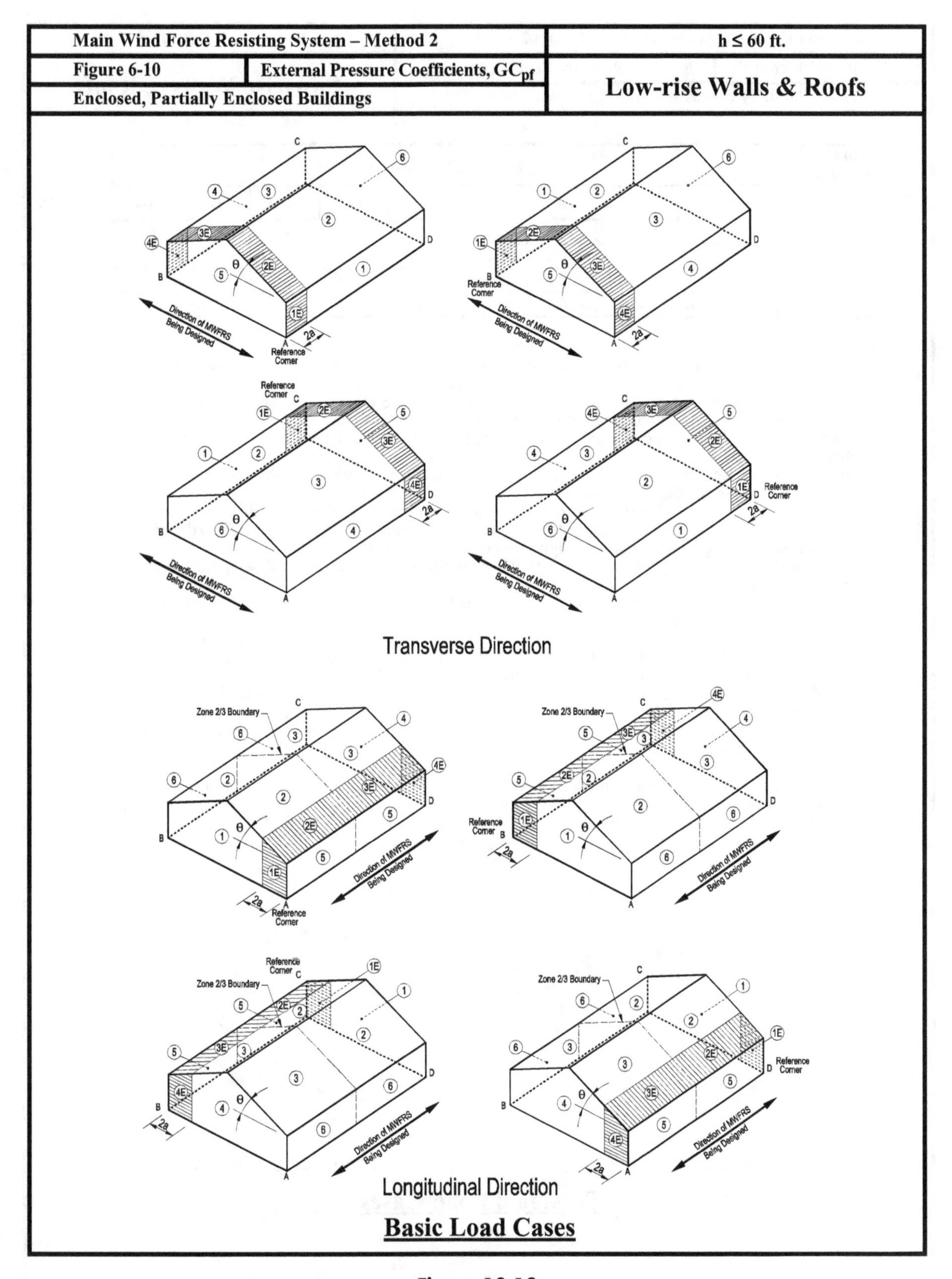
Main Wind Force Resisting System – Method 2
h ≤ 60 ft.
Figure 6-10
External Pressure Coefficients, GC_{pf}
Enclosed, Partially Enclosed Buildings
Low-rise Walls & Roofs
Reference Corner
Direction of MWFRS Being Designed
Zone 2/3 Boundary
Transverse Direction
Longitudinal Direction
Basic Load Cases

Figure 13.13

Main Wind Force Resisting System – Method 2		h ≤ 60 ft.
Figure 6-10 (cont'd)	External Pressure Coefficients, GC_{pf}	**Low-rise Walls & Roofs**
Enclosed, Partially Enclosed Buildings		

Roof Angle θ (degrees)	Building Surface									
	1	2	3	4	5	6	1E	2E	3E	4E
0-5	0.40	-0.69	-0.37	-0.29	-0.45	-0.45	0.61	-1.07	-0.53	-0.43
20	0.53	-0.69	-0.48	-0.43	-0.45	-0.45	0.80	-1.07	-0.69	-0.64
30-45	0.56	0.21	-0.43	-0.37	-0.45	-0.45	0.69	0.27	-0.53	-0.48
90	0.56	0.56	-0.37	-0.37	-0.45	-0.45	0.69	0.69	-0.48	-0.48

Notes:

1. Plus and minus signs signify pressures acting toward and away from the surfaces, respectively.
2. For values of θ other than those shown, linear interpolation is permitted.
3. The building must be designed for all wind directions using the 8 loading patterns shown. The load patterns are applied to each building corner in turn as the Reference Corner.
4. Combinations of external and internal pressures (see Figure 6-5) shall be evaluated as required to obtain the most severe loadings.
5. For the torsional load cases shown below, the pressures in zones designated with a "T" (1T, 2T, 3T, 4T) shall be 25% of the full design wind pressures (zones 1, 2, 3, 4).
   Exception: One story buildings with h less than or equal to 30 ft (9.1m), buildings two stories or less framed with light frame construction, and buildings two stories or less designed with flexible diaphragms need not be designed for the torsional load cases.
   Torsional loading shall apply to all eight basic load patterns using the figures below applied at each reference corner.
6. Except for moment-resisting frames, the total horizontal shear shall not be less than that determined by neglecting wind forces on roof surfaces.
7. For the design of the MWFRS providing lateral resistance in a direction parallel to a ridge line or for flat roofs, use θ = 0° and locate the zone 2/3 boundary at the mid-length of the building.
8. The roof pressure coefficient GC_{pf}, when negative in Zone 2 or 2E, shall be applied in Zone 2/2E for a distance from the edge of roof equal to 0.5 times the horizontal dimension of the building parallel to the direction of the MWFRS being designed or 2.5 times the eave height, h_e, at the windward wall, whichever is less; the remainder of Zone 2/2E extending to the ridge line shall use the pressure coefficient GC_{pf} for Zone 3/3E.
9. Notation:
   a: 10 percent of least horizontal dimension or 0.4h, whichever is smaller, but not less than either 4% of least horizontal dimension or 3 ft (0.9 m).
   h: Mean roof height, in feet (meters), except that eave height shall be used for θ ≤ 10°.
   θ: Angle of plane of roof from horizontal, in degrees.

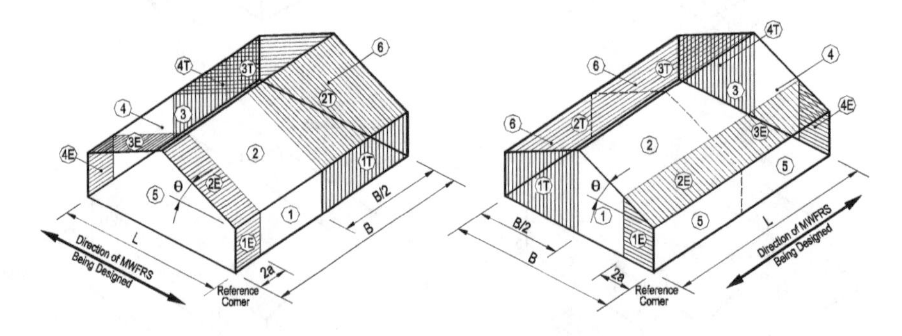

Torsional Load Cases

Figure 13.13 (continued)

Method 1—Simplified Procedure

Method 1 is applicable primarily in the design of regularly shaped low-rise buildings. In order to use this method, the following conditions must be satisfied. Although the list appears lengthy, many low-rise buildings satisfy all these criteria:

1. The building must be a simple diaphragm building, that is, a building in which both windward and leeward wind loads are transmitted through floor and roof diaphragms to the same main wind-force-resisting system. This method cannot be used to determine wind forces on buildings without diaphragms (for example, light-frame metal buildings that transmit wind forces to the main wind-force-resisting system through framing members such as girts). This also means that this method cannot be used if there is a structural separation (for example, an expansion joint) in a building.
2. The mean roof height must be 60 feet or less, and the mean roof height cannot be greater than the least horizontal dimension of the building.
3. The building must be enclosed. Buildings in *wind-borne debris regions* must have special impact-resistant glazing. Wind-borne debris regions are defined as those areas of hurricane-prone regions (1) in Hawaii or within one mile of the coast where the basic wind speed is greater than or equal to 110 miles per hour, or (2) where the basic wind speed is greater than or equal to 120 miles per hour.
4. The building must be regularly shaped; it cannot have unusual geometric irregularities.
5. The building cannot be flexible; its natural period must be one second or less.
6. The building cannot have characteristics that make it subject to across wind loading and other higher order response modes, and cannot be located where there are channeling effects of wind.
7. The building must have an approximately symmetrical cross-section in each direction and must have either a flat, gable, or hip roof with an angle less than or equal to 45 degrees.
8. The building must be exempted from torsional load cases due to building symmetry and stiffness, or have flexible diaphragms, or be free from the control of torsional load cases.

Method 2 or 3 must be used to determine wind pressures if any of these conditions are not met.

Wind pressures are tabulated in the code for a specific set of criteria. The design wind pressures on different zones of a building's surface are calculated (for any situation) by the following formula:

$$p_s = \lambda K_{zt} I p_{s30}$$

where

λ = adjustment factor for building height and exposure (ASCE 7 Figure 6-2, shown in Figure 13.14)

K_{zt} = topographic factor (defined previously)

I = importance factor (defined previously)

p_{s30} = simplified design wind pressure for Exposure B, at h = 30 feet and I = 1.0 (ASCE 7 Figure 6-2)

The pressure p_s represents the net pressures (sum of internal and external pressures) applied to the horizontal and vertical projections of the building surfaces, as illustrated in Figure 13.14. Since the buildings are enclosed, $GC_{pi} = +0.18$

and –0.18 (from ASCE 7 Figure 6-5, shown in Figure 13.15). Also note that for the horizontal pressure zones A, B, C, and D depicted in Figure 13.14, p_s is the combination of the windward and leeward net pressures (with the internal pressures canceled).

The tabulated pressures of p_{s30} in ASCE 7 Figure 6-2 have been determined using the low-rise method described in Method 2, based on exposure category B and a building not located on a hill or escarpment with a mean roof height of 30 feet. Once these pressures have been established for all zones on the surface of a building, they are multiplied by the appropriate adjustment factor for building height and exposure (λ), importance factor (I), and topographic factor (K_{zt}).

The last of the criteria of Method 1 was established because torsional loading was deemed to be too complicated for a simplified method. It essentially prevents the use of this method for buildings sensitive to torsional wind loads. The buildings listed in Note 5 of ASCE 7 Figure 6-10 are exempt from torsional load requirements because they are not sensitive to torsion. The same can be said for buildings with flexible roof and floor diaphragms. Some buildings with rigid diaphragms are also exempt, depending on the distribution of the stiffness of the main wind-force-resisting system elements.

Example #1

What is the basic wind speed in Salt Lake City?

Solution:

The exam sometimes tests candidates' familiarity with the wind speed map, which is Figure 1609 in the IBC, reproduced in Figure 13.1. Using that map, we see that the basic wind speed in Salt Lake City is *90 mph.*

Example #2

What is the static horizontal and vertical wind pressure in pounds per square foot on a flat roof fire station that is 15 feet high and located on a flat side in downtown Los Angeles, using Method 2?

The building is 50 feet by 50 feet in plan.

Solution:

Wind pressures are determined using the basic formulas presented earlier in Method 2.

On the windward wall:

$$p_w = q_z GC_p - q_i(GC_{pi})$$

where $q_z = 0.00256 K_z K_{zt} K_d V^2 I$

Use Exposure B for built-up areas. It is also safe to assume that this one-story building is enclosed and rigid. Buildings of this size will typically have a fundamental period much less than one second.

It can be seen from ASCE 7 Table 6-3 that K_z is a constant between 0 and 15 feet above the ground. Under Exposure B, Case 2 is applicable in this situation, so $K_z = 0.57$.

Since there is no mention of whether this building is located on a hill or escarpment, assume it is not: $K_{zt} = 1.0$.

From ASCE 7 Table 6-4, $K_d = 0.85$ for main wind-force-resisting systems of buildings.

The basic wind speed for Los Angeles is 85 miles per hour, which can be found in IBC Figure 1609 or ASCE 7 Figure 6-1.

Main Wind Force Resisting System – Method 1		h ≤ 60 ft.
Figure 6-2	Design Wind Pressures	Walls & Roofs
Enclosed Buildings		

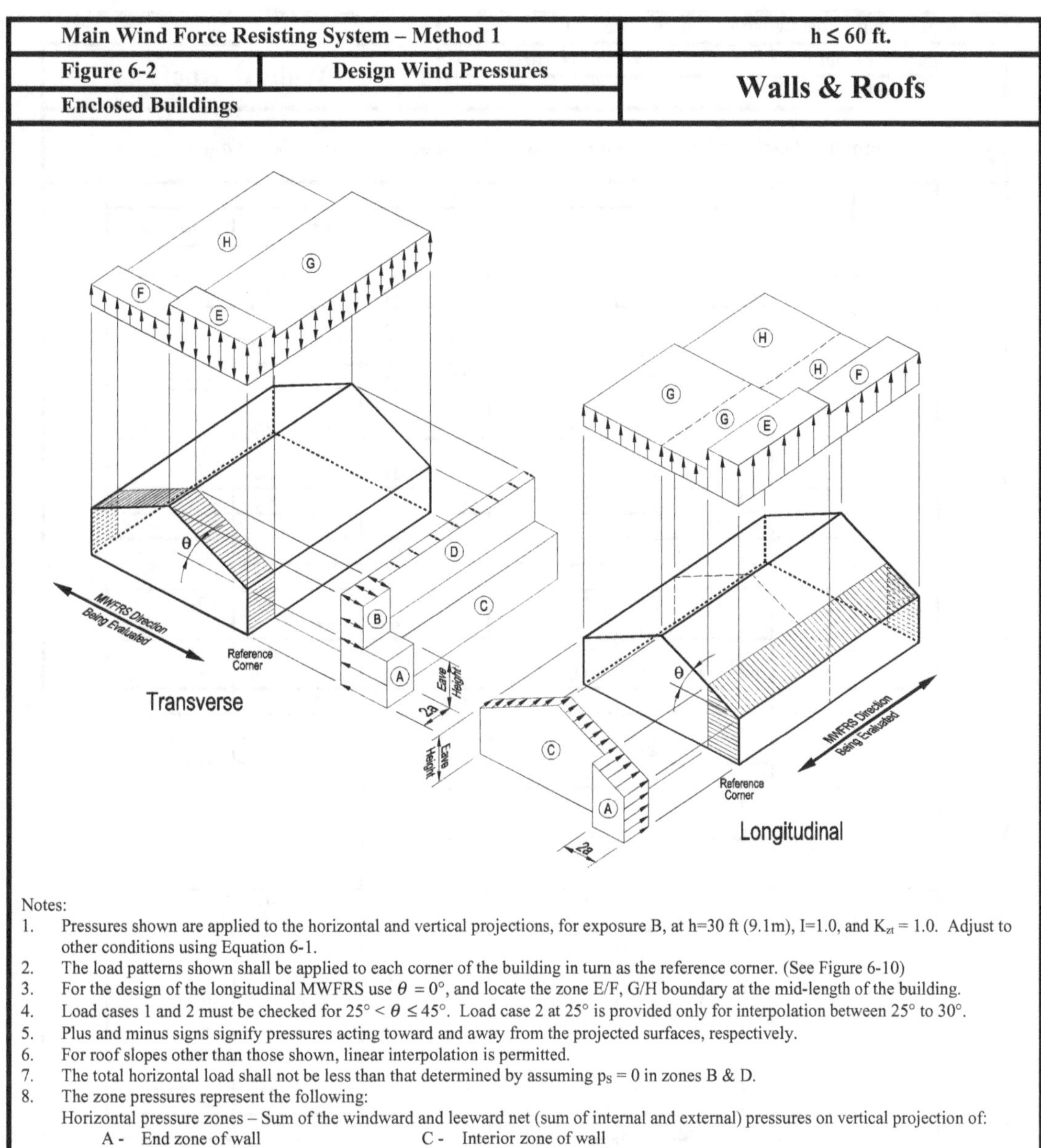

Notes:

1. Pressures shown are applied to the horizontal and vertical projections, for exposure B, at h=30 ft (9.1m), I=1.0, and K_{zt} = 1.0. Adjust to other conditions using Equation 6-1.
2. The load patterns shown shall be applied to each corner of the building in turn as the reference corner. (See Figure 6-10)
3. For the design of the longitudinal MWFRS use $\theta = 0°$, and locate the zone E/F, G/H boundary at the mid-length of the building.
4. Load cases 1 and 2 must be checked for $25° < \theta \leq 45°$. Load case 2 at 25° is provided only for interpolation between 25° to 30°.
5. Plus and minus signs signify pressures acting toward and away from the projected surfaces, respectively.
6. For roof slopes other than those shown, linear interpolation is permitted.
7. The total horizontal load shall not be less than that determined by assuming $p_S = 0$ in zones B & D.
8. The zone pressures represent the following:
   Horizontal pressure zones – Sum of the windward and leeward net (sum of internal and external) pressures on vertical projection of:
   A - End zone of wall　　C - Interior zone of wall
   B - End zone of roof　　D - Interior zone of roof
   Vertical pressure zones – Net (sum of internal and external) pressures on horizontal projection of:
   E - End zone of windward roof　　G - Interior zone of windward roof
   F - End zone of leeward roof　　H - Interior zone of leeward roof
9. Where zone E or G falls on a roof overhang on the windward side of the building, use E_{OH} and G_{OH} for the pressure on the horizontal projection of the overhang. Overhangs on the leeward and side edges shall have the basic zone pressure applied.
10. Notation:
    a: 10 percent of least horizontal dimension or 0.4h, whichever is smaller, but not less than either 4% of least horizontal dimension or 3 ft (0.9 m).
    h: Mean roof height, in feet (meters), except that eave height shall be used for roof angles <10°.
    θ: Angle of plane of roof from horizontal, in degrees.

Figure 13.14

Main Wind Force Resisting System – Method 1		h ≤ 60 ft.
Figure 6-2 (cont'd)	Design Wind Pressures	Walls & Roofs
Enclosed Buildings		

Simplified Design Wind Pressure, p_{s30} (psf) *(Exposure B at h = 30 ft., K_{zt} = 1.0, with I = 1.0)*

Basic Wind Speed (mph)	Roof Angle (degrees)	Load Case	Zones									
			Horizontal Pressures				Vertical Pressures				Overhangs	
			A	B	C	D	E	F	G	H	EOH	GOH
85	0 to 5°	1	11.5	-5.9	7.6	-3.5	-13.8	-7.8	-9.6	-6.1	-19.3	-15.1
	10°	1	12.9	-5.4	8.6	-3.1	-13.8	-8.4	-9.6	-6.5	-19.3	-15.1
	15°	1	14.4	-4.8	9.6	-2.7	-13.8	-9.0	-9.6	-6.9	-19.3	-15.1
	20°	1	15.9	-4.2	10.6	-2.3	-13.8	-9.6	-9.6	-7.3	-19.3	-15.1
	25°	1	14.4	2.3	10.4	2.4	-6.4	-8.7	-4.6	-7.0	-11.9	-10.1
		2	-------	-------	-------	-------	-2.4	-4.7	-0.7	-3.0	-------	-------
	30 to 45	1	12.9	8.8	10.2	7.0	1.0	-7.8	0.3	-6.7	-4.5	-5.2
		2	12.9	8.8	10.2	7.0	5.0	-3.9	4.3	-2.8	-4.5	-5.2
90	0 to 5°	1	12.8	-6.7	8.5	-4.0	-15.4	-8.8	-10.7	-6.8	-21.6	-16.9
	10°	1	14.5	-6.0	9.6	-3.5	-15.4	-9.4	-10.7	-7.2	-21.6	-16.9
	15°	1	16.1	-5.4	10.7	-3.0	-15.4	-10.1	-10.7	-7.7	-21.6	-16.9
	20°	1	17.8	-4.7	11.9	-2.6	-15.4	-10.7	-10.7	-8.1	-21.6	-16.9
	25°	1	16.1	2.6	11.7	2.7	-7.2	-9.8	-5.2	-7.8	-13.3	-11.4
		2	-------	-------	-------	-------	-2.7	-5.3	-0.7	-3.4	-------	-------
	30 to 45	1	14.4	9.9	11.5	7.9	1.1	-8.8	0.4	-7.5	-5.1	-5.8
		2	14.4	9.9	11.5	7.9	5.6	-4.3	4.8	-3.1	-5.1	-5.8
100	0 to 5°	1	15.9	-8.2	10.5	-4.9	-19.1	-10.8	-13.3	-8.4	-26.7	-20.9
	10°	1	17.9	-7.4	11.9	-4.3	-19.1	-11.6	-13.3	-8.9	-26.7	-20.9
	15°	1	19.9	-6.6	13.3	-3.8	-19.1	-12.4	-13.3	-9.5	-26.7	-20.9
	20°	1	22.0	-5.8	14.6	-3.2	-19.1	-13.3	-13.3	-10.1	-26.7	-20.9
	25°	1	19.9	3.2	14.4	3.3	-8.8	-12.0	-6.4	-9.7	-16.5	-14.0
		2	-------	-------	-------	-------	-3.4	-6.6	-0.9	-4.2	-------	-------
	30 to 45	1	17.8	12.2	14.2	9.8	1.4	-10.8	0.5	-9.3	-6.3	-7.2
		2	17.8	12.2	14.2	9.8	6.9	-5.3	5.9	-3.8	-6.3	-7.2
105	0 to 5°	1	17.5	-9.0	11.6	-5.4	-21.1	-11.9	-14.7	-9.3	-29.4	-23.0
	10°	1	19.7	-8.2	13.1	-4.7	-21.1	-12.8	-14.7	-9.8	-29.4	-23.0
	15°	1	21.9	-7.3	14.7	-4.2	-21.1	-13.7	-14.7	-10.5	-29.4	-23.0
	20°	1	24.3	-8.4	16.1	-3.5	-21.1	-14.7	-14.7	-11.1	-29.4	-23.0
	25°	1	21.9	3.5	15.9	3.5	-9.7	-13.2	-7.1	-10.7	-18.2	-15.4
		2	-----	-----	-----	-----	-3.7	-7.3	-1.0	-4.6	-----	-----
	30 to 45	1	19.6	13.5	15.7	10.8	1.5	-11.9	0.6	-10.3	-6.9	-7.9
		2	19.6	13.5	15.7	10.8	7.6	-5.8	6.5	-4.2	-6.9	-7.9
110	0 to 5°	1	19.2	-10.0	12.7	-5.9	-23.1	-13.1	-16.0	-10.1	-32.3	-25.3
	10°	1	21.6	-9.0	14.4	-5.2	-23.1	-14.1	-16.0	-10.8	-32.3	-25.3
	15°	1	24.1	-8.0	16.0	-4.6	-23.1	-15.1	-16.0	-11.5	-32.3	-25.3
	20°	1	26.6	-7.0	17.7	-3.9	-23.1	-16.0	-16.0	-12.2	-32.3	-25.3
	25°	1	24.1	3.9	17.4	4.0	-10.7	-14.6	-7.7	-11.7	-19.9	-17.0
		2	-------	-------	-------	-------	-4.1	-7.9	-1.1	-5.1	-------	-------
	30 to 45	1	21.6	14.8	17.2	11.8	1.7	-13.1	0.6	-11.3	-7.6	-8.7
		2	21.6	14.8	17.2	11.8	8.3	-6.5	7.2	-4.6	-7.6	-8.7
120	0 to 5°	1	22.8	-11.9	15.1	-7.0	-27.4	-15.6	-19.1	-12.1	-38.4	-30.1
	10°	1	25.8	-10.7	17.1	-6.2	-27.4	-16.8	-19.1	-12.9	-38.4	-30.1
	15°	1	28.7	-9.5	19.1	-5.4	-27.4	-17.9	-19.1	-13.7	-38.4	-30.1
	20°	1	31.6	-8.3	21.1	-4.6	-27.4	-19.1	-19.1	-14.5	-38.4	-30.1
	25°	1	28.6	4.6	20.7	4.7	-12.7	-17.3	-9.2	-13.9	-23.7	-20.2
		2	-------	-------	-------	-------	-4.8	-9.4	-1.3	-6.0	-------	-------
	30 to 45	1	25.7	17.6	20.4	14.0	2.0	-15.6	0.7	-13.4	-9.0	-10.3
		2	25.7	17.6	20.4	14.0	9.9	-7.7	8.6	-5.5	-9.0	-10.3

Unit Conversions—1.0 ft = 0.3048 m; 1.0 psf = 0.0479 kN/m^2

Figure 13.14 (continued)

Main Wind Force Resisting System – Method 1		h ≤ 60 ft.
Figure 6-2 (cont'd)	Design Wind Pressures	Walls & Roofs
Enclosed Buildings		

Simplified Design Wind Pressure, p_{s30} (psf) *(Exposure B at h = 30 ft., K_{zt} = 1.0, with I = 1.0)*

Basic Wind Speed (mph)	Roof Angle (degrees)	Load Case	Zones									
			Horizontal Pressures				Vertical Pressures				Overhangs	
			A	B	C	D	E	F	G	H	EOH	GOH
125	0 to 5°	1	24.7	-12.9	16.4	-7.6	-29.7	-16.9	-20.7	-13.1	-41.7	-32.7
	10°	1	28.0	-11.6	18.6	-6.7	-29.7	-18.2	-20.7	-14.0	-41.7	-32.7
	15°	1	31.1	-10.3	20.7	-5.9	-29.7	-19.4	-20.7	-14.9	-41.7	-32.7
	20°	1	34.3	-9.0	22.9	-5.0	-29.7	-20.7	-20.7	-15.7	-41.7	-32.7
	25°	1	31.0	5.0	22.5	5.1	-13.8	-18.8	-10.0	-15.1	-25.7	-21.9
		2	-----	-----	-----	-----	-5.2	-10.2	-1.4	-6.5	-----	-----
	30 to 45	1	27.9	19.1	22.1	15.2	2.2	-16.9	0.8	-14.5	-9.8	-11.2
		2	27.9	19.1	22.1	15.2	10.7	-8.4	9.3	-6.0	-9.8	-11.2
130	0 to 5°	1	26.8	-13.9	17.8	-8.2	-32.2	-18.3	-22.4	-14.2	-45.1	-35.3
	10°	1	30.2	-12.5	20.1	-7.3	-32.2	-19.7	-22.4	-15.1	-45.1	-35.3
	15°	1	33.7	-11.2	22.4	-6.4	-32.2	-21.0	-22.4	-16.1	-45.1	-35.3
	20°	1	37.1	-9.8	24.7	-5.4	-32.2	-22.4	-22.4	-17.0	-45.1	-35.3
	25°	1	33.6	5.4	24.3	5.5	-14.9	-20.4	-10.8	-16.4	-27.8	-23.7
		2	-------	-------	-------	-------	-5.7	-11.1	-1.5	-7.1	-------	-------
	30 to 45	1	30.1	20.6	24.0	16.5	2.3	-18.3	0.8	-15.7	-10.6	-12.1
		2	30.1	20.6	24.0	16.5	11.6	-9.0	10.0	-6.4	-10.6	-12.1
140	0 to 5°	1	31.1	-16.1	20.6	-9.6	-37.3	-21.2	-26.0	-16.4	-52.3	-40.9
	10°	1	35.1	-14.5	23.3	-8.5	-37.3	-22.8	-26.0	-17.5	-52.3	-40.9
	15°	1	39.0	-12.9	26.0	-7.4	-37.3	-24.4	-26.0	-18.6	-52.3	-40.9
	20°	1	43.0	-11.4	28.7	-6.3	-37.3	-26.0	-26.0	-19.7	-52.3	-40.9
	25°	1	39.0	6.3	28.2	6.4	-17.3	-23.6	-12.5	-19.0	-32.3	-27.5
		2[4]	-------	-------	-------	-------	-6.6	-12.8	-1.8	-8.2	-------	-------
	30 to 45	1	35.0	23.9	27.8	19.1	2.7	-21.2	0.9	-18.2	-12.3	-14.0
		2[4]	35.0	23.9	27.8	19.1	13.4	-10.5	11.7	-7.5	-12.3	-14.0
145	0 to 5°	1	33.4	-17.3	22.1	-10.3	-40.0	-22.7	-27.9	-17.6	-56.1	-43.9
	10°	1	37.7	-15.6	25.0	-9.1	-40.0	-24.5	-27.9	-18.8	-56.1	-43.9
	15°	1	41.8	-13.8	27.9	-7.9	-40.0	-26.2	-27.9	-20.0	-56.1	-43.9
	20°	1	46.1	-12.2	30.8	-6.8	-40.0	-27.9	-27.9	-21.1	-56.1	-43.9
	25°	1	41.8	6.8	30.3	6.9	-18.6	-25.3	-13.4	-20.4	-34.6	-29.5
		2[4]	-----	-----	-----	-----	-7.1	-13.7	-1.9	-8.8	-----	-----
	30 to 45	1	37.5	25.6	29.8	20.5	2.9	-22.7	1.0	-19.5	-13.2	-15.0
		2[4]	35.7	25.6	29.8	20.5	14.4	-11.3	12.6	-8.0	-13.2	-15.0
150	0 to 5°	1	35.7	-18.5	23.7	-11.0	-42.9	-24.4	-29.8	-18.9	-60.0	-47.0
	10°	1	40.2	-16.7	26.8	-9.7	-42.9	-26.2	-29.8	-20.1	-60.0	-47.0
	15°	1	44.8	-14.9	29.8	-8.5	-42.9	-28.0	-29.8	-21.4	-60.0	-47.0
	20°	1	49.4	-13.0	32.9	-7.2	-42.9	-29.8	-29.8	-22.6	-60.0	-47.0
	25°	1	44.8	7.2	32.4	7.4	-19.9	-27.1	-14.4	-21.8	-37.0	-31.6
		2[4]	-------	-------	-------	-------	-7.5	-14.7	-2.1	-9.4	-------	-------
	30 to 45	1	40.1	27.4	31.9	22.0	3.1	-24.4	1.0	-20.9	-14.1	-16.1
		2[4]	40.1	27.4	31.9	22.0	15.4	-12.0	13.4	-8.6	-14.1	-16.1
170	0 to 5°	1	45.8	-23.8	30.4	-14.1	-55.1	-31.3	-38.3	-24.2	-77.1	-60.4
	10°	1	51.7	-21.4	34.4	-12.5	-55.1	-33.6	-38.3	-25.8	-77.1	-60.4
	15°	1	57.6	-19.1	38.3	-10.9	-55.1	-36.0	-38.3	-27.5	-77.1	-60.4
	20°	1	63.4	-16.7	42.3	-9.3	-55.1	-38.3	-38.3	-29.1	-77.1	-60.4
	25°	1	57.5	9.3	41.6	9.5	-25.6	-34.8	-18.5	-28.0	-47.6	-40.5
		2[4]	-------	-------	-------	-------	-9.7	-18.9	-2.6	-12.1	-------	-------
	30 to 45	1	51.5	35.2	41.0	28.2	4.0	-31.3	1.3	-26.9	-18.1	-20.7
		2[4]	51.5	35.2	41.0	28.2	19.8	-15.4	17.2	-11.0	-18.1	-20.7

Unit Conversions—1.0 ft = 0.3048 m; 1.0 psf = 0.0479 kN/m^2

Figure 13.14 (continued)

Main Wind Force Resisting System – Method 1		$h \leq 60$ ft.
Figure 6-2 (cont'd)	Design Wind Pressures	Walls & Roofs
Enclosed Buildings		

Adjustment Factor for Building Height and Exposure, λ

Mean roof height (ft)	Exposure B	Exposure C	Exposure D
15	1.00	1.21	1.47
20	1.00	1.29	1.55
25	1.00	1.35	1.61
30	1.00	1.40	1.66
35	1.05	1.45	1.70
40	1.09	1.49	1.74
45	1.12	1.53	1.78
50	1.16	1.56	1.81
55	1.19	1.59	1.84
60	1.22	1.62	1.87

Figure 13.14 (continued)

Main Wind Force Res. Sys. / Comp and Clad. – Method 2		All Heights
Figure 6-5	Internal Pressure Coefficient, GC_{pi}	Walls & Roofs
Enclosed, Partially Enclosed, and Open Buildings		

Enclosure Classification	GC_{pi}
Open Buildings	0.00
Partially Enclosed Buildings	+0.55 -0.55
Enclosed Buildings	+0.18 -0.18

Notes:

1. Plus and minus signs signify pressures acting toward and away from the internal surfaces, respectively.
2. Values of GC_{pi} shall be used with q_z or q_h as specified in 6.5.12.
3. Two cases shall be considered to determine the critical load requirements for the appropriate condition:
   (i) a positive value of GC_{pi} applied to all internal surfaces
   (ii) a negative value of GC_{pi} applied to all internal surfaces

Figure 13.15

The importance factor I is equal to 1.15 for a fire house (Occupancy Category IV) in a non-hurricane prone region (ASCE Table 6-1).

Therefore, $q_z = 0.00256 \times 0.57 \times 1.0 \times 0.85 \times 85^2 \times 1.15 = 10.3$ psf.

The gust effect factor G is taken as 0.85.

For the windward wall, $C_p = 0.8$ from ASCE 7 Figure 6-6.

Windward pressure $p_w = 10.3 \times 0.85 \times 0.8 =$ *7.0 psf inward.*

Note that the internal pressure $q_i(GC_{pi})$ is not computed for the windward wall because it and the internal pressure on the leeward wall cancel.

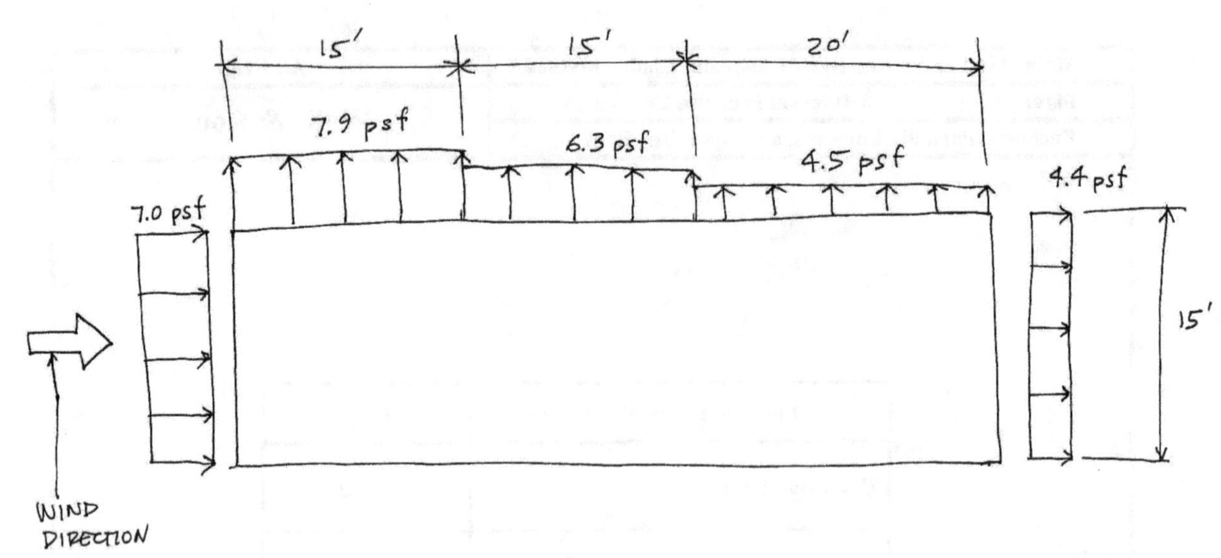

Figure 13.16

On the leeward wall:

$p_l = q_hGC_p - q_i(GC_{pi})$

where $q_h = 0.00256K_hK_{zt}K_dV^2I$

Since the building has a mean roof height of 15 feet, q_h = 10.3 psf, which is the same as on the windward wall.

The pressure coefficient C_p for the windward wall depends on the plan dimensions of the building, as shown in ASCE 7 Figure 6-6. For a square building, $C_p = -0.5$.

Leeward pressure $p_l = 10.3 \times 0.85 \times (-0.5)$ $= -4.4$ psf = *4.4 psf outward.*

On the roof:

q_h = 10.3 psf, which is the same as on the windward and leeward walls in this case.

The roof pressure coefficient is determined from ASCE 7 Figure 6-6. Since this is a flat roof, the applicable wind direction is "Normal to ridge for θ < 10 and parallel to ridge for all θ." The ratio h/L = 15/50 = 0.3 < 0.5. In the figure, two sets of C_p are given. The first set of values decreases the farther you go from the windward edge of the roof. The second set is a constant. In this example, we will use the first set of values. Also note that unlike in the horizontal direction, the internal pressure must be considered on the roof.

In general, $p_r = q_hGC_p - q_h(GC_{pi})$.

As shown in ASCE 7 Figure 6-5, $GC_{pi} = +0.18$ or -0.18 for enclosed buildings. A positive value of GC_{pi} is used, that is, internal pressure is taken to act towards the inside surface of the roof because this results in the largest pressure when the internal and external pressures are combined.

From the windward edge to h = 15 feet: $p_r = [10.3 \times 0.85 \times (-0.9)] - [10.3 \times 0.18] = -7.9 - 1.9 = -9.8$ psf = *9.8 psf outward.*

From 15 feet to 2h = 30 feet: $p_r = [10.3 \times 0.85 \times (-0.5)] - [10.3 \times 0.18] = -4.4 - 1.9 = -6.3$ psf = *6.3 psf outward.*

From 30 feet to 50 feet: p_r = [10.3 × 0.85 × (–0.3)] – [10.3 × 0.18] = –2.6 – 1.9 = –4.5 psf = *4.5 psf outward.*

A summary of the wind pressures on all of the surface of the building is given in Figure 13.16.

Example #3

If the building in Example #2 were located in an isolated desert region outside of Los Angeles, not in a *special wind region*, what would the wind pressures be?

Solution:

The only change from Example #2 is the site exposure, which is Exposure C instead of Exposure B.

On the windward wall:

$p_w = q_z GC_p - q_i(GC_{pi})$

where $q_z = 0.00256 K_z K_{zt} K_d V^2 I$

For Exposure C, K_z = 0.85 from ASCE 7 Table 6-3.

All the other quantities are the same as in Example #2.

Therefore, q_z = 0.00256 × 0.85 × 1.0 × 0.85 × 85^2 × 1.15 = 15.4 psf.

Windward pressure p_w = 15.4 × 0.85 × 0.8 = *10.5 psf inward.*

Leeward pressure p_l = 15.4 × 0.85 × (–0.5) = –6.6 psf = *6.6 psf outward.*

Roof pressure:

From the windward edge to h = 15 feet: p_r = [15.4 × 0.85 × (–0.9)] – [15.4 × 0.18] = –11.8 – 2.8 = –14.6 psf = *14.6 psf outward.*

From 15 feet to 2h = 30 feet: p_r = [15.4 × 0.85 × (–0.5)] – [15.4 × 0.18] = –6.6 – 2.8 = –9.4 psf = *9.4 psf outward.*

From 30 feet to 50 feet: p_r = [15.4 × 0.85 × (–0.3)] – [15.4 × 0.18] = –3.9 – 2.8 = –6.7 psf = *6.7 psf outward.*

If we compare these results with those of Example #2, we can see how much greater the wind pressures are for buildings on open exposed sites.

Example #4

If the building in Examples #2 and #3 is 50 feet by 50 feet in plan, and one story high, what is the total wind load resisted by the roof diaphragm?

Solution:

As in earthquake design, the walls are assumed to span vertically from the ground to the roof. Thus, one-half of the wind load is transferred from the walls into the roof diaphragm, which in turn transfers the load into the vertical shear resisting elements. The other half of the wind load is resisted at the bottom of the wall, at the ground. Thus we have:

For the urban site, wind load =

$(7.0 \text{ psf} + 4.4 \text{ psf}) \times \frac{15 \text{ feet}}{2} \times 50 \text{ feet}$
= *4,275 pounds.*

For the open site, wind load =

$(10.5 \text{ psf} + 6.6 \text{ psf}) \times \frac{15 \text{ feet}}{2} \times 50 \text{ feet}$
= *6,413 pounds.*

Example #5

For the same building as in Example #2, what would the wind pressures be if Method 1 were used instead of Method 2?

Solution:

If we use Method 1 instead of Method 2, the pressures on various surfaces of the building are computed by $p_s = \lambda K_{zt} I p_{s30}$.

The pressures p_{s30} are tabulated in ASCE 7 Figure 6-2 for different zones on the building. With a basic wind speed of V = 85 miles per hour and a roof angle of zero degrees, these pressures are read directly from the figure.

Zone A: $p_{s30} = 11.5$ psf

Zone C: $p_{s30} = 7.6$ psf

Zone E: $p_{s30} = -13.8$ psf

Zone F: $p_{s30} = -7.8$ psf

Zone G: $p_{s30} = -9.6$ psf

Zone H: $p_{s30} = -6.1$ psf

Note that pressures in Zones B and D are not applicable to buildings with flat roofs.

For Exposure B and a building with a mean roof height of 15 feet, $\lambda = 1.00$. Also, $K_{zt} = 1.0$ and I = 1.15. Therefore, the pressures are:

Zone A: $p_s = 1.00 \times 1.0 \times 1.15 \times 11.5 =$ *13.2 psf in any horizontal direction.*

Zone C: $p_s = 1.00 \times 1.0 \times 1.15 \times 7.6 =$ *8.7 psf in any horizontal direction.*

Zone E: $p_s = 1.00 \times 1.0 \times 1.15 \times (-13.8) = -15.9$ psf = *15.9 psf upward.*

Zone F: $p_s = 1.00 \times 1.0 \times 1.15 \times (-7.8) = -9.0$ psf = *9.0 psf upward.*

Zone G: $p_s = 1.00 \times 1.0 \times 1.15 \times (-9.6) = -11.0$ psf = *11.0 psf upward.*

Zone H: $p_s = 1.00 \times 1.0 \times 1.15 \times (-6.1) = -7.0$ psf = *7.0 psf upward.*

Recall that the pressures in Zones A and C are the total horizontal pressures, that is, they are the sum of the windward and leeward pressures on the building, and the pressures on the zones on the roof surface include internal pressure.

These pressures are shown in Figure 13.17.

Example #6

Rework Example #3 using Method 1.

Solution:

The tabulated values of p_{s30} remain the same as those in Example #5. However, the value of λ changes from 1.00 to 1.21 as the exposure category changes from B to C. Therefore, the pressures are determined by multiplying those from Example #5 by 1.21, as follows:

Zone A: $p_s = 1.21 \times 13.2 =$ *16.0 psf in any horizontal direction.*

Zone C: $p_s = 1.21 \times 8.7 =$ *10.5 psf in any horizontal direction.*

Zone E: $p_s = 1.21\ (-15.9) = -19.2$ psf = *19.2 psf upward.*

Zone F: $p_s = 1.21\ (-9.0) = -10.9$ psf = *10.9 psf upward.*

Zone G: $p_s = 1.21 \times (-11.0) = -13.3$ psf = *13.3 psf upward.*

Zone H: $p_s = 1.21 \times (-7.0) = -8.5$ psf = *8.5 psf upward.*

Example #7

Rework Example #4 using Method 1.

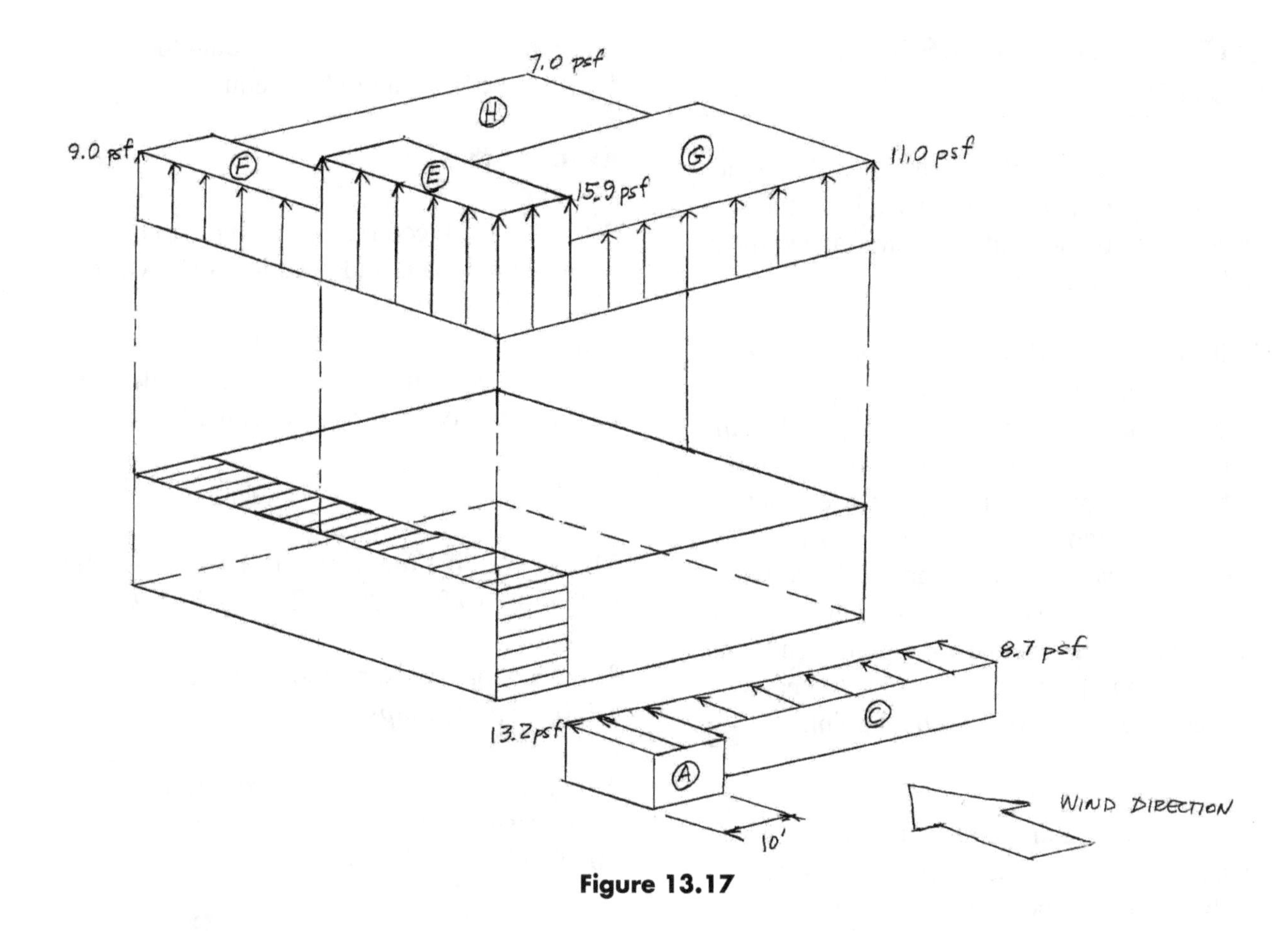

Figure 13.17

Solution:

The wind pressures from Examples #5 and #6 are used to calculate the total wind load resisted by the roof diaphragm. As shown in ASCE 7 Figure 6-2, the wind pressure over Zone A acts over a horizontal width equal to 2a, where a is 10 percent of the least horizontal dimension of the building or 40 percent of the mean roof height, whichever is smaller. It must not be taken less than 4 percent of the least horizontal dimension of the building or 3 feet.

Thus, a = 0.1 × 50 = 5 feet < 0.4 × 15 = 6 feet

2a = 10 feet

The total wind load is the sum of the pressures in Zones A and C multiplied by the respective tributary areas.

For the urban site, the total wind load =

$$\left(13.2 \times 10 \times \frac{15}{2}\right) + \left(8.7 \times 40 \times \frac{15}{2}\right) = \textit{3600\#.}$$

For the open site, the total wind =

$$\left(16.0 \times 10 \times \frac{15}{2}\right) + \left(10.5 \times 40 \times \frac{15}{2}\right) = \textit{4350\#.}$$

Comparing these results with those of Example #4, we find that the forces from Method 2 are about 1.2 and 1.5 times greater than those from Method 1.

LATERAL LOAD RESISTING SYSTEMS

The lateral load resisting systems used in wind design are the same as those used in earthquake design, namely moment-resisting frames, shear walls, and braced frames.

Unlike earthquake design, however, the type of lateral load resisting system used does not affect the magnitude of the wind load. Also, the concept of ductility, or the ability of a system to absorb energy in the inelastic range, is less important for wind design than for earthquake design. The reason is that in earthquake design, the system is expected to be stressed in the inelastic range, above the yield point, while the stresses in wind design are expected to be in the elastic range, below the yield point.

Except for these differences, the general descriptions of moment-resisting frames, shear walls, and braced frames found in Lesson Twelve for earthquake design also apply to wind design.

OVERTURNING

As with earthquake forces, buildings must be designed to resist overturning moments caused by wind forces. If strength design is used, only 90 percent of the dead load can be used to resist 1.6 times the overturning effects from the wind forces. If allowable stress design is used, only 60 percent of the dead load can be used. Put another way, the wind overturning moment cannot exceed 0.6 of the dead load resisting moment, unless the structure is anchored to resist the excess moment.

The factor of safety against overturning is equal to the dead load resisting moment divided by the wind overturning moment. The factor of safety must be greater than or equal to 1/0.6 = 1.67 for the building to be adequate.

Example #8

A two-story braced frame is subject to the service wind loads and dead loads shown in Figure 13.18. What is the overturning moment at the base? What is the dead load resisting moment? Does the structure have an adequate factor of safety against overturning?

Solution:

The overturning moment = 10 kips (10 + 10) ft. + 20 kips (10 ft.) = 200 + 200 = *400 ft.-kips.*

The dead load resisting moment = 25 kips (20 ft.) = *500 ft.-kips.*

The factor of safety against overturning = dead load resisting moment ÷ overturning moment = 500 ft.-kips ÷ 400 ft.-kips = 1.25.

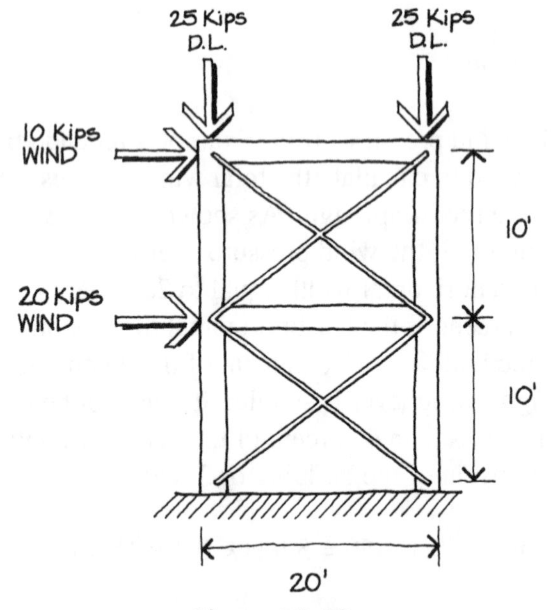

Figure 13.18

Since this is less than 1.67, *the structure does not have an adequate factor of safety against overturning.*

If the columns are adequately anchored to the foundation, the weight of the foundation may be used to increase the dead load resisting moment.

As with earthquake design, buildings with a high height-to-width ratio are most critical for overturning from wind forces. Thus, tall, slender buildings tend to overturn more than short, squat buildings, and top-heavy buildings tend to overturn more than pyramidal buildings.

DEFLECTION AND DRIFT

The deflection or drift of a building under wind loading must be limited in order to prevent damage to the brittle elements of the building and to minimize discomfort to the building's occupants. The code does not provide limits on drift between stories for wind. Limiting wind drift is complex and beyond the scope of this lesson.

As with earthquake design, care must be taken to adequately separate adjacent buildings, and nonstructural elements should be detailed to allow for movement under wind loads.

DIAPHRAGMS, COLLECTORS, AND TORSION

Wind forces are transferred to the vertical resisting elements (shear walls, braced frames, or moment-resisting frames) by diaphragms, in exactly the same way as earthquake forces. Therefore, the discussion of diaphragms in Lesson Twelve for earthquake design also applies to wind design.

Likewise, collector members may be used in wind design, just as in earthquake design, and the discussion in Lesson Twelve on this subject also applies to wind design.

As described in Lesson Twelve for earthquake design, torsion occurs in a rigid diaphragm when the center of mass does not coincide with the center of rigidity. Similarly, under wind loading, torsion occurs in a rigid diaphragm when the center of the applied wind load does not coincide with the center of rigidity. The discussion of torsion in Lesson Twelve is therefore applicable to wind design. The load cases provided in ASCE 7 Figure 6-9 include torsional moments that must be considered in design.

ELEMENTS AND COMPONENTS OF STRUCTURES

The wind pressures on components and cladding are determined from essentially the same basic formulas used to design the main force-resisting system. For low-rise buildings and buildings with a mean roof height of 60 feet or less, the following formula is used:

$$p = q_h[(GC_p) - (GC_{pi})]$$

where GC_p are the external pressure coefficients given in ASCE 7 Figures 6-11 through 6-16.

For buildings with a mean roof height greater than 60 feet, the formula for pressure is:

$$p = q_z(GC_p) - q_i(GC_{pi})$$

where GC_p are the external pressure coefficients given in ASCE 7 Figure 6-17.

The design wind pressure on the components and cladding elements of parapets is given by the following formula:

$$p = q_p(GC_p - GC_{pi})$$

where GC_p are the external pressure coefficients given in ASCE 7 Figures 6-11 through 6-17.

At any given instant, the wind pressures over the surface of the building vary greatly, and gusts cause the pressures in localized areas to exceed the average pressure on the entire building.

Furthermore, observed wind damage to buildings and wind tunnel tests indicate that very high suctions or negative pressures occur at the building's discontinuities, such as eaves, ridges, and wall corners.

Therefore, higher values for the external pressure coefficients are used in the design of components and cladding. These factors are shown in ASCE 7 Figures 6-11 through 6-17. The other factors in the pressure formula remain the same as for the design of the primary lateral-force-resisting system.

Example #9

If the building in Example #2 has CMU walls that span between the foundation and the roof, what are the wind pressures on the walls?

Solution:

Since this building has a mean roof height of less than 60 feet, the following equation can be used to determine the pressures on the CMU walls:

$$p = q_h[(GC_p) - (GC_{pi})]$$

From Example #2, $q_h = 10.3$ psf.

The external pressure coefficients are determined from ASCE 7 Figure 6-11A, shown in Figure 13.19. Pressure coefficients are needed for Zones 4 and 5, which depend on the effective wind area. The effective wind area is the span length (15 feet) multiplied by an effective width, which is equal to one-third the span length. Thus, the effective area in this example is equal to (15)(15/3) = 75 square feet.

The following external pressure coefficients are determined from Figure 6-11A for an effective area of 75 square feet:

Zone 4: $GC_p = -0.95, +0.85$
Zone 5: $GC_p = -1.09, +0.85$

According to Note 5 in the figure, these pressure coefficients are to be reduced by 10 percent for buildings with a roof slope less than or equal to 10 degrees:

Zone 4: $GC_p = -0.86, +0.77$
Zone 5: $GC_p = -0.98, +0.77$

The width of Zone 5, a, was determined from Example #7, and is equal to 5 feet.

From ASCE 7 Figure 6-5, the internal pressure coefficient GC_{pi} for an enclosed building is equal to –0.18, +0.18.

The pressures on the CMU walls are as follows:

- Corner Zone 5
  $p = 10.3(-0.98 - 0.18) =$ *–12.0 psf*
  $p = 10.3(0.77 + 0.18) =$ *9.8 psf*
- Interior Zone 4
  $p = 10.3(-0.86 - 0.18) =$ *–10.7 psf*
  $p = 10.3(0.77 + 0.18) =$ *9.8 psf*

Example #10

If the building in Example #2 has open web steel joists spaced 5 feet on center that span 50 feet between the CMU walls, what are the wind pressures on the joists?

Solution:

As in Example #9, the following equation can be used to determine the pressures on the joists:

$p = q_h[(GC_p) - (GC_{pi})]$

From Example #2, q_h = 10.3 psf.

The external pressure coefficients are determined from ASCE 7 Figure 6-11B, shown in Figure 13.20. Pressure coefficients are needed for Zones 1, 2, and 3, which depend on the effective wind area. The effective wind area is the larger of the following:

A = span × spacing of joist = 50 × 5 = 250 square feet.

A = span × (span/3) = 50 × (50/3) = 833 square feet (governs).

The following external pressure coefficients are determined from Figure 6-11B for an effective area of 833 square feet:

Zone 1: GC_p = –0.90, +0.20
Zone 2: GC_p = –1.10, +0.20
Zone 3: GC_p = –1.10, +0.20

It is evident that the corner (Zone 3) and the eave (Zone 2) have the same pressures in this example.

The width of Zones 2 and 3, a, was determined from Example #7, and is equal to 5 feet.

From ASCE 7 Figure 6-5, the internal pressure coefficient GC_{pi} for an enclosed building is equal to –0.18, +0.18.

The pressures on the joists are as follows:

- Corner Zone 3 and Eave Zone 2
  p = 10.3(–1.10 – 0.18) = *–13.2 psf*
  p = 10.3(0.20 + 0.18) = *3.9 psf*
- Interior Zone 1
  p = 10.3(–0.90 – 0.18) = *–11.1 psf*
  p = 10.3(0.20 + 0.18) = *3.9 psf*

COMBINED VERTICAL AND HORIZONTAL FORCES

All building components must be designed for load combinations using either (1) strength design or load and resistance factor design or (2) allowable stress design. The appropriate set of load combinations usually depends on the design requirements for the material specified. For example, load combinations based on strength design are used for reinforced concrete, but for steel, load combinations for strength design *or* allowable stress design are used.

Many load combinations are given for the various load types that can act on a building. The following combinations include dead loads (D), live loads (L), roof live loads (L_r), snow loads (S), and wind loads (W) only. A complete set of load combinations can be found in the IBC or ASCE 7.

Strength design load combinations:

1. $1.2D + 1.6(L_r \text{ or } S) + (f_1L \text{ or } 0.8W)$
2. $1.2D + 1.6W + f_1L + 0.5(L_r \text{ or } S)$
3. $0.9D + 1.6W$

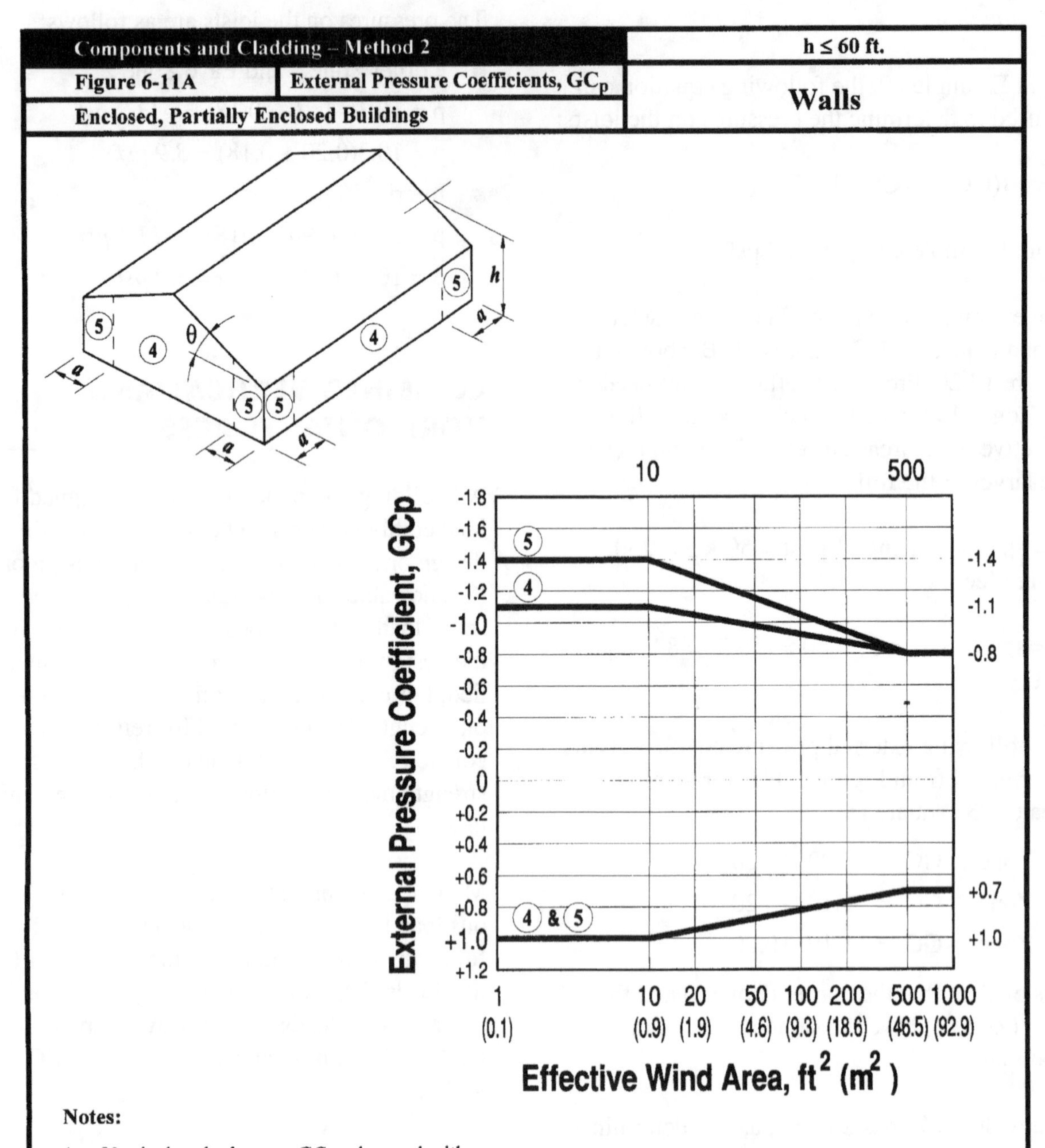

Notes:

1. Vertical scale denotes GC_p to be used with q_h.
2. Horizontal scale denotes effective wind area, in square feet (square meters).
3. Plus and minus signs signify pressures acting toward and away from the surfaces, respectively.
4. Each component shall be designed for maximum positive and negative pressures.
5. Values of GC_p for walls shall be reduced by 10% when $\theta \leq 10°$.
6. Notation:
   - a: 10 percent of least horizontal dimension or 0.4h, whichever is smaller, but not less than either 4% of least horizontal dimension or 3 ft (0.9 m).
   - h: Mean roof height, in feet (meters), except that eave height shall be used for $\theta \leq 10°$.
   - θ: Angle of plane of roof from horizontal, in degrees.

Figure 13.19

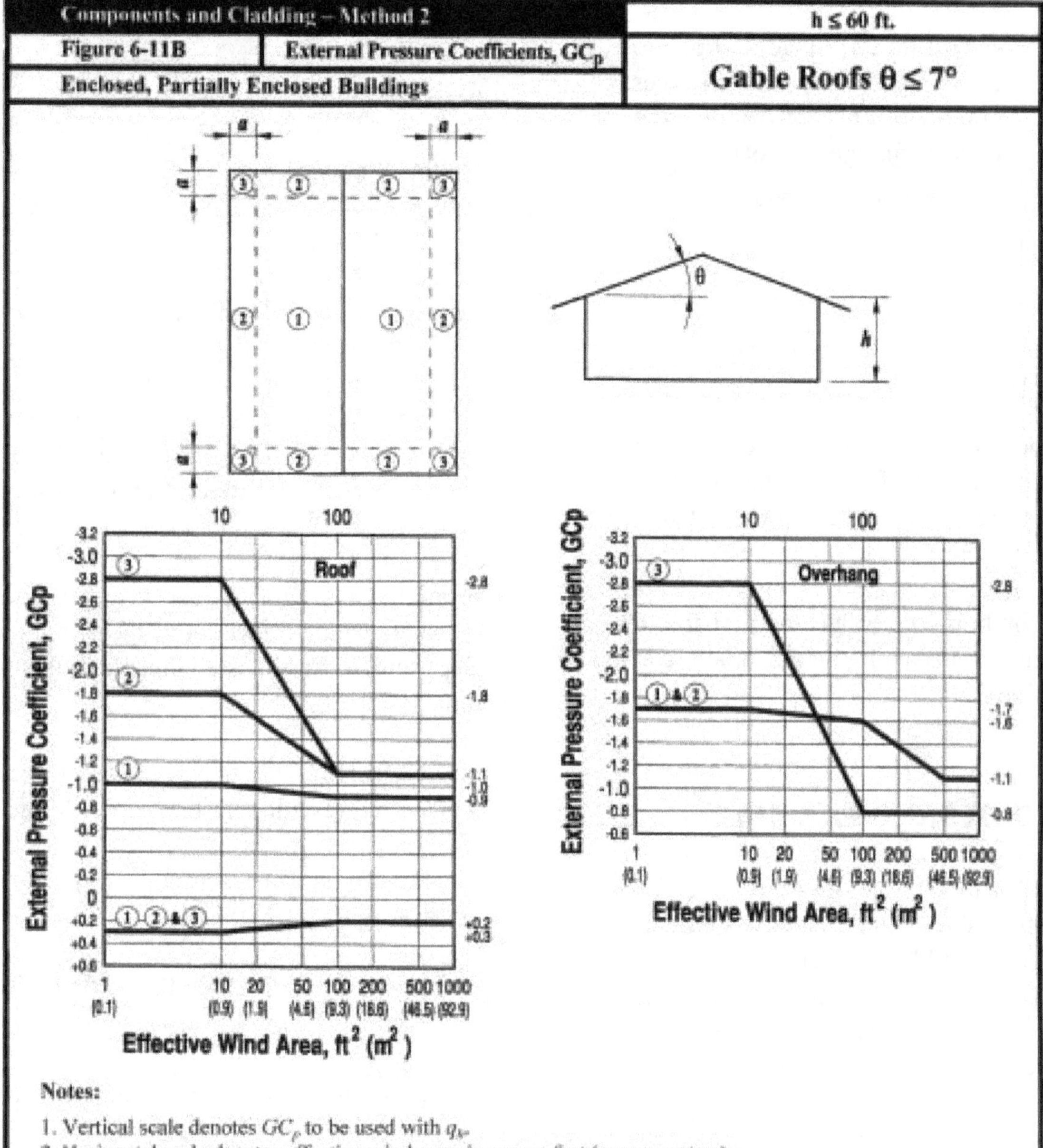

Notes:

1. Vertical scale denotes GC_p to be used with q_h.
2. Horizontal scale denotes effective wind area, in square feet (square meters).
3. Plus and minus signs signify pressures acting toward and away from the surfaces, respectively.
4. Each component shall be designed for maximum positive and negative pressures.
5. If a parapet equal to or higher than 3 ft (0.9m) is provided around the perimeter of the roof with $\theta \leq 7°$, the negative values of GC_p in Zone 3 shall be equal to those for Zone 2 and positive values of GC_p in Zones 2 and 4 shall be set equal to those for wall Zones 4 and 5 respectively n Figure 6-11A.
6. Values of GC_p for roof overhangs include pressure contributions from both upper and lower surfaces.
7. Notation:
   - a: 10 percent of least horizontal dimension or 0.4h, whichever is smaller, but not less than either 4% of least horizontal dimension or 3 ft (0.9 m).
   - h: Eave height shall be used for $\theta \leq 10°$.
   - θ: Angle of plane of roof from horizontal, in degrees.

Figure 13.20

where

f_1 = 1.0 for floors in places of public assembly, for live loads in excess of 100 pounds per square foot, and for parking garage live load

f_1 = 0.5 for other live loads

Allowable stress design load combinations:

1. D + W
2. D + 0.75W + 0.75L + 0.75(L_r or S)
3. 0.6D + W

The code also gives a set of alternate basic load combinations for allowable stress design, which were previously given in the Uniform Building Code.

Load combinations reflect the probabilities of various loads acting on the structure at the same time.

WIND DESIGN EXAMPLE

The following is a step-by-step design of a small commercial building built to resist wind forces. The building is identical to that used in the earthquake design example that starts on page 285 and is assumed to be located in an urban area where the basic wind speed is 90 mph.

As with the earthquake design example, the exam will not have a problem as lengthy or complex as this. However, careful study of this design example should help you understand the basic principles of wind design.

Wind Pressure

What are the wind pressures for wind in the north-south direction using Method 2?

Solution:

The wind pressures are determined using the basic formulas in Method 2.

On the windward wall:

$p_w = q_z G C_p - q_i(GC_{pi})$

where $q_z = 0.00256 K_z K_{zt} K_d V^2 I$

Since this building is in an urban area, it is categorized as Exposure B. It is also assumed that this one-story building is enclosed and rigid, and that it is not located on a hill or escarpment: $K_{zt} = 1.0$.

From ASCE 7 Table 6-3, K_z is a constant between 0 and 15 feet above the ground. Under Exposure B, Case 2 is applicable in this situation, so $K_z = 0.57$.

From ASCE 7 Table 6-4, $K_d = 0.85$ for main wind-force-resisting systems of buildings.

The building is not essential, high occupancy, or hazardous: I = 1.0.

Therefore, $q_z = 0.00256 \times 0.57 \times 1.0 \times 0.85 \times 90^2 \times 1.0 = 10.1$ psf

The gust effect factor G is taken as 0.85.

For the windward wall, $C_p = 0.8$ from ASCE 7 Figure 6-6.

Windward pressure $p_w = 10.1 \times 0.85 \times 0.8 =$ *6.9 psf inward.*

On the leeward wall:

$p_l = q_h G C_p - q_i(GC_{pi})$

where $q_h = 0.00256 K_h K_{zt} K_d V^2 I$

Since the building has a mean roof height of 14 feet, q_h = 10.1 psf, which is the same as on the windward wall.

The pressure coefficient C_p for the windward wall depends on the plan dimensions of the building, as shown in ASCE 7 Figure 6-6. For wind in the north-south direction, L/B = 60/150 = 0.4, so that C_p = –0.5.

Leeward pressure $p_l = 10.1 \times 0.85 \times (-0.5)$ = –4.3 psf = *4.3 psf outward.*

On the parapet:

The wind pressure on the parapet is determined from the following formula:

$$p_p = q_p GC_{pn}$$
$$q_p = 0.00256 K_z K_{zt} K_d V^2 I$$

From ASCE 7 Table 6-3, K_z = 0.59 for z = 17 feet by linear interpolation.

Thus, $q_p = 0.00256 \times 0.59 \times 1.0 \times 0.85 \times 90^2 \times 1.0 = 10.4$ psf.

For the windward parapet, GC_{pn} = 1.5, so that $p_p = 10.4 \times 1.5 =$ *15.6 psf inward.*

For the leeward parapet, GC_{pn} = –1.0, so that $p_p = 10.4 \times (-1.0) = -10.4$ psf = *10.4 psf outward.*

These pressures are shown in Figure 13.21.

Diaphragm Design for North-South Direction

Design the roof diaphragm. Determine the diaphragm shear, select a plywood diaphragm, and calculate the maximum chord force in the diaphragm.

Solution:

The north and south walls span vertically between grade and the roof diaphragm when subject to wind load, as shown in Figure 13.21.

The bottom reaction of 70 pounds per foot is resisted by the footings and soil. The pressures on the windward and leeward faces have been combined for convenience (see Figure 13.22).

The top reaction of 165 pounds per foot is resisted by the roof diaphragm, which acts as a horizontal beam spanning between the east and west shear walls, as shown in Figure 13.23.

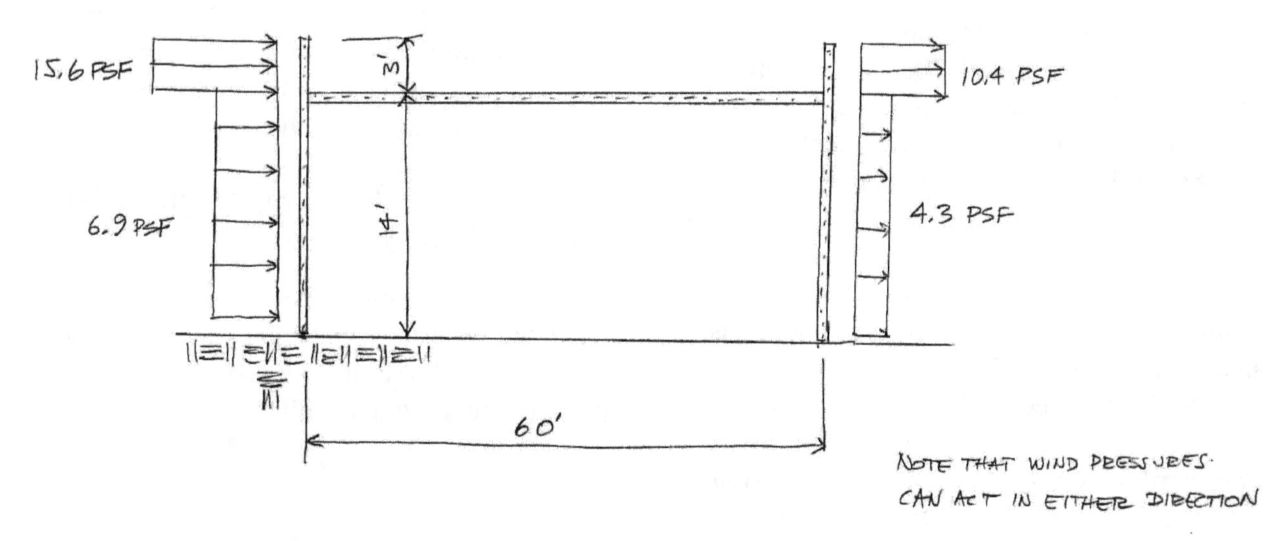

Figure 13.21

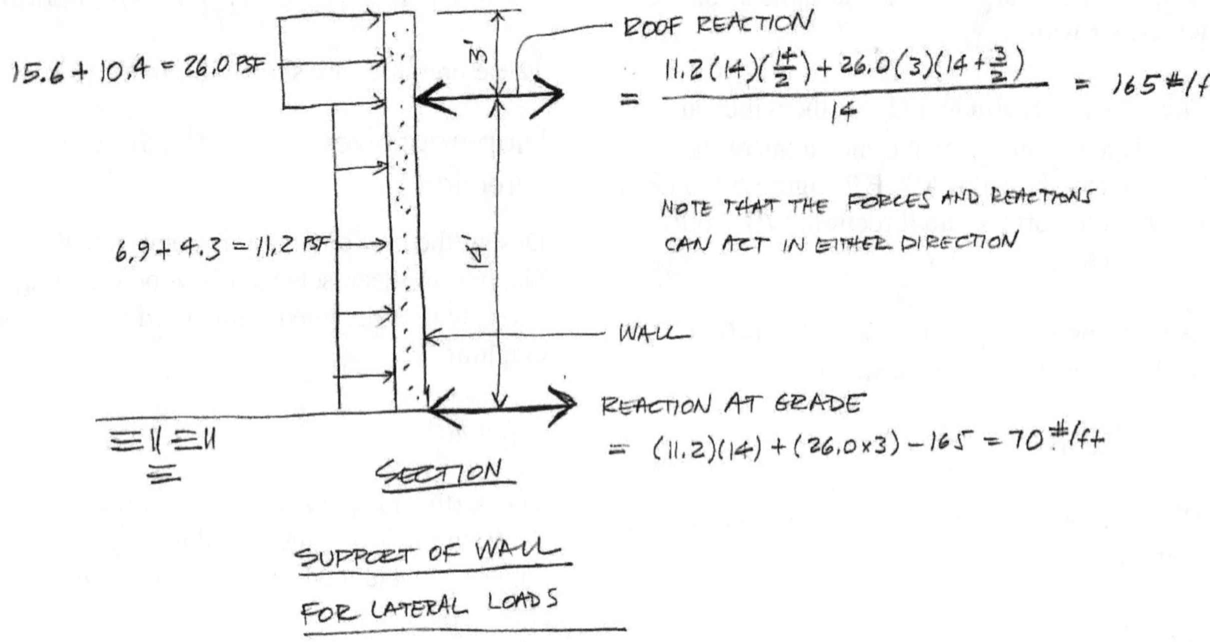

Figure 13.22

The plywood roof diaphragm is a flexible diaphragm, which acts as a simple beam. The reaction to each shear wall is therefore 165#/ft. × 150 ft./2 = 12,375#.

The diaphragm shear is equal to this reaction divided by the diaphragm width = 12,375#/60 ft. = *206#/ft.*

We select a plywood diaphragm having an allowable shear of at least 206#/ft. from IBC Table 2306.2.1, shown in Figure 12.30. Several diaphragms meet this requirement, and we select the following: *3/8" plywood sheathing, with 8d nails at 6" on center at boundaries, and continuous panel edges parallel to load, and at 6" at other panel edges, with blocking at all unsupported panel edges.* This diaphragm has an allowable shear of 240#/ft., which is greater than the 206#/ft. required.

The chord force is the tension or compression in the flanges, or chords, of the diaphragm.

Since the diaphragm acts like a uniformly loaded simple beam, its maximum moment = $wL^2/8$.

The chord force is the maximum diaphragm moment divided by the diaphragm depth

$$= \frac{165\#\text{ft.} \times (150 \text{ ft.})^2}{8 \times 60 \text{ ft.}} = 7{,}734\#.$$

As discussed in the earthquake design example, this chord force can be resisted by a continuous wood or steel member, or by reinforcing steel in the north and south walls.

Connection of Roof Diaphragm to East and West Shear Walls

Refer to the earthquake design example on page 285. What should be the maximum spacing of the anchor bolts to transfer the diaphragm shear into the walls?

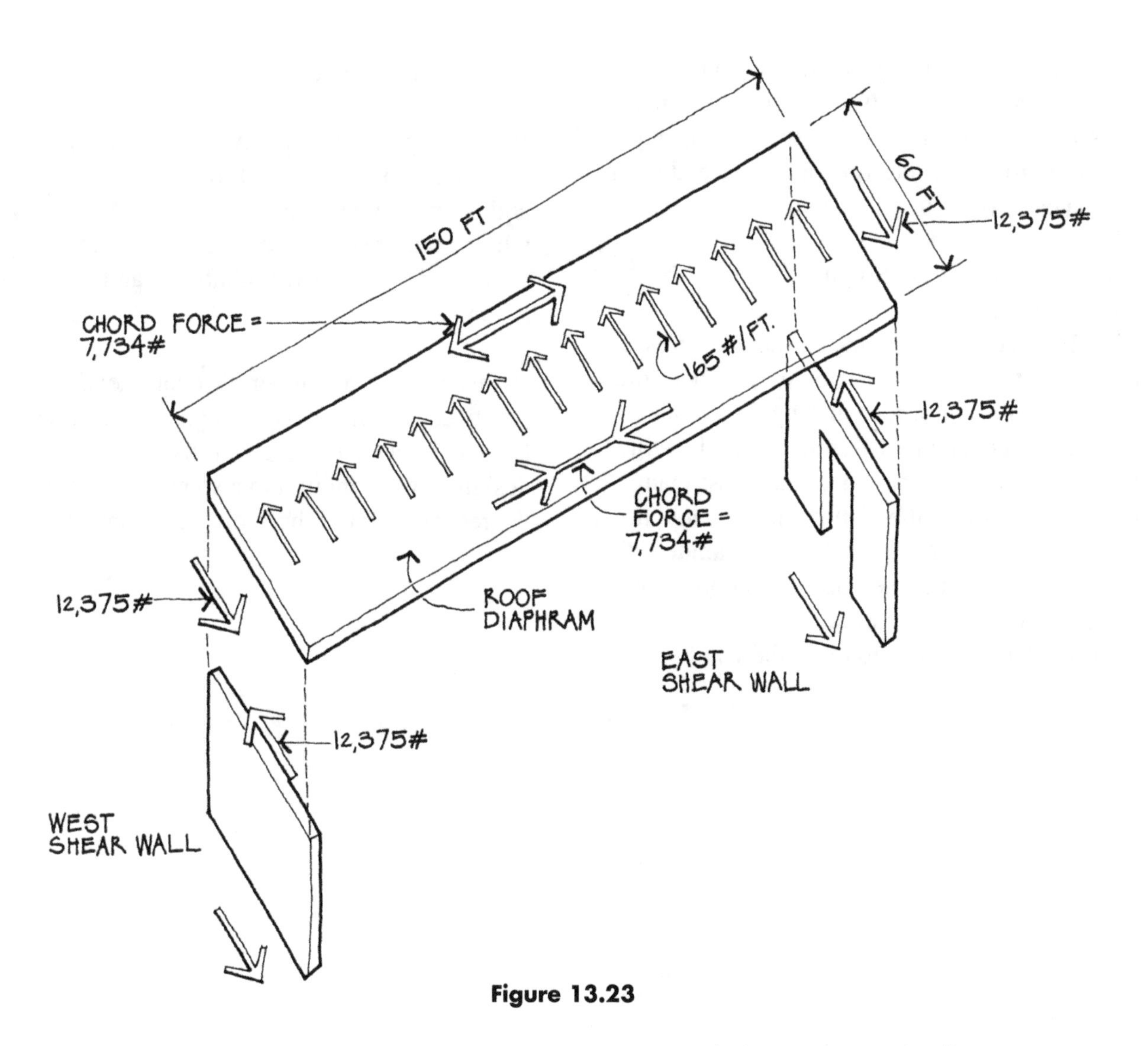

Figure 13.23

Solution:

The anchor bolts must be able to transfer the diaphragm shear of 206#/ft.

The required anchor bolt spacing is equal to $\frac{1{,}400\#/\text{bolt}}{206\#/\text{ft.}} = 6.8$ ft./bolt.

While a bolt spacing of 6 feet on center would be adequate, it is common practice to space anchor bolts at no more than *four feet on center*.

East and West Shear Walls

The east and west shear walls must be adequate to resist the shear from the diaphragm, which is 12,375#. Note that the unit shear stress is greater in the east wall because the 20-foot opening reduces the net area of the wall available to resist the shear stress.

The walls must also resist the overturning moment caused by the shear. If the resisting moment from the dead load of the wall and roof is less than 1.67 times the overturning moment, the wall must be tied to the footing to provide additional overturning resistance.

Because of the building's configuration, wind forces in the north-south direction cause greater stresses in the diaphragm and shear walls than east-west forces. Therefore, in this case, design for the east-west direction is not necessary.

Comparing this wind design with the earthquake design for the same building assigned to SDC D, we can see that the earthquake stresses are greater than the wind stresses and therefore govern the design of the building. In actual practice, once having determined that the wind load to the roof diaphragm is much less than the seismic load to the diaphragm, it would not be necessary to proceed with the remainder of the wind design; the members and connections designed to resist the earthquake forces would be more than adequate to resist the wind forces.

CONCLUSION

Wind design is very much like earthquake design. For both, we want to minimize torsion and overturning effects, avoid building shapes which could result in stress concentrations, adhere strictly to drift limitations, and provide redundancy.

In this context, a number of recommended practices for earthquake design are discussed in Lesson Twelve. These generally apply to wind design as well, though to a somewhat lesser degree, and should therefore be reviewed.

LESSON 13 QUIZ

1. Using Method 2, the wind pressure on a building is assumed to act

I. inward on the windward wall.
II. outward on the leeward wall.
III. upward on a flat roof.
IV. inward on the leeward wall.
V. downward on a flat roof.
VI. outward on the windward wall.

A. I, II, and III
B. I, III, and IV
C. II, III, and VI
D. I, II, and V

2. When designing buildings for wind forces, which of the following are considered? Check all that apply.

A. Drift
B. Liquefaction
C. Tsunamis
D. Overturning moment

3. A square building 15 feet high with Exposure C and located in an area with a basic wind speed of 90 mph should be designed for what pressure on the windward wall, using Method 2? Assume I = 1.0 and neglect internal pressure.

A. 6.2 psf
B. 10.2 psf
C. 6.7 psf
D. 14.0 psf

4. The IBC/ASCE 7 wind load coefficients account for which of the following?

I. Height of the building
II. Exposure of the site
III. Hurricanes and tornadoes
IV. Wind gusting

A. I only
B. II and III
C. I, II, and IV
D. I, II, III, and IV

5.

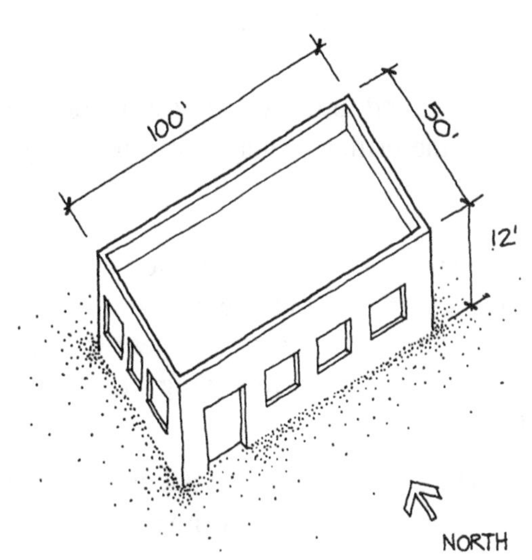

Using the Analytical Procedure, the one-story building shown above has a design wind pressure (p) on the vertical projected area of 20 psf. For wind acting in the north-south direction, what is the total wind force to be transferred from the roof diaphragm to the east or west shear wall?

A. 3 kips
B. 6 kips
C. 12 kips
D. 24 kips

6. For the building in the preceding question, what is the maximum diaphragm shear?

A. 120#/ft.
B. 240#/ft.
C. 60#/ft.
D. 480#/ft.

7. Elements and components of structures are designed for wind pressures that are

A. less than those used for the design of the primary lateral force resisting system.
B. the same as those used for the design of the primary lateral force resisting system.
C. greater than those used for the design of the primary lateral force resisting system.
D. unrelated to those used for the design of the primary lateral force resisting system.

8. A shear wall resisting wind forces is subject to an overturning moment of 400 ft.-kips. The dead load resisting moment is 500 ft.-kips. Which of the following statements is CORRECT?

A. The shear wall has an adequate factor of safety against overturning.
B. The shear wall does not have an adequate factor of safety against overturning and must, therefore, be made longer and/or heavier.
C. The shear wall does not have an adequate factor of safety against overturning and must, therefore, have sufficient anchorage to the foundation to prevent overturning.
D. None of the above statements are correct because buildings are not usually designed to resist the overturning moments caused by wind forces.

9. Which of the following methods is likely to be most economical when designing a very tall building to resist wind forces?

A. Moment-resisting frames
B. Shear walls
C. Base isolation
D. Tubular system

10. All of the following load combinations must be considered in the allowable stress design of building components EXCEPT

A. dead plus earthquake plus wind.
B. dead plus wind.
C. dead plus 3(wind)/4 plus 3(floor live)/4 plus 3(roof live or snow)/4.
D. 0.6 times dead plus wind.

LESSON FOURTEEN 14

NOTABLE BUILDINGS AND ENGINEERS

INTRODUCTION

For many years, the licensing examinations included a test in architectural history. This is no longer the case. Instead, history questions are now scattered throughout the various written tests, including the Structural Systems division.

These questions may test a candidate's knowledge of the structural systems and materials used in notable buildings, the achievements of famous engineers, as well as the ability to identify specific buildings from sketches or photographs. To help candidates prepare for such questions, we have compiled a short illustrated list of a number of buildings notable for both their structure and their architecture.

Those candidates wishing to pursue this subject further are referred to the Architectural History study guide.

Also included in this lesson are brief biographical sketches of several well-known engineers, a discussion on recent high-tech architecture resulting from the collaboration of architects and structural engineers, and a short summary of structural failures.

NOTABLE BUILDINGS

Pantheon

This great concrete structure, the largest dome of the ancient world, was built by the Romans in the year 123 AD and still stands today (Figure 14.1). Its architects intuitively understood the nature of the stresses in a concrete dome: the lower part tends to crack because of circumferential tensile stresses. Since they had no materials that were strong in tension, their solution was to make the dome walls about 20 feet thick at the bottom in order to keep the unit tensile stresses low enough to be resisted by the concrete. To reduce the weight of the dome, its underside was coffered. The Pantheon remains to this day the most extraordinary example of Roman architecture and engineering.

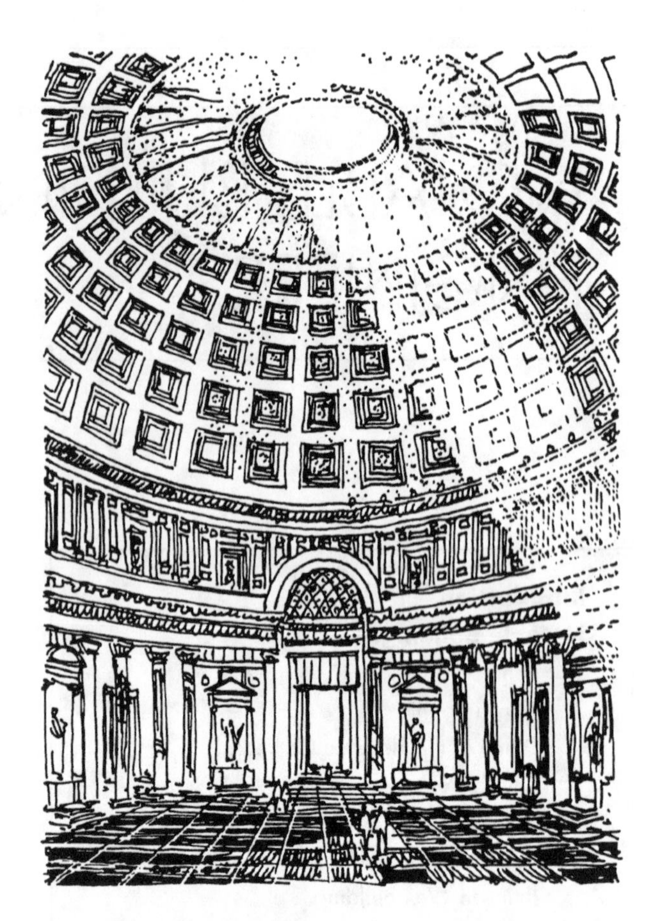

PANTHEON

Figure 14.1

Hagia Sophia

Hagia Sophia in Constantinople (now Istanbul), completed in 537 AD, is very interesting because of its architectural, engineering, historical, and religious significance (Figure 14.2). Its architects, Anthemius and Isidorus, created a magnificent edifice dedicated to the glory of both church and state. Its main dome was shallow and supported by four pillars, through pendentives and arches that rose from the pillars. The arches resisted both vertical forces and outward thrusts from the dome; unfortunately, however, the

HAGIA SOPHIA - SECTION

Figure 14.2

arches did not provide sufficient buttressing, and a portion of the dome collapsed in 558 A.D., while repairs were being made following an earthquake. The spherical dome was dismantled and rebuilt with a 20 foot greater rise, thus reducing the outward thrust. Through the years, more damage and reconstruction occurred, largely as a result of earthquakes, until in 1847, an iron tie was placed around the base of the dome. The outward thrust of the dome was now resisted by tension in the iron tie rather than by other elements of the structure. No further damage occurred, and the building still survives today, after 14 centuries.

Dome of the Florence Cathedral

The dome of Santa Maria del Fiore in Florence, designed by the great Renaissance architect Filippo Brunelleschi and completed in 1436, is a masterpiece of aesthetic and structural design (Figure 14.3). There are actually two masonry domes: a thick inner shell and a thinner outer shell. Brunelleschi understood that a dome tends to spread apart and built in a series of circumferential iron chains to act as tension rings and hold the dome in equilibrium. His design permitted the dome to be constructed without the use of any temporary shoring, a feat unparalleled in its time.

Figure 14.3

Crystal Palace

The Great Exhibition of 1851, in London, was the largest display of man's progress ever assembled up to that time. It was housed in an immense prefabricated glass and cast iron structure, known as the Crystal Palace, that was without precedent (Figure 14.4). Designed by Joseph Paxton, it was more than a third of a mile long, enclosed nearly a million square feet, and was fabricated and erected in only six months. With its lightness and transparency, it influenced subsequent iron and glass buildings, and even today's steel and glass skyscrapers.

Figure 14.4

Fallingwater

The Kaufmann House, completed in 1936 in Bear Run, Pennsylvania is popularly known as "Fallingwater" (Figure 14.5). It has become one of the best-known houses in the country and is widely considered one of the great houses of all time. Frank Lloyd Wright designed the striking three-story masonry structure, featuring six reinforced concrete terraces which cantilever over a natural waterfall and pool. Interior spaces open through large glass areas to the terraces, which offer breathtaking views of the waters below. Viewed from the exterior, Fallingwater seems to grow out of the natural site rather than being an addition to it.

FALLING WATER

Figure 14.5

Nervi's Airplane Hangars

The great Italian engineer Pier Luigi Nervi designed buildings that achieved beauty through the bold and imaginative expression of structure. His airplane hangars built for the Italian Air Force between 1936 and 1939 had lamella roofs formed by short prefabricated reinforced concrete members connected at their joints (Figure 14.6). Towards the end of World War II, the retreating German army blew up the buttresses supporting the hangar roofs to prevent the hangars from falling into Allied hands. The roofs fell 40 feet to earth, yet remained almost entirely intact, a tribute to the excellence of their design and construction.

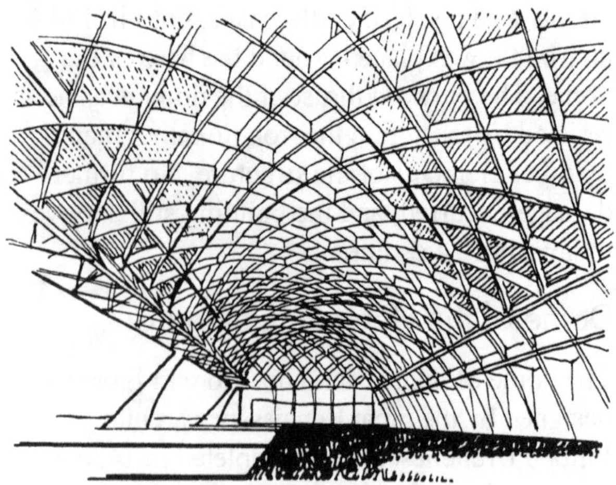

NERVI'S AIRPLANE HANGAR

Figure 14.6

Johnson Wax Building

In 1939, one of Frank Lloyd Wright's most famous structures, the Johnson Wax Building in Racine, Wisconsin was completed. The rounded exterior walls consist of horizontal bands of brick punctuated by strips of glass tubing. Perhaps the most interesting element of this building is a great work space with slender mushroom-shaped concrete columns that flare out at the top to support the roof (Figure 14.7). The structural columns, ceiling, and lighting form an integrated design and provide a space with virtually no sense of enclosure. The nearby multistoried laboratory tower, whose floors cantilever from a single central support, was built in 1950.

JOHNSON WAX

Figure 14.7

Figure 14.8

Yale University Skating Rink

In recent years, a number of buildings have been constructed in which steel cables in tension are the essential structural element. In the Yale University Skating Rink, designed by Eero Saarinen and completed in 1958, steel cables are hung from a central reinforced concrete arch (Figure 14.8). The outer ends of the cables are anchored to heavy curved perimeter walls. The roof is wood, the weight of which partially stabilizes the cables. In this building, the merger of architecture and engineering creates a unified and dramatic expression.

Palazzetto Dello Sport

In certain types of buildings, such as arenas, exhibition halls, and airplane hangars, the structure often determines the shape and character of the building. An outstanding example is the Palazzetto Dello Sport (Little Sports Palace) designed by Pier Luigi Nervi for the 1960 Rome Olympics. Its roof, a ribbed concrete shell dome, is supported by 36 Y-shaped concrete buttresses which resist the forces at the edge of the shell (Figure 14.9). In this building, the clearly expressed structure creates a pattern of unusual elegance and refinement.

Figure 14.9

Figure 14.10

Dulles International Airport

In the terminal of Dulles Airport in suburban Washington, D.C., completed in 1962, Eero Saarinen was able to depart from the conventional finger plan airport terminal by using a mobile lounge, which separated the terminal from the airplanes. He was thus able to express the character of the terminal by a single compact building. The concrete roof is supported by steel cables that are suspended between huge concrete columns that lean outward to balance the inward pull of the cables (Figure 14.10). The result is a successful marriage of architecture and structure.

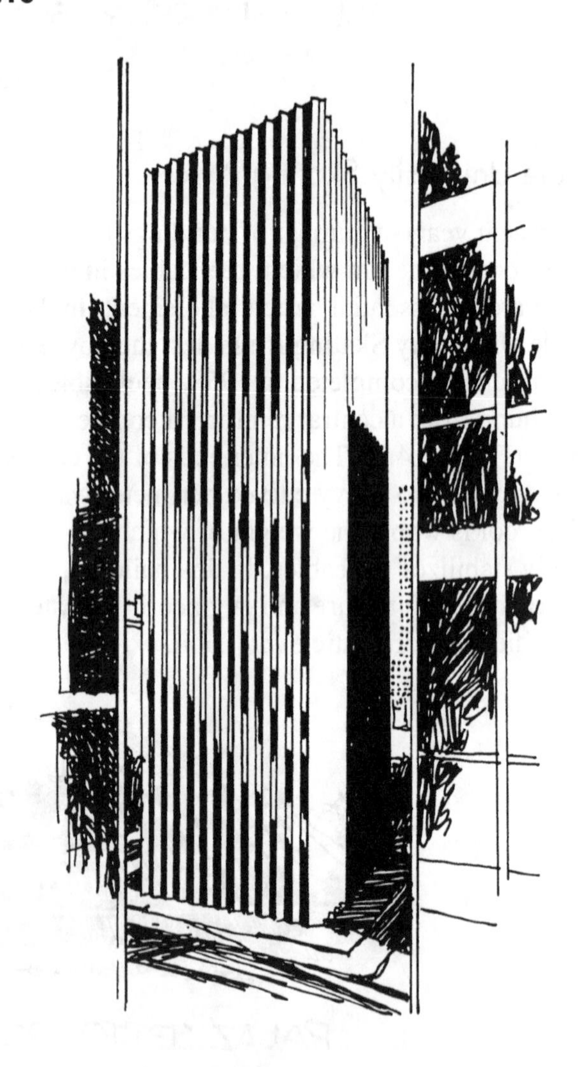

Figure 14.11

CBS Building

Most very tall buildings are framed with structural steel, but a number of interesting reinforced concrete skyscrapers have also been built. An outstanding example is the CBS Building in New York, completed in 1964, which was designed by Eero Saarinen and engineered by Paul Weidlinger. This 42-story structure resists lateral forces by both an inner core and perimeter walls, which consist of concrete piers five feet long spaced five feet apart. The building's verticality is emphasized by the granite-clad triangular piers, which extend uninterrupted from below street level to the very top of the building (Figure 14.11).

Toronto City Hall

In 1958, the Finnish architect Viljo Revell won the competition for the Toronto City Hall, completed in 1965. The design is unique among high-rise public buildings: two curved office slabs surrounding a low circular city council chamber (Figure 14.12). Although the curved shape may seem arbitrary, it is efficient structurally; each office tower is a huge curved shell that provides strength and rigidity against the overturning forces caused by wind or earthquake.

Figure 14.12

Houston Astrodome

Domes are not only the most dramatic and spectacular roof structures, they are also remarkably efficient. During the past 25 years, a number of enormous domed roofs have been built, enclosing stadiums seating up to 80,000 spectators. The first such structure, and still one of the most spectacular, is the Houston Astrodome. When it was completed in 1965, it was the largest enclosed stadium ever constructed, roofed by the largest dome ever built. Covering 9-1/2 acres, the steel lattice dome is 710 feet in diameter and rises 208 feet over the playing field (Figure 14.13). It weighs less than 30 pounds per square foot, one 20th of the weight of Brunelleschi's inner dome in Florence.

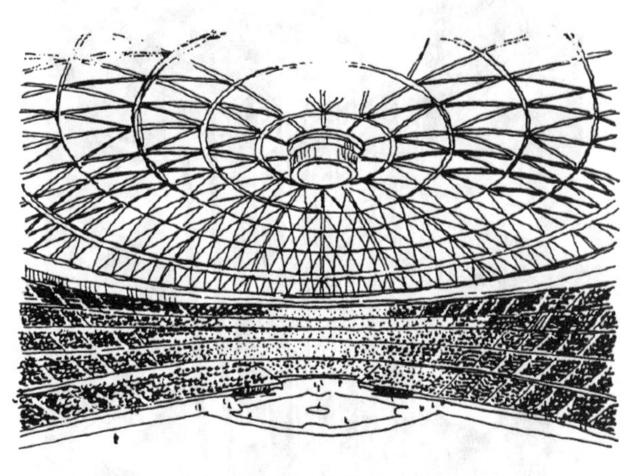

Figure 14.13

First National Bank, Chicago

High rise buildings utilize a variety of structural systems to resist lateral forces from wind or earthquake, principally rigid frames, shear walls, braced frames, and tubular systems. A special type of rigid frame sometimes used in buildings over 50 stories is called a "superframe" or "mega-frame." This consists of a very deep, stiff horizontal truss or girder wherever a mechanical floor occurs, about 15 to 20 stories apart, connected at each end to a large exterior column. The interior columns and horizontal girders at the other levels form a secondary rigid frame. This type of system was used in the 60-story First National Bank Building in Chicago, which was designed by Perkins and Will with C.F. Murphy Associates and completed in 1966. This building has a slender profile that tapers gracefully to a wider base to resist wind overturning forces more effectively (Figure 14.14).

John Hancock Building, Chicago

The familiar John Hancock Building, now a prominent part of the Chicago skyline, expresses its structure through its tapered form and enormous exposed exterior X-bracing (Figure 14.15).

Figure 14.14

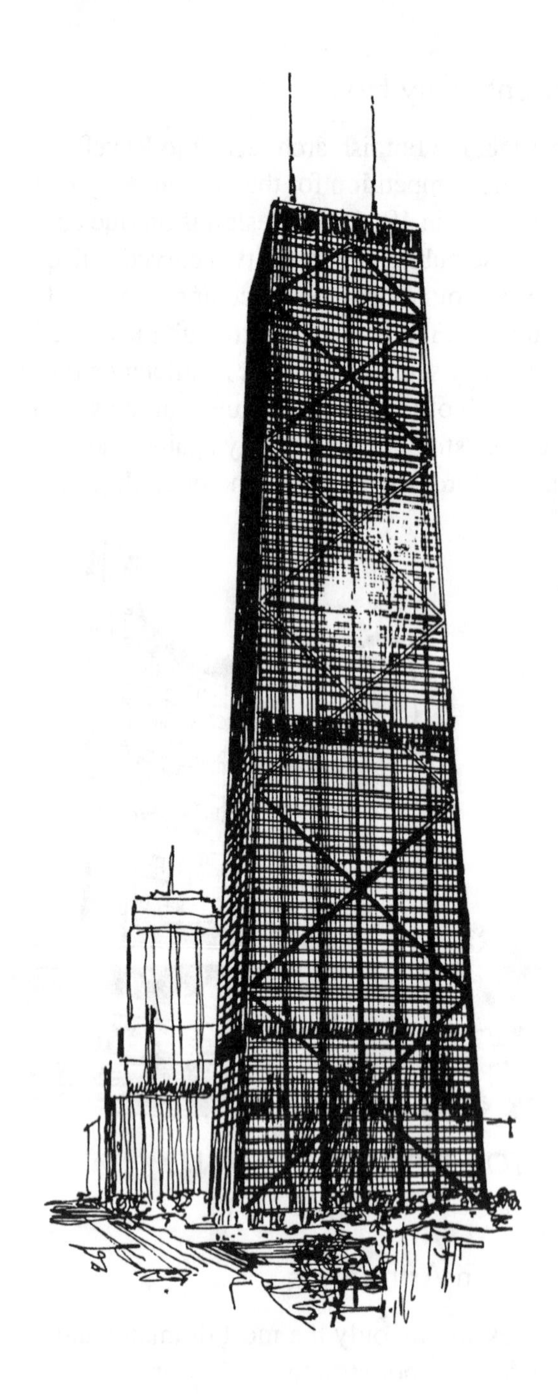

Figure 14.15

Designed by Skidmore, Owings, and Merrill and completed in 1968, "Big John" is a gigantic trussed tube, which is very efficient in resisting wind forces. The overall dimension of the building is utilized to resist overturning forces, while the truss members resist shear by direct stress rather than by bending. For a time, this 100-story multi-use structure was the tallest building in Chicago, a distinction it has since relinquished to Willis Tower.

Knights of Columbus Building

The 26-story Knights of Columbus Building in New Haven, Connecticut designed by Kevin Roche and John Dinkeloo and completed in 1969, makes a clear and powerful statement. Its four corner towers, constructed of concrete with dark brick veneer, support the main 80-foot long horizontal steel girders, which in turn support the steel floor structure (Figure 14.16). The towers also resist horizontal wind or earthquake forces by acting as huge tubes that cantilever from the foundation. Within the towers are service elements such as stairs, toilets, and mechanical shafts, and the six elevators are contained in a core at the building's center.

Figure 14.16

U.S. Pavilion at Expo '70

World's fairs and Olympic games have often been the background for exotic, state-of-the-art structures. The U.S. Pavilion at Expo '70 in Osaka, Japan, had an incredible 100,000 square foot inflatable roof made of a special vinyl membrane, with stiffening steel cables anchored to a concrete compression ring around the perimeter (Figure 14.17). The roof was designed to resist wind forces as well as the air pressure inside the pavilion. This building was a pioneering effort by engineer David Geiger in the field of pneumatic structures.

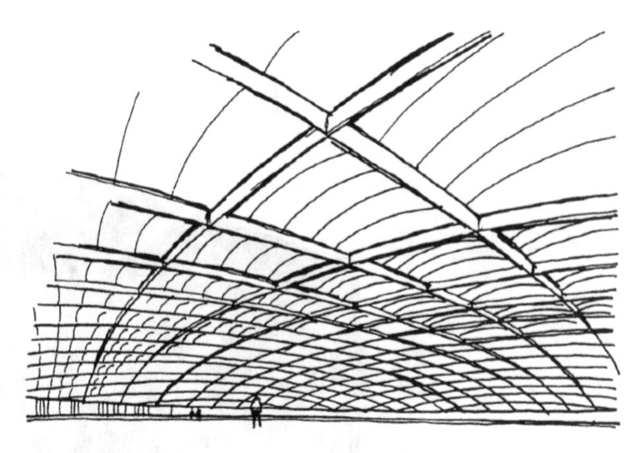

Figure 14.17

Munich Olympic Stadium

The Munich Olympics of 1972 featured three new major arenas, the largest of which was the Olympic Stadium. Its roof design, by Frei Otto, was strikingly original—an immense high-tech tent. A series of steel cable nets was stretched between steel masts that were anchored to the ground by steel cables (Figure 14.18). The net surfaces, covered with plexiglass, were shapes of double curvature for stability. The design was bold, certainly controversial, and a giant step forward for tensile structures.

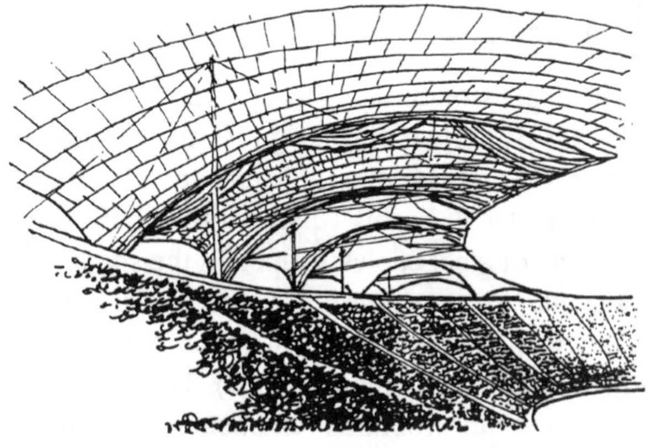

Figure 14.18

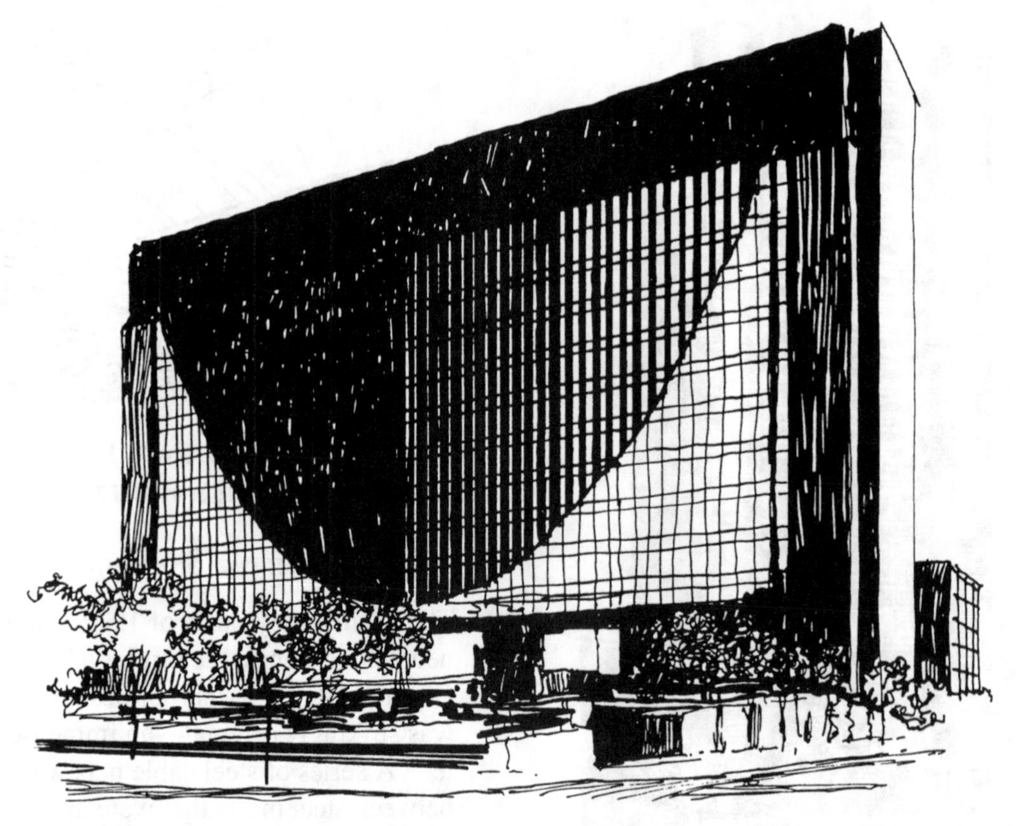

Figure 14.19

Federal Reserve Bank Building, Minneapolis

The suspension bridge concept was used by Gunnar Birkerts in his design for the Federal Reserve Bank Building in Minneapolis, completed in 1972. In this 10-story office building, two sets of steel cables, draped in the shape of a catenary, support the building's vertical load and are anchored to a pair of concrete towers 275 feet apart (Figure 14.19). The facade expressed the structure dramatically by using glass on the inside face of the mullions above the catenary and on the outside face below.

World Trade Center

The twin towers of the World Trade Center in New York, completed in 1972, were striking in scale, yet elegantly simple in design (Figure 14.20). Its statistics were staggering: 1,350 feet high, 110 stories, 9 million square feet, a working population of 50,000. As with all skyscrapers, two major obstacles that had to be overcome were the elevator system and the structural system to resist wind or earthquake forces. In this case, architect Minoru Yamasaki and his consulting engineers solved both problems imaginatively. The elevator system utilized a combination of express and local elevators, thus greatly increasing the area available for offices. The structural system comprised exterior columns only three feet apart connected by deep spandrels, so that the entire tower became an immense hollow cantilever tube. By any measure, the World Trade Center was an impressive part of the Manhattan skyline.

On September 11, 2001, the twin towers were tragically attacked by terrorists. The attack caused immense fires within the buildings and eventually compromised the buildings' steel structural system and resulted in their collapse. The strength and redundancy of the towers' innovative structural tube system did withstand the initial attack and withstood the blaze long enough for thousands of occupants to safely escape.

THE WORLD TRADE CENTER
Figure 14.20

Willis Tower

A recent innovation in the design of skyscrapers is the tubular concept, in which the structure acts like an immense, hollow, tubular column which cantilevers out of its foundation under the action of wind loads. Completed in 1976, the Willis Tower was originally named the Sears Tower. This Chicago structure is a bundle of nine tubes, each 75 feet square, placed next to each other to form a pattern of three squares in each direction (Figure 14.21). The square tubes end at varying heights, with only two of them extending the full 1,450-foot height of the building.

Designed by Skidmore, Owings, and Merrill, with Fazlur Kahn as chief engineer, the Willis Tower is one of the most notable achievements in skyscraper design and is currently the tallest building in the United States.

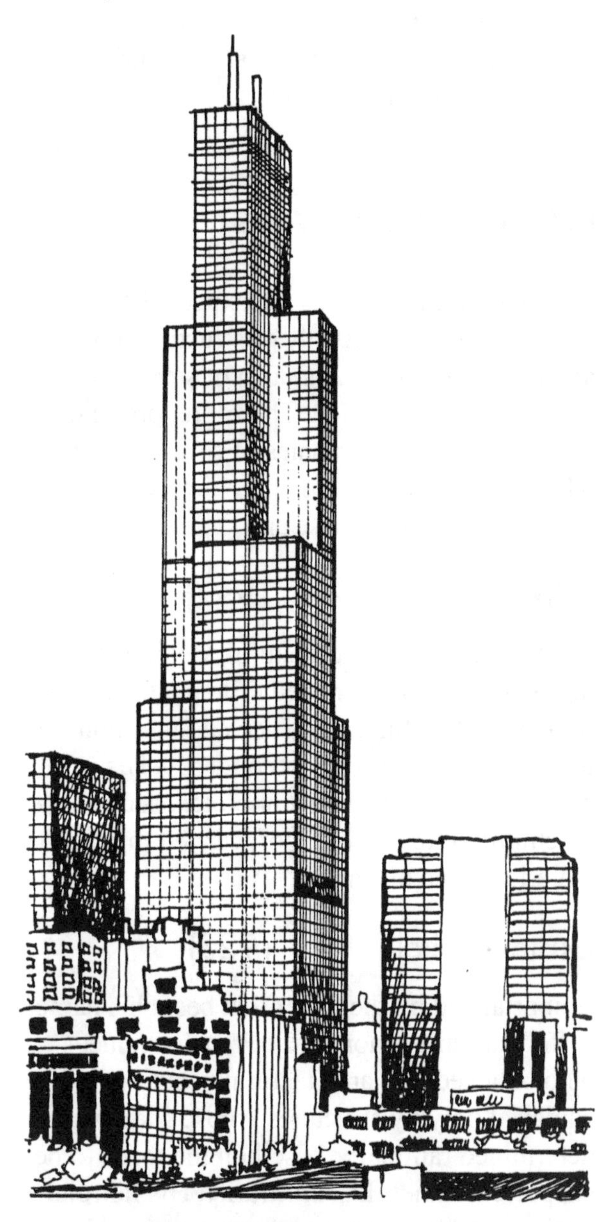

WILLIS TOWER
Figure 14.21

Taller buildings have been built in Malaysia, namely the Petronas Towers, designed by Cesar Pelli, with a measured height of 1,483 feet for each tower. The Taipei 101 building, located in Taipei, Taiwan, was completed in 2004 and reaches a height of 1,671 feet.

The Burj Dubai, recently constructed in Dubai, United Arab Emirates, was completed in 2010 and is currently the world's tallest building. The final height is 2,723 feet.

NOTABLE ENGINEERS

Architectural candidates are expected to have some knowledge not only of structural concepts and actual structures, but also of structural engineers of distinction. Therefore, presented below are brief biographies of several engineers whose talent was exceptional, transcending technology to create true art.

Felix Candela (1910–1997)

Born in Spain but a resident of Mexico for most of his life, Candela created thin-shell concrete roofs based on his experience and intuition, using mathematics only secondarily. His designs for hyperbolic paraboloids, such as the Xochimilco Restaurant roof, carried thinness of material to the ultimate.

Gustave Eiffel (1832–1923)

Immortalized by the tower that bears his name, Eiffel was an extraordinary French engineer who believed that architectural beauty could be achieved only by structures whose form was determined rationally, based on the loads to be supported, rather than arbitrarily. His railroad stations and bridges used the new materials of 19th-century technology, rolled iron and steel, and were structures of great strength and beauty.

Eugene Freyssinet (1879–1962)

The French engineer Freyssinet pioneered the development of prestressed concrete, which he used for a number of bridges that were visually and conceptually elegant. His most famous buildings were the two immense Orly Dirigible Hangars, near Paris, which were destroyed during World War II.

Fazlur Khan (1930–1982)

A Pakistani by birth, Khan was a brilliant structural engineer and a partner in the Skidmore, Owings, and Merrill firm. In his buildings, such as the John Hancock Building and Willis Tower, both in Chicago, form evolved as a result of structural ideas rather than the other way around.

Robert Maillart (1872–1940)

Maillart, a Swiss engineer, was among the first to recognize the aesthetic and technical potential of reinforced concrete. His designs for arched concrete bridges, originally accepted because of their economy, are now recognized as works of art.

Pier Luigi Nervi (1891–1979)

The great Italian contractor and engineer Nervi created soaring concrete shell roofs to house sports event, exhibitions, and aircraft. Like Maillart, his work successfully combined technical excellence with a conscious aesthetic intent.

John Roebling (1806–1869) and his son Washington Roebling (1837–1926)

The Roeblings were pioneers in the design and construction of suspension bridges. Their crowning achievement was the Brooklyn Bridge, which combined lightness with strength to create a structure of enduring beauty.

Eduardo Torroja (1899–1961)

Although trained in his native Spain as an engineer, Torroja was a great creator of architectural form. His concrete shell structures, most notably the roof of the Zarzuela racetrack grandstand in Madrid, united rational engineering design with inspired creativity.

Others

Others who are equally noteworthy include:

Othmar Ammann, the Swiss-American engineer, one of the greatest bridge designers of this century.

Benjamin Baker, designer of the Forth Bridge in Scotland.

Santiago Calatrava, Spanish-born architect and engineer, noted for engineering talent and artistic sensibility in designing contemporary bridges and buildings, including the Milwaukee Art Museum.

Horst Berger and Frei Otto, both specialists in the design of tent structures.

James Eads, whose greatest work was the Eads Bridge over the Mississippi, the first major structure built of steel.

Buckminster Fuller, the inventor of the geodesic dome.

David Geiger, pioneer designer of pneumatic structures.

William Jenney, the pioneer designer of steel frame buildings.

T.Y. Lin, the master of prestressed concrete.

Elisha Graves Otis, the inventor of the first safe passenger elevator, which helped make the skyscraper a reality.

Auguste Perret, the French architect and contractor, known as the father of reinforced concrete.

Thomas Telford, the great Scottish engineer of iron bridges.

HIGH TECH

High-tech architecture intends to reduce structure and function to the most necessary elements. "Perfection can only be achieved if nothing can be taken away," evokes the classic Modernist maxim that "less is more." The manifestation of this high-tech philosophy ranges from super-rational and spare aesthetic to dazzling structural gymnastics that realize new achievements in engineering, material science, and building systems and the architect's ability to integrate them.

Architects working in this idiom collaborate closely with engineering firms to achieve new structures that challenge expected limits of roof spans, floor cantilevers, and column supports. Evanescent cable-stayed curtain walls that use special insulating and tempered glass are commonly seen in high-tech architecture, as they allow maximum building performance for a (perceived) minimum use of material.

The equal marriage of technology, manufacturing, and architecture has always been a goal for some architects. Buckminster Fuller (b. 1895) was one of the iconic early pioneers, and the technologically rich ***Dymaxion House*** *(1945)* was one of his most famous works. Today's high-tech architecture is a design practice that attempts to effectively remove the working separation between architect and engineer. Architect-engineer teams such as Helmut Jahn and Werner Sobek or Norman Foster and Ove Arup focus on utilizing the most current structural science and materials available to them in order to produce the most high-performing and structurally daring designs possible. Designer and engineer work together from the earliest sketches to create an integrated design. Invention and progress are requirements for the high-tech architects, placing them at an edge of a constantly changing and expanding architecture.

Functionalism, rationalism, and expression of industrial production are elements linking this style to early European Modernism and the International Style. Advances in steel, insulating glass, cable structures, and concrete all allow for more dramatic forms than were possible in previous generations. Modernist movements have tried to free themselves from historicity and regionalism to pursue new horizons of form and to embrace this architectural style that addresses an increasingly global society.

While much early Modernism worked in a populist philosophy, high-tech is more frequently synonymous with large-scale projects such as airports, train stations, convention centers, and skyscrapers. Due to the high budgets of the projects and typically high square foot costs of the architecture, it is often governments, corporations, and high-end developers that commission high-tech projects. As a result, many of these works are major landmarks that affect the skyline, transportation patterns, and commerce at the scale of entire cities or countries.

Critics of the style see it as cold, soulless, and impersonal. High-tech architecture rejects a broad color palette and any extraneous elements that might distract from the overall composition of structure and skin. All-glass exteriors with steel and concrete structure, and stainless steel fittings and cladding are ubiquitous material combinations. Yet its ambition and the eternally evocative drive toward invention give high-tech architecture its own humanity.

Style/Mannerism

An early example of high-tech is Norman Foster's (b. 1935) ***Stansted Airport*** in London, built 1981-1991. Foster's consistent collaboration with Ove Arup and Partners has had a great impact on architecture. High-tech style often is a good match for transportation facilities because great efficiency is demanded for the large, open plans and building systems. The building is an efficient and simple continuous roof supported on large structural "trees." Daylight streams in from skylights in the roof as well as through the full-height glass walls at the perimeter. The transparency contributes to the successful organization of this very large building, and the visitor is able to see through the building from the roadway to the airplanes and is in no need of further orientation in the space.

Nicholas Grimshaw's (b. 1939), ***Waterloo International Terminal*** (1990–93) in London uses a long-span cable system as the main structural system. Grimshaw evokes the history and dynamics of the train shed with its immense volumes and diaphanous structures while making a serious statement about efficiency, cutting edge structural design, and responses to contingencies of a site. An asymmetrical, tapering, double cable-stayed truss supports the roof and is the main image of this project. The structure lends a sense of importance and lightness to an irregular and confined London rail station. The smaller western truss is placed on the outside of the building and supports glazing on the interior, achieving a continuous glazed interior with adequate clearance for the trains. This immense glass wall introduced daylight and also became a "public showcase for the trains." Due to the irregularity of the site and truss sizes, and movement due to climate and train vibration, standardized glass panels were designed to move over one another in response to slight movements in the structural system.

Richard Rogers' (b. 1933) ***Lloyd's Building*** (1984) in London is remarkable for the way that it boldly expresses function and the structural and building systems. The office areas are completely open while elevators, building systems, and other servant areas are clearly exposed at the rear of the building. These elements are clearly expressed on the exterior as individual elements that are hanging on the exposed structural system. This arrangement allows easy access for maintenance and allows easy upgrading of those elements that become obsolete through frequent use or advancing technology. Rogers' design recognizes that changing technology will require systems in the building to be significantly overhauled in the future. He designed even beyond the forefront of the technology of the era.

Structure

Santiago Calatrava's (b. 1951) artful interpretations of natural elements into structural forms demonstrates a perfect equilibrium of architectonic and engineering principles. In his work asymmetry, dramatic cantilevers, graceful concrete forms, and operable exteriors serve to create a sense of wonder while the interiors and the site designs create a sense of architectural order. His ***Lyon Satolas Railway Station*** (1989–94) in Lyon, France utilizes a series of soaring steel arches that fan from a single vertex to create the central light-filled hall that provides clear orientation for the visitor and an easily defined landmark for the surrounding city. The height and airiness of the central hall is in distinct contrast to the low, linear tube created by Calatrava's trademark wishbone piers that cover the train platforms below.

Calatrava also draws inspiration from natural movement and human or animal structures and form. The ***Milwaukee Art Museum's*** (2001) wing-like operable sunshades draw a large crowd each time they open and close. Located just off downtown at the end of a long street, the museum complex is a magnificently sited yet unexpected midwestern landmark. The project features a cable-stayed footbridge hung from a graceful, inclined pylon. This bridge leads to the entrance and is the main organizing axis of the project's formal composition. The sunshades qualify as public sculpture when they are open and fully extended. But the museum is more than just a mechanical curiosity: the sunshades enclose a jewel-like interior space with a stunning view of Lake Michigan.

Sustainability/Energy Conservation

One of the surprising tenets of the high-tech style is one of energy conservation. Although their palette of glass, steel, and concrete varies from that of other sustainable practitioners, these architects are working to bring the operational costs of a building down by using less energy. Also, the components of high-tech architecture are particular to a given project, and the architects frequently design these to be easily recycled or reused.

Nicolas Grimshaw's temporary ***British Pavilion*** for the 1992 Seville Exposition was entirely prefabricated and assembled in Seville, Spain. It was a showcase for the articulation of structure and the ability of a small palette of materials to capture and manipulate wind, light, and water. High-tech aesthetics and structure combined to showcase local traditions and environmentally conscious passive building systems. A water wall on one side of the building served as a sculpture and also provided passive cooling. Fabric sails and roof louvers helped to shade the building and softened the harsh sunlight. Solar panels on the roof provided energy to the Pavilion. The project was designed to be taken apart at the end of the Exposition, so all elements were discrete and easily recycled.

Helmut Jahn's ***Post Tower*** (1992) in Bonn, Germany, the headquarters for the German post office, is based on simple geometries resulting in sharp edges and elegant curves. These combine to form a handsome statement of corporate identity. The distinctive shingled façade is a double skin commonly found in European high-rise buildings. It offers a highly insulated envelope that also allows fresh air to circulate. The floor-to-ceiling glass and narrow floor plate allow a maximum amount of natural light to reach the interior. These considerations give the building a low operating cost and also give office staff a pleasant working environment. Helmut Jahn was educated in the pure Miesian style, and he typifies the high-tech architect. He collaborates with structural and environmental engineers from the beginning of the design process in a relationship Jahn refers to as "Archineering."

TALL TOWERS

Ever since the very first steel high-rises were built at the end of the 19th century, the race to construct the world's tallest buildings has been a race for civic pride and stature. The names New York or Chicago immediately call to most people's mind an image of the Empire State Building or Willis Tower. Super-tall buildings define the visual identity of the cities that welcome them, the status of corporations who finance them, and the careers of the architects who design them.

Caesar Pelli's (b. 1926) 1.6 billion-dollar ***Petronas Towers*** (1998) in Kuala Lumpur, Malaysia, emphatically demonstrates the ability of the world's tallest building to become an international cultural and commercial landmark. The tower ushered a new continent and world capital market into global consciousness. The twin 88-story towers are capped with spires that soar to 1,483 feet. The towers are linked at the 41st and 42nd floors by a double-decker skybridge. Comprising more than 8 million square feet, Petronas Towers includes a shopping mall, an interactive petroleum discovery center, an 864-seat concert hall, a mosque, and a conference center. The floor plate is based on an eight-pointed star and is said to reference Malaysian Islamic motifs. This reinforces the project's connection with national, not merely corporate, identity. The façade is reminiscent of temple spires, especially the pattern that is created by the blue glass panels that are set in aluminum frames. Other stainless steel panels help deflect the forceful Malaysian heat.

C.Y. Lee and Partners' ***Taipei 101*** (2004) in Taipei, Taiwan, at 1671 feet (509 meters), was the first building to exceed the half-kilometer mark. Holding the record with the world's tallest occupied floor and observation deck until 2010, the project was also notable for the world's fastest elevators. Based on traditional Chinese design, the rhythm of the tower is distinctive and is a departure from the orthodox, smooth Modernist line found in most high-rise buildings. The interior is organized according to the principles of feng shui.

Skidmore, Owings and Merrill's *Burj Khalifa* (2010) in Dubai, United Arab Emirates, at 828 meters (2,717 feet) is currently the world's tallest occupied floor and observation deck. The overall construction cost of Burj Khalifa was approximately $1.5 billion.

STRUCTURAL FAILURES

In recent years, a number of spectacular structural failures have occurred in the United States. Among them are the following:

1. **Hartford Civic Center Coliseum, Hartford, CT,** 1978. Steel space frame spanning 300 feet. No casualties.
2. **C.W. Post Auditorium**, Greenvale, NY, 1978. Steel and aluminum dome spanning 171 feet. No casualties.
3. **Kemper Memorial Arena,** Kansas City, MO, 1979. Steel truss roof suspended from steel space frame spanning 324 feet. No casualties.
4. **Rosemont Horizon Arena**, Rosemont, IL, 1979. Glued laminated arches spanning 290 feet. 5 dead, 19 injured.
5. **Hyatt Regency Hotel**, Kansas City, MO, 1981. Two suspended walkways. 113 dead, 186 injured. In terms of casualties, the most devastating structural collapse ever to take place in the United States.

The investigations and conferences that have taken place as a result of these failures have produced several general conclusions, which are summarized at the end of Lesson Ten.

LESSON 14 QUIZ

1. Select the CORRECT statement about the terminal of Dulles Airport.

   A. The use of a mobile lounge allowed Saarrinen to depart from the conventional finger plan airport terminal.

   B. The huge concrete piers that support the roof lean outward to express the concept of flight rather than for any structural reason.

   C. The terminal was the first to use computerized subway trains.

   D. While the terminal building is compact and efficient, it does not lend itself readily to future expansion.

2. Select the CORRECT statement about the Pantheon.

   A. Circumferential iron chains were built in to act as tension rings.

   B. An iron tie was placed around the dome's base in the 19th century to resist the outward thrust of the dome.

   C. The hoop tension in the dome is resisted by very thick concrete walls.

   D. It remains to this day a masterpiece of Greek architecture.

3. The principal structural materials utilized by Wright in Fallingwater are

   A. wood and structural steel.

   B. structural steel and cast-in-place concrete.

   C. cast-in-place concrete and masonry.

   D. precast concrete and masonry.

4. Select the CORRECT statements about the John Hancock Building in Chicago.

   A. After its completion, its windows frequently broke loose from their frames and fell to the street below.

   B. It expresses its structure through its tapered form.

   C. Its enormous exposed cross-bracing is effective in resisting wind loads.

   D. It is the tallest building in Chicago.

5. What do the Palazzetto Dello Sport and the airplane hangars built for the Italian Air Force have in common?

   I. Both were designed by Nervi.

   II. Both were outstanding examples of elegant cast-in-place concrete construction.

   III. Both had dome roofs.

   IV. Both were destroyed during World War II.

   V. Both had ribbed concrete shell roofs.

   A. I, II, IV, and V

   B. I and V

   C. II and V

   D. I, III, and IV

6. Select the CORRECT statement about the Toronto City Hall.

   A. Its two huge shells are curved arbitrarily to achieve an aesthetic effect.

   B. Each of its towers acts as a huge circular tube to resist wind forces.

   C. Its floors are suspended from immense roof trusses.

   D. It comprises two huge shells which are curved to provide resistance to wind or earthquake forces.

7. What great structure partially collapsed because its dome was insufficiently buttressed?

A. Hagia Sophia

B. Palazzetto Dello Sport

C. Santa Maria del Fiore Cathedral

D. Munich Olympic Stadium

8. Which of the following buildings BEST exemplifies the bundled tube concept?

A. First National Bank, Chicago

B. John Hancock Building, Chicago

C. Willis Tower

D. World Trade Center

9. What building has a façade that clearly expresses its catenary suspension structure?

A. CBS Building

B. Dulles International Airport

C. Johnson Wax Building

D. Federal Reserve Bank, Minneapolis

10. Which of the buildings listed below has a cable-supported roof suspended between a central reinforced concrete arch and heavy perimeter walls?

A. Italian Air Force Hangars

B. Knights of Columbus Building

C. Toronto City Hall

D. Yale University Skating Rink

Part II

The Graphic Vignette

LESSON FIFTEEN

THE NCARB SOFTWARE

15

INTRODUCTION

There is a wide variety of programs used by candidates at the firms in which they work. Therefore, an essential part of every candidate's preparation is to practice using the examination's computer tools. Candidates can download this software from the NCARB Web site (*www.ncarb.org*). This program contains tutorials and sample vignettes for all the graphic portions. Spend all the time necessary to become familiar with this material in order to develop the necessary technique and confidence. You must become thoroughly familiar with the software.

The drafting program for the graphic portions is by no means a sophisticated program. While this may frustrate candidates accustomed to advanced CAD software, it is important to recall that NCARB aimed to create an adequate drafting program that virtually anyone can use, even those with no CAD background at all.

VIGNETTE SCREEN LAYOUT

Each vignette has a number of sections and screens with which the candidate must become familiar. The first screen that appears when the vignette is opened is called the Vignette Index and starts with the Task Information Screen. Listed on this screen are all the components particular to this vignette. Each component opens a new screen when the candidate clicks on it with the mouse. A menu button appears in the upper left corner of any of these screens that returns you to the Index Screen. Also available from the Index Screen is a screen that opens the General Test Directions Screen, which gives the candidate an overview of the procedures for doing the vignettes. Here are the various screens found on the Index Screen:

- **Vignette Directions** (found on all vignettes)—describes the procedure for solving the problem
- **Program** (found on all vignettes)—describes the problem to be solved
- **Tips** (found on all vignettes)—gives advice for approaching the problem and hints about the most useful drafting tools
- **Code**—gives applicable code information if required by the vignette

- **Sections**—typically found on the Stair Design Vignette and shows a section through the space in which the stair will be located
- **Lighting Diagrams**—found on the Mechanical and Electrical Plan vignette to show light fixture distribution patterns

The beginning of the vignette lesson in this study guide provides a more detailed description of each vignette screen.

To access the actual vignette problem, press the space bar. This screen displays the problem and all the computer tools required to solve it. Toggle back and forth between the Vignette Screen and one of the screens from the Index Screen at any time by simply pressing the space bar. This is not as convenient as viewing both the drawing and, say, the printed program adjacent to each other at the same time. Thus it is a procedure that the candidate must become familiar with through practice. Also, some vignettes are too large to be displayed all at once on the screen. In this case use the scroll bars to move the screen up and down or left and right as needed. The Zoom Tool is also helpful.

THE COMPUTER TOOLS

There are two categories of computer tools found in the ARE graphic portions:

- Common Tools
- Tools specific to each vignette

The Common Tools, as the name implies, are generally present in all the tests and allow a candidate to draw lines, circles, and rectangles, adjust or move shapes, undo or erase a previously drawn object, and zoom to enlarge objects on the screen. There is also an on-screen calculator and a tool that lets you erase an entire solution and begin again.

Vignette-specific tools include additional tools that enable the candidate to turn on and off layers, rotate objects, and set elevations or roof slopes. In addition to these extra tools, each vignette also includes specific items under the draw tool required for the vignette, such as joists or skylights. Become an expert in the use of each tool.

Each tool is dependent on the mouse; there are no "shortcut" keys on the keyboard. Press the computer tool first to activate it, then select the item or items on the drawing to be affected by the tool, and then re-click the computer tool to finish the operation. Spend as much time as required to become completely familiar with this drafting program. The Common Tools section of the practice vignettes available from NCARB is particularly useful for helping you become familiar with the computer tools. Three things to note: the left mouse button activates all tools; there is no zoom wheel on the mouse, nor an associated tool on the program; and the shift key activates the Ortho Tool.

The standard computer tools and their functions are shown in Figure 15.1.

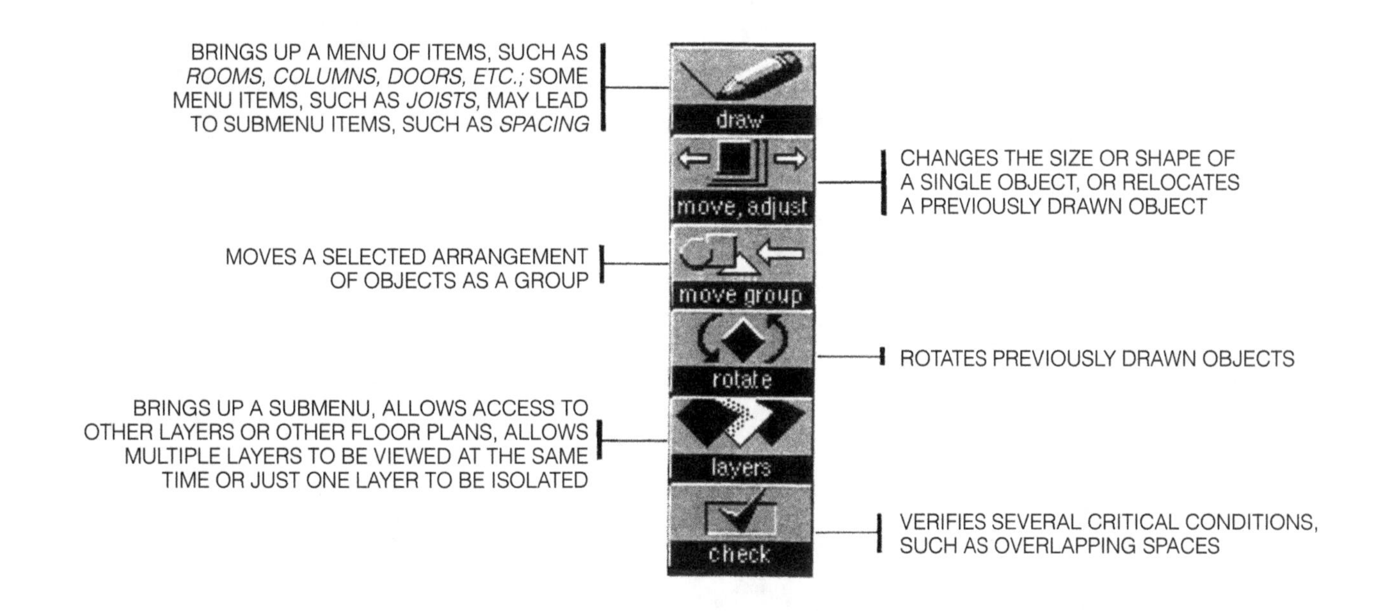
BRINGS UP A MENU OF ITEMS, SUCH AS ROOMS, COLUMNS, DOORS, ETC.; SOME MENU ITEMS, SUCH AS JOISTS, MAY LEAD TO SUBMENU ITEMS, SUCH AS SPACING
draw
CHANGES THE SIZE OR SHAPE OF A SINGLE OBJECT, OR RELOCATES A PREVIOUSLY DRAWN OBJECT
move, adjust
MOVES A SELECTED ARRANGEMENT OF OBJECTS AS A GROUP
move group
ROTATES PREVIOUSLY DRAWN OBJECTS
rotate
BRINGS UP A SUBMENU, ALLOWS ACCESS TO OTHER LAYERS OR OTHER FLOOR PLANS, ALLOWS MULTIPLE LAYERS TO BE VIEWED AT THE SAME TIME OR JUST ONE LAYER TO BE ISOLATED
layers
VERIFIES SEVERAL CRITICAL CONDITIONS, SUCH AS OVERLAPPING SPACES
check

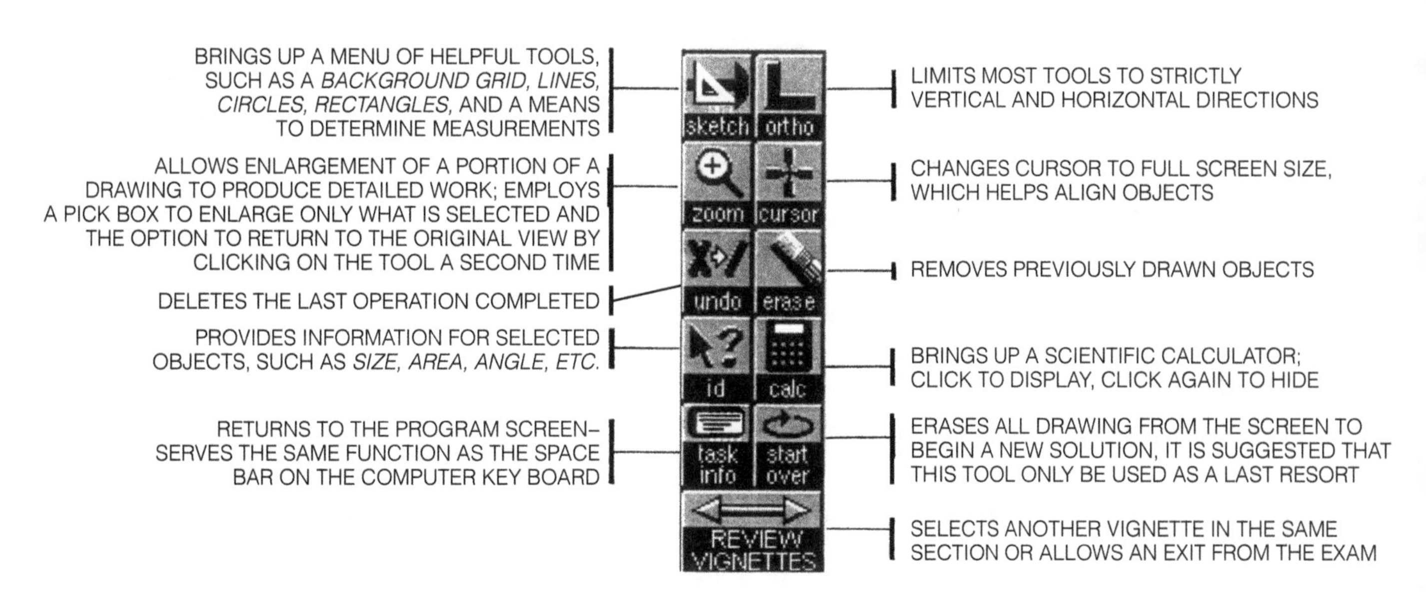
BRINGS UP A MENU OF HELPFUL TOOLS, SUCH AS A BACKGROUND GRID, LINES, CIRCLES, RECTANGLES, AND A MEANS TO DETERMINE MEASUREMENTS
sketch
ortho
LIMITS MOST TOOLS TO STRICTLY VERTICAL AND HORIZONTAL DIRECTIONS
ALLOWS ENLARGEMENT OF A PORTION OF A DRAWING TO PRODUCE DETAILED WORK; EMPLOYS A PICK BOX TO ENLARGE ONLY WHAT IS SELECTED AND THE OPTION TO RETURN TO THE ORIGINAL VIEW BY CLICKING ON THE TOOL A SECOND TIME
zoom
cursor
CHANGES CURSOR TO FULL SCREEN SIZE, WHICH HELPS ALIGN OBJECTS
REMOVES PREVIOUSLY DRAWN OBJECTS
DELETES THE LAST OPERATION COMPLETED
undo
erase
PROVIDES INFORMATION FOR SELECTED OBJECTS, SUCH AS SIZE, AREA, ANGLE, ETC.
id
calc
BRINGS UP A SCIENTIFIC CALCULATOR; CLICK TO DISPLAY, CLICK AGAIN TO HIDE
RETURNS TO THE PROGRAM SCREEN–SERVES THE SAME FUNCTION AS THE SPACE BAR ON THE COMPUTER KEY BOARD
task info
start over
ERASES ALL DRAWING FROM THE SCREEN TO BEGIN A NEW SOLUTION, IT IS SUGGESTED THAT THIS TOOL ONLY BE USED AS A LAST RESORT
REVIEW VIGNETTES
SELECTS ANOTHER VIGNETTE IN THE SAME SECTION OR ALLOWS AN EXIT FROM THE EXAM

Figure 15.1

LESSON SIXTEEN

16

TAKING THE EXAM

INTRODUCTION

Preparation for the ARE usually begins several months before taking the actual exam. The first step is to submit an application for registration with your state board or Canadian provincial association. Most, but not all, registration boards require a professional degree in architecture and completion of the Intern-Development Program (IDP) before a candidate is allowed to begin the exam process. Since the processing of educational transcripts and employment verifications may take several weeks, begin this process early. The registration board will review a candidate's application to determine whether the applicant meets the eligibility requirements.

SCHEDULING THE EXAM

The exams are available to eligible candidates at virtually any time, since test centers are open nearly every day throughout the year. However, it is the responsibility of the candidate to contact a test center to schedule an appointment. This must be done at least three days prior to the desired appointment time, but it is probably more sensible to make an appointment a month or more in advance. It is not necessary to take the test in the same jurisdiction in which you intend to be registered. Someone in San Francisco, for example, could conceivably combine the test-taking with a family visit in Philadelphia.

FINAL PREPARATION

Candidates are advised to complete all preparations the day before their appointment in order to be as relaxed as possible before the upcoming test. Avoid last minute cramming, which in most cases does more harm than good. The exams not only test design competence but also physical and emotional endurance. You must be totally prepared for the strenuous day ahead, and that requires plenty of rest and as much composure.

One of the principal ingredients for success on this exam is confidence. If you have prepared in a reasonable and realistic way, and if you have devoted the necessary time to practice, you should approach the exams with confidence.

EXAM DAY

Woody Allen once said that a large part of being successful was just showing up. That is certainly true of the licensing examination, where you must not only show up but also be on time. Get an early start on exam day and arrive at the test center at least 30 minutes before the scheduled test time. Getting an early start enables you to remain in control and maintain a sense of confidence, while arriving late creates unnecessary anxiety. If you arrive 30 minutes late, you may lose your appointment and forfeit the testing fee. Most candidates will begin their test session within one-half hour of the appointment time. You will be asked to provide a picture identification with signature and a second form of identification. For security reasons, you may also have your picture taken.

THE EXAM ROOM

Candidates are not permitted to bring anything with them into the exam room: no reference materials, no scratch paper, no drawing equipment, no food or drink, no extra sweater, no cell phones, no digital watches. You are permitted to use the restroom or retrieve a sweater from a small locker provided outside the exam room. Each testing center will have its own procedure to follow for such needs. A calculator is provided as part of the drafting program.

Scratch paper will be provided by the testing center. The candidate might wish to request graph paper, if available.

Once the candidate is seated at an assigned workstation and the test begins, the candidate must remain seated, except when authorized to leave by test center staff. When the first set of vignettes is completed, or time runs out, there is a mandatory break, during which you must leave the exam room. Photo identification will be required when you reenter the exam room for the next set of vignettes. At the conclusion of the test, staff members will collect all used scratch paper.

Exam room conditions vary considerably. Some rooms have comfortable seats, adequate lighting and ventilation, error-free computers, and a minimum of distractions. The conditions of other rooms, however, leave much to be desired. Unfortunately, there is little a candidate can do about this, unless, of course, the computer malfunctions. Staff members will try to rectify any problem within their control.

EXAM ROOM CONDUCT

NCARB has provided a lengthy list of offensive activities that may result in dismissal from a test center. Most candidates need not be concerned about these, but for those who may have entertained any of these fantasies, such conduct includes

- giving or receiving help on the test,
- using prohibited aids, such as reference material,
- failing to follow instructions of the test administrator,
- creating a disturbance,

- removing notes or scratch paper from the exam room,
- tampering with a computer, and
- taking the test for someone else.

BEGIN WITH THE PROGRAM

You can either solve the vignettes in the order they are presented or build confidence by starting with one that looks easier to you. Only you know what works best for you; the practice software should give you a sense of your preferred approach.

Every vignette solution begins with the program. Read the entire program carefully and completely and consider every requirement. During this review, identify the requirements, restrictions, limitations, code demands, and other critical clues that will influence the solution. Feel free to use scratch paper to jot down key points, data, and requirements. This will help ensure that you understand and meet all the requirements as you develop your solution.

Every vignette problem has two components: the written program and a graphic-base plan. Both components are complementary and equally important; together they completely define the problem. Candidates should not rush through a review of the program and base drawing in an attempt to begin the design sooner. It is more important to understand every constraint and to be certain that you have not overlooked any significant detail. Until you completely understand the vignette, it is pointless to continue.

GENERAL STRATEGIES

The approach to all vignette solutions is similar. Work quickly and efficiently to produce a solution that satisfies every programmatic requirement. The most important requirements are those that involve compliance with the code, such as life safety, egress, and barrier-free access.

Another important matter is design quality. Strive for an adequate solution that merely solves the problem. Exceptional solutions are not expected, nor are they necessary. You can only pass or fail this test, not win a gold medal. Produce a workable, error-free solution that is good enough to pass.

During the test session, candidates will frequently return to the program to verify element sizes, relationships, and specific restrictions. Always confirm program requirements before completing the vignette, so that oversights or omissions may be corrected while there is still time to do so. Candidates must always keep in mind the immutability of the program. That is, never—under any circumstances—modify, deviate from, or add anything to the program. Never try to *improve* the program. Lastly, taking time at the end of each section to review all the vignettes can help to eliminate small errors or omissions that could tip the balance between a passing and failing grade.

Candidates should have little trouble understanding a vignette's intent; however, the true meaning of certain details may be ambiguous and open to interpretation. Simply make a reasonable assumption and proceed with the solution.

While candidates will necessarily employ their own strategies for ordering the vignettes, what follows are some ideas others have found helpful.

1. Start with the vignette you feel most capable of completing quickly and competently. This will boost your confidence for the remaining vignettes.
2. Try to solve each problem in 10 minutes less than the allotted time. Use this additional time to review your solution.
3. Create a set of notes or a chart for each problem.

THE TIME SCHEDULE

The most critical problem on the exam is time, and you must use that fact as the organizing element around which any strategy is based. The use of a schedule is essential. During the preparation period, and especially after taking a mock exam, note the approximate amount of time that should be spent on each vignette solution. This information must then become a candidate's performance guide, and by following it faithfully, the candidate will automatically establish priorities regarding how time will be spent.

It is important to complete each vignette in approximately the time allotted. You cannot afford to dwell on a minor detail of one vignette while completely ignoring another vignette. Forget the details, do not strive for perfection, and be absolutely certain to finish the test. Even the smallest attempt at solving a vignette will add points to your total score.

Vignettes have been designed so that a reasonable solution for each of the problems can be achieved in approximately the amount of time shown in the *ARE Guidelines*. These time limits are estimates made by those who created this test. In any event, a 45-minute-long vignette may not necessarily take 45 minutes to complete. Some can be completed in 30 minutes while others may take an hour or longer. The time required depends on the complexity of the problem and your familiarity with the subject matter. Some candidates are more familiar with certain problem types than others, and since candidates' training, experience, and ability vary considerably, adjustments may have to be made to suit individual needs. It should also be noted that within each exam section, the time allotted for two vignettes may be used at the candidate's discretion. For example, in a three-vignette section that allots 150 minutes, NCARB recommeds spending one hour on one vignette and 45 minutes on the other two. However, you may actually spend 80 minutes on one problem and 35 minutes on the other two.

Candidates who are aware of the time limit are more able to concentrate on the tasks to be performed and the sequence in which they take place. You will also be able to recognize when to begin the next vignette. When the schedule tells you to stop workng on one vignette and move on to the next, you will do so, regardless of the unresolved problems that may remain. You may submit an imperfect solution, but you *will* complete the test.

Only solve what the program asks you to solve, and don't use real world knowledge, such as specific building code requirements.

TIME SCHEDULE PROBLEMS

It is always possible that a candidate will be unable to complete a certain vignette in the time allotted. What to do in that event? First, avoid this kind of trouble by adhering to a rigid time schedule, regardless of problems that may arise. Submit a solution for every vignette, even if some solutions still have problems or are incomplete.

Candidates are generally able to develop some kind of workable solution in a relatively short time. If each decision is based on a valid assumption and relies on common sense, the major elements should be readily organized into an acceptable functional arrangement. It may not be perfect, and it will certainly not be refined, but it should be good enough to proceed to the next step.

MANAGING PROBLEMS

There are other serious problems that may arise, and while each is potentially a failing error, they must be managed and resolved. Consider the following:

- The candidate has inadvertently omitted a major programmed element.
- The candidate has drawn a major element too large or too small.
- The candidate has ignored a critical adjacency or other relationship.

The corrective action for each of these issues will depend on the seriousness of the error and when the mistake is discovered. If there is time, one should rectify the design by returning to the point at which the error occurred and begin again from there. If it is late in the exam and time is running out, there may simply be insufficient time to correct the problem. In that case, continue on with the remainder of the exam and attempt to provide the most accurate solutions. The best strategy, of course, is to avoid critical mistakes in the first place, and those who concentrate and work carefully will do so.

WORKING UNDER PRESSURE

The time limit of the graphic portions creates subjective as well as real problems. This exam generates a unique psychological pressure that can be harmful to performance. While some designers thrive and do their best work under pressure, others become fearful or agitated under the same conditions. It is perfectly normal to be uneasy about this important event; although anxiety may be a common reaction, it is still uncomfortable.

Candidates should be aware that pressure is not altogether a negative influence. It may actually heighten awareness and sharpen abilities. In addition, realize that as important as this test may be, failure is not a career-ending event. Furthermore, failure is rarely an accurate measure of design ability; it simply means that you have not yet learned how to pass this difficult exam.

EXAMINATION ADVICE

Following is a short list of suggestions intended to help candidates develop their own strategies and priorities. We believe each item is important in achieving a passing score.

The *ARE Guidelines,* available from the NCARB Web site, also lists suggestions for examination preparedness.

- **Get an early start.** Begin preparation early enough to develop confidence by the time you are scheduled to take the exam. Arrive at the exam site early and be ready to go when the test begins.
- **Complete all vignettes.** Incomplete solutions risk failure. Complete every problem, even if every detail is not complete or perfect.

- **Don't modify the program.** Never add, change, improve, or omit anything from a program statement. Never assume that there is an error in the program. Verify all requirements to ensure complete compliance with every element of the program. If ambiguities exist in the program, make a reasonable assumption and complete your solution.
- **Develop a reasonable solution.** Since most vignettes generally have one preferred solution, solve the problem in the most direct and reasonable way. Never search for a unique or unconventional solution, because on this exam, creativity is not rewarded.
- **Be aware of time.** The strict time constraint compels you to be a clock-watcher. Never lose sight of how much time you are spending on each vignette. When it is time to proceed to the next problem, quit and move on to the next vignette.
- **Remain calm.** This may be easier said than done because this type of experience often creates stress in even the most self-assured candidate. Anxiety is generally related to fear of failure. However, if a candidate is well prepared, this fear may be unrealistic. Furthermore, even if the worst comes to pass and you must repeat a division, all it means is that your architectural license will be delayed for a short period of time.

LESSON SEVENTEEN

17

STRUCTURAL LAYOUT VIGNETTE

INTRODUCTION

The purpose of the Structural Layout vignette is to determine a candidate's understanding of basic structural framing concepts. This is not, however, a structural problem in the conventional sense because no structural calculations or selection of specific members is required. This vignette is a simple design problem involving structural framing. A candidate is required to develop a conceptual roof framing plan for a relatively simple one-story structure, which generally has a two-level roof. Candidates are provided with a background floor plan, on which they must superimpose an appropriate framing plan based on the floor plan and program requirements. Among the basic structural elements that a candidate must locate and arrange are beams, joists, lintels, columns, and bearing walls.

VIGNETTE INFORMATION

The Structural Layout vignette begins with the typical Task Information Screen that permits, with the click of a mouse, instant access to the **General Test Directions** that apply to all vignettes, as well as the following screens:

- **Vignette Directions**—These are instructions requiring a candidate to create a two-level roof framing solution that is structurally sound, efficient, and responsive to the program. The solution is superimposed on the background floor plans, and access to each roof level is controlled by the *layers* tool. Unless designated otherwise, all walls are assumed to be non-load-bearing.

  The directions also offer instruction on drawing joists and decking. First, one must use the *draw* icon to select both the framing direction and joist spacing. For example, the joists might be laid out in the east-west direction and 30 inches on center. Next, designate those areas containing joists by drawing their boundaries. When the framed areas are indicated, the joists within those areas appear automatically. Whereas laying out joists using pencil and paper is a relatively simple matter of drawing a series of parallel lines, the computer requires a candidate to perform the multi-step process outlined above in order to indicate the joists. The application of decking is accomplished in a similar way.

- **Program**—The program requirements are organized within several subdivisions, as follows:

  SITE / FOUNDATION describes elements of the foundation system assumed for the project. Such a requirement might read as follows: *Assume a foundation system that is adequate for all normal loads and a site with no unusual seismic or wind activity.*

  CONSTRUCTION / MATERIALS, which specifies the major construction materials and elements, for example: *Steel beams, open-web steel joists, and metal roof decking have been chosen for the roof structure.* Maximum spans for the structural elements may be given.

  GENERAL REQUIREMENTS describes specifics of the structure, such as roof heights, flat roofs, prohibited cantilevers, and certain column-free rooms.

- **Tips**—These begin by mentioning again the *layers* tool, which is used to switch between the two roof levels. Joists cannot be drawn individually; they must be selected and drawn as a unit using the *draw* tool. Finally, there is a suggestion about using the *full-screen cursor* tool, which is helpful in aligning structural elements, such as the columns supporting the ends of a beam.

If a candidate presses the space bar on the computer keyboard while viewing any of the screens described above, the **work screen** with a building plan appears. This is the plan on which the lower-level roof framing is to be superimposed. Clicking on the *layers* tool brings up the plan on which the upper-level roof framing is to be superimposed. Candidates are advised that all solutions must be drawn as two separate layers, or else they will not receive a grade.

The floor plan indicates spaces, partitions, doors, windows, and all other elements that relate to the framing, such as dotted lines representing breaks in ceilings. Key overall dimensions are also included, which may help candidates determine the scale and span of structural members.

DESIGN PROCEDURE

The Structural Layout vignette requires candidates to devise and superimpose a schematic framing plan over an existing floor plan. Candidates are not required to determine the size or spacing of members but should have some understanding of the location and arrangement of structural elements. Those who are familiar and experienced with the examination software can probably complete this vignette in less than its allotted time. There are no members to design and only four basic elements to draw: joists, bearing walls (or columns), lintels, and decking.

Solutions are evaluated on the basis of placement, orientation, and spacing of the various structural members available. In some problems, allowable spans or limits are specified, such as: *The roof deck may carry design live and dead loads on spans up to 6'-0".* In that case, space the roof joists six feet apart. In past problems where no spans or limits were specified, walls in plan were indicated so that every span, space, and building size was a multiple of the same dimension, for example five feet. If this kind of information were part of the problem, you would likely arrange joists five feet apart to correspond to the established spacing of elements.

The plan generally includes high and low roof levels in order to test a candidate's ability to visualize a structure in three dimensions and to coordinate multiple bearing levels within the same wall plane. An example of this would be a clerestory window, where the ends of two roofs at different levels are supported by members in a common wall (see Figure 17.1). You must be certain that loads from both roof levels are transferred to the ground through columns or bearing walls.

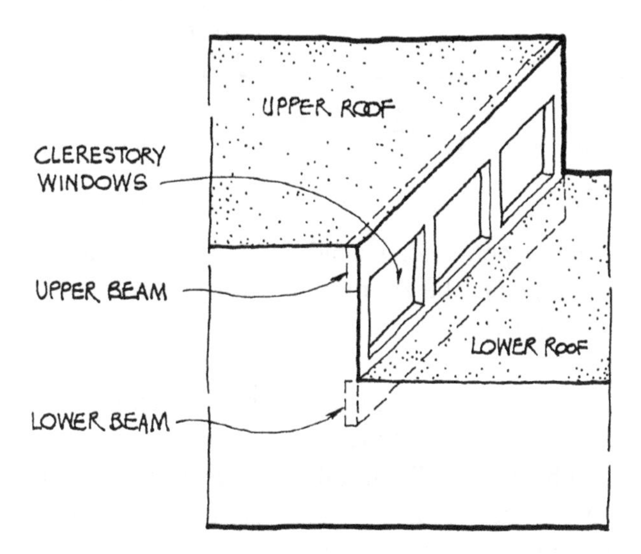

Figure 17.1

Candidates are also expected to provide support at the ends of all beams, headers, and lintels. In addition, roof decking must be supported along all edges. A quick sketch on graph paper may determine whether either bearing walls or columns will create a more suitable framing system throughout. This choice affects the efficiency of the layout, which is an important criterion on which a candidate's solution is evaluated. If you use bearing walls, then few, if any, columns should appear in plan. On the other hand, if columns are used, the solution should employ columns in preference to bearing walls. It is also possible to employ columns in one area and bearing walls in another. Again, efficiency is the key ingredient.

SOLVING THE PROBLEM

Any approach to solving a Structural Layout vignette must recognize that vertical building loads are transferred downward, through a succession of bearing members, from the roof of a structure to its foundation. Therefore, candidates should begin their solutions at the roof level. The first determination generally involves the spacing of joists, which is usually the same as the maximum span of the roof decking. A Structural Layout program may describe the roof deck as *ribbed metal decking with an allowable span of six feet.* Move to the next level of support and provide open-web steel joists spaced at six-foot centers.

Once the joist spacing and direction of framing are determined, candidates must decide if beams are required to support the joists. In most cases, this will not be an issue because joists will generally span the short dimension of spaces. However, look for a statement or clue in the program, such as specific span limits or spaces, which are described as clear-spanned and free of columns. Without such clues, candidates must use good judgment and provide supports if they seem appropriate.

Because of the scale of the Structural Layout vignette, together with the screen size and scope of past problems, it is unlikely that your maximum span will require complicated long-span girders or truss systems. Remember, the point of this exercise is to determine your understanding of basic structural framing concepts for a simple structure.

Figure 17.2

Once you have determined the basic structural layout, you must then be certain that each supporting member bears on a column, bearing wall, header, or lintel. Incidentally, although the terms are often used interchangeably, headers and lintels are not the same. A header is a member that crosses and supports the end of a joist or rafter by transferring its load to parallel joists or rafters (see the left portion of Figure 17.2). A lintel, on the other hand, is a horizontal member spanning an opening that carries the load of the wall above the opening (see the right portion of Figure 17.2). Thus, a member spanning an exterior window opening could be either a header or a lintel, depending on whether there is a portion of wall above the member.

If a solution results in a forest of columns, reconsider the overall solution. Possibly the beam spans are too short, perhaps the beams are too close together, or it may be an indication that bearing walls are a more appropriate solution to the problem. Bearing walls may be preferable to columns for one reason alone: aligning columns across a plan requires the use of the full-screen cursor tool. This tool requires some experience, a steady hand, and constant reference to the zoom tool to verify a column's position.

It should be noted that some past Structural Layout vignettes have included both columns and bearing walls. In other words, even in those cases where bearing walls were used predominantly, some columns were required to complete the solution.

FINAL LAYOUT

As time allows, candidates may try alternative framing schemes to determine the most efficient layout. One may find that turning the principal supporting beams 90 degrees will produce a more orderly or suitable layout. On the other hand, running principal framing members in the shortest direction is usually the best way to go. It takes a distinct design ability to produce a clear and logical framing layout, and the most desirable scheme may take more than one attempt.

Before completing the final drawing, verify that allowable spans have not been exceeded, that all horizontal roof or floor areas are supported by joists or beams, and that the ends of all supporting members bear on columns or bearing walls.

Using a computer program to devise a structural layout is efficient and in many ways faster than drawing with pencil and paper. For example, when determining the joist spacing and the area covered by joists, a simple click of the mouse places every joist in its correct position. Likewise, after determining the area covered by roof decking, a mouse click instantly covers the area with a color and texture.

KEY COMPUTER TOOLS

- **Drawing Tool**—Central to this vignette is the orientation and spacing of structural elements found in the drawing tool submenu.
- **Layer Tool**—Key to this vignette because the candidate is required to draw the structural elements on two separate layers.
- **Ortho Tool**—This tool helps in lining up drawing components.
- **Full Screen Cursor**—Along with the Ortho Tool, the Full Screen Cursor helps the candidate keep elements lined up accurately.

VIGNETTE STRUCTURAL LAYOUT

Develop an upper-level and lower-level roof framing system over the partial floor plans of the small museum shown. Your design should be structurally sound, efficient, and accommodate all programmed conditions and requirements. Show the layout of columns and/or load-bearing walls, the placement of beams and lintels, and the location of all roof joists. Indicate the span direction of decking with an arrow.

Program Requirements

- The soils and foundation system are adequate for all loads, including seismic and wind.
- Use structural steel sections for columns, beams, and open-web joists.
- All portions of the roof framing are flat and do not extend beyond the walls shown.
- The Exhibit Room roof height is 20 feet; all other areas have a roof height of 14 feet.
- The structure must accommodate a clerestory window located along the south wall of the Exhibit Room.
- Columns may be located within walls, including those with conventional windows or clerestory windows.
- Any wall shown on the floor plans may be designated as a bearing wall.
- Lintels are required at all openings in bearing walls.
- The Exhibit Room must remain column-free.
- Show only lower roof framing on lower level plan and only upper roof framing on upper level plan.

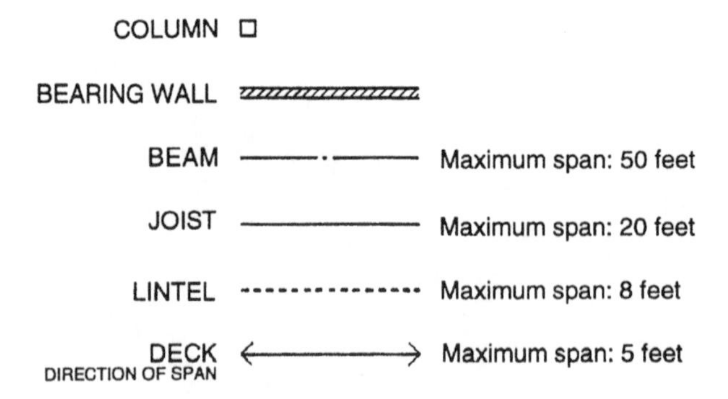

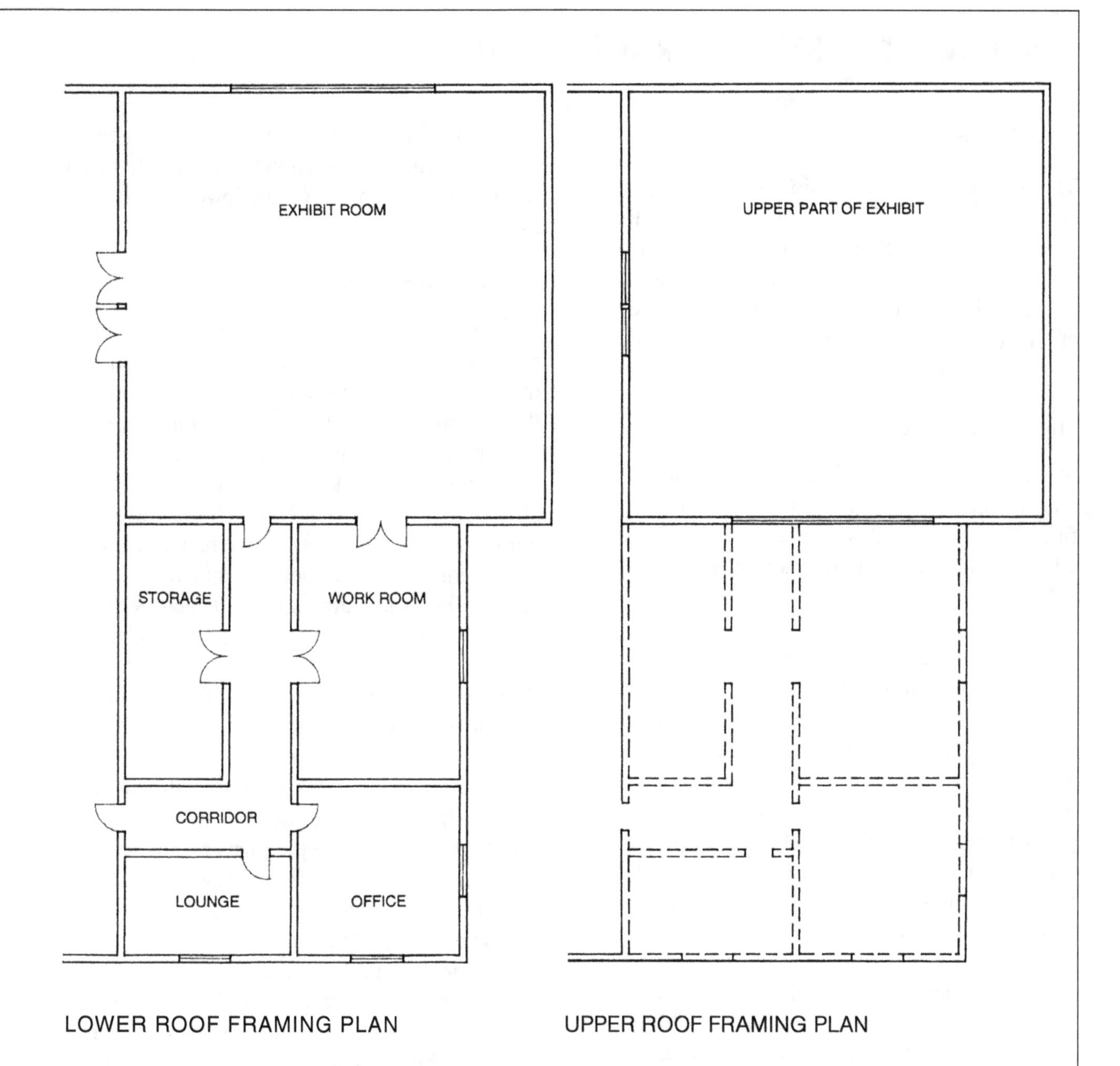
EXHIBIT ROOM
STORAGE
WORK ROOM
CORRIDOR
LOUNGE
OFFICE
UPPER PART OF EXHIBIT
LOWER ROOF FRAMING PLAN
UPPER ROOF FRAMING PLAN
NORTH

VIGNETTE STRUCTURAL LAYOUT

Introduction

The purpose of this discussion is to instruct candidates in an efficient method of solving this problem. It is a method that develops logically, one step at a time, and it works. Regardless of the design approach, every system must be efficient because time does not permit endless design variations.

The Exam Sheet

Shown are the Structural Layout program and plans, which were preprinted on the sheet furnished to candidates. The original sheet was 12" × 18" in size and drawn at a scale of 1/16" = 1'-0".

The program describes requirements for a small museum, although the building type is completely irrelevant. The framing layout would be exactly the same whether this were a museum, school, or mortuary. The program requirements follow a conventional and logical pattern. This is a steel-framed structure that uses structural steel for columns, beams, and open-web joists. The roof levels are flat, and there are no overhangs.

No columns are permitted within the Exhibit Room, which is the large space with the higher roof. The foundation is adequate for all loads, columns may be located within solid or window walls, and clerestory windows are located along the south wall of the Exhibit Room. Finally, the two roof heights are given.

The program directs you to draw the necessary structural members on the two floor plans shown, using the graphic symbols shown in the legend. The right side of the sheet contains the two partial floor plans of the museum. The upper roof framing plan is where the Exhibit Room framing will be shown, and all other roof framing will be shown on the lower roof framing plan.

Framing Analysis

Before developing a roof framing plan, review the program and plans to identify every requirement and restriction. In this case, you know that the building is framed in structural steel, and the roof framing consists of open-web steel joists. Since the decking over the joists has an allowable span of five feet, it is logical to space our open-web steel joists at five feet on center. Incidentally, this five-foot joist spacing should result in an exact number of joists across the entire building. You also know that beams must be used to clear-span the column-free Exhibit Room because the space scales 50 feet square, and joists are only able to span 20 feet.

The east wall of the corridor is exactly 20 feet from both east and west exterior walls. You should therefore frame the joists for the lower roof area in the east-west direction because the maximum allowable joist span is 20 feet. While the lower roof joists could have been framed in the north-south direction, the 30-foot maximum span in that direction would have required an intermediate beam in the Work Room and Storage. This does not appear to be as good a solution.

Framing the Small Offices

Begin at the most southerly wall of the building and lay out joists at five-foot intervals spanning east-west, while moving north across the entire lower roof area. End joists at both the north and south walls complete this level's roof framing. Select the east wall of the corridor as a line of bearing because it divides the lower roof area into two sections that are each 20 feet wide. That bearing line could be a bearing wall or a line of columns, but you can decide that later. It is enough to know now that all structural members are neatly spaced at five-foot centers.

Supporting the Joists

The joists covering the lower roof area may be supported by bearing walls wherever possible or with beams and columns (Figure 17.3). The only way to know which is more efficient is to make quick sketches of both layouts.

In the first sketch, using bearing walls, the east and west walls in the lower roof area are mostly solid, as is most of the east corridor wall and the door openings in the corridor's east wall, because none of these openings exceed the 8'-0" limit for a lintel span (see the top portion of 17.3). The wall areas that are not solid, such as at the windows in the Work Room and Office, may be spanned by lintels.

The alternative sketch shows the same joist layout, but in this case, 50-foot-long beams support the joists (Figure 17.4). One each is placed at the east and west exterior walls and along the east side of the corridor. No columns are necessary, except at the beam ends, because no beam exceeds the maximum allowable 50-foot length. Select the bearing-wall system because it is more direct and economical than using beams and columns.

Framing the Exhibit Room

The 20-foot-long joists cannot span the 50-foot-square Exhibit Room in either direction without additional supporting beams. In this case, use two 50-foot-long beams spaced equally, that is, 16'-8" apart (3 spaces @ 16'-8" = 50'-0"). Spacing the beams unequally would have been acceptable, but the logic and simplicity of equal spacing across the room is preferable.

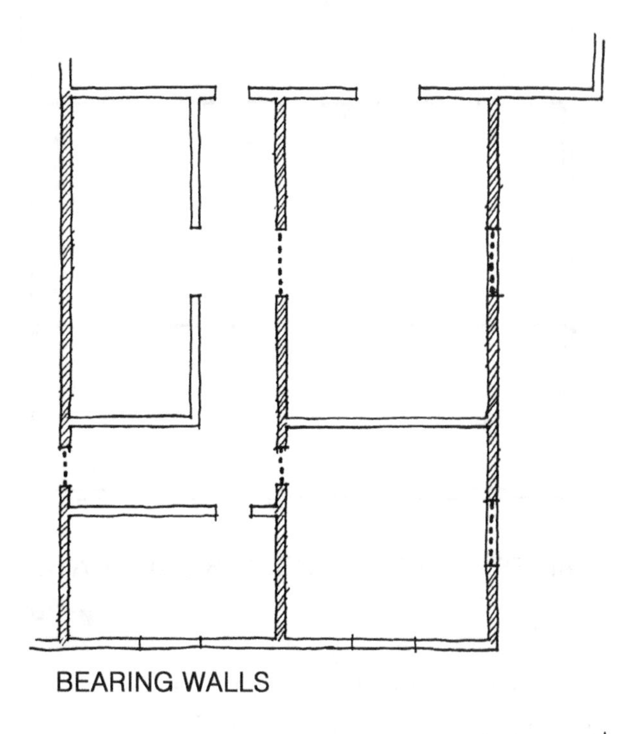

BEARING WALLS

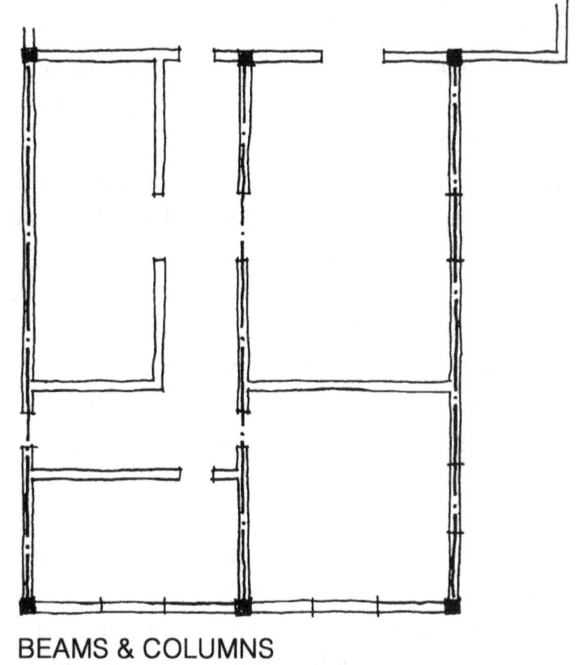

BEAMS & COLUMNS

Figure 17.3

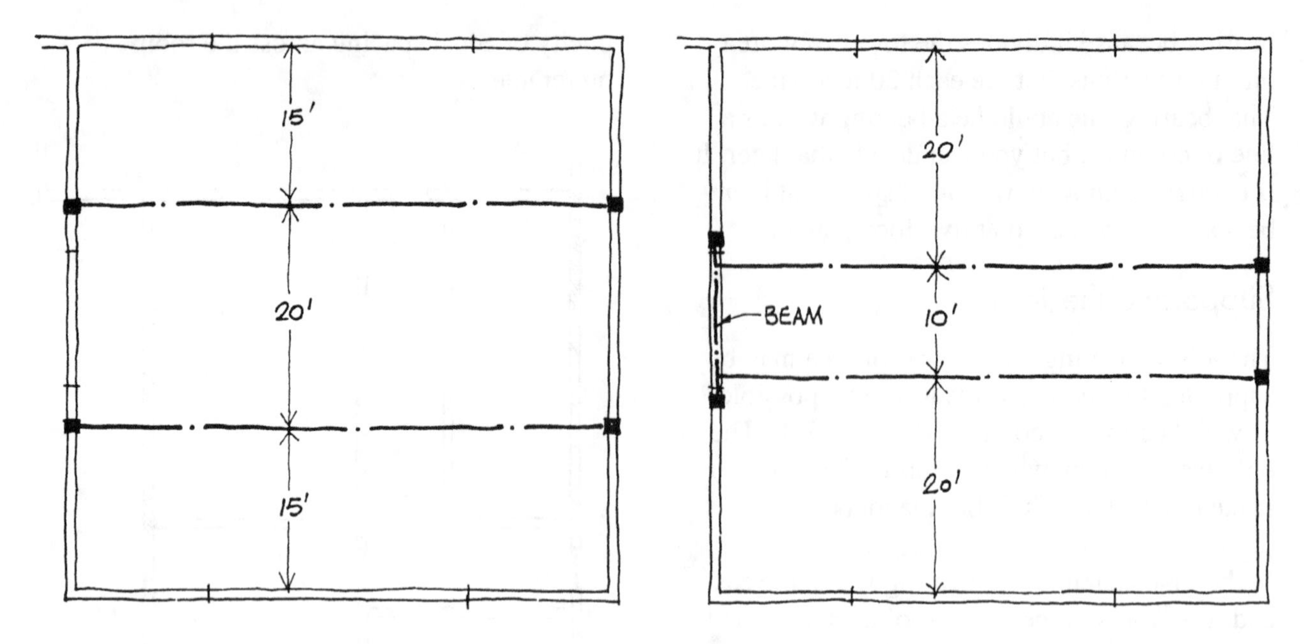

ALTERNATIVE SPACING OF BEAMS IN A 50' x 50' ROOM

Figure 17.4

Now decide on the direction of the beams, and consequently, the direction of the joists. There are three wall openings that influence this decision. At the north side of the Exhibit Room is a 24-foot-wide section of windows, at the west side is a 12-foot width of doors, and at the south side of the room is another 24-foot opening of clerestory windows. If beams were run in the north-south direction, columns supporting at least one beam would fall within the glass areas at both the north and south sides. Although the program states that columns are permitted within window walls, it is always preferable to avoid this situation if possible.

The other possibility is to run the beams in the east-west direction, where equally-spaced beams may be placed to avoid the door openings. This seems to be a more desirable solution. With nearly half of both the north and south walls glazed, it seems logical to place supporting beams at those ends of the Exhibit Room. The alternative would have been to use the end segments of those walls as bearing walls and use beams to span over the windows, since the window openings exceed the allowable span for lintels. The result of this would be a mixture of bearing walls and columns supporting beams, which is not as simple or efficient as using only beams and columns.

Again we space the joists at 5'-0" centers, which is the maximum span of the decking, and we run them across the upper roof area in the north-south direction.

Supporting the Beams

The next decision is how to support the four 50-foot-long beams that carry the upper roof loads. You could make the east and west walls of the Exhibit Room bearing walls, and that would be an acceptable solution. On the other hand, there are only four concentrated points of support along the length of each wall. A column placed at those points, that is, at each beam end, seems to be somewhat more efficient. Select this solution and indicate columns on *both* the upper and lower roof framing plans.

Remember, columns must be carried vertically through each level to the foundation below.

Final Drawing

At this point you have resolved every issue in this vignette, and now you must complete the final drawing. Everything that is drawn is superimposed over the preprinted floor plans using the graphic designations found in the legend. The actual amount of required drawing is small, and notes are not required. The graphic designations of the structural elements are self-explanatory; don't forget to indicate the framing direction of the decking. It is always important to produce a solution that is as complete and professional as possible. A suggested solution is provided in Figure 17.5.

As illustrated in this example, there may be alternative solutions to vignette problems. More than one arrangement may solve the problem directly and efficiently. Therefore, the choice between a beam-and-column scheme and a bearing-wall scheme is rarely a pass or fail decision. Candidates should use good logical and architectural sense to solve the problem directly, quickly, and with confidence, striving for efficiency in the structural layout vignette.

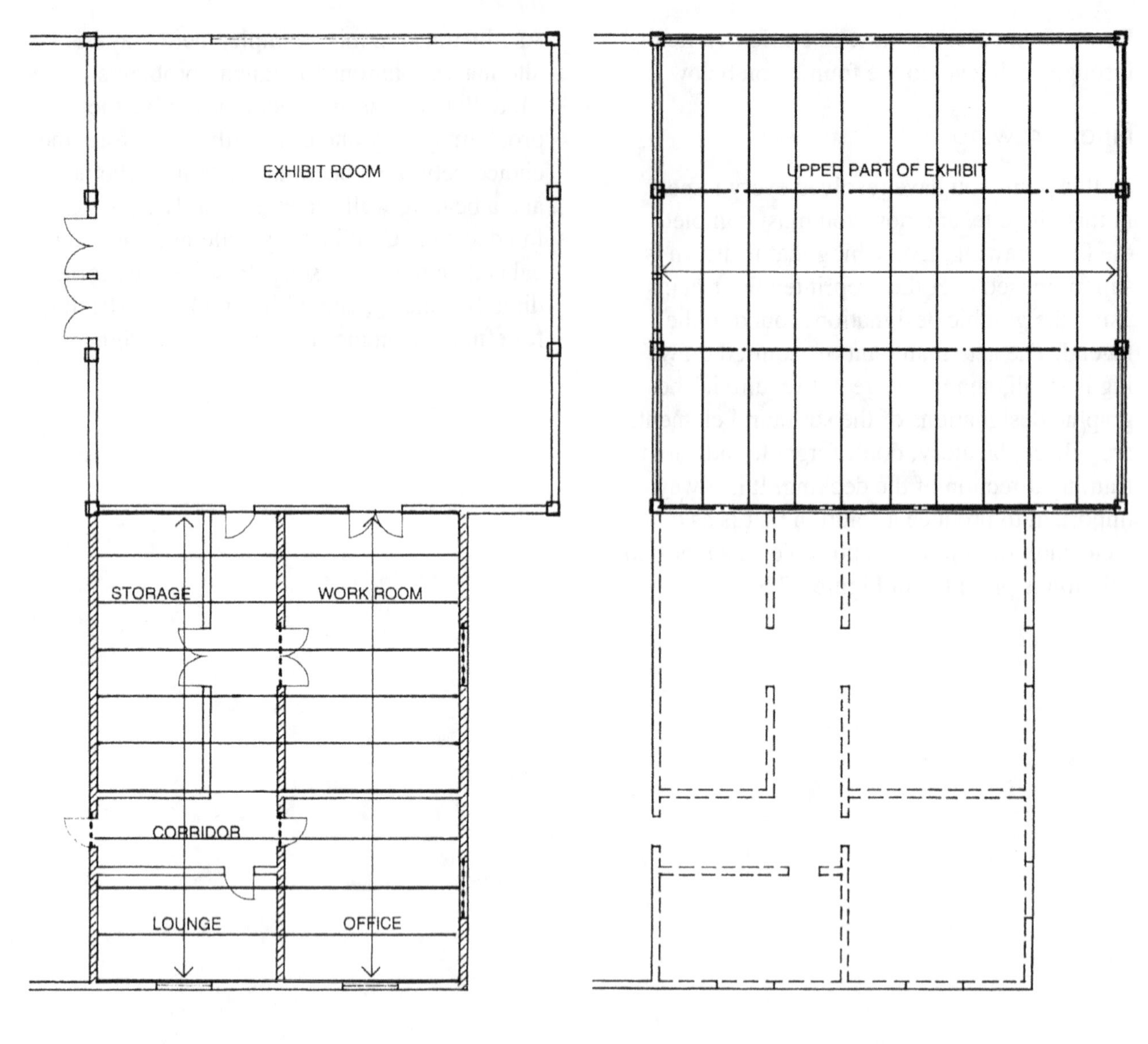

STRUCTURAL LAYOUT VIGNETTE SUGGESTED SOLUTION

Figure 17.5

APPENDIX

This appendix contains several tables used or referenced by the International Building Code. Tables from the National Design Specifications and the ACI Building Code are included as well. As discussed earlier in the Introduction of this book, the ARE only requires you to familiarize yourself with any one of the codes in use today. Any exam question involving codes will provide you with the table or equation needed to answer it correctly.

Tables 1604.3, 1907.7.1, and 1804.2 are used courtesy of the International Code Council (ICC). For more information about the ICC codes and related services, please contact ICC customer service at 800-786-4452.

TABLE 1604.3—DEFLECTION LIMITS[A,B,C,H,I]

TABLE 1604.3
DEFLECTION LIMITS [a, b, c, h, i]

CONSTRUCTION	L	*W* f	*D* + *L* d,g
Roof members: [e]			
Supporting plaster ceiling	*l* /360	*l* /360	*l* /240
Supporting nonplaster ceiling	*l* /240	*l* /240	*l* /180
Not supporting ceiling	*l* /180	*l* /180	*l* /120
Members supporting screen surfaces			
Floor members	*l* /360	—	*l* /240
Exterior walls and interior partitions:			
With brittle finishes	—	*l* /240	—
With flexible finishes	—	*l* /120	—
Farm buildings	—	—	*l* /180
Greenhouses	—	—	*l* /120

For SI: 1 foot = 304.8 mm.

a. For structural roofing and siding made of formed metal sheets, the total load deflection shall not exceed *l*/60. For secondary roof structural members supporting formed metal roofing, the live load deflection shall not exceed *l* /150. For secondary wall members supporting formed metal siding, the design wind load deflection shall not exceed *l* /90. For roofs, this exception only applies when the metal sheets have no roof covering.

b. Interior partitions not exceeding 6 feet in height and flexible, folding and portable partitions are not governed by the provisions of this section. The deflection criterion for interior partitions is based on the horizontal load defined in Section 1607.13 .

c. See Section 2403 for glass supports.

d. For wood structural members having a moisture content of less than 16 percent at time of installation and used under dry conditions, the deflection resulting from *L* + 0.5 *D* is permitted to be substituted for the deflection resulting from *L* + *D* .

e. The above deflections do not ensure against ponding. Roofs that do not have sufficient slope or camber to assure adequate drainage shall be investigated for ponding. See Section 1611 for rain and ponding requirements and Section 1503.4 for roof drainage requirements.

f. The wind load is permitted to be taken as 0.7 times the "component and cladding" loads for the purpose of determining deflection limits herein.

g. For steel structural members, the dead load shall be taken as zero.

h. For aluminum structural members or aluminum panels used in roofs or walls of sunroom additions or patio covers, not supporting edge of glass or aluminum sandwich panels, the total load deflection shall not exceed *l* /60. For aluminum sandwich panels used in roofs or walls of sunroom additions or patio covers, the total load deflection shall not exceed *l* /120.

i. For cantilever members, *l* shall be taken as twice the length of the cantilever

TABLE 4A—REFERENCE DESIGN VALUES FOR VISUALLY GRADED DIMENSION LUMBER

Table 4A (Cont.) Reference Design Values for Visually Graded Dimension Lumber (2" - 4" thick)[1,2,3]

(All species except Southern Pine — see Table 4B) (Tabulated design values are for normal load duration and dry service conditions. See NDS 4.3 for a comprehensive description of design value adjustment factors.)

USE WITH TABLE 4A ADJUSTMENT FACTORS

Species and commercial grade	Size classification	Design values in pounds per square inch (psi)							Grading Rules Agency
		Bending F_b	Tension parallel to grain F_t	Shear parallel to grain F_v	Compression perpendicular to grain $F_{c\perp}$	Compression parallel to grain F_c	Modulus of Elasticity E	E_{min}	
BEECH-BIRCH-HICKORY									
Select Structural		1,450	850	195	715	1,200	1,700,000	620,000	
No.1		1,050	600	195	715	950	1,600,000	580,000	
No.2	2" & wider	1,000	600	195	715	750	1,500,000	550,000	
No.3		575	350	195	715	425	1,300,000	470,000	NELMA
Stud	2" & wider	775	450	195	715	475	1,300,000	470,000	
Construction		1,150	675	195	715	1,000	1,400,000	510,000	
Standard	2" - 4" wide	650	375	195	715	775	1,300,000	470,000	
Utility		300	175	195	715	500	1,200,000	440,000	
COTTONWOOD									
Select Structural		875	525	125	320	775	1,200,000	440,000	
No.1		625	375	125	320	625	1,200,000	440,000	
No.2	2" & wider	625	350	125	320	475	1,100,000	400,000	
No.3		350	200	125	320	275	1,000,000	370,000	NSLB
Stud	2" & wider	475	275	125	320	300	1,000,000	370,000	
Construction		700	400	125	320	650	1,000,000	370,000	
Standard	2" - 4" wide	400	225	125	320	500	900,000	330,000	
Utility		175	100	125	320	325	900,000	330,000	
DOUGLAS FIR-LARCH									
Select Structural		1,500	1,000	180	625	1,700	1,900,000	690,000	
No.1 & Btr		1,200	800	180	625	1,550	1,800,000	660,000	
No.1	2" & wider	1,000	675	180	625	1,500	1,700,000	620,000	
No.2		900	575	180	625	1,350	1,600,000	580,000	WCLIB
No.3		525	325	180	625	775	1,400,000	510,000	WWPA
Stud	2" & wider	700	450	180	625	850	1,400,000	510,000	
Construction		1,000	650	180	625	1,650	1,500,000	550,000	
Standard	2" - 4" wide	575	375	180	625	1,400	1,400,000	510,000	
Utility		275	175	180	625	900	1,300,000	470,000	
DOUGLAS FIR-LARCH (NORTH)									
Select Structural		1,350	825	180	625	1,900	1,900,000	690,000	
No.1 & Btr		1,150	750	180	625	1,800	1,800,000	660,000	
No.1/No.2	2" & wider	850	500	180	625	1,400	1,600,000	580,000	
No.3		475	300	180	625	825	1,400,000	510,000	NLGA
Stud	2" & wider	650	400	180	625	900	1,400,000	510,000	
Construction		950	575	180	625	1,800	1,500,000	550,000	
Standard	2" - 4" wide	525	325	180	625	1,450	1,400,000	510,000	
Utility		250	150	180	625	950	1,300,000	470,000	
DOUGLAS FIR-SOUTH									
Select Structural		1,350	900	180	520	1,600	1,400,000	510,000	
No.1		925	600	180	520	1,450	1,300,000	470,000	
No.2	2" & wider	850	525	180	520	1,350	1,200,000	440,000	
No.3		500	300	180	520	775	1,100,000	400,000	WWPA
Stud	2" & wider	675	425	180	520	850	1,100,000	400,000	
Construction		975	600	180	520	1,650	1,200,000	440,000	
Standard	2" - 4" wide	550	350	180	520	1,400	1,100,000	400,000	
Utility		250	150	180	520	900	1,000,000	370,000	
EASTERN HEMLOCK-BALSAM FIR									
Select Structural		1,250	575	140	335	1,200	1,200,000	440,000	
No.1		775	350	140	335	1,000	1,100,000	400,000	
No.2	2" & wider	575	275	140	335	825	1,100,000	400,000	
No.3		350	150	140	335	475	900,000	330,000	NELMA
Stud	2" & wider	450	200	140	335	525	900,000	330,000	NSLB
Construction		675	300	140	335	1,050	1,000,000	370,000	
Standard	2" - 4" wide	375	175	140	335	850	900,000	330,000	
Utility		175	75	140	335	550	800,000	290,000	

TABLE 4D—REFERENCE DESIGN VALUES FOR VISUALLY GRADED TIBERS (5" × 5" AND LARGER)

Table 4D Reference Design Values for Visually Graded Timbers (5" x 5" and larger)[1,3]

(Tabulated design values are for normal load duration and dry service conditions, unless specified otherwise. See NDS 4.3 for a comprehensive description of design value adjustment factors.)

USE WITH TABLE 4D ADJUSTMENT FACTORS

Species and commercial grade	Size classification	Design values in pounds per square inch (psi)							Grading Rules Agency
		Bending F_b	Tension parallel to grain F_t	Shear parallel to grain F_v	Compression perpendicular to grain $F_{c\perp}$	Compression parallel to grain F_c	Modulus of Elasticity E	E_{min}	
ALASKA CEDAR									
Select Structural	Beams and Stringers	1,400	675	155	525	925	1,200,000	440,000	WCLIB
No.1		1,150	475	155	525	775	1,200,000	440,000	
No.2		750	300	155	525	500	1,000,000	370,000	
Select Structural	Posts and Timbers	1,300	700	155	525	975	1,200,000	440,000	
No.1		1,050	575	155	525	850	1,200,000	440,000	
No.2		625	350	155	525	600	1,000,000	370,000	
BALDCYPRESS									
Select Structural	5"x5" and Larger	1,150	750	200	615	1,050	1,300,000	470,000	SPIB
No.1		1,000	675	200	615	925	1,300,000	470,000	
No.2		625	425	175	615	600	1,000,000	370,000	
BALSAM FIR									
Select Structural	Beams and Stringers	1,350	900	125	305	950	1,400,000	510,000	NELMA NSLB
No.1		1,100	750	125	305	800	1,400,000	510,000	
No.2		725	350	125	305	500	1,100,000	400,000	
Select Structural	Posts and Timbers	1,250	825	125	305	1,000	1,400,000	510,000	
No.1		1,000	675	125	305	875	1,400,000	510,000	
No.2		575	375	125	305	400	1,100,000	400,000	
BEECH-BIRCH-HICKORY									
Select Structural	Beams and Stringers	1,650	975	180	715	975	1,500,000	550,000	NELMA
No.1		1,400	700	180	715	825	1,500,000	550,000	
No.2		900	450	180	715	525	1,200,000	440,000	
Select Structural	Posts and Timbers	1,550	1,050	180	715	1,050	1,500,000	550,000	
No.1		1,250	850	180	715	900	1,500,000	550,000	
No.2		725	475	180	715	425	1,200,000	440,000	
COAST SITKA SPRUCE									
Select Structural	Beams and Stringers	1,150	675	115	455	775	1,500,000	550,000	NLGA
No.1		950	475	115	455	650	1,500,000	550,000	
No.2		625	325	115	455	425	1,200,000	440,000	
Select Structural	Posts and Timbers	1,100	725	115	455	825	1,500,000	550,000	
No.1		875	575	115	455	725	1,500,000	550,000	
No .2		525	350	115	455	500	1,200,000	440,000	
DOUGLAS FIR-LARCH									
Dense Select Structural	Beams and Stringers	1,900	1,100	170	730	1,300	1,700,000	620,000	WCLIB
Select Structural		1,600	950	170	625	1,100	1,600,000	580,000	
Dense No.1		1,550	775	170	730	1,100	1,700,000	620,000	
No.1		1,350	675	170	625	925	1,600,000	580,000	
No.2		875	425	170	625	600	1,300,000	470,000	
Dense Select Structural	Posts and Timbers	1,750	1,150	170	730	1,350	1,700,000	620,000	
Select Structural		1,500	1,000	170	625	1,150	1,600,000	580,000	
Dense No.1		1,400	950	170	730	1,200	1,700,000	620,000	
No.1		1,200	825	170	625	1,000	1,600,000	580,000	
No.2		750	475	170	625	700	1,300,000	470,000	
Dense Select Structural	Beams and Stringers	1,900	1,100	170	730	1,300	1,700,000	620,000	WWPA
Select Structural		1,600	950	170	625	1,100	1,600,000	580,000	
Dense No.1		1,550	775	170	730	1,100	1,700,000	620,000	
No.1		1,350	675	170	625	925	1,600,000	580,000	
No.2 Dense		1,000	500	170	730	700	1,400,000	510,000	
No.2		875	425	170	625	600	1,300,000	470,000	
Dense Select Structural	Posts and Timbers	1,750	1,150	170	730	1,350	1,700,000	620,000	
Select Structural		1,500	1,000	170	625	1,150	1,600,000	580,000	
Dense No.1		1,400	950	170	730	1,200	1,700,000	620,000	
No.1		1,200	825	170	625	1,000	1,600,000	580,000	
No.2 Dense		850	550	170	730	825	1,400,000	510,000	
No.2		750	475	170	625	700	1,300,000	470,000	

TABLE 5A—DESIGN VALUES FOR STRUCTURAL GLUED LAMINATED SOFTWOOD TIMBER

Table 5A Expanded - Reference Design Values for Structural Glued Laminated Softwood Timber Combinations[1]

(Members stressed primarily in bending) (Tabulated design values are for normal load duration and dry service conditions.) Table 5A Expanded is an expanded list of structural glued laminated timber combinations from AITC 117 and APA-EWS Y117 that meet the requirements of each stress class in the new structural glued laminated timber stress class system. Table 5A Expanded is provided to allow easy conversion from the old combination symbols system to the new stress class system.

Use with Table 5A Adjustment Factors

Combination Symbol	Species Outer/Core	Bending About X-X Axis (Loaded Perpendicular to Wide Faces of Laminations): Extreme Fiber in Bending, Tension Zone Stressed in Tension[2] F_{bx}^{+} (psi)	Extreme Fiber in Bending, Compression Zone Stressed in Tension F_{bx}^{-} (psi)	Compression Perpendicular to Grain, Tension Face $F_{c\perp x}$ (psi)	Compression Perpendicular to Grain, Compression Face $F_{c\perp x}$ (psi)	Shear Parallel to Grain (Horizontal) F_{vx}[3] (psi)	Modulus of Elasticity E_x (10^6 psi)	Modulus of Elasticity for Beam and Column Stability $E_{x\,min}$ (10^6 psi)	Bending About Y-Y Axis (Loaded Parallel to Wide Faces of Laminations): Extreme Fiber in Bending F_{by} (psi)	Compression Perpendicular to Grain $F_{c\perp y}$ (psi)	Shear Parallel to Grain (Horizontal) F_{vy}[4] (psi)	Modulus of Elasticity E_y (10^6 psi)	Modulus of Elasticity for Beam and Column Stability $E_{y\,min}$ (10^6 psi)	Axially Loaded: Tension Parallel to Grain F_t (psi)	Compression Parallel to Grain F_c (psi)	Modulus of Elasticity E_{axial} (10^6 psi)	Fasteners: Specific Gravity for Fastener Design G, Top or Bottom Face	Specific Gravity for Fastener Design G, Side Face
16F-1.3E		1,600	925	315		195	1.3	0.67	800	315	170	1.1	0.57	675	925	1.2	0.42	
16F-V6	DF/DF	1,600	1,600	560	560	265	1.5	0.78	1,450	560	230	1.5	0.78	900	1,550	1.6	0.5	0.5
16F-E2	HF/HF	1,600	1,050	375	375	215	1.3	0.67	1,200	375	190	1.3	0.67	825	1,200	1.4	0.43	0.43
16F-E3	DF/DF	1,600	1,200	560	560	215	1.6	0.83	1,450	560	190	1.5	0.78	925	1,600	1.6	0.5	0.5
16F-E6	DF/DF	1,600	1,600	560	560	265	1.6	0.83	1,550	560	230	1.5	0.78	975	1,600	1.6	0.5	0.5
16F-E7	HF/HF	1,600	1,600	375	375	215	1.4	0.73	1,450	375	190	1.4	0.73	1,000	1,450	1.5	0.43	0.43
16F-V2	SP/SP	1,600	1,350	650	650	300	1.4	0.73	1,450	650	260	1.4	0.73	975	1,350	1.5	0.55	0.55
16F-V3	SP/SP	1,600	1,450	740	740	300	1.4	0.73	1,450	650	260	1.4	0.73	925	1,400	1.4	0.55	0.55
16F-V5	SP/SP	1,600	1,600	650	650	300	1.4	0.73	1,750	650	260	1.4	0.73	1,000	1,500	1.5	0.55	0.55
16F-E1	SP/SP	1,600	1,250	650	650	300	1.6	0.83	1,750	650	260	1.5	0.78	1,050	1,550	1.6	0.55	0.55
16F-E3	SP/SP	1,600	1,600	650	650	300	1.6	0.83	1,750	650	260	1.5	0.78	1,100	1,600	1.6	0.55	0.55
20F-1.5E		2,000	1,100	425		210	1.5	0.78	800	315	185	1.2	0.62	725	925	1.3	0.42	
20F-V3	DF/DF	2,000	1,450	650	560	265	1.6	0.83	1,450	560	230	1.5	0.78	975	1,550	1.6	0.5	0.5
20F-V7	DF/DF	2,000	2,000	650	650	265	1.6	0.83	1,450	560	230	1.6	0.83	1,000	1,600	1.6	0.5	0.5
20F-V9	HF/HF	2,000	2,000	500	500	215	1.5	0.78	1,350	375	190	1.4	0.73	975	1,400	1.5	0.43	0.43
20F-V12	AC/AC	2,000	1,400	560	560	265	1.5	0.78	1,250	470	230	1.4	0.73	900	1,500	1.4	0.46	0.46
20F-V13	AC/AC	2,000	2,000	560	560	265	1.5	0.78	1,250	470	230	1.4	0.73	925	1,550	1.5	0.46	0.46
20F-E2	HF/HF	2,000	1,400	500	500	215	1.6	0.83	1,200	375	190	1.4	0.73	925	1,350	1.5	0.43	0.43
20F-E3	DF/DF	2,000	1,200	560	560	265	1.7	0.88	1,450	560	230	1.6	0.83	1,000	1,600	1.7	0.5	0.5
20F-E6	DF/DF	2,000	2,000	560	560	265	1.7	0.88	1,550	560	230	1.6	0.83	1,100	1,650	1.7	0.5	0.5
20F-E7	HF/HF	2,000	2,000	500	500	215	1.6	0.83	1,450	375	190	1.4	0.73	1,050	1,450	1.5	0.43	0.43
20F-V2	SP/SP	2,000	1,550	740	650	300	1.5	0.78	1,450	650	260	1.4	0.73	975	1,350	1.5	0.55	0.55
20F-V3	SP/SP	2,000	1,450	650	650	300	1.5	0.78	1,750	650	260	1.4	0.73	1,050	1,400	1.5	0.55	0.55
20F-V5	SP/SP	2,000	2,000	740	740	300	1.6	0.83	1,450	650	260	1.4	0.73	1,050	1,500	1.5	0.55	0.55
20F-E1	SP/SP	2,000	1,250	650	650	300	1.7	0.88	1,750	650	260	1.5	0.78	1,050	1,550	1.6	0.55	0.55
20F-E3	SP/SP	2,000	2,000	650	650	300	1.7	0.88	1,900	650	260	1.5	0.78	1,150	1,650	1.6	0.55	0.55
24F-1.7E		2,400	1,450	500		210	1.7	0.88	1,050	315	185	1.3	0.67	775	1,000	1.4	0.42	
24F-V5	DF/HF	2,400	1,600	650	650	215	1.7	0.88	1,200	375	190	1.5	0.78	1,150	1,450	1.6	0.5	0.43
24F-V10	DF/HF	2,400	2,400	650	650	215	1.8	0.93	1,450	375	190	1.5	0.78	1,100	1,550	1.6	0.5	0.43
24F-E11	HF/HF	2,400	2,400	500	500	215	1.8	0.93	1,550	375	190	1.5	0.78	1,150	1,550	1.6	0.43	0.43
24F-E15	HF/HF	2,400	1,600	500	500	215	1.8	0.93	1,200	375	190	1.5	0.78	975	1,500	1.6	0.43	0.43
24F-V1	SP/SP	2,400	1,750	740	650	300	1.7	0.88	1,450	650	260	1.5	0.78	1,100	1,550	1.6	0.55	0.55
24F-V4[5]	SP/SP	2,400	1,450	740	650	210	1.7	0.88	1,050	470	185	1.3	0.67	875	1,000	1.5	0.55	0.43
24F-V5	SP/SP	2,400	2,400	740	740	300	1.7	0.88	1,750	650	260	1.5	0.78	1,150	1,650	1.6	0.55	0.55

TABLE 9.5 (A)—MINIMUM THICKNESS OF NONPRESTRESSED BEAMS OR ONE-WAY SLABS UNLESS DEFLECTIONS ARE CALCULATED*

TABLE 9.5(a)—MINIMUM THICKNESS OF NONPRESTRESSED BEAMS OR ONE-WAY SLABS UNLESS DEFLECTIONS ARE CALCULATED

	Minimum thickness, h			
	Simply supported	One end continuous	Both ends continuous	Cantilever
Member	Members not supporting or attached to partitions or other construction likely to be damaged by large deflections.			
Solid one-way slabs	$\ell/20$	$\ell/24$	$\ell/28$	$\ell/10$
Beams or ribbed one-way slabs	$\ell/16$	$\ell/18.5$	$\ell/21$	$\ell/8$

Notes:
Values given shall be used directly for members with normalweight concrete ($w_c = 145$ lb/ft^3) and Grade 60 reinforcement. For other conditions, the values shall be modified as follows:
a) For structural lightweight concrete having unit weight, w_c, in the range 90-120 lb/ft^3, the values shall be multiplied by $(1.65 - 0.005w_c)$ but not less than 1.09.
b) For f_y other than 60,000 psi, the values shall be multiplied by $(0.4 + f_y/100,000)$.

From ACI 318/318R-05, "Building Code Requirements for Structural Concrete and Commentary," courtesy American Concrete Institute.

TABLE 1806.2—PRESUMPTIVE LOAD-BEARING VALUES

TABLE 1806.2
PRESUMPTIVE LOAD-BEARING VALUES

CLASS OF MATERIALS	VERTICAL FOUNDATION PRESSURE (psf)	LATERAL BEARING PRESSURE (psf/f below natural grade)	LATERAL SLIDING RESISTANCE	
			Coefficient of friction[a]	Cohesion (psf)[b]
1. Crystalline bedrock	12,000	1,200	0.70	-
2. Sedimentary and foliated rock	4,000	400	0.35	-
3. Sandy gravel and/or gravel (GW and GP)	3,000	200	0.35	-
4. Sand, silty sand, clayey sand, silty gravel and clayey gravel (SW, SP, SM, SC, GM and GC)	2,000	150	0.25	-
5. Clay, sandy clay, silty clay, clayey silt, silt and sandy silt (CL, ML, MH and CH)	1,500	100	-	130

For SI: 1 pound per square foot = 0.0479 kPa, 1 pound per square foot per foot = 0.157 kPa/m.
a. Coefficient to be multiplied by the dead load.
b. Cohesion value to be multiplied by the contact area, as limited by Section 1806.3.2.

TABLE 601
FIRE-RESISTANCE RATING REQUIREMENTS FOR BUILDING ELEMENTS (hours)

BUILDING ELEMENT	TYPE I		TYPE II		TYPE III		TYPE IV	TYPE V	
	A	B	A[e]	B	A[e]	B	HT	A[e]	B
Structural frame[a]	3[b]	2[b]	1	0	1	0	HT	1	0
Bearing walls Exterior[g] Interior	 3 3[b]	 2 2[b]	 1 1	 0 0	 2 1	 2 0	 2 1/HT	 1 1	 0 0
Nonbearing walls and partitions Exterior	See Table 602								
Nonbearing walls and partitions Interior[f]	0	0	0	0	0	0	See Section 602.4.6	0	0
Floor construction Including supporting beams and joists	2	2	1	0	1	0	HT	1	0
Roof construction Including supporting beams and joists	$1^1/_2$[c]	1[c, d]	1[c, d]	0[c, d]	1[c, d]	0[c, d]	HT	1[c, d]	0

For SI: 1 foot = 304.8 mm.

a. The structural frame shall be considered to be the columns and the girders, beams, trusses and spandrels having direct connections to the columns and bracing members designed to carry gravity loads. The members of floor or roof panels which have no connection to the columns shall be considered secondary members and not a part of the structural frame.
b. Roof supports: Fire-resistance ratings of structural frame and bearing walls are permitted to be reduced by 1 hour where supporting a roof only.
c. Except in Group F-1, H, M and S-1 occupancies, fire protection of structural members shall not be required, including protection of roof framing and decking where every part of the roof construction is 20 feet or more above any floor immediately below. Fire-retardant-treated wood members shall be allowed to be used for such unprotected members.
d. In all occupancies, heavy timber shall be allowed where a 1-hour or less fire-resistance rating is required.
e. An approved automatic sprinkler system in accordance with Section 903.3.1.1 shall be allowed to be substituted for 1-hour fire-resistance-rated construction, provided such system is not otherwise required by other provisions of the code or used for an allowable area increase in accordance with Section 506.3 or an allowable height increase in accordance with Section 504.2. The 1-hour substitution for the fire resistance of exterior walls shall not be permitted.
f. Not less than the fire-resistance rating required by other sections of this code.
g. Not less than the fire-resistance rating based on fire separation distance (see Table 602).

GLOSSARY

The following glossary defines a number of structural terms, many of which have appeared on past exams. While this list is by no means complete, it comprises much of the terminology with which candidates should be familiar. You are therefore encouraged to review these definitions as part of your preparation for the structural exams.

A

Acceleration The rate of change of velocity, usually expressed as a fraction or percentage of g, the acceleration of gravity.

Accelerograph A seismological instrument that is normally inoperative, but becomes activated when subject to strong earth motion, records the earth motion, and then shuts off.

Active Pressure The pressure exerted by retained earth against a retaining wall.

Admixture A prepared substance added to concrete to alter or achieve certain characteristics.

Aggregate The chemically inert element of concrete, usually consisting of sand, gravel, and/or other granular material.

Air-Supported Structure A membrane enclosing a pressurized occupied space, which must be held down to its foundation.

Allowable Stress Same as Working Stress.

Allowable Stress Design The design method used for most reinforced concrete design until the middle of the 1960s. Still used for steel design. Largely replaced by strength design.

Arch A curved structure in which the internal stresses are essentially compression.

Axial Load A longitudinal load that acts at the centroid of a member and perpendicular to its cross-section.

B

Balloon Framing A method of framing wood stud walls, in which the studs are continuous for the full height of the building, which is usually two stories.

Base Isolation A method of isolating a structure from the ground by specially designed bearing and dampers that absorb earthquake forces.

Beam A structural member that supports loads perpendicular to its longitudinal axis.

Bearing Wall A wall that supports any vertical load in addition to its own weight.

Bending Moment The algebraic sum of the moments of all forces that are on one side of a given cross-section of a beam.

Braced Frame A vertical truss used to resist lateral forces.

C

C A standard designation for a structural steel American Standard channel.

Cable Roof A curved structure in which the internal stresses are pure tension.

Camber A curve built into a structural member to compensate for deflection.

Cantilever Beam A beam that is restrained against rotation at one end and free at the other.

Cantilever Footing An exterior column footing joined by a concrete beam to an interior column footing.

Cantilever Wall A retaining wall in which the stem, heel, and toe act as cantilever slabs.

Catenary The shape assumed by a cable when the only load acting on it is its own weight.

Centroid The point in a cross-section where all of the area may be considered concentrated without affecting the moment of the area about any axis.

Chord A perimeter member of a truss.

Coefficient of Thermal Expansion The ratio of unit strain to temperature change, which is constant for a given material.

Collector A member used to collect seismic load from a diaphragm and deliver it to a shear resisting element.

Column A member, usually vertical, which is subject primarily to axial compressive load.

Column Footing A spread footing, generally square or rectangular in plan, used to support a single column.

Combined Footing A footing supporting two or more columns.

Component One of two or more forces that will produce the same effect on a body as a given force.

Composite Beam A steel beam and a concrete slab connected so that they act together as a single structural unit to resist bending stresses.

Composite Deck Steel floor decking with embossed ridges, bonded to a concrete slab so that they act together as a single structural unit.

Compression Stress that tends to shorten a member or crush it.

Compressive Reinforcement Reinforcing steel embedded in the compression face of a reinforced concrete beam.

Concentrated Load A load that acts at one point on a structure.

Concrete A mixture of fine and coarse aggregates, portland cement, and water.

Concurrent Describing the condition when the lines of action of several forces pass through a common point.

Continuous Beam A beam that rests on more than two supports.

Core Test A compression test of hardened concrete that has been cut from the structure.

Counterfort Wall A retaining wall in which the stem and base are connected at intervals by transverse walls called counterforts.

Couple Two forces equal in magnitude, but opposite in direction, and acting at some distance from each other.

Creep Continued deformation of a structural member with time, with no increase of load.

Curing Maintaining concrete at the proper moisture and temperature after it is cast.

Curvature Factor A factor used to modify the allowable unit stress in bending for the curved portion of glued laminated members.

Cylinder Test A test to determine the compressive strength of concrete.

D

Dead Load The vertical load due to the weight of all permanent structural and nonstructural components of a building, such as walls, floors, roofs, and fixed service equipment.

Deflection The movement of a beam from its original location when load is applied to it.

Deformation See Strain.

Diaphragm The horizontal floor or roof system that distributes lateral forces caused by wind or earthquake, by functioning as a horizontal girder.

Dome A roof structure whose shape is that of an arch rotated about its vertical axis to form a curved surface.

Drift The horizontal movement of a structure when subject to wind or earthquake force.

Drilled Caisson An end-bearing pile, the bottom of which may be belled, which is constructed by pouring concrete into a drilled shaft.

Drilled Pile A vertical shaft drilled into the ground and filled with concrete, which supports building loads by skin friction.

Ductility As used in earthquake design, the ability of structural systems and materials to deform and absorb energy, without failure or collapse.

E

E A symbol for modulus of elasticity.

Eccentric Load A longitudinal load that acts at a distance from a member's centroid, thereby producing bending moment in addition to axial stress.

Elastic Limit The unit stress for a material, below which Hooke's Law applies.

End-Bearing Caisson See Drilled Caisson.

Engineering News Formula A dynamic formula used to determine the capacity of driven piles.

Epicenter The projection of the focus on the ground surface.

Equilibrant A force equal in magnitude to the resultant, but opposite in direction and on the same line of action as the resultant.

Equilibrium A state of rest due to balanced forces and balanced moments.

Euler's Equation A basic equation that applies to all columns and gives the maximum stress a slender column can resist without failing by sudden buckling.

Expansive Soil A fine-grained cohesive soil which undergoes large volume changes with changes in moisture content.

F

Factor of Safety The ratio of the ultimate strength of a material to its working stress.

Fault The boundary between adjacent rock plates along which movement may take place during an earthquake.

Fillet Weld A weld placed in the right angle formed by lapping or intersecting plates and generally subject to shear stress.

Fixed End Beam A beam that is restrained (fixed) against rotation at both ends.

Flat Plate A flat slab without column capitals or drop panels.

Flat Slab A concrete slab reinforced in two directions that brings its load directly to supporting columns without any beams or girders, usually requiring column capitals (widened tops of columns) and drop panels (thickened slab around columns).

Flexure Bending.

Focus The location in the earth's crust where rock slippage begins during an earthquake.

Folded Plate A structural roof system consisting of inclined planes that support each other and function as deep beams.

Force A push or pull exerted on an object. The description of a force includes its magnitude, direction, and point of application.

Friction Pile See Drilled Pile.

Frost Line The maximum depth of frost penetration in the ground expected in a given area.

G

Gabled Frame A frame consisting of two columns and two inclined beams that meet at the ridge, in which the joint between each column and beam is rigid.

Girder A main beam that supports secondary beams.

Glued Laminated Beam An assembly of laminations of lumber in which the grain of all the laminations is approximately parallel longitudinally. The laminations are bonded with adhesives and fabricated in accordance with certain accepted standards.

Gravity Wall A retaining wall that depends entirely on its own weight to resist the pressure of the retained earth and provide stability.

Groove Weld A weld placed between two butting pieces of metal to be joined.

Grout A high slump concrete, consisting of portland cement, sand, hydrated lime, water, and sometimes pea gravel.

H

Hooke's Law The physical law that states that up to a certain unit stress, called the elastic limit, unit stress is directly proportional to unit strain.

Hoop A horizontal member that extends around the circumference of a dome.

Hydrostatic Pressure The pressure exerted by a liquid against every surface it contacts.

Hyperbolic Paraboloid A thin shell saddle-shaped surface formed by moving a vertical parabola with downward curvature along and perpendicular to another parabola with upward curvature.

Hypocenter Same as Focus.

I

I (1) A symbol for moment of inertia. (2) The importance factor used in earthquake or wind design.

Impact Hammer Test A nondestructive test of hardened concrete to determine its approximate strength.

J

Joist One of a series of small, closely-spaced beams used to support floor, ceiling, or roof loads.

Joist Girder A shop-fabricated steel truss that supports evenly-spaced steel joists along its top chord.

K

K An effective length factor used in the design of structural steel columns.

Kelly Ball Test A method of measuring the workability of fresh concrete.

Kip A unit of force or weight equal to 1,000 pounds.

Ksi An abbreviation for kips per square inch.

L

L A standard designation for a structural steel angle.

Lamella A roof structure comprising a series of parallel arches, skewed to the axes of the building, which are intersected by another series of skewed arches, so that they interact with each other.

Lateral Load Any horizontal load on a building, including the load from wind or earthquake.

Lift Slab A flat plate cast at grade around columns and then lifted to position with hydraulic jacks.

Line of Action A line parallel to and in line with a force.

Lintel A structural member placed over an opening and supporting construction above.

Liquefaction Transformation of soil into a liquefied state, similar to quicksand, as a result of earthquake vibrations.

Live Load The vertical load caused by the use and occupancy of a building, not including wind, earthquake, or dead loads.

Load A force applied to a body.

M

M A symbol for bending moment.

Mat Foundation A large footing under an entire building, which distributes the building load over the entire area.

Membrane A thin sheet that can resist tension, but cannot resist compression, bending, or shear.

Meridian A curved line on the surface of a dome, usually circular, which is formed by the intersection of a vertical plane with the dome, when the plane passes through the top of the dome.

Method of Joints An analytical method for determining the forces in the members of a truss.

Method of Sections An analytical method for determining the forces in the members of a truss.

Modified Mercalli Scale A scale used to measure the intensity of an earthquake.

Modulus of Elasticity Within the elastic limit, the constant ratio of the unit stress in a material to the corresponding unit strain.

Modulus of Rupture The unit bending stress calculated from the flexure formula, for the maximum bending moment resisted by a beam before rupture.

Moment The tendency of a force to cause rotation about a given point or axis.

Moment Diagram A graphic representation of the value of the bending moment at any point along a beam.

Moment of Inertia The sum of the products obtained by multiplying each unit of area by the square of its distance to the neutral axis.

Moment-Resisting Frame A frame with rigid joints, in which the members and joints are capable of resisting vertical and horizontal forces primarily by flexure.

N

Natural Period of Vibration The time it takes for a structure to go through one complete back-and-forth motion under the action of dynamic loads.

Negative Moment Bending moment that produces tension in the upper part of a beam and compression in the lower part.

Neutral Axis The line on a beam cross-section that has zero bending stress when the beam is loaded.

Nonbearing Wall A wall that supports no vertical load other than its own weight.

O

Open Web Steel Joist A shop-fabricated, lightweight steel truss used to span between main members or bearing walls and support roof or floor loads.

Overhanging Beam A beam that rests on two or more supports and has one or both ends projecting beyond the support.

P

P-Delta Effect The secondary effect on frame members produced by vertical loads acting on a building frame that is laterally displaced by earthquake loads.

Pile An underground wood, concrete, or steel member, usually vertical, and usually driven into place, which is used to support building loads.

Plate Girder An assembly of steel plates, or plates and angles, which are fastened together to form an integral member.

Plate Tectonics The theory that explains earthquake phenomena.

Platform Framing A method of framing wood stud walls in which the studs are one story in height and the floor joists bear on the top plates of the wall below.

Point of Inflection The point in a beam or other flexural member where the bending moment changes sign and has a value of zero.

Portland Cement The finely-ground material used as the binder for structural concrete.

Positive Moment Bending moment that produces compression in the upper part of a beam and tension in the lower part.

Posttensioning A method of prestressing concrete in which the concrete is cast and then the steel tendons stressed by jacking.

Prestressed Concrete Concrete that is permanently loaded so as to cause stresses opposite in direction from those caused by dead and live loads.

Pretensioning A method of prestressing concrete in which the tensile force is put into high strength steel wires before the concrete is cast.

Proctor Test A laboratory compaction test to determine the optimum moisture content and density for a soil.

Psf An abbreviation for pounds per square foot.

Psi An abbreviation for pounds per square inch.

Punching Shear Two-way shear that occurs in a flat slab, spread footing, or pile cap.

Purlin A regularly spaced roof beam that spans between girders or trusses.

R

r A symbol for radius of gyration.

R A numerical coefficient used in earthquake design.

Radiographic Inspection Nondestructive testing of welded joints using x-rays and gamma rays.

Radius of Gyration A term used in column design to describe the distribution of cross sectional area in a beam around its centroidal axis.

Reactions Forces acting at the supports of a structure that hold the structure in equilibrium.

Redundancy The ability of part of a structural system to redistribute loads to other parts of the system.

Redundant Member Any truss member not necessary for stability.

Reinforced Brick Masonry A type of wall construction consisting of brick units, usually two tiers, with a solidly grouted space between in which vertical and horizontal reinforcing bars are placed.

Reinforced Concrete Block Masonry A type of wall construction consisting of hollow concrete masonry units, with certain cells continuously filled with grout in which reinforcing bars are embedded.

Resolving Forces Replacing a force with two or more other forces (components) that will produce the same effect on a body as the original force.

Response Spectrum A curve that shows the maximum acceleration of a series of idealized structures when subject to an earthquake.

Resultant One force that will produce the same effect as two or more other forces.

Retaining Wall A wall that resists the lateral pressure of retained earth or other material.

Richter Scale A logarithmic scale used to measure earthquake magnitudes.

Rigid Frame Same as Moment-Resisting Frame.

Rigidity Same as Stiffness.

S

S (1) A symbol for section modulus. (2) A standard designation for a structural steel I beam, also known as an American Standard beam.

Section Modulus The ratio of the moment of inertia of a beam (I) to the distance from its neutral axis to the most remote fiber (c). Thus, section modulus (S) = I/c.

Shear Stress that tends to make two members, or two parts of a member, slide past each other.

Shear Diagram A graphic representation of the value of the vertical shear at any point along a beam.

Shear Wall A wall designed to resist lateral forces parallel to itself caused by wind or earthquake.

Simple Beam A beam that rests on a support at each end.

Size Factor A factor used to reduce the allowable bending stress for wood beams deeper than 12 inches.

Slenderness Ratio The ratio l/r used in column design, where l is the length and r is the radius of gyration.

Slump Test A test for mixed concrete to determine consistency and workability.

Space Frame A series of trusses that intersect in a consistent grid pattern and are rigidly connected at their points of intersection.

Spiral Column A reinforced concrete column, usually square or round, containing longitudinal reinforcing bars enclosed by a closely-spaced continuous steel spiral.

Statical Moment The product of an area and the distance from the centroid of the area to a given axis.

Statically Determinate Describing a structure whose reactions can be determined from the equations of equilibrium.

Statically Indeterminate Describing a structure whose reactions cannot be found from the equations of equilibrium only, but requires additional equations.

Stiffness Resistance to deformation.

Stirrup A vertical steel bar, usually U-shaped, used to reinforce a reinforced concrete beam where the shear stresses are excessive.

Strain The deformation (change in size) of a body caused by external forces.

Strap Footing Same as Cantilever Footing.

Strength Design The method generally used for reinforced concrete design, formerly called ultimate strength design. Used in steel design since the 1980s.

Stress An internal force in a body that resists an external force.

Stress Diagram A graphical method for determining the forces in the members of a truss.

Stressed Skin A structural system consisting of spaced members solidly sheathed on one or both sides, in which the sheathing forms the flanges and resists flexure while the spaced members comprise the webs and resist shear.

Stud Wall A wall consisting of small closely-spaced members (studs) usually sheathed on both faces with a wall material.

Surcharge Increased earth pressure against a retaining wall caused by vertical load behind the wall or a sloping ground surface.

T

T-Beam A reinforced concrete beam consisting of a portion of the slab and the integrally constructed beam, which act together.

Tapered Girder A plate girder having a tapered profile, usually varying from minimum depth at the supports to maximum depth at midspan.

Tension Stress that tends to stretch a member or pull it apart.

Thin Shell A structure with a curved surface that supports load by tension, compression, and shear in the plane of its surface, but which is too thin to resist bending stresses.

Tied Column A reinforced concrete column, usually square or rectangular, containing longitudinal reinforcing bars and separate lateral ties.

Tilt-Up Wall A reinforced concrete wall that is precast at the job site, usually in a flat position, and later tilted up and set into place.

Torsion The rotation caused in a rigid diaphragm by lateral load from wind or earthquake, when the center of mass does not coincide with the center of rigidity.

Tributary Area The floor or roof area supported by an individual structural member.

Truss A jointed structure designed to support vertical or horizontal loads and composed generally of straight members forming a number of triangles.

Trussed Rafter A prefabricated lightweight wood truss used to support roof loads for dwellings and other small structures.

Tsunami An ocean wave produced by displacements of the ocean bottom as a result of earthquake or volcanic activity.

Tubular System A structural system used in tall buildings, consisting of closely-spaced columns at the perimeter connected by deep spandrel beams, which acts like a tube that cantilevers from the ground when subject to lateral wind or earthquake loads.

U

Ultimate Strength The maximum unit stress that can be developed in a material.

Ultrasonic Testing Nondestructive testing of welded joints using high-frequency sound waves.

Uniformly Distributed Load A beam loading of constant magnitude per unit of length.

V

V (1) A symbol for vertical shear. (2) The total lateral earthquake force at the base of a structure.

Vault A series of arches placed side-by-side to form a continuous structure.

Vertical Shear The algebraic sum of the forces that are on one side of a given cross-section of a beam.

Vibratory Compactor A machine used primarily for the compaction of granular soils, such as sands.

Vierendeel Truss A truss with no diagonals.

Volume Factor A factor used to reduce the allowable bending stress for glued laminated beams, based on width, depth, and span.

W

W (1) The total dead load used in earthquake design. (2) A standard designation for a structural steel wide flange shape.

Wall Footing A continuous spread footing supporting a uniformly loaded wall.

Water-Cement Ratio The ratio of water to cement in a concrete mix, the main factor that determines concrete strength.

Web Members The interior members of a truss, which connect the chords.

Wind Bent A frame used to resist lateral forces from wind.

Workability The ease with which concrete can be placed and consolidated in the forms.

Working Stress The maximum unit stress permissible in a structural member.

Y

Yield Point The unit stress at which a material deforms with no increase in load.

Z

Z Plastic section modulus of a structural steel section.

Part III

ARE Strategic Planning

LESSON EIGHTEEN

18

ARE STRATEGIC PLANNING

GOAL SETTING

Setting specific goals in order to reach an accomplishment is like identifying a destination prior to leaving. If you know exactly where you are going, you are able to focus on how you will reach your destination, stay on course, and get back on course if any obstacles get in the way.

Goals also give you something against which to measure your progress, to see if what you are doing is moving you closer or farther from your ultimate accomplishment. If your actions are moving you closer to your goal, then keep doing what you're doing. If your actions are moving you away from you goal, then you will need to adjust what you are doing to get back on track.

Your Specific Licensure Goals

In what state, or states, do you want to practice architecture?

When are you going to be eligible to start taking ARE exams?

By what date do you want to be licensed?

GETTING STARTED

Select the order in which you will take the exams.

There are numerous opinions on what is the optimum order for taking the exams. Recently, one Kaplan candidate passionately argued for ordering exams based on pass rates with a bell curve model, where the easiest exams were taken first and last and the hardest exams were taken in the middle. Our research suggests many candidates choose exam order based primarily on personal convenience and availability of study materials, seminars, and study partners. Occasionally, a candidate will insist on a systematic but whimsical approach, such as taking the exams in alphabetical order.

Kaplan recommends carefully considering several factors to determine the order that is best for you, including the following:

- *Personal areas of strength*—Many candidates start with subjects they are familiar with, or strongest in, in order to become familiar with the exam format first.
- *Timing concerns*—Leaving the hardest exams until the end can be problematic if the candidate does not pass and has to wait six months before retaking the exam.
- *Grouping of exams with closely related content*—While Kaplan strongly recommends delineating which exam a candidate is studying for at any given time, there is a benefit in studying related subject areas sequentially. Studying should be easier as the candidate progresses and builds on previous knowledge.

Sample exam order strategy.

Following is a sample order of common exam divisions, as well as an explanation of why this order might be considered:

1. Schematic Design
2. Structural Systems
3. Building Design & Construction Systems
4. Building Systems
5. Construction Documents & Services
6. Site Planning & Design
7. Programming, Planning & Practice

This order starts with a division this candidate thinks will be "easier" in order for him to become familiar with taking the exams. After that, the candidate has front-loaded the divisions he feels will be more "challenging" in order to allow for the possibility of retaking one of these exams and still meeting the overall licensure target date. Divisions with related content, such as Site Planning & Design and Programming, Planning & Practice, are taken sequentially. Programming, Planning & Practice was intentionally left for the end as the content touches so many disparate general topics the candidate found it disorienting, particularly as he had been out of school for a long time. Building Design & Construction Systems also includes information from other exams; however, this candidate thought the content was easy enough to take sooner.

The downside of this order is that Schematic Design, the first division, does not allow a student to become familiar with the multiple-choice exam format prior to the harder divisions. This candidate decided to compensate by spending additional time reviewing for multiple-choice section exam format prior to the second division.

Alternative sample exam order strategy.

Following is an alternate exam order strategy and rationale based around a different candidate's strengths and weaknesses:

1. Site Planning & Design
2. Programming, Planning & Practice
3. Construction Documents & Services
4. Structural Systems
5. Building Systems
6. Building Design & Construction Systems
7. Schematic Design

This order also starts with an exam the candidate believes will be "easier," so she has time to become more familiar with the exam format and the general process of taking these exams. This order also groups related subject areas sequentially. However, this candidate, as a recent graduate, decided to take Programming, Planning & Practice and Construction Documents & Services earlier, as much of this information was still fresh in her mind. Memorization and quickly navigating the multiple-choice format were her strengths, so she also elected to put the vignette-heavy exams later to allow more time for her to become familiar with the NCARB software.

The downside of this order is putting Building Design & Construction Systems, a division with historically low pass rates, so close to the end and increasing the chances the candidate will be delayed in reaching her licensure date.

YOUR DIVISION ORDER

Division	Order
Building Design & Construction Systems	
Building Systems	
Construction Documents & Services	
Programming, Planning & Practice	
Schematic Design	
Site Planning & Design	
Structural Systems	

This order makes sense for you because...

Set aside scheduled study time.

This is one of the biggest challenges for busy professionals, but one that is often necessary for success.

Establish a routine.

Unfortunately, ad hoc studying is often incomplete and uneven studying. A little organization and dedication can go a long way.

Identify areas for improvement.

You should not just review areas you know you are weak in. You need to at least test yourself on your perceived areas of strength. There are things you think you know, but do not. You need to find those areas for improvement, too.

Increase your speed.

This exam, like most exams, is timed. Speed counts. It is not just a matter of knowing the material, but of recalling it and applying it in the exam format quickly. Speed is something that should be planned for and practiced.

Don't forget to learn something.

Too many people try to cram or rely on educational tools with poor instructional design. Most people forget exam preparation should involve actual learning about the exam format, the best way to approach problems, personal areas of strength and weakness and of course the content being tested.

Sample Daily Breakdown

Daily study routine: 1 hour morning + 1 hour lunch + 2 hours evening = 4 hours/day

Sample Study Schedule

Week	Activity	Hours
1	Diagnostic quizzes and test banks	20
2	Multiple-choice areas of weakness	20
3	Multiple-choice areas of weakness and NCARB sample problems	20
4	Seminar review	16
5	Flashcards and mock exams	20
6	NCARB software and vignette sample problems	20
7	NCARB software and vignette sample problems	20
8	Multiple-choice and vignette areas of weakness	15
9	Exam	

It is important to remember that the study tools used often have different strengths and weaknesses from an educational standpoint. These strengths and weaknesses should be considered when developing your study plan.

Seminars and study groups.

The time constraints of most seminars and study groups do not allow for an in-depth review of all the content. Instead, they are great opportunities to ask others about the exam format, the best way to approach problems, and how to address personal areas of strength and weakness.

Reference books.

The value of reading or rereading about *all* of the exam content topics in depth varies significantly by student. However, reference books are almost always extremely helpful once an area of weakness has been identified.

Quizzes, test banks, and problems and solutions books.

These allow you the opportunity to pinpoint areas where you need improvement.

Mock final exams and NCARB sample problems.

These are designed to simulate the ARE in a way that can teach you about the exam format and allow you to practice for speed as well as accuracy.

Flashcards and terminology and definitions mobile applications.

These are great for ad hoc studying to ensure the main concepts are understood and can be recalled quickly. They should always be supplemented with study tools that review the exam format and the exam required concept application.

NCARB vignette software.

You need to learn how to use this software prior to taking the exam. You do not have time to learn to use it during the exam. Many people do not find the software very intuitive. Practice, practice, practice!

Sample Study Breakdown

To ensure that your studying is appropriate for the exam, consider which portion of your preparation will be "active studying," such as information gathering, reading, and memorization, and which will be "active practice," such as diagnostic testing, mock examinations, flashcards, and so on.

- 60 percent active study—information gathering, reading, and memorization
- 40 percent active practice—flashcards, mock exams, and answering approaches

MULTIPLE-CHOICE QUESTION CONSIDERATIONS

It helps to recognize that there are distinct question types within the multiple-choice part of the exams. These include:

- Checking all that apply
- Comparative analysis
- Terminology and definitions
- Diagrams, principles, materials, definitions, and details
- Scenarios
- Questions with partially correct answers that appear to be correct but have minor errors
- Selecting the most accurate statement: code, public health, safety, welfare, and so on

Historically, the most common question type has been comparative analysis. "Check all that apply" questions are newer.

Many candidates fall victim to questions seeking the "best" answer. In these cases, it may appear at first glance as though several choices are correct. Remember the importance of reviewing the question carefully and interpreting the language literally. Consider the following example:

1. Which of these cities is located on the east coast of the United States?

   A. Boston

   B. Philadelphia

   C. Washington, DC

   D. Atlanta

At first glance, it may appear that all of the cities could be correct answers. However, if you interpret the question literally, you'll identify the critical phrase as "on the east coast." Although each of the cities listed is arguably an "eastern" city, only Boston sits on the Atlantic coast. All the other choices are located in the eastern part of the country but are not coastal cities.

GENERAL TEST-TAKING TIPS

- Arrive in plenty of time—you do not need the additional stress.
- Bring layered clothing—the room temperature at testing facilities can vary, and you should not let that distract you.
- Breathe slowly and deeply—this will help you remain calm, focused, and confident.

MULTIPLE-CHOICE TIPS

Pace yourself.

Each division allows candidates at least one minute per question. You should be able to comfortably read and reread each question and fully understand what is being asked before answering.

Read carefully.

Begin each question by reading it carefully and fully reviewing the choices, eliminating those that are obviously incorrect. Interpret language literally, and keep an eye out for negatively worded questions.

Decide on a strategy.

Have a strategy in mind for how you will ensure that you answer all the questions in the allotted time. Some people answer every question before moving on, while others use the three-pass method.

Review difficult questions.

The exam allows candidates to review and change answers within the time limit. Utilize this feature to mark troubling questions for review after completing the rest of the exam.

Use your instincts.

If you continue to be unsure of the answer after returning to the question a second time, it is usually best to stick with your first guess. Unless you have misread the question, statistics show that the first answer you choose is usually correct.

Answer every question.

All unanswered questions are considered incorrect, so answer every question. If you are unsure of the correct answer, select your best guess and/or mark the question for later review. If unfamiliar with terminology, look for something similar.

Be cautious of wording.

Words such as *never* or *always* suggest a situation where the conditions "always" or "never" exist, which rarely happens in the practice of architecture. Look for strong words such as *guarantee* or *eliminate.*

Use scratch paper liberally.

Sketching helps many candidates think more clearly. Do not hesitate to sketch out a vignette strategy if it will improve your solution and still meet the time restrictions.

Double-check your math calculations.

This is a good idea for the complex formulas and the simple problems as well. Small miscalculations can be easily avoided with a little extra attention.

Utilize reference materials available to you.

Some divisions include reference materials accessible through an on-screen icon. These materials include formulas and other reference content that may prove helpful when answering questions in these divisions. Note, however, that candidates may *not* bring reference material with them to the testing center.

The Three-Pass Method

Pass #1. Read all questions, answering only those you can answer quickly. This gives you an opportunity to see all questions and gauge time-management requirements.

Pass #2. Answer all remaining questions, except for those where you clearly do not have the knowledge base.

Pass #3. Guess at remaining questions and try to eliminate any nonviable answers to increase your odds. Remember that all unanswered questions are marked incorrect and count against you.

VIGNETTE CONSIDERATIONS

Style doesn't count.

Vignettes are graded on their conformance with program requirements and instructions. Don't waste time creating aesthetically pleasing solutions and adding unnecessary design elements.

Pace yourself, and don't panic.

Speed is often important, but each vignette allows candidates ample time to complete a solution within the time allotted. It is typically very helpful to already have a strategy for how you will approach each vignette type.

Read carefully.

Review instructions and requirements. Quickly make a list of program and code requirements to check your work against as you proceed through the vignette.

STRUCTURAL SYSTEMS

EXAM OVERVIEW

According to NCARB, the Structural Systems division covers "the identification and incorporation of general structural and lateral force principles in the design and construction of buildings."

SITE PLANNING & DESIGN		
Intro Time	:15	
MC Testing Time	**3:30**	**125 items**
Scheduled Break	:15	
Intro Time	:15	
Graphic Testing Time	**1:00**	**Structural Layout**
Exit Questionnaire	:15	
Scheduled Appointment Time	**5:30**	

Multiple-Choice Topic Areas

Multiple-choice topic areas will likely include the following:

1. Structural principles.

You will need to be able to apply general structural principles to the design and construction of buildings. You will need to be able to apply lateral forces principles to the design and construction of buildings to resist seismic forces and wind forces. You also will need to be able to apply lateral forces principles to the design and construction of buildings. Consider the following topics:

- Forces
  - Magnitude and direction
  - Concentrated load
  - Distributed load
- Moments
- Connections
  - Roller
  - Pin
  - Cantilever
- Free body diagrams
- Static equilibrium
  - PVM
  - Diagrams
- Internal stresses
  - Varies by shape and material
  - Moment of inertia (statical moment)
- Strain
  - Deformation of the body
  - Varies by material
  - Modulus of elasticity (like a spring)
- Vertical forces
  - Components
    - Foundations
    - Columns
    - Beams
    - Bearing walls
    - Floors and roofs
  - Types
    - Dead load
    - Live load
    - Other loads
- Lateral Forces
  - Wind loads
    - Uniformly distributed
  - Seismic loads
    - Linearly distributed
  - Base shear
  - Overturning moment
  - Shear load transfer
    - Diaphragm
    - Shear Wall
    - Brace
- Special conditions
  - Trusses
  - Footings
  - Retaining walls

2. Materials and technology.

You will need to be able to analyze the implications of design decisions in the selection of systems, materials, and construction details related to general structural design. You will need to be able to analyze the implications of design decisions in the selection of systems, materials, and construction details related to seismic forces and wind forces. You will need to be able to analyze the implications of design

Notes

decisions in the selection of systems, materials, and construction details related to lateral forces. Consider the following topics:

- Wood
  - General properties
  - Reactive behavior
- Steel
  - General properties
  - Reactive behavior
- Reinforced concrete
  - Reinforcing (asymmetrical, heterogeneous)
  - Reactive behavior
- Assemblies
  - Walls (shear, bearing, retaining, etc.)
  - Floors and roofs
  - Connections

3. Codes and regulations.

You will need to be able to incorporate building codes, specialty codes, and other regulatory requirements in the design of general structural systems. You will need to be able to incorporate building codes, specialty codes, and other regulatory requirements related to seismic forces and wind forces. Consider the following topics.

- Allowable stress tables
- IBC
  - Seismic design
    - Seismic ground motion values
    - Seismic design category
    - Seismic response coefficient
  - Design loads
    - Live loads
    - Dead loads
    - Others
  - Allowable deflection

Tip.

- Do not just know how to calculate problems. Understand the concepts so you know how to apply the formulas.

Structural Layout Vignette

This problem tests the understanding of basic structural framing concepts through the development of a conceptual framing plan for a one-story building with a multilevel roof.

The information typically provided is *a background floor plan along with the program requirements.*

The goal is to complete the roof framing by locating basic structural elements, including beams, joists, lintels, columns, and bearing walls.

The solution is analyzed for compliance with program and code requirements, in addition to the general efficiency of the layout.

1. Strategy.

- Sketch a 3-D view of the plan, starting with the highest roof framing and sketching the load-bearing horizontal beams and vertical columns or bearing walls for each subsequent level.
- Make sure that each end of a beam is supported on the level below it.
- Create a chart of typical spans for types of structural elements.

2. Analysis.

- Analyze floor plan to determine best use of columns or bearing walls.

Notes

- Understand program requirements for placement of structural elements, and clerestory window.
- Look for the maximum span for decking, which equals the joist spacing.

3. Draw upper framing plan.

- Draw joists for upper framing plan (high space).
- Joists should span in short direction and be supported at ends by beams or bearing walls. Joists are drawn automatically at ends.
- Draw beams and/or bearing walls to support upper framing plan joists.
- Draw the upper framing on the *upper framing plan layer*.
- Use the Draw tool to indicate that a wall is a bearing wall.
- Draw beam and intermediate columns over clerestory window.
- Make sure that all ends of beams are supported!
- Draw roof deck boundary. Deck direction is perpendicular to joists.

4. Draw lower framing plan.

- Use the same process as for upper, but pick up all loads from upper framing.
- Draw on *lower framing plan layer*.
- Avoid drawing columns in areas required to be column-free.
- Draw lintels in all openings in bearing walls (windows, doors, etc.).

5. Review.

- Compare both framing plans to ensure that all loads are picked up!

Notes

YOUR STRUCTURAL SYSTEMS SELF-STUDY PLAN

Your target test date is:____________

Your study breakdown.

____ percent active study—information gathering, reading, memorization

____ percent active practice—flashcards, mock exams, answering approaches

Your daily breakdown.

Daily study routine ___ hour(s) morning +

___ hour(s) lunch + ___ hour(s) evening =

___ hours/day

Your study tools.

Your study schedule.

Week	Activity	Hours

Preparation plan.

My areas to improve.

My areas of strength.

My exam strategy.

Multiple-choice strategy.

Vignette strategy.

BIBLIOGRAPHY

The Architect Registration Examination (ARE) is closed book; no reference material is permitted for any part of the exam. If you need any reference material for the structural test, you will be able to access it during the test.

We suggest that prior to taking the test, you obtain some of the reference books that may be excerpted in order to become more familiar with them and thereby save valuable time at the exam.

The reference books that have often had sections reproduced for reference during the test include the *AISC Manual*, the *International Building Code*, and the *Standard Specifications for Steel Joists*.

These reference books are listed below, along with other useful books for those candidates wishing to do additional studying or review.

ACI Code (Building Code Requirements for Reinforced Concrete ACI 318)
American Concrete Institute
P.O. Box 9094
Farmington Hills, MI 48333

Architects and Earthquakes
AIA Research Corporation
1735 New York Ave., NW
Washington, DC 20006

Manual of Steel Construction
Thirteenth edition
American Institute of Steel Construction
One East Wacker Drive
Chicago, IL 60601

National Design Specification for Wood Construction
American Forest & Paper Association
1111 19th Street, NW
Washington, DC 20036

Parker Handbooks
Wiley
1. Simplified Design of Structural Steel
2. Simplified Mechanics and Strength of Materials
3. Simplified Design of Reinforced Concrete
4. Simplified Engineering for Architects and Builders

Standard Specifications for Steel Joists
Steel Joist Institute
1205 48th Ave. North
Myrtle Beach, SC 29577

Structure in Architecture
Salvadori and Heller
Prentice-Hall

Timber Construction Manual
American Institute of Timber Construction
Centennial, CO
Wiley

Minimum Design Loads for Building and Other Structures
American Society of Civil Engineers
1801 Alexander Bell Drive
Reston, VA 20191

International Building Code
International Code Council
500 New Jersey Ave., NW
6th Floor
Washington, DC 20001

QUIZ ANSWERS

Lesson One

1. **A** See pages 4 and 6 for the definitions of concurrent, non-concurrent, and components. A vector is any quantity that has both magnitude and direction, such as a force.

2. **Strain** See pages 15 and 17 for definitions of stress and strain. Other related topics include yielding, which is explained on page 18, and elasticity. Elasticity is the property of a material that enables it to deform when a force is applied to it and to return to its original size and shape when the force is removed. Candidates should be familiar with all of these terms.

3. **C** Modulus of elasticity, yield point, and factor of safety are explained on pages 17 to 19. Moment of inertia is one of the factors that determine a beam's resistance to deflection and is independent of the beam's material.

4. **C** P = 10,000# L = 10 ft. = 120"
   $A = \pi d^2/4 = \pi\ (1)^2/4 = 0.785$ in.^{2}
   E = 29,000,000 psi

   From page 17,
   $\Delta = PL/AE$
   $= \dfrac{10{,}000(120)}{0.785(29{,}000{,}000)}$
   = 0.053"

5. **B** See pages 9 and 12.

6. **D** If a load acts through a body's center of gravity, the body has no tendency to rotate, but it tends to translate (move) in the direction of the applied force (I and II are correct). If movement cannot take place, as in building structures, stress will be developed in the body (III is incorrect). IV is also incorrect, since the body has no tendency to rotate.

7. **C** The unit stress (f) is equal to the load (P) divided by the cross-sectional area (A).
   f = P/A
   = 20,000#/10" × 10"
   = 200 psi

8. **D** See page 18.

9. **B** See page 17.

10. **D** From page 14
    $I = bd^3/12$
    $= 8(12)^3/12$
    = 1,152 in.^{4}

Lesson Two

1. **C** The maximum stress that a column can resist without failing by buckling depends on the column's slenderness ratio l/r, where l is the unbraced length of the column and r is the radius of gyration of the column cross-section.

2. **B** See page 30.

3. **A and C** The maximum flexural stress in a beam is equal to M/S, where M is the bending moment and S is the section modulus of the beam. Therefore A and C are the correct choices.

4. **A** See page 37.

5. **D** See page 31.

6. **B** The reaction at each end of a uniformly loaded simple beam is equal to wL/2 (see Figure 2.24).

   Thus, wL/2 = 2,400
   w(24/2) = 2,400
   w = 2,400/12 = 200 lbs. per foot, or 0.2 kips per foot

7. **D** The deflection formula for a uniformly loaded simple beam is

   $\Delta = \dfrac{5wL^4}{384EI}$ (see Figure 2.24).

   For the 20-foot span, Δ = 0.50".

For the 25-foot span, all the terms in the formula remain the same, except the length L. Therefore, the deflection varies only with L^4.

$\frac{(25)^4}{(20)^4} = 2.44$

$0.50 \times 2.44 = 1.2207"$

8. C I is correct (see page 34).

II is incorrect. The point of inflection occurs some distance from the middle support.

III is correct (see page 24).

IV is correct (see page 34).

V is correct. The moment diagram for any beam supporting a uniform load is a parabola.

9. C From page 38, the maximum horizontal shear stress $= \frac{3}{2}\frac{V}{bd}$

$= \frac{3}{2} \times \frac{6{,}000}{8 \times 12} = 93.8$ psi

Since the vertical and horizontal shear stresses are equal, this is also the maximum vertical shear stress.

10. B See page 36.

Lesson Three

1. D See page 49.

2. B, C, and D See page 44 for a complete explanation.

3. A Most lumber used structurally is softwood, which comes from coniferous, evergreen trees. See page 44.

4. B $F_b = 2{,}200$ psi
$S = bd^2/6 = 6.75 \times (24)^2/6$
$= 648$ in.^{3}
$M = F_b \times S = 2{,}200 \times 648$
$= 1{,}425{,}600$ in.-lbs.
$= 1{,}425.6$ in.-kips

5. A II, IV, and V are correct. See page 55.

6. D $S = M/F_b = (30{,}000 \times 12)/1{,}000 \times 1.25$
$= 288.0$ in.^{3}

The factor 12 in the numerator converts the moment from ft.-lbs to in.-lbs. The 1.25 factor in the denominator provides for a 25 percent increase in the design value for bending, which is permitted for a wood member supporting roof live load.

7. B II and III are correct. See page 52.

8. D See page 54.

9. B The flexural stress varies inversely with the section modulus ($f_b = M/S$). For a given moment, therefore, the greater the section modulus, the lower the flexural stress. S for a 4 × 10 = 49.91 in.^{3}, and S for a 4 × 12 = 73.83 in.^{3} (see Figure 3.2). Therefore, by ratio, the flexural stress in the 4 × 12 is (49.91/73.83) × 1,250 = 845.0 psi.

10. C From IBC Table 1604.3, the deflection limits for sawn wood members are L/360 for live load only and L/240 for live load plus dead load. L/360 = 22 × 12/360 = 0.73" for live load L/240 = 22 × 12/240 = 1.10" for total load

Lesson Four

1. B See page 71.

2. C See the table in Figure 4.17.

3. A We compute the slenderness ratio Kl/r, using the smaller value of r to obtain the larger Kl/r value.

Kl/r = 20 × 12/2.04 = 117.6

We enter the table in Figure 4.17 with Kl/r = 118 (the closest whole number) and read $F_{cr}/\Omega = 10.8$ ksi.

$P = F_{cr}/\Omega \times A = 10.8$ kips/in.$^2 \times 11.7$ in.^{2}
$= 126.4$ kips, or approximately 126 kips.

4. **B** See page 62.

5. **D** See page 61.

6. **Wide flange** See page 62 for a complete explanation.

7. **B** First compute the moment = $1{,}250(30)^2/8$ = 140,625 ft.-lbs. Enter the chart on page 72 with an unbraced length of 30/3 = 10 ft. Locate the intersection of the moment coordinate (140.6 ft.-kips) and the unbraced length coordinate (10 ft.). Any beam whose curve lies above and to the right of this intersection is satisfactory, and the first such curve which is a solid line represents the lightest section which satisfies the design criteria, in this case a W16 × 40.

8. **B** M = 140.6 ft.-kips

   Review the table in Figure 4.3 on page 65 and choose a W18 × 35, which has a design flexural strength M_p/Ω = 166 ft.-kips > 140.6 ft-kips. This is the lightest section that is able to resist the maximum applied bending moment.

9. **C** See page 73.

10. **B** See page 73.

Lesson Five

1. **C** See page 87.

2. **A** See page 87.

3. **B** See page 86.

4. **D** See pages 98-99.

5. **D** From page 93,
   $\rho_{min.} = 0.0033$
   $A_s = \rho_{min.}\,bd = 0.0033 \times 12 \times 24$
   $= 0.95$ sq. in.

   From the table in Figure 5.4, the minimum flexural reinforcement is 2-#7

   ($A_s = 0.60 \times 2 = 1.20$ in^2)

6. **B** See pages 99 through 101.

7. **A** From page 92,
   $M_u = ØA_sf_y\,(d\text{-}a/2)$
   $= 0.90 \times 4 \times 0.79 \times 60\,(46\text{-}9.2/2)$
   = 7,064.5 in.-kips, or 7,064.5/12
   = 588.7 ft.-kips

8. **C** See pages 97 and 98.

9. **D** See pages 97 and 98.

10. **A, B, C, and D** All the available answers except choice E are correct. See page 94 for a complete explanation.

Lesson Six

1. **D** See page 114. Weep holes are not used in basement walls, since this would cause water to flow into the basement (A). Regardless of the type of backfill used, basement walls are designed to resist earth pressure (B).

2. **D** I, II, and V are incorrect (see page 113). III and IV are correct (see pages 112 to 114). The total earth pressure = $30 \times h \times h/2 = 30\,h^2/2$.

3. **D** See pages 116 and 117.

4. **Tilt-up** See page 112. Tilt-up walls may be used as shear walls, and may be as thin as 4 inches under certain conditions. They are cast on site and lifted into place—often spanning between footing pads and acting as a very deep beam to support roof loads.

5. **C** See page 108.

6. **B** See pages 109 to 111.

7. **B** See page 111. Additional reinforcing bars are placed at openings in reinforced concrete walls, rather than angle lintels (C). Reinforced concrete walls generally have conventional reinforcing bars and are not prestressed (D).

8. **D** See pages 114 and 116.

9. **C** The soil pressure at the base = 30 × 10 = 300 lbs./ft.

   The total earth pressure = 300 × 10/2 = 1,500 lbs.

   The bending moment at the base = 1,500 × 10/3 = 5,000 ft.-lbs.

10. **A** I and IV are true (see pages 109 through 111). II is false; reinforcing bars cannot stabilize a wall against buckling. III is also false. Grout and mortar are different and may not be used interchangeably.

Lesson Seven

1. **D** See page 121.
2. **A** From Table 7-3 in Figure 7.21, the value of each 3/4" A325-SC STD bolt in double shear is 14.8 kips. The total allowable load for 4 bolts is 14.8 × 4 = 59.2 kips. From Tables 7-5 and 7-6 in Figures 7.24 and 7.25, the bearing value of each bolt on the 1" plate is 52.2 kips. Since this is greater than 14.8 kips, it doesn't govern.
3. **D** See Figure 7.28.
4. **A** See page 131.
5. **C** See pages 123 and 124.
6. **B** See page 131.
7. **Butt** See page 138 for a complete explanation of weld types. Candidates should be familiar with different types of weld joints and be able to recognize the symbols associated with them.
8. **A** From Table 11-F in Figure 7.9, each 7/8" bolt has a value in double shear parallel to the grain in a 3-1/2" main member with a 1.5" side member of 3,180 lbs. The total safe load P is 3,180 × 2 = 6,360 lbs.
9. **D** The allowable load on the 1/4" fillet weld = 0.707 × 0.25 × 18.0 = 3.18 kips per lineal inch (see page 137). Since there is a total of 12 inches of weld (3 inches each side of each angle), the total allowable load P = 3.18 × 12 = 38.16 kips.
10. **A** Wood fasteners are strongest when the load is parallel to the grain (0° to grain).

Lesson Eight

1. **D** See page 148.
2. **B** See page 151.
3. **B** See page 147.
4. **D** See page 147.
5. **B and C** Drilled piles transmit their load to the soil by skin friction and belled caissons transmit their load by end bearing. See pages 155 to 158 for a complete description of different types of piles and caissons.
6. **D** See page 152.
7. **A** Locate the resultant of the two column loads by taking moments about A. Distance of the resultant from column A

   $$= \frac{P_B \times 20 \text{ ft.}}{P_B + P_A}$$

   $$= \frac{245 \text{ kips} \times 20 \text{ ft.}}{(245 \text{ kips} + 200 \text{ kips})}$$

   = 11.0 ft. from A

   The footing length should be 2(11 ft. + 3 ft.) = 28 feet.

   Required footing area

   $$= \frac{245 \text{ kips} + 200 \text{ kips} + 40 \text{ kips}}{4.0 \text{ ksf}}$$

   = 485/4 = 121.25 sq. ft.

   Required footing width = 121.25/28 = 4.33 ft.

   Use footing 28'-0" × 4' – 4"

8. B See page 152.

9. A Soil bearing value = 5,000 psf

Weight of concrete footing = 150(2) = 300 psf

Weight of earth over footing = 1(100) = 100 psf

Net soil bearing value = 5,000 – 300 – 100 = 4,600 psf

Required bearing area = (200 kips + 150 kips)/4.6 ksf

= 76.09 sq. ft.

Required D = $\sqrt{76.09}$ = 8.72 ft.

Use 8'–9"

10. C See pages 146–149.

Lesson Nine

1. A and D Composite beams and T-beams both consist of flat and deep elements acting together to resist bending. See page 170 for further discussion.

2. B Flat plate, flat slab, and waffle slab are two-way systems (see page 174). Rigid frame is a one-way system (see page 172).

3. A See page 170.

4. C See pages 170 and 171.

5. D Stiffness and strength are not the same. Stiffness means resistance to deflection, while flexural strength is the ability to resist bending stress.

6. B A and D are incorrect (see page 172). C is also incorrect, since the base of a rigid frame has moment only if the base is fixed. If the base is pinned, there is no moment.

7. D A is incorrect. Doubling the spans increases the structural cost, not the total construction cost, about 25 percent. B is also incorrect; unit structural costs for tall buildings may be as much as double those of short buildings, because of the lateral resistance required for wind or earthquake loads. The most expensive element in cast-in-place concrete construction is formwork (C is incorrect).

8. D See page 171.

9. C II and III are correct and I is incorrect (see page 162). IV is incorrect, because the allowable live load reduction is .08(250 – 150) = 8%.

10. D The systems described in choices A, B, and C are likely to be efficient (see pages 173 and 174), but for the waffle slab system to be efficient, the bays should be approximately square.

Lesson Ten

1. A See page 183.

2. C See page 187.

3. Flanges Since most of the material of a steel beam or girder is in the flanges, at the maximum possible distance from the neutral axis, it has a large section modulus, and therefore it is an efficient bending member.

4. C All of the statements are correct except C.

5. B Joists are generally made of steel with a yield point of 50 ksi (II is incorrect), and may have a number of different profiles (V is incorrect). All of the other statements are correct.

6. D See page 195.

7. A See page 201.

8. A See page 204.

9. D See page 207.

10. D See page 215.

Lesson Eleven

1. **D** See page 223.
2. **A** Isolate joint D

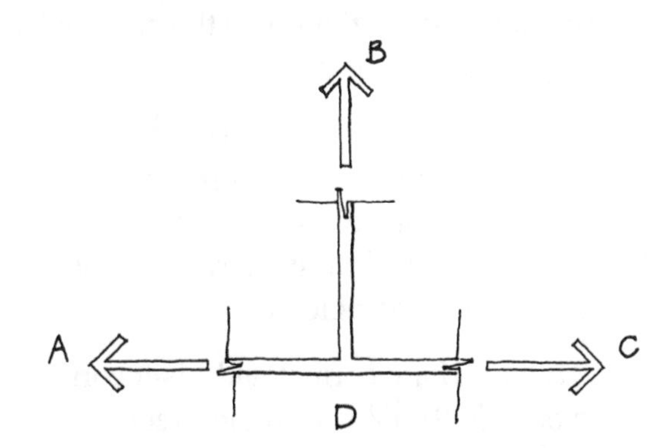

$\Sigma V = 0$

Since there is no external load applied at D, $F_{BD} = 0$. The 5 kip loads between A and B and between B and C cause bending stresses in members AB and BC, in addition to the axial compressive stresses caused by truss action.

3. **A** See page 224. B and C are incorrect; a truss can support load between panel points and can resist horizontal load. D is also incorrect; if the axial force acts toward the joint, the member is in compression.
4. **B** See page 223.
5. **C** See page 230.
6. **B, D and E**

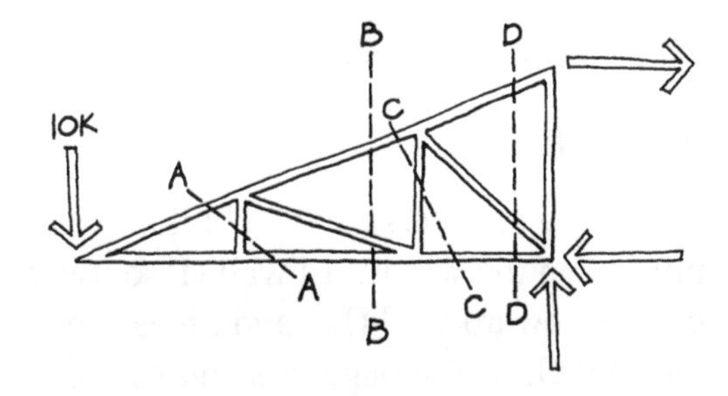

Cut section A-A through the truss as shown and take moments about the left end of the truss.

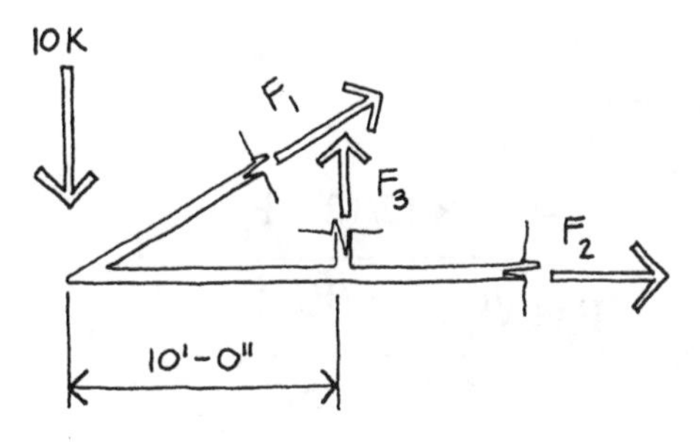

The lines of action of the chord members and the 10 kip applied load all intersect at the left end of the truss, and their moment arms are therefore equal to zero.

$\Sigma M = 0$

$(10 \text{ kips} \times 0) + (F_1 \times 0) + (F_2 \times 0)$
$- (F_3 \times 10 \text{ ft.}) = 0$

$F_3 = 0$ (the force in member 3 is zero)

Similary, if we cut sections B-B, C-C, and D-D through the truss, we can see that the force in every diagonal and vertical web member is zero (D and E). As with all cantilever structures, the top of the truss is in tension and the bottom is in compression (B).

7. **D** The design is based on the lower r value, which may be either $r_{x\text{-}x}$ or $r_{y\text{-}y}$.
8. **C** Since the loads are symmetrical, each reaction is equal to one half of the total load on the truss = (10 + 10 + 10 + 10 + 10)/2 = 25 kips. We cut a section through the first truss panel.

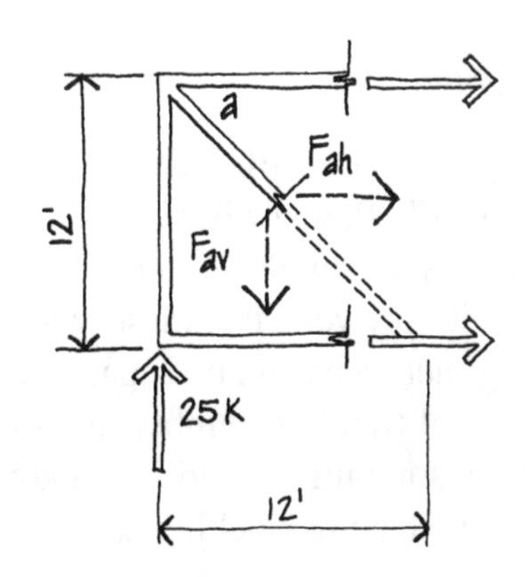

$$\Sigma V = 0$$
$$25 - F_{av} = 0$$
$$F_{av} = 25$$
$$F_{ah} = F_{av} \times 12/12 = F_{av}$$
$$F_a = \sqrt{(F_{av})^2 + (F_{ah})^2}$$
$$= \sqrt{(25)^2 + (25)^2}$$
$$= 35.4 \text{ kips tension}$$

9. **A** Cut a section that cuts member b, the bottom chord, and a vertical, as shown below. Take moments about point 0.

$\Sigma M = 0$

25 kips (12 ft. + 12 ft.) – 10 kips(12 ft.) + F_b(12 ft.) = 0

$F_b = -(600 - 120)/12$

$= -40$ kips (compression)

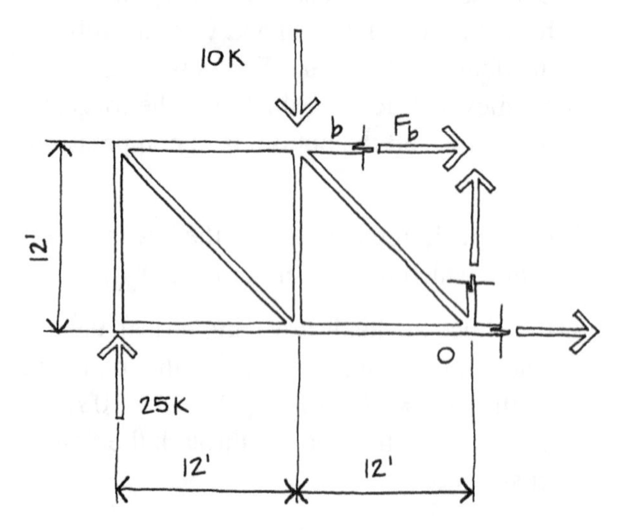

Lesson Twelve

1. **D** If you refer to IBC Table 2306.2.1 in Figure 12.30, you will note that the allowable shear in plywood diaphragms varies with plywood thickness, width of framing members, and nail spacing. The direction of framing members does not affect the strength of a blocked diaphragm, that is, one having blocking at the plywood panel edges.

2. **B** Moment-resisting frames are the most ductile lateral load resisting system, and are therefore most effective in resisting earthquakes, particularly for tall buildings. Locating these frames at the building's perimeter tends to minimize the potentially damaging effects of torsion.

3. **D** A, B, and C are correct (see pages 253–254 and 268). D is incorrect; torsion is the rotation in a diaphragm that occurs when the center of mass does not coincide with the center of rigidity.

4. **D** II and III are correct and I is incorrect (see page 239). IV is also incorrect, since live loads are vertical gravity loads unrelated to earthquake loads.

5. **D** The system described is system B.5, which, according to Table 12.2-1 in Figure 12.19, has an R value of 6.

6. **B**

7. **C** One of the factors in the formula for the period of a structure (T) is the height of the building (see page 249). The taller the building, the longer the period.

8. **A, B, and C** See pages 255–256.

9. **B** The intent of earthquake provisions is to make buildings capable of resisting major earthquakes without collapse, although some structural and nonstructural damage may occur (see page 242).

10. **A** See pages 238–239.

Lesson Thirteen

1. **A** The wind pressure on a building acts inward on the windward wall (I), outward on the leeward wall (II), and upward on a flat roof (III). See page 298 and ASCE 7 Figure 6-6 in Figure 13.8.

2. **A and D** Drift and overturning moment (A and D) are considered in wind design,

while liquefaction and tsunamis (B and C) are phenomena associated with earthquakes.

3. **B** On the windward wall: $p_w = q_z G C_p$ (neglecting internal pressure)

   $q_z = 0.00256 K_z K_{zt} K_d V^2 I$

   For Exposure C, $K_z = 0.85$ (ASCE 7 Table 6-3).

   $K_{zt} = 1.0$ for a building not located on a hill or escarpment.

   $K_d = 0.85$ for main wind-force-resisting systems of buildings (ASCE 7 Table 6-4).

   $q_z = 0.00256 \times 0.85 \times 1.0 \times 0.85 \times 90^2 \times 1.0 = 15.0$ psf

   $G = 0.85$

   For windward wall, $C_p = 0.8$ (ASCE 7 Figure 6-6).

   $p_w = 15.0 \times 0.85 \times 0.8 = 10.2$ psf

4. **C** The IBC wind load coefficients account for all of the factors shown, except tornadoes. (I, II, and IV are correct, III is incorrect.)

5. **B** For wind load, the north and south walls span vertically between the ground and the roof diaphragm. The total wind load delivered to the roof diaphragm is therefore 20 psf × 12 ft./2 × 100 ft. = 12,000#. Half of this load goes into the east wall and half to the west wall, or 6,000# to each wall.

6. **A** The maximum diaphragm shear is equal to the wind load transferred to each end wall divided by the width of the diaphragm, or 6,000 #/50 ft. = 120#/ft.

7. **C** See page 326.

8. **C** The factor of safety against overturning is equal to the dead load resisting moment divided by the overturning moment = 500 ft.-kips ÷ 400 ft.-kips = 1.25. This is inadequate, since the code requires a factor of safety of at least 1.67. Although the factor of safety could be increased by making the wall longer and/or heavier (B), this is not generally feasible. A better solution is to provide sufficient anchorage from the wall to the foundation (C), so that the dead load of the foundation can be used to help resist the overturning moment.

9. **D** The tubular concept is discussed in Lesson Twelve, since tubular systems are used to resist lateral loads from either earthquake or wind. For very tall buildings, this system is often economical (see page 282).

10. **A** The load combinations shown in B, C, and D must be considered in the design of building components. The combination shown in A is not considered in the design, since the likelihood that the earthquake load will be acting on the structure during a windstorm is very remote.

Lesson Fourteen

1. **A** See page 342.

2. **C** See page 338. A refers to the Dome of the Florence Cathedral, B to Hagia Sophia, and D to the Parthenon.

3. **C** See page 340.

4. **B and C** See pages 343 to 344. A refers to the John Hancock Building in Boston and D refers to the Willis Tower.

5. **B** See pages 340 to 341. Both structures were designed by Nervi (I), had ribbed concrete shell roofs (V), and consisted of prefabricated elements (II is incorrect).

III and IV are also incorrect. Only the Palazzetto has a dome roof, and the Palazzetto was not built until long after World War II.

6. D See page 343.

7. A See page 338.

8. C See page 347.

9. D See page 346.

10. D See page 341.

EXAMINATION DIRECTIONS

The examination on the following pages should be taken when you have completed your study of all the lessons in this course. It is designed to simulate the Structural Systems division of the Architect Registration Examination. Many questions are intentionally difficult in order to reflect the pattern of questions you may expect to encounter on the actual examination.

You will also notice that the subject matter for several questions has not been covered in the course material. This situation is inevitable and, thus, should provide you with practice in making an educated guess. A few questions may appear ambiguous, trivial, or simply unfair. This too, unfortunately, reflects the actual experience of the exam and should prepare you for the worst you may encounter.

Answers and complete explanations will be found on the pages following the examination to permit self-grading. **Do not look at these answers until you have completed the entire exam**. Once the examination is completed and graded, your weaknesses will be revealed, and you are urged to do further study in those areas.

Please observe the following directions:

1. If a question requires you to refer to a table or chart in the course, it will so state. Otherwise, please do not use any reference material.
2. Allow about 1-1/2 hours to answer all questions. Time is definitely a factor to be seriously considered.
3. Read all questions *carefully* and mark the appropriate answer on the answer sheet provided.
4. Answer all questions, even if you must guess. Do not leave any questions unanswered.
5. If time allows, review your answers, but do not arbitrarily change any answer.
6. Turn to the answers only after you have completed the entire examination.

GOOD LUCK!

EXAMINATION ANSWER SHEET

Directions: Read each question and its lettered answers. When you have decided which answer is correct, blacken the corresponding space on this sheet. After completing the exam, you may grade yourself; complete answers and explanations will be found on the pages following the examination.

1. ⒶⒷⒸⒹⒺⒻ
2. ⒶⒷⒸⒹⒺⒻ
3. ⒶⒷⒸⒹⒺⒻ
4. ⒶⒷⒸⒹⒺⒻ
5. ⒶⒷⒸⒹⒺⒻ
6. ⒶⒷⒸⒹⒺⒻ
7. ⒶⒷⒸⒹⒺⒻ
8. ⒶⒷⒸⒹⒺⒻ
9. ⒶⒷⒸⒹⒺⒻ
10. ⒶⒷⒸⒹⒺⒻ
11. ⒶⒷⒸⒹⒺⒻ
12. ⒶⒷⒸⒹⒺⒻ
13. ⒶⒷⒸⒹⒺⒻ
14. ⒶⒷⒸⒹⒺⒻ
15. ⒶⒷⒸⒹⒺⒻ
16. ⒶⒷⒸⒹⒺⒻ
17. ⒶⒷⒸⒹⒺⒻ
18. ⒶⒷⒸⒹⒺⒻ
19. ⒶⒷⒸⒹⒺⒻ
20. ⒶⒷⒸⒹⒺⒻ
21. ____________
22. ⒶⒷⒸⒹⒺⒻ
23. ⒶⒷⒸⒹⒺⒻ
24. ⒶⒷⒸⒹⒺⒻ
25. ⒶⒷⒸⒹⒺⒻ
26. ⒶⒷⒸⒹⒺⒻ
27. ⒶⒷⒸⒹⒺⒻ
28. ⒶⒷⒸⒹⒺⒻ
29. ⒶⒷⒸⒹⒺⒻ
30. ⒶⒷⒸⒹⒺⒻ
31. ⒶⒷⒸⒹⒺⒻ
32. ⒶⒷⒸⒹⒺⒻ
33. ⒶⒷⒸⒹⒺⒻ
34. ⒶⒷⒸⒹⒺⒻ
35. ⒶⒷⒸⒹⒺⒻ
36. ⒶⒷⒸⒹⒺⒻ
37. ⒶⒷⒸⒹⒺⒻ
38. ⒶⒷⒸⒹⒺⒻ
39. ⒶⒷⒸⒹⒺⒻ
40. ⒶⒷⒸⒹⒺⒻ
41. ⒶⒷⒸⒹⒺⒻ
42. ⒶⒷⒸⒹⒺⒻ
43. ⒶⒷⒸⒹⒺⒻ
44. ⒶⒷⒸⒹⒺⒻ
45. ⒶⒷⒸⒹⒺⒻ
46. ____________
47. ⒶⒷⒸⒹⒺⒻ
48. ⒶⒷⒸⒹⒺⒻ
49. ⒶⒷⒸⒹⒺⒻ
50. ⒶⒷⒸⒹⒺⒻ
51. ⒶⒷⒸⒹⒺⒻ
52. ⒶⒷⒸⒹⒺⒻ
53. ⒶⒷⒸⒹⒺⒻ
54. ____________
55. ⒶⒷⒸⒹⒺⒻ
56. ⒶⒷⒸⒹⒺⒻ
57. ⒶⒷⒸⒹⒺⒻ
58. ⒶⒷⒸⒹⒺⒻ
59. ⒶⒷⒸⒹⒺⒻ
60. ⒶⒷⒸⒹⒺⒻ
61. ⒶⒷⒸⒹⒺⒻ
62. ⒶⒷⒸⒹⒺⒻ
63. ⒶⒷⒸⒹⒺⒻ
64. ⒶⒷⒸⒹⒺⒻ
65. ⒶⒷⒸⒹⒺⒻ
66. ⒶⒷⒸⒹⒺⒻ
67. ⒶⒷⒸⒹⒺⒻ
68. ⒶⒷⒸⒹⒺⒻ
69. ⒶⒷⒸⒹⒺⒻ
70. ⒶⒷⒸⒹⒺⒻ
71. ⒶⒷⒸⒹⒺⒻ
72. ⒶⒷⒸⒹⒺⒻ
73. ⒶⒷⒸⒹⒺⒻ
74. ⒶⒷⒸⒹⒺⒻ
75. ⒶⒷⒸⒹⒺⒻ
76. ⒶⒷⒸⒹⒺⒻ
77. ⒶⒷⒸⒹⒺⒻ
78. ____________
79. ⒶⒷⒸⒹⒺⒻ
80. ⒶⒷⒸⒹⒺⒻ
81. ⒶⒷⒸⒹⒺⒻ
82. ⒶⒷⒸⒹⒺⒻ
83. ⒶⒷⒸⒹⒺⒻ
84. ⒶⒷⒸⒹⒺⒻ
85. ⒶⒷⒸⒹⒺⒻ
86. ⒶⒷⒸⒹⒺⒻ
87. ⒶⒷⒸⒹⒺⒻ
88. ⒶⒷⒸⒹⒺⒻ
89. ⒶⒷⒸⒹⒺⒻ
90. ⒶⒷⒸⒹⒺⒻ
91. ⒶⒷⒸⒹⒺⒻ
92. ⒶⒷⒸⒹⒺⒻ
93. ⒶⒷⒸⒹⒺⒻ
94. ⒶⒷⒸⒹⒺⒻ
95. ⒶⒷⒸⒹⒺⒻ
96. ⒶⒷⒸⒹⒺⒻ
97. ⒶⒷⒸⒹⒺⒻ
98. ⒶⒷⒸⒹⒺⒻ

FINAL EXAMINATION

1. The shear diagram for a beam is

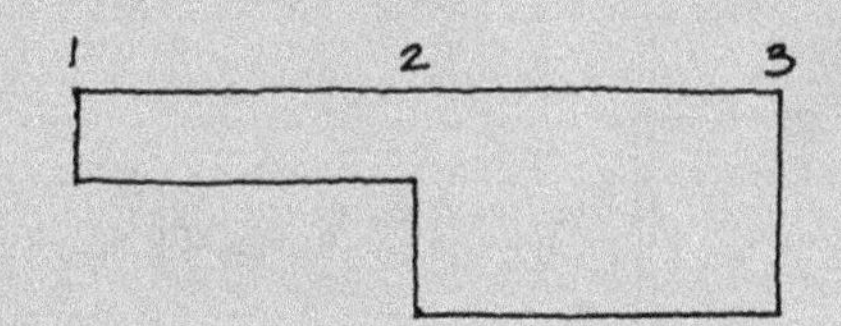

The corresponding moment diagram is

A.

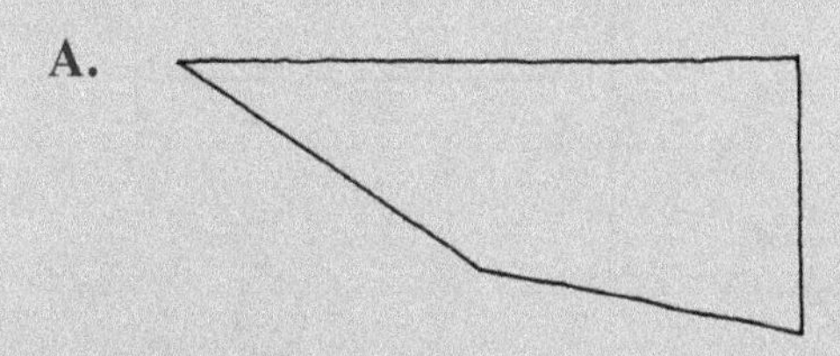

B.

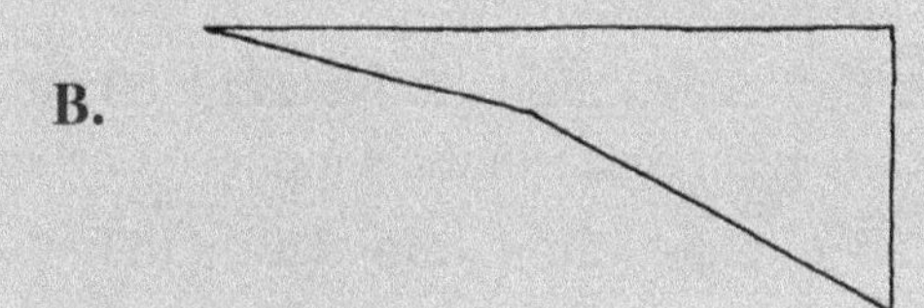

C.

D.

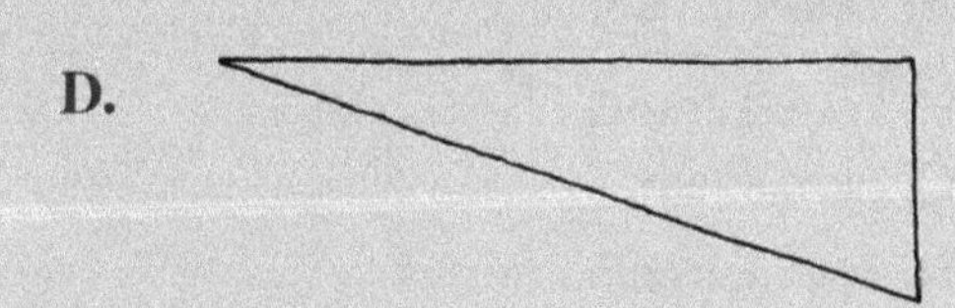

2. A five-inch concrete slab increases in temperature 40°F. If it is restrained against movement, what internal stress is developed in the concrete? Assume that the coefficient of expansion is 0.0000065 and the modulus of elasticity is 3,000,000 psi.

A. 780 psi compression

B. 390 psi tension

C. 390 psi compression

D. 7,800 psi compression

3. A reinforced concrete beam with b = 14" and d = 30" is subject to an ultimate moment M_u equal to 400 ft.-kips. If f_y = 60,000 psi and f'_c = 3,000 psi, what should the reinforcement be? Assume a = 0.2 d, and use the table in Figure 5.4.

A. 3-#8 **C.** 2-#11

B. 3-#9 **D.** 3-#10

4. Select the CORRECT statement about reinforced concrete columns.

A. Reinforced concrete columns are normally designed for axial compression only.

B. The area of longitudinal reinforcement must be at least 1 percent of the cross-sectional area of the column.

C. Spiral and tied columns are both designed with the same strength reduction factor Ø.

D. Tied columns generally cost more than spiral columns.

5. A structural steel beam spans 30 feet and supports a uniformly distributed load of 2,000 pounds per lineal foot. The maximum permissible deflection of the beam is one inch. What is the required I for the beam, if the value of E is 29,000,000 psi? Use the formula $\Delta = 5\ wL^4/384\ EI$.

A. 251 $in.^4$ **C.** 1,508 $in.^4$

B. 873 $in.^4$ **D.** 1,257 $in.^4$

6. In the concrete retaining walls shown below, the proper placement of reinforcing steel is shown in

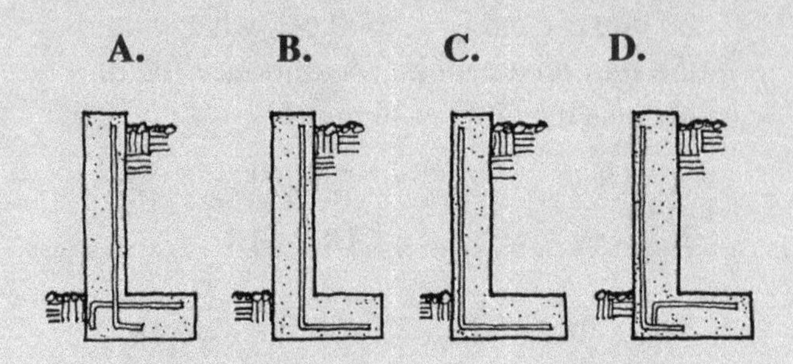

7. The impact hammer test is used to determine the

A. compressive strength of concrete cylinders.

B. bearing capacity of compacted fill.

C. workability of concrete.

D. strength of hardened concrete.

8. A wood column is used to support a concentric load of 100 kips. The column is 24 feet long and its ends are restrained against lateral movement. Assume that the column has a high l/d value and will tend to fail by buckling. The adjusted axial stress capacity has been determined, after a few trials, to be approximately 768 psi. What is the minimum size column that may be used?

A. 8 × 8

B. 10 × 10

C. 12 × 12

D. 14 × 14

9. The detail shown below indicates what type of weld?

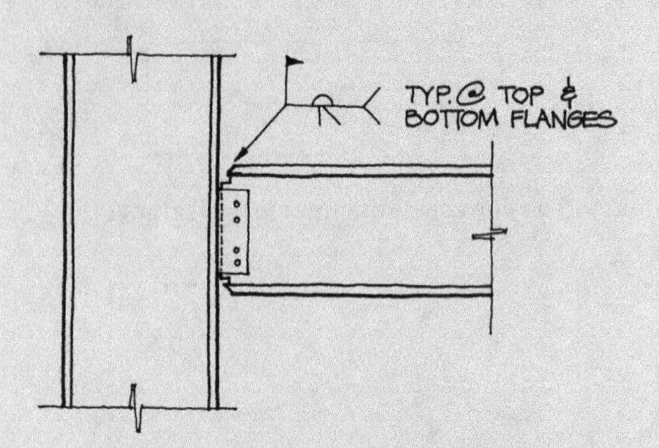

A. Overhead fillet weld to both flanges, made in the field

B. Full penetration groove weld to both flanges, made in the shop

C. Full penetration groove weld to both flanges, made in the field

D. Slot weld to both flanges

10. The advantages of continuous beams over simple beams include which of the following? Check all that apply.

A. Less deflection

B. Lower moment

C. Less shear

D. Simpler calculations

E. Simpler construction details

11. If the wind speed doubles, the wind pressure

A. doubles.

B. increases fourfold.

C. remains the same.

D. decreases by one half.

12. Which of the following statements about live load is CORRECT? Check all that apply.

A. Live load is not combined with wind or earthquake loads in the structural design of buildings.

B. Live load is the load superimposed by the use and occupancy of the building, not including the wind load, earthquake load, or dead load.

C. The allowable stresses may be increased 25 percent for structural members supporting dead load plus live load.

D. In allowable stress design, roof live load is assumed to act concurrently with wind, snow, or earthquake loads only.

13. A seven-story, 80-foot-tall building assigned to SDC D has special moment-resisting frames (reinforced concrete). What is the seismic base shear as a percentage of the building weight? Use S_{DS} = 1.11g; S_{D1} = 0.64g; and S_1 = 0.74g.

A. 6.9%

B. 9.6%

C. 11.0%

D. 18.3%

14. Floor plans of four reinforced concrete buildings are shown below. Which would experience the greatest amount of torsion when resisting lateral forces?

A.

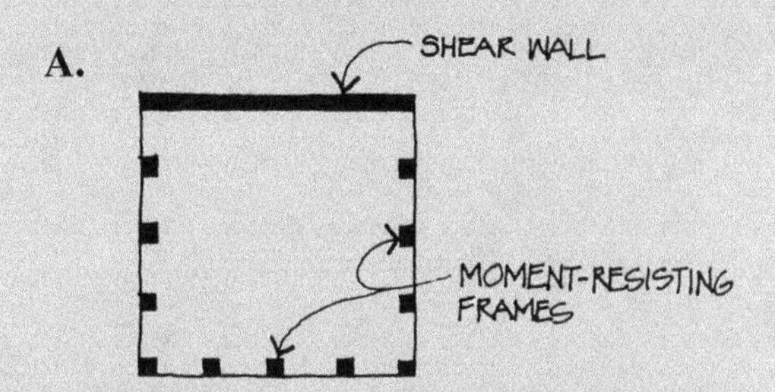

B.

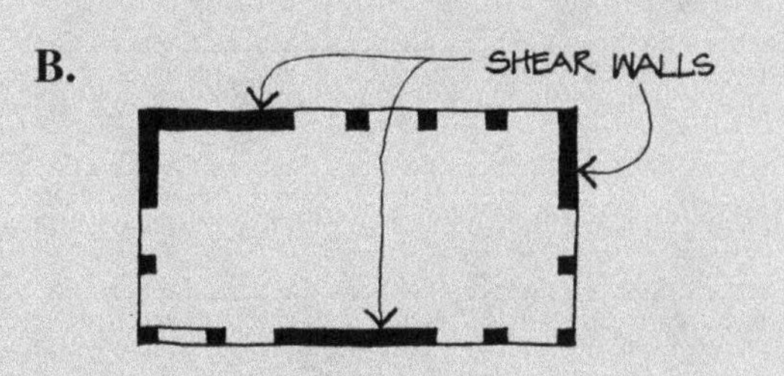

C.

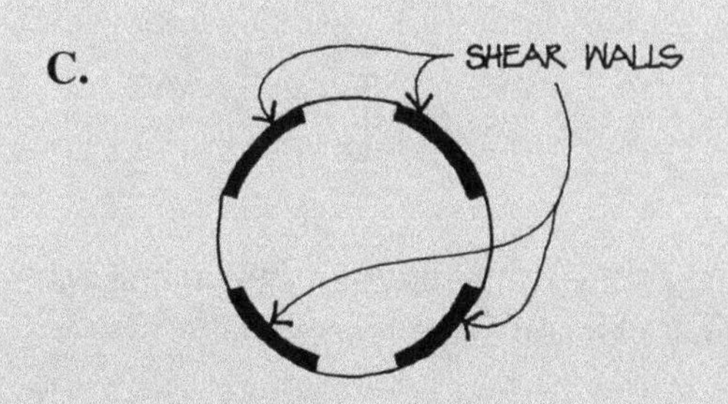

D.

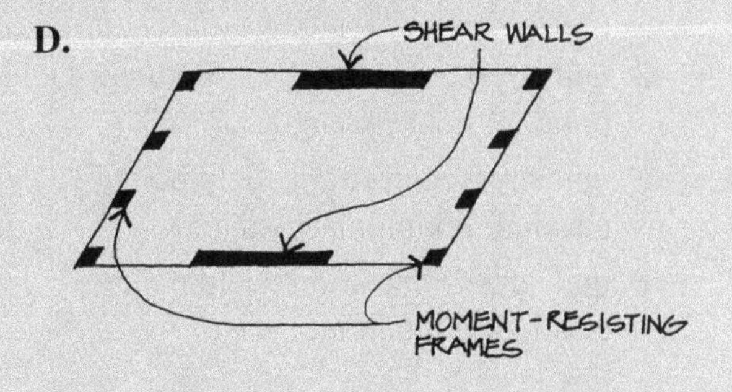

15.

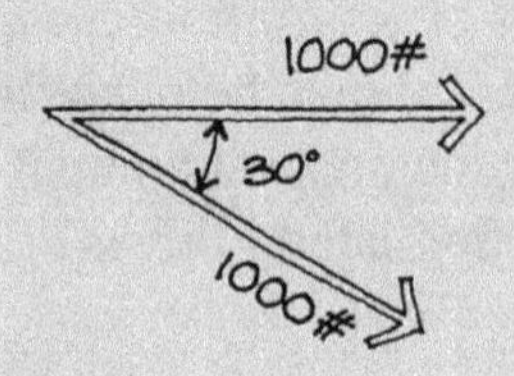

The resultant of the two forces shown above is

A.

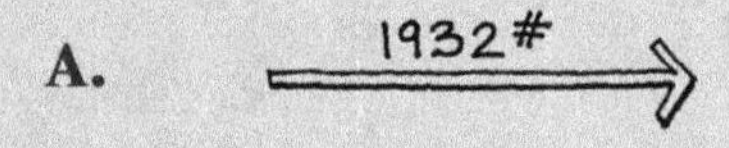

B.

C.

D.

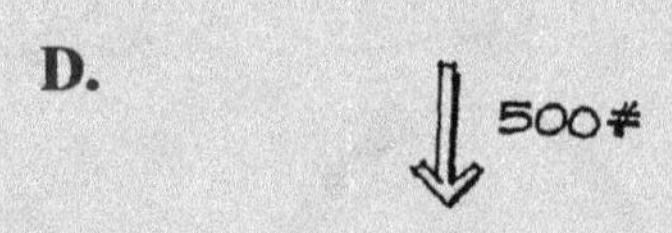

16. In the stress-strain diagram shown below, A, B, C, D, and E are respectively

A. unit strain, unit stress, elastic limit, yield point, ultimate strength.

B. unit stress, unit strain, elastic limit, yield point, ultimate strength.

C. unit stress, unit strain, yield point, elastic limit, ultimate strength.

D. unit stress, unit strain, yield point, elastic limit, failure.

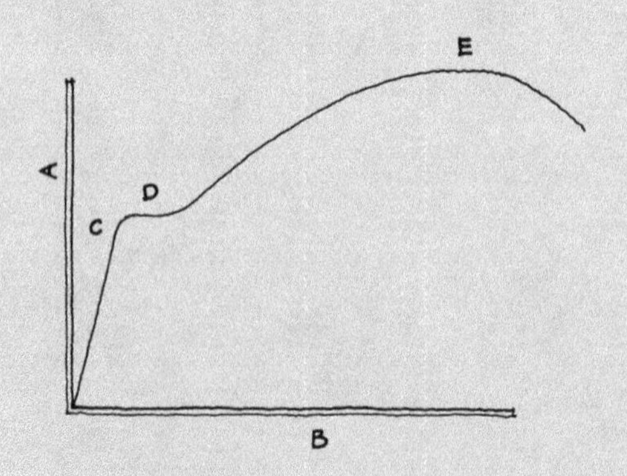

17. Which of the three-span beams shown below are statically determinate when subject to a uniform load?

I.

II.

III.

A. I only

B. II only

C. III only

D. II and III

18. During the construction documents phase of a project, it is determined that the calculated deflection of a steel beam is excessive. Which of the following is the best way to remedy this situation?

A. Modify the deflection limitation.

B. Use a beam with a greater section modulus.

C. Use a beam with a higher yield strength.

D. Use a beam with a greater moment of inertia.

19. Which of the following connectors are generally NOT used in a moment-resisting steel frame?

A. ASTM A325-SC high-strength bolts

B. ASTM A307 machine bolts

C. Full penetration groove welds

D. Fillet welds

20. A W 24 × 68 beam (web thickness = 7/16") is connected to a W 27 × 102 girder with 2 angles 4 × 3-1/2 × 3/8 with 5-3/4" A325 high-strength bolts (slip-critical) in each leg. The beams are A992 steel, and all connection material is ASTM A36 steel. What is the capacity of the connection? You may use the table in Figure 7.33.

A. 73.8 kips **C.** 46.2 kips
B. 79.5 kips **D.** 111.6 kips

21. A thick reinforced concrete slab that distributes load to a group of piles is called a _________ .

22. The load capacity of a structural steel column is a function of Kl/r. Which of the following statements are CORRECT?

I. K is a constant determined by the yield strength (F_y) of the structural steel.
II. K is a constant determined by the degree of fixity at the ends of the column.
III. K is a constant that depends on the buckled shape of the column at failure.
IV. An increase in the value of K results in a decrease in the column load capacity.

A. II only **C.** II, III, and IV
B. III only **D.** I, II, III, and IV

23. The support for a handrail is subject to a horizontal load of 120# as shown below. What are the horizontal forces in bolts A and B?

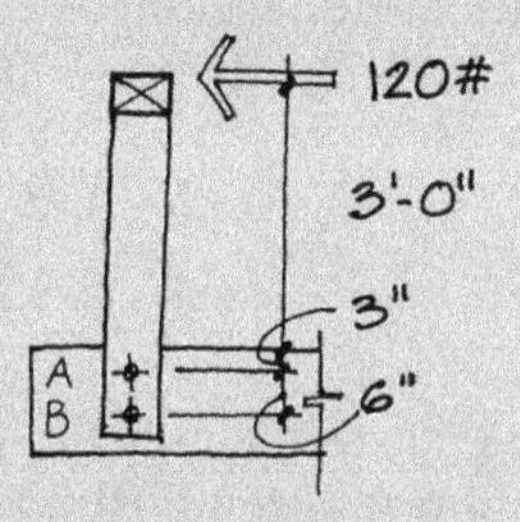

A. A = 780#, B = 780#
B. A = 900#, B = 900#
C. A = 900#, B = 780#
D. A = 780#, B = 900#

24. Which of the following factors affects the shear capacity of a reinforced concrete beam?

A. The ultimate 28-day strength of the concrete.
B. The grade, size, and spacing of the longitudinal tension reinforcing.
C. The size of aggregate used in the concrete mix.
D. The span length of the beam.

25. A steel bar two inches in diameter and 20 feet long resists a tensile load of 50,000 pounds. How much does it elongate? Assume E = 29,000,000 psi.

A. 0.011" **C.** 0.110"
B. 0.132" **D.** 0.032"

26. A concrete bearing wall supports a uniform load of 6 kips per lineal foot, including the weight of the wall. If the wall footing is supported on clayey sand, what is the minimum required footing width? You may refer to IBC Table 1806.2 in the appendix.

A. 1'-6"
B. 2'-0"
C. 3'-0"
D. 6'-4"

27. Building A has a fundamental period of vibration of one second. Building B has a fundamental period of vibration of two seconds. All other factors for the two buildings are equal. Select the CORRECT statement.

A. Building A has a greater seismic force.
B. Building B has a greater seismic force.
C. Both buildings have equal seismic forces.
D. The information given is insufficient to determine which building has the greater seismic force.

28.

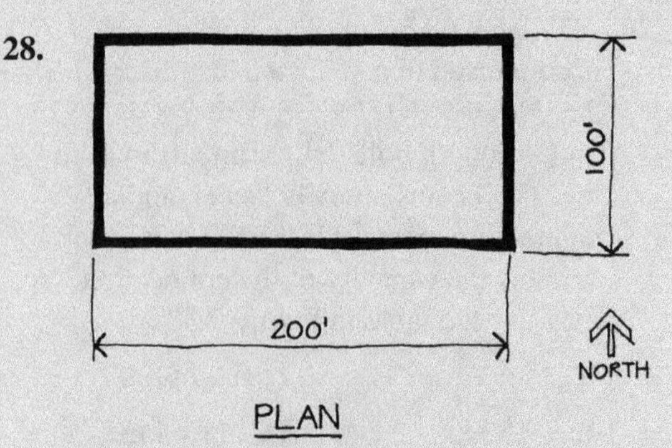

PLAN

The building shown above has rigid horizontal diaphragms that transfer seismic forces to vertical shear walls at the building's perimeter. The center of mass coincides with the center of rigidity. In the seismic design of the building, horizontal torsional moments

A. are ignored.
B. must be provided for by assuming the mass displaced 10 feet for north-south loads and 5 feet for east-west loads.
C. must be provided for by assuming the mass displaced 5 feet in each direction.
D. must be provided for by assuming the mass displaced 10 feet in each direction.

29. Select the CORRECT statements about damage from earthquake forces. Check all that apply.

A. Tall buildings are particularly susceptible to damage from slow rocking motions of an earthquake.

B. Damage to elevators in tall buildings is often significant.

C. One-story wood frame dwellings perform comparatively well, and substantially better than two-story dwellings.

D. Heavy damage from earthquakes may be expected where soft alluvial soils meet firmer soils.

E. Major damage and hazard to life is generally from ground shaking rather than surface faulting.

30. Which of the following buildings best expresses the structural concept of the "bundled tube?"

A. Willis Tower

B. John Hancock Building

C. CBS Building

D. Knights of Columbus Building

31. What is the maximum bending moment that can be resisted by a 4 × 6 Select Structural beam of Douglas fir-larch? Use Table 4-A in the Appendix and the table in Figure 3.2.

A. 25,593 ft.-lbs.

B. 5,864 ft.-lbs.

C. 2,326 ft.-lbs.

D. 2,206 ft.-lbs.

32. Which of the following statements about steel columns is INCORRECT?

A. Wide flange columns are ideal for resisting buckling.

B. The maximum slenderness ratio permitted is 200.

C. The K factor is used to convert the actual length of the column to the effective length.

D. The buckling tendency of a column is independent of its yield point value.

33.

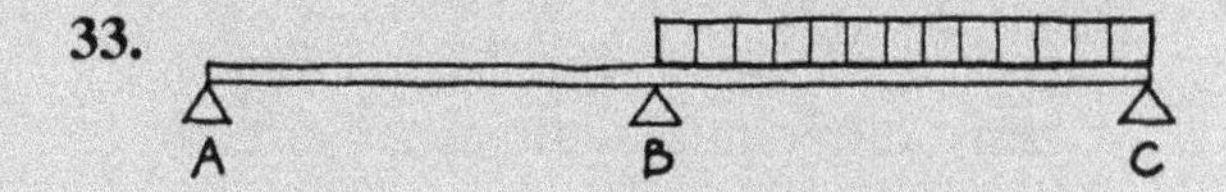

Select the CORRECT statements about the three-span continuous beam shown above, neglecting the weight of the beam.

A. The moment at B is negative.

B. The reaction at A is downward.

C. The moment in span AB is always negative.

D. The moment in span AB changes from positive to negative.

E. The moment in span BC is always negative.

F. The moment in span BC changes from positive to negative.

34. 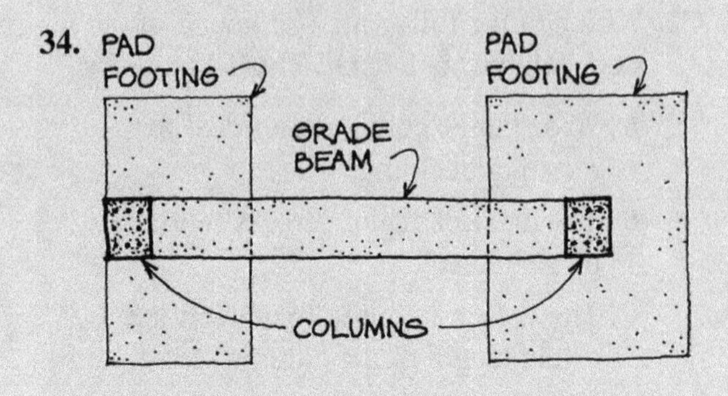

The foundation shown above in plan is usually used to

A. resist differential settlement caused by poor soil conditions.

B. resist eccentricity caused by a column located immediately adjacent to a property line.

C. span utility trenches between pad footings.

D. provide fixity for rigid frame footings.

35. Select the CORRECT statement about steel column base plates.

A. They fix the column against rotation.

B. They cannot resist any bending.

C. They are designed to transmit shear to the foundation by friction.

D. They are designed to spread the column load so that the bearing pressure on the foundation is not excessive.

36. Select the INCORRECT statement about wood screws.

A. Wood screws have greater holding power than nails of the same diameter.

B. The allowable withdrawal value for a screw inserted into the end grain of a timber is the same as for a screw inserted into side grain.

C. Wood screws may not be driven with a hammer.

D. Connections in which wood screws are subject to withdrawal loads should be avoided if possible.

37. The principal reason that high-early-strength cement is used is because

A. it is less expensive than other cements.

B. early removal of forms shortens the construction period.

C. Type III cement is not readily available.

D. it increases the ultimate moment capacity of reinforced concrete members.

38. A steel beam spans 24 feet and supports a load of 1,200 pounds per foot. Neglecting the weight of the beam, what is the required plastic section modulus? Assume ASTM A992 steel and full lateral support.

A. 34.6 in^3

B. 36.0 in^3

C. 180.0 in^3

D. 103.7 in^3

39. Which of the following statements is INCORRECT? For a body to be in static equilibrium,

A. the sum of the forces acting in any direction is equal to zero.

B. the sum of the moments about any axis is equal to zero.

C. the line of action of the resultant force passes through the centroid.

D. the resultant force is equal to zero.

40. A steel bar 1-1/2 inches square resists a tensile force of 40,000 pounds. What is the unit tensile stress in the bar?

A. 17,778 psi **C.** 22,635 psi

B. 1,778 psi **D.** 5,659 psi

41.

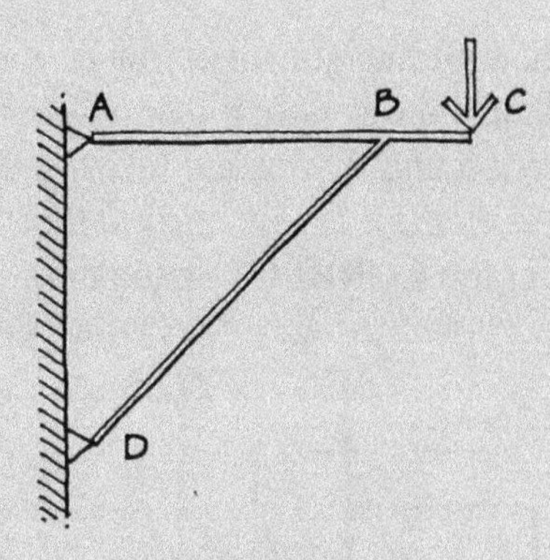

Which of the illustrations below best represents the components of the reactions acting on the structure shown above?

A.

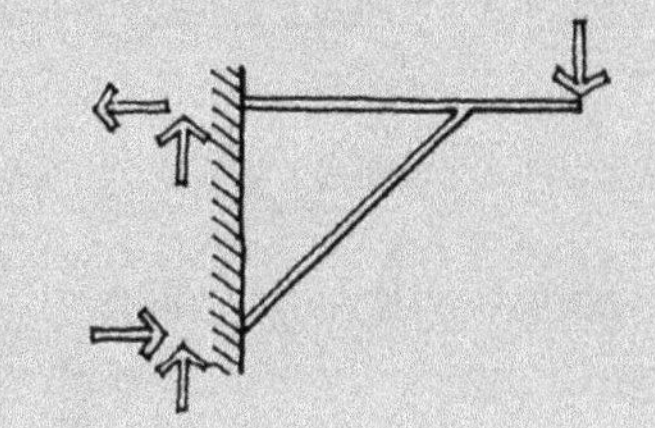

B.

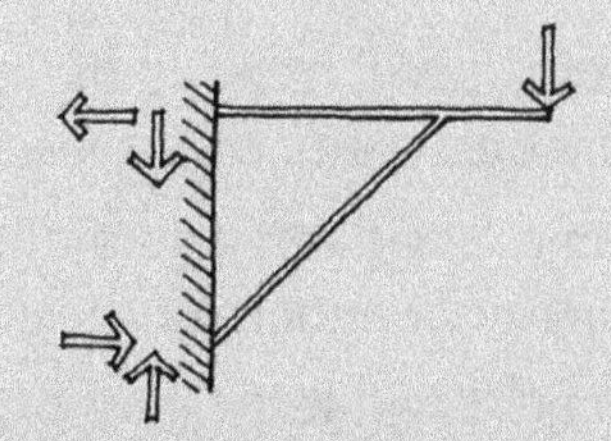

C.

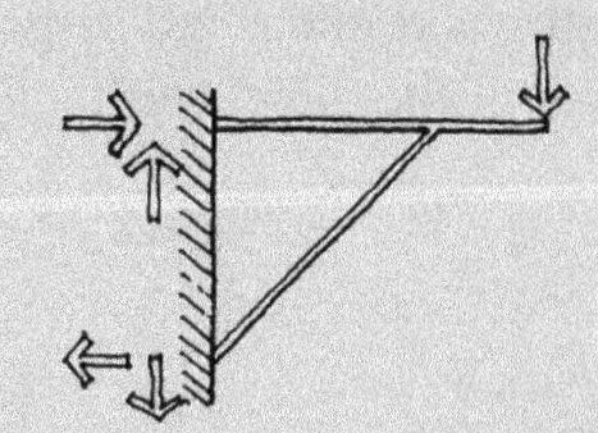

D.

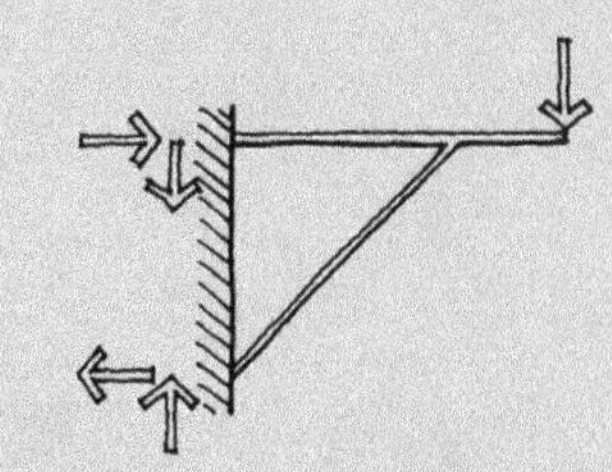

42. In the building shown below in plan, earthquake forces in the north-south direction and torsion are resisted by bents 1 and 2. Bents 2 are more rigid than bents 1. Select the CORRECT statement.

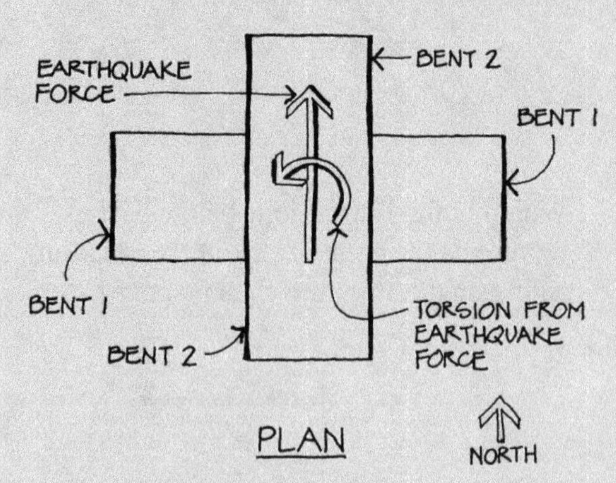

A. Bents 1 will resist most of the direct earthquake shear and bents 2 will resist most of the torsional earthquake shear.

B. Bents 1 will resist all of the direct and torsional earthquake shear.

C. Bents 2 will resist all of the direct and torsional earthquake shear.

D. Bents 1 will resist most of the torsional earthquake shear and bents 2 will resist most of the direct earthquake shear.

43. The braced frame below resists a wind force of 6 kips as shown. Assuming that the braces can resist tension only, what are the forces in braces 1 and 2?

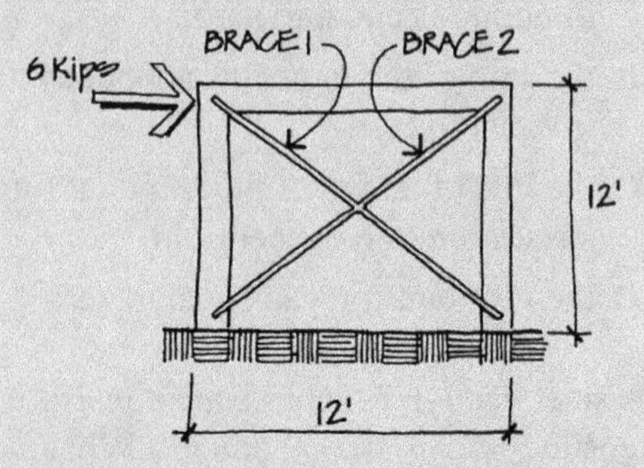

A. Brace 1 = 6.0 kips tension, brace 2 = zero

B. Brace 1 = zero, brace 2 = 6.0 kips tension

C. Brace 1 = 8.5 kips tension, brace 2 = zero

D. Brace 1 = zero, brace 2 = 8.5 kips tension

44. In the braced frame of the previous question, what is the force in the right column caused by the wind?

A. 6 kips compression

B. 6 kips tension

C. 8.5 kips compression

D. 8.5 kips tension

45. Select the CORRECT statement.

A. All concrete has approximately the same modulus of elasticity.

B. In a properly designed reinforced concrete beam, the deflection under load remains constant.

C. Doubling the amount of reinforcing steel in a reinforced concrete beam will reduce the deflection approximately by half.

D. The modulus of elasticity of concrete increases as the concrete strength increases.

46. The ratio of unit stress to unit strain for a material is a measure of its ____________.

47. Select the INCORRECT statement about strength design of reinforced concrete.

A. The code provisions assure that yielding of the tensile steel will occur before failure of the concrete.

B. The depth of the rectangular stress block (a) is less than the distance from the extreme compression fiber to the neutral axis.

C. The concrete stress at the extreme compression fiber, at failure, is equal to f'_c.

D. For a singly reinforced beam, the steel is assumed to resist all the tensile stresses and the concrete is assumed to resist all the compressive stresses.

48. Select the CORRECT statements about beam behavior.

A. The difference of bending moments between two sections of a beam is equal to the area of the shear diagram between those two sections.

B. The vertical shear at any section of a beam is the algebraic sum of all the forces acting on the beam.

C. A statically indeterminate beam is one for which the equations of equilibrium do not apply.

D. The maximum unit shear stress in a beam occurs at the neutral axis.

E. The most critical section of a beam is the section where the vertical shear is maximum.

49. The calculations for the design of a retaining wall show that the dead load resisting moment is 1.25 times the overturning moment from earth pressure. Select the MOST CORRECT statement.

A. The design is adequate.

B. A key integral with the footing should be added.

C. The footing should be made wider.

D. The amount of reinforcing steel should be increased.

50. A 4" nominal member is connected to two 2" nominal members by two 3/4" bolts as shown below. All members are Douglas fir-larch. What is the safe load P that can be transferred by the bolts? Use Table 11F in the National Design Specification in Figure 7.9.

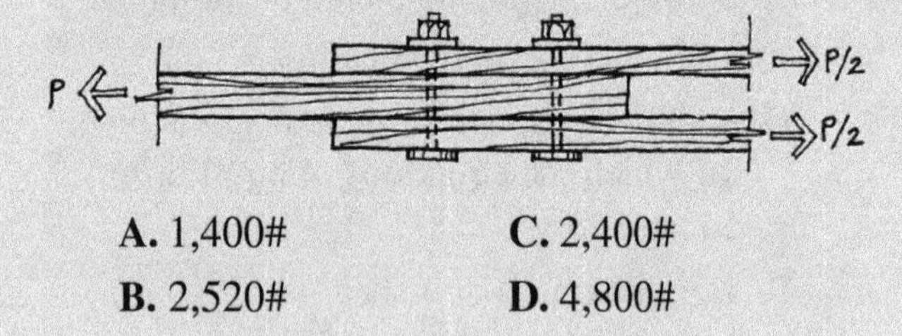

A. 1,400# **C.** 2,400#

B. 2,520# **D.** 4,800#

51. Select the INCORRECT statement about glued laminated members.

A. All the laminations are required to be of the same species and grade.

B. With proper quality control, the adhesive is at least as strong as the wood.

C. The individual laminations are thin enough to be readily seasoned before fabrication.

D. The members may be varied in cross-section along their length in accordance with strength requirements.

52. A beam supports the loads shown below. What is the reaction at R_2, neglecting the weight of the beam?

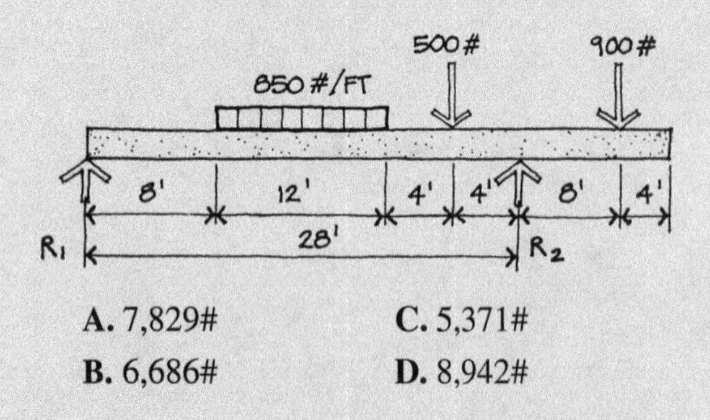

A. 7,829# **C.** 5,371#

B. 6,686# **D.** 8,942#

53. Select the INCORRECT statements.

I. The greatest drying shrinkage in wood members occurs in the longitudinal direction.

II. Glued laminated members generally shrink less than sawn timber members.

III. Shrinkage in wood is generally greater in heavier pieces than in lighter pieces of the same species.

IV. Shrinkage is usually greater in hardwoods than in softwoods.

V. Expansion of wood members caused by temperature change is an important consideration in the detailing of such members.

VI. Where wood joists are supported from a steel beam by joist hangers, the tops of the joists when installed should be above the top of the beam to provide for shrinkage of the joists.

A. I, IV, V, and VI

B. III, IV, and V

C. I and V

D. II, III, IV, and VI

54. Small diameter reinforcing bars used in concrete beams to provide shear reinforcement are called ______________________.

55. The calculations for the design of a column footing show that the footing is overstressed in shear. How should the design be changed?

A. The footing should be made larger in plan.

B. The amount of reinforcing steel should be increased.

C. Stirrups should be provided.

D. The thickness of the footing should be increased.

56. What is the ratio of the effective length of a column to its least radius of gyration called?

A. Poisson's ratio
B. Young's modulus
C. Slenderness ratio
D. Section modulus

57. What test is used to measure the workability of concrete?

A. Slump test
B. Air content test
C. Mortar bar test
D. Core test

58. What laboratory compaction test is used to determine a soil's optimum moisture content and density?

A. Impact hammer test
B. Proctor test
C. Cylinder test
D. Kelly ball test

59. A steel beam spans 30 feet and supports a load of 1,200 pounds per foot. The beam's compression flange is laterally braced at its ends and midspan only. Using ASTM A992 steel and neglecting the weight of the beam, what is the most economical beam which may be used? You may use the chart in Figure 4.13.

A. W14 × 43 **C.** W18 × 46
B. W16 × 45 **D.** W21 × 50

60. Select the CORRECT statement about basement walls.

A. Reinforcing steel is not required unless the wall exceeds eight feet in height.
B. If the wall is properly drained, it need not be designed to resist earth pressure.
C. If the backfill is placed before the first floor is constructed, the wall must be temporarily braced to resist earth pressure.
D. The wall must be designed to resist overturning caused by earth pressure.

61. An ordinary roof member supports a 500 square foot tributary area. If the roof pitch is 3 in 12, the reduced live load that may be taken on the roof is

A. 20 psf. **C.** 12 psf.
B. 14 psf. **D.** 7 psf.

62. A 24K8 open web steel joist spans 32 feet and supports a load of 300 pounds per lineal foot. How many rows of bridging are required? You may use the table in Figure 10.10.

A. 1 **C.** 3
B. 2 **D.** 4

63. In earthquake design of buildings, torsion

A. occurs only in buildings which are circular in plan.
B. occurs when the center of resistance of the primary lateral load resisting system is not coincident with the center of mass of the building.
C. is most significant when the lateral load resisting system is a dual system of shear walls and moment-resisting frames.
D. must be considered in buildings with flexible diaphragms.

64. Wind pressures used in the design of buildings

A. do not account for the effects of hurricanes.

B. increase when there are buildings or surface irregularities close to the site.

C. act both inward and outward on wall surfaces.

D. are the same for wall elements and wall corners.

65. In earthquake design, which of the following are examples of plan irregularities?

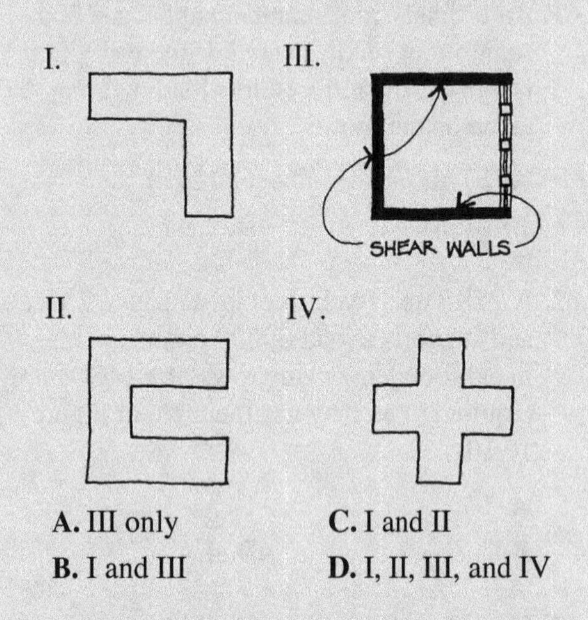

A. III only

B. I and III

C. I and II

D. I, II, III, and IV

66. Your client has proposed that an existing large warehouse, designed and built under the requirements of the building code, be converted to a meeting hall for 200 persons. To meet requirements, you must show that the building's lateral force resisting system is adequate to resist how much more lateral force than its original design forces?

A. No additional lateral force

B. 15 percent more wind force only

C. 15 percent more wind force, 25 percent more seismic force

D. 25 percent more seismic force only

67. In the detail shown below, what is the purpose of anchor A?

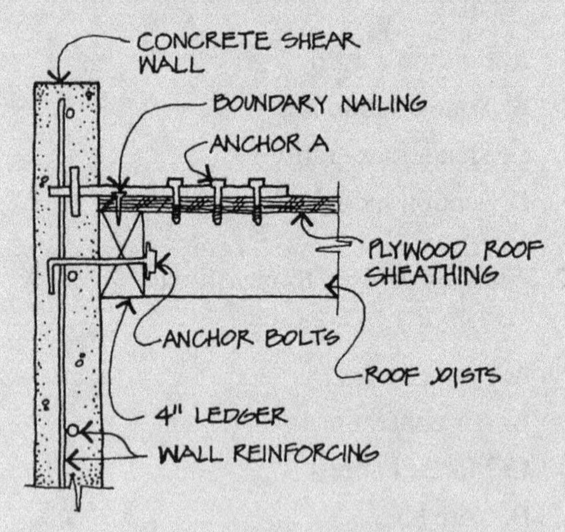

A. Transfers vertical load from roof joists to wall.

B. Transfers lateral load from roof diaphragm to wall.

C. Anchors wall to roof diaphragm for seismic or wind forces perpendicular to wall.

D. Connects plywood panels together.

68. Which of the following are advantages of a Vierendeel truss over a conventional truss?

I. Less deflection

II. May have larger openings within the depth of the truss

III. Less material

IV. Simpler joint details

V. Ability to have load applied to either top or bottom chord

A. All of the above

B. II, IV, and V

C. II only

D. III and IV

69. The roof of a circular building is supported by a series of radial arch-shaped ribs, which connect to a ring at the center of the roof and to another ring at the perimeter. Select the CORRECT statement.

A. Both rings are in tension.

B. Both rings are in compression.

C. The center ring is in tension and the perimeter ring is in compression.

D. The center ring is in compression and the perimeter ring is in tension.

70.

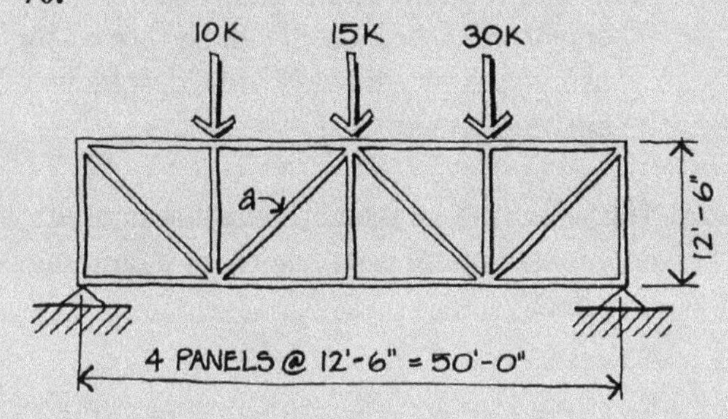

For the truss shown above, what is the axial force in member a?

A. 31.8 kips tension

B. 17.7 kips tension

C. 31.8 kips compression

D. 17.7 kips compression

71. A waffle slab is

I. a one-way system, with beams between the columns in two directions.

II. a two-way system, with beams between the columns in two directions.

III. a two-way system without beams.

IV. generally formed with long adjustable steel pans.

V. generally formed with removable dome-shaped steel forms.

A. I and IV **C.** III and V

B. II and V **D.** II and IV

72. Select the CORRECT statements about a three-hinged gabled frame.

I. It is statically indeterminate.

II. Its supports permit rotation.

III. The moment at the center is zero.

IV. It is generally subject to greater moments than a rectangular rigid frame.

V. The maximum moment occurs at the intersection of the column and the sloping beam.

A. II, III, and V **C.** I, III, and IV

B. All of the above **D.** II and V

73.

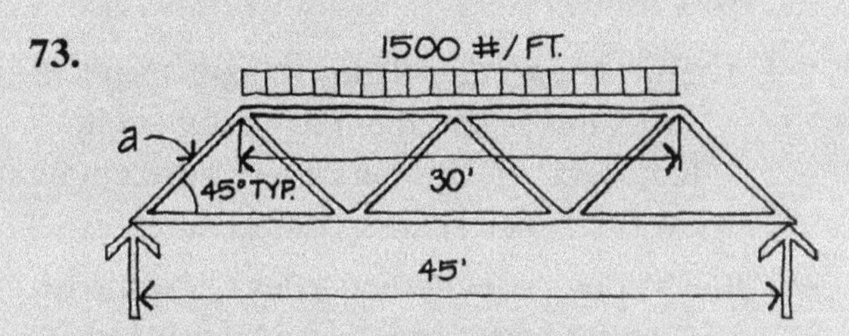

For the truss shown above, what is the axial force in member a?

A. 22.5 kips compression

B. 45.0 kips compression

C. 31.8 kips compression

D. 15.9 kips compression

74. In general, concentrated loads are best resisted by a(n)

A. arch.

B. cable structure.

C. truss.

D. dome.

75. A floor is framed with open web steel joists spaced three feet on center supporting a uniform dead load plus live load of 100 pounds per square foot, including the joist weight. The joists span 72 feet. Which of the following is the lightest joist which may be used, neglecting deflection? You may use the tables in Figures 10.13 and 10.14.

A. 36 LH10 **C.** 40 LH10

B. 36 LH11 **D.** 40 LH11

76. There have been several collapses of long span structures in the recent past. In this regard, which of the following statements is NOT true?

A. The lack of redundancy in long span structures makes them more vulnerable to general collapse in case of accidental overload than short span structures.

B. The potential consequences of failure of long span structures are greater than for conventional structures.

C. The structural systems used for long span structures often surpass the frontier of existing knowledge, and this is a contributing cause of failure.

D. Systems that adequately brace short spans may not be stiff enough for long span structures.

77. Select the CORRECT statement about dome construction.

A. Since a dome is essentially an arch rotated about its vertical axis, the stiffness of a dome is about the same as that of an arch.

B. The largest domes built to date are geodesic domes.

C. Framed domes are frequently constructed with a center ring to simplify erection and alleviate congestion of framing members.

D. A thin shell dome can easily be constructed, because it can be formed by using straight timber boards, despite its curved surface.

78. The weight of all the permanent structural and nonstructural components of a building is called ________________________.

79.

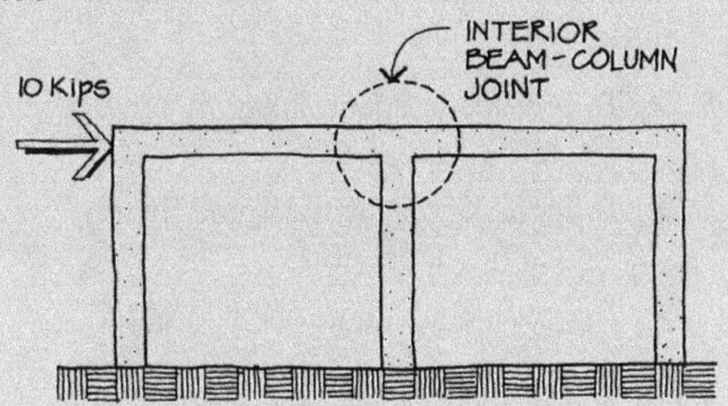

A wind load of 10 kips is resisted by a moment-resisting frame as shown above. Which of the choices below correctly shows the internal forces in the members at the interior beam-column joint? (T = tension, C = compression.)

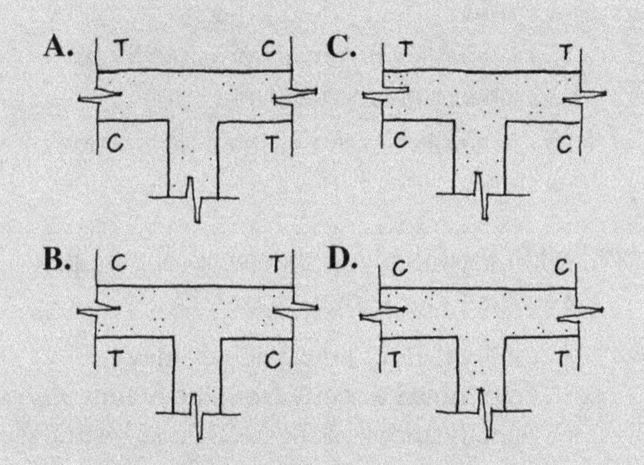

80. In designing a building for earthquake forces, which of the following are NOT considered?

I. The building's orientation

II. The subsoil conditions

III. The geographical location of the building

IV. The dead weight of the building

V. The type of lateral load resisting system

A. I and II

B. I, II, and III

C. II, III, IV, and V

D. I only

81. For the water tank supported on the four-legged braced frame shown below, what is the maximum vertical force at the bottom of column A due to gravity and seismic loads? V = 0.3W.

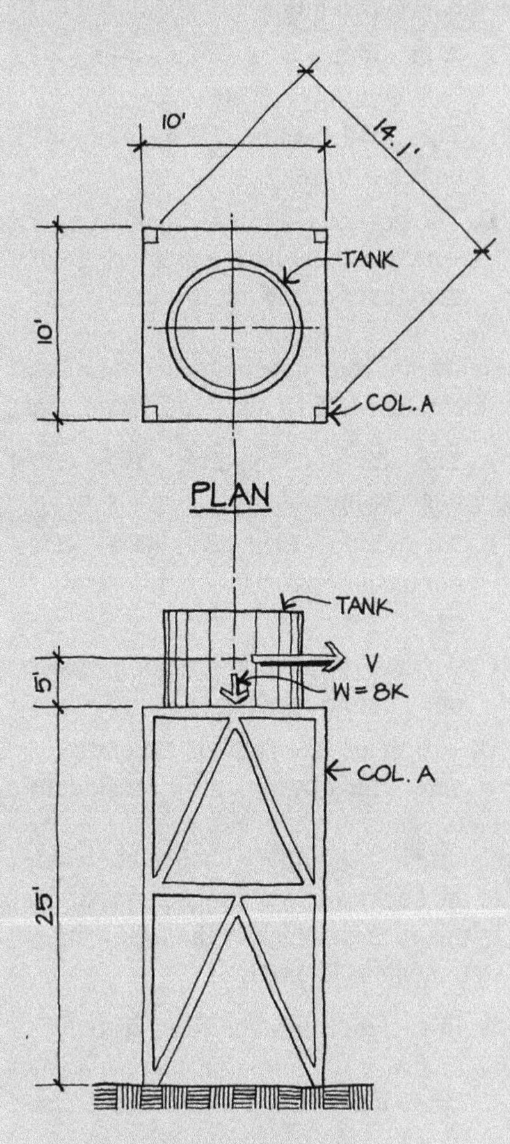

A. 11.2 kips

B. 7.1 kips

C. 5.6 kips

D. 5.1 kips

82. Select the CORRECT statement about membrane structures.

A. A membrane can only resist tension, compression, or shear in its own plane, but no bending.

B. A membrane adapts to changing loads by changing its shape.

C. The classic example of a membrane in nature is the egg.

D. The occupied space in an air-supported structure must have a pressure greater than that outside the space.

83. Below are four statements about arches. Select the MOST CORRECT statement.

A. The supports of an arch may be either fixed or hinged.

B. An arch under balanced loading does not have horizontal reactions at its bases.

C. An arch is able to resist compression only, but no bending.

D. The shape of an arch is generally determined by structural considerations.

84. Candidates should have some knowledge of the background of long span roof design. In this regard, which of the following statements is CORRECT?

A. In ancient times, the Romans built arches and domes that spanned distances approaching those of modern times.

B. The first roof spanning 300 feet was built about 50 years ago.

C. The size of an arena roof depends partially on the size of its central floor space, and the indoor sport that requires the largest such space is hockey.

D. Long span roofs are used almost exclusively to enclose theaters, arenas, and stadiums.

85. What is the main reason for tapering a girder?

A. To provide adequate bending strength where it is needed.

B. To increase the stiffness and thereby reduce the deflection.

C. To provide adequate shear strength.

D. To provide for roof drainage.

86. Select the CORRECT statement about the Parthenon.

A. Its structure is ideally suited to its material.

B. It had the largest dome of the ancient world.

C. Its great span is made possible by its arched construction.

D. Its members were adapted from wood to stone.

87. All of the following statements about open web steel joists are true, EXCEPT

A. the load carrying capacities may be determined directly from the Standard Specifications of the Steel Joist Institute.

B. bridging may be omitted if the joist span is less than 32 feet.

C. open web steel joists may be used in spans greater than 100 feet.

D. when selecting steel joists, consideration must be given to both load carrying capacity and deflection.

88. Select the CORRECT statements about trusses. Check all that apply.

A. The stress in the web members is a function of the depth of the truss.

B. Trusses are generally expensive to fabricate, but usually require less material than competitive systems.

C. Truss members are most often designed for axial stress only.

89.

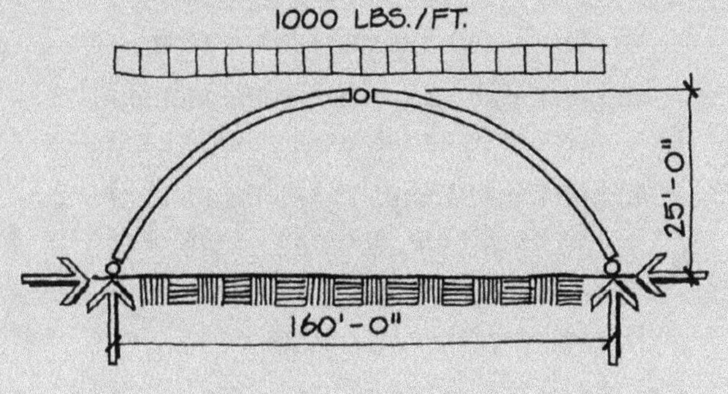

What is the horizontal thrust at each end of the three-hinged arch shown above?

A. 40 kips **C.** 128 kips

B. 80 kips **D.** 256 kips

90. A series of inclined trusses that intersect in a grid pattern and are connected at their points of intersection is known as a

A. lamella.

B. Vierendeel truss.

C. rigid frame.

D. space frame.

91. In multi-story rigid frames made of cast-in-place concrete, continuity at the beam-column joints is provided by

A. welding the reinforcing bars.

B. shear keys.

C. high strength bolts.

D. the monolithic nature of the concrete and proper lapping of the reinforcing bars.

92.

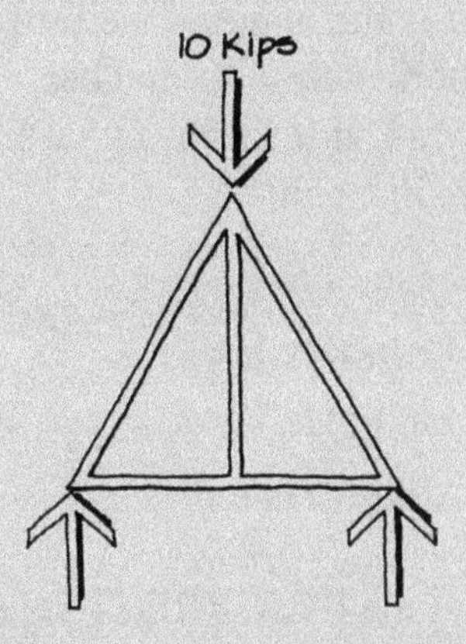

What is the force in the vertical member of the truss shown above?

A. 10 kips **C.** 3.3 kips

B. 5 kips **D.** zero

93. What Roman structure has a dome that was the longest-spanning unreinforced concrete dome until the 20th century?

A. Palazzetto Dello Sport

B. Pantheon

C. Parthenon

D. Santa Maria del Fiore Cathedral

94. Which of the following engineers was known for his roof structures that typically utilized diagonal precast concrete ribs?

A. Eugene Freyssinet

B. Robert Maillart

C. Fazlur Khan

D. Pier Luigi Nervi

95. What is the name of the immense prefabricated glass and cast iron structure that was built for the Great Exhibition of 1851, in London?

A. Crystal Palace
B. Machinery Hall
C. Market Hall
D. Exhibition Hall

96. Select the INCORRECT statement.

A. Nonstructural walls can drastically affect the performance of a building in an earthquake.
B. A building with many partitions in the upper stories and a relatively open first story may be expected to perform well in an earthquake.
C. Reinforced concrete columns subject to earthquake forces often fail because of inadequate ties.
D. An irregular floor plan tends to increase torsional stresses from earthquake forces.

97. Select the INCORRECT statement about rigid frames.

A. A rigid frame can resist both gravity and wind loads.
B. Steel or reinforced concrete may be used in rigid frame construction.
C. A rigid frame is more rigid than a braced frame of comparable dimensions.
D. The joints of a rigid frame are capable of resisting bending moment.

98.

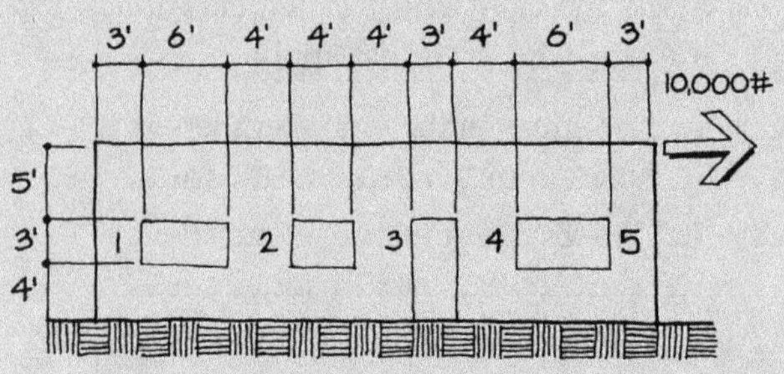

In the wall shown above,

A. piers 2, 3, and 4 resist 2,300# each and piers 1 and 5 resist 1,500# each.
B. piers 1 and 5 resist 2,300# each and piers 2, 3, and 4 resist 1,800# each.
C. piers 1 and 5 resist 1,500# each, pier 2 resists 3,000#, and piers 3 and 4 resist 2,000# each.
D. each pier resists 2,000#.

The examination answers and explanations will be found on the following pages.

Do not look at the answers until you have completed the exam.

EXAMINATION ANSWERS

1. **B** The slope of the moment diagram is equal to the value of the shear diagram (see page 29). Between points 1 and 2, the value of the shear diagram is constant and negative, and therefore the slope of the moment diagram is constant and negative. Between 2 and 3, the shear diagram is constant and has a greater negative value than between 1 and 2. Therefore, the moment diagram has a constant negative slope between 2 and 3, which is greater than the slope between 1 and 2. The only moment diagram that meets these criteria is correct answer B.

2. **A** Since the slab is restrained against movement, one can consider that it expands because of the increase in temperature, and then its supports apply a force to it to compress it back to its original length. Expansion Δ due to temperature change = nLt = shortening Δ due to load = PL/AE
nLt = PL/AE
P/A = nEt, where
P/A is internal stress
n is coefficient of expansion
E is modulus of elasticity
t is change of temperature

   Thus, P/A = 0.0000065 × 3,000,000 × 40 = 780 psi compression (answer A).

3. **D** To solve this problem, we use the formula on page 92.

$$M_u = ØA_s f_y(d - a/2)$$

$$A_s = \frac{M_u}{Øf_y(d - a/2)}$$

$$a = 0.2 \times d = 0.2 \times 30 = 6.0"$$

$$A_s = \frac{400{,}000 \times 12}{0.90(60{,}000)(30 - 6/2)}$$

$$= 3.29 \text{ sq. in.}$$

   From the table in Figure 5.4, we select 3-#10 bars (correct answer D), which have an area of 1.27 × 3 = 3.81 sq. in. All of the other answers are inadequate.

4. **B** Statement A is incorrect because the IBC requires that columns must be designed for a minimum amount of eccentricity, which is equivalent to bending moment, even if the load theoretically is axial. C is also incorrect; the strength reduction factor Ø is equal to 0.70 for spiral columns and 0.65 for tied columns, because of the greater toughness of spiral columns. Spiral columns are usually more expensive than tied columns because the spiral reinforcement costs more to fabricate (incorrect answer D). Only B is correct, in accordance with the IBC.

5. **D** Questions involving deflection sometimes appear on the exam. In that case, the necessary deflection formulas will probably be provided. However, it may be advisable to memorize the formula for the maximum deflection of a uniformly loaded simple beam ($\Delta = 5wL^4/384EI$), in case the formula is not provided or is difficult to locate. Since wl = W, the total load on the beam, the formula becomes $\Delta = 5\ WL^3/384\ EI$

   where
   Δ = 1.0"
   W = 2,000#/ft. × 30 ft.
   $L^3 = (30 \text{ ft.} \times 12 \text{ in./ft.})^3$
   E = 29,000,000 psi

$$1.0 = \frac{5(2{,}000)(30)(30 \times 12)^3}{384(29{,}000{,}000)(I)}$$

$$I = \frac{5(2{,}000)(30)(30 \times 12)^3}{384(29{,}000{,}000)(1.0)}$$

$$= 1{,}257 \text{ in}^4 \text{ (answer D).}$$

6. **A** Reinforcing steel is placed in the tension face of a reinforced concrete structure, i.e., the face that elongates or stretches under load. Thus, by visualizing

the shape of the deflected structure, one is able to determine where the reinforcing steel should be placed. The retaining wall is acted on by the loads shown, causing it to deflect as indicated by the dashed line. The face of the wall closest to the earth is in tension, as well as the upper face of the footing. Therefore, the placement of reinforcing steel shown in diagram A is correct.

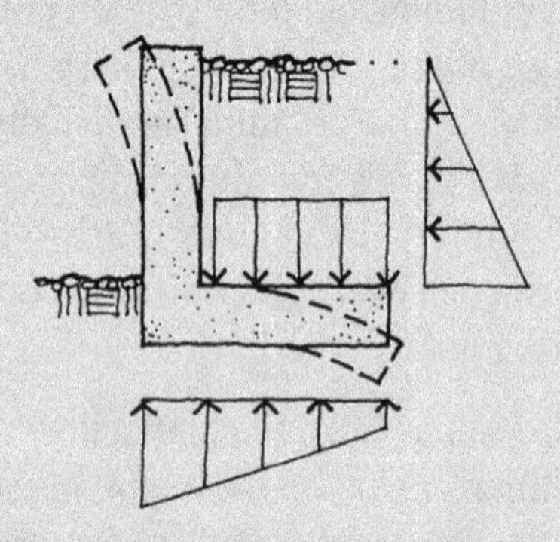

7. D The impact hammer test is a simple nondestructive test that is used to determine the approximate strength of hardened concrete. D is the correct answer.

8. C Required column area = 100,000/768 = 130.2 square inches. $\sqrt{130.2} = 11.41"$ The smallest column which may be used is therefore a 12 × 12, answer C.

9. C The weld symbol along the horizontal line indicates the type of weld; in this case indicates a full penetration groove weld. A darkened flag ( ) indicates a field weld, as opposed to one made in the shop. Thus, this weld symbol indicates a full penetration groove weld to both flanges made in the field (C is correct, B is incorrect). A fillet weld (A) is designated by the symbol , and a slot weld (D) is indicated by the symbol .

10. A and B Continuous beams generally have smaller deflections (A) and lower maximum moments (B) than comparable simple beams. However, the maximum shear in continuous beams may be greater (C is incorrect), the calculations are more complex (D is incorrect), and the construction details are often more complex as well (E is incorrect).

11. B The wind pressure varies as the square of the velocity, and therefore doubling the wind velocity increase the wind pressure fourfold.

12. B The correct definition of live load is given in answer B. Floor live load is assumed to act concurrently with wind or earthquake loads (A is incorrect). C is incorrect; there is usually no basis for increasing the allowable stresses for a member supporting dead plus live load. D is also incorrect because in allowable stress design, roof live load is assumed to act concurrently with wind loads *or* earthquake loads, not just wind loads.

13. B The seismic base shear $V = C_s W$ where $C_s = S_{DS}/(R/I)$

For a special reinforced concrete moment frame, R = 8 from ASCE 7 Table 12.2-1 (system C.5).

The building is not a high occupancy, essential, or hazardous facility, so from ASCE 7 Table 11.5-1, I = 1.0.

Thus, $C_s = 1.11/(8/1.0) = 0.139$.

Also, C_s need not exceed the following:

$C_s = S_{D1}/T(R/I)$
$T = C_t(h_n)^x$

where $C_t = 0.016$ and x = 0.9 from ASCE 7 Table 12.8-2 for concrete moment-resisting frames.

$T = 0.016(80)^{0.9} = 0.83$ seconds

Thus, $C_s = (0.64 \times 1)/(0.83 \times 8) = 0.096$.

Minimum C_s = 0.01 or $0.5S_1/(R/I)$ = (0.5 × 0.74 × 1)/8 = 0.05 (governs).

Therefore, C_s = 0.096 and V = 0.096W, or 9.6% of W.

14. A Torsional forces on a building result when lateral forces act on a building that is not symmetrical with respect to rigidity. In the four plans in this question, the lateral forces from wind or earthquake act through the center of gravity of the building, which is located at the center of the building's plan. The center of the resistance to these lateral forces is called the center of rigidity. If the building is symmetrical with respect to resistance, that is, if opposite sides of the building are equally rigid, then the center of rigidity will coincide with the center of gravity (plans B, C, and D), and torsional forces will be small. If the building is not symmetrical with respect to resistance, that is, if two opposite sides have very different rigidities, then the center of resistance will be located near the stiffer wall, resulting in larger torsional forces. This is the case in plan A, where the upper shear wall is much stiffer than the lower moment-resisting frame.

15. B To solve this problem we resolve the 1,000# force acting at 30° with the horizontal into its horizontal and vertical components. Its horizontal component = 1,000 cos 30° = 866# to the right, and its vertical component = 1,000 sin 30° = 500# downward. The sum of the horizontal forces is 1,000 + 866 = 1,866# to the right. The only vertical force is 500# downward.

The resultant = $\sqrt{1{,}866^2 + 500^2}$ = 1932#. Tan θ = 500/1,866, from which θ = 15°. The correct answer is therefore B.

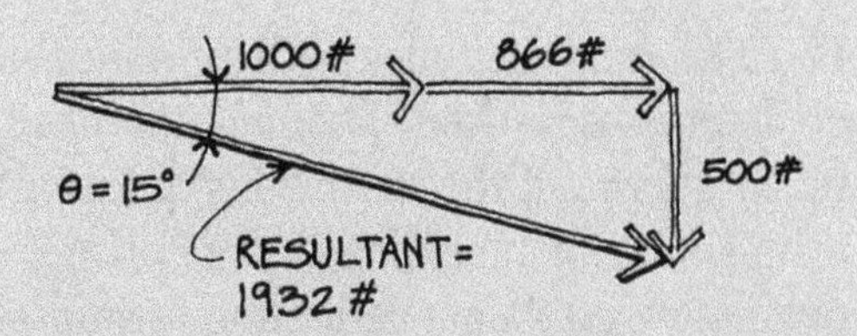

This question can also be answered without performing a single calculation, but simply by using a little logic. Since the two forces have the same magnitude and are 30° apart, their resultant must be 15° from each force, larger than each force, and in the same general direction as the forces (downward and to the right). The only answer that fits is B. The lesson here is that it pays to look at the question before you start punching numbers on the calculator.

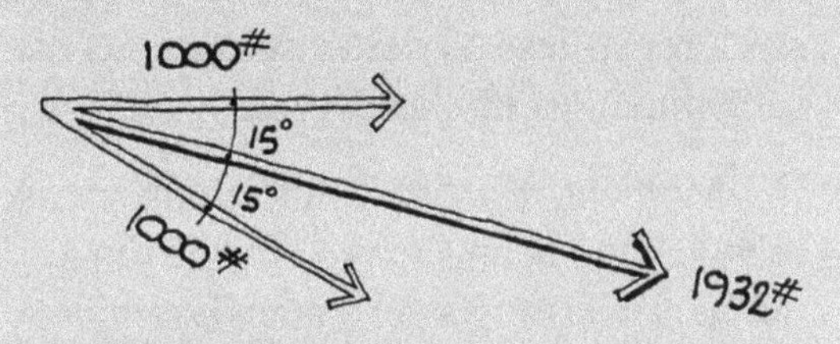

16. B The stress-strain diagram shown is typical for steel, and the correct identification is given in answer B.

17. D When we say that a beam, or any other structure, is *statically determinate*, we mean that its reactions can be determined using the basic laws of statics ($\Sigma F_x = 0$, $\Sigma F_y = 0$, $\Sigma M = 0$). Examples include simple beams and overhanging beams on two supports. Conversely, the reactions of a *statically indeterminate* structure cannot be determined from statics alone, but additional methods must be used. A multi-span beam continuous over one or more supports (I) is statically indeterminate. Beam II is statically determinate: its center span is a simple beam supported by overhanging end spans. III is also statically determinate: its end spans are simple beams supported by the overhanging center span.

18. D The exam includes both numerical and conceptual questions, although there are usually more of the latter. This question tests your understanding of the concept of deflection. All beam deflection formulas are essentially in the form $\Delta = KL^3/EI$, where K is a constant that depends on the load and the loading condition. Thus, to *reduce the deflection*, one must *increase the moment of inertia I* (correct answer D). We would not arbitrarily change the deflection limitation to accommodate our design (A is incorrect). B is also incorrect: while a beam with a greater section modulus has a greater ability to resist flexural stress, it does not necessarily have a greater value of I. Finally, C is also incorrect, because all steel, regardless of yield strength, has the same modulus of elasticity E.

19. B Moment-resisting connections are used to attach the beams to the columns in a moment-resisting steel frame. Examples of such connections are shown in Figure 7.32. In a moment-resisting steel frame, the moments in the beams produce tension in one beam flange and compression in the other flange. To transfer these forces from the beam flanges to the columns, the connection must have adequate strength and must not allow any slippage. For these reasons, the beam flanges are usually attached directly to the columns with full penetration groove welds (C), or with cover plates welded to the columns and welded (D) or bolted with high-strength slip-critical bolts (A) to the beam flanges. ASTM A307 machine bolts are generally *not* used for moment connections because they rely on bearing against the steel to develop their strength, and in so doing allow some movement in the connection. B is therefore the correct answer.

20. A From Figure 7.33, read the value directly from the upper portion of Table 10-1 for bolt and angle available strength = 73.8 kips. The available strength of the web of the W24 × 68 must also be checked. Assuming a 1.25" by 1.75" cope in the top flange of the W24 × 68, the available beam web strength is 216 kips per inch, which is obtained from the lower portion of Table 10-1. For the 7/16" web thickness, the available strength is $216 \times 7/16 = 94.5$ kips > 73.8 kips. Thus, the capacity of the connection is 73.8 kips (answer A).

21. Pile cap The bearing capacity of a single pile is generally inadequate to support the load of a column; therefore, a group of piles must be used. In the usual arrangement, the column bears on a concrete pile cap, which in turn distributes the column load to the group of piles.

22. C Questions testing your understanding of structural behavior often appear on the exam. Structural steel columns fail by buckling. The ratio Kl/r is a measure of the buckling tendency of a steel column; the larger the value of Kl/r, the greater the tendency of the column to buckle, resulting in a lower column capacity. In this ratio, K is a constant determined by the degree of fixity at the ends of the column (II), and the buckled shape of the column at failure (III), and is found in Figure 4.25. An increase in the value of K results in a higher value of Kl/r, a greater tendency of the column to buckle, and consequently a lower column load capacity (IV). The value of K is not related to the yield strength (F_y) of the structural steel (I is incorrect).

23. C This problem is solved by using the basic equations of statics: $\Sigma F_x = 0$, $\Sigma F_y = 0$, $\Sigma M = 0$. We take moments about bolt A.

$\Sigma M = 0$
$-120\# \times 39" + B \times 6" = 0$
$B = 120 \times 39/6 = 780\#$
$\Sigma F_x = 0$
$A - 780\# - 120\# = 0$
$A = 900\#$

Therefore the correct answer is C.

24. A Excessive shear stresses in reinforced concrete beams cause diagonal cracks to form in the beam. Resistance to shear stresses is provided by the concrete and by shear reinforcement, usually in the form of stirrups. The shear resistance of the concrete is determined by the ultimate 28-day strength of the concrete (A), and the width and depth of the beam. The longitudinal tension reinforcing (B) does *not* resist shear stresses. Likewise, the aggregate size (C) and the beam span (D) have no effect on the beam's shear capacity.

25. B The elongation $\Delta = PL/AE$ (see page 17).

$$\Delta = \frac{50{,}000\#(20 \text{ ft.})(12 \text{ in./ft.})}{\frac{\pi}{4}(2 \text{ in.})^2 (29{,}000{,}000\#/\text{in.}^2)} = 0.132"$$

26. C The formula for foundation pressure is $F = P/A$, where F is the foundation pressure, P is the load on the foundation, and A is the area of the footing. To solve for the required area of a footing, this formula can be rearranged so that $A = P/F$. In this problem P is given as 6 kips/ft. (6,000 lbs./ft.). For a wall footing, A = width (W) × 1.0 ft. You are also given the foundation soil type as clayey sand, and from the table in the appendix, F is 2,000 psf. Substituting in the formula

$$W = P/F = \frac{6{,}000 \text{ lbs./ft.}}{2{,}000 \text{ lbs./ft.}^2} = 3.0 \text{ ft.}$$

27. A In general, the longer the fundamental period, the lower the acceleration and the resulting seismic force. Conversely, the shorter the period, the greater the seismic force. The correct answer is therefore A.

28. B Refer to ASCE 7 Section 12.8.4. To account for various uncertainties, accidental torsion must be provided for by assuming the mass displaced from the calculated center of mass in each direction a distance equal to 5 percent of the building dimension perpendicular to the direction of the force. Thus, for north-south loads, 0.05 × 200 = 10 feet, and for east-west loads, 0.05 × 100 = 5 feet.

29. A, B, C, D, and E All of the statements are true. Tall buildings have a long fundamental period of vibration, and therefore tend to be in resonance with the long period of a slow rocking motion (A). Elevator cables and the containment of elevator counterweights in tall buildings were typically damaged in the San Fernando earthquake of 1971 (B). The typical one-story wood frame dwelling is particularly suited to survive strong shaking without serious danger to occupants, although such houses may be damaged. Two-story houses typically do not perform as well as one-story houses (C). In the San Fernando earthquake, intensified damage occurred at the discontinuity between soft alluvial soils and the much firmer soils of the foothills (D). Most of the deaths in recent earthquakes were due to ground shaking, rather than surface faulting (E). Building directly on known active faults should be avoided, but it is more important to design structures to resist ground shaking with minimum hazard to occupants and without excessive damage.

30. A In the "tube" concept, the entire structure acts like an immense, hollow tubular column which cantilevers out of its foundation under the action of wind loads. The Willis Tower (correct answer A) consists of a "bundle" of nine tubes that terminate at varying heights, thus expressing the "bundled tube" concept visually.

31. D You are asked to find the maximum moment that this beam can resist. Using the formula $F_b = M/S$, we can rearrange the terms so that $M = F_bS$. We enter Figure 3.2 with the given beam size (4 × 6) and read the value of the section modulus S as 17.65 in.^{3}. From Table 4A, for a 4 × 6 Select Structural, F_b (the design bending stress) is 1,500 psi.

$M = F_bS = 1{,}500\#/\text{in.}^2 \times 17.65 \text{ in.}^3$
$= 26{,}475 \text{ in.-lbs.} \times 1 \text{ ft./12 in.}$
$= 2{,}206 \text{ ft.-lbs.}$ (D).

32. A All of the statements are true except A, which is therefore the correct answer to this question. The ideal column to resist buckling is one whose radius of gyration is the same in both directions, such as a pipe column, and not a wide flange section.

33. A, B, C, and F. The moment at the interior supports of a continuous beam is always negative when the beam supports normal downward vertical loads. Therefore, A is correct. If we isolate span AB we can see that the reaction at A is downward (B is correct), and the moment in span AB is always negative (C is correct).

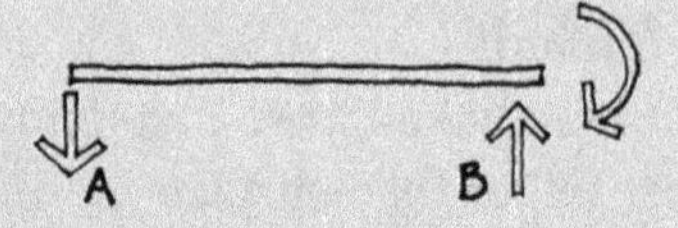

The moment in span BC varies as shown below.

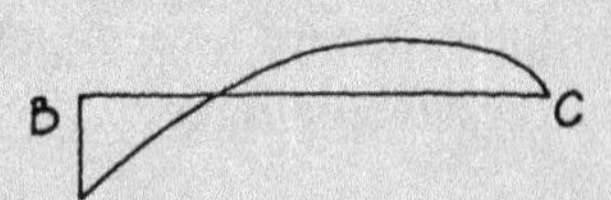

Thus, statement F is also correct.

34. B For concentrated column loads, the most common type of footing is a square pad centered beneath the column, which results in a uniform soil pressure on the footing. However, if the column is located adjacent to a property line, the footing cannot be centered beneath the column, and the eccentricity between the center of the column and the center of the footing results in a non-uniform soil pressure distribution. This could result in undesirable footing settlement, or an uneconomically large footing. To avoid these problems, a cantilevered footing as shown in the sketch is often used, in which the grade beam between the two pad footings resists the bending caused by the footing eccentricity.

35. D Steel columns generally bear on steel base plates, which spread the column load over the foundation, so that the allowable bearing pressure is not exceeded (correct statement D). This action causes bending in the plate (B is incorrect). A column with a typical base plate is usually considered to be pin-ended, not fixed against rotation (A is incorrect). Any shear in the column is usually transmitted to the foundation by anchor bolts, not friction (incorrect statement C).

36. B Although connections that subject wood screws to withdrawal loads are permissible, it is better to avoid them if possible (correct statement D). Loading wood screws in withdrawal from the end grain of wood is not permitted at all (B is the incorrect statement and therefore the correct answer to this question). Both A and C are true statements.

37. B There are five types of portland cement, which are described on page 86. Type III (high-early-strength) is generally used to speed the construction process (B). Such cement is more costly than normal cements

(A is incorrect), but achieves high strength in one to three days. A given concrete strength can be obtained with any type of cement, and all types of cement are readily available.

38. A The moment in a simple beam supporting a uniform load is equal to $wL^2/8$. Thus, $M = 1{,}200(24)^2/8 = 86{,}400$'#. We convert this to inch-pounds by multiplying by 12 = 86,400(12) = 1,036,800"#. The required plastic section modulus $Z = M\Omega/F_y = 1{,}036{,}800 \times 1.67/50{,}000 = 34.6\ \text{in}^3$ (correct answer A).

39. C For a body to be in equilibrium, it must have no unbalanced force acting on it (A and D are correct), and no unbalanced moment (B is correct). C is incorrect, and therefore the answer to this question, because if there is a resultant force on a body, it will move and not be in equilibrium.

40. A Unit stress is equal to $P/A = 40{,}000\#/(1.5\ \text{in.})^2 = 17{,}778$ psi.

41. B By taking moments about D, we determine that A_H must act to the left. For equilibrium, D_H must therefore act to the right. If we draw a free body diagram of member ABC and take moments about B, we can determine that A_V acts downward. D_V must therefore act upward. The components of the reactions are shown correctly in answer B.

42. D Bents 1, being furthest from the center of the building, will resist most of the shear from torsion. Bents 2, being stiffer, will resist most of the direct shear. The correct answer is therefore D.

43. D A braced frame is essentially a vertical truss which resists horizontal loads. We cut a section through the braces, as shown in the right column.

$\Sigma H = 0$

$6 - F_{1H} - F_{2H} = 0$

The internal force in brace 1 acts toward the upper left joint, which means that the brace is in compression. But the braces can only resist tension, and therefore brace 1 is ineffective and has zero force.

Since $6 - F_{1H} - F_{2H} = 0$

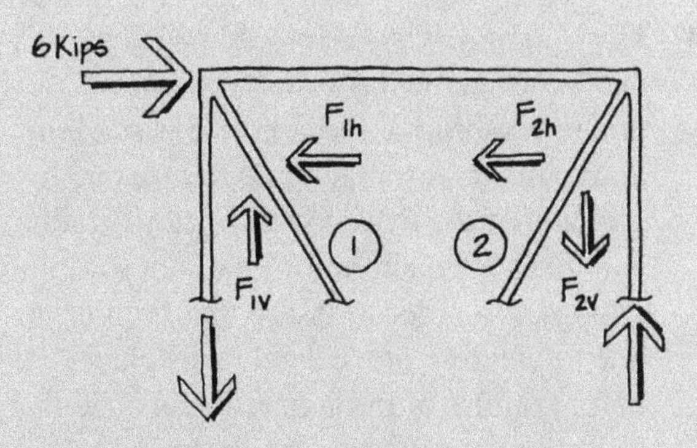

$6 - 0 - F_{2H} = 0$

$F_{2H} = 6$ kips

Since the brace is at 45° to the horizontal, its force is $F_{2H} \div \cos 45° = 6 \div 0.707 = 8.5$ kips.

44. A Take moments about the bottom of the left column. 6 kips (12 ft.) – force in right column (12 ft.) = 0. Force in right column = 6 kips. Since the foundation acts upward on the column, the column is in compression.

45. D Unlike structural steel, which has a constant value of modulus of elasticity, the modulus of elasticity of concrete varies with its strength and unit weight (A is incorrect, D is correct). Reinforced concrete beams continue to deflect after they reach their initial deflection (B is incorrect). C is also incorrect; the amount of reinforcing steel does not have a great effect on the deflection of a reinforced concrete beam.

46. Stiffness Candidates should know and understand the definitions of unit stress, unit strain, modulus of elasticity, and stiffness. Unit stress is the stress per unit of area and is measured in pounds per square inch. Unit strain is the total stretching or shortening of the member divided by its original length. The ratio of unit stress to unit strain is called the modulus of elasticity, E. The higher the E value of a material, the greater its stiffness, or resistance to deformation.

47. C You may wish to refer to pages 91 and 92. Statement A is correct; failure due to crushing of the concrete is sudden and without warning, while failure due to yielding of the tensile steel is more gradual and gives adequate warning of approaching collapse. In order to assure that failure due to yielding of the steel takes place before failure of the concrete, the code sets a limit on the net tensile strain in the reinforcement. Statement B is also correct; the depth of the rectangular stress block (a) is equal to $\beta_1 c$, where β_1 is 0.85 up to $f'_c = 4{,}000$ and c is the distance from the extreme compression fiber to the neutral axis. Statement D is also correct. Only statement C is incorrect, and therefore the answer to this question. The concrete stress at failure is equal to $0.85\ f'_c$, not f'_c.

48. A and D A and D are the only correct statements. Statement B is incorrect, because the vertical shear is the algebraic sum of the forces on one side of the section under consideration. C is also incorrect; the equations of equilibrium apply to every structure in equilibrium. A statically indeterminate beam is one that cannot be solved from the equations of equilibrium alone, but requires additional equations. Finally, E is also incorrect, because the critical section for bending stress is where the bending moment is maximum, which occurs where the vertical shear passes through zero.

49. C The dead load resisting moment must be at least 1.5 times the overturning moment caused by earth pressure (A is incorrect). The simplest solution is to widen the base (correct answer C). Choices B and D do not affect the dead load resisting moment.

50. D This is a double shear joint. According to Table 11F of the National Design Specification, which is reproduced in Figure 7.9, the design load for each 3/4" bolt in double shear parallel to grain when the thickness of the main member is 3-1/2" (nominal 4") and the side member is 1-1/2" (2" nominal) is 2,400 pounds. The total load which can be transferred by the two bolts is therefore 2,400 × 2 = 4,800 pounds (correct answer D).

51. A All of the statements are true, except for A, which is therefore the correct answer. Lower grade material is often used for the less highly stressed inner laminations, while the outer laminations are comprised of higher grade material.

52. B In order to determine the reaction at R_2, we take moments about R_1.

$\Sigma M_{R1} = 0$
850#/ft.(12 ft.)(8 ft. + 12 ft./2)
+ 500#(8 ft.+ 12 ft. + 4 ft.) + 900#(28 ft. + 8 ft.) – R_2(28 ft.) = 0
R_2 = 187,200 ft.-lbs. ÷ 28 ft. = 6,686#

Therefore the correct answer is B.

53. C Statement I is incorrect, since the greatest shrinkage occurs across the grain, and shrinkage parallel to the grain (longitudinal direction of the member) is negligible. Statement II is correct, since the individual laminations are relatively thin and can easily be seasoned, reducing shrinkage. Statements III and IV are also correct. Statement V is incorrect; in most designs the expansion of wood caused by temperature change is negligible. Statement VI is correct. If the tops of the joists and beam are flush, and the joists shrink, the floor over the beam will be higher than that over the joists, resulting in a bump in the floor. The incorrect statements are I and V, and the correct answer is therefore C.

54. Stirrups Stirrups are small diameter bars, usually #3 or #4, generally U-shaped, placed vertically in a reinforced concrete beam to provide reinforcement for shear stress.

55. D Two-way shear, which is often critical in the design of single column footings, is virtually unaffected by the footing size (A is incorrect). Shear stress is unrelated to the amount of reinforcing steel (B is also incorrect). While stirrups could theoretically be provided to carry part of the shear stress (C), the simpler solution is to increase the thickness of the footing (correct answer D).

56. C The ratio of a column's effective length (l or Kl) to its least radius of gyration (r) is called the *slenderness ratio* (correct answer C) and is used in the design of columns.

57. A The *workability* of concrete—the ease with which it can be placed and consolidated—is an important quality that is usually measured in the field by the *slump test* (correct answer A), or occasionally, by the *Kelly ball test.*

58. B Properly compacted fill is sometimes used for the support of buildings. Where this is anticipated, typical soil samples are subjected to a laboratory compaction test, called the *Proctor test*, to determine the soil's optimum moisture content and density (correct answer B).

59. A The maximum moment = $wL^2/8$ = $1{,}200(30)^2/8$ = 135,000 ft.-lbs. = 135 ft.-kips. We enter the chart in Figure 4.13 and locate the intersection of the moment coordinate (135 ft.-kips) and the unbraced length coordinate (30/2 = 15 ft.). The first solid line beam curve at this intersection is that of W14 × 43 (answer A).

60. C Basement walls are usually designed to resist earth pressure by spanning between the basement floor and the first floor (B and D are incorrect). Therefore, if the first floor is not in place when the backfill is placed, temporary bracing is required (C is correct). A is incorrect: in most cases, the bending moment in the wall requires the use of reinforcing steel.

61. B From IBC Table 1607.1, the roof live load L_o for an ordinary flat, pitched, or curved roof is equal to 20 psf.

Since the tributary area A_t is equal to 500 square feet, reduction factor R_1 is determined by the following formula:

$R_1 = 1.2 - 0.001A_t = 1.2 - 0.001(500) = 0.70$

With a roof pitch of 3 in 12, F = 3. Thus, reduction factor R_2 = 1 for F less than or equal to 4.

Therefore, the live load reduction = 1 – (0.70 × 1.0) = 1 – 0.70 = 0.30, or 30 percent. The design roof live load may be reduced from 20 psf to 0.70 × 20 = 14 psf, which is greater than the minimum permitted reduced live load of 12 psf (correct answer B).

62. B The required number of rows of bridging for open web steel joists is a function of the chord size and span only, not the load. From the table in Figure 10.10 for K series joists, for chord size #8 (the last digit in the joist designation) and span of 20 to 33 feet, we read 2 rows of bridging (correct answer B).

63. B Torsional forces result when lateral forces act on a building that has a rigid diaphragm, usually of concrete or steel, and in which the center of resistance of the lateral load resisting system (the center of rigidity) is not located at the center of mass of the building (B is correct, D is incorrect). Torsion can occur in any shape building (A is incorrect) and with any type of lateral load resisting system (C is incorrect).

64. C Wind pressures used in IBC design take into account the high wind speeds of hurricanes (A is incorrect). Buildings or surface irregularities interfere with the wind, and thus decrease the wind pressures (B is incorrect). C is correct: wind creates both positive inward pressures and negative outward pressures on wall surfaces. D is incorrect, because the wind pressures at corners and discontinuities are higher than those on typical wall elements.

65. D All four are examples of plan irregularities. I, II, and IV are plans with reentrant corners, and III is an example of torsional irregularity caused by unsymmetrical rigidities. Regular and symmetrical structures have better seismic response characteristics than irregular structures. Therefore, irregular structures in earthquake-prone areas should be avoided if possible.

66. A The formulas for both wind and seismic forces include an importance factor (I), which depends on the occupancy or use of the building. In this case, both the original warehouse occupancy and the meeting hall occupancy have an importance factor I of 1.0. Therefore, no additional design lateral force is required. See Table 1604.5. Certain high-density, essential, or hazardous facilities have an importance factor greater than 1.0, protecting such facilities against increased wind or seismic forces.

67. C Walls must be anchored to all floors and roofs that provide lateral support for the wall. The anchorage for masonry or concrete walls must resist the design seismic or wind forces perpendicular to the wall, or a minimum force of 280 pounds per lineal foot of wall. Anchor A in the detail serves this function (correct answer C). Vertical load is transferred from the roof joists to the ledger, usually by joist hangers, and from the ledger to the wall by anchor bolts. Lateral load is transferred from the roof diaphragm to the ledger by boundary nails, and from the ledger to the wall by anchor bolts. Plywood panels may be connected together by being nailed to blocking at the panel edges.

68. C A Vierendeel truss is one without diagonals. Therefore, large openings, such as doors and windows, may be made within the depth of a Vierendeel truss without conflicting with diagonal members (II is correct). However, Vierendeel trusses tend to have greater deflection (I is incorrect) and more material (III is incorrect) than conventional triangulated trusses. While the joints of a Vierendeel truss may appear simple, they are not necessarily so, since they must be designed to transfer moment (IV is incorrect). V is also incorrect, since many conventional trusses may have load applied to either top or bottom chord. Since only II is true, C is the correct answer.

69. D

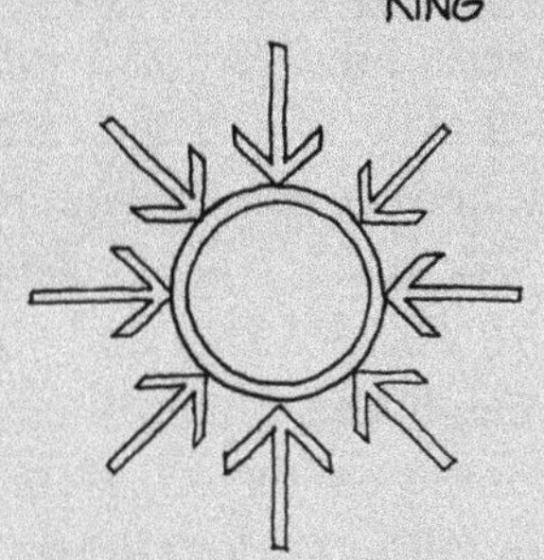

PLAN OF CENTER RING

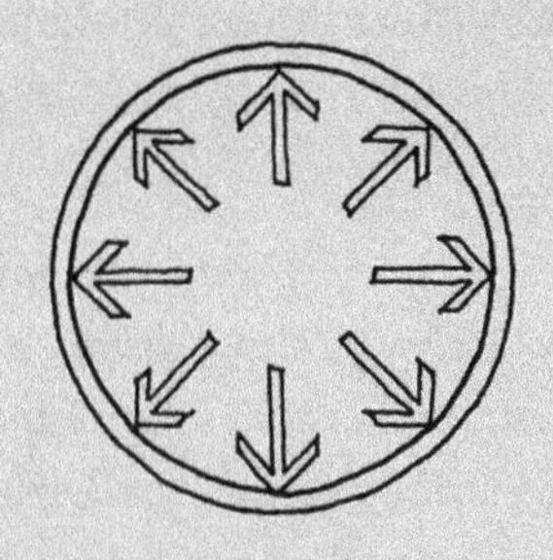

PLAN OF PERIMETER RING

As with all arches, each arch rib thrusts against its supports. These thrusts compress the center ring and tend to open up the perimeter ring, stressing it in tension. The correct answer is D.

70. D We first compute the left reaction by taking moments about the right end of the truss.

$\Sigma M = 0$
R_L(50ft) – 10 kips (37.5 ft.) – 15 kips (25 ft.) – 30 kips(12.5 ft.) = 0
R_L = 22.5 kips

We cut a section through member a, as shown in the next column.

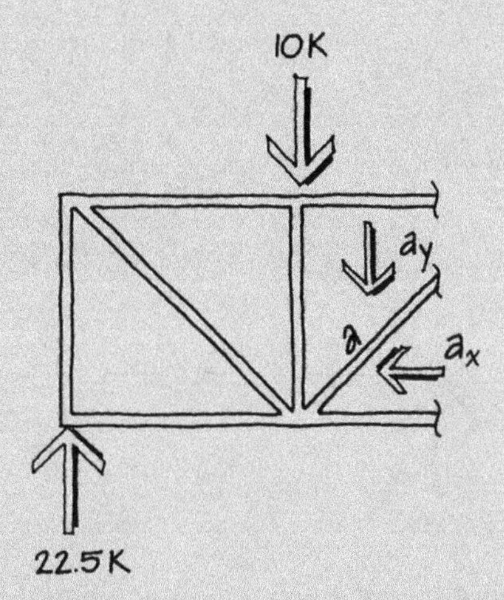

$\Sigma F_y = 0$
22.5 kips – 10 kips – a_y = 0
a_y = 12.5 kips
a_x = 12.5 kips, since member a is at 45°.

Therefore $a = \sqrt{12.5^2 + 12.5^2}$ = 17.7 kips compression (correct answer D).

71. C A waffle slab is a two-way ribbed slab that generally does not require any beams (III is correct, I and II are incorrect). It is usually formed using dome-shaped steel forms which are removed after the concrete has hardened (V is correct, IV is incorrect). Since only III and V are correct, C is the answer.

72. A A three-hinged gabled frame, or arch, is a statically determinate structure (I is incorrect). The hinges at the supports and center permit rotation and therefore can develop no moment (II and III are correct). The maximum moment occurs at the haunch, the intersection of the column and the sloping beam (correct statement V), but this moment is generally less than in a rectangular rigid frame (incorrect statement IV). Since II, III, and V are correct, A is the answer.

73. C

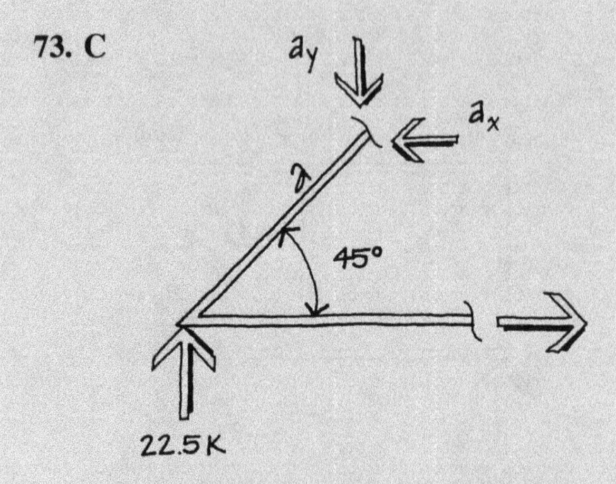

By symmetry, each reaction is equal to 1,500 × 30/2 = 22,500# = 22.5 kips.

Isolating the left support, as shown above, $\Sigma F_y = 0$

$22.5 \text{ kips} - a_y = 0$

$a_y = 22.5 \text{ kips}$

Since member a is at 45°, $a_x = a_y = 22.5$ kips.

The force in member a = $\sqrt{22.5^2 + 22.5^2}$

= 31.8 kips.

74. C The information given in this question is very sparse; we are not told what kind of arch or truss or the magnitude of the concentrated loads. So we have to use our best judgment to arrive at the answer. In general, concentrated loads tend to induce bending moment in a structure and therefore we must determine which of the four systems is best able to resist bending moment. That system is the truss (correct answer C), which resists concentrated loads and the resulting moments by developing compressive stresses in the top chord, tensile stresses in the bottom chord, and tension or compression in the web members.

75. C First, we convert the load in pounds per square foot to pounds per lineal foot, by multiplying the square foot load by the joist spacing. w = 100 × 3 = 300 #/ft. From the load table in Figures 10.13 and 10.14, we select 40LH10, which can safely support 305 #/ft. and weighs 21 #/ft. The 36LH10 (A) is inadequate, since it can only support =295 #/ft. The 36LH11 and 40LH11 (B and D) can both support the load, but they weigh more than the 40LH10. Generally, we would also check deflection, but in this case, the problem statement specifically told us to neglect deflection.

76. C The AIA Long Span Building Panel, convened to review the design and construction of long span structures in the light of several failures, published its report in 1981. The following were among its findings:

(A) Many long span structures...have fewer alternative paths of load resistance. This lack of redundancy makes the long span structure more vulnerable to general collapse in case of accidental overload or member weakness.

(B) ...the potential consequences of failure ... points to a need for special care in design ...

(D) ... some roof decks which may be used to adequately brace short spans may not be stiff enough to adequately brace long span structures. However, the panel also concluded that ...none of the basic structural systems...could be defined as surpassing the frontier of existing knowledge. The incorrect statement, and therefore the answer to this question, is C.

77. C The first part of A is correct since a dome can be thought of as an arch rotated about its vertical axis. But the rest of A is not correct; most domes are stiffened by circumferential rings or hoops, making them much stiffer than arches. B is also incorrect; while a number of large geodesic domes have been built, the largest domes are generally framed in some other manner. D is another incorrect statement; a hyperbolic paraboloid, not a dome, can be formed using straight timber boards. C is therefore the only correct statement.

78. Dead load

79. A We isolate the joint in question, as shown below. Since the wind load acts to the right, the internal force in the column must act to the left. That causes a clockwise moment in the column, which must be balanced by counterclock-wise moments in the beams. The shear in the left beam therefore acts down, while that in the right beam acts up. The face of the beam or column that tends to open up because of the moments caused by these forces is in tension (T), while the face tending to close or crush is in compression (C), as shown. A is therefore the correct answer.

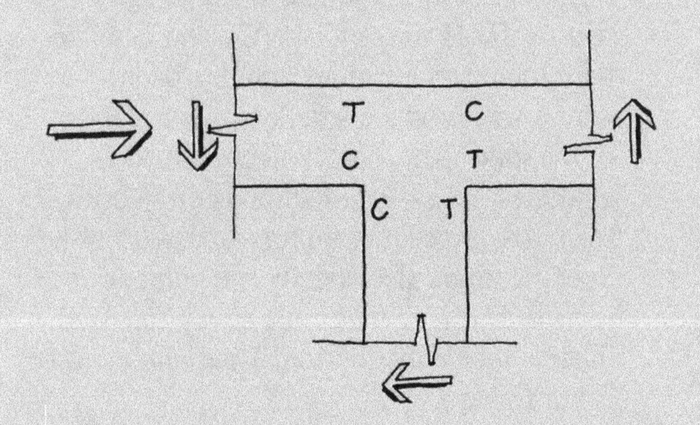

80. D The total lateral earthquake force on a structure is equal to C_sW. In this formula, C_s is the seismic response coefficient, which depends on the subsoil conditions at the site (II), the geographical location of the building (III), and the type of lateral load resisting system (V). W is the total dead load (IV). The building's orientation (I) is not considered in earthquake design, making D the correct answer.

81. B Questions involving overturning moment from wind or earthquake forces have appeared on recent exams, and candidates are therefore advised to become familiar with this concept. For the water tank in this question, the gravity load on each of the four legs is one-quarter of the total load or 8 kips/4 = 2 kips. The seismic force on the structure is given as V = 0.3 W = 0.3 × 8 kips = 2.4 kips. The overturning moment on the structure is then 2.4 kips × (25 ft. + 5 ft.) = 72 ft.-kips. If the seismic force acts parallel to the main axes of the structure, all four legs resist the overturning moment and the vertical force due to overturning on each leg is 72 ft.-kips/10 ft. × 2 legs = 3.6 kips (up or down). However, the seismic force may act in any direction, and if it acts along a 45° diagonal, then only two legs resist the overturning moment. In that case, the vertical force on two diagonal legs, due to the overturning moment, is 72 ft.-kips/14.1 ft. = 5.1 kips (up or down). The maximum vertical force due to both gravity and seismic overturning is 2 kips + 5.1 kips = 7.1 kips (correct answer B).

82. B A shell, not a membrane, can resist tension, compression, or shear. A membrane can only resist tension (incorrect statement A). B is correct; like a cable structure, a membrane changes its shape as the load changes. The egg is a shell, not a membrane (C is incorrect); an example of a membrane in nature is the soap bubble. D is also incorrect. Although many pneumatic type structures have habitable space that is under pressure, double membrane roofs and pillow roofs are pressurized only within the membrane roof, not the occupied space.

83. A Since the supports of an arch may be either fixed against rotation or hinged, A is correct. B is incorrect, because all arches have horizontal reactions at their bases. While a so-called "pure" arch is stressed only in compression, and compression is always the primary stress in an arch, virtually all arches are designed to resist some bending as well (C is incorrect). D is also incorrect, because the shape of an arch is frequently determined by aesthetics, rather than structural considerations.

84. C Although the Romans were the master builders of antiquity, they simply did not have the technology that is available today. The largest dome built by the Romans was the Pantheon, a masterpiece of ancient architecture and engineering. Its inside span is 142 feet, compared to 710 feet for the Houston Astrodome, and the surface area of the Astrodome is 25 times greater than that of the Pantheon (incorrect statement A). B is also incorrect, since a number of roofs spanning up to 300 feet and greater were built during the 19th century to enclose railroad stations, factories, and armories. C is the correct statement: the 85-foot by 200-foot ice surface required for hockey is larger than for any other common indoor sport. Finally, D is incorrect. Although long span roofs are widely used for theaters, arenas, and stadiums, they are also used for a variety of other applications, including hangars, factories, stores, and airport terminal buildings.

85. A All of the reasons given have some validity; stiffness (B), shear strength (C), and roof drainage (D) are all factors that must be considered. However, the primary reason for tapering a girder is to provide only as much bending resistance as is needed at each point along the girder's length, rather than providing a constant, maximum bending resistance for the full length of the girder.

86. D Architectural history questions have appeared in many parts of the exam, including the structural tests. This question test your knowledge concerning one of the masterpieces of Greek architecture. The Parthenon was adapted from the wood construction of early Greek temples to marble (correct statement D). Its post-and-beam structure is appropriate for wood, but not for stone, whose low tensile strength limits its ability to span as a beam (A is incorrect). Later, the Romans exploited the great compressive strength of stone and concrete and spanned great distances using vaults, domes, and arches (C is incorrect). B refers to the Pantheon, the great Roman structure, not the Parthenon.

87. B The load capacities for open web steel joists are tabulated in the Standard Specifications of the Steel Joist Institute (A). These tables include spans of up to 144 feet for DLH series joists (C), and tabulate both the load capacities and the loads which will produce a deflection of 1/360 of the joist span. This deflection load is tabulated because deflection is often critical for joists, especially on long spans (D). The Specifications also specify the number of rows of bridging that must always be used in steel joist construction. Thus statement B is not true.

88. B and C B is correct; the number of members to be joined, as well as the complexity of the joints, usually makes truss fabrication expensive, but this is often offset by the economy of material. C is also correct; truss members are usually subject to axial stress only, unless they support loads between the panel points. A is incorrect; the stress in the chord members depends on the depth, but the stress in the web members is a function of their slope, not the truss depth.

89. C To calculate the horizontal thrust, we isolate the left half of the arch.

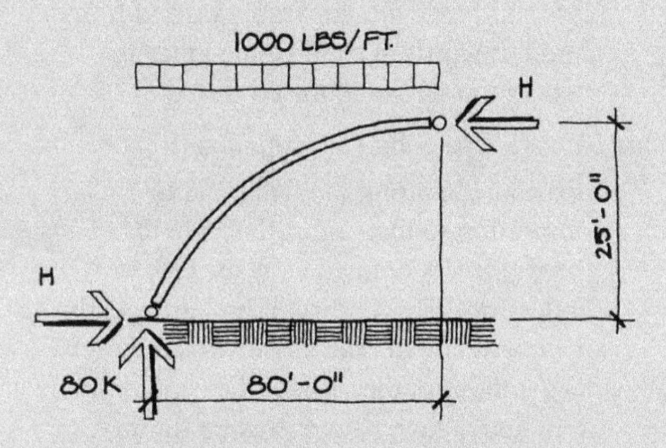

By symmetry, each vertical reaction is equal to one-half of the total load on the arch = 1,000 #/ft. × 160 ft./2 = 80,000# = 80 kips.

Take moments about the center hinge, where the moment is equal to zero.

$\Sigma M = 0$

H = [80 kips(80 ft.) – 1.0 kips/ft. (80 ft.)(80 ft./2)] ÷ 25 ft.
= (6,400 – 3,200)/25
= 3,200/25 = 128 kips (answer C).

90. D Candidates should be familiar with some of the terms used for long span structural systems. The lamella (A) is a series of parallel arches that are skewed with respect to the axes of a building and that intersect another series of skewed arches. A Vierendeel truss (B) is a truss without diagonals. A rigid frame (C) is a structure in which the joints between the beams and columns are rigid and capable of transferring bending moment. The correct answer is space frame (D), a three-dimensional, statically indeterminate system used to span large areas.

91. D Cast-in-place concrete is appropriate for multi-story rigid frames because continuity at the joints is easily provided by the monolithic nature of the concrete and lapping of the reinforcing bars (correct answer D). Steel rigid frames require high strength bolts (C) or welding to achieve joint rigidity. Although concrete frames may use shear keys (B) or welded reinforcing bars (A), these are not the primary means of providing continuity.

92. D To solve this problem, we isolate the joint at the intersection of the vertical and bottom chord.

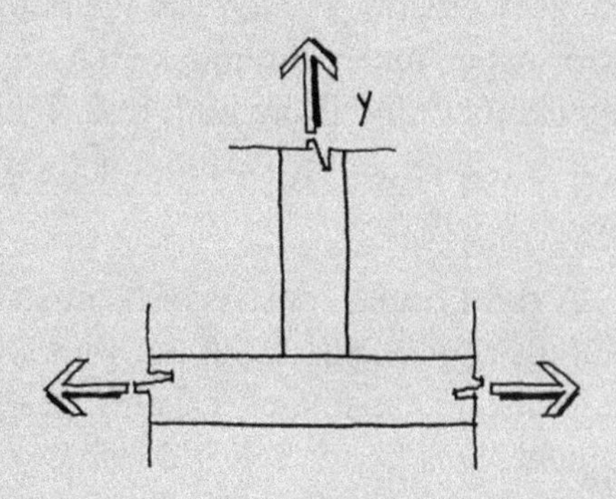

We call the force in the vertical member y and apply the equation of equilibrium $\Sigma F_y = 0$. Since the only vertical force acting at the joint is y, y = 0.

93. B The Romans regularly used concrete in their structures. The concrete, of course, was unreinforced, and therefore limited to compression structures, such as domes. The Pantheon remains today one of the greatest achievements of Roman architecture; its dome, spanning 142 feet, remained the longest-spanning unreinforced concrete dome until it was surpassed in 1913.

94. D Pier Luigi Nervi was an Italian engineer who successfully merged mathematics and aesthetics. His concrete domes and barrel shells, generally using diagonal precast ribs, were structures of exceptional lightness, economy, and beauty.

95. A The Great Exhibition of 1851, in London, was housed in the Crystal Palace, an immense prefabricated glass and cast iron structure. Because of its lightness and transparency, it strongly influenced subsequent steel and glass buildings.

96. B Statement A is true. Although nonstructural walls are assumed to provide no resistance to a building's earthquake forces, they can in fact affect the way the building performs in an earthquake. Statement B is false. The sudden change from the stiff upper stories to the flexible first story creates problems in the building's performance under earthquake loading. Statements C and D are both true. The only incorrect statement is B, and it is therefore the correct answer.

97. C A rigid frame consists of beams and columns rigidly connected together so that the joints are capable of resisting bending moment (D is correct). Rigid frames can efficiently resist both vertical and horizontal forces (correct statement A). Their greatest use is in multi-story construction, and they may be constructed of either structural steel or reinforced concrete (B is correct). Statement C is the answer we are seeking. Despite its name, a rigid frame is generally less rigid than a comparable braced frame. In other words, a rigid frame tends to deflect more than a braced frame.

98. A The 10,000# lateral load will be distributed among the five piers in proportion to their rigidities, where the rigidity is defined as resistance to deflection. The rigidity of a pier depends on its height and depth. The lower the height-to-depth ratio, the greater the rigidity. Piers 2, 3, and 4 each have the same rigidity, based on a height of three feet and a depth of four feet, and therefore a height-to-depth ratio of 3/4 = 0.75. Piers 1 and 5 have equal rigidities based on a height of three feet and a depth of three feet, and a height-to-depth ratio of 3/3 = 1.0. The rigidity of piers 2, 3, and 4 is therefore greater than that of piers 1 and 5. Thus, the lateral loads resisted by piers 2, 3, and 4 are equal, and the lateral loads resisted by piers 1 and 5 are equal, but smaller than those resisted by piers 2, 3, and 4. Only answer A meets these criteria.

INDEX